**MECHANICAL
METALLURGY**

**McGRAW-HILL
BOOK COMPANY**
New York
St. Louis
San Francisco
Auckland
Düsseldorf
Johannesburg
Kuala Lumpur
London
Mexico
Montreal
New Delhi
Panama
Paris
São Paulo
Singapore
Sydney
Tokyo
Toronto

**GEORGE E. DIETER**

*Professor of Engineering*
*Carnegie-Mellon University*

# Mechanical Metallurgy

**SECOND EDITION**

This book was set in Times Roman.
The editors were B. J. Clark and Michael Gardner;
the production supervisor was Dennis J. Conroy.
The drawings were done by J & R Services, Inc.
Kingsport Press, Inc., was printer and binder.

**Library of Congress Cataloging in Publication Data**

Dieter, George Ellwood.
    Mechanical metallurgy.

    (McGraw-Hill series in materials science and engineering)
    Includes index.
    1. Strength of materials. 2. Physical metallurgy. 3. Metal-work. I. Title.
TA405.D53    1976        620.1'6'3       75-41430
ISBN 0-07-016891-1

**MECHANICAL
METALLURGY**

234567890KPKP7832109876

# CONTENTS

# PREFACE TO THE SECOND EDITION

In the 12 years since the first edition of "Mechanical Metallurgy" at least 25 text-books dealing with major segements of the book have appeared in print. For example, at least 10 books dealing with the mechanics of metalworking have been published during this period. However, none of these books have dealt with the entire scope of mechanical metallurgy, from an understanding of the continuum description of stress and strain through crystalline and defect mechanisms of flow and fracture, and on to a consideration of the major mechanical property tests and the basic metalworking processes.

Important advances have been made in understanding the mechanical behavior of solids in the period since the first edition. The dislocation theory of plastic deformation has become well established, with excellent experimental verification for most of the theory. These advances have led to a better understanding of the strengthening mechanisms in solids. Developments such as fracture mechanics have matured to a high level of technical sophistication and engineering usefulness. An important development during this period has been the "materials science movement" in which crystalline solids, metals, ceramics, and polymers are considered as a group whose properties are controlled by basic structural defects common to all classes of crystalline solids.

In this revision of the book the emphasis is as before. The book is intended for the senior first-year graduate-student level. Extensive revisions have been made to up-date material, to introduce new topics which have emerged as important areas, and to clarify sections which have proven difficult for students to understand. In some sections advanced material intended primarily for graduate students has been set in smaller type. The problems have been extensively revised and expanded, and a solutions manual has been prepared. Two new chapters, one dealing with the mechanical properties of polymers and the other with the machining of metals, have been added, while the chapters of statistical methods and residual stresses have been deleted. In total, more than one-half of the book has been rewritten completely.

GEORGE E. DIETER

# PREFACE TO THE FIRST EDITION

Mechanical metallurgy is the area of knowledge which deals with the behavior and response of metals to applied forces. Since it is not a precisely defined area, it will mean different things to different persons. To some it will mean mechanical properties of metals or mechanical testing, others may consider the field restricted to the plastic working and shaping of metals, while still others confine their interests to the more theoretical aspects of the field, which merge with metal physics and physical metallurgy. Still another group may consider that mechanical metallurgy is closely allied with applied mathematics and applied mechanics. In writing this book an attempt has been made to cover, in some measure, this great diversity of interests. The objective has been to include the entire scope of mechanical metallurgy in one fairly comprehensive volume.

The book has been divided into four parts. Part One, Mechanical Fundamentals, presents the mathematical framework for many of the chapters which follow. The concepts of combined stress and strain are reviewed and extended into three dimensions. Detailed consideration of the theories of yielding and an introduction to the concepts of plasticity are given. No attempt is made to carry the topics in Part One to the degree of completion required for original problem solving. Instead, the purpose is to acquaint metallurgically trained persons with the mathematical language encountered in some areas of mechanical metallurgy. Part Two, Metallurgical Fundamentals, deals with the structural aspects of plastic deformation and fracture. Emphasis is on the atomistics of flow and fracture and the way in which metallurgical structure affects these processes. The concept of the dislocation is introduced early in Part Two and is used throughout to provide qualitative explanations for such phenomena as strain hardening, the yield point, dispersed phase hardening, and fracture. A more mathematical treatment of the properties of dislocations is given in a separate chapter. The topics covered in Part Two stem from physical metallurgy. However, most topics are discussed in greater detail and with a different emphasis than when they are first covered in the usual undergraduate course in physical metallurgy. Certain topics that are more physical metallurgy than mechanical metallurgy are included to provide continuity and the necessary background for readers who have not studied modern physical metallurgy.

Part Three, Applications to Materials Testing, deals with the engineering aspects of the common testing techniques of mechanical failure of metals. Chapters are devoted to the tension, torsion, hardness, fatigue, creep, and impact tests. Others take up the important subjects of residual stresses and the statistical analysis of mechanical-property data. In Part Three emphasis is placed on the interpretation of the tests and on the effect of metallurgical variables on mechanical behavior rather than on the procedures for conducting the tests. It is assumed that the actual performance of these tests will be covered in a concurrent laboratory course or in a separate course. Part Four, Plastic Forming of Metals, deals with the common mechanical processes for producing useful metal shapes. Little emphasis is given to the descriptive aspects of this subject, since this can best be covered by plant trips and illustrated lectures. Instead, the main attention is given to the mechanical and metallurgical factors which control each process such as forging, rolling, extrusion, drawing, and sheet-metal forming.

This book is written for the senior or first-year graduate student in metallurgical or mechanical engineering, as well as for practicing engineers in industry. While most universities have instituted courses in mechanical metallurgy or mechanical properties, there is a great diversity in the material covered and in the background of the students taking these courses. Thus, for the present there can be nothing like a standardized textbook on mechanical metallurgy. It is hoped that the breadth and scope of this book will provide material for these somewhat diverse requirements. It is further hoped that the existence of a comprehensive treatment of the field of mechanical metallurgy will stimulate the development of courses which cover the total subject.

Since this book is intended for college seniors, graduate students, and practicing engineers, it is expected to become a part of their professional library. Although there has been no attempt to make this book a handbook, some thought has been given to providing abundant references to the literature on mechanical metallurgy. Therefore, more references are included than is usual in the ordinary textbook. References have been given to point out derivations or analyses beyond the scope of the book, to provide the key to further information on controversial or detailed points, and to emphasize important papers which are worthy of further study. In addition, a bibliography of general references will be found at the end of each chapter. A collection of problems is included at the end of the volume. This is primarily for the use of the reader who is engaged in industry and who desires some check on his comprehension of the material.

The task of writing this book has been mainly one of sifting and sorting facts and information from the literature and the many excellent texts on specialized aspects of this subject. To cover the breadth of material found in this book would require parts of over 15 standard texts and countless review articles and individual contributions. A conscientious effort has been made throughout to give credit to original sources. For the occasional oversights that may have developed during the " boiling-down process " the author offers his apologies. He is indebted to many authors and publishers who consented to the reproduction of illustrations. Credit is given in the captions of the illustrations.

Finally, the author wishes to acknowledge the many friends who advised him in this work. Special mention should be given to Professor A. W. Grosvenor, Drexel Institute of Technology, Dr. G. T. Horne, Carnegie Institute of Technology, Drs. T. C. Chilton, J. H. Faupel, W. L. Phillips, W. I. Pollock, and J. T. Ransom of the du Pont Company, and Dr. A. S. Nemy of the Thompson-Ramo-Wooldridge Corp.

GEORGE E. DIETER

$A$ = Area; amplitude

$a$ = Linear distance; crack length

$a_0$ = Interatomic spacing

$B$ = Constant; specimen thickness

$b$ = Width or breadth

$\mathbf{b}$ = Burgers vector of a dislocation

$C$ = Generalized constant; specific heat

$C_{ij}$ = Elastic coefficients

$c$ = Length of Griffith crack

$D$ = Diameter, grain diameter

$E$ = Modulus of elasticity for axial loading (Young's modulus)

$e$ = Conventional, or engineering, linear strain

$\exp$ = Base of natural logarithms ($=2.718$)

$F$ = Force per unit length on a dislocation line

$G$ = Modulus of elasticity in shear (modulus of rigidity)

$\mathscr{G}$ = Crack-extension force

$H$ = Activation energy

$h$ = Distance, usually in thickness direction

$(h, k, l) = $ Miller indices of a crystallographic plane
$I = $ Moment of inertia
$J = $ Invariant of the stress deviator; polar moment of inertia
$K = $ Strength coefficient
$K_f = $ Fatigue-notch factor
$K_t = $ Theoretical stress-concentration factor
$K_{Ic} = $ Fracture toughness
$k = $ Yield stress in pure shear
$L = $ Length
$l, m, n = $ Direction cosines of normal to a plane
$\ln = $ Natural logarithm
$\log = $ Logarithm to base 10
$M_B = $ Bending moment
$M_T = $ Torsional moment, torque
$m = $ Strain-rate sensitivity
$N = $ Number of cycles of stress or vibration
$n = $ Strain-hardening exponent
$n' = $ Generalized constant in exponential term
$P = $ Load or external force
$Q = $ Activation energy
$p = $ Pressure
$q = $ Reduction in area; plastic-constraint factor; notch sensitivity index in fatigue
$R = $ Radius of curvature; stress ratio in fatigue; gas constant
$r = $ Radial distance
$S = $ Total stress on a plane before resolution into normal and shear components
$S_{ij} = $ Elastic compliance
$s = $ Engineering stress
$T = $ Temperature
$T_m = $ Melting point
$t = $ Time; thickness
$t_r = $ Time for rupture
$U = $ Elastic strain energy
$U_0 = $ Elastic strain energy per unit volume
$u, v, w = $ Components of displacement in $x$, $y$, and $z$ directions
$[uvw] = $ Miller indices for a crystallographic direction
$V = $ Volume
$v = $ Velocity
$W = $ Work
$Z = $ Zener-Hollomon parameter

$\alpha = $ Linear coefficient of thermal expansion; phase angle
$\alpha, \beta, \theta, \phi = $ Generalized angles

$\Gamma$ = Line tension of a dislocation

$\gamma$ = Shear strain

$\Delta$ = Volume strain or cubical dilatation; finite change

$\delta$ = Deformation or elongation; deflection; logarithmic decrement; Kronecker delta

$\epsilon$ = General symbol for strain; natural or true strain

$\bar{\epsilon}$ = Significant, or effective, true strain

$\dot{\epsilon}$ = True-strain rate

$\dot{\epsilon}_s$ = Minimum creep rate

$\eta$ = Efficiency; coefficient of viscosity

$\theta$ = Dorn time-temperature parameter

$\kappa$ = Bulk modulus or volumetric modulus of elasticity

$\lambda$ = Lamé's constant; interparticle spacing

$\mu$ = Lode's stress parameter; coefficient of friction

$\nu$ = Poisson's ratio; Lode's strain parameter

$\rho$ = Density

$\sigma$ = Normal stress; true stress

$\sigma_0$ = Yield stress or yield strength

$\sigma_0'$ = Yield stress in plane strain

$\bar{\sigma}$ = Significant, or effective, true stress

$\sigma_1, \sigma_2, \sigma_3$ = Principal stresses

$\sigma'$ = Stress deviator

$\sigma''$ = Hydrostatic component of stress

$\sigma_a$ = Alternating, or variable, stress

$\sigma_m$ = Average principal stress; mean stress

$\sigma_r$ = Range of stress

$\sigma_u$ = Ultimate tensile strength

$\sigma_w$ = Working stress

$\tau$ = Shearing stress; relaxation time

$\Phi$ = Airy stress function

$\psi$ = Specific damping capacity

MECHANICAL
METALLURGY

# PART ONE

# Mechanical Fundamentals

# 1

## INTRODUCTION

### 1-1  SCOPE OF THIS BOOK

*Mechanical metallurgy* is the area of metallurgy which is concerned primarily with the response of metals to forces or loads. The forces may arise from the use of the metal as a member or part in a structure or machine, in which case it is necessary to know something about the limiting values which can be withstood without failure. On the other hand, the objective may be to convert a cast ingot into a more useful shape, such as a flat plate, and here it is necessary to know the conditions of temperature and rate of loading which minimize the forces that are needed to do the job.

Mechanical metallurgy is *not* a subject which can be neatly isolated and studied by itself. It is a combination of many disciplines and many approaches to the problem of understanding the response of materials to forces. On the one hand is the approach used in strength of materials and in the theories of elasticity and plasticity, where a metal is considered to be a homogeneous material whose mechanical behavior can be rather precisely described on the basis of only a very few material constants. This approach is the basis for the rational design of structural members and machine parts. The topics of strength of materials, elasticity, and plasticity are treated in Part One of this book from a more generalized point of view than is usually considered in a first course in strength of materials. The material in Chaps. 1 to 3 can be considered the mathematical framework on which much of the remainder of the book rests. For students of engineering who have had an advanced course in strength of materials or machine design, it probably will be possible to skim rapidly over these chapters. However, for most students of metal-lurgy and for practicing engineers in industry, it is worth spending the time to become familiar with the mathematics presented in Part One.

The theories of strength of materials, elasticity, and plasticity lose much of their power when the structure of the metal becomes an important consideration and it can no longer be considered a homogeneous medium. Examples of this are in the high-temperature behavior of metals, where the metallurgical structure may continuously change with time, or in the ductile-to-brittle transition, which occurs in carbon steel. The determination of the relationship between mechanical behavior and structure (as detected chiefly with microscopic and x-ray techniques) is the main responsibility of the mechanical metallurgist. When mechanical behavior is understood in terms of metallurgical structure, it is generally possible to improve the mechanical properties or at least to control them. Part Two of this book is concerned with the metallurgical fundamentals of the mechanical behavior of metals. Metallurgical students will find that some of the material in Part Two has been covered in a previous course in physical metallurgy, since mechanical metallurgy is part of the broader field of physical metallurgy. However, these subjects are considered in greater detail than is usually the case in a first course in physical metallurgy. In addition, certain topics which pertain more to physical metallurgy than mechanical metallurgy have been included in order to provide continuity and to assist nonmetallurgical students who may not have had a course in physical metallurgy.

The last three chapters of Part Two are concerned primarily with atomistic concepts of the flow and fracture of metals. Many of the developments in these areas have been the result of the alliance of the solid-state physicist with the metallurgist. This has been an area of great progress. The introduction of trans-mission electron microscopy has provided an important experimental tool for verifying theory and guiding analysis. A body of basic dislocation theory is pre-sented which is useful for understanding the mechanical behavior of crystalline solids.

Basic data concerning the strength of metals and measurements for the routine control of mechanical properties are obtained from a relatively small number of standardized mechanical tests. Part Three, Applications to Materials Testing, considers each of the common mechanical tests, not from the usual standpoint of testing techniques, but instead from the consideration of what these tests tell about the service performance of metals and how metallurgical variables affect the results of these tests. Much of the material in Parts One and Two has been utilized in Part Three. It is assumed that the reader either has completed a conventional course in materials testing or will be concurrently taking a laboratory course in which familiarization with the testing techniques will be acquired.

Part Four considers the metallurgical and mechanical factors involved in forming metals into useful shapes. Attempts have been made to present mathematical analyses of the principal metalworking processes, although in certain cases this has not been possible, either because of the considerable detail required or because the analysis is beyond the scope of this book. No attempt has been made to include the extensive specialized technology associated with each metalworking process, such as rolling or extrusion, although some effort has been made to give a general impression of the mechanical equipment required and to familiarize the reader with the specialized vocabulary of the metalworking field. Major emphasis has been placed on presenting a fairly simplified picture of the forces involved in each process and of how geometrical and metallurgical factors affect the forming loads and the success of the metalworking process.

## 1-2 STRENGTH OF MATERIALS—BASIC ASSUMPTIONS

Strength of materials is the body of knowledge which deals with the relation between internal forces, deformation, and external loads. In the general method of analysis used in strength of materials the first step is to assume that the member is in equilibrium. The equations of static equilibrium are applied to the forces acting on some part of the body in order to obtain a relationship between the external forces acting on the member and the internal forces resisting the action of the external loads. Since the equations of equilibrium must be expressed in terms of forces acting external to the body, it is necessary to make the internal resisting forces into external forces. This is done by passing a plane through the body at the point of interest. The part of the body lying on one side of the cutting plane is removed and replaced by the forces it exerted on the cut section of the part of the body that remains. Since the forces acting on the "free body" hold it in equilibrium, the equations of equilibrium may be applied to the problem.

The internal resisting forces are usually expressed by the *stress*[1] acting over a certain area, so that the internal force is the integral of the stress times the differential area over which it acts. In order to evaluate this integral, it is necessary to know the distribution of the stress over the area of the cutting plane. The stress distribution is arrived at by observing and measuring the strain distribution

---

[1] For present purposes *stress* is defined as force per unit area. The companion term *strain* is defined as the change in length per unit length. More complete definitions will be given later.

in the member, since stress cannot be physically measured. However, since stress is proportional to strain for the small deformations involved in most work, the determination of the strain distribution provides the stress distribution. The expression for the stress is then substituted into the equations of equilibrium, and they are solved for stress in terms of the loads and dimensions of the member.

Important assumptions in strength of materials are that the body which is being analyzed is continuous, homogeneous, and isotropic. A *continuous body* is one which does not contain voids or empty spaces of any kind. A body is *homogeneous* if it has identical properties at all points. A body is considered to be *isotropic* with respect to some property when that property does not vary with direction or orientation. A property which varies with orientation with respect to some system of axes is said to be *anisotropic*.

While engineering materials such as steel, cast iron, and aluminum may appear to meet these conditions when viewed on a gross scale, it is readily apparent when they are viewed through a microscope that they are anything but homogeneous and isotropic. Most engineering metals are made up of more than one phase, with different mechanical properties, such that on a micro scale they are heterogeneous. Further, even a single-phase metal will usually exhibit chemical segregation, and therefore the properties will not be identical from point to point. Metals are made up of an aggregate of crystal grains having different properties in different crystallographic directions. The reason why the equations of strength of materials describe the behavior of real metals is that, in general, the crystal grains are so small that, for a specimen of any macroscopic volume, the materials are statistically homogeneous and isotropic. However, when metals are severely deformed in a particular direction, as in rolling or forging, the mechanical properties may be anisotropic on a macro scale. Other examples of anisotropic properties are fiber-reinforced composite materials and single crystals. Lack of continuity may be present in porous castings or powder metallurgy parts and, on an atomic level, at defects such as vacancies and dislocations.

## 1-3 ELASTIC AND PLASTIC BEHAVIOR

Experience shows that all solid materials can be deformed when subjected to external load. It is further found that up to certain limiting loads a solid will recover its original dimensions when the load is removed. The recovery of the original dimensions of a deformed body when the load is removed is known as *elastic behavior*. The limiting load beyond which the material no longer behaves elastically is the *elastic limit*. If the elastic limit is exceeded, the body will experience a permanent set or deformation when the load is removed. A body which is permanently deformed is said to have undergone *plastic deformation*.

For most materials, as long as the load does not exceed the elastic limit, the deformation is proportional to the load. This relationship is known as Hooke's law; it is more frequently stated as *stress is proportional to strain*. Hooke's law requires that the load-deformation relationship should be linear. However, it does not necessarily follow that all materials which behave elastically will have a linear stress-strain relationship. Rubber is an example of a material with a non-linear stress-strain relationship that still satisfies the definition of an elastic material.

Elastic deformations in metals are quite small and require very sensitive instruments for their measurement. Ultrasensitive instruments have shown that the elastic limits of metals are much lower than the values usually measured in engineering tests of materials. As the measuring devices become more sensitive, the elastic limit is decreased, so that for most metals there is only a rather narrow range of loads over which Hooke's law strictly applies. This is, however, primarily of academic importance. Hooke's law remains a quite valid relationship for engineering design.

## 1-4 AVERAGE STRESS AND STRAIN

As a starting point in the discussion of stress and strain, consider a uniform cylindrical bar which is subjected to an axial tensile load (Fig. 1-1). Assume that two gage marks are put on the surface of the bar in its unstrained state and that $L_0$ is the gage length between these marks. A load $P$ is applied to one end of the bar, and the gage length undergoes a slight increase in length and decrease in diameter. The distance between the gage marks has increased by an amount $\delta$, called the deformation. The *average linear strain e* is the ratio of the change in length to the original length.

$$e = \frac{\delta}{L_0} = \frac{\Delta L}{L_0} = \frac{L - L_0}{L_0} \qquad (1\text{-}1)$$

Strain is a dimensionless quantity since both $\delta$ and $L_0$ are expressed in units of length.

**Fig. 1-1** Cylindrical bar subjected to axial load.

**Fig. 1-2** Free-body diagram for Fig. 1-1.

Figure 1-2 shows the free-body diagram for the cylindrical bar shown in Fig. 1-1. The external load $P$ is balanced by the internal resisting force $\int \sigma \, dA$, where $\sigma$ is the stress normal to the cutting plane and $A$ is the cross-sectional area of the bar. The equilibrium equation is

$$P = \int \sigma \, dA \qquad (1\text{-}2)$$

If the stress is distributed uniformly over the area $A$, that is, if $\sigma$ is constant, Eq. (1-2) becomes

$$P = \sigma \int dA = \sigma A$$

$$\sigma = \frac{P}{A} \qquad (1\text{-}3)$$

In general, the stress will not be uniform over the area $A$, and therefore Eq. (1-3) represents an *average stress*. For the stress to be absolutely uniform, every longitudinal element in the bar would have to experience exactly the same strain, and the proportionality between stress and strain would have to be identical for each element. The inherent anisotropy between grains in a polycrystalline metal rules out the possibility of complete uniformity of stress over a body of macroscopic size. The presence of more than one phase also gives rise to nonuniformity of stress on a microscopic scale. If the bar is not straight or not centrally loaded, the strains will be different for certain longitudinal elements and the stress will not be uniform. An extreme disruption in the uniformity of the stress pattern occurs when there is an abrupt change in cross section. This results in a stress raiser or stress concentration (see Sec. 2-16).

In engineering practice, the load is usually measured in pounds and the area in square inches, so that stress has units of pounds per square inch (psi). Since it is common for engineers to deal with loads in the thousands of pounds, an obvious simplification is to work with units of 1,000 lb, called *kips*. The stress may be expressed in units of kips per square inch (ksi). (1 ksi = 1,000 psi.) In scientific work stresses are often expressed in units of kilograms per square millimeter or dynes per square centimeter. (1 kg/mm$^2$ = 9.81 × 10$^7$ dyn/cm$^2$.)

In the International System of Units (SI), an updated version of the metric system, the official unit of stress is the newton per square meter, N/m$^2$. One newton per square meter has been designated a *pascal* (Pa). In the SI system the unit of force is the newton, derived through Newton's first law from the basic units kg-m/s$^2$. Since the newton per square meter represents a very small stress (1 N/m$^2$ = 0.000145 psi), it is more common to express stress in units of meganewtons per square meter, MN/m$^2$. 1 MN/m$^2$ = 10$^6$ N/m$^2$ = 145 psi.

Below the elastic limit Hooke's law can be considered valid, so that the average stress is proportional to the average strain,

$$\frac{\sigma}{e} = E = \text{constant} \qquad (1\text{-}4)$$

The constant $E$ is the *modulus of elasticity*, or *Young's modulus*.

## 1-5  TENSILE DEFORMATION OF DUCTILE METAL

The basic data on the mechanical properties of a ductile metal are obtained from a tension test, in which a suitably designed specimen is subjected to increasing axial load until it fractures. The load and elongation are measured at frequent intervals during the test and are expressed as average stress and strain according to the equations in the previous section. (More complete details on the tension test are given in Chap. 9.)

The data obtained from the tension test are generally plotted as a stress-strain diagram. Figure 1-3 shows a typical stress-strain curve for a metal such as aluminum or copper. The initial linear portion of the curve $OA$ is the elastic region within which Hooke's law is obeyed. Point $A$ is the elastic limit, defined as the greatest stress that the metal can withstand without experiencing a permanent strain when the load is removed. The determination of the elastic limit is quite tedious, not at all routine, and dependent on the sensitivity of the strain-measuring instrument. For these reasons it is often replaced by the *proportional limit*, point $A'$. The proportional limit is the stress at which the stress-strain curve deviates from linearity. The slope of the stress-strain curve in this region is the modulus of elasticity.

For engineering purposes the limit of usable elastic behavior is described by the *yield strength*, point $B$. The yield strength is defined as the stress which will produce a small amount of permanent deformation, generally equal to a strain

**Fig. 1-3**  Typical tension stress-strain curve.

of 0.002. In Fig. 1-3 this permanent strain, or offset, is *OC*. Plastic deformation begins when the elastic limit is exceeded. As the plastic deformation of the specimen increases, the metal becomes stronger (strain hardening) so that the load required to extend the specimen increases with further straining. Eventually the load reaches a maximum value. The maximum load divided by the original area of the specimen is the *ultimate tensile strength*. For a ductile metal the diameter of the specimen begins to decrease rapidly beyond maximum load, so that the load required to continue deformation drops off until the specimen fractures. Since the average stress is based on the original area of the specimen, it also decreases from maximum load to fracture.

## 1-6 DUCTILE VS. BRITTLE BEHAVIOR

The general behavior of materials under load can be classified as ductile or brittle depending upon whether or not the material exhibits the ability to undergo plastic deformation. Figure 1-3 illustrates the tension stress-strain curve of a ductile material. A completely brittle material would fracture almost at the elastic limit (Fig. 1-4*a*), while a brittle metal, such as white cast iron, shows some slight measure of plasticity before fracture (Fig. 1-4*b*). Adequate ductility is an important engineering consideration, because it allows the material to redistribute localized stresses. When localized stresses at notches and other accidental stress concentrations do not have to be considered, it is possible to design for static situations on the basis of average stresses. However, with brittle materials, localized stresses continue to build up when there is no local yielding. Finally, a crack forms at one or more points of stress concentration, and it spreads rapidly over the section. Even if no stress concentrations are present in a brittle material, fracture will still occur suddenly because the yield stress and tensile strength are practically identical.

It is important to note that brittleness is not an absolute property of a metal. A metal such as tungsten, which is brittle at room temperature, is ductile at an elevated temperature. A metal which is brittle in tension may be ductile under

**Fig. 1-4** (*a*) Stress-strain curve for completely brittle material (ideal behavior); (*b*) stress-strain curve for brittle metal with slight amount of ductility.

hydrostatic compression. Furthermore, a metal which is ductile in tension at room temperature can become brittle in the presence of notches, low temperature, high rates of loading, or embrittling agents such as hydrogen.

## 1-7 WHAT CONSTITUTES FAILURE?

Structural members and machine elements can fail to perform their intended functions in three general ways:

*1.* Excessive elastic deformation
*2.* Yielding, or excessive plastic deformation
*3.* Fracture

An understanding of the common types of failure is important in good design because it is always necessary to relate the loads and dimensions of the member to some significant material parameter which limits the load-carrying capacity of the member. For different types of failure, different significant parameters will be important.

Two general types of excessive elastic deformation may occur: (1) excessive deflection under conditions of stable equilibrium, such as the deflection of beam under gradually applied loads; (2) sudden deflection, or *buckling*, under conditions of unstable equilibrium.

Excessive elastic deformation of a machine part can mean failure of the machine just as much as if the part completely fractured. For example, a shaft which is too flexible can cause rapid wear of the bearing, or the excessive deflection of closely mating parts can result in interference and damage to the parts. The sudden buckling type of failure may occur in a slender column when the axial load exceeds the Euler critical load or when the external pressure acting against a thin-walled shell exceeds a critical value. Failures due to excessive elastic deformation are controlled by the modulus of elasticity, not by the strength of the material. Generally, little metallurgical control can be exercised over the elastic modulus. The most effective way to increase the stiffness of a member is usually by changing its shape and increasing the dimensions of its cross section.

Yielding, or excessive plastic deformation, occurs when the elastic limit of the metal has been exceeded. Yielding produces permanent change of shape, which may prevent the part from functioning properly any longer. In a ductile metal under conditions of static loading at room temperature yielding rarely results in fracture, because the metal strain hardens as it deforms, and an increased stress is required to produce further deformation. Failure by excessive plastic deformation is controlled by the yield strength of the metal for a uniaxial condition of loading. For more complex loading conditions the yield strength is still the significant parameter, but it must be used with a suitable failure criterion (Sec. 3-4). At temperatures significantly greater than room temperature metals no longer exhibit strain hardening. Instead, metals can continuously deform at constant stress in a time-dependent yielding known as *creep*. The failure criterion under creep conditions is complicated by the fact that stress is not proportional to strain and the further fact that the mechanical properties of the material may change appreciably during service. This complex phenomenon will be considered in greater detail in Chap. 13.

The formation of a crack which can result in complete disruption of continuity of the member constitutes fracture. A part made from a ductile metal which is loaded statically rarely fractures like a tensile specimen, because it will first fail by excessive plastic deformation. However, metals fail by fracture in three general ways: (1) sudden brittle fracture; (2) fatigue, or progressive fracture; (3) delayed fracture. In the previous section it was shown that a brittle material fractures under static loads with little outward evidence of yielding. A sudden brittle type of fracture can also occur in ordinarily ductile metals under certain conditions. Plain carbon structural steel is the most common example of a material with a ductile-to-brittle transition. A change from the ductile to the brittle type of fracture is promoted by a decrease in temperature, an increase in the rate of loading, and the presence of a complex state of stress due to a notch. This problem is considered in Chap. 14.

Most fractures in machine parts are due to *fatigue*. Fatigue failures occur in parts which are subjected to alternating, or fluctuating, stresses. A minute crack starts at a localized spot, generally at a notch or stress concentration, and gradually spreads over the cross section until the member breaks. Fatigue failure occurs without any visible sign of yielding at *nominal* or average stresses that are well below the tensile strength of the metal. Fatigue failure is caused by a critical *localized* tensile stress which is very difficult to evaluate, and therefore design for fatigue failure is based primarily on empirical relationships using nominal stresses. Fatigue of metals is discussed in greater detail in Chap. 12.

One common type of delayed fracture is *stress-rupture* failure, which occurs when a metal has been statically loaded at an elevated temperature for a long period of time. Depending upon the stress and the temperature there may be no yielding prior to fracture. A similar type of delayed fracture, in which there is no warning by yielding prior to failure, occurs at room temperature when steel is statically loaded in the presence of hydrogen.

All engineering materials show a certain variability in mechanical properties, which in turn can be influenced by changes in heat treatment or fabrication. Further, uncertainties usually exist regarding the magnitude of the applied loads, and approximations are usually necessary in calculating the stresses for all but the most simple member. Allowance must be made for the possibility of accidental loads of high magnitude. Thus, in order to provide a margin of safety and to protect against failure from unpredictable causes, it is necessary that the allowable stresses be smaller than the stresses which produce failure. The value of stress for a particular material used in a particular way which is considered to be a safe stress is usually called the *working stress* $\sigma_w$. For static applications the working stress of ductile metals is usually based on the yield strength $\sigma_0$ and for brittle metals on the ultimate tensile strength $\sigma_u$. Values of working stress are established by local and federal agencies and by technical organizations such as the American Society of Mechanical Engineers (ASME). The working stress may be considered as either the yield strength or the tensile strength divided by a number called the *factor of safety.*

$$\sigma_w = \frac{\sigma_0}{N_0} \quad \text{or} \quad \sigma_w = \frac{\sigma_u}{N_u} \qquad (1\text{-}5)$$

where $\sigma_w$ = working stress, psi

$\quad \sigma_0$ = yield strength, psi

$\quad \sigma_u$ = tensile strength, psi

$\quad N_0$ = factor of safety based on yield strength

$\quad N_u$ = factor of safety based on tensile strength

The value assigned to the factor of safety depends on an estimate of all the factors discussed above. In addition, careful consideration should be given to the consequences, which would result from failure. If failure would result in loss of life, the factor of safety should be increased. The type of equipment will also influence the factor of safety. In military equipment, where light weight may be a prime consideration, the factor of safety may be lower than in commercial equipment. The factor of safety will also depend on the expected type of loading. For static loading, as in a building, the factor of safety would be lower than in a machine, which is subjected to vibration and fluctuating stresses.

## 1-8  CONCEPT OF STRESS AND THE TYPES OF STRESSES

Stress is defined as force per unit area. In Sec. 1-4 the stress was considered to be uniformly distributed over the cross-sectional area of the member. However, this is not the general case. Figure 1-5a represents a body in equilibrium under the action of external forces $P_1, P_2, \ldots, P_5$. There are two kinds of external forces which may act on a body: surface forces and body forces. Forces distributed over the surface of the body, such as hydrostatic pressure or the pressure exerted by one body on another, are called *surface forces.* Forces distributed over the volume of a body, such as gravitational forces, magnetic forces, or inertia forces (for a body in motion), are called *body forces.* The two most common types of

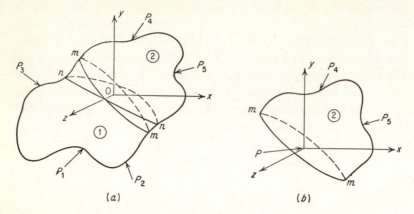

**Fig. 1-5** (a) Body in equilibrium under action of external forces $P_1, \ldots, P_5$; (b) forces acting on parts.

body forces encountered in engineering practice are centrifugal forces due to high-speed rotation and forces due to temperature differential over the body (thermal stress).

In general the force will not be uniformly distributed over any cross section of the body illustrated in Fig. 1-5a. To obtain the stress at some point $O$ in a plane such as $mm$, part 1 of the body is removed and replaced by the system of external forces on $mm$ which will retain each point in part 2 of the body in the same position as before the removal of part 1. This is the situation in Fig. 1-5b. We then take an area $\Delta A$ surrounding the point $O$ and note that a force $\Delta P$ acts on this area. If the area $\Delta A$ is continuously reduced to zero, the limiting value of the ratio $\Delta P/\Delta A$ is the stress at the point $O$ on plane $mm$ of body 2.

$$\lim_{\Delta A \to 0} \frac{\Delta P}{\Delta A} = \sigma \qquad (1\text{-}6)$$

The stress will be in the direction of the resultant force $P$ and will generally be inclined at an angle to $\Delta A$. The same stress at point $O$ in plane $mm$ would be obtained if the free body were constructed by removing part 2 of the solid body. However, the stress will be different on any other plane passing through point $O$, such as the plane $nn$.

It is inconvenient to use a stress which is inclined at some arbitrary angle to the area over which it acts. The total stress can be resolved into two components, a *normal stress* $\sigma$ perpendicular to $\Delta A$, and a *shearing stress* (or shear stress) $\tau$ lying in the plane $mm$ of the area. To illustrate this point, consider Fig. 1-6. The force $P$ makes an angle $\theta$ with the normal $z$ to the plane of the area $A$. Also, the plane

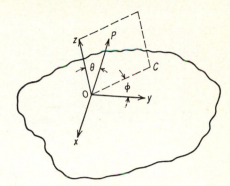

**Fig. 1-6**  Resolution of total stress into its components.

containing the normal and $P$ intersects the plane $A$ along a dashed line that makes an angle $\phi$ with the $y$ axis. The normal stress is given by

$$\sigma = \frac{P}{A} \cos \theta \qquad (1\text{-}7)$$

The shear stress in the plane acts along the line $OC$ and has the magnitude

$$\tau = \frac{P}{A} \sin \theta \qquad (1\text{-}8)$$

This shear stress may be further resolved into components parallel to the $x$ and $y$ directions lying in the plane.

$x$ direction $\qquad\qquad\qquad\qquad\qquad\qquad \tau = \frac{P}{A} \sin \theta \sin \phi \qquad (1\text{-}9)$

$y$ direction $\qquad\qquad\qquad\qquad\qquad\qquad \tau = \frac{P}{A} \sin \theta \cos \phi \qquad (1\text{-}10)$

Therefore, in general a given plane may have one normal stress and two shear stresses acting on it.

## 1-9  CONCEPT OF STRAIN AND THE TYPES OF STRAIN

In Sec. 1-4 the average linear strain was defined as the ratio of the change in length to the original length of the same dimension.

$$e = \frac{\delta}{L_0} = \frac{\Delta L}{L_0} = \frac{L - L_0}{L_0}$$

where $e$ = average linear strain
$\quad\quad \delta$ = deformation

**Fig. 1-7**  Shear strain.

By analogy with the definition of stress at a point, the strain at a point is the ratio of the deformation to the gage length as the gage length approaches zero.

Rather than referring the change in length to the original gage length, it often is more useful to define the strain as the change in linear dimension divided by the instantaneous value of the dimension.

$$\varepsilon = \int_{L_0}^{L_f} \frac{dL}{L} = \ln \frac{L_f}{L_0} \qquad (1\text{-}11)$$

The above equation defines the *natural*, or *true*, *strain*. True strain, which is useful in dealing with problems in plasticity and metal forming, will be discussed more fully in Chap. 3. For the present it should be noted that for the very small strains for which the equations of elasticity are valid the two definitions of strain give identical values.

Not only will the elastic deformation of a body result in a change in length of a linear element in the body, but it may also result in a change in the initial angle between any two lines. The angular change in a right angle is known as *shear strain*. Figure 1-7 illustrates the strain produced by the pure shear of one face of a cube. The angle at $A$, which was originally 90°, is decreased by the application of a shear stress by a small amount $\theta$. The shear strain $\gamma$ is equal to the displacement $a$ divided by the distance between the planes, $h$. The ratio $a/h$ is also the tangent of the angle through which the element has been rotated. For the small angles usually involved, the tangent of the angle and the angle (in radians) are equal. Therefore, shear strains are often expressed as angles of rotation.

$$\gamma = \frac{a}{h} = \tan \theta = \theta \qquad (1\text{-}12)$$

## BIBLIOGRAPHY

Crandall, S. H., and N. C. Dahl (eds.): "An Introduction to the Mechanics of Solids," McGraw-Hill Book Company, New York, 1959.

Drucker, D. C.: "Introduction to Mechanics of Deformable Solids," McGraw-Hill Book Company, New York, 1967.

Freudenthal, A.: "Mechanics of Solids," John Wiley & Sons, Inc., New York, 1966.

Gillam, E.: "Materials Under Stress," Butterworths & Co. (Publishers), Ltd., London, 1969.

Housner, G. W., and T. Vreeland: "The Analysis of Stress and Deformation," The Macmillan Company, New York, 1966.

Polakowski, N. H., and E. J. Ripling: "Strength and Structure of Engineering Materials," Prentice-Hall, Inc., Englewood Cliffs, N.J., 1964.

# 2
## STRESS AND STRAIN RELATIONSHIPS FOR ELASTIC BEHAVIOR

## 2-1 INTRODUCTION

The purpose of this chapter is to present the mathematical relationships for expressing the stress and strain at a point and the relationships between stress and strain in a solid which obeys Hooke's law. While part of the material covered in this chapter is a review of information generally covered in strength of materials, the subject is extended beyond this point to a consideration of stress and strain in three dimensions and to an introduction to the theory of elasticity. The material included in this chapter is important for an understanding of most of the pheno-menological aspects of mechanical metallurgy, and for this reason it should be given careful attention by those readers to whom it is unfamiliar. In the space available for this subject it has not been possible to carry it to the point where extensive problem solving is possible. The material covered here should, however, provide a background for intelligent reading of the more mathematical literature in mechanical metallurgy.

It should be recognized that the equations describing the state of stress or strain in a body are applicable to any solid continuum, whether it be an elastic or plastic solid or a viscous fluid. Indeed, this body of knowledge is often called *continuum mechanics.* The equations relating stress and strain are called *constitutive equations* because they depend on the material behavior. In this chapter we shall only consider the constitutive equations for an elastic solid.

## 2-2 DESCRIPTION OF STRESS AT A POINT

As described in Sec. 1-8, it is often convenient to resolve the stresses at a point into normal and shear components. In the general case the shear components are at arbitrary angles to the coordinate axes, so that it is convenient to resolve each shear stress further into two components. The general case is shown in Fig. 2-1. Stresses acting normal to the faces of the elemental cube are identified by the subscript which also identifies the direction in which the stress acts; that is $\sigma_x$ is the normal stress acting in the $x$ direction. Since it is a normal stress, it must act on the plane perpendicular to the $x$ direction. By convention, values of normal stresses greater than zero denote tension; values less than zero indicate compression. All the normal stresses shown in Fig. 2-1 are tensile.

Two subscripts are needed for describing shearing stresses. The first subscript indicates the plane in which the stress acts and the second the direction in which the stress acts. Since a plane is most easily defined by its normal, the first subscript refers to this normal. For example, $\tau_{yz}$ is the shear stress on the plane perpendicular to the $y$ axis in the direction of the $z$ axis. $\tau_{yx}$ is the shear stress on a plane normal to the $y$ axis in the direction of the $x$ axis.

A shear stress is positive if it points in the positive direction on the positive face of a unit cube. (It is also positive if it points in the negative direction on the

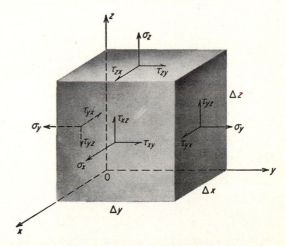

**Fig. 2-1** Stresses acting on an elemental unit cube.

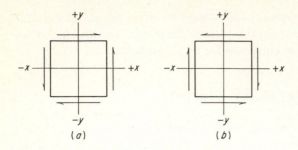

**Fig. 2-2** Sign convention for shear stress. (*a*) Positive; (*b*) negative.

negative face of a unit cube.) All of the shear stresses in Fig. 2-2*a* are positive shear stresses regardless of the type of normal stresses that are present. A shear stress is negative if it points in the negative direction of a positive face of a unit cube and vice versa. The shearing stresses shown in Fig. 2-2*b* are all negative stresses.

The notation for stress given above is the one used by Timoshenko[1] and most American workers in the field of elasticity. However, many other notations have been used, some of which are given below.

$$\sigma_x \quad \sigma_{11} \quad X_x \quad \widehat{xx} \quad p_{xx}$$
$$\sigma_y \quad \sigma_{22} \quad Y_y \quad \widehat{yy} \quad p_{yy}$$
$$\sigma_z \quad \sigma_{33} \quad Z_z \quad \widehat{zz} \quad p_{zz}$$
$$\tau_{xy} \quad \sigma_{12} \quad X_y \quad \widehat{xy} \quad p_{xy}$$
$$\tau_{yz} \quad \sigma_{23} \quad Y_z \quad \widehat{yz} \quad p_{yz}$$
$$\tau_{zx} \quad \sigma_{31} \quad Z_x \quad \widehat{zx} \quad p_{zx}$$

It can be seen from Fig. 2-1 that nine quantities must be defined in order to establish the state of stress at a point. They are $\sigma_x$, $\sigma_y$, $\sigma_z$, $\tau_{xy}$, $\tau_{xz}$, $\tau_{yx}$, $\tau_{yz}$, $\tau_{zx}$, and $\tau_{zy}$. However, some simplification is possible. If we assume that the areas of the faces of the unit cube are small enough so that the change in stress over the face is negligible, by taking the summation of the moments of the forces about the $z$ axis it can be shown that $\tau_{xy} = \tau_{yx}$.

$$(\tau_{xy} \, \Delta y \, \Delta z) \, \Delta x = (\tau_{yx} \, \Delta x \, \Delta z) \, \Delta y \qquad (2\text{-}1)$$

$$\therefore \tau_{xy} = \tau_{yx}$$

[1] S. P. Timoshenko, and J. N. Goodier, "Theory of Elasticity," 2d ed., McGraw-Hill Book Company, New York, 1951.

and in like manner

$$\tau_{xz} = \tau_{zx} \qquad \tau_{yz} = \tau_{zy}$$

Thus, the state of stress at a point is completely described by six components: three normal stresses and three shear stresses, $\sigma_x$, $\sigma_y$, $\sigma_z$, $\tau_{xy}$, $\tau_{xz}$, $\tau_{yz}$.

## 2-3 STATE OF STRESS IN TWO DIMENSIONS (PLANE STRESS)

Many problems can be simplified by considering a two-dimensional state of stress. This condition is frequently approached in practice when one of the dimensions of the body is small relative to the others. For example, in a thin plate loaded in the plane of the plate there will be no stress acting perpendicular to the surface of the plate. The stress system will consist of two normal stresses $\sigma_x$ and $\sigma_y$ and a shear stress $\tau_{xy}$. A stress condition in which the stresses are zero in one of the primary directions is called *plane stress*.

Figure 2-3 illustrates a thin plate with its thickness normal to the plane of the paper. In order to know the state of stress at point $O$ in the plate, we need to be able to describe the stress components at $O$ for any orientation of the axes through the point. To do this, consider an oblique plane normal to the plane of the paper at an angle $\theta$ between the $x$ axis and the outward normal to the oblique plane. Let the normal to this plane be the $x'$ direction and the direction lying in the oblique plane the $y'$ direction. It is assumed that the plane shown in Fig. 2-3 is an infinitesimal distance from $O$ and that the element is so small that variations in stress over the sides of the element can be neglected. The stresses acting on the oblique plane are the normal stress $\sigma$ and the shear stress $\tau$. The direction cosines between $x'$ and the $x$ and $y$ axes are $l$ and $m$, respectively. From the geometry of Fig. 2-3, $l = \cos\theta$ and $m = \sin\theta$. If $A$ is the area of the oblique plane, the areas of the sides of the element perpendicular to the $x$ and $y$ axes are $Al$ and $Am$.

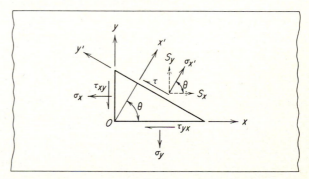

**Fig. 2-3**   Stress on oblique plane (two dimensions).

Let $S_x$ and $S_y$ denote the $x$ and $y$ components of the total stress acting on the inclined face. By taking the summation of the *forces* in the $x$ direction and the $y$ direction, we obtain

$$S_x A = \sigma_x Al + \tau_{xy} Am$$
$$S_y A = \sigma_y Am + \tau_{xy} Al$$

or

$$S_x = \sigma_x \cos \theta + \tau_{xy} \sin \theta$$
$$S_y = \sigma_y \sin \theta + \tau_{xy} \cos \theta$$

The components of $S_x$ and $S_y$ in the direction of the normal stress $\sigma$ are

$$S_{xN} = S_x \cos \theta \quad \text{and} \quad S_{yN} = S_y \sin \theta$$

so that the normal stress acting on the oblique plane is given by

$$\sigma_{x'} = S_x \cos \theta + S_y \sin \theta$$
$$\sigma_{x'} = \sigma_x \cos^2 \theta + \sigma_y \sin^2 \theta + 2\tau_{xy} \sin \theta \cos \theta \qquad (2\text{-}2)$$

The shearing stress on the oblique plane is given by

$$\tau_{x'y'} = S_y \cos \theta - S_x \sin \theta$$
$$\tau_{x'y'} = \tau_{xy}(\cos^2 \theta - \sin^2 \theta) + (\sigma_y - \sigma_x) \sin \theta \cos \theta \qquad (2\text{-}3)$$

The stress $\sigma_{y'}$ may be found by substituting $\theta + \pi/2$ for $\theta$ in Eq. (2-2), since $\sigma_{y'}$ is orthogonal to $\sigma_{x'}$.

$$\sigma_{y'} = \sigma_x \cos^2 (\theta + \pi/2) + \sigma_y \sin^2 (\theta + \pi/2) + 2\tau_{xy} \sin (\theta + \pi/2) \cos (\theta + \pi/2)$$

and since $\sin (\theta + \pi/2) = \cos \theta$ and $\cos (\theta + \pi/2) = -\sin \theta$, we obtain

$$\sigma_{y'} = \sigma_x \sin^2 \theta + \sigma_y \cos^2 \theta - 2\tau_{xy} \sin \theta \cos \theta \qquad (2\text{-}4)$$

Equations (2-2) to (2-4) are the transformation of stress equations which give the stresses in an $x'y'$ coordinate system if the stresses in an $xy$ coordinate system and the angle $\theta$ are known.

To aid in computation, it is often convenient to express Eqs. (2-2) to (2-4) in terms of the double angle $2\theta$. This can be done with the following identities:

$$\cos^2 \theta = \frac{\cos 2\theta + 1}{2}$$

$$\sin^2 \theta = \frac{1 - \cos 2\theta}{2}$$

$$2 \sin \theta \cos \theta = \sin 2\theta$$

$$\cos^2 \theta - \sin^2 \theta = \cos 2\theta$$

The transformation of stress equations now become

$$\sigma_{x'} = \frac{\sigma_x + \sigma_y}{2} + \frac{\sigma_x - \sigma_y}{2}\cos 2\theta + \tau_{xy}\sin 2\theta \qquad (2\text{-}5)$$

$$\sigma_{y'} = \frac{\sigma_x + \sigma_y}{2} - \frac{\sigma_x - \sigma_y}{2}\cos 2\theta - \tau_{xy}\sin 2\theta \qquad (2\text{-}6)$$

$$\tau_{x'y'} = \frac{\sigma_y - \sigma_x}{2}\sin 2\theta + \tau_{xy}\cos 2\theta \qquad (2\text{-}7)$$

It is important to note that $\sigma_{x'} + \sigma_{y'} = \sigma_x + \sigma_y$. Thus the sum of the normal stresses on two perpendicular planes is an *invariant* quantity, that is, it is independent of orientation or angle $\theta$.

Equations (2-2) and (2-3) and their equivalents, Eqs. (2-5) and (2-7), describe the normal stress and shear stress on any plane through a point in a body

**Fig. 2-4** Variation of normal stress and shear stress on oblique plane with angle $\theta$.

subjected to a plane-stress situation. Figure 2-4 shows the variation of normal stress and shear stress with $\theta$ for the biaxial-plane-stress situation given at the top of the figure. Note the following important facts about this figure:

1. The maximum and minimum values of normal stress on the oblique plane through point $O$ occur when the shear stress is zero.
2. The maximum and minimum values of both normal stress and shear stress occur at angles which are 90° apart.
3. The maximum shear stress occurs at an angle halfway between the maximum and minimum normal stresses.
4. The variation of normal stress and shear stress occurs in the form of a sine wave, with a period of $\theta = 180°$. These relationships are valid for any state of stress.

For any state of stress it is always possible to define a new coordinate system which has axes perpendicular to the planes on which the maximum normal stresses act and on which no shearing stresses act. These planes are called the *principal planes*, and the stresses normal to these planes are the *principal stresses*. For two-dimensional plane stress there will be two principal stresses $\sigma_1$ and $\sigma_2$ which occur at angles that are 90° apart (Fig. 2-4). For the general case of stress in three dimensions there will be three principal stresses $\sigma_1$, $\sigma_2$, and $\sigma_3$. According to convention, $\sigma_1$ is the algebraically greatest principal stress, while $\sigma_3$ is the algebraically smallest stress. The directions of the principal stresses are the *principal axes* 1, 2, and 3. Although in general the principal axes 1, 2, and 3 do not coincide with the cartesian-coordinate axes $x$, $y$, $z$, for many situations that are encountered in practice the two systems of axes coincide because of symmetry of loading and deformation. The specification of the principal stresses and their direction provides a convenient way of describing the state of stress at a point.

Since by definition a principal plane contains no shear stress, its angular relationship with respect to the $xy$ coordinate axes can be determined by finding the values of $\theta$ in Eq. (2-3) for which $\tau_{x'y'} = 0$.

$$\tau_{xy}(\cos^2 \theta - \sin^2 \theta) + (\sigma_y - \sigma_x) \sin \theta \cos \theta = 0$$

$$\frac{\tau_{xy}}{\sigma_x - \sigma_y} = \frac{\sin \theta \cos \theta}{\cos^2 \theta - \sin^2 \theta} = \frac{\frac{1}{2}(\sin 2\theta)}{\cos 2\theta} = \frac{1}{2} \tan 2\theta$$

$$\tan 2\theta = \frac{2\tau_{xy}}{\sigma_x - \sigma_y} \qquad (2\text{-}8)$$

Since $\tan 2\theta = \tan (\pi + 2\theta)$, Eq. (2-8) has two roots, $\theta_1$ and $\theta_2 = \theta_1 + n\pi/2$. These roots define two mutually perpendicular planes which are free from shear.

Equation (2-5) will give the principal stresses when values of $\cos 2\theta$ and $\sin 2\theta$ are substituted into it from Eq. (2-8). The values of $\cos 2\theta$ and $\sin 2\theta$ are

**Fig. 2-5** Method of establishing direction of $\sigma_1$.

found from Eq. (2-8) by means of the pythagorean relationships.

$$\sin 2\theta = \pm \frac{\tau_{xy}}{[(\sigma_x - \sigma_y)^2/4 + \tau_{xy}^2]^{\frac{1}{2}}}$$

$$\cos 2\theta = \pm \frac{(\sigma_x - \sigma_y)/2}{[(\sigma_x - \sigma_y)^2/4 + \tau_{xy}^2]^{\frac{1}{2}}}$$

Substituting these values into Eq. (2-5) results in the expression for the maximum and minimum principal stresses for a two-dimensional (biaxial) state of stress.

$$\left. \begin{array}{l} \sigma_{max} = \sigma_1 \\ \sigma_{min} = \sigma_2 \end{array} \right\} = \frac{\sigma_x + \sigma_y}{2} \pm \left[ \left( \frac{\sigma_x - \sigma_y}{2} \right)^2 + \tau_{xy}^2 \right]^{\frac{1}{2}} \qquad (2\text{-}9)$$

The direction of the principal planes is found by solving for $\theta$ in Eq. (2-8), taking special care to establish whether $2\theta$ is between 0 and $\pi/2$, $\pi$, and $3\pi/2$, etc. Figure 2-5 shows a simple way to establish the direction of the largest principal stress $\sigma_1$. $\sigma_1$ will lie between the algebraically largest normal stress and the shear diagonal. To see this intuitively, consider that if there were no shear stresses, then $\sigma_x = \sigma_1$. If only shear stresses act, then a normal stress (the principal stress) would exist along the shear diagonal. If both normal and shear stresses act on the element, then $\sigma_1$ lies between the influences of these two effects.

To find the maximum shear stress we return to Eq. (2-7). We differentiate the expression for $\tau_{x'y'}$ and set this equal to zero.

$$\frac{d\tau_{x'y'}}{d\theta} = (\sigma_y - \sigma_x) \cos 2\theta - 2\tau_{xy} \sin 2\theta = 0$$

$$\tan 2\theta_s = \frac{\sigma_y - \sigma_x}{2\tau_{xy}} = -\frac{\sigma_x - \sigma_y}{2\tau_{xy}} \qquad (2\text{-}10)$$

Comparing this with the angle at which the principal planes occur, Eq. (2-8), $\tan 2\theta_n = 2\tau_{xy}/(\sigma_x - \sigma_y)$, we see that $\tan 2\theta_s$ is the negative reciprocal of $\tan 2\theta_n$. This means that $2\theta_s$ and $2\theta_n$ are orthogonal, and that $\theta_s$ and $\theta_n$ are separated in space by 45°. The magnitude of the maximum shear stress is found by substituting Eq. (2-10) into Eq. (2-7).

$$\tau_{max} = \pm \left[ \left( \frac{\sigma_x - \sigma_y}{2} \right)^2 + \tau_{xy}^{\,2} \right]^{\frac{1}{2}} \qquad (2\text{-}11)$$

## 2-4 MOHR'S CIRCLE OF STRESS—TWO DIMENSIONS

A very useful graphical method for representing the state of stress at a point on an oblique plane through the point was suggested by O. Mohr. The transformation of stress equations, Eqs. (2-5) and (2-7), can be rearranged to give

$$\sigma_{x'} - \frac{\sigma_x + \sigma_y}{2} = \frac{\sigma_x - \sigma_y}{2} \cos 2\theta + \tau_{xy} \sin 2\theta$$

$$\tau_{y'x'} = \frac{\sigma_y - \sigma_x}{2} \sin 2\theta + \tau_{xy} \cos 2\theta$$

We can solve for $\sigma_{x'}$ in terms of $\tau_{x'y'}$ by squaring each of these equations and adding

$$\left( \sigma_{x'} - \frac{\sigma_x + \sigma_y}{2} \right)^2 + \tau_{x'y'}^{\,2} = \left( \frac{\sigma_x - \sigma_y}{2} \right)^2 + \tau_{xy}^{\,2} \qquad (2\text{-}12)$$

Equation (2-12) is the equation of a circle of the form $(x - h)^2 + y^2 = r^2$. Thus, Mohr's circle is a circle in $\sigma_{x'}$, $\tau_{x'y'}$ coordinates with a radius equal to $\tau_{max}$ and the center displaced $(\sigma_x + \sigma_y)/2$ to the right of the origin.

In working with Mohr's circle there are only a few basic rules to remember. An angle of $\theta$ on the physical element is represented by $2\theta$ on Mohr's circle. The same sense of rotation (clockwise or counterclockwise) should be used in each case. A different convention to express shear stress is used in drawing and interpreting Mohr's circle. This convention is that a shear stress causing a clockwise rotation about any point in the physical element is plotted above the horizontal axis of the Mohr's circle. A point on Mohr's circle gives the magnitude and direction of the normal and shear stresses on any plane in the physical element.

Figure 2-6a illustrates the plotting and use of Mohr's circle for a particular stress state shown at the upper left. Normal stresses are plotted along the x axis, shear stresses along the y axis. The stresses on the planes normal to the x and y axes are plotted as points A and B. The intersection of the line AB with the σ axis determines the center of the circle. At points D and E the shear stress is zero, so these points represent the values of the principal stresses. The angle between $\sigma_x$ and $\sigma_1$ on Mohr's circle is $2\theta$. Since this angle is measured counterclockwise on Mohr's circle on the physical element, $\sigma_1$ acts counterclockwise from the x axis at an angle $\theta$ (see sketch, upper right). The stresses on any other plane whose

**Fig. 2-6** (*a*) Mohr's circle for two-dimensional state of stress; (*b*) Mohr's circle using pole for normals.

normal makes an angle of $\theta$ with the $x$ axis could be found from Mohr's circle in the same way.

A particularly simple method[1] of determining the stresses on any plane from Mohr's circle is by determining the point called the *pole for normals*. The stresses on the $x$ and $y$ planes and the greatest principal plane are shown again in Fig. 2-6*b*. The pole is determined by the line through point $A$ parallel to the $x$ axis and the line through point $B$ parallel to the $y$ axis of the physical element. Note that a line parallel to $\sigma_1$ is now inclined to the $x$ axis at an angle $\theta$ instead of $2\theta$. This is so because angular rotation about the pole is half the angular rotation about the center of the circle. Therefore, rotation about the pole of normals is exactly the angular rotation on the physical element. For example, if we want to

[1] D. C. Drucker, "Introduction to Mechanics of Deformable Solids," pp. 226–228, McGraw-Hill Book Company, New York, 1967.

know the stresses on a plane whose *normal* is 30° counterclockwise from $\sigma_1$, we have only to draw the line from the pole of normals and scale off the normal and shear stresses acting on that plane.

## 2-5 STATE OF STRESS IN THREE DIMENSIONS

The general three-dimensional state of stress consists of three unequal principal stresses acting at a point. This is called a *triaxial state of stress*. If two of the three principal stresses are equal, the state of stress is known as *cylindrical*, while if all three principal stresses are equal, the state of stress is said to be *hydrostatic*, or *spherical*.

The determinaton of the principal stresses for a three-dimensional state of stress in terms of the stresses acting on an arbitrary cartesian-coordinate system is an extension of the method described in Sec. 2-3 for the two-dimensional case. Figure 2-7 represents an elemental free body similar to that shown in Fig. 2-1 with a diagonal plane *JKL* of area *A*. The plane *JKL* is assumed to be a principal plane cutting through the unit cube. $\sigma$ is the principal stress acting normal to the plane *JKL*. Let *l, m, n* be the direction cosines of $\sigma$, that is, the cosines of the angles between $\sigma$ and the *x, y,* and *z* axes. Since the free body in Fig. 2-7 must be in equilibrium, the forces acting on each of its faces must balance. The components of $\sigma$ along each of the axes are $S_x$, $S_y$, and $S_z$.

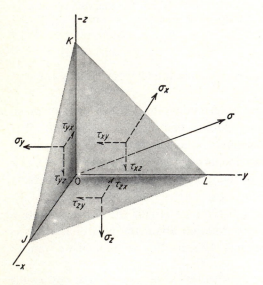

**Fig. 2-7** Stresses acting on elemental free body.

$$S_x = \sigma l \qquad\qquad S_y = \sigma m \qquad\qquad S_z = \sigma n$$

$$\text{Area } KOL = Al \qquad \text{Area } JOK = Am \qquad \text{Area } JOL = An$$

Taking the summation of the forces in the $x$ direction results in

$$\sigma Al - \sigma_x Al - \tau_{yx} Am - \tau_{zx} An = 0$$

which reduces to

$$(\sigma - \sigma_x)l - \tau_{yx} m - \tau_{zx} n = 0 \qquad (2\text{-}13a)$$

Summing the forces along the other two axes results in

$$-\tau_{xy} l + (\sigma - \sigma_y)m - \tau_{zy} n = 0 \qquad (2\text{-}13b)$$

$$-\tau_{xz} l - \tau_{yz} m + (\sigma - \sigma_z)n = 0 \qquad (2.13c)$$

Equations (2-13) are three homogeneous linear equations in terms of $l$, $m$, and $n$. The only nontrivial solution can be obtained by setting the determinant of the coefficients of $l$, $m$, and $n$ equal to zero, since $l$, $m$, and $n$ cannot all be zero.

$$\begin{vmatrix} \sigma - \sigma_x & -\tau_{yx} & -\tau_{zx} \\ -\tau_{xy} & \sigma - \sigma_y & -\tau_{zy} \\ -\tau_{xz} & -\tau_{yz} & \sigma - \sigma_z \end{vmatrix} = 0$$

Solution of the determinant results in a cubic equation in $\sigma$.

$$\sigma^3 - (\sigma_x + \sigma_y + \sigma_z)\sigma^2 + (\sigma_x \sigma_y + \sigma_y \sigma_z + \sigma_x \sigma_z - \tau_{xy}^2 - \tau_{yz}^2 - \tau_{xz}^2)\sigma$$

$$- (\sigma_x \sigma_y \sigma_z + 2\tau_{xy} \tau_{yz} \tau_{xz} - \sigma_x \tau_{yz}^2 - \sigma_y \tau_{xz}^2 - \sigma_z \tau_{xy}^2) = 0 \qquad (2\text{-}14)$$

The three roots of Eq. (2-14) are the three principal stresses $\sigma_1$, $\sigma_2$, and $\sigma_3$. To determine the direction, with respect to the original $x$, $y$, $z$ axes, in which the principal stresses act, it is necessary to substitute, $\sigma_1$, $\sigma_2$, and $\sigma_3$ each in turn into the three equations of Eq. (2.13). The resulting equations must be solved simultaneously for $l$, $m$, and $n$ with the help of the auxiliary relationship $l^2 + m^2 + n^2 = 1$.

Note that there are three combinations of stress components in Eq. (2-14) that make up the coefficients of the cubic equation. Since the values of these coefficients determine the principal stresses, they obviously do not vary with changes in the coordinate axes. Therefore, they are invariant coefficients.

$$\sigma_x + \sigma_y + \sigma_z = I_1$$

$$\sigma_x \sigma_y + \sigma_y \sigma_z + \sigma_x \sigma_z - \tau_{xz}^2 - \tau_{xz}^2 - \tau_{yz}^2 = I_2$$

$$\sigma_x \sigma_y \sigma_z + 2\tau_{xy} \tau_{yz} \tau_{xz} - \sigma_x \tau_{yz}^2 - \sigma_y \tau_{xz}^2 - \sigma_z \tau_{xy}^2 = I_3$$

The first invariant of stress $I_1$ has been seen before for the two-dimensional state of stress. It states the useful relationship that the sum of the normal stresses for any orientation in the coordinate system is equal to the sum of the normal stresses for any other orientation. For example

$$\sigma_x + \sigma_y + \sigma_z = \sigma_{x'} + \sigma_{y'} + \sigma_{z'} = \sigma_1 + \sigma_2 + \sigma_3 \qquad (2\text{-}15)$$

In the discussion above we developed the equation for the stress on a particular oblique plane, a principal plane in which there is no shear stress. Let us now develop the equations for the normal and shear stress on *any* oblique plane whose normal has the direction cosines *l*, *m*, *n* with the *x*, *y*, *z* axes. We can use Fig. 2-7 once again if we realize that for this general situation the total stress on the plane *S* will not be coaxial with the normal stress, and that $S^2 = \sigma^2 + \tau^2$. Once again the total stress can be resolved into components $S_x$, $S_y$, $S_z$, so that

$$S^2 = S_x^2 + S_y^2 + S_z^2 \qquad (2\text{-}16)$$

Taking the summation of the *forces* in the *x*, *y*, and *z* directions, we arrive at the expressions for the orthogonal components of the total stress:

$$S_x = \sigma_x l + \tau_{yx} m + \tau_{zx} n \qquad (2\text{-}17a)$$
$$S_y = \tau_{xy} l + \sigma_y m + \tau_{zy} n \qquad (2\text{-}17b)$$
$$S_z = \tau_{xz} l + \tau_{yz} m + \sigma_z n \qquad (2\text{-}17c)$$

To find the normal stress $\sigma$ on the oblique plane, it is necessary to determine the components of $S_x$, $S_y$, $S_z$ in the direction of the normal to the oblique plane. Thus,

$$\sigma = S_x l + S_y m + S_z n$$

or, after substituting from Eqs. (2-17) and simplifying with $\tau_{xy} = \tau_{yx}$, etc.

$$\sigma = \sigma_x l^2 + \sigma_y m^2 + \sigma_z n^2 + 2\tau_{xy} lm + 2\tau_{yz} mn + 2\tau_{zx} nl \qquad (2\text{-}18)$$

The magnitude of the shear stress on the oblique plane can be found from $\tau^2 = S^2 - \sigma^2$. To get the magnitude and direction of the two shear stress components lying in the oblique plane it is necessary to resolve the stress components $S_x$, $S_y$, $S_z$ into the $y'$ and $z'$ directions lying in the oblique plane.[1] This development will not be carried out here because the pertinent equations can be derived more easily by the methods given in Sec. 2-6.

Since plastic flow involves shearing stresses, it is important to identify the planes on which the *maximum* or *principal shear stresses* occur. In our discussion of the two-dimensional state of stress we saw that $\tau_{max}$ occurred on a plane halfway between the two principal planes. Therefore it is easiest to define the principal shear planes in terms of the three principal axes 1, 2, 3. From $\tau^2 = S^2 - \sigma^2$ it can be shown that

$$\tau^2 = (\sigma_1 - \sigma_2)^2 l^2 m^2 + (\sigma_1 - \sigma_3)^2 l^2 n^2 + (\sigma_2 - \sigma_3)^2 m^2 n^2 \qquad (2\text{-}19)$$

where *l*, *m*, *n* are the direction cosines between the normal to the oblique plane and the principal axes.

[1] P. C. Chou and N. J. Pagano, "Elasticity," p. 24, D. Van Nostrand Company, Inc., Princeton, N.J., 1967.

The principal shear stresses occur for the following combinations of direction cosines that bisect the angle between two of the three principal axes:

$$
\begin{array}{cccc}
l & m & n & \tau \\[4pt]
0 & \pm\sqrt{\dfrac{1}{2}} & \pm\sqrt{\dfrac{1}{2}} & \tau_1 = \dfrac{\sigma_2 - \sigma_3}{2} \\[12pt]
\pm\sqrt{\dfrac{1}{2}} & 0 & \pm\sqrt{\dfrac{1}{2}} & \tau_2 = \dfrac{\sigma_1 - \sigma_3}{2} \\[12pt]
\pm\sqrt{\dfrac{1}{2}} & \pm\sqrt{\dfrac{1}{2}} & 0 & \tau_3 = \dfrac{\sigma_1 - \sigma_2}{2}
\end{array}
\tag{2-20}
$$

Since according to convention $\sigma_1$ is the algebraically greatest principal normal stress and $\sigma_3$ is the algebraically smallest principal stress, $\tau_2$ has the largest value of shear stress and it is called the *maximum shear stress* $\tau_{max}$.

$$
\tau_{max} = \frac{\sigma_1 - \sigma_3}{2}
\tag{2-21}
$$

The maximum shear stress is important in theories of yielding and metal-forming operations. Figure 2-8 shows the planes of the principal shear stresses for a cube

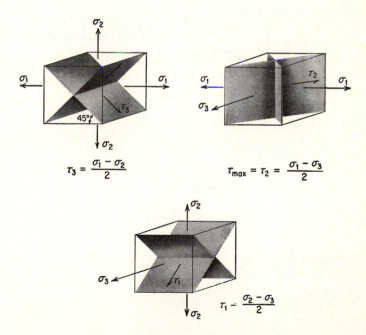

$$
\tau_3 = \frac{\sigma_1 - \sigma_2}{2}
$$

$$
\tau_{max} = \tau_2 = \frac{\sigma_1 - \sigma_3}{2}
$$

$$
\tau_1 = \frac{\sigma_2 - \sigma_3}{2}
$$

**Fig. 2-8** Planes of principal shear stresses.

whose faces are the principal planes. Note that for each pair of principal stresses there are two planes of principal shear stress, which bisect the directions of the principal stresses.

## 2-6   STRESS TENSOR

Many aspects of the analysis of stress, such as the equations for the transformation of the stress components from one set of coordinate axes to another coordinate system or the existence of principal stresses, become simpler when it is realized that stress is a second-rank tensor quality. Many of the techniques for manipulating second-rank tensors do not require a deep understanding of tensor calculus, so it is advantageous to learn something about the properties of tensors.

We shall start with the consideration of the transformation of a vector (a first-rank tensor) from one coordinate system to another. Consider the vector $\mathbf{S} = S_1 i_1 + S_2 i_2 + S_3 i_3$, when the unit vectors $i_1, i_2, i_3$ are in the directions $x_1, x_2, x_3$. (In accordance with convention and convenience in working with tensor quantities, the coordinate axes will be designated $x_1, x_2$, etc., where $x_1$ is equivalent to our previous designation $x$, $x_2$ is equivalent to the old $y$, etc.) $S_1, S_2, S_3$ are the components of $\mathbf{S}$ referred to the axes $x_1, x_2, x_3$. We now want to find the components of $\mathbf{S}$ referred to the $x_1', x_2', x_3'$ axes, Fig. 2-9. $S_1'$ is obtained by resolving $S_1, S_2, S_3$ along the new direction $x_1'$.

$$S_1' = S_1 \cos (x_1 x_1') + S_2 \cos (x_2 x_1') + S_3 \cos (x_3 x_1')$$

or
$$S_1' = a_{11} S_1 + a_{12} S_2 + a_{13} S_3 \tag{2-22a}$$

where $a_{11}$ is the direction cosine between $x_1'$ and $x_1$, $a_{12}$ is the direction cosine between $x_1'$ and $x_2$, etc. Similarly,

$$S_2' = a_{21} S_1 + a_{22} S_2 + a_{23} S_3 \tag{2-22b}$$
$$S_3' = a_{31} S_1 + a_{32} S_2 + a_{33} S_3 \tag{2-22c}$$

Fig. 2-9   Transformation of axes for a vector.

We note that the leading suffix for each direction cosine in each equation is the same, so we could write these equations as

$$S_1' = \sum_{j=1}^{3} a_{1j} S_j \qquad S_2' = \sum_{j=1}^{3} a_{2j} S_j \qquad S_3' = \sum_{j=1}^{3} a_{3j} S_j$$

These three equations could be combined by writing

$$S_i' = \sum_{j=1}^{3} a_{ij} S_j (i = 1, 2, 3) = a_{i1} S_1 + a_{i2} S_2 + a_{i3} S_3 \qquad (2\text{-}23)$$

Still greater brevity is obtained by writing Eq. (2-23) in the Einstein suffix notation

$$S_i' = a_{ij} A_j \qquad (2\text{-}24)$$

The suffix notation is a very useful way of compactly expressing the systems of equations usually found in continuum mechanics. In Eq. (2-24) it is understood that when a suffix occurs twice in the same term (in this case the suffix $j$), it indicates *summation* with respect to that suffix. Unless otherwise indicated, the summation of the other index is from 1 to 3.

In the above example, $i$ is a *free suffix* and it is understood that in the expanded form there is one equation for each value of $i$. The repeated index is called a *dummy suffix*. Its only purpose is to indicate summation. Exactly the same three equations would be produced if some other letter were used for the dummy suffix, for example, $S_i' = a_{ir} S_r$ would mean the same thing as Eq. (2-24).

We saw in Sec. 2-5 that the complete determination of the state of stress at a point in a solid requires the specification of nine components of stress on the orthogonal faces of the element at the point. A vector quantity only requires the specification of three components. Obviously, stress is more complicated than a vector. Physical quantities that transform with coordinate axes in the manner of Eq. (2-18) are called *tensors* of the *second rank*. Stress, strain, and many other physical quantities are second-rank tensors. A scalar quantity, which remains unchanged with transformation of axes, requires only a single number for its specification. Scalars are tensors of zero rank. Vector quantities require three components for their specification, so they are tensors of the first rank. The number of components required to specify a quantity is $3^n$, where $n$ is the rank of the tensor.[1] The elastic constant that relates stress with strain in an elastic solid is a fourth-rank tensor with 81 components in the general case.

The product of two vectors **A** and **B** having components $(A_1, A_2, A_3)$ and $(B_1, B_2, B_3)$ results in a second-rank tensor $T_{ij}$. The components of this tensor can be displayed as a $3 \times 3$ matrix.

---

[1] A more precise relationship is $N = k^n$, where $N$ is the number of components required for the description of a tensor of the $n$th rank in a $k$-dimensional space. For a two dimensional space only four components are required to describe a second-rank tensor.

$$T_{ij} = \begin{vmatrix} T_{11} & T_{12} & T_{13} \\ T_{21} & T_{22} & T_{23} \\ T_{31} & T_{32} & T_{33} \end{vmatrix} = \begin{vmatrix} A_1 B_1 & A_1 B_2 & A_1 B_3 \\ A_2 B_1 & A_2 B_2 & A_2 B_3 \\ A_3 B_1 & A_3 B_2 & A_3 B_3 \end{vmatrix}$$

On transformation of axes the vector components become $(A'_1, A'_2, A'_3)$ and $(B'_1, B'_2, B'_3)$. We wish to find the relationship between the nine components of $T_{ij}$ and the nine components of $T'_{ij}$ after the transformation of axes.

$$A'_i = a_{ij} A_j \qquad B'_k = a_{kl} B_l$$
$$A'_i B'_k = (a_{ij} A_j)(a_{kl} B_l)$$

or
$$T'_{ik} = a_{ij} a_{kl} T_{jl} \tag{2-25}$$

Since stress is a second-rank tensor, the components of the stress tensor can be written as

$$\sigma_{ij} = \begin{vmatrix} \sigma_{11} & \sigma_{12} & \sigma_{13} \\ \sigma_{21} & \sigma_{22} & \sigma_{23} \\ \sigma_{31} & \sigma_{32} & \sigma_{33} \end{vmatrix} = \begin{vmatrix} \sigma_x & \tau_{xy} & \tau_{xz} \\ \tau_{yx} & \sigma_y & \tau_{yz} \\ \tau_{zx} & \tau_{zy} & \sigma_z \end{vmatrix}$$

The transformation of the stress tensor $\sigma_{ij}$ from the $x_1, x_2, x_3$ system of axes to the $x'_1, x'_2, x'_3$ axes is given by

$$\sigma_{kl} = a_{ki} a_{lj} \sigma_{ij} \tag{2-26}$$

where $i$ and $j$ are dummy suffixes and $k$ and $l$ are free suffixes. To expand the tensor equation, we first sum over $j = 1, 2, 3$.

$$\sigma_{kl} = a_{ki} a_{l1} \sigma_{i1} + a_{ki} a_{l2} \sigma_{i2} + a_{ki} a_{l3} \sigma_{i3}$$

Now summing over $i = 1, 2, 3$

$$\begin{aligned} \sigma_{kl} = &\, a_{k1} a_{l1} \sigma_{11} + a_{k1} a_{l2} \sigma_{12} + a_{k1} a_{l3} \sigma_{13} \\ &+ a_{k2} a_{l1} \sigma_{21} + a_{k2} a_{l2} \sigma_{22} + a_{k2} a_{l3} \sigma_{23} \\ &+ a_{k3} a_{l1} \sigma_{31} + a_{k3} a_{l2} \sigma_{32} + a_{k3} a_{l3} \sigma_{33} \end{aligned} \tag{2-27}$$

For each value of $k$ and $l$ there will be an equation similar to (2-27). Thus, to find the equation for the normal stress in the $x'_1$ direction, let $k = 1$ and $l = 1$

$$\begin{aligned} \sigma_{11} = &\, a_{11} a_{11} \sigma_{11} + a_{11} a_{12} \sigma_{12} + a_{11} a_{13} \sigma_{13} \\ &+ a_{12} a_{11} \sigma_{21} + a_{12} a_{12} \sigma_{22} + a_{12} a_{13} \sigma_{23} \\ &+ a_{13} a_{11} \sigma_{31} + a_{13} a_{12} \sigma_{32} + a_{13} a_{13} \sigma_{33} \end{aligned}$$

The reader should verify that this reduces to Eq. (2-18) when recast in the symbolism of Sec. 2-5.

Similarly, if we want to determine the shear stress on the $x'$ plane in the $z'$ direction, that is, $\tau_{x'z'}$, let $k = 1$ and $l = 3$

$$\begin{aligned} \sigma_{13} = &\, a_{11} a_{31} \sigma_{11} + a_{11} a_{32} \sigma_{12} + a_{11} a_{33} \sigma_{13} \\ &+ a_{12} a_{31} \sigma_{21} + a_{12} a_{32} \sigma_{22} + a_{12} a_{33} \sigma_{23} \\ &+ a_{13} a_{31} \sigma_{31} + a_{13} a_{32} \sigma_{32} + a_{13} a_{33} \sigma_{33} \end{aligned}$$

It is perhaps worth emphasizing again that it is immaterial what letters are used for subscripts in tensor notation. Thus, the transformation of a second-rank tensor could just as well be written as $T'_{st} = a_{sp} a_{tq} T_{pq}$, where $T_{pq}$ are the components in the original unprimed axes and $T'_{st}$ are the components referred to the new primed axes. The transformation law for a third-rank tensor is written as

$$T'_{stv} = a_{sp} a_{tq} a_{vr} T_{pqr}$$

The material presented so far in this section is really little more than tensor notation. However, even with the minimal topics that have been discussed we have gained a powerful shorthand method for writing the often unwieldy equations of continuum mechanics. (The student will find that this will greatly ease the problem of remembering equations.) We have also gained a useful technique for transforming a tensor quantity from one set of axes to another. There are only a few additional facts about tensors that we need to consider. The student interested in pursuing this topic further is referred to a number of applications-oriented texts on cartesian tensors.[1]

A useful quantity in tensor theory is the Kronecker delta $\delta_{ij}$. The Kronecker delta is a second-rank unit isotropic tensor, that is, it has identical components in any coordinate system.

$$\delta_{ij} = \begin{vmatrix} 1 & 0 & 0 \\ 0 & 1 & 0 \\ 0 & 0 & 1 \end{vmatrix} = \begin{cases} 1 & i = j \\ 0 & i \neq j \end{cases} \qquad (2\text{-}28)$$

Multiplication of a tensor or products of tensors by $\delta_{ij}$ result in a reduction of two in the rank of the tensor. This is called *contraction* of the tensor. The rule is stated here without proof, but examples are given so we can make use of this operation in subsequent discussions. Consider the product of two second-rank tensors $A_{pq} B_{vw}$. This multiplication would produce a fourth-rank tensor, nine equations each with nine terms. If we multiply the product by $\delta_{qw}$, it is reduced to a second-rank tensor.

$$A_{pq} B_{vw} \delta_{qw} = A_{pq} B_{vq}$$

The "rule" is, replace $w$ by $q$ and drop $\delta_{qw}$. The process of contraction can be repeated several times. Thus, $A_{pq} B_{vw} \delta_{qw} \delta_{pv}$ reduces to $A_{pq} B_{vq} \delta_{pv}$ on the first contraction, and then to $A_{pq} B_{pq}$, which is a zero-rank tensor (scalar).

If we apply contraction to the second-rank stress tensor

$$\sigma_{ij} \delta_{ij} = \sigma_{ii} = \sigma_{11} + \sigma_{22} + \sigma_{23} = I_1$$

we obtain the first invariant of the tensor (a scalar).

[1] L. G. Jaeger, "Cartesian Tensors in Engineering Science," Pergamon Press, New York, 1966.

The invariants of the stress tensor may be determined readily from the matrix of its components. Since $\sigma_{12} = \sigma_{21}$, etc., the stress tensor is a *symmetric tensor*.

$$\sigma_{ij} = \begin{vmatrix} \sigma_{11} & \sigma_{12} & \sigma_{13} \\ \sigma_{12} & \sigma_{22} & \sigma_{23} \\ \sigma_{13} & \sigma_{23} & \sigma_{33} \end{vmatrix}$$

The first invariant is the trace of the matrix, i.e., the sum of the main diagonal terms

$$I_1 = \sigma_{11} + \sigma_{22} + \sigma_{33}$$

The second invariant is the sum of the principal minors. A minor of an element of a matrix is the determinant of the next lower order which remains when the row and column in which the element stands are suppressed. Thus, taking each of the principal (main diagonal) terms in order and suppressing that row and column we have

$$I_2 = \begin{vmatrix} \sigma_{22} & \sigma_{23} \\ \sigma_{23} & \sigma_{33} \end{vmatrix} + \begin{vmatrix} \sigma_{11} & \sigma_{13} \\ \sigma_{13} & \sigma_{33} \end{vmatrix} + \begin{vmatrix} \sigma_{11} & \sigma_{12} \\ \sigma_{12} & \sigma_{22} \end{vmatrix}$$

Finally, the third invariant is the determinant of the entire matrix of the components of the stress tensor.

As an example of the advantages of compression and concept that are provided by tensor notation we shall derive again the equations for principal stress that were developed in Sec. 2-5. The reader is warned that it is easy to lose the physical significance in the mathematical manipulation. It is a basic theorem of tensor theory that there is some orientation of the coordinate axes such that the components of a symmetric tensor of rank 2 will all be equal to zero for $i \neq j$. This is equivalent to stating that the concepts of principal stress and principal axes are inherent in the tensor character of stress.

The three force summation equations, Eqs. (2-17), can be written as

$$\sigma_{nj} = a_{ni}\sigma_{ij} \qquad (2\text{-}29)$$

where the suffix $n$ is used to denote that we are dealing with the angles to the normal of an oblique plane. If we let the oblique plane be a principal plane and let the normal stress on it be $\sigma_p$, then we can write

$$\sigma_{nj} = a_{pj}\sigma_p \qquad (2\text{-}30)$$

Combining Eqs. (2-29) and (2-30)

$$(a_{ni}\sigma_{ij} - a_{pj}\sigma_p) = 0 \qquad (2\text{-}31)$$

But, $a_{pj} = a_{pi}\delta_{ji}$ (replace $i$ by $j$ and drop $\delta_{ji}$)

$$a_{ni}\sigma_{ij} - \sigma_p a_{pi}\delta_{ji} = 0$$

However, $a_{ni} = a_{pi}$, since the principal stress lies in the direction of the normal to the oblique plane, so

$$(\sigma_{ij} - \sigma_p\delta_{ji})a_{pi} = 0 \qquad (2\text{-}32)$$

Expanding Eq. (2-32) will give the three equations (2-13), since $a_{p1} = l$, $a_{p2} = m$, etc., and $\delta_{ji} = 0$ when $j \neq i$. For Eq. (2-32) to have a nontrivial solution in $a_{pi}$ the determinant of the coefficients must vanish, resulting in

$$|\sigma_{ij} - \sigma_p \delta_{ji}| = \begin{vmatrix} \sigma_x - \sigma_p & \tau_{xy} & \tau_{xz} \\ \tau_{yx} & \sigma_y - \sigma_p & \tau_{yz} \\ \tau_{zx} & \tau_{zy} & \sigma_z - \sigma_p \end{vmatrix} = 0$$

which yields the cubic equation Eq. (2-14). The coefficients of this equation in tensor notation are

$$I_1 = \sigma_{ii}$$
$$I_2 = \tfrac{1}{2}(\sigma_{ik}\sigma_{ki} - \sigma_{ii}\sigma_{kk})$$
$$I_3 = \tfrac{1}{6}(2\sigma_{ij}\sigma_{jk}\sigma_{ki} - 3\sigma_{ij}\sigma_{ji}\sigma_{kk} + \sigma_{ii}\sigma_{jj}\sigma_{kk})$$

The fact that only dummy subscripts appear in these equations indicates the scalar nature of the invariants of the stress tensor.

## 2-7 MOHR'S CIRCLE—THREE DIMENSIONS

The discussion given in Sec. 2-4 of the representation of a two-dimensional state of stress by means of Mohr's circle can be extended to three dimensions. Figure 2-10 shows how a triaxial state of stress, defined by the three principal stresses, can be represented by three Mohr's circles. It can be shown[1] that all possible stress conditions within the body fall within the shaded area between the circles in Fig. 2-10.

**Fig. 2-10**   Mohr's circle representation of a three-dimensional state of stress.

[1] A. Nadai, "Theory of Flow and Fracture of Solids," 2d ed., pp. 96–98, McGraw-Hill Book Company, New York, 1950.

While the only physical significance of Mohr's circle is that it gives a geometrical representation of the equations that express the transformation of stress components to different sets of axes, it is a very convenient way of visualizing the state of stress. Figure 2-11 shows Mohr's circle for a number of common states of stress. Note that the application of a tensile stress $\sigma_2$ at right angles to an existing tensile stress $\sigma_1$ (Fig. 2-11c) results in a decrease in the principal shear stress on

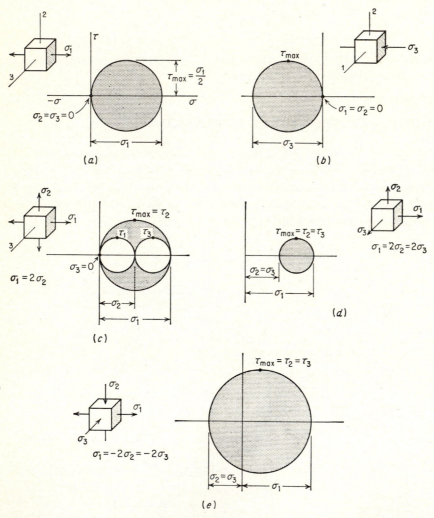

**Fig. 2-11** Mohr's circles (three-dimensional) for various states of stress. (a) Uniaxial tension; (b) uniaxial compression; (c) biaxial tension; (d) triaxial tension (unequal); (e) uniaxial tension plus biaxial compression.

two of the three sets of planes on which a principal shear stress acts. However, the maximum shear stress is not decreased from what it would be for uniaxial tension, although if only the two-dimensional Mohr's circle had been used, this would not have been apparent. If a tensile stress is applied in the third principal direction (Fig. 2-11d), the maximum shear stress is reduced appreciably. For the limiting case of equal triaxial tension (hydrostatic tension) Mohr's circle reduces to a point, and there are no shear stresses acting on any plane in the body. The effectiveness of biaxial- and triaxial-tension stresses in reducing the shear stresses results in a considerable decrease in the ductility of the material, because plastic deformation is produced by shear stresses. Thus, brittle fracture is invariably associated with triaxial stresses developed at a notch or stress raiser. However, Fig. 2-11e shows that, if compressive stresses are applied lateral to a tensile stress, the maximum shear stress is larger than for the case of either uniaxial tension or compression. Because of the high value of shear stress relative to the applied tensile stress the material has an excellent opportunity to deform plastically without fracturing under this state of stress. Important use is made of this fact in the plastic working of metals. For example, greater ductility is obtained in drawing wire through a die than in simple uniaxial tension because the reaction of the metal with the die will produce lateral compressive stresses.

## 2-8  DESCRIPTION OF STRAIN AT A POINT

The displacement of points in a continuum may result from rigid-body translation, rotation, and deformation. The deformation of a solid may be made up of *dilatation*, change in volume, or *distortion*, change in shape. Situations involving translation and rotation are usually treated in the branch of mechanics called *dynamics*. Small deformations are the province of elasticity theory, while larger deformations are treated in the disciplines of plasticity and hydrodynamics. The equations developed in this section are basically geometrical, so that they apply to all types of continuous media.

Consider a solid body in fixed coordinates $x, y, z$ (Fig. 2-12). Let a combination of deformation and movement displace point $Q$ to $Q'$ with new coordinates $x + u, y + v, z + w$. The components of the displacement are $u, v, w$. The displacement of $Q$ is the vector $\mathbf{u}_Q = f(u, v, w)$. If the displacement vector is constant for all particles in the body then there is no strain. However, in general, $u_i$ is different from particle to particle so that displacement is a function of distance, $u_i = f(x_i)$. For elastic solids and small displacements, $u_i$ is a linear function of $x_i$, homogeneous displacements, and the displacement equations are linear. However, for other materials the displacement may not be linear with distance, which leads to cumbersome mathematical relationships.

**Fig. 2-12**  Displacement of point $Q$.

**Fig. 2-13**  One-dimensional strain.

To start our discussion of strain, consider a simple one-dimensional case (Fig. 2-13). In the undeformed state points $A$ and $B$ are separated by a distance $dx$. When a force is applied in the $x$ direction $A$ moves to $A'$ and $B$ moves to $B'$. Since displacement $u$, in this one-dimensional case, is a function of $x$, $B$ is displaced slightly more than $A$ since it is further from the fixed end. The normal strain is given by

$$e_x = \frac{\Delta L}{L} = \frac{A'B' - AB}{AB} = \frac{dx + \dfrac{\partial u}{\partial x} dx - dx}{\partial x} = \frac{\partial u}{\partial x} \qquad (2\text{-}33)$$

For this one-dimensional case, the displacement is given by $u = e_x x$. To generalize this to three dimensions, each of the components of the displacement will be linearly related to each of the three initial coordinates of the point.

$$u = e_{xx} x + e_{xy} y + e_{xz} z$$
$$v = e_{yx} x + e_{yy} y + e_{yz} z$$
$$w = e_{zx} x + e_{zy} y + e_{zz} z \qquad (2\text{-}34)$$

or

$$u_i = e_{ij} x_j \qquad (2\text{-}35)$$

The coefficients relating displacement with the coordinates of the point in the body are the components of the relative displacement tensor. Three of these terms can be identified readily as the normal strains.

$$e_{xx} = \frac{\partial u}{\partial x} \qquad e_{yy} = \frac{\partial v}{\partial y} \qquad e_{zz} = \frac{\partial w}{\partial z} \qquad (2\text{-}36)$$

However, the other six coefficients require further scrutiny.

**Fig. 2-14** Angular distortion of an element.

Consider an element in the $xy$ plane which has been distorted by shearing stresses (Fig. 2-14). The element has undergone angular distortion. The displacement of points along the line $AD$ is parallel to the $x$ axis, but this component of displacement increases in proportion to the distance out along the $y$ axis. Thus, referring to Eq. (2-34)

$$e_{xy} = \frac{DD'}{DA} = \frac{\partial u}{\partial y} \qquad (2\text{-}37)$$

Similarly, for the angular distortion of the $x$ axis

$$e_{yx} = \frac{BB'}{AB} = \frac{\partial v}{\partial x} \qquad (2\text{-}38)$$

These shear displacements are positive when they rotate a line from one positive axis towards another positive axis. By similar methods the rest of the components of the displacement tensor can be seen to be

$$e_{ij} = \begin{vmatrix} e_{xx} & e_{xy} & e_{xz} \\ e_{yx} & e_{yy} & e_{yz} \\ e_{zx} & e_{zy} & e_{zz} \end{vmatrix} = \begin{vmatrix} \dfrac{\partial u}{\partial x} & \dfrac{\partial u}{\partial y} & \dfrac{\partial u}{\partial z} \\ \dfrac{\partial v}{\partial x} & \dfrac{\partial v}{\partial y} & \dfrac{\partial v}{\partial z} \\ \dfrac{\partial w}{\partial x} & \dfrac{\partial w}{\partial y} & \dfrac{\partial w}{\partial z} \end{vmatrix} \qquad (2\text{-}39)$$

In general, displacement components such as $e_{xy}$, $e_{yx}$, etc., produce both *shear strain* and *rigid-body rotation*. Figure 2-15 illustrates several cases. Since we need to identify that part of the displacement that results in strain, it is important to break the displacement tensor into a strain contribution and a rotational contribution. Fortunately, a basic postulate of tensor theory states that any second-rank tensor can be decomposed into a symmetric tensor and an antisymmetric (skew-symmetric) tensor.

$$e_{ij} = \tfrac{1}{2}(e_{ij} + e_{ji}) + \tfrac{1}{2}(e_{ij} - e_{ji}) \qquad (2\text{-}40)$$

or

$$e_{ij} = \varepsilon_{ij} + \omega_{ij} \qquad (2\text{-}41)$$

**Fig. 2-15** Some examples of displacement with shear and rotation. (*a*) Pure shear without rotation; (*b*) pure rotation without shear; (*c*) simple shear.

where $\qquad \varepsilon_{ij} = \dfrac{1}{2}\left(\dfrac{\partial u_i}{\partial x_j} + \dfrac{\partial u_j}{\partial x_i}\right) \qquad$ and is called the *strain tensor*

$$\omega_{ij} = \dfrac{1}{2}\left(\dfrac{\partial u_i}{\partial x_j} - \dfrac{\partial u_j}{\partial x_i}\right) \qquad \text{and is called the } \textit{rotation tensor}$$

$$
\varepsilon_{ij} = \begin{vmatrix} \varepsilon_{xx} & \varepsilon_{xy} & \varepsilon_{xz} \\ \varepsilon_{yx} & \varepsilon_{yy} & \varepsilon_{yz} \\ \varepsilon_{zx} & \varepsilon_{zy} & \varepsilon_{zz} \end{vmatrix} = \begin{vmatrix} \dfrac{\partial u}{\partial x} & \dfrac{1}{2}\left(\dfrac{\partial u}{\partial y} + \dfrac{\partial v}{\partial x}\right) & \dfrac{1}{2}\left(\dfrac{\partial u}{\partial z} + \dfrac{\partial w}{\partial x}\right) \\ \dfrac{1}{2}\left(\dfrac{\partial u}{\partial y} + \dfrac{\partial v}{\partial x}\right) & \dfrac{\partial v}{\partial y} & \dfrac{1}{2}\left(\dfrac{\partial v}{\partial z} + \dfrac{\partial w}{\partial y}\right) \\ \dfrac{1}{2}\left(\dfrac{\partial u}{\partial z} + \dfrac{\partial w}{\partial x}\right) & \dfrac{1}{2}\left(\dfrac{\partial v}{\partial z} + \dfrac{\partial w}{\partial y}\right) & \dfrac{\partial w}{\partial z} \end{vmatrix} \qquad (2\text{-}42)
$$

$$
\omega_{ij} = \begin{vmatrix} \omega_{xx} & \omega_{xy} & \omega_{xz} \\ \omega_{yx} & \omega_{yy} & \omega_{yz} \\ \omega_{zx} & \omega_{zy} & \omega_{zz} \end{vmatrix} = \begin{vmatrix} 0 & \dfrac{1}{2}\left(\dfrac{\partial u}{\partial y} - \dfrac{\partial v}{\partial x}\right) & \dfrac{1}{2}\left(\dfrac{\partial u}{\partial z} - \dfrac{\partial w}{\partial x}\right) \\ \dfrac{1}{2}\left(\dfrac{\partial v}{\partial x} - \dfrac{\partial u}{\partial y}\right) & 0 & \dfrac{1}{2}\left(\dfrac{\partial v}{\partial z} - \dfrac{\partial w}{\partial y}\right) \\ \dfrac{1}{2}\left(\dfrac{\partial w}{\partial x} - \dfrac{\partial u}{\partial z}\right) & \dfrac{1}{2}\left(\dfrac{\partial w}{\partial y} - \dfrac{\partial v}{\partial z}\right) & 0 \end{vmatrix} \qquad (2\text{-}43)
$$

Note that $\varepsilon_{ij}$ is a symmetric tensor since $\varepsilon_{ij} = \varepsilon_{ji}$, that is, $\varepsilon_{xy} = \varepsilon_{xz}$, etc. $\omega_{ij}$ is an antisymmetric tensor since $\omega_{ij} = -\omega_{ji}$, that is, $\omega_{xy} = -\omega_{yx}$. If $\omega_{ij} = 0$, the deformation is said to be irrotational.

By substituting Eq. (2-41) into Eq. (2-35), we get the general displacement equations

$$u_i = \varepsilon_{ij} x_j + \omega_{ij} x_j \qquad (2\text{-}44)$$

Earlier in Sec. 1-9 the shear strain $\gamma$ was defined as the total angular change from a right angle. Referring to Fig. 2-15a, $\gamma = e_{xy} + e_{yx} = \varepsilon_{xy} + \varepsilon_{yx} = 2\varepsilon_{xy}$. This definition of shear strain, $\gamma_{ij} = 2\varepsilon_{ij}$, is called the *engineering shear strain*.

$$\gamma_{xy} = \frac{\partial u}{\partial y} + \frac{\partial v}{\partial x}$$

$$\gamma_{xz} = \frac{\partial w}{\partial x} + \frac{\partial u}{\partial z} \qquad (2\text{-}45)$$

$$\gamma_{yz} = \frac{\partial w}{\partial y} + \frac{\partial v}{\partial z}$$

This definition of shear strain commonly is used in engineering elasticity. However, the shear strain defined according to Eq. (2-45) *is not a tensor quantity*.

Because of the obvious advantages in the transformation of tensors by the methods discussed in Sec. 2-6, it is profitable to use the strain tensor as defined by Eq. (2-42). Since the strain tensor is a second-rank tensor, it has all of the properties that have been described earlier for stress. Thus, the strain tensor may be transformed from one set of coordinate axes to a new system of axes by

$$\varepsilon_{kl} = a_{ki} a_{lj} \varepsilon_{ij} \qquad (2\text{-}46)$$

For simplicity, equations for strain analogous with those for stress can be written directly by substituting $\varepsilon$ for $\sigma$ and $\gamma/2$ for $\tau$. Thus, the normal strain on an oblique plane is given by

$$\varepsilon = \varepsilon_x l^2 + \varepsilon_y m^2 + \varepsilon_z n^2 + \gamma_{xy} lm + \gamma_{yz} mn + \gamma_{xz} ln$$

[Compare the above with Eq. (2-18).]

In complete analogy with stress, it is possible to define a system of coordinate axes along which there are no shear strains. These axes are the principal strain axes. For an isotropic body the directions of principal strains coincide with principal stress directions.[1] An element oriented along one of the principal strain axes will undergo pure extension or contraction without any rotation or shear strain. The three principal shear strains are the roots of the cubic equation

$$\varepsilon^3 - I_1 \varepsilon^2 + I_2 \varepsilon - I_3 = 0 \qquad (2\text{-}47)$$

where
$$I_1 = \varepsilon_x + \varepsilon_y + \varepsilon_z$$
$$I_2 = \varepsilon_x \varepsilon_y + \varepsilon_y \varepsilon_z + \varepsilon_z \varepsilon_x - \tfrac{1}{4}(\gamma_{xy}^2 + \gamma_{zx}^2 + \gamma_{yz}^2)$$
$$I_3 = \varepsilon_x \varepsilon_y \varepsilon_z + \tfrac{1}{4}\gamma_{xy}\gamma_{zx}\gamma_{yz} - \tfrac{1}{4}(\varepsilon_x \gamma_{yz}^2 + \varepsilon_y \gamma_{zx}^2 + \varepsilon_z \gamma_{xy}^2)$$

---

[1] For a derivation of this point see C. T. Wang, "Applied Elasticity," pp. 26–27, McGraw-Hill Book Company, New York, 1953.

The directions of the principal strains are obtained from the three equations analogous to Eqs. (2-13)

$$2l(\varepsilon_x - \varepsilon) + m\gamma_{xy} + n\gamma_{xz} = 0$$

$$l\gamma_{xy} + 2m(\varepsilon_y - \varepsilon) + n\gamma_{yz} = 0$$

$$l\gamma_{xz} + m\gamma_{yz} + 2n(\varepsilon_z - \varepsilon) = 0$$

Continuing the analogy between stress and strain equations, the equations for the *principal shearing strains* can be obtained from Eq. (2-20).

$$\gamma_1 = \varepsilon_2 - \varepsilon_3$$

$$\gamma_{max} = \gamma_2 = \varepsilon_1 - \varepsilon_3 \qquad (2\text{-}48)$$

$$\gamma_3 = \varepsilon_1 - \varepsilon_2$$

In general, the deformation of a solid involves a combination of volume change and change in shape. Therefore, we need a way to determine how much of the deformation is due to these contributions. The *volume strain,* or cubical dilatation, is the change in volume per unit volume. Consider a rectangular parallelepiped with edges $dx$, $dy$, $dz$. The volume in the strained condition is $(1 + \varepsilon_x)(1 + \varepsilon_y)(1 + \varepsilon_z)\, dx\, dy\, dz$, since only normal strains result in volume change. The volume strain $\Delta$ is

$$\Delta = \frac{(1 + \varepsilon_x)(1 + \varepsilon_y)(1 + \varepsilon_z)\, dx\, dy\, dz - dx\, dy\, dz}{dx\, dy\, dz}$$

$$= (1 + \varepsilon_x)(1 + \varepsilon_y)(1 + \varepsilon_z) - 1$$

which for small strains, after neglecting the products of strains, becomes

$$\Delta = \varepsilon_x + \varepsilon_y + \varepsilon_z \qquad (2\text{-}49)$$

Note that the volume strain is equal to the first invariant of the strain tensor, $\Delta = \varepsilon_x + \varepsilon_y + \varepsilon_z = \varepsilon_1 + \varepsilon_2 + \varepsilon_3$. We can also define $(\varepsilon_x + \varepsilon_y + \varepsilon_z)/3$ as the *mean strain* or the hydrostatic (spherical) component of strain.

$$\varepsilon_m = \frac{\varepsilon_x + \varepsilon_y + \varepsilon_z}{3} = \frac{\varepsilon_{kk}}{3} = \frac{\Delta}{3} \qquad (2\text{-}50)$$

That part of the strain tensor which is involved in shape change rather than volume change is called the *strain deviator* $\varepsilon'_{ij}$. To obtain the deviatoric strains, we simply subtract $\varepsilon_m$ from each of the normal strain components. Thus,

$$\varepsilon'_{ij} = \begin{vmatrix} \varepsilon_x - \varepsilon_m & \varepsilon_{xy} & \varepsilon_{xz} \\ \varepsilon_{yx} & \varepsilon_y - \varepsilon_m & \varepsilon_{yz} \\ \varepsilon_{zx} & \varepsilon_{zy} & \varepsilon_z - \varepsilon_m \end{vmatrix} = \begin{vmatrix} \dfrac{2\varepsilon_x - \varepsilon_y - \varepsilon_z}{3} & \varepsilon_{xy} & \varepsilon_{xz} \\ \varepsilon_{yx} & \dfrac{2\varepsilon_y - \varepsilon_z - \varepsilon_x}{3} & \varepsilon_{yz} \\ \varepsilon_{zx} & \varepsilon_{zy} & \dfrac{2\varepsilon_z - \varepsilon_x - \varepsilon_y}{3} \end{vmatrix}$$

$$(2\text{-}51)$$

The division of the total strain tensor into deviatoric and dilatational strains is given in tensor notation by

$$\varepsilon_{ij} = \varepsilon'_{ij} + \varepsilon_m = \left(\varepsilon_{ij} - \frac{\Delta}{3}\delta_{ij}\right) + \frac{\Delta}{3}\delta_{ij} \qquad (2\text{-}52)$$

For example, when $\varepsilon_{ij}$ are the principal strains, $(i = j)$, the strain deviators are $\varepsilon'_{11} = \varepsilon_{11} - \varepsilon_m$, $\varepsilon'_{22} = \varepsilon_{22} - \varepsilon_m$, $\varepsilon'_{33} = \varepsilon_{33} - \varepsilon_m$. These strains represent elongations or contractions along the principal axes that change the shape of the body at constant volume.

## 2-9 MOHR'S CIRCLE OF STRAIN

Except in a few cases involving contact stresses, it is not possible to measure stress directly. Therefore, experimental measurements of stress are actually based on measured strains and are converted to stresses by means of Hooke's law and the more general relationships which are given in Sec. 2-11. The most universal strain-measuring device is the bonded-wire resistance gage, frequently called the SR-4 strain gage.[1] These gages are made up of several loops of fine wire or foil of special composition, which are bonded to the surface of the body to be studied. When the body is deformed, the wires in the gage are strained and their electrical resistance is altered. The change in resistance, which is proportional to strain, can be accurately determined with a simple Wheatstone-bridge circuit. The high sensitivity, stability, comparative ruggedness, and ease of application make resistance strain gages a very powerful tool for strain determination.

For practical problems of experimental stress analysis it is often important to determine the principal stresses. If the principal directions are known, gages can be oriented in these directions and the principal stresses determined quite readily. In the general case the direction of the principal strains will not be known, so that it will be necessary to determine the orientation and magnitude of the principal strains from the measured strains in arbitrary directions. Because no stress can act perpendicular to a free surface, strain-gage measurements involve a two-dimensional state of stress. The state of strain is completely determined if $\varepsilon_x$, $\varepsilon_y$, and $\gamma_{xy}$ can be measured. However, strain gages can make only direct readings of linear strain, while shear strains must be determined indirectly. Therefore, it is the usual practice to use three strain gages separated at fixed angles in the form of a "rosette," as in Fig. 2-16. Strain-gage readings at three values of $\theta$ will give three simultaneous equations similar to Eq. (2-53) which can be solved

---

[1] For a treatment of strain gages and other techniques of experimental stress analysis see J. W. Dally, and W. F. Riley, "Experimental Stress Analysis," McGraw-Hill Book Company, New York, 1965.

**Fig. 2-16** Typical strain-gage rosettes. (a) Rectangular; (b) delta.

for $\varepsilon_x$, $\varepsilon_y$, and $\gamma_{xy}$. The two-dimensional version of Eq. (2.47) can then be used to determine the principal strains.

$$\varepsilon_\theta = \varepsilon_x \cos^2 \theta + \varepsilon_y \sin^2 \theta + \gamma_{xy} \sin \theta \cos \theta \qquad (2\text{-}53)$$

A more convenient method of determining the principal strains from strain-gage readings than the solution of three simultaneous equations in three unknowns is the use of Mohr's circle. In constructing a Mohr's circle representation of strain, values of linear normal strain $\varepsilon$ are plotted along the $x$ axis, and the shear strain divided by 2 is plotted along the $y$ axis. Figure 2-17 shows the Mohr's circle construction[1] for the generalized strain-gage rosette illustrated at the top of the figure. Strain-gage readings $\varepsilon_a$, $\varepsilon_b$, and $\varepsilon_c$ are available for three gages situated at arbitrary angles $\alpha$ and $\beta$. The objective is to determine the magnitude and orientation of the principal strains $\varepsilon_1$ and $\varepsilon_2$.

1. Along an arbitrary axis $X'X'$ lay off vertical lines $aa$, $bb$, and $cc$ corresponding to the strains $\varepsilon_a$, $\varepsilon_b$, and $\varepsilon_c$.
2. From any point on the line $bb$ (middle strain gage) draw a line $DA$ at an angle $\alpha$ with $bb$ and intersecting $aa$ at point $A$. In the same way, lay off $DC$ intersecting $cc$ at point $C$.
3. Construct a circle through $A$, $C$, and $D$. The center of this circle is at $O$, determined by the intersection of the perpendicular bisectors to $CD$ and $AD$.
4. Points $A$, $B$, and $C$ on the circle give the values of $\varepsilon$ and $\gamma/2$ (measured from the new $x$ axis through $O$) for the three gages.
5. Values of the principal strains are determined by the intersection of the circle with the new $x$ axis through $O$. The angular relationship of $\varepsilon_1$ to the gage $a$ is one-half the angle $AOP$ on the Mohr's circle ($AOP = 2\theta$).

[1] G. Murphy, *J. Appl. Mech.*, vol. 12, p. A209, 1945; F. A. McClintock, *Proc. Soc. Exp. Stress Anal.*, vol. 9, p. 209, 1951.

**Fig. 2-17**  Mohr's circle for determination of principal strains.

## 2-10  HYDROSTATIC AND DEVIATOR COMPONENTS OF STRESS

Having introduced the concept that the strain tensor can be divided into a hydro-static or mean strain and a strain deviator, it is important to consider the physical significance of a similar operation on the stress tensor. The total stress tensor can be divided into a *hydrostatic* or *mean stress tensor* $\sigma_m$, which involves only pure tension or compression, and a *deviator stress tensor* $\sigma'_{ij}$, which represents the shear stresses in the total state of stress (Fig. 2-18). In direct analogy with the situation for strain, the hydrostatic component of the stress tensor produces only elastic volume changes and does not cause plastic deformation. Experiment shows that the yield stress of metals is independent of hydrostatic stress, although the fracture strain is strongly influenced by hydrostatic stress. Because the stress deviator involves the shearing stresses, it is important in causing plastic deformation. In Chap. 3 we shall see that the stress deviator is useful in formulating theories of yielding.

Total stress   =   Hydrostatic stress   +   Stress deviator

**Fig. 2-18** Resolution of total stress into hydrostatic stress and stress deviator.

The hydrostatic or mean stress is given by

$$\sigma_m = \frac{\sigma_{kk}}{3} = \frac{\sigma_x + \sigma_y + \sigma_z}{3} = \frac{\sigma_1 + \sigma_2 + \sigma_3}{3} \qquad (2\text{-}54)$$

The decomposition of the stress tensor is given by

$$\sigma_{ij} = \sigma'_{ij} + \tfrac{1}{3}\delta_{ij}\sigma_{kk} \qquad (2\text{-}55)$$

Therefore,

$$\sigma'_{ij} = \sigma_{ij} - \sigma_m \delta_{ij} \qquad (2\text{-}56)$$

$$
\sigma'_{ij} =
\begin{vmatrix}
\dfrac{2\sigma_x - \sigma_y - \sigma_z}{3} & \tau_{xy} & \tau_{xz} \\[2ex]
\tau_{yx} & \dfrac{2\sigma_y - \sigma_z - \sigma_x}{3} & \tau_{yz} \\[2ex]
\tau_{zx} & \tau_{zy} & \dfrac{2\sigma_z - \sigma_x - \sigma_y}{3}
\end{vmatrix}
\qquad (2\text{-}57)
$$

It can be seen readily that the stress deviator involves shear stresses. For example, referring $\sigma'_{ij}$ to a system of principal axes,

$$\sigma'_1 = \frac{2\sigma_1 - \sigma_2 - \sigma_3}{3} = \frac{(\sigma_1 - \sigma_2) + (\sigma_1 - \sigma_3)}{3}$$

$$\sigma'_1 = \frac{2}{3}\left(\frac{\sigma_1 - \sigma_2}{2} + \frac{\sigma_1 - \sigma_3}{2}\right) = \frac{2}{3}(\tau_3 + \tau_2) \qquad (2\text{-}58)$$

where $\tau_3$ and $\tau_2$ are principal shearing stresses.

Since $\sigma'_{ij}$ is a second-rank tensor, it has principal axes. The principal values of the stress deviator are the roots of the cubic equation[1]

$$(\sigma')^3 - J_1(\sigma')^2 - J_2\,\sigma' - J_3 = 0 \qquad (2\text{-}59)$$

where $J_1$, $J_2$, $J_3$ are the invariants of the deviator stress tensor. $J_1$ is the sum of the principal terms in the diagonal of the matrix of components of $\sigma'_{ij}$.

$$J_1 = (\sigma_x - \sigma_m) + (\sigma_y - \sigma_m) + (\sigma_z - \sigma_m) = 0 \qquad (2\text{-}60)$$

$J_2$ can be obtained from the sum of the principal minors of $\sigma'_{ij}$.

$$
\begin{aligned}
J_2 &= \tau_{xy}{}^2 + \tau_{yz}{}^2 + \tau_{xz}{}^2 - \sigma'_x\sigma'_y - \sigma'_y\sigma'_z - \sigma'_z\sigma'_x \\
&= \tfrac{1}{6}[(\sigma_x - \sigma_y)^2 + (\sigma_y - \sigma_z)^2 + (\sigma_z - \sigma_x)^2 + 6(\tau_{xy}{}^2 + \tau_{yz}{}^2 + \tau_{xz}{}^2)]
\end{aligned} \qquad (2\text{-}61)
$$

The third invariant $J_3$ is the determinant of Eq. (2-57).

## 2-11  ELASTIC STRESS-STRAIN RELATIONS

Up till now our discussion of stress and strain has been perfectly general and applicable to any continuum. Now, if we want to relate the stress tensor with the strain tensor, we must introduce the properties of the material. Equations of this nature are called *constitutive equations*. In this chapter we shall consider only constitutive equations for elastic solids. Moreover, initially we shall only consider isotropic elastic solids.

In Chap. 1 we saw that elastic stress is linearly related to elastic strain by means of the modulus of elasticity (Hooke's law).

$$\sigma_x = E\varepsilon_x \qquad (2\text{-}62)$$

where $E$ is the modulus of elasticity in tension or compression. While a tensile force in the $x$ direction produces an extension along that axis, it also produces a contraction in the transverse $y$ and $z$ directions. The transverse strain has been found by experience to be a constant fraction of the strain in the longitudinal direction. This is known as *Poisson's ratio*, denoted by the symbol $v$.

$$\varepsilon_y = \varepsilon_z = -v\varepsilon_x = -\frac{v\sigma_x}{E} \qquad (2\text{-}63)$$

Only the absolute value of $v$ is used in calculations. Poisson's ratio is 0.25 for a perfectly isotropic elastic material, but for most metals the values[2] of $v$ are closer to 0.33.

To develop the stress-strain relations for a three-dimensional state of stress, consider a unit cube subjected to normal stresses $\sigma_x$, $\sigma_y$, $\sigma_z$ and shearing stresses

---

[1] Note that we use a negative sign for the coefficient of $\sigma'$. Compare with Eq. (2-14).

[2] W. Koster and H. Franz, *Metall. Rev.*, vol. 6, pp. 1–55, 1961.

$\tau_{xy}$, $\tau_{yz}$, $\tau_{zx}$. Because the elastic stresses are small and the material is isotropic, we can assume that normal stress $\sigma_x$ does not produce shear strain on the $x$, $y$, or $z$ planes and that a shear stress $\tau_{xy}$ does not produce normal strains on the $x$, $y$, or $z$ planes. We can then apply the principle of superposition[1] to determine the strain produced by more than one stress component. For example, the stress $\sigma_x$ produces a normal strain $\varepsilon_x$ and two transverse strains $\varepsilon_y = -\nu\varepsilon_x$ and $\varepsilon_z = -\nu\varepsilon_x$. Thus,

| Stress | Strain in the $x$ Direction | Strain in the $y$ Direction | Strain in the $z$ Direction |
|--------|--------|--------|--------|
| $\sigma_x$ | $\varepsilon_x = \dfrac{\sigma_x}{E}$ | $\varepsilon_y = -\dfrac{\nu\sigma_x}{E}$ | $\varepsilon_z = -\dfrac{\nu\sigma_x}{E}$ |
| $\sigma_y$ | $\varepsilon_x = -\dfrac{\nu\sigma_y}{E}$ | $\varepsilon_y = \dfrac{\sigma_y}{E}$ | $\varepsilon_z = -\dfrac{\nu\sigma_y}{E}$ |
| $\sigma_z$ | $\varepsilon_x = -\dfrac{\nu\sigma_z}{E}$ | $\varepsilon_y = -\dfrac{\nu\sigma_z}{E}$ | $\varepsilon_z = \dfrac{\sigma_z}{E}$ |

By superposition of the components of strain in the $x$, $y$, and $z$ directions

$$\varepsilon_x = \frac{1}{E}[\sigma_x - \nu(\sigma_y + \sigma_z)]$$

$$\varepsilon_y = \frac{1}{E}[\sigma_y - \nu(\sigma_z + \sigma_x)] \qquad (2\text{-}64)$$

$$\varepsilon_z = \frac{1}{E}[\sigma_z - \nu(\sigma_x + \sigma_y)]$$

The shearing stresses acting on the unit cube produce shearing strains.

$$\tau_{xy} = G\gamma_{xy} \qquad \tau_{yz} = G\gamma_{yz} \qquad \tau_{xz} = G\gamma_{xz} \qquad (2\text{-}65)$$

The proportionality constant $G$ is the *modulus of elasticity in shear*, or the *modulus of rigidity*. Values of $G$ are usually determined from a torsion test.

We have seen that the stress-strain equations for an *isotropic* elastic solid involve three constants, $E$, $G$, and $\nu$. Typical values of these constants for a number of metals are given in Table 2-1.

Still another elastic constant is the *bulk modulus* or the *volumetric modulus of elasticity $K$*. The bulk modulus is the ratio of the hydrostatic pressure to the dilatation that it produces

$$K = \frac{\sigma_m}{\Delta} = \frac{-p}{\Delta} = \frac{1}{\beta} \qquad (2\text{-}66)$$

where $-p$ is the hydrostatic pressure and $\beta$ is the compressibility.

---

[1] The principle of superposition states that two strains may be combined by direct superposition. The order of application has no effect on the final strain of the body.

**Table 2-1  TYPICAL ROOM-TEMPERATURE VALUES OF ELASTIC CONSTANTS FOR ISOTROPIC MATERIALS**

| Material | Modulus of elasticity, $10^{-6}$ psi | Shear modulus, $10^{-6}$ psi | Poisson's ratio |
|---|---|---|---|
| Aluminum alloys | 10.5 | 4.0 | 0.31 |
| Copper | 16.0 | 6.0 | 0.33 |
| Steel (plain carbon and low-alloy) | 29.0 | 11.0 | 0.33 |
| Stainless steel (18-8) | 28.0 | 9.5 | 0.28 |
| Titanium | 17.0 | 6.5 | 0.31 |
| Tungsten | 58.0 | 22.8 | 0.27 |

Many useful relationships may be derived between the elastic constants $E$, $G$, $v$, $K$. For example, if we add up the three equations (2-64),

$$\varepsilon_x + \varepsilon_y + \varepsilon_z = \frac{1 - 2v}{E}(\sigma_x + \sigma_y + \sigma_z)$$

The term on the left is the volume strain $\Delta$, and the term on the right is $3\sigma_m$.

$$\Delta = \frac{1 - 2v}{E} 3\sigma_m$$

or

$$K = \frac{\sigma_m}{\Delta} = \frac{E}{3(1 - 2v)} \qquad (3\text{-}67)$$

Another important relationship is the expression relating $E$, $G$, and $v$. This equation is usually developed in a first course in strength of materials.[1]

$$G = \frac{E}{2(1 + v)} \qquad (2\text{-}68)$$

Many other relationships can be developed between these four isotropic elastic constants. For example,

$$E = \frac{9K}{1 + 3K/G} \qquad v = \frac{1 - 2G/3K}{2 + 2G/3K}$$

$$G = \frac{3(1 - 2v)K}{2(1 + v)} \qquad K = \frac{E}{9 - 3E/G}$$

---

[1] For a geometric development see D. C. Drucker, "Introduction to Mechanics of Deformable Solids," pp. 64–65, McGraw-Hill Book Company, New York, 1967. For a derivation based on isotropy and transformation of axes see Chou and Pagano, op. cit., pp. 58–59.

Equations (2-64) and (2-65) may be expressed succinctly in tensor notation

$$\varepsilon_{ij} = \frac{1+v}{E}\,\sigma_{ij} - \frac{v}{E}\,\sigma_{kk}\,\delta_{ij} \qquad (2\text{-}69)$$

For example, if $i = j = x$,

$$\varepsilon_{xx} = \frac{1+v}{E}\,\sigma_{xx} - \frac{v}{E}(\sigma_{xx} + \sigma_{yy} + \sigma_{zz})(1)$$

$$= \frac{1}{E}\,[\sigma_{xx} - v(\sigma_{yy} + \sigma_{zz})]$$

If $i = x$ and $j = y$,

$$\varepsilon_{xy} = \frac{\gamma_{xy}}{2} = \frac{1+v}{E}\,\tau_{xy} - \frac{v}{E}\,\sigma_{kk}(0)$$

where

$$\frac{1+v}{E} = \frac{1}{2G} \qquad \text{and} \qquad \gamma_{xy} = \frac{1}{G}\,\tau_{xy}$$

## 2-12  CALCULATION OF STRESSES FROM ELASTIC STRAINS

Since for small elastic strains there is no coupling between the expressions for normal stress and strain and the equations for shear stress and shear strain, it is possible to invert Eqs. (2-64) and (2-65) to solve for stress in terms of strain. From Eq. (2-64),

$$\sigma_x + \sigma_y + \sigma_z = \frac{E}{1-2v}(\varepsilon_x + \varepsilon_y + \varepsilon_z) \qquad (2\text{-}70)$$

$$\varepsilon_x = \frac{1+v}{E}\,\sigma_x - \frac{v}{E}(\sigma_x + \sigma_y + \sigma_z) \qquad (2\text{-}71)$$

Substitution of Eq. (2-70) into Eq. (2-71) gives

$$\sigma_x = \frac{E}{1+v}\,\varepsilon_x + \frac{vE}{(1+v)(1-2v)}(\varepsilon_x + \varepsilon_y + \varepsilon_z) \qquad (2\text{-}72)$$

or in tensor notation

$$\sigma_{ij} = \frac{E}{1+v}\,\varepsilon_{ij} + \frac{vE}{(1+v)(1-2v)}\,\varepsilon_{kk}\,\delta_{ij} \qquad (2\text{-}73)$$

Upon expansion, Eq. (2-73) gives three equations for normal stress and six equations for shear stress. Equation (2-72) is often written in a briefer form by letting

$$\frac{vE}{(1+v)(1-2v)} = \lambda \qquad \text{Lamé's constant}$$

and noting that $\Delta = \varepsilon_x + \varepsilon_y + \varepsilon_z$.

$$\sigma_x = 2G\varepsilon_x + \lambda\Delta \qquad (2\text{-}74)$$

The stresses and the strains can be broken into deviator and hydrostatic components. The deviatoric response (distortion) is related to the stress deviator by

$$\sigma'_{ij} = \frac{E}{1+v}\,\varepsilon'_{ij} = 2G\varepsilon'_{ij} \qquad (2\text{-}75)$$

while the relationship between hydrostatic stress and mean strain is

$$\sigma_{ii} = \frac{E}{1-2v}\,\varepsilon_{kk} = 3K\varepsilon_{kk} \qquad (2\text{-}76)$$

For a case of *plane stress* ($\sigma_3 = 0$), two simple and useful equations relating stress to strain may be obtained by solving simultaneously two of the equations of (2-64).

$$\sigma_1 = \frac{E}{1-v^2}\,(\varepsilon_1 + v\varepsilon_2)$$

$$\sigma_2 = \frac{E}{1-v^2}\,(\varepsilon_2 + v\varepsilon_1) \qquad (2\text{-}77)$$

A situation of plane stress exists typically in a thin sheet loaded in the plane of the sheet or a thin-wall tube loaded by internal pressure where there is no stress normal to a free surface.

Another important situation is *plane strain* ($\varepsilon_3 = 0$), which occurs typically when one dimension is much greater than the other two, as in a long rod or a cylinder with restrained ends. Some type of physical restraint exists to limit the strain in one direction, so

$$\varepsilon_3 = \frac{1}{E}[\sigma_3 - v(\sigma_1 + \sigma_2)] = 0$$

but

$$\sigma_3 = v(\sigma_1 + \sigma_2)$$

Therefore, a stress exists even though the strain is zero. Substituting this value into Eq. (2-64), we get

$$\varepsilon_1 = \frac{1}{E}[(1-v^2)\sigma_1 - v(1+v)\sigma_2]$$

$$\varepsilon_2 = \frac{1}{E}[(1-v^2)\sigma_2 - v(1+v)\sigma_1] \qquad (2\text{-}78)$$

$$\varepsilon_3 = 0$$

## 2-13 STRAIN ENERGY

The *elastic strain energy* $U$ is the energy expended by the action of external forces in deforming an elastic body. Essentially all the work performed during elastic deformation is stored as elastic energy, and this energy is recovered on the release of the applied forces. Energy (or work) is equal to a force multiplied by the distance over which it acts. In the deformation of an elastic body, the force and deformation increase linearly from initial values of zero so that the average energy is equal to one-half of their product. This is also equal to the area under the load-deformation curve.

$$U = \tfrac{1}{2}P\delta$$

For an elemental cube that is subjected to only a tensile stress along the $x$ axis, the elastic strain energy is given by

$$dU = \tfrac{1}{2}P \, du = \tfrac{1}{2}(\sigma_x A)(\varepsilon_x \, dx)$$
$$= \tfrac{1}{2}(\sigma_x \varepsilon_x)(A \, dx) \tag{2-79}$$

Equation (2-79) describes the total elastic energy absorbed by the element. Since $A \, dx$ is the volume of the element, the *strain energy per unit volume* or strain energy density $U_0$ is given by

$$U_0 = \frac{1}{2}\sigma_x \varepsilon_x = \frac{1}{2}\frac{\sigma_x^{\,2}}{E} = \frac{1}{2}\varepsilon_x^{\,2}E \tag{2-80}$$

Note that the lateral strains which accompany deformation in simple tension do not enter into the expression for strain energy because forces do not exist in the direction of the lateral strains.

By the same type of reasoning, the strain energy per unit volume of an element subjected to *pure shear* is given by

$$U_0 = \frac{1}{2}\tau_{xy}\gamma_{xy} = \frac{1}{2}\frac{\tau_{xy}^{\,2}}{G} = \frac{1}{2}\gamma_{xy}^{\,2}G \tag{2-81}$$

The elastic strain energy for a general three-dimensional stress distribution may be obtained by superposition.

$$U_0 = \tfrac{1}{2}(\sigma_x \varepsilon_x + \sigma_y \varepsilon_y + \sigma_z \varepsilon_z + \tau_{xy}\gamma_{xy} + \tau_{xz}\gamma_{xz} + \tau_{yz}\gamma_{yz}) \tag{2-82}$$

or in tensor notation

$$U_0 = \tfrac{1}{2}\sigma_{ij}\varepsilon_{ij} \tag{2-83}$$

Substituting the equations of Hooke's law [Eqs. (2-64) and (2-65)] for the strains in Eq. (2-82) results in an expression for strain energy per unit volume expressed solely in terms of the stress and the elastic constants

$$U_0 = \frac{1}{2E}(\sigma_x^{\,2} + \sigma_y^{\,2} + \sigma_z^{\,2}) - \frac{\nu}{E}(\sigma_x \sigma_y + \sigma_y \sigma_z + \sigma_x \sigma_z)$$

$$+ \frac{1}{2G}(\tau_{xy}^{\,2} + \tau_{xz}^{\,2} + \tau_{yz}^{\,2}) \tag{2-84}$$

Also, by substituting Eqs. (2-74) into Eq. (2-82), the stresses are eliminated, and the strain energy is expressed in terms of strains and the elastic constants

$$U_0 = \tfrac{1}{2}\lambda\Delta^2 + G(\varepsilon_x^2 + \varepsilon_y^2 + \varepsilon_z^2) + \tfrac{1}{2}G(\gamma_{xy}^2 + \gamma_{xz}^2 + \gamma_{yz}^2) \qquad (2\text{-}85)$$

It is interesting to note that the derivative of $U_0$ with respect to any strain component gives the corresponding stress component. For example,

$$\frac{\partial U_0}{\partial \varepsilon_x} = \lambda\Delta + 2G\varepsilon_x = \sigma_x \qquad (2\text{-}86)$$

In the same way, $\partial U_0 / \delta\sigma_x = \varepsilon_x$. Methods of calculation using strain energy to arrive at stresses and strains are powerful tools in elasticity analysis. Some of the better known techniques are Castigliano's theorem, the theorem of least work, and the principal of virtual work.

## 2-14  ANISOTROPY OF ELASTIC BEHAVIOR

Up to this point we have considered elastic behavior from a simple phenomenological point of view, i.e., Hooke's law was presented as a well-established empirical law and our attention was directed at developing useful relationships between stress and strain in an isotropic elastic solid. In this section we consider the fact that the elastic constants of a crystal vary markedly with orientation. However, first it is important to discuss briefly the nature of the elastic forces between atoms.

When a force is applied to a crystalline solid, it either pulls the atoms apart or pushes them together. The applied force is resisted by the forces of attraction or repulsion between the atoms. A convenient way to look at this is with an energy-distance diagram (Fig. 2-19), which represents the interaction energy (potential energy) between two atoms as they are separated by a distance $a$. When the external force is zero, the atoms are separated by a distance equal to the equilibrium spacing $a = a_0$. For small applied forces, the atoms will find a new equilibrium spacing $a$ at which the external and internal forces are balanced. The displacement of the atom is $u = a - a_0$. Since force is the derivative of potential

**Fig. 2-19**  Interaction energy vs. separation between atoms.

energy with distance [compare Eq. (2-86)], the force to produce a given equilibrium displacement is

$$P = \frac{d\phi(u)}{du} \qquad (2\text{-}87)$$

where $\phi(u)$ is the interaction bond energy at a displacement $u$. Thus, the force on a bond is a function of displacement $u$. For each displacement there is a characteristic value of force $P(u)$. Moreover, the deformation of the bonds between atoms is reversible. When the displacement returns to some initial value $u_1$ after being extended to $u_2$ the force returns to its previous value $P(u_1)$.

In an elastic solid the bond energy is a continuous function of displacement.[1] Thus, we can express $\phi(u)$ as a Taylor series

$$\phi(u) = \phi_0 + \left(\frac{d\phi}{du}\right)_0 u + \frac{1}{2}\left(\frac{d^2\phi}{du^2}\right)_0 u^2 + \cdots \qquad (2\text{-}88)$$

where $\phi_0$ is the energy at $u = 0$ and the differential coefficients are measured at $u = 0$. Since the force is zero when $a = a_0$, $d\phi/du = 0$

$$\phi(u) = \phi_0 + \frac{1}{2}\left(\frac{d^2\phi}{du^2}\right)_0 u^2$$

$$P = \frac{d\phi(u)}{du} = \left(\frac{d^2\phi}{du^2}\right)_0 u \qquad (2\text{-}89)$$

The coefficient $(d^2\phi/du^2)_0$ is the curvature of the energy-distance curve at $u = a_0$. Since it is independent of $u$, the coefficient is a constant, and Eq. (2-89) is equivalent to $P = ku$, which is Hooke's law in its original form. When Eq. (2-89) is expressed in terms of stress and strain, the coefficient is directly proportional to the elastic constant of the material. It has the same value for both tension and compression since it is independent of the sign of $u$. Thus, we have shown that the elastic constant is determined by the sharpness of curvature of the minimum in the energy-distance curve. It is therefore a basic property of the material, not readily changed by heat treatment or defect structure, although it would be expected to decrease with increasing temperature. Moreover, since the binding forces will be strongly affected by distance between atoms, the elastic constants will vary with direction in the crystal lattice.

In the generalized case[2] Hooke's law may be expressed as

$$\varepsilon_{ij} = S_{ijkl}\,\sigma_{kl} \qquad (2\text{-}90)$$

---

[1] This development follows that given by A. H. Cottrell, "The Mechanical Properties of Matter," pp. 84–85, John Wiley & Sons, Inc., New York, 1964.

[2] An excellent text that deals with the anisotropic properties of crystals in tensor notation is J. F. Nye, "Physical Properties of Crystals," Oxford University Press, London, 1957. For a treatment of anisotropic elasticity see R. F. S. Hearmon, "An Introduction to Applied Anisotropic Elasticity," Oxford University Press, London, 1961. A fairly concise but complete discussion of crystal elasticity is given by S. M. Edelglass, "Engineering Materials Science," pp. 277–301, The Ronald Press Company, New York, 1966.

and

$$\sigma_{ij} = C_{ijkl}\varepsilon_{kl} \qquad (2\text{-}91)$$

where $S_{ijkl}$ is the *compliance tensor* and $C_{ijkl}$ is the *elastic stiffness* (often called just the elastic constants). Both $S_{ijkl}$ and $C_{ijkl}$ are fourth-rank tensor quantities. If we expanded Eq. (2-90) or (2-91), we would get nine equations, each with nine terms, 81 constants in all. However, we know that both $\varepsilon_{ij}$ and $\sigma_{ij}$ are symmetric tensors, that is, $\varepsilon_{ij} = \varepsilon_{ji}$, which immediately leads to appreciable simplification. Thus, we can write

$$\varepsilon_{ij} = S_{ijkl}\sigma_{kl} \qquad \text{or} \qquad \varepsilon_{ij} = S_{ijlk}\sigma_{lk}$$

and since

$$S_{ijkl}\sigma_{kl} = S_{ijlk}\sigma_{lk}$$

$$\sigma_{kl} = \sigma_{lk} \qquad \text{and} \qquad S_{ijkl} = S_{ijlk}$$

Also, we could write

$$\varepsilon_{ij} = S_{ijkl}\sigma_{kl} = \varepsilon_{ji} = S_{jikl}\sigma_{kl}$$

$$S_{ijkl} = S_{jikl}$$

Therefore, because of the symmetry of the stress and strain tensors, only 36 of the components of the compliance tensor are independent and distinct terms. The same is true of the elastic stiffness tensor.

Expanding Eq. (2-91) and taking into account the above relationships gives equations like

$$\sigma_{11} = C_{1111}\varepsilon_{11} + C_{1122}\varepsilon_{22} + C_{1133}\varepsilon_{33} + C_{1123}(2\varepsilon_{23}) + C_{1113}(2\varepsilon_{13}) + C_{1112}(2\varepsilon_{12})$$

$$\cdots \cdots \cdots \cdots$$

$$\sigma_{23} = C_{2311}\varepsilon_{11} + C_{2322}\varepsilon_{22} + C_{2333}\varepsilon_{33} + C_{2323}(2\varepsilon_{23}) + C_{2313}(2\varepsilon_{13}) + C_{2312}(2\varepsilon_{12})$$

$$\cdots \cdots \cdots \cdots$$

$$(2\text{-}92)$$

These equations show that, in contrast to the situation for an isotropic elastic solid, Eq. (2-72), for an anisotropic elastic solid both normal strains and shear strains are capable of contributing to a normal stress.

In expanding Eq. (2-90), we express the shearing strains by the more conventional engineering shear strain $\gamma = 2\varepsilon$.

$$\varepsilon_{11} = S_{1111}\sigma_{11} + S_{1122}\sigma_{22} + S_{1133}\sigma_{33} + 2S_{1123}\sigma_{23} + 2S_{1113}\sigma_{13} + 2S_{1112}\sigma_{12}$$

$$\cdots \cdots \cdots \cdots$$

$$\gamma_{23} = 2\varepsilon_{23} = 2S_{2311}\sigma_{11} + 2S_{2322}\sigma_{22} + 2S_{2333}\sigma_{33} + 4S_{2323}\sigma_{23}$$
$$+ 4S_{2313}\sigma_{13} + 4S_{2312}\sigma_{12}$$

$$\cdots \cdots \cdots \cdots$$

$$(2\text{-}93)$$

The usual convention for designating components of elastic compliance and elastic stiffness uses only two subscripts instead of four. This is called the *contracted notation*. The subscripts simply denote the row and column in the matrix of components in which they fall.

$$\sigma_{11} = C_{11}\varepsilon_{11} + C_{12}\varepsilon_{22} + C_{13}\varepsilon_{33} + C_{14}\gamma_{23} + C_{15}\gamma_{13} + C_{16}\gamma_{12}$$

$$\cdots\cdots\cdots\cdots\cdots\cdots\cdots\cdots\cdots\cdots\cdots\cdots\cdots$$

$$\cdots\cdots\cdots\cdots\cdots\cdots\cdots\cdots\cdots\cdots\cdots\cdots\cdots \tag{2-94}$$

$$\sigma_{23} = C_{41}\varepsilon_{11} + C_{42}\varepsilon_{22} + C_{43}\varepsilon_{33} + C_{44}\gamma_{23} + C_{45}\gamma_{13} + C_{46}\gamma_{12}$$

$$\cdots\cdots\cdots\cdots\cdots\cdots\cdots\cdots\cdots\cdots\cdots\cdots\cdots$$

$$\cdots\cdots\cdots\cdots\cdots\cdots\cdots\cdots\cdots\cdots\cdots\cdots\cdots$$

and

$$\varepsilon_{11} = S_{11}\sigma_{11} + S_{12}\sigma_{22} + S_{13}\sigma_{33} + S_{14}\sigma_{23} + S_{15}\sigma_{13} + S_{16}\sigma_{12}$$

$$\cdots\cdots\cdots\cdots\cdots\cdots\cdots\cdots\cdots\cdots\cdots\cdots\cdots$$

$$\cdots\cdots\cdots\cdots\cdots\cdots\cdots\cdots\cdots\cdots\cdots\cdots\cdots \tag{2-95}$$

$$\gamma_{23} = S_{41}\sigma_{11} + S_{42}\sigma_{22} + S_{43}\sigma_{33} + S_{44}\sigma_{23} + S_{45}\sigma_{13} + S_{46}\sigma_{12}$$

$$\cdots\cdots\cdots\cdots\cdots\cdots\cdots\cdots\cdots\cdots\cdots\cdots\cdots$$

$$\cdots\cdots\cdots\cdots\cdots\cdots\cdots\cdots\cdots\cdots\cdots\cdots\cdots$$

By comparing coefficients in Eqs. (2-92) and (2-94) and Eqs. (2-93) and (2-95) we note, for example, that

$$C_{2322} = C_{42} \qquad C_{1122} = C_{12}$$
$$S_{1122} = C_{12} \qquad 2S_{2311} = C_{41} \qquad 4S_{2323} = S_{44}$$

The elastic stiffness constants are defined by equations like

$$C_{11} = \frac{\Delta\sigma_{11}}{\Delta\varepsilon_{11}} \qquad \text{all } \varepsilon_{ij} \text{ constant except } \varepsilon_{11}$$

Unfortunately, a measurement such as this is difficult to do experimentally since the specimen must be constrained mechanically to prevent strains such as $\varepsilon_{23}$. It is much easier to experimentally determine the coefficients of the elastic compliance from equations of the type

$$S_{11} = \frac{\Delta\varepsilon_{11}}{\Delta\sigma_{11}} \qquad \text{all } \sigma_{ij} \text{ constant except } \sigma_{11}$$

If the components of $S_{ij}$ have been determined experimentally, then the components of $C_{ij}$ can be determined by matrix inversion.

At this stage we have 36 independent constants, but further reduction in the number of independent constants is possible. By using the relationship given in Eq. (2-86), we can show that the constants are symmetrical, that is, $C_{ij} = C_{ji}$.

For example,

$$\frac{\partial U}{\partial \varepsilon_{11}} = \sigma_{11} = C_{11}\varepsilon_{11} + C_{12}\varepsilon_{22} + C_{13}\varepsilon_{33} + C_{14}\gamma_{23} + C_{15}\gamma_{13} + C_{16}\gamma_{12}$$

$$\frac{\partial^2 U}{\partial \varepsilon_{11}\, \partial \varepsilon_{22}} = C_{12}$$

$$\frac{\partial U}{\partial \varepsilon_{22}} = \sigma_{22} = C_{21}\varepsilon_{11} + C_{22}\varepsilon_{22} + C_{23}\varepsilon_{33} + C_{24}\gamma_{23} + C_{25}\gamma_{13} + C_{26}\gamma_{12}$$

$$\frac{\partial^2 U}{\partial \varepsilon_{22}\, \partial \varepsilon_{11}} = C_{12}$$

$$\therefore \quad \frac{\partial^2 U}{\partial \varepsilon_{11}\, \partial \varepsilon_{22}} = \frac{\partial^2 U}{\partial \varepsilon_{22}\, \partial \varepsilon_{11}} = C_{12} = C_{21}$$

In general, $C_{ij} = C_{ji}$ and $S_{ij} = S_{ji}$. Now, we start with 36 constants $C_{ij}$, but of these there are six constants where $i = j$. This leaves 30 constants where $i \neq j$, but only one-half of these are independent constants since $C_{ij} = C_{ji}$. Therefore, for the general anisotropic linear elastic solid there are $30/2 + 6 = 21$ independent elastic constants.

   Further simplification in the number of independent elastic constants results from the consideration of the symmetry of different crystal lattices. The simplest and most important case is the cubic crystal structure. Figure 2-20 shows that a 90° rotation about one of the axes produces an orientation that is physically identical with the original orientation. For a 90° rotation about the $z$ axis the direction cosines are:

|     | $x$ | $y$ | $z$ |
| --- | --- | --- | --- |
| $x'$ | $a_{x'x} = 0$ | $a_{x'y} = 1$ | $a_{x'z} = 0$ |
| $y'$ | $a_{y'x} = -1$ | $a_{y'y} = 0$ | $a_{y'z} = 0$ |
| $z'$ | $a_{z'x} = 0$ | $a_{z'y} = 0$ | $a_{z'z} = 1$ |

Using these direction cosines in the equation for the transformation of the strain tensor, Eq. (2-46), the strain components in the $x'$, $y'$, $z'$ axes are

$$\varepsilon_{x'x'} = \varepsilon_{yy} \qquad \varepsilon_{x'y'} = -\varepsilon_{xy} \qquad \text{or } \tfrac{1}{2}\gamma_{x'y'} = -\tfrac{1}{2}\gamma_{xz}$$

$$\varepsilon_{y'y'} = \varepsilon_{xx} \qquad \varepsilon_{y'z'} = -\varepsilon_{xz}$$

$$\varepsilon_{z'z'} = \varepsilon_{zz} \qquad \varepsilon_{x'z'} = \varepsilon_{yz}$$

The components of stress in the $x'$, $y'$, $z'$ axes are found from Eq. (2-27) to be

$$\sigma_{x'x'} = \sigma_{yy} \qquad \tau_{x'y'} = -\tau_{xy}$$

$$\sigma_{y'y'} = \sigma_{xx} \qquad \tau_{y'z'} = -\tau_{xz}$$

$$\sigma_{z'z'} = \sigma_{zz} \qquad \tau_{x'z'} = \tau_{yz}$$

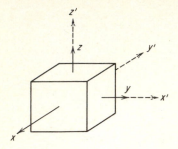

**Fig. 2-20** Symmetry rotation of cubic crystal about the z axis.

The normal stress in the $x'$ direction must equal the normal stress in the original $y$ direction. From Eq. (2-94)

$$\sigma_{x'x'} = C_{11}\varepsilon_{x'x'} + C_{12}\varepsilon_{y'y'} + C_{13}\varepsilon_{z'z'} + C_{14}\gamma_{y'z'} + C_{15}\gamma_{x'z'} + C_{16}\gamma_{x'y'}$$
$$= C_{11}\varepsilon_{yy} + C_{12}\varepsilon_{xx} + C_{13}\varepsilon_{zz} - \gamma_{xz} + C_{14}C_{15}\gamma_{yz} - C_{16}\gamma_{xy}$$

$$\sigma_{yy} = C_{21}\varepsilon_{xx} + C_{22}\varepsilon_{yy} + C_{23}\varepsilon_{zz} + C_{24}\gamma_{yz} + C_{25}\gamma_{xz} + C_{26}\gamma_{xy}$$

Since $\sigma_{x'x'} = \sigma_{yy}$,

$$C_{11}\varepsilon_{yy} + C_{12}\varepsilon_{xx} + C_{13}\varepsilon_{zz} - C_{14}\gamma_{xz} + C_{15}\gamma_{yz} - C_{16}\gamma_{xz}$$
$$= C_{21}\varepsilon_{xx} + C_{22}\varepsilon_{yy} + C_{23}\varepsilon_{zz} + C_{24}\gamma_{yz} + C_{25}\gamma_{xz} + C_{26}\gamma_{xy}$$

By making a term-by-term comparison of this equation, we conclude that $C_{11} = C_{22}$; $C_{12} = C_{21}$; $C_{13} = C_{23}$; $C_{15} = C_{24}$ and $C_{14} = C_{16} = C_{25} = C_{26} = 0$ since $-C_{14}\gamma_{xz} = C_{25}\gamma_{xz}$, etc. From the relationship $\sigma_{y'y'} = \sigma_{xx}$, we get the results $C_{11} = C_{22}$; $C_{12} = C_{21}$; $C_{23} = C_{13}$ and $C_{14} = C_{25} = 0$ (from above); $C_{15} = C_{24} = 0$; $C_{16} = C_{26} = 0$, since $-C_{14}\gamma_{yz} = C_{25}\gamma_{xz}$, etc.

By using the relationships between shear stresses, we eliminate additional constants. Thus, $\tau_{y'z'} = -\tau_{xz}$ results in $C_{41} = -C_{52} = 0$; $C_{42} = -C_{51} = 0$; $C_{43} = -C_{53} = 0$; $C_{45} = -C_{54} = 0$; $C_{44} = C_{55}$; $C_{46} = C_{56}$. From $\tau_{x'z'} = \tau_{yz}$, we get the additional relation that $C_{46} = -C_{56} = 0$.

Thus, by the 90° rotation about the $z$ axis, we have found that $C_{11} = C_{22}$; $C_{12} = C_{21}$; $C_{13} = C_{23}$; $C_{44} = C_{55}$, and that the following constants are zero: $C_{14}$, $C_{15}$, $C_{16}$, $C_{24}$, $C_{25}$, $C_{26}$, $C_{41}$, $C_{42}$, $C_{43}$, $C_{45}$, $C_{46}$, $C_{51}$, $C_{52}$, $C_{53}$, $C_{54}$, $C_{56}$. By now considering in turn 90° rotations about the $x$ and $y$ axes, it will be found that the elastic constants for a cubic crystal obey the relationships

$$C_{11} = C_{22} = C_{33}$$
$$C_{12} = C_{13} = C_{23}(=C_{21} = C_{31} = C_{32})$$
$$C_{44} = C_{55} = C_{66}$$

with all other constants equal to zero. Thus, for cubic symmetry the matrix of 36 elastic constants reduces to

$$C_{ij} = \begin{bmatrix} C_{11} & C_{12} & C_{12} & 0 & 0 & 0 \\ C_{12} & C_{11} & C_{12} & 0 & 0 & 0 \\ C_{12} & C_{12} & C_{11} & 0 & 0 & 0 \\ 0 & 0 & 0 & C_{44} & 0 & 0 \\ 0 & 0 & 0 & 0 & C_{44} & 0 \\ 0 & 0 & 0 & 0 & 0 & C_{44} \end{bmatrix}$$

**Table 2-2  STIFFNESS AND COMPLIANCE CONSTANTS FOR CUBIC CRYSTALS**

| Metal | $C_{11}$ | $C_{12}$ | $C_{44}$ | $S_{11}$ | $S_{12}$ | $S_{44}$ |
|---|---|---|---|---|---|---|
| Aluminum | 10.82 | 6.13 | 2.85 | 1.57 | −0.57 | 3.51 |
| Copper | 16.84 | 12.14 | 7.54 | 1.49 | −0.62 | 1.33 |
| Iron | 23.70 | 14.10 | 11.60 | 0.80 | −0.28 | 0.86 |
| Tungsten | 50.10 | 19.80 | 15.14 | 0.26 | −0.07 | 0.66 |

Stiffness constants in units of $10^{11}$ dyn/cm$^2$.
Compliance in units of $10^{-12}$ cm$^2$/dyn.

For a cubic crystal the 21 constants reduce to only three independent elastic constants. A similar matrix of coefficients would hold for the elastic compliances of a cubic metal. Some typical values of these constants for cubic metals are given in Table 2-2. Note that the elastic stiffness constants $C_{ij}$ are not simply the reciprocals of $S_{ij}$ but must be determined by the inversion of the $S_{ij}$ matrix. For a cubic crystal structure

$$C_{11} = \frac{S_{11} + S_{12}}{(S_{11} - S_{12})(S_{11} + 2S_{12})}$$

$$C_{12} = \frac{-S_{12}}{(S_{11} - S_{12})(S_{11} + 2S_{12})} \tag{2-96}$$

$$C_{44} = \frac{1}{S_{44}}$$

Similar equations for converting stiffnesses to the compliances can be obtained simply by writing $C$ for $S$ and $S$ for $C$.

We have seen that the number of independent elastic constants decreases with increasing symmetry of the crystal structure:

| Crystal Structure | Number of Independent Constants |
|---|---|
| Triclinic | 21 |
| Monoclinic | 13 |
| Orthorhombic | 9 |
| Tetragonal | 6 |
| Hexagonal | 5 |
| Cubic | 3 |
| Isotropic | 2 |

An isotropic solid has only two independent elastic constants. If we work with elastic compliances and consider the independent constants to be $S_{11}$ and $S_{12}$, then we can arrive at an equivalence between the compliances and the more common elastic moduli.

**From Eq. (2-95)**

**From Eq. (2-64)**

$$\varepsilon_x = S_{11}\sigma_x + S_{12}\sigma_y + S_{12}\sigma_z \qquad \varepsilon_x = \frac{1}{E}[\sigma_x - v(\sigma_y + \sigma_z)]$$

$$\varepsilon_y = S_{12}\sigma_x + S_{11}\sigma_y + S_{12}\sigma_z \qquad \varepsilon_y = \frac{1}{E}[\sigma_y - v(\sigma_x + \sigma_z)]$$

$$\varepsilon_z = S_{12}\sigma_x + S_{12}\sigma_y + S_{11}\sigma_z \qquad \varepsilon_z = \frac{1}{E}[\sigma_z - v(\sigma_x + \sigma_y)]$$

$$\gamma_{yz} = S_{44}\tau_{yz} \qquad \gamma_{yz} = \frac{1}{G}\tau_{yz}$$

$$\gamma_{xz} = S_{44}\tau_{xz} \qquad \gamma_{xz} = \frac{1}{G}\tau_{xz}$$

$$\gamma_{xy} = S_{44}\tau_{xy} \qquad \gamma_{xy} = \frac{1}{G}\tau_{xy}$$

It is apparent by comparing the two sets of equations that

$$S_{11} = \frac{1}{E} \qquad S_{12} = -\frac{v}{E} \qquad S_{44} = \frac{1}{G}$$

Since $S_{11}$ and $S_{12}$ are the independent constants, their relationship to $S_{44}$ can be obtained from Eq. (2-68)

$$G = \frac{E}{2(1 + v)} = \frac{1}{2(1/E + v/E)}$$

$$G = \frac{1}{S_{44}} = \frac{1}{2(S_{11} - S_{12})} \qquad (2\text{-}97)$$

or

$$S_{44} = 2(S_{11} - S_{12})$$

Comparable equations relating the elastic stiffness constants can be developed from Eqs. (2-95) and (2-74).

$$C_{12} = \lambda \qquad \text{Lamé's constant}$$

$$C_{11} = 2G + \lambda \qquad (2\text{-}98)$$

$$C_{44} = \tfrac{1}{2}(C_{11} - C_{12})$$

The technical elastic moduli $E$, $v$, and $G$ are usually measured by direct static measurements in the tension or torsion tests. However, where more precise measurements are required or where measurements are required in small single-crystal specimens cut along specified directions, dynamic techniques using measurement of frequency or elapsed time are frequently employed. Dynamic measurements involve very small atomic displacements and low stresses compared with static modulus measurements. The velocity of propagation of a displacement down a cylindrical-crystal specimen is given by

$$v_x = \frac{\omega\lambda}{2\pi}\sqrt{\frac{E_x}{\rho}} \qquad (2\text{-}99)$$

where $\omega$ is the natural frequency of vibration of a stress pulse of wavelength $\lambda$ in a crystal of density $\rho$. Dynamic techniques consist of measuring either the natural frequency of vibration or the elapsed time for an ultrasonic pulse to travel down the specimen and return. Because the strain cycles produced in dynamic testing occur at high rates, there is very little time for heat transfer to take place. Thus, dynamic measurements of elastic constants are obtained under adiabatic conditions, while static elastic measurements are obtained under essentially isothermal conditions. There is a small difference between adiabatic and isothermal elastic moduli.[1]

$$E_{adi} = \frac{E_{iso}}{1 - \dfrac{E_{iso}\,T\alpha^2}{9c}} \qquad (2\text{-}100)$$

where $\alpha$ is the volume coefficient of thermal expansion and $c$ is the specific heat. Since the specific heat of a solid is large compared to a gas, the difference between adiabatic and isothermal moduli is not great and can be ignored for practical purposes.

An important class of engineering materials are *orthotropic* materials. An orthotropic material has three mutually perpendicular planes of elastic symmetry. Examples of orthotropic materials are wood, plywood, glass-fiber-reinforced polymer composites, and highly textured cold-rolled metal sheet. The constitutive equations for an orthotropic material can be derived from Eq. (2-94) by the procedure described earlier for cubic symmetry. If the orthogonal $x$, $y$, $z$ axes are the axes of orthotropic symmetry, then a $180°$ rotation about the $z$ axis, followed by another $180°$ rotation about the $x$ axis reduces the matrix of elastic stiffness to

$$C_{ij} = \begin{bmatrix} C_{11} & C_{12} & C_{13} & 0 & 0 & 0 \\ C_{12} & C_{22} & C_{23} & 0 & 0 & 0 \\ C_{13} & C_{23} & C_{33} & 0 & 0 & 0 \\ 0 & 0 & 0 & C_{44} & 0 & 0 \\ 0 & 0 & 0 & 0 & C_{55} & 0 \\ 0 & 0 & 0 & 0 & 0 & C_{66} \end{bmatrix}$$

A similar matrix exists for the elastic compliances. Thus, the stress-strain equations for an orthotropic material are given by

$$\begin{aligned} \varepsilon_x &= S_{11}\sigma_x + S_{12}\sigma_y + S_{13}\sigma_z \\ \varepsilon_y &= S_{12}\sigma_x + S_{22}\sigma_y + S_{23}\sigma_z \\ \varepsilon_z &= S_{13}\sigma_x + S_{23}\sigma_y + S_{33}\sigma_z \\ \varepsilon_{yz} &= S_{44}\tau_{yz} \\ \varepsilon_{xz} &= S_{55}\tau_{xz} \\ \varepsilon_{xy} &= S_{66}\tau_{xy} \end{aligned} \qquad (2\text{-}101)$$

[1] For a derivation of Eq. (2-100) see S. M. Edelglass, op. cit., pp. 294–297.

To relate Eq. (2-101) to the technical elastic moduli, consider a situation where the only stress is applied parallel to the $x$ axis of symmetry.

$$\varepsilon_x = S_{11}\sigma_x \qquad \varepsilon_y = S_{12}\sigma_x \qquad \varepsilon_z = S_{13}\sigma_x$$

From the first relationship it is apparent that $S_{11} = 1/E_x$.

$$S_{12} = \frac{\varepsilon_y}{\sigma_x} = \varepsilon_y \frac{S_{11}}{\varepsilon_x} = \frac{\varepsilon_y}{\varepsilon_x} S_{11} = -\frac{\nu_{yx}}{E_x}$$

and

$$S_{13} = \frac{\nu_{zx}}{E_x}$$

By applying stresses in the $y$ and $z$ axes, we could develop similar relationships. It can also be shown[1] that $\nu_{yx}E_x = \nu_{xy}E_y$, etc. Therefore, the constitutive equations for an orthotropic material involve nine independent constants

$$\varepsilon_x = \frac{1}{E_x}\sigma_x - \frac{\nu_{yx}}{E_y}\sigma_y - \frac{\nu_{zx}}{E_z}\sigma_z$$

$$\varepsilon_y = -\frac{\nu_{xy}}{E_x}\sigma_x + \frac{1}{E_y}\sigma_y - \frac{\nu_{zy}}{E_z}\sigma_z \qquad (2\text{-}102)$$

$$\varepsilon_z = -\frac{\nu_{xz}}{E_x}\sigma_x - \frac{\nu_{yz}}{E_y}\sigma_y + \frac{1}{E_z}\sigma_z$$

$$\gamma_{yz} = \frac{1}{G_{yz}}\tau_{yz} \qquad \gamma_{xz} = \frac{1}{G_{xz}}\tau_{xz} \qquad \gamma_{xy} = \frac{1}{G_{xy}}\tau_{xy}$$

## 2-15  THEORY OF ELASTICITY

The mathematical theory of elasticity entails a more detailed consideration of the stresses and strains in a loaded member than is required by the ordinary methods of analysis of strength of materials. The solutions arrived at by strength of materials are usually made mathematically simpler by the assumption of a strain distribution in the loaded member which satisfies the physical situation but may not be mathematically rigorous. In the theory of elasticity no simplifying assumptions are made concerning the strain distribution.

As with strength of materials, the first requirement for a solution is to satisfy the conditions of equilibrium. Figure 2-21 illustrates the forces acting on an element of the body for a plane-stress situation. Taking the summation of forces in the $x$ and $y$ directions results in

$$\Sigma P_x = \frac{\partial \sigma_x}{\partial x} + \frac{\partial \tau_{xy}}{\partial y} = 0$$

$$\Sigma P_y = \frac{\partial \sigma_y}{\partial y} + \frac{\partial \tau_{xy}}{\partial x} + \rho g = 0 \qquad (2\text{-}103)$$

[1] B. W. Shaffer, *SPE Trans.*, October 1964, pp. 267–276.

**Fig. 2-21**  Forces acting on volume element.

The term $\rho g$ arises from the consideration of the weight of the body, where $\rho$ is the mass per unit volume and $g$ is the acceleration of gravity. The above equations constitute the *equations of equilibrium* for plane stress. For a three-dimensional stress system there will be three equations, each containing one partial derivative of the normal stress and two partial derivatives of the shear stresses.

Equations (2-103) must be satisfied at all points throughout the body. Note that these equilibrium equations do not give a relationship between the stresses and the external loads. Instead, they give the rate of change of the stresses at any point in the body. However, the relationship between stress and external load must be such that at the boundary of the body the stresses become equal to the surface forces per unit area; i.e., it must satisfy the boundary conditions.

One of the important requirements of the theory of elasticity is that the deformation of each element must be such that continuity is preserved. Physically, this means that the displacements must vary smoothly throughout the body so that no gaps occur in the material. This requires that the displacement components $u_i$ must be continuous, single-valued functions of the coordinates $x_i$. Certain relationships between the strain components must be satisfied to meet this requirement. These relationships are the *equations of compatibility*.

For example, for the two-dimensional case we have

$$(a) \quad \varepsilon_x = \frac{\partial u}{\partial x} \qquad (b) \quad \varepsilon_y = \frac{\partial v}{\partial y} \qquad (c) \quad \gamma_{xy} = \frac{\partial u}{\partial y} + \frac{\partial v}{\partial x}$$

There must be a definite relationship between the three strain components because they are expressed in terms of two displacements $u$ and $v$. Differentiating Eq. (a) twice with respect to $y$, Eq. (b) twice with respect to $x$, and Eq. (c) with respect to both $x$ and $y$ results in

$$\frac{\partial^2 \varepsilon_x}{\partial y^2} + \frac{\partial^2 \varepsilon_y}{\partial x^2} = \frac{\partial^2 \gamma_{xy}}{\partial x\,\partial y} \qquad (2\text{-}104)$$

Further manipulation leads to

$$2\frac{\partial^2 \varepsilon_x}{\partial y\, \partial z} = \frac{\partial}{\partial x}\left(-\frac{\partial \gamma_{yz}}{\partial x} + \frac{\partial \gamma_{xz}}{\partial y} + \frac{\partial \gamma_{xy}}{\partial z}\right) \qquad (2\text{-}105)$$

There are each two other compatibility equations of the types given by Eqs. (2-104) and (2-105). These may be obtained most easily by permuting the subscripts $x$, $y$, and $z$.

Basically, the solution of a problem with the theory of elasticity involves the determination of the displacements $u_i$, the strains $\varepsilon_{ij}$ and the stresses $\sigma_{ij}$ throughout a body subjected to a specified external force system. Thus, in general there are 15 unknowns, but there are in turn 15 equations which relate these unknowns in the $x$, $y$, $z$ coordinate system. These equations are: the six equations relating strains with displacements, Eq. (2-42); the six constitutive equations relating strain with stress, Eqs. (2-64) and (2-65); and the three equilibrium equations (2-103). These field equations must be satisfied at all points inside an elastic solid in equilibrium. Once these equations are satisfied the compatibility equations will be satisfied automatically. The mathematical requirements for a general solution of a three-dimensional problem are quite formidable. However, mathematical techniques for obtaining solutions have been well established, particularly for two-dimensional problems. For problems with complicated geometry and loading systems finite-difference techniques and relaxation methods are useful approaches.

The equations of equilibrium and the equations of compatibility have been presented here within the context of the theory of elasticity. However, it should be emphasized that these are general equations applicable to any continuum. But, with a plastic solid or a viscoelastic material the constitutive equations are more complex and complete solutions to problems are even more difficult.

## 2-16 STRESS CONCENTRATION

A geometrical discontinuity in a body, such as a hole or a notch, results in a nonuniform stress distribution at the vicinity of the discontinuity. At some region near the discontinuity the stress will be higher than the average stress at distances removed from the discontinuity. Thus, a *stress concentration* occurs at the discontinuity, or *stress raiser*. Figure 2-22a shows a plate containing a circular hole which is subjected to a uniaxial load. If the hole were not present, the stress would be uniformly distributed over the cross section of the plate and it would be equal to the load divided by the cross-sectional area of the plate. With the hole present, the distribution is such that the axial stress reaches a high value at the edges of the hole and drops off rapidly with distance away from the hole.

**Fig. 2-22** Stress distributions due to (*a*) circular hole and (*b*) elliptical hole.

The stress concentration is expressed by a theoretical stress-concentration factor $K_t$. Generally $K_t$ is described as the ratio of the maximum stress to the nominal stress based on the net section, although some workers use a value of nominal stress based on the entire cross section of the member in a region where there is no stress concentrator.

$$K_t = \frac{\sigma_{max}}{\sigma_{nominal}} \qquad (2\text{-}106)$$

In addition to producing a stress concentration, a notch also creates a localized condition of biaxial or triaxial stress. For example, for the circular hole in a plate subjected to an axial load, a radial stress is produced as well as a longitudinal stress. From elastic analysis,[1] the stresses produced in an infinitely wide plate containing a circular hole and axially loaded can be expressed as

$$\sigma_r = \frac{\sigma}{2}\left(1 + \frac{a^2}{r^2}\right) + \frac{\sigma}{2}\left(1 + 3\frac{a^4}{r^4} - 4\frac{a^2}{r^2}\right)\cos 2\theta$$

$$\sigma_\theta = \frac{\sigma}{2}\left(1 + \frac{a^2}{r^2}\right) - \frac{\sigma}{2}\left(1 + 3\frac{a^4}{r^4}\right)\cos 2\theta \qquad (2\text{-}107)$$

$$\tau = -\frac{\sigma}{2}\left(1 - 3\frac{a^4}{r^4} + 2\frac{a^2}{r^2}\right)\sin 2\theta$$

Examination of these equations shows that the maximum stress occurs at point *A* when $\theta = \pi/2$ and $r = a$. For this case

$$\sigma_\theta = 3\sigma = \sigma_{max} \qquad (2\text{-}108)$$

[1] Timoshenko and Goodier, *op. cit.*, pp. 78–81.

where $\sigma$ is the uniform tensile stress applied at the ends of the plate.  The theoretical stress-concentration factor for a plate with a circular hole is therefore equal to 3. Further study of these equations shows that $\sigma_\theta = -\sigma$ for $r = a$ and $\theta = 0$.  Therefore, when a tensile stress is applied to the plate, a compressive stress of equal magnitude exists at the edge of the hole at point $B$ in a direction perpendicular to the axis of loading in the plane of the plate.

(a)

(b)

(c)

**Fig. 2-23**  Theoretical stress-concentration factors for different geometrical shapes. (*After G. H. Neugebauer, Prod. Eng. (NY), vol. 14, pp. 82–87, 1943.*)

Another interesting case for which an analytical solution for the stress concentration is available[1] is the case of a small elliptical hole in a plate. Figure

[1] C. E. Inglis, *Trans. Inst. Nav. Archit.*, pt. 1, pp. 219–230, 1913.

2-22*b* shows the geometry of the hole. The maximum stress at the ends of the hole is given by the equation

$$\sigma_{max} = \sigma \left(1 + 2\frac{a}{b}\right) \qquad (2\text{-}109)$$

Note that, for a circular hole ($a = b$), the above equation reduces to Eq. (2-108). Equation (2-109) shows that the stress increases with the ratio $a/b$. Therefore, a very narrow hole, such as a crack, normal to the tensile direction will result in a very high stress concentration.

Mathematical complexities prevent the calculation of elastic stress-concentration factors in all but the simplest geometrical cases. Much of this work has been compiled by Neuber,[1] who has made calculations for various types of notches. Stress-concentration factors for practical problems are usually determined by experimental methods.[2] Photoelastic analysis[3] of models is the most widely used technique. This method is especially applicable to plane-stress problems, although it is possible to make three-dimensional photoelastic analyses. Figure 2-23 shows typical curves for the theoretical stress-concentration factor of certain machine elements that were obtained by photoelastic methods. Much of the information on stress concentrations in machine parts has been collected by Peterson.[4]

The effect of a stress raiser is much more pronounced in a brittle material than in a ductile material. In a ductile material, plastic deformation occurs when the yield stress is exceeded at the point of maximum stress. Further increase in load produces a local increase in strain at the critically stressed region with little increase in stress. Because of strain hardening, the stress increases in regions adjacent to the stress raiser, until, if the material is sufficiently ductile, the stress distribution becomes essentially uniform. Thus, a ductile metal loaded statically will not develop the full theoretical stress-concentration factor. However, redistribution of stress will not occur to any extent in a brittle material, and therefore a stress concentration of close to the theoretical value will result. Although stress raisers are not usually dangerous in ductile materials subjected to static loads, appreciable stress-concentration effects will occur in ductile materials under fatigue conditions of alternating stresses. Stress raisers are very important in the fatigue failure of metals and will be discussed further in Chap. 12.

[1] H. Neuber, "Theory of Notch Stresses," English translation, J. W. Edwards, Publisher, Incorporated, Ann Arbor, Mich., 1946.
[2] M. Hetenyi, "Handbook on Experimental Stress Analysis," John Wiley & Sons, Inc., New York, 1950.
[3] M. M. Frocht, "Photoelasticity," John Wiley & Sons, Inc., New York, 1955.
[4] R. E. Peterson, "Stress-Concentration Factors," John Wiley & Sons, Inc., New York, 1974.

# BIBLIOGRAPHY

Chou, P. C., and N. J. Pagano: "Elasticity," D. Van Nostrand Company, Inc., Princeton, N.J., 1967.

Jaeger, J. C.: "Elasticity, Fracture, and Flow," 2d ed., Methuen & Co., Ltd., London, 1962.

Love, A. E. H.: "A Treatise on the Mathematical Theory of Elasticity," 4th ed., Dover Publications, Inc., New York, 1949.

Reid, C. N.: "Deformation Geometry for Materials Scientists," Pergamon Press, New York, 1973.

Suh, N. P., and A. P. L. Turner: "Elements of the Mechanical Behavior of Solids," McGraw-Hill Book Company, New York, 1974.

Timoshenko, S. P., and J. N. Goodier: "Theory of Elasticity," 2d ed., McGraw-Hill Book Company, New York, 1951.

Wang, C. T.: "Applied Elasticity," McGraw-Hill Book Company, New York, 1953.

# 3

# ELEMENTS OF THE THEORY OF PLASTICITY

## 3-1 INTRODUCTION

The theory of plasticity deals with the behavior of materials at strains where
Hooke's law is no longer valid. A number of aspects of plastic deformation make
the mathematical formulation of a theory of plasticity more difficult than the
description of the behavior of an elastic solid. For example, plastic deformation is
not a reversible process like elastic deformation. Elastic deformation depends only
on the initial and final states of stress and strain, while the plastic strain depends on
the loading path by which the final state is achieved. Moreover, in plastic deformation there is no easily measured constant relating stress to strain as with Young's
modulus for elastic deformation. The phenomenon of strain hardening is difficult
to accommodate within the theory of plasticity without introducing considerable
mathematical complexity. Also, several aspects of real material behavior, such as
plastic anisotropy, elastic hysteresis, and the Bauschinger effect (see Sec. 3-2) cannot
be treated easily by plasticity theory. Nevertheless, the theory of plasticity has been
one of the most active areas of continuum mechanics, and considerable progress
has been made in developing a theory which can solve important engineering
problems.

The theory of plasticity is concerned with a number of different types of problems. From the viewpoint of design, plasticity is concerned with predicting the maximum load which can be applied to a body without causing excessive yielding. The yield criterion must be expressed in terms of stress in such a way that it is valid for all states of stress. The designer is also concerned with plastic deformation in problems where the body is purposely stressed beyond the yield stress into the plastic region. For example, plasticity must be considered in designing for processes such as autofrettage, shrink fitting, and the overspeeding of rotor disks. The consideration of small plastic strains allows economies in building construction through the use of the theory of limit design.

The analysis of large plastic strains is required in the mathematical treatment of the plastic forming of metals. This aspect of plasticity will be considered in Part Four. It is very difficult to describe, in a rigorous analytical way, the behavior of a metal under these conditions. Therefore, certain simplifying assumptions are usually necessary to obtain a tractable mathematical solution.

Another aspect of plasticity is concerned with acquiring a better under-standing of the mechanism of the plastic deformation of metals. Interest in this field is centered on the imperfections in crystalline solids. The effect of metallurgical variables, crystal structure, and lattice imperfections on the deformation behavior are of chief concern. This aspect of plasticity is considered in Part Two.

## 3-2   THE FLOW CURVE

The stress-strain curve obtained by uniaxial loading, as in the ordinary tension test, is of fundamental interest in plasticity when the curve is plotted in terms of true stress $\sigma$ and true strain $\varepsilon$. True stress and true strain are discussed in the next section. The purpose of this section[1] is to describe typical stress-strain curves for real metals and to compare them with the theoretical flow curves for ideal materials.

The true stress-strain curve for a typical ductile metal, such as aluminum, is illustrated in Fig. 3-1$a$. Hooke's law is followed up to some yield stress $\sigma_0$. (The value of $\sigma_0$ will depend upon the accuracy with which strain is measured.) Beyond $\sigma_0$, the metal deforms plastically. Most metals strain-harden in this region, so that increases in strain require higher values of stress than the initial yield stress $\sigma_0$. However, unlike the situation in the elastic region, the stress and strain are not related by any simple constant of proportionality. If the metal is strained to point $A$, when the load is released the total strain will immediately decrease from $\varepsilon_1$ to $\varepsilon_2$ by an amount $\sigma/E$. The strain decrease $\varepsilon_1 - \varepsilon_2$ is the *recoverable elastic strain*.

[1] See Chap. 9 for a more complete discussion of the mathematics of the true stress-strain curve.

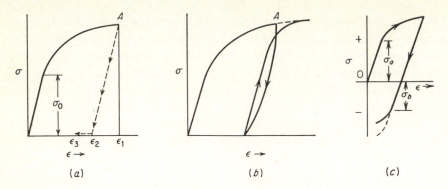

**Fig. 3-1** Typical true stress-strain curves for a ductile metal.

However, the strain remaining is not all permanent plastic strain. Depending upon the metal and the temperature, a small amount of the plastic strain $\varepsilon_2 - \varepsilon_3$ will disappear with time. This is known as *anelastic behavior*. Generally the anelastic strain is neglected in mathematical theories of plasticity.

Usually the stress-strain curve on unloading from a plastic strain will not be exactly linear and parallel to the elastic portion of the curve (Fig. 3-1*b*). Moreover, on reloading the curve will generally bend over as the stress approaches the original value of stress from which it was unloaded. With a little additional plastic strain the stress-strain curve becomes a continuation of what it would have been had no unloading taken place. The hysteresis behavior resulting from unloading and loading from a plastic strain is generally neglected in plasticity theories.

If a specimen is deformed plastically beyond the yield stress in one direction, e.g., in tension, and then after unloading to zero stress it is reloaded in the opposite direction, e.g., in compression, it is found that the yield stress on reloading is less than the original yield stress. Referring to Fig. 3-1*c*, $\sigma_b < \sigma_a$. This dependence of the yield stress on loading path and direction is called the *Bauschinger effect*. The Bauschinger effect is commonly ignored in plasticity theory, and it is usual to assume that the yield stress in tension and compression are the same.

A true stress-strain curve is frequently called a *flow curve* because it gives the stress required to cause the metal to flow plastically to any given strain. Many attempts have been made to fit mathematical equations to this curve. The most common is a power expression of the form

$$\sigma = K\varepsilon^n \qquad (3\text{-}1)$$

where $K$ is the stress at $\varepsilon = 1.0$ and $n$, the strain-hardening coefficient, is the slope of a log-log plot of Eq. (3-1). This equation can be valid only from the beginning of plastic flow to the maximum load at which the specimen begins to neck down.

**Fig. 3-2**  Idealized flow curves. (*a*) Rigid ideal plastic material; (*b*) ideal plastic material with elastic region; (*c*) piecewise linear (strain-hardening) material.

Even the simple mathematical expression for the flow curve that is given by Eq. (3-1) can result in considerable mathematical complexity when it is used with the equations of the theory of plasticity. Therefore, in this field it is common practice to devise idealized flow curves which simplify the mathematics without deviating too far from physical reality. Figure 3-2a shows the flow curve for a *rigid, perfectly plastic* material. For this idealized material, a tensile specimen is completely rigid (zero elastic strain) until the axial stress equals $\sigma_0$, whereupon the material flows plastically at a constant flow stress (zero strain hardening). This type of behavior is approached by a ductile metal which is in a highly cold worked condition. Figure 3-2b illustrates the flow curve for a perfectly plastic material with an elastic region. This behavior is approached by a material such as plain carbon steel which has a pronounced yield-point elongation (see Sec. 6-5). A more realistic approach is to approximate the flow curve by two straight lines corresponding to the elastic and plastic regions (Fig. 3-2c). This type of curve results in somewhat more complicated mathematics.

## 3-3   TRUE STRESS AND TRUE STRAIN

The engineering stress-strain curve does not give a true indication of the deformation characteristics of a material because it is based entirely on the original dimensions of the specimen, and these dimensions change continuously during the test. Also in metalworking processes, such as wiredrawing, the workpiece undergoes appreciable change in cross-sectional area. Thus, measures of stress and strain which are based on the instantaneous dimensions are needed. Since dimensional changes are small in elastic deformation, it was not necessary to make this distinction in the previous chapter.

Equation (1-1) describes the conventional concept of unit linear strain namely, the change in length referred to the original unit length.

$$e = \frac{\Delta L}{L_0} = \frac{1}{L_0} \int_{L_0}^{L} dL$$

This definition of strain is satisfactory for elastic strains where $\Delta L$ is very small. However, in plastic deformation the strains are frequently large, and during the extension the gage length changes considerably. Ludwik[1] first proposed the definition of true strain, or natural strain, $\varepsilon$, which obviates this difficulty. In this definition of strain the change in length is referred to the instantaneous gage length, rather than to the original gage length.

$$\varepsilon = \sum \frac{L_1 - L_0}{L_0} + \frac{L_2 - L_1}{L_1} + \frac{L_3 - L_2}{L_2} + \cdots \qquad (3\text{-}2)$$

or

$$\varepsilon = \int_{L_0}^{L} \frac{dL}{L} = \ln \frac{L}{L_0} \qquad (3\text{-}3)$$

The relationship between true strain and conventional linear strain[2] follows from Eq. (1-1).

$$e = \frac{\Delta L}{L_0} = \frac{L - L_0}{L_0} = \frac{L}{L_0} - 1$$

$$e + 1 = \frac{L}{L_0}$$

$$\varepsilon = \ln \frac{L}{L_0} = \ln (e + 1) \qquad (3\text{-}4)$$

Values of true strain and conventional linear strain are given for comparison:

| True strain $\varepsilon$ | 0.01 | 0.10 | 0.20 | 0.50 | 1.0 | 4.0 |
|---|---|---|---|---|---|---|
| Conventional Strain $e$ | 0.01 | 0.105 | 0.22 | 0.65 | 1.72 | 53.6 |

Thus, the two measures of strain give nearly identical results up to a strain of 0.1.

The advantage of using true strain should be apparent from the following example: Consider a uniform cylinder which is extended to twice its

---

[1] P. Ludwik, "Elemente der technologischen Mechanik," Springer-Verlag OHG, Berlin, 1909.

[2] The reader is warned against confusion in notation for various measures of strain. We in general shall use $\varepsilon$ for strain. The true strain will be implied unless otherwise specified. For elastic strains ($\varepsilon < 0.01$) the numerical values of $\varepsilon$ and $e$ are identical and on occasion we may use $e$ for linear strain, especially where we want to designate a small elastic strain. In other texts true strain is sometimes denoted by $\delta$ or $\bar{\varepsilon}$.

original length. The linear strain is then $e = (2L_0 - L_0)/L_0 = 1.0$, or a strain of 100 percent. To achieve the same amount of negative linear strain in compression, the cylinder would have to be squeezed to zero thickness. However, intuitively we should expect that the strain produced in compressing a cylinder to half its original length would be the same as, although opposite in sign to, the strain produced by extending the cylinder to twice its length. If true strain is used, equivalence is obtained for the two cases. For extension to twice the original length, $\varepsilon = \ln (2L_0/L_0) = \ln 2$. For compression to half the original length, $\varepsilon = \ln [(L_0/2)/L_0] = \ln \frac{1}{2} = -\ln 2$.

Another advantage of working with true strain is that the total true strain is equal to the sum of the incremental true strains. This can be seen from the following example. Consider a rod initially 2.0 in long that is elongated in three increments, each increment being a conventional strain of $e = 0.1$.

| Increment | Length of Rod | |
|-----------|---------------|---|
| 0 | 2.00 | |
| 1 | 2.20 | $e_{0-1} = 0.2/2.0 = 0.1$ |
| 2 | 2.42 | $e_{1-2} = 0.22/2.20 = 0.1$ |
| 3 | 2.662 | $e_{2-3} = 0.242/2.42 = 0.1$ |

We note that the total conventional strain $e_{0-3} = 0.662/2.0 = 0.331$ is not equal to $e_{0-1} + e_{1-2} + e_{2-3}$. However, if we use true strain, the sum of the increments equals the total strain.

$$\varepsilon_{0-1} + \varepsilon_{1-2} + \varepsilon_{2-3} = \ln \frac{2.2}{2.0} + \ln \frac{2.42}{2.2} + \ln \frac{2.662}{2.42} = \ln \frac{2.662}{2.0} = \varepsilon_{0-3}$$

One of the basic characteristics of plastic deformation is that a metal is essentially incompressible. The density changes measured on metals after large plastic strain are less than 0.1 percent. Therefore, as a good engineering approximation we can consider that the *volume* of a solid *remains constant* during plastic deformation.

In Sec. 2-8 we determined the volume strain by considering a cube of initial volume $dx\, dy\, dz$ which when deformed had a volume $dx\, (1 + e_x)\, dy\, (1 + e_y)\, dz\, (1 + e_z)$. The volume strain $\Delta$ is given by

$$\Delta = \frac{\Delta V}{V} = \frac{(1 + e_x)(1 + e_y)(1 + e_z)\, dx\, dy\, dz - dx\, dy\, dz}{dx\, dy\, dz}$$

or 

$$\Delta = (1 + e_x)(1 + e_y)(1 + e_z) - 1$$

When we previously determined $\Delta$ for small elastic strains, it was permissible to neglect products of strains compared to the strain itself, but this is no longer possible when the larger plastic strains are being considered. Since the volume change is zero for plastic deformation,

$$\Delta + 1 = 0 + 1 = (1 + e_x)(1 + e_y)(1 + e_z)$$

or

$$\ln 1 = 0 = \ln (1 + e_x) + \ln (1 + e_y) + \ln (1 + e_z)$$

But $\varepsilon_x = \ln (1 + e_x)$, etc.

$$\therefore \varepsilon_x + \varepsilon_y + \varepsilon_z = \varepsilon_1 + \varepsilon_2 + \varepsilon_3 = 0 \qquad (3\text{-}5)$$

Equation (3-5) represents the first invariant of the strain tensor when strain is expressed as true strain. It is a very useful relationship in plasticity problems. Note particularly that Eq. (3-5) is not valid for elastic strains since there is an appreciable volume change relative to the magnitude of the elastic strains. Thus, if we add up the three equations for Hooke's law (2-64)

$$\Delta = e_x + e_y + e_z = \frac{1 - 2v}{E} (\sigma_x + \sigma_y + \sigma_z)$$

we see that $\Delta$ can be zero only if $v = \frac{1}{2}$. This result can be interpreted that Poisson's ratio is equal to $\frac{1}{2}$ for a plastic material for which $\Delta = 0$.

Because of constancy of volume $A_0 L_0 = AL$, and Eq. (3-3) can be written in terms of either length or area.

$$\varepsilon = \ln \frac{L}{L_0} = \ln \frac{A_0}{A} \qquad (3\text{-}6)$$

*True stress* is the load at any instant divided by the cross-sectional area over which it acts. The *engineering stress*, or conventional stress, is the load divided by the original area. In considering elastic behavior it was not necessary to make this distinction, but in certain problems in plasticity, particularly when dealing with the mathematics of the tension test (Chap. 9), it is important to distinguish between these two definitions of stress. True stress will be denoted by the familiar symbol $\sigma$, while engineering stress will be denoted by $s$.

True stress

$$\sigma = \frac{P}{A} \qquad (3\text{-}7)$$

Engineering stress

$$s = \frac{P}{A_0} \qquad (3\text{-}8)$$

The true stress may be determined from the engineering stress as follows:

$$\sigma = \frac{P}{A} = \frac{P}{A_0} \frac{A_0}{A}$$

But, by the constancy-of-volume relationships

$$\frac{A_0}{A} = \frac{L}{L_0} = e + 1$$

$$\therefore \sigma = \frac{P}{A_0}(e+1) = s(e+1) \qquad (3\text{-}9)$$

## 3-4 YIELDING CRITERIA FOR DUCTILE METALS

The problem of deducing mathematical relationships for predicting the conditions at which plastic yielding begins when a material is subjected to any possible combination of stresses is an important consideration in the field of plasticity. In uniaxial loading, as in a tension test, macroscopic plastic flow begins at the yield stress $\sigma_0$. It is expected that yielding under a situation of combined stresses can be related to some particular combination of principal stresses. There is at present no theoretical way of calculating the relationship between the stress components to correlate yielding for a three-dimensional state of stress with yielding in the uniaxial tension test.

The yielding criteria are essentially empirical relationships. However, a yield criterion must be consistent with a number of experimental observations, the chief of which is that pure hydrostatic pressure does not cause yielding in a continuous solid.[1] As a result of this, the hydrostatic component of a complex state of stress does not influence the stress at which yielding occurs. Therefore, we look for the stress deviator to be involved with yielding. Moreover, for an isotropic material, the yield criterion must be independent of the choice of axes, i.e., it must be an invariant function. These considerations lead to the conclusion that the yield criteria must be some function of the invariants of the stress deviator. At present there are two generally accepted criteria for predicting the onset of yielding in ductile metals.

### VON MISES' OR DISTORTION-ENERGY CRITERION

Von Mises (1913) proposed that yielding would occur when the second invariant of the stress deviator $J_2$ exceeded some critical value.

$$J_2 = k^2 \qquad (3\text{-}10)$$

where $J_2 = \frac{1}{6}[(\sigma_1 - \sigma_2)^2 + (\sigma_2 - \sigma_3)^2 + (\sigma_3 - \sigma_1)^2]$.

---

[1] A significant influence of hydrostatic or mean stress of modest values on yielding has been observed in glassy polymers such as PMMA. S. S. Sternstein and L. Ongchin, *Polym. Prepr. Am. Chem. Soc. Div. Polym. Chem.*, September 1969.

To evaluate the constant $k$ and relate it to yielding in the tension test, we realize that at yielding in uniaxial tension $\sigma_1 = \sigma_0$, $\sigma_2 = \sigma_3 = 0$

$$\sigma_0{}^2 + \sigma_0{}^2 = 6k^2$$

$$\sigma_0 = \sqrt{3k} \qquad (3\text{-}11)$$

Substituting Eq. (3-11) in Eq. (3-10) results in the usual form of the von Mises' yield criterion

$$\sigma_0 = \frac{1}{\sqrt{2}} [(\sigma_1 - \sigma_2)^2 + (\sigma_2 - \sigma_3)^2 + (\sigma_3 - \sigma_1)^2]^{\frac{1}{2}} \qquad (3\text{-}12)$$

or from Eq. (2-61)

$$\sigma_0 = \frac{1}{\sqrt{2}} [(\sigma_x - \sigma_y)^2 + (\sigma_y - \sigma_z)^2 + (\sigma_z - \sigma_x)^2 + 6(\tau_{xy}{}^2 + \tau_{yz}{}^2 + \tau_{xz}{}^2)]^{\frac{1}{2}} \qquad (3\text{-}13)$$

Equation (3-12) or (3-13) predicts that yielding will occur when the differences of stresses on the right side of the equation exceed the yield stress in uniaxial tension $\sigma_0$.

To identify the constant $k$ in Eq. (3-10), consider the state of stress in pure shear, as is produced in a torsion test.

$$\sigma_1 = -\sigma_3 = \tau \qquad \sigma_2 = 0$$

at yielding $\qquad \sigma_1{}^2 + \sigma_1{}^2 + 4\sigma_1{}^2 = 6k^2$

$$\therefore \sigma_1 = k$$

so that $k$ represents the yield stress in pure shear (torsion). Therefore, the von Mises' criterion predicts that the yield stress in torsion will be less than in uniaxial tension according to

$$k = \frac{1}{\sqrt{3}} \sigma_0 = 0.577\sigma_0 \qquad (3\text{-}14)$$

To summarize, note that the von Mises' yield criterion implies that yielding is not dependent on any particular normal stress or shear stress, but instead, yielding depends on a function of all three values of principal shearing stress. Since the yield criterion is based on differences of normal stresses, $\sigma_1 - \sigma_2$, etc., the criterion is independent of the component of hydrostatic stress. Since the von Mises' yield criterion involves squared terms, the result is independent of the sign of the individual stresses. This is an important advantage since it is not necessary to know which are the largest and smallest principal stresses in order to use this yield criterion.

Von Mises originally proposed this criterion because of its mathematical simplicity. Subsequently, other workers have attempted to give it physical meaning.

Hencky (1924) showed that Eq. (3-12) was equivalent to assuming that yielding occurs when the *distortion energy* reaches a critical value. The distortion energy is that part of the total strain energy per unit volume that is involved in change of shape as opposed to a change in volume.

The fact that the total strain energy can be split into a term depending on change of volume and a term depending on distortion can be seen by expressing Eq. (2-84) in terms of principal stresses.

$$U_0 = \frac{1}{2E}[\sigma_1^2 + \sigma_2^2 + \sigma_3^2 - 2v(\sigma_1\sigma_2 + \sigma_2\sigma_3 + \sigma_1\sigma_3)] \qquad (3\text{-}15)$$

or expressing in terms of the invariants of the stress tensor

$$U_0 = \frac{1}{2E}[I_1^2 - 2I_2(1+v)] \qquad (3\text{-}16)$$

This equation is more meaningful if we express it in terms of the bulk modulus (volume change) and the shear modulus (distortion). From Sec. 2-11,

$$E = \frac{9GK}{3K+G} \qquad v = \frac{3K-2G}{6K+2G}$$

Substituting into Eq. (3-16)

$$U_0 = \frac{I_1^2}{18K} + \frac{1}{6G}(I_1^2 - 3I_2) \qquad (3\text{-}17)$$

Equation (3-17) is important because it shows that the total strain energy can be split into a term depending on change of volume and a term depending on distortion.

$$(U_0)_{\text{distortion}} = \frac{1}{6G}(\sigma_1^2 + \sigma_2^2 + \sigma_3^2 - \sigma_1\sigma_2 - \sigma_2\sigma_3 - \sigma_1\sigma_3)$$

or

$$(U_0)_{\text{distortion}} = \frac{1}{12G}[(\sigma_1 - \sigma_2)^2 + (\sigma_2 - \sigma_3)^2 + (\sigma_3 - \sigma_1)^2] \qquad (3\text{-}18)$$

For a uniaxial state of stress, $\sigma_1 = \sigma_0$, $\sigma_2 = \sigma_3 = 0$

$$(U_0)_{\text{distortion}} = \frac{1}{12G}2\sigma_0^2$$

or

$$\sigma_0 = \frac{1}{\sqrt{2}}[(\sigma_1 - \sigma_2)^2 + (\sigma_2 - \sigma_3)^2 + (\sigma_3 - \sigma_1)^2] \qquad (3\text{-}19)$$

Another physical interpretation given to the von Mises' yield criterion is that it represents the critical value of the octahedral shear stress (see Sec. 3-6). This is the shear stress on the octahedral planes which make equal angles with the principal axes. Still another interpretation is that it represents the mean square of the shear stress averaged over all orientations in the solid.[1]

## MAXIMUM-SHEAR-STRESS OR TRESCA CRITERION

This yield criterion assumes that yielding occurs when the maximum shear stress reaches the value of the shear stress in the uniaxial-tension test. From Eq. (2-21), the maximum shear stress is given by

$$\tau_{max} = \frac{\sigma_1 - \sigma_3}{2} \qquad (3\text{-}20)$$

where $\sigma_1$ is the algebraically largest and $\sigma_3$ is the algebraically smallest principal stress.

For uniaxial tension, $\sigma_1 = \sigma_0$, $\sigma_2 = \sigma_3 = 0$, and the shearing yield stress $\tau_0$ is equal to $\sigma_0/2$. Substituting in Eq. (3-20),

$$\tau_{max} = \frac{\sigma_1 - \sigma_3}{2} = \tau_0 = \frac{\sigma_0}{2}$$

Therefore, the maximum-shear-stress criterion is given by

$$\sigma_1 - \sigma_3 = \sigma_0 \qquad (3\text{-}21)$$

For a state of pure shear, $\sigma_1 = -\sigma_3 = k$, $\sigma_2 = 0$, the maximum-shear-stress criterion predicts that yielding will occur when

$$\sigma_1 - \sigma_3 = 2k = \sigma_0$$

or

$$k = \frac{\sigma_0}{2}$$

so that the maximum-shear-stress criterion may be written

$$\sigma_1 - \sigma_3 = \sigma_1' - \sigma_3' = 2k \qquad (3\text{-}22)$$

We note that the maximum-shear-stress criterion is less complicated mathematically than the von Mises' criterion, and for this reason it is often used in engineering design. However, the maximum-shear criterion does not take into consideration the intermediate principal stress. It suffers from the major difficulty that it is necessary to know in advance which are the maximum and minimum principal stresses. Moreover, the general form of the maximum-shear-stress criterion, Eq. (3-23), is far more complicated than the von Mises' criterion, Eq. (3-10), and for this reason the von Mises' criterion is preferred in most theoretical work.

$$4J_2{}^3 - 27J_3{}^2 - 36k^2J_2{}^2 + 96k^4J_2 - 64k^6 = 0 \qquad (3\text{-}23)$$

---

[1] See G. Sines, "Elasticity and Strength," pp. 54–56, Allyn and Bacon, Inc., Boston, 1969.

## 3-5  COMBINED STRESS TESTS

The conditions for yielding under states of stress other than uniaxial and torsion loading can be studied conveniently with thin-wall tubes. Axial stress can be combined with torsion to produce various combinations of shear stress to normal stress intermediate between the values obtained separately in tension and torsion. Alternatively, a hydrostatic pressure may be introduced to produce a circumferential hoop stress in the tube.[1]

For the stresses shown in Fig. 3-3, from Eq. (2-9) the principal stresses are

$$\sigma_1 = \frac{\sigma_x}{2} + \left(\frac{\sigma_x^2}{4} + \tau_{xy}^2\right)^{1/2}$$

$$\sigma_2 = 0 \tag{3-24}$$

$$\sigma_3 = \frac{\sigma_x}{2} - \left(\frac{\sigma_x^2}{4} + \tau_{xy}^2\right)^{1/2}$$

Therefore, the maximum-shear-stress criterion of yielding is given by

$$\left(\frac{\sigma_x}{\sigma_0}\right)^2 + 4\left(\frac{\tau_{xy}}{\sigma_0}\right)^2 = 1 \tag{3-25}$$

and the distortion-energy theory of yielding is expressed by

$$\left(\frac{\sigma_x}{\sigma_0}\right)^2 + 3\left(\frac{\tau_{xy}}{\sigma_0}\right)^2 = 1 \tag{3-26}$$

Both equations define an ellipse. Figure 3-4 shows that the experimental results[2] agree best with the distortion-energy theory.

Fig. 3-3  Combined tension and torsion in a thin-walled tube.

[1] See for example S. S. Hecker, *Metall. Trans.*, vol. 2, pp. 2077–2086, 1971. A unique method for determining the yield locus of a flat sheet has been presented by D. Lee and W. A. Backofen, *Trans. Metall. Soc. AIME*, vol. 236, pp. 1077–1084, 1966. This method is well suited for studying the anisotropy of rolled sheet.
[2] G. I. Taylor and H. Quinney, *Proc. R. Soc. London Ser. A.*, vol. 230A, pp. 323–362, 1931.

**Fig. 3-4** Comparison between maximum-shear-stress theory and distortion-energy (von Mises') theory.

For the plane-stress condition, the distortion-energy theory of yielding can be expressed mathematically by

$$\sigma_1^2 + \sigma_2^2 - \sigma_1\sigma_2 = \sigma_0^2 \qquad (3\text{-}27)$$

This is the equation of an ellipse whose major semiaxis is $\sqrt{2}\sigma_0$ and whose minor semiaxis is $\sqrt{\tfrac{2}{3}}\sigma_0$.

A convenient way of comparing yielding criteria for a two-dimensional state of stress is with a plot of yield locus such as Fig. 3-5. Note that the maximum-shear-stress theory and the distortion-energy theory predict the same yield stress for conditions of uniaxial stress and balanced biaxial stress ($\sigma_1 = \sigma_2$). The greatest divergence between the two theories occurs for a state of pure shear ($\sigma_1 = -\sigma_2$). We have already seen that for pure shear the von Mises' yield criterion predicts $k = \sigma_0/\sqrt{3}$, while the maximum-shear-stress criterion predicts $k = \sigma_0/2$. The von Mises' criterion is $2/\sqrt{3} = 1.155$ greater than the yield stress predicted by the maximum-shear-stress criterion. This difference of about 15 percent is the greatest divergence between the two yield criteria.

A very sensitive method of differentiating between the two yield criteria is the procedure adopted by Lode of determining the effect of the intermediate principal stress on yielding. According to the maximum-shear-stress law, there should be no effect of the value of the intermediate stress $\sigma_2$. Thus, $(\sigma_1 - \sigma_3)/\sigma_0 = 1$. For the distortion-energy theory, to account for the influence of the intermediate principal stress, Lode introduced the parameter $\mu$, called *Lode's stress parameter*.

$$\mu = \frac{2\sigma_2 - \sigma_3 - \sigma_1}{\sigma_1 - \sigma_3} \qquad (3\text{-}28)$$

Solving this equation for $\sigma_2$ and eliminating $\sigma_2$ from Eq. (3-12) results in

$$\frac{\sigma_1 - \sigma_3}{\sigma_0} = \frac{2}{(3 + \mu^2)^{\frac{1}{2}}} \qquad (3\text{-}29)$$

**Fig. 3-5** Comparison of yield criteria for plane stress.

Experimental data plot much better against Eq. (3-29) than against the maximum-shear-stress equation, indicating that the intermediate principal stress has an influence on yielding.

Another contribution of Lode was the introduction of a strain parameter $\nu$.

$$\nu = \frac{2\,\Delta\epsilon_2 - \Delta\epsilon_3 - \Delta\epsilon_1}{\Delta\epsilon_1 - \Delta\epsilon_3} \qquad (3\text{-}30)$$

where $\Delta\varepsilon$ is a finite increment of strain. A plot of $\mu$ against $\nu$ should yield a straight line at 45° to the axis if the metal behaves according to the Levy–von Mises equations of plasticity (see Sec. 3-9). Most metals show some slight but systematic deviation from Lode's relationship $\mu = \nu$.

The yielding criteria considered so far assume that the material is isotropic. While this may be the case at the start of plastic deformation, it certainly is no longer a valid assumption after the metal has undergone appreciable plastic deformation. Moreover, most fabricated metal shapes have anisotropic properties, so that it is likely that the tubular specimens used for basic studies of yield criteria incorporate some degree of anisotropy. Certainly the von Mises' criterion as formulated in Eq. (3-12) would not be valid for a highly oriented cold-rolled sheet or a fiber-reinforced composite material.

Hill[1] has formulated the von Mises' yield criterion for an anisotropic material having orthotropic symmetry.

$$F(\sigma_y - \sigma_z)^2 + G(\sigma_z - \sigma_x)^2 + H(\sigma_x - \sigma_z)^2 + 2L\tau_{yz}^2 + 2M\tau_{zx}^2$$
$$+ 2N\tau_{xy}^2 = 1$$

[1] R. Hill, *Proc. R. Soc. London, Ser. B*, vol. 193, pp. 281–297, 1948.

where $F$, $G$, ..., $N$ are constants defining the degree of anisotropy. For principal axes of orthotropic symmetry

$$F(\sigma_2 - \sigma_3)^2 + G(\sigma_3 - \sigma_1)^2 + H(\sigma_1 - \sigma_2)^2 = 1 \qquad (3\text{-}31)$$

If $X$ is the yield stress in the 1 direction, $Y$ is the yield stress in the 2 direction, $Z$ is the yield stress in the 3 direction, then by substituting into Eq. (3-31) we can evaluate the constants by

$$G + H = \frac{1}{X^2} \qquad H + F = \frac{1}{Y^2} \qquad F + G = \frac{1}{Z^2}$$

Lubahn and Felgar[1] give detailed plasticity calculations for anisotropic behavior.

On a plane-stress yield locus, such as Fig. 3-5, anisotropic yielding results in distortion of the yield ellipse. Figure 3-6 shows the yield locus for highly textured titanium alloy sheet.[2] Note that the experimentally determined curve is non-symmetric when compared with the ideal isotropic curve.

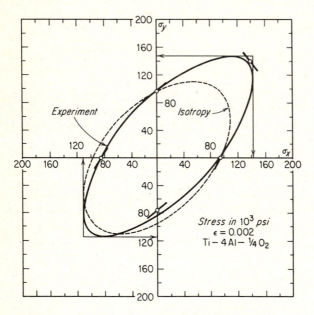

**Fig. 3-6** Yield locus for textured titanium-alloy sheet. (*From D. Lee and W. A. Backofen, Trans. Metall. Soc. AIME, vol. 236, p. 1083, 1966. By permission of the publishers.*)

[1] J. D. Lubahn and R. P. Felgar, "Plasticity and Creep of Metals," chap. 13, John Wiley & Sons, Inc., New York, 1961.

[2] These curves were obtained with the method of D. Lee and W. A. Backofen, op. cit.

An important aspect of yield anisotropy is *texture hardening*.[1]  Consider a highly textured sheet that is fabricated into a thin-wall pressure vessel, so that the thickness stress $\sigma_3$ is negligible.  From Eq. (3-31)

$$F\sigma_2{}^2 + G\sigma_1{}^2 + H(\sigma_1{}^2 - 2\sigma_1\sigma_2 + \sigma_2{}^2) = 1$$

$$(G + H)\sigma_1{}^2 + (F + H)\sigma_2{}^2 - 2H\sigma_1\sigma_2 = 1$$

or

$$\left(\frac{\sigma_1}{X}\right)^2 + \left(\frac{\sigma_2}{Y}\right)^2 - 2HXY\left(\frac{\sigma_1}{X}\frac{\sigma_2}{Y}\right) = 1$$

For simplicity, we shall assume that the yield stresses in the plane of the sheet are equal, that is, $X = Y$.  Thus,

$$\sigma_1{}^2 + \sigma_2{}^2 - 2HY^2\sigma_1\sigma_2 = Y^2$$

and

$$G = F = \frac{1}{2Z^2} \qquad HY^2 = 1 - \frac{1}{2}\left(\frac{Y}{Z}\right)^2$$

However, the yield stress in the thickness direction of the sheet, $Z$, is a difficult property to measure.  This problem can be circumvented by measuring the $R$ value, the ratio of the width strain to the thickness strain

$$R = \frac{\ln(w_0/w)}{\ln(t_0/t)}$$

Since $\left(\dfrac{Z}{Y}\right)^2 = \tfrac{1}{2}(1 + R)$, the equation for the yield locus can be written as

$$\sigma_1{}^2 + \sigma_2{}^2 - \frac{2R}{1 + R}\sigma_1\sigma_2 = Y^2$$

High through-thickness yield stress $Z$ results in low-thickness strain and a high value of $R$.  The extent of strengthening from the texture effect can be seen from Fig. 3-6.  For a spherical pressure vessel $\sigma_1 = \sigma_2$.  Thus, by moving out a 45° line on Fig. 3-6, we see that the resistance to yielding increases markedly with increased $R$.

The relationships that have been developed for yield criteria, Eqs. (3-12) and (3-21), can be represented geometrically by a cylinder oriented at equal angles to the $\sigma_1$, $\sigma_2$, $\sigma_3$ axes (Fig. 3-7).  A state of stress which gives a point inside of the cylinder represents elastic behavior.  Yielding begins when the state of stress reaches the surface of the cylinder, which is called the *yield surface*.  The radius of the cylinder $MN$ is the stress deviator.  Since the axis of the cylinder $OM$ makes equal angles with the principal stress axes, $l = m = n = 1/\sqrt{3}$, and from Eq. (2-18), $\sigma = (\sigma_1 + \sigma_2 + \sigma_3)/3 = \sigma_m$.  Therefore, the axis of the cylinder is the hydrostatic

[1] W. A. Backofen, W. F. Hosford, Jr., and J. J. Burke, *ASM Trans Q.*, vol. 55, p. 264, 1962.

**Fig. 3-7**  Yield surface for von Mises' criterion.

component of stress. Since plastic deformation is not influenced by hydrostatic stress, the generator of the yield surface is a straight line parallel to $OM$, so that the radius of the cylinder is constant. As plastic deformation occurs we can consider that the yield surface expands outward, maintaining its same geometric shape.

The yield surface shown in Fig. 3-7 is a circular cylinder if it represents the von Mises' yield criterion. If a plane is passed through this surface parallel to the $\sigma_3$ axis, it intersects on the $\sigma_1\sigma_2$ plane as an ellipse (see Fig. 3-5). The yield surface for the maximum-shear-stress criterion is a hexagonal cylinder. It should be noted that although the yield surface is an important concept in plasticity theory, there is no extensive body of experimental data on the shape of the surface. There is some work[1] which indicates that the yield surface is not a cylinder of uniform radius.

## 3-6  OCTAHEDRAL SHEAR STRESS AND SHEAR STRAIN

The octahedral stresses are a particular set of stress functions which are important in the theory of plasticity. They are the stresses acting on the faces of a three-dimensional octahedron which has the geometric property that the faces of the planes make equal angles with each of the three principal directions of stress. For such a geometric body, the angle between the normal to one of the faces and the nearest principal axis is 54°44′, and the cosine of this angle is $1/\sqrt{3}$. This is equivalent to {111} plane in an fcc crystal lattice.

The stress acting on each face of the octahedron can be resolved[2] into a normal octahedral stress $\sigma_{oct}$ and an octahedral shear stress lying in the octahedral

---

[1] L. W. Hu, J. Markowitz, and T. A. Bartush, *Exp. Mech.* vol. 6, pp. 58–65, 1966.
[2] A. Nadai, "Theory of Flow and Fracture of Solids," 2d ed., vol. I, pp. 99–105, McGraw-Hill Book Company, New York, 1950.

plane, $\tau_{oct}$. The normal octahedral stress is equal to the hydrostatic component of the total stress.

$$\sigma_{oct} = \frac{\sigma_1 + \sigma_2 + \sigma_3}{3} = \sigma_m \qquad (3\text{-}32)$$

The octahedral shear stress $\tau_{oct}$ is given by

$$\tau_{oct} = \tfrac{1}{3}[(\sigma_1 - \sigma_2)^2 + (\sigma_2 - \sigma_3)^2 + (\sigma_3 - \sigma_1)^2]^{1/2} \qquad (3\text{-}33)$$

Since the normal octahedral stress is a hydrostatic stress, it cannot produce yielding in solid materials. Therefore, the octahedral shear stress is the component of stress responsible for plastic deformation. In this respect, it is analogous to the stress deviator.

If it is assumed that a critical octahedral shear stress determines yielding, the failure criterion can be written as

$$\tau_{oct} = \frac{1}{3}[(\sigma_1 - \sigma_2)^2 + (\sigma_2 - \sigma_3)^2 + (\sigma_3 - \sigma_1)^2]^{1/2} = \frac{\sqrt{2}}{3}\sigma_0$$

or

$$\sigma_0 = \frac{1}{\sqrt{2}}[(\sigma_1 - \sigma_2)^2 + (\sigma_2 - \sigma_3)^2 + (\sigma_3 - \sigma_1)^2]^{1/2} \qquad (3\text{-}34)$$

Since Eq. (3-34) is identical with the equation already derived for the distortion-energy theory, the two yielding theories give the same results. In a sense, the octahedral theory can be considered the *stress equivalent* of the distortion-energy theory. According to this theory, the octahedral shear stress corresponding to yielding in uniaxial stress is given by

$$\tau_{oct} = \frac{\sqrt{2}}{3}\sigma_0 = 0.471\sigma_0 \qquad (3\text{-}35)$$

Octahedral strains are referred to the same three-dimensional octahedron as the octahedral stresses. The octahedral linear strain is given by

$$\tau_{oct} = \frac{\varepsilon_1 + \varepsilon_2 + \varepsilon_3}{3} \qquad (3\text{-}36)$$

Octahedral shear strain is given by

$$\gamma_{oct} = \tfrac{2}{3}[(\varepsilon_1 - \varepsilon_2)^2 + (\varepsilon_2 - \varepsilon_3)^2 + (\varepsilon_3 - \varepsilon_1)^2]^{1/2} \qquad (3\text{-}37)$$

## 3-7 INVARIANTS OF STRESS AND STRAIN

It is frequently useful to simplify the representation of a complex state of stress or strain by means of invariant functions of stress and strain. If the plastic stress-strain curve (the flow curve) is plotted in terms of invariants of stress and strain, approximately the same curve will be obtained regardless of the state of stress. For example, the flow curves obtained in a uniaxial-tension test and a biaxial-torsion test of a thin tube with internal pressure will coincide when the curves are plotted in terms of invariant stress and strain functions.

Nadai[1] has shown that the octahedral shear stress and shear strain are invariant functions which describe the flow curve independent of the type of test. Other frequently used invariant functions are the *effective*, or *significant*, stress and strain. These quantities are defined by the following equations for the case where the coordinate axes correspond to the principal directions:

Effective or significant stress:

$$\bar{\sigma} = \frac{\sqrt{2}}{2} [(\sigma_1 - \sigma_2)^2 + (\sigma_2 - \sigma_3)^2 + (\sigma_3 - \sigma_1)^2]^{\frac{1}{2}} \qquad (3\text{-}38)$$

Effective or significant plastic strain:

$$\bar{\varepsilon} = \frac{\sqrt{2}}{3} [(\varepsilon_1 - \varepsilon_2)^2 + (\varepsilon_2 - \varepsilon_3)^2 + (\varepsilon_3 - \varepsilon_1)^2]^{\frac{1}{2}} \qquad (3\text{-}39)$$

Note that both effective stress and effective strain reduce to the axial normal component of stress and strain for a tensile test. These values are also related to the octahedral shearing stress and strain, as can be seen by comparing Eqs. (3-33) and (3-37) with the above equations.

$$\tau_{oct} = \frac{\sqrt{2}}{3} \bar{\sigma} \qquad \gamma_{oct} = \sqrt{2}\,\bar{\varepsilon} \qquad (3\text{-}40)$$

Drucker[2] has pointed out that there are a large number of different functions of stress and strain which might serve as invariant stress and strain parameters. For example, he shows that combined stress data for aluminum-alloy tubes show better agreement when the equivalent shearing stress $\tau_{eq}$, defined below, is plotted against octahedral shearing strain instead of $\tau_{oct}$ being plotted against $\gamma_{oct}$.

$$\tau_{eq} = \tau_{oct}\left(1 - 2.25\frac{J_3^2}{J_2^3}\right)^{\frac{1}{6}} \qquad (3\text{-}41)$$

where $J_2$ and $J_3$ are invariants of stress deviator. There appears to be no theoretical or experimental justification for choosing invariant stress and strain parameters other than the closeness of agreement with the data and mathematical convenience

[1] A. Nadai, *J. Appl. Phys.*, vol. 8, p. 205, 1937.
[2] D. C. Drucker, *J. Appl. Mech.*, vol. 16, pp. 349–357, 1949.

## 3-8 PLASTIC STRESS-STRAIN RELATIONS

Having discussed the relationships between stress state and plastic yielding, it is now necessary to consider the relations between stress and strain in plastic deformation. In the elastic region the strains are uniquely determined by the stresses through Hooke's law without regard to how the stress state was achieved. This is not the case for plastic deformation. In the plastic region the strains in general are not uniquely determined by the stresses but depend on the entire history of loading. Therefore, in plasticity it is necessary to determine the differentials or *increments of plastic strain* throughout the loading path and then obtain the total strain by integration or summation. As a simple example, consider a rod 1 in long extended in tension to $1\frac{1}{4}$ in and then compressed to the original 1 in length.

On the basis of total deformation

$$\varepsilon = \int_{1}^{1\frac{1}{4}} \frac{dL}{L} + \int_{1\frac{1}{4}}^{1} \frac{dL}{L} = 0$$

However, on an incremental basis

$$\varepsilon = \int_{1}^{1\frac{1}{4}} \frac{dL}{L} + \int_{1\frac{1}{4}}^{1} -\frac{dL}{L} = 2 \ln 1\frac{1}{4} = 0.445$$

For the particular class of loading paths in which all the stresses increase in the same ratio, *proportional loading*, i.e.,

$$\frac{d\sigma_1}{\sigma_1} = \frac{d\sigma_2}{\sigma_2} = \frac{d\sigma_3}{\sigma_3}$$

the plastic strains are independent of the loading path and depend only on the final state of stress.

There are two general categories of plastic stress-strain relationships. *Incremental* or *flow theories* relate the stresses to the plastic strain increments. *Deformation* or *total strain theories* relate the stresses to the total plastic strain. Deformation theory simplifies the solution of plasticity problems, but the plastic strains in general cannot be considered independent of loading path.[1]

## LEVY-MISES EQUATIONS (IDEAL PLASTIC SOLID)

Levy (1871) and later von Mises (1913) proposed that the increments of *plastic strain* are related to the *stress deviators* by

$$\frac{d\varepsilon_x}{\sigma'_x} = \frac{d\varepsilon_y}{\sigma'_y} = \frac{d\varepsilon_z}{\sigma'_z} = \frac{d\varepsilon_{yz}}{\tau_{yz}} = \frac{d\varepsilon_{xz}}{\tau_{xz}} = \frac{d\varepsilon_{xy}}{\tau_{xy}} = d\lambda \qquad (3\text{-}42)$$

---

[1] Arguments to show that deformation theories of plasticity should be valid for loading paths other than proportional loading have been given by B. Budiansky, *J. Appl. Mech.*, vol. 26, no. 2, pp. 259–264, 1959.

where $d\lambda$ is a nonnegative constant which may vary throughout the loading path. These equations are based on the assumption that the principal axes of strain increment coincide with the principal stress axes. Since elastic strains are not considered in Eq. (3-42), they apply only for an ideal (rigid) plastic solid in which the elastic strain is small compared with the plastic strain. Equations (3-42) are often written as[1]

$$\frac{d\varepsilon_x - d\varepsilon_y}{\sigma'_x - \sigma'_y} = \frac{d\varepsilon_y - d\varepsilon_z}{\sigma'_y - \sigma'_z} = \frac{d\varepsilon_z - d\varepsilon_x}{\sigma'_z - \sigma'_x} = \frac{d\varepsilon_{yz}}{\tau_{yz}} = \frac{d\varepsilon_{xz}}{\tau_{xz}} = \frac{d\varepsilon_{xy}}{\tau_{xy}} = d\lambda \qquad (3\text{-}43)$$

By using Eq. (2-57), the Levy–Mises equations can be written in terms of the actual stresses

$$d\varepsilon_x = \tfrac{2}{3}\, d\lambda[\sigma_x - \tfrac{1}{2}(\sigma_y + \sigma_z)]$$

$$d\varepsilon_y = \tfrac{2}{3}\, d\lambda[\sigma_y - \tfrac{1}{2}(\sigma_z + \sigma_x)]$$

$$d\varepsilon_z = \tfrac{2}{3}\, d\lambda[\sigma_z - \tfrac{1}{2}(\sigma_x + \sigma_y)] \qquad (3\text{-}44)$$

$$d\varepsilon_{yz} = d\lambda\tau_{yz}$$

$$d\varepsilon_{xz} = d\lambda\tau_{xz}$$

$$d\varepsilon_{xy} = d\lambda\tau_{xy}$$

To evaluate the constant $d\lambda$, we utilize the invariant plastic strain, the effective strain, $\bar{\varepsilon}$.

$$d\bar{\varepsilon} = \frac{\sqrt{2}}{3}\,[(d\varepsilon_x - d\varepsilon_y)^2 + (d\varepsilon_y - d\varepsilon_z)^2 + (d\varepsilon_z - d\varepsilon_x)^2 + 6(d\varepsilon_{xy})^2$$

$$+\, 6(d\varepsilon_{yz})^2 + 6(d\varepsilon_{xz})^2]^{1/2} \qquad (3\text{-}45)$$

Substituting Eqs. (3-44) into Eq. (3-45) results in

$$d\bar{\varepsilon} = \frac{2}{3}\, d\lambda \left\{ \frac{1}{\sqrt{2}}\,[(\sigma_x - \sigma_y)^2 + (\sigma_y - \sigma_z)^2 + (\sigma_z - \sigma_x)^2 + 6(\tau_{xy}{}^2 + \tau_{yz}{}^2 + \tau_{xz}{}^2)]^{1/2} \right\}$$

$$d\bar{\varepsilon} = \frac{2}{3}\, d\lambda\bar{\sigma}$$

---

[1] These equations also may be developed by assuming that Lode's stress parameter equals Lode's strain parameter. The condition $\mu = \nu$ is called the *von Mises' condition*.

So that

$$d\varepsilon_x = \frac{d\bar{\varepsilon}}{\bar{\sigma}} [\sigma_x - \tfrac{1}{2}(\sigma_y + \sigma_z)]$$

$$d\varepsilon_y = \frac{d\bar{\varepsilon}}{\bar{\sigma}} [\sigma_y - \tfrac{1}{2}(\sigma_z + \sigma_x)]$$

$$d\varepsilon_z = \frac{d\bar{\varepsilon}}{\bar{\sigma}} [\sigma_z - \tfrac{1}{2}(\sigma_x + \sigma_y)]$$

$$d\varepsilon_{yz} = \frac{3}{2} \frac{d\bar{\varepsilon}}{\bar{\sigma}} \tau_{yz}$$

(3-46)

$$d\varepsilon_{xz} = \frac{3}{2} \frac{d\bar{\varepsilon}}{\bar{\sigma}} \tau_{xz}$$

$$d\varepsilon_{xy} = \frac{3}{2} \frac{d\bar{\varepsilon}}{\bar{\sigma}} \tau_{xy}$$

The term $d\bar{\varepsilon}/\bar{\sigma}$ can be evaluated from the effective stress–effective strain curve. Equations (3-46) show a strong resemblance to the equations for Hooke's law, Eqs. (2-64) and (2-65). In place of a fixed Young's modulus, the plasticity equations have a variable plastic modulus $\bar{\sigma}/d\bar{\varepsilon}$, and Poisson's ratio for the plasticity equations is $\tfrac{1}{2}$.

## PRANDTL-REUSS EQUATIONS (ELASTIC-PLASTIC SOLID)

The Levy–Mises equations can only be applied to problems of large plastic deformation because they neglect elastic strains. To treat the important, but more difficult problems in the elastic-plastic region it is necessary to consider both elastic and plastic components of strain. These equations were proposed by Prandtl (1925) and Reuss (1930).

The total strain increment is the sum of an elastic strain increment $de^E$ and a plastic strain increment $de^P$.

$$d\varepsilon_{ij} = de_{ij}{}^E + d\varepsilon_{ij}{}^P \qquad (3\text{-}47)$$

From Eqs. (2-52) and (2-69), the elastic strain increment is given by

$$de_{ij}{}^E = \left( de_{ij} - \frac{de_{kk}}{3} \delta_{ij} \right) + \frac{de_{kk}}{3} \delta_{ij} = \frac{1+v}{E} d\sigma_{ij} - \frac{v}{E} \sigma_{kk} \delta_{ij}$$

or

$$de_{ij}{}^E = \frac{1+v}{E} d\sigma'_{ij} + \frac{1-2v}{E} \frac{d\sigma_{kk}}{3} \delta_{ij} \qquad (3\text{-}48)$$

The plastic strain increment is given by the Levy–Mises equations, which can be written as

$$d\varepsilon_{ij}{}^{P} = \frac{3}{2}\frac{d\bar{\varepsilon}}{\bar{\sigma}}\sigma'_{ij} \qquad (3\text{-}49)$$

Thus, the stress, strain relations for an elastic-plastic solid are given by

$$d\varepsilon_{ij} = \frac{1+v}{E}d\sigma'_{ij} + \frac{1+2v}{E}\frac{d\sigma_{kk}}{3}\delta_{ij} + \frac{3}{2}\frac{d\bar{\varepsilon}}{\bar{\sigma}}\sigma'_{ij} \qquad (3\text{-}50)$$

## SOLUTION OF PLASTICITY PROBLEMS

The Levy–Mises and Prandtl-Reuss equations provide relations between the increments of plastic strain and the stresses. The basic problem is to calculate the next increment of plastic strain for a given state of stress when the loads are increased incrementally. If all of the increments of strain are known, then the total plastic strain is simply determined by summation. To do this we have available a set of plastic stress-strain relationships, either Eqs. (3-46) or (3-50), a yield criterion, and a basic relationship for the flow behavior of the material in terms of a curve of $\bar{\sigma}$ vs. $\bar{\varepsilon}$. In addition, a complete solution also must satisfy the equations of equilibrium, the strain-displacement relations, and the boundary conditions. The reader is referred to the several excellent texts on plasticity listed at the end of this chapter for examples of detailed solutions.[1] Although the incremental nature of plasticity solutions in the past has resulted in much labor and infrequent application of the available techniques, the current widespread use of digital computers should make plasticity analysis of engineering problems more commonplace. A numerical technique[2] involving successive approximations is very useful for solving elastic-plastic problems in which strain hardening is considered.

## 3-9  TWO-DIMENSIONAL PLASTIC FLOW—SLIP-LINE FIELD THEORY

In many practical problems, such as rolling and strip drawing, all displacements can be considered to be limited to the $xy$ plane, so that strains in the $z$ direction can be neglected in the analysis. This is known as a condition of *plane strain*. When a problem is too difficult for an exact three-dimensional solution, a good indication of the stresses often can be obtained by consideration of the analogous plane-strain problem.

---

[1] A number of plasticity problems are worked out in great detail in Lubahn and Felgar op. cit., Chaps 8 and 9.

[2] A. Mendelson and S. S. Manson, *NASA Tech. Rept.* R.28, 1959. See also A. Mendelson, "Plasticity: Theory and Application," chap. 9, The Macmillan Company, New York, 1968.

**Fig. 3-8**  Methods of developing plastic constraint.

Since a plastic material tends to deform in all directions, to develop a plane-strain condition it is necessary to constrain flow in one direction. Constraint can be produced by an external lubricated barrier, such as a die wall (Fig. 3-8a), or it can arise from a situation where only part of the material is deformed and the rigid (elastic) material outside the plastic region prevents the spread of deformation (Fig. 3-8b).

If the plane-strain deformation occurs on planes parallel to the $xy$ plane, then

$$\varepsilon_z = \varepsilon_{xz} = \varepsilon_{yz} = 0 \quad \text{and} \quad \tau_{xz} = \tau_{yz} = 0$$

Since $\tau_{xz} = \tau_{yz} = 0$, it follows that $\sigma_z$ is a principal stress. From the Levy–Mises equations, Eq. (3-44)

$$d\varepsilon_z = 0 = \tfrac{2}{3}\, d\lambda [\sigma_z - \tfrac{1}{2}(\sigma_x + \sigma_y)]$$

and

$$\sigma_z = \frac{\sigma_x + \sigma_y}{2} \tag{3-51}$$

Note that although the strain is zero in the $z$ direction, a restraining stress acts in this direction.

Equation (3-51) could just as well have been written in terms of the principal stresses $\sigma_3 = (\sigma_1 + \sigma_2)/2$.

This principal stress will be intermediate between $\sigma_1$ and $\sigma_2$, so that the maximum-shear-stress yield criterion is given by

$$\sigma_1 - \sigma_2 = \sigma_0 = 2k \tag{3-52}$$

where $k$ is the yield stress in pure shear.

If the value for the intermediate principal stress $\sigma_3$ is substituted into the von Mises' yield criterion, Eq. (3-12) it reduces to

$$\sigma_1 - \sigma_2 = \frac{2}{\sqrt{3}}\, \sigma_0 \tag{3-53}$$

However, for the von Mises' yield criterion $\sigma_0 = \sqrt{3}\,k$ so that Eq. (3-53) becomes

$$\sigma_1 - \sigma_2 = 2k \qquad (3\text{-}54)$$

Thus, for a state of plane strain the maximum-shear-stress and von Mises' yield criteria are equivalent. It can be considered that two-dimensional plastic flow will begin when the shear stress reaches a critical value of $k$.

Slip-line field theory is based on the fact that any general state of stress in plane strain consists of *pure shear* plus a *hydrostatic pressure*. We could show this by applying the equations for transformation of stress from one set of axes to another, Eqs. (2-5) to (2-7), but it perhaps is more instructive to see this diagrammatically. In Fig. 3-9, let the state of stress consist of $\sigma_1 = -Q$, $\sigma_3 = -P$, and

**Fig. 3-9** Demonstration that a state of stress in plane strain may be expressed as the sum of a hydrostatic stress and pure shear.

$\sigma_2 = (-P - Q)/2$.  The maximum shear stress is given by

$$\tau_{max} = \sigma_1 - \sigma_3 = 2k$$
$$-Q + P = 2k$$

or in Fig. 3-9b,

$$P = Q + 2k$$

But we can write the state of stress in Fig. 3-9b as in Fig. 3-9c, which in turn can be written as the sum of a hydrostatic pressure and a biaxial state of stress Fig. 3-9d. The latter is the stress state in pure torsion, which for planes rotated by 45° consists of pure shear stresses.  Thus, a general state of stress in plane strain can be decomposed into a hydrostatic state of stress $p$ (in this case compression) and a state of pure shear $k$.  The components of the stress tensor for plane strain are

$$\sigma_{ij} = \begin{vmatrix} p & k & 0 \\ k & p & 0 \\ 0 & 0 & p \end{vmatrix}$$

Mohr's circle representation for the state of stress given in Fig. 3-9 is shown in Fig. 3-10.  If $\sigma_1 = -Q$ and $\sigma_3 = -P$, then $\sigma_2 = (-Q - P)/2 = -p$.  This follows because

$$p = \sigma_m = \frac{\sigma_1 + \sigma_2 + \sigma_3}{3} = -\frac{1}{3}\left(Q + \frac{Q}{2} + \frac{P}{2} + P\right)$$

$$\therefore p = \frac{Q + P}{2} = -\sigma_2$$

Also, the radius of Mohr's circle is $\tau_{max} = k$, where $k$ is the yield stress in pure shear. Thus, using Fig. 3-10, we can express the principal stresses

$$\sigma_1 = -p + k$$
$$\sigma_2 = -p$$
$$\sigma_3 = -p - k$$

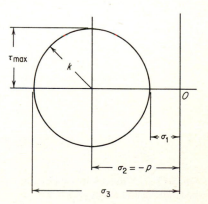

**Fig. 3-10**  Mohr's circle representation of stresses
in Fig. 3-9a.

The slip-line field theory for plane strain allows the determination of stresses in a plastically deformed body when the deformation is not uniform throughout the body. In addition to requiring plane-strain conditions, the theory assumes an isotropic, homogeneous, rigid ideal plastic material. For such a non-strain-hardening material $k$ is everywhere constant but $p$ may vary from point to point. The state of stress at any point can be determined if we can find the magnitude of $p$ and the direction of $k$. The lines of maximum shear stress occur in two orthogonal directions $\alpha$ and $\beta$. These lines of maximum shear stress are called *slip lines* and have the property that shear strain is a maximum and linear strain is zero tangent to their directions. The slip lines give the direction of $p$ at any point and the changes in magnitude of $p$ are deduced from the rotation of the slip line between one point and another in the field. It should be noted that the slip lines referred to in this section are geometric constructions which define the characteristic directions of the hyperbolic partial differential equations for the stress under plane-strain conditions. These slip lines bear no relationship to the slip lines observed under the microscope on the surface of a plastically deformed metal.

To arrive at the equations for calculating stress through the use of slip-line fields, we must now relate the stresses on a physical body in the $xy$ coordinate system to $p$ and $k$. Figure 3-11$b$ shows the Mohr's circle representation of the stress state given in Fig. 3-11$a$. The stresses may be expressed as

$$\sigma_x = -p - k \sin 2\phi$$
$$\sigma_y = -p - (-k \sin 2\phi) = -p + k \sin 2\phi$$
$$\sigma_z = -p$$
$$\tau_{xy} = k \cos 2\phi$$

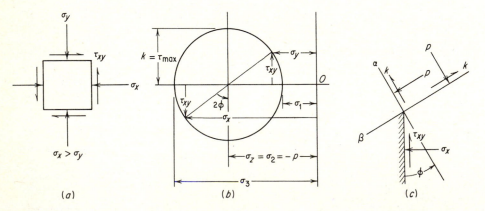

**Fig. 3-11** (a) Stress state on physical body; (b) Mohr's circle for (a); (c) relationship of physical body and $\alpha$ and $\beta$ slip lines.

where $2\phi$ is a counterclockwise angle on Mohr's circle from the *physical* $x$ plane to the first plane of maximum shear stress. This plane of maximum shear stress is known as an $\alpha$ slip line. The relationship between the stress state on the physical body and the $\alpha$ and $\beta$ slip lines is given in Fig. 3-11c.

The variation of hydrostatic pressure $p$ with change in direction of the slip lines is given by the *Hencky equations*

$$p + 2k\phi = \text{constant along an } \alpha \text{ line} \qquad \text{(3-55)}$$

$$p - 2k\phi = \text{constant along a } \beta \text{ line}$$

These equations are developed[1] from the equilibrium equations in plane strain. The use of the Hencky equations will be illustrated with the example of the indentation of a thick block with a flat frictionless punch. The slip-line field shown in Fig. 3-12 was first suggested by Prandtl[2] in 1920. At the free surface on the frictionless interface between the punch and the block the slip lines meet the surface at 45° (see Prob. 3-14). We could construct the slip-line field by starting with triangle $AFB$, but we would soon see that if all plastic deformation were restricted to this region, the metal could not move because it would be surrounded by rigid (elastic) material. Therefore, the plastic zone described by the slip-line field must be extended along the free surface to $AH$ and $BD$.

To determine the stresses from the slip-line field, we start with a simple point such as $D$. Since $D$ is on a free surface, there is no stress normal to this surface.

$$\sigma_y = 0 = -p + k \sin 2\phi$$

and

$$\sigma_x = -p - k \sin 2\phi = -p - p = -2p$$

**Fig. 3-12**   Slip-line field for frictionless indentation with a flat punch.

---

[1] See for example W. Johnson and P. B. Mellor, "Plasticity for Mechanical Engineers," pp. 263–265, D. Van Nostrand Company, Inc., Princeton, N.J., 1962.

[2] A different slip-line field was later suggested by R. Hill. Although the slip field is different, it leads to the same value of indentation pressure. This illustrates the fact that slip-line field solutions are not necessarily unique.

**Fig. 3-13**   (*a*) Stresses at point $D$; (*b*) Mohr's circle.

The stresses at point $D$ are shown in Fig. 3-13. From the Mohr's circle we learn that $p = k$. In order to use the Hencky equations we need to know whether the slip line through $D$ is an $\alpha$ or $\beta$ line. This is done most simply from the following sign convention:

> For a counterclockwise rotation about the point of intersection of two slip lines, starting from an $\alpha$-line the direction of the algebraically highest principal stress $\sigma_1$ is crossed before a $\beta$ line is crossed.

Applying this convention, we see that the slip line from $D$ to $E$ is an $\alpha$ line. Thus, the first Hencky equation applies,

$$p + 2k\phi = C_1$$

and if we use $DE$ as the reference direction so $\phi = 0$,

$$p = C_1 = k$$

Because $DE$ is straight $p$ is constant from $D$ to $E$ and

$$p_D = p_E = k$$

Between $E$ and $F$ the tangent to the $\alpha$ slip line rotates through $\pi/2$ rad. Since the tangent to the $\alpha$ line rotates *clockwise*, $d\phi = -\pi/2$. If we write the Hencky equation in differential form, for clarity

$$dp + 2k d\phi = 0$$

or

$$(p_F - p_E) + 2k(\phi_F - \phi_E) = 0$$

$$p_F - k + 2k\left(-\frac{\pi}{2} - 0\right) = 0$$

$$p_F = k(\pi + 1)$$

Note that the pressure at $F'$ is the same as at $F$ because the slip line is straight and that the value of $p$ under the punch face at $G$ is also the same. (We stayed away from $A$ and $B$ at the punch edges because these are points of pressure discontinuity.) To find the punch pressure required to indent the block, it is necessary to convert the hydrostatic pressure at the punch interface into the vertical stress $\sigma_y$.

$$p_F = p_{F'} = p_G = k(\pi + 1)$$

$$\sigma_y = -p_G + k \sin 2\phi$$

From Fig. 3-11c, recall that the angle $\phi$ is measured by the counterclockwise angle from the physical $x$ axis to the $\alpha$ line.

$$\sigma_y = -k(\pi + 1) + k \sin 2\left(\frac{3\pi}{4}\right)$$

$$\sigma_y = -k\pi - k - k = -2k\left(1 + \frac{\pi}{2}\right) \tag{3-56}$$

If we trace out other slip lines, we shall find in the same way that the normal compressive stress under the punch is $2k(1 + \pi/2)$, and the pressure is uniform.

Since $k = \sigma_0/\sqrt{3}$,

$$\sigma_y = \frac{2\sigma_0}{\sqrt{3}}\left(1 + \frac{\pi}{2}\right) \approx 3\sigma_0 \tag{3-57}$$

This shows that the yield pressure for the indentation of a thick block with a narrow punch is nearly three times the stress required for the yielding of a cylinder in frictionless compression. This increase in flow stress is a *geometrical constraint* resulting from the localized deformation under the narrow punch.

The example described above is one of the simplest situations that involves slip-line fields. In the general case the slip-line field selection must also satisfy certain velocity conditions to assure equilibrium. Prager[1] and Thomsen[2] have given general procedures for constructing slip-line fields. However, there is no easy method of checking the validity of a solution. Partial experimental verification of theoretically determined slip-line fields has been obtained for mild steel by etching techniques[3] which delineate the plastically deformed regions. Highly localized plastic regions can be delineated by an etching technique in Fe-3 % Si steel.[4]

[1] W. Prager, *Trans. R. Inst. Technol. Stockholm*, no. 65, 1953.
[2] E. G. Thomsen, *J. Appl. Mech.*, vol 24, pp 81–84, 1957.
[3] B. Hundy, *Metallurgia*, vol 49, no. 293, pp 109–118, 1954.
[4] G. T. Hahn, P. N. Mincer and A. R. Rosenfield, *Exp. Mech.*, vol. 11, pp. 248–253, 1971.

## BIBLIOGRAPHY

Calladine, C. R.: "Engineering Plasticity," Pergamon Press Inc., New York, 1969.

Hill, R.: "The Mathematical Theory of Plasticity," Oxford University Press, New York, 1950.

Johnson, W., and P. B. Mellor: "Engineering Plasticity," Van Nostrand Reinhold Company, New York, 1973.

Johnson, W., R. Sowerby, and J. B. Haddow: "Plane-Strain Slip-Line Fields," Elsevier Publishing Company, New York, 1970.

Mendelson, A.: "Plasticity: Theory and Application," The Macmillan Company, New York, 1968.

Nadai, A.: "Theory of Flow and Fracture of Solids," 2d ed., vol. I, McGraw-Hill Book Company, New York, 1950; vol. II, 1963.

Prager, W., and P. G. Hodge: "Theory of Perfectly Plastic Solids," John Wiley & Sons, New York, 1951.

# PART TWO

# Metallurgical Fundamentals

# PLASTIC DEFORMATION
# OF SINGLE CRYSTALS

## 4-1  INTRODUCTION

The previous three chapters have been concerned with the phenomenological description of the elastic and plastic behavior of metals. It has been shown that formal mathematical theories have been developed for describing the mechanical behavior of metals based upon the simplifying assumptions that metals are homogeneous and isotropic. That this is not true should be obvious to anyone who has examined the structure of metals under a microscope. However, for fine-grained metals subjected to static loads within the elastic range the theories are perfectly adequate for design. Within the plastic range the theories describe the observed behavior, although not with the precision which is frequently desired. For conditions of dynamic and impact loading we are forced, in general, to rely heavily on experimentally determined data. As the assumption that we are dealing with an isotropic homogeneous medium becomes less tenable, our ability to predict the behavior of metals under stress by means of the theories of elasticity and plasticity decreases.

   Following the discovery of the diffraction of x-rays by metallic crystals by Von Laue in 1912 and the realization that metals were fundamentally composed of atoms arranged in specific geometric lattices there have been a great many investigations of the relationships between atomic structure and the plastic behavior of metals. Much of the fundamental work on the plastic deformation of metals has been performed with single-crystal specimens, so as to eliminate the complicating effects of grain boundaries and the restraints imposed by neighboring grains and second-phase particles. Techniques for preparing single crystals have been described in a number of sources.[1-4]

    The basic mechanisms of plastic deformation in single crystals will be discussed in this chapter. The dislocation theory, which plays such an important part in modern concepts of plastic deformation, will be introduced in this chapter to the extent needed to provide a qualitative understanding. A more detailed consideration of dislocation theory will be found in Chap. 5. Using dislocation theory as the main tool, consideration will be given to the strengthening mechanisms in polycrystalline solids in Chap. 6. Primary consideration will be given to tensile deformation. The fundamental deformation behavior in creep and fatigue will be covered in chapters in Part Three specifically devoted to these subjects. This part closes with a chapter on the fundamental aspects of fracture (Chap. 7) and a chapter on time-dependent deformation in metals and polymers (Chap. 8).

## 4-2  CONCEPTS OF CRYSTAL GEOMETRY

X-ray diffraction analysis shows that the atoms in a metal crystal are arranged in a regular, repeated three-dimensional pattern. The atom arrangement of metals is most simply portrayed by a crystal lattice in which the atoms are visualized as hard balls located at particular locations in a geometrical arrangement.

    The most elementary crystal structure is the simple cubic lattice (Fig. 4-1). This is the type of structure cell found for ionic crystals, such as NaCl and LiF, but not for any of the metals. Three mutually perpendicular axes are arbitrarily placed through one of the corners of the cell. Crystallographic planes and directions will be specified with respect to these axes in terms of *Miller indices*. A crystallographic plane is specified in terms of the length of its intercepts on the three axes, measured from the origin of the coordinate axes. To simplify the crystallographic formulas, the reciprocals of these intercepts are used. They are reduced to a lowest common denominator to give the Miller indices $(hkl)$ of the plane. For example, the plane $ABCD$ in Fig. 4-1 is parallel to the $x$ and $z$ axes and intersects the $y$ axis

[1] R. W. K. Honeycombe, *Metall. Rev.*, vol. 4, no. 13, pp. 1–47, 1969.
[2] A. N. Holden, *Trans. Am. Soc. Met.*, vol. 42, pp. 319–346, 1950.
[3] W. D. Lawson and S. Nielsen, "Preparation of Single Crystals," Academic Press, Inc., New York, 1958.
[4] J. J. Gilman (ed.), "The Art and Science of Growing Crystals," John Wiley & Sons, Inc., New York, 1963.

**Fig. 4-1**   Simple cubic structure.

at one interatomic distance $a_0$. Therefore, the indices of the plane are $1/\infty$, $1/1$, $1/\infty$, or $(hkl) = (010)$. Plane *EBCF* would be designated as the $(\bar{1}00)$ plane, since the origin of the coordinate system can be moved to *G* because every point in a space lattice has the same arrangement of points as every other point. The bar over one of the integers indicates that the plane intersects one of the axes in a negative direction. There are six crystallographically equivalent planes of the type (100), any one of which can have the indices (100), (010), (001), $(\bar{1}00)$, $(0\bar{1}0)$, $(00\bar{1})$ depending upon the choice of axes. The notation {100} is used when they are to be considered as a group, or *family of planes*.

Crystallographic directions are indicated by integers in brackets: $[uvw]$. Reciprocals are not used in determining directions. As an example, the direction of the line *FD* is obtained by moving out from the origin a distance $a_0$ along the *x* axis and moving an equal distance in the positive *y* direction. The indices of this direction are then [110]. A family of crystallographically equivalent directions would be designated $\langle uvw \rangle$. For the cubic lattice only, a direction is always perpendicular to the plane having the same indices.

Many of the common metals have either a body-centered cubic (bcc) or face-centered cubic (fcc) crystal structure. Figure 4-2*a* shows a body-centered cubic structure cell with an atom at each corner and another atom at the body center of the cube. Each corner atom is surrounded by eight adjacent atoms, as is the atom located at the center of the cell. Therefore, there are two atoms *per structure cell* for the body-centered cubic structure $(\frac{8}{8} + 1)$. Typical metals which have this crystal structure are alpha iron, columbium, tantalum, chromium, molybdenum, and tungsten. Figure 4-2*b* shows the structure cell for a face-centered cubic crystal structure. In addition to an atom at each corner, there is an atom at the center of each of the cube faces. Since these latter atoms belong to two unit cells, there are four atoms per structure cell in the face-centered cubic structure $(\frac{8}{8} + \frac{6}{2})$. Aluminum, copper, gold, lead, silver, and nickel are common face-centered cubic metals.

(100) – HADG
(110) – HBCG
(111) – GEC
(112) – GJC

(100) – HADG
(110) – HBCG
(111) – GEC
(112) – GJC

(a)    (b)

**Fig. 4-2** (a) Body-centered cubic structure; (b) face-centered cubic structure.

The third common metallic crystal structure is the hexagonal close-packed (hcp) structure[1] (Fig. 4-3). In order to specify planes and directions in the hcp structure, it is convenient to use the Miller-Bravais system with four indices of the type $(hkil)$. These indices are based on four axes; the three axes $a_1, a_2, a_3$ are 120° apart in the basal plane, and the vertical $c$ axis is normal to the basal plane. These axes and typical planes in the hcp crystal structure are given in Fig. 4-3. The third index is related to the first two by the relation $i = -(h + k)$.

The face-centered cubic and hexagonal close-packed structures can both be built up from a stacking of close-packed planes of spheres. Figure 4-4 shows that there are two ways in which the spheres can be stacked. The first layer of spheres is arranged so that each sphere is surrounded by and just touching six other spheres. This corresponds to the solid circles in Fig. 4-4. A second layer of close-packed spheres can be placed over the bottom layer so that the centers of the atoms in the second plane cover one-half the number of valleys in the bottom layer (dashed circles in Fig. 4-4). There are two ways of adding spheres to give a third close-packed plane. Although the spheres in the third layer must fit into the valleys in the second plane, they may lie either over the valleys not covered in the first plane (the dots in Fig. 4-4) or directly above the atoms in the first plane (the crosses in Fig. 4-4). The first possibility results in a stacking sequence $ABCABC\ldots$, which is found for the {111} planes of an fcc structure. The other possibility results in the stacking sequence $ABAB\ldots$, which is found for the (0001) basal plane of the hcp structure. For the ideal hcp packing, the ratio of $c/a$ is $\sqrt{\frac{8}{3}}$, or 1.633. Table 4-1 shows that actual hcp metals deviate from the ideal $c/a$ ratio.

[1] A detailed review of the crystallography and deformation in hcp metals is given by P. G. Partridge, *Metall. Rev.*, no. 118 and *Met. Mater.* vol. 1, no. 11, pp. 169–194, 1967.

Basal plane (0001) – *ABCDEF*
Prism plane (10$\bar{1}$0) – *FEJH*
Pyramidal planes
  Type I, Order 1 (10$\bar{1}$1) – *GHJ*
  Type I, Order 2 (10$\bar{1}$2) – *KJH*
  Type II, Order 1 (11$\bar{2}$1) – *GHL*
  Type II, Order 2 (11$\bar{2}$2) – *KHL*
Digonal axis [11$\bar{2}$0] – *FGC*

**Fig. 4-3**  Hexagonal close-packed structure.

**Fig. 4-4**  Stacking of close packed spheres.

The fcc and hcp structures are both close-packed structures. Seventy-four percent of the volume of the unit cell is occupied by atoms, on a hard sphere model, in the fcc and hcp structures. This is contrasted with 68 percent packing for a bcc unit cell and 52 percent of the volume occupied by atoms in the simple cubic unit cell.

Plastic deformation is generally confined to the low-index planes, which have a higher density of atoms per unit area than the high-index planes. Table 4-2 lists

**Table 4-1   AXIAL RATIOS OF SOME HEXAGONAL METALS**

| Metal | $c/a$ |
| --- | --- |
| Be | 1.567 |
| Ti | 1.587 |
| Mg | 1.623 |
| Ideal hcp | 1.633 |
| Zn | 1.856 |
| Cd | 1.886 |

**Table 4-2  ATOMIC DENSITY OF LOW-INDEX PLANES**

| Crystal Structure | Plane | Atomic Density, Atoms Per Unit Area | Distance Between Planes |
|---|---|---|---|
| Face-centered cubic | Octahedral {111} | $4/\sqrt{3}\, a_0^2$ | $a_0/\sqrt{3}$ |
| | Cube {100} | $2/a_0^2$ | $a_0/2$ |
| | Dodecahedral {110} | $2/\sqrt{2}\, a_0^2$ | $a_0/2\sqrt{2}$ |
| Body-centered cubic | Dodecahedral {110} | $2/\sqrt{2}\, a_0^2$ | $a_0/\sqrt{2}$ |
| | Cube {100} | $1/a_0^2$ | $a_0/2$ |
| | Octahedral {111} | $1/\sqrt{3}\, a_0^2$ | $a_0/2\sqrt{3}$ |
| Hexagonal close-packed | Basal {0001} | $2/\sqrt{3}\, a_0^2$ | $c$ |

the atomic density per unit area for the common low-index planes. Note that the planes of greatest atomic density also are the most widely spaced planes for the crystal structure.

## 4-3  LATTICE DEFECTS

Real crystals deviate from the perfect periodicity that was assumed in the previous section in a number of important ways. While the concept of the perfect lattice is adequate for explaining the *structure-insensitive* properties of metals, for a better understanding of the *structure-sensitive* properties it has been necessary to consider a number of types of lattice defects. The description of the structure-sensitive properties then reduces itself largely to describing the behavior of these defects.

| Structure-insensitive | Structure-sensitive |
|---|---|
| Elastic constants | Electrical conductivity |
| Melting point | Semiconductor properties |
| Density | Yield stress |
| Specific heat | Fracture strength |
| Coefficient of thermal expansion | Creep strength |

As is suggested by the above brief tabulation, practically all the mechanical properties are structure-sensitive properties. Only since the realization of this fact, in relatively recent times, have really important advances been made in understanding the mechanical behavior of materials.

The term *defect*, or *imperfection*, is generally used to describe any deviation from an orderly array of lattice points. When the deviation from the periodic arrangement of the lattice is localized to the vicinity of only a few atoms it is called a *point defect*, or *point imperfection*. However, if the defect extends through microscopic regions of the crystal, it is called a *lattice imperfection*. Lattice imperfections may be divided into *line defects* and *surface*, or *plane, defects*. Line defects obtain their name because they propagate as lines or as a two-dimensional net in the crystal. The edge and screw dislocations that are discussed in this section are the common line defects encountered in metals. Surface defects arise from the clustering of line defects into a plane. Low-angle boundaries and grain boundaries are surface defects (see Chap. 5). The *stacking fault* between two close-packed regions of the crystal that have alternate stacking sequences (Sec. 4-10) and twinned regions of a crystal (Sec. 4-9) are other examples of surface defects. It is important to note that even at the places where the long-range periodicity of the crystal structure breaks down, as at dislocations and stacking faults, it does so only in certain well-defined ways. Thus, the defects in crystals have regular and reproducible structures and properties.

## POINT DEFECTS

Figure 4-5 illustrates three types of point defects. A *vacancy*, or vacant lattice site,[1] exists when an atom is missing from a normal lattice position (Fig. 4-5a). In pure metals, small numbers of vacancies are created by thermal excitation, and these are thermodynamically stable at temperatures greater than absolute zero. At equilibrium, the fraction of lattices that are vacant at a given temperature is given approximately by the equation

$$\frac{n}{N} = e^{-E_s/kT} \qquad (4\text{-}1)$$

where $n$ is the number of vacant sites in $N$ sites and $E_s$ is the energy required to move an atom from the interior of a crystal to its surface. Table 4-3 illustrates how the fraction of vacant lattice sites in a metal increases rapidly with temperature.

(a)        (b)        (c)

**Fig. 4-5**  Point defects. (a) Vacancy; (b) interstitial; (c) impurity atom.

[1] A .C. Damask and G. J. Dienes," Point Defects in Metals," Gordon and Breach, Science Publishers, Inc., New York. 1963.

**Table 4-3   EQUILIBRIUM VACANCIES IN A METAL**

| Temperature, °C | Approximate Fraction of Vacant Lattice Sites |
|---|---|
| 500 | $1 \times 10^{-10}$ |
| 1000 | $1 \times 10^{-5}$ |
| 1500 | $5 \times 10^{-4}$ |
| 2000 | $3 \times 10^{-3}$ |

$$E_s \approx 1 \text{ ev}$$

By rapid quenching from close to the melting point, it is possible to trap in a greater than equilibrium number of vacancies at room temperature. Higher than equilibrium concentrations of vacancies can also be produced by extensive plastic deformation (cold-work) or as the result of bombardment with high-energy nuclear particles. When the density of vacancies becomes relatively large, it is possible for them to cluster together to form voids.

An atom that is trapped inside the crystal at a point intermediate between normal lattice positions is called an *interstitial atom*, or interstitialcy (Fig. 4-5b). The interstitial defect occurs in pure metals as a result of bombardment with high-energy nuclear particles (radiation damage), but it does not occur frequently as a result of thermal activation.

The presence of an *impurity atom* at a lattice position (Fig. 4-5c] or at an interstitial position results in a local disturbance of the periodicity of the lattice, the same as for vacancies and interstitials.

It is important to realize that no material is completely pure. Most commercially "pure" materials contain usually 0.01 to 1 percent impurities, while ultrapurity materials, such as germanium and silicon crystals for transistors, contain purposely introduced foreign atoms on the order of one part in $10^{10}$. In alloys, foreign atoms are added usually in the range 1 to 50 percent to impart special properties.

## LINE DEFECTS—DISLOCATIONS

The most important two-dimensional, or line, defect is the *dislocation*. The dislocation is the defect responsible for the phenomenon of slip, by which most metals deform plastically. Therefore, one way of thinking about a dislocation is to consider that it is the region of localized lattice disturbance separating the slipped and unslipped regions of a crystal. In Fig. 4-6, *AB* represents a dislocation lying in the slip plane, which is the plane of the paper. It is assumed that slip is advancing to the right. All the atoms above area *C* have been displaced one atomic distance in the slip direction; the atoms above *D* have not yet slipped. *AB* is then the boundary

**Fig. 4-6** A dislocation in a slip plane.

between the slipped and unslipped regions. It is shown shaded to indicate that for a few atomic distances on each side of the dislocation line there is a region of atomic disorder in which the slip distance is between zero and one atomic spacing. As the dislocation moves, slip occurs in the area over which it moves. In the absence of obstacles, a dislocation can move easily on the application of only a small force; this helps explain why real crystals deform much more readily than would be expected for a crystal with a perfect lattice. Not only are dislocations important for explaining the slip of crystals, but they are also intimately connected with nearly all other mechanical phenomena such as strain hardening, the yield point, creep, fatigue, and brittle fracture.

The two basic types of dislocations are the edge dislocation and the screw dislocation. The simplest type of dislocation, which was originally suggested by Orowan, Polanyi, and Taylor, is called the *edge dislocation*, or Taylor-Orowan dislocation. Figure 4-7 shows the slip that produces an edge dislocation for an element of crystal having a simple cubic lattice. Slip has occurred in the direction of the slip vector over the area *ABCD*. The boundary between the right-hand slipped part of the crystal and the left-hand part which has not yet slipped is the line *AD*, the edge dislocation. Note that the parts of the crystal above the slip plane have been displaced, in the direction of slip, with respect to the part of the crystal

**Fig. 4-7** Edge dislocation produced by slip in a simple cubic lattice. Dislocation lies along *AD*, perpendicular to slip direction. Slip has occurred over area *ABCD*. (*From W. T. Read, Jr., "Dislocation in Crystals," p. 2, McGraw-Hill Book Company, New York, 1953.*)

below the slip plane by an amount indicated by the shaded area in Fig. 4-7. All points in the crystal which were originally coincident across the slip plane have been displaced relative to each other by this same amount. The amount of displacement is equal to the *Burgers vector* **b** of the dislocation. A defining characteristic of an edge dislocation is that its Burgers vector is always perpendicular to the dislocation line.

Although the exact arrangement of atoms along *AD* is not known, it is generally agreed that Fig. 4-8 closely represents the atomic arrangement in a plane normal to the edge dislocation *AD*. The plane of the paper in this figure corresponds to a (100) plane in a simple cubic lattice and is equivalent to any plane parallel to the front face of Fig. 4-7. Note that the lattice is distorted in the region of the dislocation. There is one more vertical row of atoms above the slip plane than below it. The atomic arrangement results in a compressive stress above the slip plane and a tensile stress below the slip plane. An edge dislocation with the extra plane of atoms above the slip plane, as in Fig. 4-8, by convention is called a *positive edge dislocation* and is frequently indicated by the symbol ⊥. If the extra plane of atoms lies below the slip plane, the dislocation is a negative edge dislocation, ⊤.

A pure edge dislocation can glide or slip in a direction perpendicular to its length. However, it may move vertically by a process known as *climb*, if diffusion of atoms or vacancies can take place at an appreciable rate. Consider Fig. 4-8. For the edge dislocation to move upward (positive direction of climb), it is necessary to remove the extra atom directly over the symbol ⊥ or to add a vacancy to this spot. One such atom would have to be removed for every atomic spacing that the dislocation climbs. Conversely, if the dislocation moved down, atoms would have to be added. Atoms could be removed from the extra plane of atoms by the extra atom interacting with a lattice vacancy. Atoms are added to the extra plane for negative climb by the diffusion of an atom from the surrounding crystal, creating

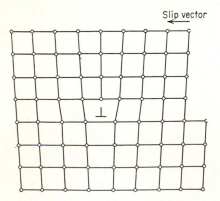

Slip vector

Fig. 4-8  Atomic arrangement in a plane normal to an edge dislocation. (*From W. T. Read, Jr., "Dislocations in Crystals," p. 3, McGraw-Hill Book Company, New York, 1953.*)

**Fig. 4-9** Slip that produces a screw dislocation in a simple cubic lattice. Dislocation lies along *AD*, parallel to slip direction. Slip has occurred over the area *ABCD*. (*From W. T. Read, Jr., " Dislocations in Crystals," p. 15, McGraw-Hill Book Company, New York, 1953.*)

a vacancy. Since movement by climb is diffusion-controlled, motion is much slower than in glide and less likely except at high temperatures.

The second basic type of dislocation is the *screw*, or Burgers, dislocation. Figure 4-9 shows a simple example of a screw dislocation. The upper part of the crystal to the right of *AD* has moved relative to the lower part in the direction of the slip vector. No slip has taken place to the left of *AD*, and therefore *AD* is a dislocation line. Thus, the dislocation line is parallel to its Burgers vector, or slip vector, and by definition this must be a screw dislocation. Consider the trace of a circuit around the dislocation line, on the front face of the crystal. Starting at *X* and completing a counterclockwise circuit, we arrive at *X'*, one atomic plane behind that containing *X*. In making this circuit we have traced the path of a right-handed screw. Every time a circuit is made around the dislocation line, the end point is displaced one plane parallel to the slip plane in the lattice. Therefore, the atomic planes are arranged around the dislocation in a spiral staircase or screw.

The arrangement of atoms (in two dimensions) around a screw dislocation in a simple cubic lattice is shown in Fig. 4-10. In this figure we are looking down on the slip plane in Fig. 4-9. The open circles represent atoms just above the slip plane, and the solid circles are atoms just below the slip plane. A screw dislocation does not have a preferred slip plane, as an edge dislocation has, and therefore the motion of a screw dislocation is less restricted than the motion of an edge dislocation. However, movement by climb is not possible with a screw dislocation.

For the present, the discussion of dislocations will be limited to the geometrical concepts presented in this section. After a more complete discussion of the plastic deformation of single crystals and polycrystalline specimens, we shall return to a detailed discussion of dislocation theory in Chap. 5. Among the topics covered will be the effect of crystal structure on dislocation geometry, the experimental evidence for dislocations, and the interaction between dislocations.

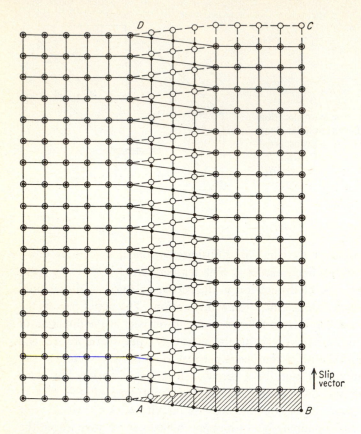

**Fig. 4-10** Atomic arrangement around the screw dislocation shown in Fig. 4-9. The plane of the figure is parallel to the slip plane. *ABCD* is the slipped area, and *AD* is the screw dislocation. Open circles represent atoms in the atomic plane just above the slip plane, and the solid circles are atoms in the plane just below the slip plane. (*From W. T. Read, Jr., " Dislocations in Crystals," p. 17, McGraw-Hill, Book Company, New York, 1953.*)

## 4-4 DEFORMATION BY SLIP

The usual method of plastic deformation in metals is by the sliding of blocks of the crystal over one another along definite crystallographic planes, called *slip planes*. As a very crude approximation, the slip, or glide, of a crystal can be considered analogous to the distortion produced in a deck of cards when it is pushed from one end. Figure 4-11 illustrates this classical picture of slip. In Fig. 4-11*a*, a shear stress is applied to a metal cube with a top polished surface. Slip occurs when the shear stress exceeds a critical value. The atoms move an integral number of atomic distances along the slip plane, and a step is produced in the polished surface (Fig. 4-11*b*). When we view the polished surface from above with a microscope,

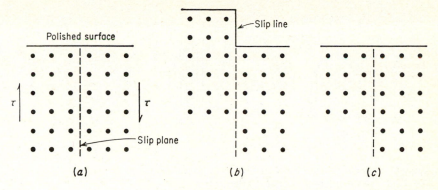

Fig. 4-11  Schematic drawing of classical idea of slip.

the step shows up as a line, which we call a *slip line*. If the surface is then repolished after slip has occurred, so that the step is removed, the slip line will disappear (Fig. 4-11*c*).

Because of the translational symmetry of a crystal lattice, the crystal structure is perfectly restored after slip has taken place provided that the deformation was uniform. Each atom in the slipped part of the crystal moves forward the same integral number of lattice spacings. Note that slip lines are due to changes in surface elevation and that the surface must be suitably prepared for microscopic observation prior to deformation if the slip lines are to be observed. Figure 4-12 shows straight slip lines in copper.

Fig. 4-12  Straight slip lines in copper (500X). (*Courtesy W. L. Phillips.*)

Fig. **4-13** Schematic drawing of the fine structure of a slip band. (*a*) Small deformation; (*b*) large deformation.

The fine structure of slip lines has been studied at high magnification by means of the electron microscope. What appears as a line, or at best a narrow band at 1,500 diameters' magnification in the optical microscope can be resolved by the electron microscope as discrete slip lamellae at 20,000 diameters, shown schematically in Fig. 4-13.

The fact that a single crystal remains a single crystal after homogeneous plastic deformation imposes limitations on the way in which plastic deformation may occur. Slip occurs most readily in specific directions on certain crystallographic planes. Generally the slip plane is the plane of greatest atomic density (Table 4-2) and the slip direction is the closest-packed direction within the slip plane. Since the planes of greatest atomic density are also the most widely spaced planes in the crystal structure, the resistance to slip is generally less for these planes than for any other set of planes. The slip plane together with the slip direction establishes the *slip system*.

In the hexagonal close-packed metals, the only plane with high atomic density is the basal plane (0001). The digonal axes $\langle 11\bar{2}0 \rangle$ are the close-packed directions. For zinc, cadmium, magnesium, and cobalt slip occurs on the (0001) plane in the $\langle 11\bar{2}0 \rangle$ directions.[1] Since there is only one basal plane per unit cell and three $\langle 11\bar{2}0 \rangle$ directions, the hcp structure possesses three slip systems. The limited number of slip systems is the reason for the extreme orientation dependence and low ductility in hcp crystals.

In the face-centered cubic structure, the {111} octahedral planes and the $\langle 110 \rangle$ directions are the close-packed systems. There are eight {111} planes in the fcc unit cell. However, the planes at opposite faces of the octahedron are parallel

---

[1] Zirconium and titanium, which have low $c/a$ ratios, slip primarily on the prism and pyramidal planes in the $\langle 11\bar{2}0 \rangle$ direction.

to each other,[1] so that there are only four *sets* of octahedral planes. Each {111} plane contains three ⟨110⟩ directions (the reverse directions being neglected). Therefore, the fcc lattice has 12 possible slip systems.

The bcc structure is not a close-packed structure like the fcc or hcp structures. Accordingly, there is no one plane of predominant atomic density, as (111) in the fcc structure and (0001) in the hcp structure. The {110} planes have the highest atomic density in the bcc structure, but they are not greatly superior in this respect to several other planes. However, in the bcc structure the ⟨111⟩ direction is just as close-packed as the ⟨110⟩ and ⟨1120⟩ directions in the fcc and hcp structures. Therefore, the bcc metals obey the general rule that the slip direction is the close-packed direction, but they differ from most other metals by not having a definite single slip plane. Slip in bcc metals is found to occur on the {110}, {112}, and {123} planes, while the slip direction is always the [111] direction. There are 48 possible slip systems, but since the planes are not so close-packed as in the fcc structure, higher shearing stresses are usually required to cause slip.

Slip in bcc alpha iron has been particularly well studied.[2] It has been concluded that the slip plane in alpha iron may occupy any position in the [111] zone, its position being determined by the orientation of the stress axis with respect to the crystal axis and the variation in the shearing strengths of the planes in the slip zone. These studies have shown that observed deviations from the low-index planes {110}, {112}, and {123} are real effects, which supports the belief that slip in alpha iron is noncrystallographic. Further evidence for noncrystallographic slip is the fact that slip lines in alpha iron are wavy[3] (Fig. 4-14).

Certain metals show additional slip systems with increased temperature. Aluminum deforms on the {110} plane at elevated temperature, while in magnesium the {10$\bar{1}$1} pyramidal plane plays an important role in deformation by slip above 225°C. In all cases the slip direction remains the same when the slip plane changes with temperature.

---

[1] In a cubic structure the angle between two planes $(h_1 k_1 l_1)$ and $(h_2 k_2 l_2)$ is given by

$$\cos \theta = \frac{h_1 h_2 + k_1 k_2 + l_1 l_2}{(h_1{}^2 + k_1{}^2 + l_1{}^2)^{1/2}(h_2{}^2 + k_2{}^2 + l_2{}^2)^{1/2}}$$

Thus, for example, the (111) and ($\bar{1}\bar{1}\bar{1}$) planes are parallel, since cos $\theta = 1$. Also, in a cubic structure, the direction [*hkl*] is perpendicular to the (*hkl*) plane. Therefore, when the slip direction [*uvw*] lies in the slip plane (*hkl*), $\theta = 90°$ and cos $\theta = 0 = hu + kv + lw$. For example the (11$\bar{1}$) plane contains the slip directions [101], and [1$\bar{1}$0].

[2] F. L. Vogel and R. M. Brick, *Trans. AIME*, vol. 197, p. 700, 1958; R. P. Steijn and R. M. Brick, *Trans. Am. Soc. Met.*, vol. 46, pp, 1406–1448, 1954; J. J. Cox, G. T. Horne, and R. F. Mehl, *Trans. Am. Soc. Met.*, vol. 49, pp. 118–131, 1957.

[3] J. R. Low and R. W. Guard, *Acta Metall.*, vol. 7, pp. 171–179, 1959, have shown that curved slip lines are produced in iron by screw components of the dislocation loop but that the slip lines are straight when viewed normal to the edge component of the dislocation.

**Fig. 4-14** Wavy slip lines in alpha iron (150X). (*Courtesy J. J. Cox.*)

Slip in an ionic crystal, such as NaCl, or a covalently bonded ceramic, such as MgO, imposes the further consideration that the interactions between like and unlike atoms must be satisfied. For example, in MgO the resistance to slip in the [100] direction is high because ions of like charge are brought into close contact (Fig. 4-15a). Slip in the [110] direction (Fig. 4-15b) does not require any nearest neighbor ions of the same polarity to come in close contact during glide so that large electrostatic forces do not develop. Thus, slip in the [110] direction is preferred in many ionic crystals. The necessity to account for the electrostatic forces between ions of different species places a strong restriction on the plastic deformation of ionic solids. In these materials there are far fewer directions of low critical shear stress than in cubic metals.

**Fig. 4-15** Deformation of MgO. (*a*) Slip in the [100] direction; (*b*) slip in the [110] direction. Symbols: □ Mg$^{++}$, O, O$^{--}$. (*From L. H. Van Vlack, "Physical Ceramics for Engineers," Addison-Wesley Publishing Company, Inc., Reading, Mass., 1964, p. 132. By permission of the publishers.*)

## 4-5 SLIP IN A PERFECT LATTICE

If slip is assumed to occur by the translation of one plane of atoms over another, it is possible to make a reasonable estimate[1] of the shear stress required for such a movement in a perfect lattice. Consider two planes of atoms subjected to a homogeneous shear stress (Fig. 4-16). The shear stress is assumed to act in the slip plane along the slip direction. The distance between atoms in the slip directions is $b$, and the spacing between adjacent lattice planes is $a$. The shear stress causes a displacement $x$ in the slip direction between the pair of adjacent lattice planes. The shearing stress is initially zero when the two planes are in coincidence, and it is also zero when the two planes have moved one identity distance $b$, so that point 1 in the top plane is over point 2 on the bottom plane. The shearing stress is also zero when the atoms of the top plane are midway between those of the bottom plane, since this is a symmetry position. Between these positions each atom is attracted toward the nearest atom of the other row, so that the shearing stress is a periodic function of the displacement.

As a first approximation, the relationship between shear stress and displacement can be expressed by a sine function

$$\tau = \tau_m \sin \frac{2\pi x}{b} \qquad (4\text{-}2)$$

(a)

(b)

**Fig. 4-16** (a) Shear displacement of one plane of atoms over another atomic plane; (b) variation of shearing stress with displacement in slip direction.

[1] J. Frenkel, *Z. Phys.*, vol. 37, p. 572, 1926.

where $\tau_m$ is the amplitude of the sine wave and $b$ is the period. At small values of displacement, Hooke's law should apply.

$$\tau = G\gamma = \frac{Gx}{a} \qquad (4\text{-}3)$$

For small values of $x/b$, Eq. (4-2) can be written

$$\tau \approx \tau_m \frac{2\pi x}{b} \qquad (4\text{-}4)$$

Combining Eqs. (4-3) and (4-4) provides an expression for the maximum shear stress at which slip should occur.

$$\tau_m = \frac{G}{2\pi} \frac{b}{a} \qquad (4\text{-}5)$$

As a rough approximation, $b$ can be taken equal to $a$, with the result that the theoretical shear strength of a perfect crystal is approximately equal to the shear modulus divided by $2\pi$.

$$\tau_m = \frac{G}{2\pi} \qquad (4\text{-}6)$$

The shear modulus for metals is in the range $10^6$ to $10^7$ psi ($10^{11}$ to $10^{12}$ dyn/cm$^2$). Therefore, Eq. (4-6) predicts that the theoretical shear stress will be in the range $10^5$ to $10^6$ psi, while actual values of the shear stress required to produce plastic deformation in metal single crystals are in the range $10^2$ to $10^4$ psi. Even if more refined calculations are used to correct the sine-wave assumption, the value of $\tau_m$ cannot be made equal to the observed shear stress. Tyson,[1] using a computer solution of the interatomic force equations, predicted $\tau_m = G/16$ for an fcc metal, $G/8$ for an NaCl structure, and $G/4$ for a covalently bonded diamond structure. Since the theoretical shear strength of metal crystals is at least 100 times greater than the observed shear strength, it must be concluded that a mechanism other than bodily shearing of planes of atoms is responsible for slip. In the next section it is shown that dislocations provide such a mechanism.

## 4-6 SLIP BY DISLOCATION MOVEMENT

The concept of the dislocation was first introduced to explain the discrepancy between the observed and theoretical shear strengths of metals. For the dislocation concept to be valid it is necessary to show (1) that the motion of a dislocation through a crystal lattice requires a stress far smaller than the theoretical shear

---

[1] W. R. Tyson, *Philos. Mag.*, vol. 14, pp. 925–936, 1966.

**Fig. 4-17**  (a) Atom movements near dislocation in slip; (b) movement of an edge dislocation.

stress, and (2) that the movement of the dislocation produces a step, or slip band, at the free surface.

In a perfect lattice all atoms above and below the slip plane are in minimum energy positions. When a shear stress is applied to the crystal, the same force opposing the movement acts on all the atoms. This is the model for slip presented in Fig. 4-16. When there is a dislocation in the crystal, the atoms well away from the dislocation are still in the minimum energy positions but at the dislocation only a small movement of the atoms is required. Referring to Fig. 4-17a, the extra plane of atoms at the edge dislocation initially is at 4. Under the action of the shear stress, a very small movement of atoms to the right will allow this half plane to line up with the half plane 5', at the same time cutting the half plane 5 from its neighbors below the slip plane. By this process the edge dislocation line has moved from its initial position between planes 4' and 5' to a new position between planes 5' and 6'. Since the atoms around the dislocation are symmetrically placed on opposite sides of the extra half plane, equal and opposite forces oppose and assist the motion. Thus, in a first approximation there is no net force on the dislocation and the stress required to move the dislocation is zero. The continuation of this process under the stresses shown in Fig. 4-17 moves the dislocation to the right. When the extra half plane of atoms reaches a free surface (Fig. 4-17b), it results in a slip step of one Burgers vector, or one atomic distance for the simple cubic lattice.

A particularly instructive way of looking at slip by dislocation motion has been proposed by Cottrell.[1] Consider that plastic deformation is the transition from an unslipped to a slipped state (Fig. 4-18a). Since the process is opposed by an energy barrier $\Delta E$, in order to facilitate the process it is logical to assume that the

[1] A. H. Cottrell, "An Introduction to Metallurgy," pp. 266–269, Edward Arnold (Publishers) Ltd. London, 1967.

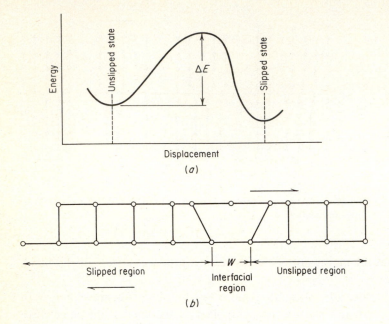

**Fig. 4-18**   (*a*) Energy change from unslipped to slipped state; (*b*) stages in growth of slipped region.

material will not all make the transition simultaneously. To minimize the energetics of the process, the slipped material will grow at the expense of the unslipped region by the advance of an interfacial region (Fig. 4-18*b*). The interfacial region is a *dislocation*. To minimize the energy for the transition, we expect the interface thickness *w* to be narrow. The length *w* is the *width* of the dislocation. The smaller the width of the dislocation, the lower is the interfacial energy, but the wider the dislocation, the lower is the elastic energy of the crystal because then the atomic spacing in the slip direction is closer to its equilibrium spacing. Thus, the equilibrium width of the dislocation is determined by a balance between these two opposing energy changes.

The dislocation width is important because it determines the force required to move a dislocation through the crystal lattice. This force is called the *Peierls-Nabarro force*. The Peierls stress is the shear stress required to move a dislocation through a crystal lattice in a particular direction

$$\tau_p \approx \frac{2G}{1-v} e^{-2\pi w/b} \approx \frac{2G}{1-v} e^{-[2\pi a/(1-v)b]} \qquad (4\text{-}7)$$

where *a* is the distance between slip planes and *b* is the distance between atoms in the slip direction. Note that the dislocation width appears in the exponential term in Eq. (4-7), so that the Peierls stress will be very sensitive to the atomic positions at the core of a dislocation. These are not known with any high degree of accuracy and, since Eq. (4-7) was derived for the sinusoidal force-distance law that has only limited validity, the equation cannot be used for precise calculations. However, it is accurate enough to show that the stress needed to move a dislocation in a metal is quite low.[1]

In spite of these limitations, the Peierls equation has important conceptual value. It shows that materials with wide dislocations will require a low stress to move the dislocations. Physically, this is so because when the dislocation is wide, the highly distorted region at the core of the dislocation is not localized on any particular atom in the crystal lattice. In ductile metals the dislocation width is of the order of 10 atomic spacings. However, in ceramic materials with directional covalent bonds, the interfacial energy is high and the dislocations are narrow and relatively immobile. This fact combined with the restriction on slip systems imposed by the requirements of electrosatic forces results in the low degree of plasticity of ceramic materials. Ceramics become more ductile at high temperatures because thermal activation helps the dislocations overcome the energy barrier.

The fact that slip occurs in close-packed directions means that $b$ is minimized, and from Eq. (4-7) the Peierls stress will be lower. If $a < b$, as would occur for closely spaced but loosely packed planes, the Peierls stress would be high. Thus, Eq. (4-7) provides a basis for the observation that slip occurs most readily on close-packed planes in the close-packed directions. When the crystal structure is complex, without highly close-packed planes and directions, dislocations tend to be immobile. This causes brittleness and high hardness of intermetallic compounds.

It is instructive to calculate the shear strain that results from the motion of a dislocation through a single crystal. The dislocation produces a slip offset of length $b$, so that the shear strain ($\gamma$ from Fig. 4-19a) is $\gamma = b/h$. If the crystal is 1 cm high, the passage of a single dislocation produces a shear strain of only $3 \times 10^{-8}$. It is obvious that the motion of many dislocations is involved in producing strains of engineering significance. The total macroscopic plastic strain is the sum of all the small strains due to a very large number of individual dislocations. Thus if three dislocations on three parallel planes moved through the crystal, the shear strain would be $\gamma = 3b/h$.

Now consider the case where a dislocation has moved part way through the crystal along the slip plane (Fig. 4-19b). Since $b$ is very small compared with $L$ or $h$, the displacement $\delta_i$ for a dislocation at an intermediate position between $x_i = 0$ and $x_i = L$ would be proportional to the fractional displacement $x_i/L$.

$$\delta_i = \frac{x_i b}{L}$$

[1] Recent concepts about the Peierls stress have been given by Rosenfield et al., "Dislocation Dynamics," McGraw-Hill Book Company, New York, 1968.

**Fig. 4-19** (a) Shear strain associated with passage of a single dislocation through a crystal; (b) shear strain due to motion of dislocation part way through a crystal.

The total displacement of the top of the crystal relative to the bottom for many dislocations on many slip planes will be

$$\Delta = \sum \delta_i = \frac{b}{L} \sum_i^N x_i \qquad (4\text{-}8)$$

where $N$ is the total number of dislocations that have moved in the volume of the crystal.

The macroscopic shear strain is

$$\gamma = \frac{\Delta}{h} = \frac{b}{hL} \sum_1^N x_i \qquad (4\text{-}9)$$

If the average distance that dislocations have moved is $\bar{x}$, where

$$\bar{x} = \frac{\sum_1^N x_i}{N}$$

then

$$\gamma = \frac{bN\bar{x}}{hL} \qquad (4\text{-}10)$$

This equation can best be written in terms of the *dislocation density* $\rho$

$$\gamma = b\rho\bar{x} \qquad (4\text{-}11)$$

where $\rho = N/hL$. The dislocation density is the total length of dislocation line per unit volume or, alternatively, the number of dislocation lines that cut through a unit cross-sectional area. Either way, $\rho$ has units of number of dislocations per square centimeter. The argument used in deriving Eq. (4-11) is applicable to both edge and screw dislocations. However, such highly simplified motion by parallel dislocations rarely occurs in real metals, so that Eq. (4-11) needs to be modified to account for complexities in geometry and dislocation configuration.

It is often useful to express Eq. (4-11) in terms of shear-strain rate.

$$\dot{\gamma} = \frac{d\gamma}{dt} = b\rho \frac{d\bar{x}}{dt} = b\rho\bar{v} \qquad (4\text{-}12)$$

where $\bar{v}$, the average dislocation velocity, is a quantity which can be measured experimentally in many systems. From Eq. (4-12) we see that if we want to describe macroscopic plastic deformation in terms of dislocation behavior we need to know (1) the crystal structure in order to evaluate $b$, (2) the number of mobile dislocations $\rho$, and (3) the average dislocation velocity $\bar{v}$. The quantities $\rho$ and $\bar{v}$ will depend on stress, time, temperature, and prior thermomechanical history.

Measurements of the velocity of dislocation motion[1] in a number of ionic crystals and metals have shown that the velocity is a very strong function of the shear stress in the slip plane. This is given by the equation $v = A\tau^{m'}$, where $m'$ is a constant varying from about 1.5 to 40 for different materials. There is a critical stress required to start dislocations moving, with small increases in stress leading to large increases in dislocation velocity. Above about 10 cm/s the velocity increases less rapidly and tends toward an upper limit close to the shear wave velocity of the material.

## 4-7 CRITICAL RESOLVED SHEAR STRESS FOR SLIP

The extent of slip in a single crystal depends on the magnitude of the shearing stress produced by external loads, the geometry of the crystal structure, and the orientation of the active slip planes with respect to the shearing stresses. Slip begins when the shearing stress on the slip plane in the slip direction reaches a threshold value called the *critical resolved shear stress*. This value[2] is really the single-crystal equivalent of the yield stress of an ordinary stress-strain curve. The value of the critical resolved shear stress depends chiefly on composition and temperature.

The fact that different tensile loads are required to produce slip in single crystals of different orientation can be rationalized by a critical resolved shear stress; this was first recognized by Schmid.[3] To calculate the critical resolved shear stress from a single crystal tested in tension, it is necessary to know, from

---

[1] W. G. Johnston and J. H., Gilman, *J. Appl. Phys.*, vol. 30, p. 129, 1959.
[2] In practice it is very difficult to determine the stress at which the first slip bands are produced. In most cases, the critical shear stress is obtained by the intersection of the extrapolated elastic and plastic regions of the stress-strain curve.
[3] E. Schmid, *Z. Elektrochem.*, vol. 37, 447, 1931.

**Fig. 4-20** Diagram for calculating critical resolved shear stress.

x-ray diffraction, the orientation with respect to the tensile axis of the plane on which slip first appears and the slip direction. Consider a cylindrical single crystal with cross-sectional area $A$ (Fig. 4-20). The angle[1] between the *normal* to the slip plane and the tensile axis is $\phi$, and the angle which the slip direction makes with the tensile axis is $\lambda$. The area of the slip plane inclined at the angle $\phi$ will be $A/\cos \phi$, and the component of the axial load acting in the slip plane in the slip direction is $P \cos \lambda$. Therefore, the critical resolved shear stress is given by

$$\tau_R = \frac{P \cos \lambda}{A/\cos \phi} = \frac{P}{A} \cos \phi \cos \lambda \qquad (4\text{-}13)$$

Equation (4-13) gives the shear stress resolved on the slip plane in the slip direction. This shear stress is a maximum when $\phi = \lambda = 45°$, so that $\tau_R = \frac{1}{2}P/A$. If the tension axis is normal to the slip plane ($\lambda = 90°$) or if it is parallel to the slip plane ($\phi = 90°$), the resolved shear stress is zero. Slip will not occur for these extreme orientations since there is no shear stress on the slip plane. Crystals close to these orientations tend to fracture rather than slip.

Table 4-4 gives values of critical resolved shear stress for a number of metals. The importance of small amounts of impurities in increasing the critical resolved shear stress is shown by the data for silver and copper. Alloying-element additions have even a greater effect, as shown by the data for gold-silver alloys in Fig. 4-21. Note that a large increase in the resistance to slip is produced by alloying gold and silver even though these atoms are very much alike in size and electronega-

---

[1] Note that $\phi$ and $\lambda$ are complementary angles only in the special case when the slip direction is in the plane defined by the stress axis and the normal to the slip plane.

**Table 4-4  ROOM-TEMPERATURE SLIP SYSTEMS AND CRITICAL
RESOLVED SHEAR STRESS FOR METAL SINGLE CRYSTALS**

| Metal | Crystal structure | Purity, % | Slip plane | Slip direction | Critical shear stress, g/mm$^2$ | Ref. |
|-------|-------------------|-----------|------------|----------------|--------------------------------|------|
| Zn | hcp | 99.999 | (0001) | [11$\bar{2}$0] | 18 | a |
| Mg | hcp | 99.996 | (0001) | [1120] | 77 | b |
| Cd | hcp | 99.996 | (0001) | [11$\bar{2}$0] | 58 | c |
| Ti | hcp | 99.99 | (1010) | [11$\bar{2}$0] | 1,400 | d |
|    |     | 99.9  | (1010) | [11$\bar{2}$0] | 9,190 | d |
| Ag | fcc | 99.99 | (111) | [110] | 48 | e |
|    |     | 99.97 | (111) | [110] | 73 | e |
|    |     | 99.93 | (111) | [110] | 131 | e |
| Cu | fcc | 99.999 | (111) | [110] | 65 | e |
|    |     | 99.98 | (111) | [110] | 94 | e |
| Ni | fcc | 99.8 | (111) | [110] | 580 | e |
| Fe | bcc | 99.96 | (110) | [111] | 2,800 | f |
|    |     |       | (112) |       |       |   |
|    |     |       | (123) |       |       |   |
| Mo | bcc | ... | (110) | [111] | 5,000 | g |

[a] D. C. Jillson, *Trans. AIME*, vol. 188, p. 1129, 1950.
[b] E. C. Burke and W. R. Hibbard, Jr., *Trans. AIME*, vol. 194, p. 295, 1952.
[c] E. Schmid, "International Conference on Physics," vol. 2, Physical Society, London. 1935.
[d] A. T. Churchman, *Proc. R. Soc. London Ser. A*, vol. 226A, p. 216, 1954.
[e] F. D. Rosi, *Trans., AIME*, vol. 200, p. 1009, 1954.
[f] J. J. Cox, R. F. Mehl, and G. T. Horne, *Trans. Am. Soc. Met.*, vol. 49, p. 118, 1957.
[g] R. Maddin and N. K. Chen, *Trans. AIME*, vol. 191, p. 937, 1951.

**Fig. 4-21**  Variation of critical resolved shear stress with
composition in silver-gold-alloy single crystals. (*From G.
Sachs and J. Weerts, Z. Phys., vol. 62, p. 473, 1930.*)

tivity, and hence they form a solid solution over the complete range of composition. In solid solutions, where the solute atoms differ considerably in size from the solvent atoms, an even greater increase in critical resolved shear stress would be observed.

The magnitude of the critical resolved shear stress of a crystal is determined by the interaction of its population of dislocations with each other and with defects such as vacancies, interstitials, and impurity atoms. This stress is, of course, greater than the stress required to move a single dislocation, but it is appreciably lower than the stress required to produce slip in a perfect lattice. On the basis of this reasoning, the critical resolved shear stress should decrease as the density of defects decreases, provided that the total number of imperfections is not zero. When the last dislocation is eliminated, the critical resolved shear stress should rise abruptly to the high value predicted for the shear strength of a perfect crystal. Experimental evidence for the effect of decreasing defect density is shown by the fact that the critical resolved shear stress of soft metals can be reduced to less than one-third by increasing the purity. At the other extreme, micron-diameter single crystal filaments, or whiskers, can be grown essentially dislocation-free. Tensile tests[1] on these filaments have given strengths which are approximately equal to the calculated strength of a perfect crystal.

The ratio of the resolved shear stress to the axial stress is called the *Schmid factor m*. For a single crystal loaded in tension or compression along its axis, $m = \cos \phi \cos \lambda$. Hartley and Hirth[2] have presented graphical methods of determining $m$ for any crystal orientation and slip system.

It is observed experimentally that a single crystal will slip when the resolved shear stress on the slip plane reaches a critical value. This behavior, known as Schmid's law, is best demonstrated with hcp metals where the limited number of slip systems allows large differences in orientation between the slip plane and the tensile axis (see Prob. 4-8). In fcc metals the high symmetry results in so many equivalent slip systems that it is possible to get a variation in the yield stress of only about a factor of 2 because of differences in the orientation of the slip plane with the tensile axis.

## 4-8 DEFORMATION OF SINGLE CRYSTALS

Most studies of the mechanical properties of single crystals are made by subjecting the crystal to simple uniaxial tension. In the ordinary tension test, the movement of the crosshead of the testing machine constrains the specimen at the grips since

[1] S. S. Brenner, *J. Appl. Phys.*, vol. 27, pp. 1484–1491, 1956.
[2] C. S. Hartley and J. P. Hirth, *Trans. Metall. Soc. AIME*, vol. 223, pp. 1415–1419, 1965.

**Fig. 4-22**  (*a*) Tensile deformation of single crystal without constraint; (*b*) rotation of slip planes due to constraint.

(*a*)  (*b*)

the grips must remain in line.  Therefore, the specimen is not permitted to deform freely by uniform glide on every slip plane along the gage length of the specimen, as is pictured in Fig. 4-22*a*.  Instead, the slip planes rotate toward the tensile axis since the tensile axis of the specimen remains fixed, as in Fig. 4-22*b*.

Since plastic flow occurs by slip on certain planes in particular directions, the measured increase in length of the specimen for a given amount of slip will depend on the orientations of the slip plane and direction with the specimen axis.[1] The fundamental measure of plastic strain in a single crystal is the crystallographic glide strain $\gamma$.  Glide strain is the relative displacement of two parallel slip planes separated at a unit distance.  The equations relating glide strain with specimen extension can be derived from Fig. 4-23.  As the single crystal elongates, the slip direction rotates toward the tensile axis.  For simplicity in Fig. 4-23, the glide elements are kept fixed and the tensile axis is rotated as the crystal elongates from $L_0$ to $L_1$.  The two cases are equivalent geometrically.  Moreover, for simplicity the orientation of the slip plane is given by the angle $\chi$ between the axis of the glide ellipse and the tensile axis rather than the angle $\phi$ between the normal to the glide ellipse (slip plane) and the tensile axis.  With this selection of angles, $\tau_R = P/A \sin \chi \cos \lambda$.

From triangle $ABB'$, using the law of sines, we can see that

$$\frac{L_0}{\sin \lambda_1} = \frac{L_1}{\sin (180 - \lambda_0)} = \frac{L_1}{\sin \lambda_0}$$

[1] E. Schmid and W. Boas, "Plasticity of Crystals," English translation, pp. 58–60, F. A. Hughes & Co., London, 1950.

**Fig. 4-23** Extension of a single crystal.

From triangles $ABC$ and $AB'C$

$$AC = L_0 \sin \chi_0 = L_1 \sin \chi_1$$

The *glide strain* is defined as the total amount of slip divided by the thickness of the glide packet

$$\lambda = \frac{BB'}{AC}$$

Again, from the law of sines

$$BB' = \frac{L_1 \sin (\lambda_0 - \lambda_1)}{\sin \lambda_0}$$

**Fig. 4-24** Typical single-crystal stress-strain curves.

Substitution in the expression for glide strain, and after considerable trigonometric manipulation, results in

$$\gamma = \frac{1}{\sin \chi_0} \left\{ \left[ \left( \frac{L_1}{L_0} \right)^2 - \sin^2 \lambda_0 \right]^{1/2} - \cos \lambda_0 \right\} \qquad (4\text{-}14)$$

Thus, the glide shear strain may be determined from the initial orientation of the slip plane and slip direction ($\chi_0$ and $\lambda_0$) and the extension of the specimen $L_1/L_0$. This analysis assumes that slip occurs on only a single slip system.

If the orientation of the glide elements can be determined during or after deformation, the glide strain may be determined from

$$\gamma = \frac{\cos \lambda_1}{\sin \chi_1} - \frac{\cos \lambda_0}{\sin \chi_0} \qquad (4\text{-}15)$$

The fundamental way to present single-crystal data is by plotting *resolved shear stress vs. glide shear strain*. Figure 4-24 shows that there are important differences between metals. Typically, fcc metals exhibit greater strain hardening than hcp metals. Additional details will be given in Sec. 4-14.

## 4-9 DEFORMATION OF FACE-CENTERED CUBIC CRYSTALS

Because fcc crystals have high symmetry and 12 potential slip systems, there is a wide choice of slip systems. The slip plane will not have to undergo much rotation before the resolved shear stress becomes high on another {111} ⟨110⟩ slip system. The initial operative slip system, the primary slip system, will be the one with the highest Schmid factor, $m = \sin \chi \cos \lambda$. The primary system will depend on the orientation of the crystal relative to the tensile stress axis.

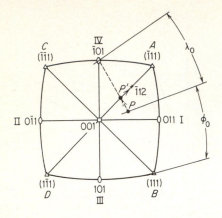

Fig. 4-25 Standard [001] stereographic projection for a cubic crystal.

The relationship between the stress axis and the 12 possible slip systems is best shown on a stereographic projection[1] (Fig. 4-25), where each of the unit triangles defines a region in which a particular slip system operates. There are four $\langle 111 \rangle$ poles *ABCD* representing the normals to the octahedral {111} slip planes. Slip directions are indicated I through IV. For a specimen axis at *P*, the slip system *B*IV will be operative. $\phi_0$ and $\lambda_0$ are given by the great circles through *B-P-IV*. We can use the stereographic plot to follow the rotation of the slip system toward the tensile axis. As the specimen elongates, $\lambda$ decreases and $\phi$ increases. However, it is more convenient to consider that the slip system remains fixed and the specimen axis rotates.

As the specimen elongates, the specimen axis eventually reaches the [001] − [$\bar{1}$11] boundary at *P′*. Now the resolved shear stress is equal on the *primary slip system* and the *conjugate slip system* ($\bar{1}\bar{1}1$) [011]. At this point deformation proceeds on the two slip systems simultaneously to produce *duplex slip* or *multiple slip*.

Under the microscope conjugate slip appears as another set of intersecting slip lines. The fact that slip can occur equally on both slip systems indicates that latent strain hardening must have occurred on the conjugate system when only the primary system was acting. The specimen axis rotates along the [001] − [$\bar{1}$11] boundary to the [$\bar{1}$12] pole, which is midway between the two operative slip directions [$\bar{1}$01] and [011]. When the specimen axis reaches [$\bar{1}$12] it stays at that orientation until the specimen necks down and fractures. Thus, the introduction of duplex slip interrupts the free rotation of the glide system and leads to lower ductility at fracture[2] than in the case of hcp crystals where easy glide on only a

---

[1] For a description of stereographic projection see C. S. Barrett and T. B. Massalski, "Structure of Metals," 3rd ed., chap. 2, McGraw-Hill Book Company, New York, 1966; A. Kelly and G. W. Groves, "Crystallography and Crystal Defects," chap. 2, Addison-Wesley Publishing Company, Inc., Reading, Mass., 1970.
[2] Duplex slip results in high ductility in polycrystalline material for reasons discussed in Chap. 5.

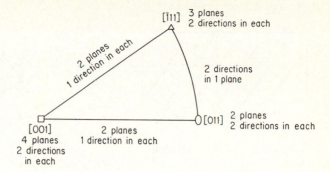

**Fig. 4-26** Operative slip systems along boundaries of stereo-graphic triangle.

single slip system occurs out to fracture. For duplex slip on primary and conjugate slip systems, the glide shear strain is given by[1]

$$\gamma = \sqrt{6} \ln \frac{1 + \sqrt{2} \cot \beta}{1 + \sqrt{2} \cot \beta_0} \qquad (4\text{-}16)$$

where $\beta$ is the angle between the stress axis and the $[001]$-$[\bar{1}11]$ boundary.

Crystals whose axes lie at orientations along the boundaries of the stereographic triangle represent a special situation because the critical resolved shear stress will be the same on more than one slip system. Therefore, plastic deformation will begin on more than one slip plane and they will initially deform by duplex slip. Figure 4-26 shows the number of operative slip systems in a cubic crystal at these orientations. Deformation by duplex slip results in a high degree of strain hardening because of interaction between dislocations on two intersecting slip systems. This is shown in Fig. 4-24 where Mg and Zn deform on a single slip system (because of the hcp geometry) while the stress-strain curves for Al and Cu are for crystals oriented for duplex slip.

## 4-10 DEFORMATION BY TWINNING

The second important mechanism by which metals deform is the process known as twinning.[2] Twinning results when a portion of the crystal takes up an orientation that is related to the orientation of the rest of the untwinned lattice in a definite,

[1] D. K. Bowen and J. W. Christian, *Philos. Mag.*, vol. 12, pp. 369–378, 1965.
[2] For complete reviews on this subject see E. O. Hall, "Twinning and Diffusionless Transformations in Metals," Butterworth & Co. (Publishers), Ltd., London, 1954, R. W. Cahn, *Adv. Phys.*, vol. 3, pp. 363–445, 1954, and S. Mahagan and D. F. Williams, *Int. Metall. Rev.*, vol. 18, pp. 43–61, June 1973. A good presentation of the crystallography of twinning is given by R. E. Reed-Hill, "Physical Metallurgy Principles," chap. 15, D. Van Nostrand Company, Inc., Princeton, N.J., 1964.

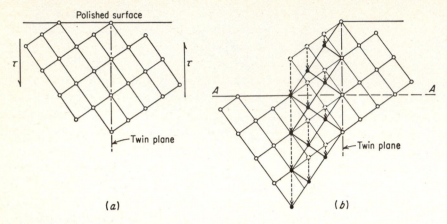

**Fig. 4-27** Classical picture of twinning.

symmetrical way. The twinned portion of the crystal is a mirror image of the parent crystal. The plane of symmetry between the two portions is called the *twinning plane*. Figure 4-27 illustrates the classical atomic picture of twinning. Figure 4-27a represents a section perpendicular to the surface in a cubic lattice with a low-index plane parallel to the paper and oriented at an angle to the plane of polish. The twinning plane is perpendicular to the paper. If a shear stress is applied, the crystal will twin about the twinning plane (Fig. 4-27b]. The region to the right of the twinning plane is undeformed. To the left of this plane, the planes of atoms have sheared in such a way as to make the lattice a mirror image across the twin plane. In a simple lattice such as this, each atom in the twinned region moves by a homogeneous shear a distance proportional to its distance from the twin plane. In Fig. 4-27b, open circles represent atoms which have not moved, dashed circles indicate the original positions in the lattice of atoms which change position, and solid circles are the final positions of these atoms in the twinned region. Note that the twin is visible on the polished surface because of the change in elevation produced by the deformation and because of the difference in crystallographic orientation between the deformed and undeformed regions. If the surface were polished down to section $AA$, the difference in elevation would be eliminated but the twin would still be visible after etching because it possesses a different orientation from the untwinned region.

It should be noted that twinning differs from slip in several specific respects. In slip, the orientation of the crystal above and below the slip plane is the same after deformation as before, while twinning results in an orientation difference across the twin plane. Slip is usually considered to occur in discrete multiples of the atomic spacing, while in twinning the atom movements are much less than an atomic distance. Slip occurs on relatively widely spread planes, but in the twinned region of a crystal every atomic plane is involved in the deformation.

Twins may be produced by mechanical deformation or as the result of annealing following plastic deformation. The first type are known as *mechanical twins*; the latter are called *annealing twins*. Mechanical twins are produced in bcc or hcp metals under conditions of rapid rate of loading (shock loading) and decreased temperature. Face-centered cubic metals are not ordinarily considered to deform by mechanical twinning, although gold-silver alloys twin fairly readily when deformed at low temperature, and mechanical twins have been produced in copper by tensile deformation at 4°K and by shock loading. Twins can form in a time as short as a few microseconds, while for slip there is a delay time of several milliseconds before a slip band is formed. Under certain conditions, twins can be heard to form with a click or loud report [tin cry]. If twinning occurs during a tensile test, it produces serrations in the the stress-strain curve.

Twinning occurs in a definite direction on a specific crystallographic plane for each crystal structure. Table 4-5 lists the common twin planes and twin directions. It is not known whether or not there is a critical resolved shear stress for twinning. However, twinning is not a dominant deformation mechanism in metals which possess many possible slip systems. Twinning generally occurs when the slip systems are restricted or when something increases the critical resolved shear stress so that the twinning stress is lower than the stress for slip. This explains the occurrence of twinning at low temperatures or high strain rates in bcc and fcc metals or in hcp metals at orientations which are unfavorable for basal slip.

The lattice strains needed to produce a twin configuration in a crystal are small, so that the amount of gross deformation that can be produced by twinning is small. For example,[1] the maximum extension which it is possible to produce in a zinc crystal when the entire crystal is converted into a twin on the {1012} plane is only 7.39 percent. The important role of twinning in plastic deformation comes not from the strain produced by the twinning process but from the fact that orient-

Table 4-5  TWIN PLANES AND TWIN DIRECTIONS

| Crystal structure | Typical examples | Twin plane | Twin direction |
|---|---|---|---|
| bcc | $\alpha$-Fe, Ta | (112) | [111] |
| hcp | Zn, Cd, Mg, Ti | (10$\bar{1}$2) | [$\bar{1}$011] |
| fcc | Ag, Au, Cu | (111) | [112] |

[1] Barrett, *op. cit.*, p. 384.

**Fig. 4-28** Microstructures of twins. (a) Neumann bands in iron; (b) mechanical twins produced in zinc by polishing; (c) annealing twins in gold-silver alloy.

ation changes resulting from twinning may place new slip systems in a favorable orientation with respect to the stress axis so that additional slip can take place. Thus, twinning is important in the overall deformation of metals with a low number of slip systems, such as the hcp metals. However, it should be understood that only a relatively small fraction of the total volume of a crystal is reoriented by twinning, and therefore hcp metals will, in general, possess less ductility than metals with a greater number of slip systems.

Figure 4-28 shows some metallographic features of twins in several different systems. Figure 4-28a is an example of mechanical twins in iron (Neumann bands). Note that the width of the twins can be readily resolved at rather low magnification. The boundaries of the twins etch at about the same rate as grain boundaries, indicating that they are rather high-energy boundaries. Figure 4-28b shows the broad, lense-shaped twins commonly found in hcp metals. Note that twins do not extend beyond a grain boundary. Figure 4-28c shows annealing twins in an fcc gold-silver alloy. Annealing twins are usually broader and with straighter sides than mechanical twins. The energy of annealing twin boundaries is about 5 percent of the average grain-boundary energy. Most fcc metals form annealing twins. Their presence in the microstructure is a good indication that the metal has been given mechanical deformation prior to annealing, since it is likely that they grow from twin nuclei produced during deformation.

A process closely related to twinning is the formation of a *martensite* region by a diffusionless shear transformation. Although both processes produce a local region of new lattice orientation, the basic difference is that in a martensite plate the crystal structure is different from the parent crystal. The driving force for twinning is the applied shear stress, while in the martensite transformation, the driving force is the free energy difference between the parent crystal and the martensitic phase. This thermodynamic driving force may be assisted by the applied shear stress.

## 4-11   STACKING FAULTS

In an earlier section, it was shown that the atomic arrangement on the {111} plane of an fcc structure and the {0001} plane of an hcp structure could be obtained by the stacking of close-packed planes of spheres. For the fcc structure, the stacking sequence of the planes of atoms is given by *ABC ABC ABC*. For the hcp structure, the stacking sequence is given by *AB AB AB*.

Errors, or faults, in the stacking sequence can be produced in most metals by plastic deformation.[1] Slip on the {111} plane in an fcc lattice produces a deformation stacking fault by the process shown in Fig. 4-29*b*. Slip has occurred between

Fig. 4-29   Faulted structures. (*a*) Fcc packing; (*b*) deformation fault in fcc; (*c*) twin fault in fcc; (*d*) hcp packing.

[1] Very precise x-ray diffraction measurements are needed to detect the presence of stacking faults. For example, see B. E. Warren and E. P. Warekois, *Acta Metall.* vol. 3 p. 473, 1955.

an *A* and a *B* layer, moving each atom layer above the slip plane one identity distance to the right. The stacking sequence then becomes *ABC A¦CAB*. Comparison of this faulted stacking sequence (Fig. 4-29*b*) with the stacking sequence for an hcp structure without faults (Fig. 4-29*d*) shows that the deformation stacking fault contains four layers of an hcp sequence. Therefore, the formation of a stacking fault in an fcc metal is equivalent to the formation of a thin hcp region. Another way in which a stacking fault could occur in an fcc metal is by the sequence[1] shown in Fig. 4-29*c*. The stacking sequence *ABC¦ACB¦CA* is called an extrinsic, or twin, stacking fault. The three layers *ACB* constitute the twin. Thus, stacking faults in fcc metals can also be considered as submicroscopic twins of nearly atomic thickness. The reason why mechanical twins of microscopically resolvable width are not formed readily when fcc metals are deformed is that the formation of stacking faults is so energetically favorable.

The situation for the hcp structure is somewhat different from that found in fcc metals. Figure 4-29*d* shows that, on going from an *A* layer to a *B* layer, if we continue in a straight line we will not come to another atom on the next *A* layer. However, slip can occur between two of the planes so that the stacking sequence becomes *ABABA¦CBCBC*. As a result, four layers of atoms *BACB* are in the straight-line fcc stacking order. Thus, a stacking fault in an hcp metal is equivalent to the formation of a thin fcc region. It is more difficult to form stacking faults in a bcc lattice than in the close-packed fcc and hcp structures. The possibility of stacking faults in the {112} planes has been investigated theoretically and demonstrated by x-ray diffraction.[2] Stacking faults have been observed with thin-film electron microscopy in columbium.[3]

Stacking faults occur most readily in fcc metals, and they have been most extensively studied for this crystal structure. For example, it is now known that differences in the deformation behavior of fcc metals can be related to differences in stacking-fault behavior. From the point of view of dislocation theory, a stacking fault in an fcc metal can be considered to be an *extended dislocation* consisting of a thin hexagonal region bounded by partial dislocations[4] (Fig. 4-30). The nearly parallel dislocations tend to repel each other, but this is balanced by the surface tension of the stacking fault pulling them together. The lower the *stacking-fault energy*, the greater the separation between the partial dislocations and the wider the stacking fault. Stacking-fault energies in fcc metals have been estimated on the assumption that the stacking-fault energy is equal to twice the energy of a

[1] C. N. J. Wagner, *Acta Metall.*, vol. 5, pp. 427–434, 1957.
[2] P. B. Hirsch and H. M. Otte, *Acta Crystallog.*, vol. 10, pp. 447–453, 1957; O. J. Guenter and B. E. Warren, *J. Appl. Phys.*, vol. 29, pp. 40–48, 1958.
[3] A. Fourdeux and A. Berghezen, *J. Inst. Met.*, vol. 89, pp. 31–32, 1960–1961.
[4] Partial dislocations will be considered in more detail in Chap. 5. The splitting of dislocations into separated partials has been observed with the electron miscroscope in many metals.

**Fig. 4-30** Schematic model of a stacking fault.

coherent boundary of an annealing twin. On this basis, the stacking-fault energies for copper, nickel, and aluminum are approximately 40, 80, and 200 erg/cm². Since the lower the energy of the twin boundary, the greater the tendency for the formation of annealing twins, the estimates of stacking-fault energy are in qualitative agreement with metallographic observations of the frequency of occurrence of annealing twins; e.g., aluminum rarely shows annealing twins. X-ray work has shown that the energy of stacking faults in brass decreases with zinc content, and this is in agreement with the fact that alpha brass forms a greater number of annealing twins than copper.

Stacking faults enter into the plastic deformation of metals in a number of ways. Metals with wide stacking faults strain-harden more rapidly, twin easily on annealing, and show a different temperature dependence of flow stress from metals with narrow stacking faults. This topic will be considered in greater detail in Chap. 5.

## 4-12 DEFORMATION BANDS AND KINK BANDS

In homogeneous deformation of a crystal results in regions of different orientation called *deformation bands*. When slip occurs without restraint in a perfectly homogeneous fashion, the slip lines are removed by subsequent polishing of the surface. Deformation bands, however, can be observed even after repeated polishing and etching because they represent regions of different crystallographic orientation. In single crystals, deformation bands several millimeters wide may occur, while in polycrystalline specimens microscopic observation is needed to see them. The tendency for the formation of deformation bands is greater in polycrystalline specimens because the restraints imposed by the grain boundaries make it easy for orientation differences to arise in a grain during deformation. Deformation bands generally appear irregular in shape but are elongated in the direction of principal strain. The outline of the bands is generally indistinct and poorly defined, indicating a general fading out of the orientation difference. Deformation bands have been observed in both fcc and bcc metals, but not in hcp metals.

Consideration of the equation for critical resolved shear stress shows that it

**Fig. 4-31** Kink band.

will be difficult to deform a hexagonal crystal when the basal plane is nearly parallel to the crystal axis. Orowan[1] found that if a cadmium crystal of this orientation were loaded in compression, it would deform by a localized region of the crystal suddenly snapping into a tilted position with a sudden shortening of the crystal. The buckling, or *kinking*, behavior is illustrated in Fig. 4-31. The horizontal lines represent basal planes, and the planes designated $p$ are the kink planes at which the orientation suddenly changes. Distortion of the crystal is essentially confined to the kink band. Further study of kink bands by Hess and Barrett[2] showed that they can be considered to be a simple type of deformation band. Kink bands have also been observed in zinc crystals tested in tension, where a nonuniform distribution of slip can produce a bending moment which can cause kink formation.

## 4-13 MICROSTRAIN BEHAVIOR

The ordinary engineering yield strength determined at an offset strain of 0.002 represents a stress at which a very great number of dislocations have already moved. From a dislocation viewpoint this is a very microscopic measurement of initial yielding. Therefore, considerable experimentation has gone into developing reliable techniques for work in the preyield microstrain region at plastic strains less than 0.001. Mechanical property studies on single crystals at plastic strains of the order of $10^{-5}$ are important in providing data for studying basic dislocation behavior. The key to these studies[3] is a technique for loading with near-perfect alignment and a transducer, such as a linear variable differential transformer (LVDT), which provides reproducible strain measurements at a level of $10^{-6}$.

The shape of the stress-strain curve on repeated loading to higher stresses in the microplasticity regime is shown in Fig. 4-32. At very low stresses a load-unload cycle produces a single straight line with the material exhibiting ideal elastic behavior (Fig. 4-32a). Deviation from ideal elastic behavior occurs at some higher

[1] E. Orowan, *Nature*, vol. 149, p. 643, 1942.
[2] J. A. Hess and C. S. Barrett, *Trans. Metall. Soc. AIME*, vol. 185, p. 599, 1949.
[3] A. Lawley and J. D. Meakin, "Microplasticity," C. J. McMahon (ed.), John Wiley & Sons, Inc., New York, 1968.

**Fig. 4-32**  Types of stress-strain curves found in microplasticity regime.

stress $\tau_E$, the *true elastic limit* (Fig. 4-32b). Above $\tau_E$ the load-unload cycle produces a closed mechanical hysteresis loop which closely approximates a parallelogram. The initial high elastic modulus corresponds well with dynamic measurements of modulus by high-frequency methods. The lower modulus corresponds to the relaxed modulus characteristic of usual static tension tests. Stress cycling above some value $\tau_A$ causes permanent plastic deformation since the hysteresis loop does not close on unloading (Fig. 4-32d). This stress is generally known as the anelastic limit $\tau_A$. Finally, at much higher stresses gross yielding occurs. Typical values for molybdenum are: $\tau_E = 0.05 \text{ kg/mm}^2$, $\tau_A = 0.5 \text{ kg/mm}^2$, $\tau_0$ (macroyield) $= 5 \text{ kg/mm}$.

## 4-14  STRAIN HARDENING OF SINGLE CRYSTALS

One of the chief characteristics of the plastic deformation of metals is the fact that the shear stress required to produce slip continuously increases with increasing shear strain. The increase in the stress required to cause slip because of previous plastic deformation is known as *strain hardening*, or *work hardening*. An increase in flow stress of over 100 percent from strain hardening is not unusual in single crystals of ductile metals.

Strain hardening is caused by dislocations interacting with each other and with barriers which impede their motion through the crystal lattice. Hardening due to dislocation interaction is a complicated problem because it involves large groups of dislocations, and it is difficult to specify group behavior in a simple mathematical way. It is known that the number of dislocations in a crystal increases

**Fig. 4-33** Effect of complete reversal of slip direction on stress-strain curve. (*From E. H. Edwards, J. Washburn, and E. R. Parker, Trans. AIME, vol. 197, p. 1526, 1953.*)

with strain over the number present in the annealed crystal. The dislocation density of a good annealed crystal is $10^5$ to $10^6$ cm$^{-2}$, while the observed dislocation density in cold-worked metal is $10^{10}$ to $10^{12}$ cm$^{-2}$. Dislocation multiplication can arise from condensation of vacancies, by regeneration under applied stress from existing dislocations by either the Frank-Read mechanism or a multiple cross-slip mechanism (see Chap. 5) or by emission of dislocations from a high-angle grain boundary.

One of the eariest dislocation concepts to explain strain hardening was the idea that dislocations pile up on slip planes at barriers in the crystal. The pile-ups produce a *back stress* which opposes the applied stress on the slip plane. The existence of a back stress was demonstrated experimentally by shear tests on zinc single crystals.[1] Zinc crystals are ideal for crystal-plasticity experiments because they slip only on the basal plane, and hence complications due to duplex slip are easily avoided. In Fig. 4-33, the crystal is strained to point $O$, unloaded, and then reloaded in the direction opposite to the original slip direction. Note that on reloading the crystal yields at a lower shear stress than when it was first loaded. This is because the back stress developed as a result of dislocations piling up at barriers during the first loading cycle is aiding dislocation movement when the direction of slip is reversed. Furthermore, when the slip direction is reversed, dislocations of opposite sign could be created at the same sources that produced the dislocations responsible for strain in the first slip direction. Since dislocations of opposite sign attract and annihilate each other, the net effect would be a further softening of the lattice. This explains the fact that the flow curve in the reverse direction lies below the curve for continued flow in the original direction. The lowering of the yield stress when deformation in one direction is followed by deformation in the opposite direction is called the *Bauschinger effect*.[2] While all metals exhibit a Bauschinger effect, it may not always be of the magnitude shown here for zinc crystals. Moreover, the flow curve after reversal of direction does not fall below the original flow curve for all metals.

[1] E. H. Edwards, J. Washburn, and E. R. Parker, *Trans. AIME*, vol. 197, p. 1525, 1953.
[2] J. Bauschinger, *Zivilingur.*, vol. 27, pp. 289–347, 1881.

The existence of back stress and its importance to strain hardening in metals having been established, the next step is to identify the barriers to dislocation motion in single crystals. Microscopic precipitate particles and foreign atoms can serve as barriers, but other barriers which are effective in pure single crystals must be found. Such barriers arise from the fact that glide dislocations on intersecting slip planes may combine with one another to produce a new dislocation that is not in a slip direction. The dislocation of low mobility that is produced by a dislocation reaction is called a *sessile dislocation*. Since sessile dislocations do not lie on the slip plane of low shear stress, they act as a barrier to dislocation motion until the stress is increased to a high enough level to break down the barrier. The most important dislocation reaction, which leads to the formation of sessile dislocations, is the formation of Lomer Cottrell barriers in fcc metals by slip on intersecting {111} planes.

Another mechanism of strain hardening, in addition to that due to the back stress resulting from dislocation pile-ups at barriers, is believed to occur when dislocations moving in the slip plane cut through other dislocations intersecting the active slip plane. The dislocations threading through the active slip plane are often called a *dislocation forest*, and this strain-hardening process is referred to as the *intersection* of a forest of dislocations.

Figure 4-34a shows that dislocation intersection results in a small step or jog in the dislocation line. Jogs on a dislocation restrict its motion so that they

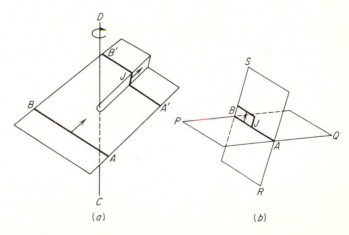

(a)                                     (b)

**Fig. 4-34** Formation of a jog $J$ (a) by a dislocation cutting through a screw dislocation as it glides from $AB$ to $A'B'$; (b) by part of a screw dislocation line $AB$ cross slipping from the primary slip plane $PQ$ into the plane $RS$. (*From A. H. Cottrell, "The Mechanical Properties of Matter," John Wiley & Sons, Inc., New York, 1964. By permission of the publishers.*)

contribute to strain hardening. Jogs are also formed by a screw dislocation *cross slipping* from the primary slip plane to another plane which contains the common slip direction (Fig. 4-34*b*).

The phenomenon of *cross slip* is restricted to screw dislocations. Since the line of a screw dislocation and its Burgers vector are parallel, this does not define a specific plane as with an edge dislocation (where **b** is perpendicular to the dislocation line). To a screw dislocation, all directions around its axis look the same, and it can glide on any plane as long as it moves parallel to its original orientation. The slip plane of a screw dislocation can be any plane containing the dislocation, and it can cross slip from one plane to another just so long as both planes contain a common slip direction. This permits screw dislocations (or screw components of mixed dislocations) to detour around obstacles and barriers. If cross slip of screw dislocations could not occur, dislocation motion would be impeded early in the deformation process and the rate of strain hardening (slope of stress-strain curve) would be very high and fracture would occur at low values of strain. The formation of jogs in screw dislocations impedes their motion and may even lead to the formation of vacancies and interstitials if the jogs are forced to move nonconservatively. Jogs in edge dislocations do not impede their motion. All these processes require an increased expenditure of energy, and therefore they contribute to hardening.

Strain hardening due to a dislocation cutting process arises from short-range forces occurring over distances less than 5 to 10 interatomic distances. This hardening can be overcome at finite temperatures with the help of thermal fluctuations, and therefore it is temperature- and strain-rate-dependent. On the other hand, strain hardening arising from dislocation pile-up at barriers occurs over longer distances, and therefore it is relatively independent of temperature and strain rate. Accordingly, data on the temperature and strain-rate dependence of strain hardening can be used[1] to determine the relative contribution of the two mechanisms.

When the stress-strain curves for single crystals are plotted as resolved shear stress vs. shear strain, certain generalizations can be made for all fcc metals. Following the notation proposed by Seeger,[2] the flow curve for pure-metal single crystals can be divided into three stages (Fig. 4-35).

Stage I, the region of *easy glide*, is a stage in which the crystal undergoes little strain hardening. During easy glide, the dislocations are able to move over relatively large distances without encountering barriers. The low strain hardening produced during this stage implies that most of the dislocations escape from the crystal at

[1] Z. S. Basinski, *Philos. Mag.*, vol. 4, ser. 8, pp. 393–432, 1959. For an extensive review see H. Conrad, *J. Met.*, pp. 582–588, July 1964.
[2] A. Seeger, in "Dislocations and Mechanical Properties of Crystals," John Wiley & Sons, Inc., New York, 1957.

**Fig. 4-35** Generalized flow curve for fcc single crystals.

the surface. During easy glide, slip always occurs on only one slip system. For this reason, stage I slip is sometimes called *laminar flow*.

Stage II is a nearly linear part of the flow curve where strain hardening increases rapidly. In this stage, slip occurs on more than one set of planes. The length of the active slip lines decreases with increasing strain, which is consistent with the formation of a greater number of Lomer-Cottrell barriers with increasing strain. During stage II, the ratio of the strain-hardening coefficient (the slope of the curve) to the shear modulus is nearly independent of stress and temperature, and approximately independent of crystal orientation and purity. The fact that the slope of the flow curve in stage II is nearly independent of temperature agrees with the theory that assumes the chief strain-hardening mechanism to be piled-up groups of dislocations.

As a result of slip on several slip systems, lattice irregularities are formed. Dislocation tangles begin to develop and these eventually result in the formation of a dislocation cell structure consisting of regions almost free of dislocations surrounded by material of high dislocation density (about five times the average dislocation density). Although the heterogeneity of dislocation distribution makes precise measurements difficult, measurements over a wide range of systems show that the average dislocation density in stage II correlates with resolved shear stress according to

$$\tau = \tau_0 + \alpha Gb\rho^{\frac{1}{2}} \qquad (4\text{-}17)$$

where $\tau_0$ is the shear stress needed to move a dislocation in the absence of other dislocations and $\alpha$ is a numerical constant which varies from 0.3 to 0.6 for different fcc and bcc metals.

Stage III is a region of decreasing rate of strain hardening. The processes occurring during this stage are often called *dynamical recovery*. In this region of the flow curve, the stresses are high enough so that dislocations can take part in processes that are suppressed at lower stresses. Cross slip is believed to be the main process by which dislocations, piled up at obstacles during stage II, can escape and reduce the internal-strain field. The stress at which stage III begins, $\tau_3$, is strongly temperature-dependent. Also, the flow stress of a crystal strained into stage III is more temperature-dependent than if it had been strained only into stage II. This temperature dependence suggests that the intersection of forests of dislocations is the chief strain-hardening mechanism in stage III.

The curve shown in Fig. 4-35 represents a general behavior for fcc metals. Certain deviations from a three-stage flow curve have been observed. For example, metals with a high stacking-fault energy, like aluminum, usually show only a very small stage II region at room temperature because they can deform so easily by cross slip. The shape and magnitude of a single-crystal flow curve, particularly during the early stages, depends upon the purity of the metal, the orientation of the crystal, the temperature at which it is tested, and the rate at which it is strained. The easy-glide region is much more prominent in hcp crystals than in fcc metals. A region of easy glide in the flow curve is favored by slip on a single system, high purity, low temperature, absence of surface oxide films, an orientation favorable for simple slip, and a method of testing which minimizes extraneous bending stresses. Figure 4-36 shows that crystal orientation can have a very strong effect on the flow curve of fcc single crystals. When the tensile axis is parallel to a ⟨011⟩ direction, one slip system is carrying appreciably more shear stress than any other and the flow curve shows a relatively large region of easy glide. When the tensile

**Fig. 4-36**  Effect of specimen orientation on the shape of the flow curve for fcc single crystals.

axis is close to a $\langle 100 \rangle$ or $\langle 111 \rangle$ direction, the stress on several slip systems is not very different and the flow curves show rapid rates of strain hardening.

The value of the resolved shear stress at a given shear strain decreases with increasing temperature. With increasing temperature both stage I and stage II decrease in extent until at high temperature the stress-strain curve shows entirely parabolic stage III behavior.

## BIBLIOGRAPHY

Barrett, C. S., and T. B. Massalski, "Structure of Metals," 3rd ed., McGraw-Hill Book Company, New York, 1966.

Clarebrough, L. M., and M. E. Hargreaves: Work Hardening of Metals, in "Progress in Metal Physics," vol. 8, Pergamon Press, Ltd., London, 1959.

Maddin, R., and N. K. Chen: Geometric Aspects of the Plastic Deformation of Metal Single Crystals, in "Progress in Metal Physics," vol. 5, Pergamon Press, Ltd., London, 1954.

Nabarro, F. R. N., Z. S., Basinski, and D. B., Holt: Plasticity of Pure Single Crystals, *Adv. Phys.*, vol. 13, pp. 193–323, 1964.

Reid, C. N.: "Deformation Geometry for Materials Scientists," Pergamon Press, New York, 1973.

Schmid, E., and W. Boas: "Plasticity of Crystals," English translation, F. A. Hughes & Co., London, 1950.

Tegart, W. J. McG., "Elements of Mechanical Metallurgy," The Macmillan Company, New York, 1968.

# 5

# DISLOCATION THEORY

## 5-1  INTRODUCTION

A dislocation is the linear lattice defect that is responsible for nearly all aspects of the plastic deformation of metals. This concept was introduced in Chap. 4, where the geometry of edge and screw dislocations was presented for the case of a simple cubic lattice. It was shown that the existence of a dislocationlike defect is necessary to explain the low values of yield stress observed in real crystals. A general picture of the factors which impede dislocation motion and lead to strain hardening was also given.

This chapter is intended to present a more complete treatment of dislocation theory. Techniques for observing dislocations in metals are discussed. The effect on dislocation behavior of considering real fcc, bcc, or hcp crystal structures are considered. The origin of dislocations and the mechanisms for their multiplication are discussed. Interaction of dislocations with other dislocations, vacancies, and foreign atoms is discussed in some detail. The object of this chapter is the presentation of the basic geometric and mathematical relationships which describe dislocation behavior. These relationships will be used to explain mechanical behavior and strengthening mechanisms in subsequent chapters of this book.

## 5-2 OBSERVATION OF DISLOCATIONS

The concept of the dislocation was proposed independently by Taylor, Orowan, and Polanyi[1] in 1934, but the idea lay relatively undeveloped until the end of World War II. There followed a period of approximately 10 years in which the theory of dislocation behavior was developed extensively and applied to practically every aspect of the plastic deformation of metals. Because there were no really reliable methods for detecting dislocations in real materials, it was necessary to build much of this theory on the basis of indirect observations of dislocation behavior. However, in the last 10 years extensive research has developed a variety of techniques for observing and studying dislocations in real materials. These studies leave no doubt that dislocations exist, and more importantly, they have provided experimental verification for most of the theoretical concepts of dislocation theory.

The resolving power of the best electron microscope would have to be improved by a factor of 5 to 10 in order to observe directly the distortion of the individual lattice planes around a dislocation in a metal crystal.[2] Practically all the experimental techniques for detecting dislocations utilize the strain field around a dislocation to increase its effective size. These experimental techniques can be roughly classified into two categories, those involving chemical reactions with the dislocation, and those utilizing the physical changes at the site of a dislocation.[3] Chemical methods include etch-pit techniques and precipitation techniques. Methods based on the physical structure at a dislocation site include transmission electron microscopy of thin films and x-ray diffraction techniques.

The simplest chemical technique is the use of an etchant which forms a pit at the point where a dislocation intersects the surface. Etch pits are formed at dislocation sites because the strain field surrounding the dislocation causes preferential chemical attack. A great deal of information about dislocation behavior in the ionic crystal LiF has been obtained in this way by Gilman and Johnston.[4]

---

[1] G. I. Taylor, *Proc. R. Soc. London*, vol. 145A, p. 362, 1934; E. Orowan, *Z. Phys.*, vol. 89, pp. 605, 614, 634, 1934; M. Polanyi, *Z. Phys.*, vol. 89, p. 660, 1934.

[2] It has been possible by means of an electron microscope to observe this lattice distortion in an organic crystal of platinum phthalocyanine, which has a very large lattice spacing (12 A) [J. W. Menter, *Proc. R. Soc. London Ser. A*; vol. 236A, p. 119, 1956]. An indication of the lattice distortion at a dislocation in metals has been obtained by making use of the magnification resulting from moiré patterns produced by electron transmission through two thin overlapping crystals with slightly different orientations or lattice spacings. See G. A. Bassett, J. W. Menter, and D. W. Pashley, *Proc. R. Soc. London Ser. A*, vol. 246A, p. 345, 1958.

[3] Several excellent reviews of experimental techniques have been published. See P. B. Hirsch., *Metall. Rev.*, vol. 4, no, 14, pp. 101–140, 1959; J. Nutting, Seeing Dislocations, in "The Structure of Metals," Institution of Metallurgists, Interscience Publishers, Inc., New York, 1959; S. Amelinckx, The Direct Observation of Dislocations, *Solid State Phys. Suppl.* 6, 1964.

[4] J. J. Gilman and W. G. Johnston, in "Dislocations and Mechanical Properties of Crystals," John Wiley & Sons, Inc., New York, 1957.

**Fig. 5-1** Etch pits on slip bands in alpha brass crystals (5,000X). (*From J. D. Meakin and H. G. F. Wilsdorf, Trans. Metall. Soc. AIME, vol. 218, p. 740, 1960.*)

Important information about dislocations in metals has also been obtained with etch-pit techniques. Figure 5-1 shows the excellent resolution obtainable from etch-pit studies on alpha brass.[1] Pits only 500 Å apart have been resolved. In the region of heavy slip shown in this electron micrograph the dislocation density is $10^{10}$ cm$^{-2}$.

In metals, etch-pit formation at dislocations appears to be dependent on purity.[2] Because of solute segregation to the dislocation, the region around the dislocation becomes anodic to the surrounding metal, and consequently preferential etching occurs at the dislocation. Figure 6-4 shows an etch-pit structure in an iron-silicon alloy which was made visible by diffusion of carbon atoms to the dislocations. Etch-pit techniques are useful because they can be used with bulk samples.

---

[1] J. D. Meakin and H. G. F. Wilsdorf, *Trans. Metall. Soc. AIME*, vol. 218, pp. 737–745, 1960.
[2] A summary of etch-pit techniques in metals is given by L. C. Lowell, F. L. Vogel and J. H. Wernick, *Met. Prog.*, vol. 75, pp. 96–96D, 1959.

In certain systems it may be possible to distinguish between edge and screw dislocations and between positive and negative edge dislocations. The technique can also be used to study the movement of dislocations. However, care must be taken to ensure that pits are formed only at dislocation sites and that all dislocations intersecting the surface are revealed. Because the etch pits have a finite size and are difficult to resolve when they overlap, the etch-pit technique is limited generally to crystals with a low dislocation density of about $10^6$ cm$^{-2}$.

A similar method of detecting dislocations is to form a visible precipitate along the dislocation lines. Usually a small amount of impurity is added to form the precipitate after suitable heat treatment. The procedure is called "decoration" of dislocations. This technique was first used by Hedges and Mitchell[1] to decorate dislocations in AgBr with photolytic silver. It has since been used with many other ionic crystals,[2] such as AgCl, NaCl, KCl, and CaF$_2$. With these optically transparent crystals this technique has the advantage that it shows the internal structure of the dislocation lines. Figure 5-2 shows a hexagonal network of dislocations in

**Fig. 5-2**  Hexagonal network of dislocations in NaCl detected by a decoration technique. (*From S. Amelinckx, in: "Dislocations and Mechanical Properties of Crystals," John Wiley & Sons, Inc., New York, 1957. By permission of the publishers.*)

[1] J. M. Hedges and J. W. Mitchell, *Philos. Mag.*, vol. 44, p. 223, 1953.
[2] S. Amelinckx, in "Dislocations and Mechanical Properties of Crystals," John Wiley & Sons, Inc., New York, 1957.

a NaCl crystal which was made visible by decoration. Although dislocation decoration has not been used extensively with metals, some work has been done along these lines with the Al-Cu precipitation-hardening system and with silicon crystals.

The most powerful method available today for the detection of dislocations in metals is transmission electron microscopy of thin foils.[1] Thin sheet, less than 1 mm thick, is thinned after deformation by electropolishing to a thickness of about 1,000 Å. At this thickness the specimen is transparent to electrons in the electron microscope. Although the crystal lattice cannot be resolved, individual dislocation lines can be observed because the intensity of the diffracted electron beam is altered by the strain field of the dislocation. The width of the diffraction image of a dislocation in a thin foil is about 100 Å, so that this technique can be used at dislocation densities up to about $10^{11}$ cm$^{-2}$. By means of this technique it has been possible to observe dislocation networks (Fig. 5-3), stacking faults, dislocation pile-up at grain boundaries (Fig. 6-1), Lomer-Cottrell barriers, and many other structural features of dislocation theory. Dislocation movement has been observed by generating thermal stresses in the thin film with the electron beam or by using special straining fixtures in the electron microscope.

Transmission electron microscopy is the most powerful and universally applicable technique available for studying dislocations in solids. By application of

**Fig. 5-3** Dislocation network in cold-worked aluminum (32,500X). (*From P. B. Hirsch, R. W. Horne, and M. J. Whelan, Philos. Mag., ser. 8, vol. 1, p. 677, 1956.*)

---

[1] G. Thomas, "Transmission Electron Microscopy of Metals," John Wiley & Sons, Inc., New York, 1962.

the kinematic[1] and dynamic[2] theories of electron diffraction, it is possible to make detailed analysis of the images to determine the number of dislocations, their Burgers vectors, and the slip planes on which they lie. However, this technique is not without disadvantages. Since only a miniscule volume of material is examined with thin films, great care must be exerted to obtain a representative sample. It is possible to alter the defect structure during sectioning and polishing to a thin film, and dislocation structures may relax in a very thin foil. The greatest defect of transmission electron microscopy is that it is not very effective in detecting long-range stresses, nor does it give very much information about slip-line lengths or surface step heights.

The dislocation structure of a crystal can be detected by x-ray microscopy. The most common techniques are the Berg-Barrett reflection method[3] and the Lang topography method.[4] Unfortunately, the limiting resolution of these techniques is about $10^5$ dislocations per square centimeter. Important features of plastically deformed metal such as the mean size of coherently diffracting domains (crystallite size), the microstrains within these domains, and the probability of lattice faulting can be determined by detailed analysis of the breadth and shape of x-ray diffraction peaks.[5] X-ray peak profile studies are indirect and statistical, and depend on models of the dislocation structure to relate their results to dislocation behavior.

The ultimate technique for observing defect structure is the field-ion microscope.[6] Since its resolution is 2 to 3 Å, individual atoms may be distinguished. Thus, it is the only experimental technique for directly observing vacancies. At its present state of development, field-ion microscopy is limited to metals with a strong binding force, such as W, Mo, or Pt. Since the specimen must be a fine wire bent into a sharp tip, the method is limited in flexibility and examines only a very small area of the surface.

## 5-3 BURGERS VECTOR AND THE DISLOCATION LOOP

The Burgers vector **b** is the vector which defines the magnitude and direction of slip. Therefore, it is the most characteristic feature of a dislocation. It has already been shown that for a *pure* edge dislocation the Burgers vector is perpendicular to the dislocation line, while for a *pure* screw dislocation the Burgers vector is parallel

[1] P. B. Hirsch, A. Howie, and M. J. Whelan, *Philos. Trans. R. Soc. London Ser. A*, vol. A252, pp. 499–529, 1960.

[2] A. Howie and M. J. Whelan, *Proc. R. Soc. London Ser. A*, vol. A263, pp. 217–237, 1961; *Proc. R. Soc. London Ser. A*, vol. A267, pp. 206–230, 1962.

[3] C. S. Barrett, *Trans. Metall. Soc. AIME*, vol. 161, pp. 15–64, 1945.

[4] A. R. Lang, *J. Appl. Phys.*, vol. 30, pp. 1748–1755, 1959.

[5] B. E. Warren, X-ray Studies of Deformed Metals, *Prog. Met. Phys.*, vol. 8, pp. 147–202, 1958.

[6] E. W. Muller, "Direct Observation of Imperfections in Crystals," Interscience Publishers, Inc., New York, 1962.

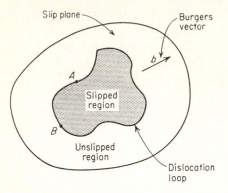

**Fig. 5-4** Dislocation loop lying in a slip plane (schematic).

to the dislocation line. Actually, dislocations in real crystals are rarely straight lines and rarely lie in a single plane. In general, a dislocation will be partly edge and partly screw in character. As shown by Figs. 5-2 and 5-3, dislocations will ordinarily take the form of curves or loops, which in three dimensions form an interlocking dislocation network. In considering a dislocation loop in a slip plane any small segment of the dislocation line can be resolved into edge and screw components. For example, in Fig. 5-4, the dislocation loop is pure screw at point *A* and pure edge at point *B*, while along most of its length it has mixed edge and screw components. Note, however, that the Burgers vector is the same along the entire dislocation loop. If this were not so, part of the crystal above the slipped region would have to slip by a different amount relative to another part of the crystal and this would mean that another dislocation line would run across the slipped region.

A convenient way of defining the Burgers vector of a dislocation is with a *Burgers circuit*. Consider the positive edge dislocation shown in Fig. 5-5a. If we start at a lattice point and imagine a clockwise path traced from atom to atom an equal distance in each direction, we find that at the finish of the path the circuit

**Fig. 5-5** Burgers circuits. (*a*) Around positive edge dislocation; (*b*) around a right-handed screw dislocation.

**Fig. 5-6**  Definition of unit tangent vector.

does not close. The closure failure from finish to start is the Burgers vector **b** of the dislocation. (If we had made the Burgers circuit around the dislocation in the anticlockwise direction, the direction of the Burgers vector would have been in the opposite sense.) Moreover, if we traverse a Burgers circuit about the screw dislocation shown in Fig. 5-5*b*, we would find the closure error pointing out of the front face of the crystal. This is a right-handed screw dislocation since in traversing the circuit around the dislocation line, we advance the helix one atomic plane into the crystal.

We now need to be able to distinguish between a positive and negative Burgers vector. To do this we define the unit tangent vector **t** which is tangent at every point to the dislocation line (Fig. 5-6). The selection of the positive direction of **t** is arbitrary, but once it is selected the sign of the Burgers vector is fixed without ambiguity. With the positive direction of the tangent vector selected, sight down the dislocation line in the positive direction and make a clockwise Burgers circuit around the dislocation line to determine the magnitude and sense of **b** measured from finish to start of the circuit.[1] From this definition it follows that two straight parallel dislocations are identical if either (1) their tangent vectors point in the same direction and their Burgers vectors are of the same magnitude and direction, or (2) their tangent vectors point in opposite directions and their Burgers vectors are equal in magnitude but point in opposite directions. Also, according to this convention, the Burgers vector of a left-handed screw dislocation points in the same direction as its tangent vector (Fig. 5-7*a*), while the Burgers vector of a right-handed screw dislocation points in the direction opposite to its tangent vector (Fig. 5-7*b*).

**Fig. 5-7** Convention for screw dislocations. (*a*) Left-handed screw dislocation; (*b*) right-handed screw dislocation.

---

[1] All authorities on dislocation theory do not agree on the definition of the Burgers vector. Sometimes a convention is adopted which is equivalent to taking the Burgers circuit in a counterclockwise direction, resulting in a Burgers vector opposite in sense to that given by the above definition.

We note that the tangent vector of an edge dislocation and its Burgers vector define a plane, the slip plane. The direction of motion of an edge dislocation is in the direction of **b**, which is normal to **t**. The position of the extra half plane of atoms can be found with respect to the Burgers vector by making a 90° counterclockwise rotation from the direction of **b** while sighting down the dislocation in the positive direction. (This establishes whether an edge dislocation is positive or negative.) Since for a screw dislocation **b** and **t** are parallel, no specific plane is defined. The screw dislocation moves in a direction normal to **b**, but it is free to slip on any plane which contains its Burgers vector. Since screw dislocations (or screw components of mixed dislocations) are free to wander out of their original planes of motion, they can take part in cross slip and in complicated reactions with other dislocations leading to the formation of complex dislocation networks.

The process of *cross slip* is illustrated in Fig. 5-8, which will also serve as an application of the rules discussed above. In Fig. 5-8$a$ a small loop of dislocation line with $\mathbf{b} = a_0/2[\bar{1}01]$ is moving on a (111) plane in an fcc crystal. Let the **t** vector be counterclockwise in the plane of the loop, so that the dislocation loop is pure positive edge at $w$ and pure negative edge at $y$ (using the rule of 90° counterclockwise rotation). At $x$ the dislocation is a right-handed screw (since **b** and **t** are parallel), while at $z$ the dislocation loop is a pure left-handed screw dislocation. At some stage (Fig. 5-8$b$), the shear stress causing expansion of the loop tends to move the dislocation on the intersecting $(1\bar{1}1)$ plane. Since the dislocation is pure screw at $z$, it is free to move on this plane. In Fig. 5-8$c$ the loop has expanded on the second plane, while in Fig. 5-8$d$ *double cross slip* has taken place as the loop

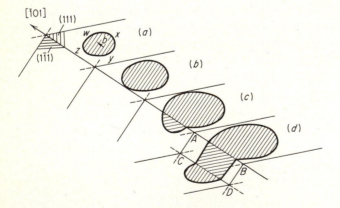

**Fig. 5-8** Cross slip in a face-centered cubic crystal. (*From D. Hull, "Introduction to Dislocations," p. 56, Pergamon Press, New York, 1965. By permission of the publishers.*)

glides back onto the original (111) plane. Note that during the glide of the dislocation on the cross-slip plane only the screw component of the loop has moved.

Because a dislocation represents the boundary between the slipped and unslipped region of a crystal, topographic considerations require that it either must be a closed loop or else must end at the free surface of a crystal or at a grain boundary. In general, a dislocation line cannot end inside of a crystal. The exception is at a *node*, where three or four dislocation lines meet. At a node two dislocations with Burgers vectors $\mathbf{b}_1$ and $\mathbf{b}_2$ combine to produce a resultant dislocation $\mathbf{b}_3$. The vector $\mathbf{b}_3$ is given by the vector sum of $\mathbf{b}_1$ and $\mathbf{b}_2$.

Since the periodic force field of the crystal lattice requires that atoms must move from one equilibrium position to another, it follows that the Burgers vector must always connect one equilibrium lattice position with another. Therefore, the crystal structure will determine the possible Burgers vectors. A dislocation with a Burgers vector equal to one lattice spacing is said to be a *dislocation of unit strength*. Because of energy considerations dislocations with strengths larger than unity are generally unstable and dissociate into two or more dislocations of lower strength. The criterion for deciding whether or not dissociation will occur is based on the fact that the strain energy of a dislocation is proportional to the square of its Burgers vector. Therefore, the dissociation reaction $\mathbf{b}_1 \rightarrow \mathbf{b}_2 + \mathbf{b}_3$ will occur when $b_1{}^2 > b_2{}^2 + b_3{}^2$, but not if $b_1{}^2 < b_2{}^2 + b_3{}^2$.

Dislocations with strengths less than unity are possible in close-packed lattices where the equilibrium positions are not the edges of the structure cell. A Burgers vector is specified by giving its components along the axes of the crystallographic structure cell. Thus, the Burgers vector for slip in a cubic lattice from a cube corner to the center of one face has the components $a_0/2$, $a_0/2$, 0. The Burgers vector is $[a_0/2 \ \ a_0/2 \ \ 0]$, or as generally written $\mathbf{b} = (a_0/2)[110]$. The strength of a dislocation with Burgers vector $a_0[uvw]$ is $|b| = a_0[u^2 + v^2 + w^2]^{1/2}$. For example, the magnitude of the Burgers vector given above is $|b| = a_0/\sqrt{2}$.

In adding Burgers vectors, each of the corresponding components are added separately. Thus $b_1 + b_2 = a_0[110] + a_0[211] = a_0[321]$. In adding or subtracting components common unit vectors must be used. Thus $a_0/3[112] + a_0/6[11\bar{1}]$ must be expressed as $a_0/6[224] + a_0/6[11\bar{1}] = a_0/6[333] = a_0/2[111]$.

A dislocation of unit strength, or *unit dislocation*, has a minimum energy when its Burgers vector is parallel to a direction of closest atomic packing in the lattice. This agrees with the experimental observation that crystals almost always slip in the close-packed directions. A unit dislocation of this type is also said to be a *perfect dislocation* because translation equal to one Burgers vector produces an identity translation. For a perfect dislocation there is perfect alignment of atom planes above and below the slip plane within the dislocation loop. A unit dislocation parallel to the slip direction cannot dissociate further unless it becomes an *imperfect dislocation*, where a translation of one Burgers vector does not result in an identity translation. A stacking fault is produced by the dissociation of a unit dislocation into two imperfect dislocations. For a stacking fault to be stable, the decrease in energy due to dissociation must be greater than the increase in interfacial energy of the faulted region.

## 5-4 DISLOCATIONS IN THE FACE-CENTERED CUBIC LATTICE

Slip occurs in the fcc lattice on the {111} plane in the ⟨110⟩ direction. The shortest lattice vector is $(a_0/2)$ [110], which connects an atom at a cube corner with a neighboring atom at the center of a cube face. The Burgers vector is therefore $(a_0/2)$ [110].

However, consideration of the atomic arrangement on the {111} slip plane shows that slip will not take place so simply. Figure 5-9 represents the atomic packing on a close-packed (111) plane. It has already been shown that the {111} planes are stacked in a sequence $ABC\ ABC$ .... The vector $\mathbf{b}_1 = (a_0/2)$ [10$\bar{1}$] defines one of the observed slip directions. However, if the atoms are considered as hard spheres,[1] it is easier for an atom on a type $B$ plane to move along a zigzag path $\mathbf{b}_2 + \mathbf{b}_3$ in the valleys instead of moving over the hump that lies in the path of the vector $\mathbf{b}_1$. The dislocation reaction is given by

$$\mathbf{b}_1 \rightarrow \mathbf{b}_2 + \mathbf{b}_3$$

$$\frac{a_0}{2}\ [10\bar{1}] \rightarrow \frac{a_0}{6}\ [2\bar{1}\bar{1}] + \frac{a_0}{6}\ [11\bar{2}]$$

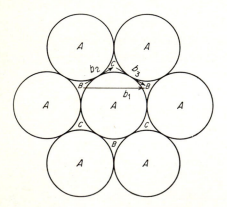

Fig. 5-9 Slip in a close-packed (111) plane in an fcc lattice. (*After A. H. Cottrell, "Dislocations and Plastic Flow in Crystals," p. 73, Oxford University Press, New York, 1953. By permission of the publishers.*)

[1] F. C. Thompson and W. E. W. Millington, *J. Iron Steel Inst. London*, vol. 109, p. 67 1924; C. H. Mathewson, *Trans. AIME*, vol. 32, p. 38, 1944.

To check this reaction, the summation of the $x$, $y$, $z$ components of the right-hand side of the equation must add up to the $x$, $y$, $z$ components of the original dislocation.

$$x \text{ component} \quad \tfrac{1}{2} = \tfrac{2}{6} + \tfrac{1}{6}$$

$$y \text{ component} \quad 0 = -\tfrac{1}{6} + \tfrac{1}{6}$$

$$z \text{ component} \quad -\tfrac{1}{2} = -\tfrac{1}{6} - \tfrac{2}{6}$$

The above reaction is energetically favorable since there is a decrease in strain energy proportional to the change $a_0{}^2/2 \to a_0{}^2/3$.

| Original dislocation | Product of reaction | |
|---|---|---|
| $\lvert b_1 \rvert = a_0[\tfrac{1}{4} + 0 + \tfrac{1}{4}]^{1/2}$ | $\lvert b_2 \rvert = a_0[\tfrac{4}{36} + \tfrac{1}{36} + \tfrac{1}{36}]^{1/2}$ | |
| $\lvert b_1 \rvert = \dfrac{\sqrt{2}}{2} a_0$ | $\lvert b_2 \rvert = \dfrac{a_0}{\sqrt{6}}$ | $\lvert b_3 \rvert = \dfrac{a_0}{\sqrt{6}}$ |
| $b_1{}^2 = \dfrac{a_0{}^2}{2}$ | $> \quad b_2{}^2 = \dfrac{a_0{}^2}{6}$ | $b_3{}^2 = \dfrac{a_0{}^2}{6}$ |
| $b_1{}^2$ | $> \quad b_2{}^2 + b_3{}^2$ | |

Slip by this two-stage process creates a stacking fault $ABCAC \vert ABC$ in the stacking sequence. As Fig. 5-10 shows, the dislocation with Burgers vector $\mathbf{b}_1$ has been dissociated into two *partial dislocations* $\mathbf{b}_2$ and $\mathbf{b}_3$. This dislocation reaction

**Fig. 5-10** Dissociation of a dislocation into two partial dislocations.

was suggested by Heidenreich and Shockley,[1] and therefore this dislocation arrangement is often known as *Shockley partials*, since the dislocations are imperfect ones which do not produce complete lattice translations. Figure 5-10 represents the situation looking down on (111) along [11$\bar{1}$]. *AB* represents the perfect dislocation line having the full slip vector $b_1$. This dissociates according to the above reaction into partial dislocations with Burgers vectors $b_2$ and $b_3$. The combination of the two partials *AC* and *AD* is known as an *extended dislocation*. The region between them is a stacking fault representing a part of the crystal which has undergone slip intermediate between full slip and no slip. Because $b_2$ and $b_3$ are at a 60° angle, there will be a repulsive force between them (Sec. 6-9). However, the surface tension of the stacking fault tends to pull them together. The partial dislocations will settle at an equilibrium separation determined primarily by the stacking-fault energy. As was discussed in Sec. 4-11, the stacking-fault energy can vary considerably for different fcc metals and alloys and this in turn can have an important influence on their deformation behavior.

Dissociation of unit dislocations is independent of the character (edge, screw, or mixed) of the dislocation. However, unlike the unextended screw dislocation, the extended screw dislocation defines a specific slip plane, the {111} plane of the fault, and it will be constrained to move in this plane. The partial dislocations move as a unit maintaining the equilibrium width of the faulted region. Because of this restriction to a specific slip plane, an extended screw dislocation cannot cross slip unless the partial dislocations recombine into a perfect dislocation. Constrictions in the stacking fault ribbon which permit cross slip are possible (Fig. 4-30), but this requires energy. The greater the width of the stacking fault (or the lower the stacking-fault energy) the more difficult it is to produce constrictions in the stacking faults. This explains why cross slip is quite prevalent in aluminum, which has a very narrow stacking-fault ribbon, while it is not observed usually in copper, which has a wide stacking-fault ribbon.

Stacking faults can be readily detected by transmission electron microscopy. The nature of the dislocation network in fcc metals changes with the stacking-fault energy. Austenitic stainless steel, with a stacking-fault energy around 13 erg/cm$^2$, shows dislocation networks only along slip planes, even for large deformations. Gold, copper, and nickel, where the energy is about 30, 40, and 80 erg/cm$^2$, respectively, show the dislocations arranged in complex three-dimensional networks at low strains. This changes into poorly developed subboundaries at higher deformations. Aluminum, with a stacking-fault energy of 200 erg/cm$^2$, shows almost perfect subboundaries. This picture of a graded transition in the way the dislocations are arranged is in agreement with the influence of the stacking-fault energy on the ability of a metal to undergo cross slip. Cross slip is very difficult in stainless

[1] R. D. Heidenreich and W. Shockley, " Report on Strength of Solids," p. 37, Physical Society, London, 1948.

steel, even at high strains, so that the dislocations are confined to the slip planes. In gold, copper, and nickel, cross slip is possible, but probably only at highly stressed regions. Therefore, cross slip of screw dislocations occurs, and at high strains they try to form low-angle boundary networks to lower their strain energy. In aluminum, cross slip is very prevalent, and screw dislocations can easily arrange themselves into a network of low-angle boundaries.

Frank[1] pointed out that another type of partial dislocation can exist in the fcc lattice. Figure 5-11 illustrates a set of (111) planes viewed from the edge. The center part of the middle $A$ plane is missing. An edge dislocation is formed in this region with a Burgers vector $(a_0/3)[111]$. This is called a *Frank partial dislocation*. Its Burgers vector is perpendicular to the central stacking fault. Since glide must be restricted to the plane of the stacking fault and the Burgers vector is normal to this plane, the Frank partial dislocation cannot move by glide. For this reason it is called a *sessile dislocation*. A sessile dislocation can move only by the diffusion of atoms or vacancies to or from the fault, i.e., by the process of climb. Because climb is not a likely process at ordinary temperatures, sessile dislocations provide obstacles to the movement of other dislocations. Dislocations which glide freely over the slip plane, such as perfect dislocations or Shockley partials, are called *glissile*. A method by which a missing row of atoms can be created in the (111) plane is by the condensation of a disk of vacancies on that plane. Evidence for the collapse of disks of vacancies in aluminum has been obtained by transmission electron microscopy.[2]

Sessile dislocations are produced in the fcc lattice by the glide of dislocations on intersecting {111} planes during duplex slip. The sessile dislocation produced by the reaction is called a *Lomer-Cottrell barrier*. Consider two perfect dislocations $a_0/2[\bar{1}10]$ and $a_0/2[101]$ lying in different {111} planes and both parallel to the line of intersection of the {111} planes (Fig. 5-12). These dislocations attract each other and move toward their line of intersection. Lomer[3] suggested that they react according to

$$\frac{a_0}{2}[101] + \frac{a_0}{2}[\bar{1}10] \rightarrow \frac{a_0}{2}[011]$$

**Fig. 5-11** A Frank partial dislocation or sessile dislocation. (*After A. H. Cottrell, " Dislocations and Plastic Flow in Crystals," p. 75, Oxford University Press, New York, 1953. By permission of the publishers.*)

[1] F. C. Frank, *Proc. Phys. Soc. London*, vol. 62A, p. 202, 1949.
[2] P. B. Hirsch, J. Silcox, R. E. Smallman, and K. H. Westmacott, *Philos. Mag.*, vol. 3, p. 897, 1958.
[3] W. M. Lomer, *Philos. Mag.*, vol. 42, p. 1327, 1951.

**Fig. 5-12** Lomer-Cottrell barrier.

to produce a new dislocation of reduced energy. This new dislocation lies parallel to the line of intersection of the initial slip planes in the (100) plane bisecting the slip planes. Its Burgers vector lying in the (100) plane is normal to the line of intersection so it is a pure edge dislocation. Since (100) is not a close-packed slip plane in the fcc lattice, this dislocation will not glide freely. However, it is not a true sessile dislocation in the sense of the Frank partial because it is not an imperfect dislocation.

Cottrell[1] showed that the product of Lomer's reaction could be made strictly immobile if it is considered that dislocations on the {111} planes of an fcc metal are normally dissociated into partials. The leading partial dislocations on each slip plane will interact with each other in a reaction of the type

$$\frac{a_0}{6}[\bar{1}2\bar{1}] + \frac{a_0}{6}[1\bar{1}2] \rightarrow \frac{a_0}{6}[011]$$

Like before, the new dislocation $a_0/6[011]$ lies parallel to the line of intersection of the slip plane and has a pure edge character in the (100) plane. The barrier is triangular in shape with this dislocation at the apex and two stacking faults bounded by partial dislocations on the slip planes. The dislocation is sessile because its Burgers vector does not lie in either of the planes of its stacking faults.

Lomer-Cottrell barriers can be overcome at high stresses and/or temperatures A mathematical analysis of the stress required to break down a barrier either by slip on the (100) plane or by reaction back into the dislocations from which the barrier formed has been given by Stroh.[2] However, it has been shown[3] that for the important case of screw dislocations piled up at Lomer-Cottrell barriers, the screw dislocations can escape the pile-up by cross slip before the stress is high

[1] A. H. Cottrell, *Philos. Mag.*, vol. 43, p. 645, 1952.
[2] A. N. Stroh, *Philos. Mag.*, vol. 1, sec, 8, p. 489, 1956.
[3] A. Seeger, J. Diehl, S. Mader, and R. Rebstook, *Philos. Mag.*, vol. 2, p. 323, 1957.

enough to collapse the barrier. While the formation of Lomer-Cottrell barriers is an important mechanism in the strain hardening of fcc metals, they do not constitute the chief contribution to strain hardening.

Because of the multiplicity of slip systems in the fcc lattice, a large number of dislocation reactions of the types discussed above are possible. These have been worked out in detail by Hirth.[1] Also, the Thompson tetrahedron[2] is a useful geometrical method for visualizing the geometry of these reactions.

## 5-5  DISLOCATIONS IN THE HEXAGONAL CLOSE-PACKED LATTICE

The basal plane of the hcp lattice is a close-packed plane with the stacking sequence *ABABAB* .... Slip occurs on the basal plane (0001) in the $\langle 11\bar{2}0 \rangle$ direction (Fig. 4-3). The smallest unit vector for the hcp structure has a length $a_0$ and lies in the close-packed $\langle 11\bar{2}0 \rangle$ direction. Therefore, the Burgers vector is $a_0[11\bar{2}0]$. Dislocations in the basal plane can reduce their energy by dissociating into Shockley partials according to the reaction

$$a_0[11\bar{2}0] \rightarrow a_0[10\bar{1}0] + a_0[01\bar{1}0]$$

The stacking fault produced by this reaction lies in the basal plane, and the extended dislocation which forms it is confined to glide in this plane.

## 5-6  DISLOCATIONS IN THE BODY-CENTERED CUBIC LATTICE

Slip occurs in the $\langle 111 \rangle$ direction in the bcc lattice. The shortest lattice vector extends from an atom corner to the atom at the center of the unit cube. Therefore, the Burgers vector is $(a_0/2)[111]$. It will be recalled that slip lines in iron have been found to occur on {110}, {112}, and {123}, although in other bcc metals slip appears to occur predominantly on the {110} planes. While the slip plane is normally (110), it is important to realize that three {110}-type planes intersect in a [111] direction. Thus, screw dislocations with a $(a_0/2)[111]$ Burgers vector may move at random on the {111} planes with a high resolved shear stress. This is the origin of the wavy and poorly defined slip lines in iron (Fig. 4-14).

Extended dislocations are not commonly observed in bcc metals as they are in fcc and hcp metals. Although dislocation reactions involving partials have been suggested,[3] there are no well-established reactions that have been substantiated by extensive experimentation.

[1] J. P. Hirth, *J. Appl. Phys.*, vol. 32, pp. 700–706, 1961.
[2] N. Thompson, *Proc. Phys. Soc. London Ser. B*, vol. B66, p. 481, 1953.
[3] For a summary see J. P. Hirth and J. Lothe, "Theory of Dislocations," pp. 344–353, McGraw-Hill Book Company, New York, 1968.

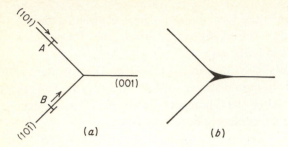

**Fig. 5-13** Slip on intersecting (110) planes. (*From A. H. Cottrell, Trans. AIME, vol. 212, p. 196, 1958.*)

However, Cottrell[1] has suggested a dislocation reaction which appears to lead to the formation of immobile dislocations in the bcc lattice. This reaction has been shown to be one of the mechanisms for producing a crack nucleus for brittle fracture. It is also a mechanism for producing $a_0[001]$ dislocation networks that are observed in iron. In Fig. 5-13, dislocation $A$ with Burgers vector $(a_0/2)[\bar{1}\bar{1}1]$ is gliding on the (101) plane, while dislocation $B$ with $(a_0/2)[111]$ is gliding on the intersecting $(10\bar{1})$ slip plane. The two dislocations come together and react to lower their strain energy by providing a pure edge dislocation which lies on the (001) plane.

$$\frac{a_0}{2}[\bar{1}\bar{1}1] + \frac{a_0}{2}[111] \rightarrow a_0[001]$$

Since the (001) is not a close-packed slip plane in the bcc lattice, the dislocation is immobile. Moreover, the (001) plane is the cleavage plane along which brittle fracture occurs.

## 5-7 STRESS FIELDS AND ENERGIES OF DISLOCATIONS

A dislocation is surrounded by an elastic stress field that produces forces on other dislocations and results in interaction between dislocations and solute atoms. For the case of a perfect dislocation a good approximation of the stress field can be obtained from the mathematical theory of elasticity for continuous media. However the equations obtained are not valid close to the core of the dislocation line. The equations given below apply to straight edge and screw dislocations in an isotropic crystal.[2] The stress around a straight dislocation will be a good approximation to

---

[1] A. H. Cottrell, *Trans. Metall. Soc. AIME*, vol. 212, p. 192, 1958.
[2] For derivations see F. R. N. Nabarro, *Adv. Phys.*, vol. 1, no. 3, pp. 271–395, 1952; W. T. Read, Jr., "Dislocations in Crystals," pp. 114–123, McGraw-Hill Book Company, New York, 1953; J. D. Eshelby, *Brit J. Appl. Phys.*, vol. 17, pp. 1131–1135, 1966.

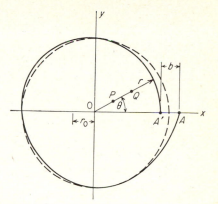

**Fig. 5-14** Deformation of a circle containing an edge dislocation. The unstrained circle is shown by a dashed line. The solid line represents the circle after the dislocation has been introduced.

that around a curved dislocation at distances that are small compared with the radius of curvature. Appreciably greater complexity results from the consideration of a crystal with anisotropic elastic constants.[1]

Figure 5-14 represents the cross section of a cylindrical piece of elastic material containing an edge dislocation running through point $O$ parallel to the $z$ axis (normal to the plane of the figure). The original undistorted cylinder without a dislocation is shown by the dashed line. The dislocation was produced by making a radial cut along the plane $y = 0$ (line $OA$), sliding the cut surfaces along each other the distance $AA'$, and joining them back together again. This sequence of operations[2] produces a positive edge dislocation running along the $z$ axis with a strain field identical with that around a dislocation model such as that of Fig. 4-8. Since the dislocation line is parallel to the $z$ axis, strains in that direction are zero and the problem can be treated as one in plane strain.

For the case of a straight edge dislocation in an elastically isotropic material the stresses, in terms of three orthogonal coordinate axes, are given by the following equations. The notation is the same as that used in Chaps. 1 and 2.

$$\sigma_x = -\tau_0 \frac{by(3x^2 + y^2)}{x^2 + y^2} \tag{5-1}$$

$$\sigma_y = \tau_0 \frac{by(x^2 - y^2)}{(x^2 + y^2)^2} \tag{5-2}$$

$$\sigma_z = v(\sigma_x + \sigma_y) = \frac{2\tau_0 \, vy}{x^2 + y^2} \tag{5-3}$$

---

[1] J. D. Eshelby, W. T. Read, and W. Shockley, *Acta Metall.*, vol. 1, pp. 351–359. 1953.

[2] It is interesting that this problem was analyzed by Volterra in 1907, long before the concept of dislocations was originated. The mathematical details may be found in A. E. H. Love, "A Treatise on the Mathematical Theory of Elasticity," pp. 221–228, Cambridge University Press, New York, 1934.

where

$$\tau_0 = \frac{G}{2\pi(1-v)}$$

$$\tau_{xy} = \tau_0 \frac{bx(x^2 - y^2)}{(x^2 + y^2)^2} \tag{5-4}$$

$$\tau_{xz} = \tau_{yz} = 0 \tag{5-5}$$

The largest normal stress $\sigma_x$ is along the $x$ axis, and is compressive above the slip plane and tensile below the slip plane. The shear stress $\tau_{xy}$ is a maximum in the slip plane, i.e., when $y = 0$.

For polar coordinates, the equations are

$$\sigma_r = \sigma_\theta = \frac{-\tau_0 b \sin \theta}{r} \tag{5-6}$$

$$\tau_{r\theta} = \tau_{\theta r} = \tau_0 \frac{b \cos \theta}{r} \tag{5-7}$$

$\sigma_r$ acts in the radial direction, while $\sigma_\theta$ acts in a plane perpendicular to $r$. Note that the stresses vary inversely with distance from the dislocation line. Since the stress becomes infinite at $r = 0$, a small cylindrical region $r = r_0$ around the dislocation line must be excluded from the analysis. Estimates of $r_0$ indicate that it is on the order of 5 to 10 Å.

A straight screw dislocation in an isotropic medium has complete cylindrical symmetry. For a rectangular-coordinate system only two components of stress are not equal to zero.

$$\tau_{xz} = -\frac{Gb}{2\pi} \frac{y}{x^2 + y^2} \tag{5-8}$$

$$\tau_{yz} = \frac{Gb}{x} \frac{2\pi}{x^2 + y^2} \tag{5-9}$$

Since there is no extra half plane of atoms in a screw dislocation, there are no tensile or compressive normal stresses. The stress field is simply one of shear. The radial symmetry of this stress field is apparent when the shear stress is expressed in a polar-coordinate system.

$$\tau_{\theta z} = \frac{Gb}{2\pi r} \tag{5-10}$$

The strain field around an edge dislocation in a silicon crystal has been observed[1] by means of polarized infrared radiation. The variation in intensity is in

---

[1] W. L. Bond and J. Andrus, *Phys. Rev.*, vol. 101, p. 1211, 1956.

agreement with what would be expected from the equations for a stress field around an edge dislocation in an isotropic medium.

. The strain energy involved in the formation of an edge dislocation can be estimated from the work involved in displacing the cut $OA$ in Fig. 5-14 a distance $b$ along the slip plane.

$$U = \frac{1}{2} \int_{r_0}^{r_1} \tau_{r\theta} b \, dr = \frac{1}{2} \int_{r_0}^{r_1} \tau_0 b^2 \cos\theta \frac{dr}{r} \qquad (5\text{-}11)$$

But $\cos\theta = 1$ along the slip plane $y = 0$, so that the strain energy is given by

$$U = \frac{Gb^2}{4\pi(1-v)} \ln\frac{r_1}{r_0} \qquad (5\text{-}12)$$

In the same way, the strain energy of a screw dislocation is given by

$$U = \frac{1}{2} \int_{r_0}^{r_1} \tau_{\theta z} b \, dr = \frac{Gb^2}{4\pi} \ln\frac{r_1}{r_0} \qquad (5\text{-}13)$$

Note that, in accordance with our assumption up to this point, the strain energy per unit length of dislocation is proportional to $Gb^2$.

The total strain energy of a dislocation is the sum of the elastic strain energy [Eq. (5-12) or (5-13)] plus the energy of the core of the dislocation. Although estimates of the core energy are quite approximate, quantum-mechanical calculations indicate that the core energy is about one-fifteenth of the total energy. As a good approximation, the core energy is added to the elastic strain energy by taking $r_0 \approx b$, so that the total energy per unit length of screw dislocation is given by

$$U_t = \frac{Gb^2}{4\pi} \ln\frac{r_1}{b} \qquad (5\text{-}14)$$

Typical values for an annealed crystal are $r_1 = 10^{-5}$ cm and $b = 2 \times 10^{-8}$ cm. Since the natural logarithm of a large number changes slowly, for this range of values $\ln(r_1/b) \approx 2\pi$, and the dislocation energy per unit length (ignoring small differences in Eqs. (5-12) and (5-13) simplifies to

$$U = \frac{Gb^2}{2} \qquad (5\text{-}15)$$

The strain energy of a dislocation is about 8 eV for each atom plane threaded by the dislocation (see Prob. 5-8) while the core energy is the order of 0.5 eV per atom plane. This large, positive strain energy means that the free energy of a crystal is increased by the introduction of a dislocation. Since nature tries to minimize the free energy of a system, a crystal with dislocations is thermodynamically unstable and will try to lower its free energy by the elimination of dislocations in a process such as annealing. This situation should be contrasted with that of point defects (vacancies), which are thermodynamically stable crystal defects. At a given temperature there is an equilibrium concentration of vacancies given by Eq. (4-1).

## 5-8  FORCES ON DISLOCATIONS

When an external force of sufficient magnitude is applied to a crystal, the dislocations move and produce slip. Thus, there is a force acting on a dislocation line which tends to drive it forward. Figure 5-15 shows a dislocation line moving in the direction of its Burgers vector under the influence of a uniform shear stress $\tau$. An element of the dislocation line $ds$ is displaced in the direction of slip normal to $ds$ by an amount $dl$. The area swept out by the line element is then $ds\, dl$. This corresponds to an average displacement of the crystal above the slip plane to the crystal below the slip plane of an amount $(ds\, dl/A)b$, where $A$ is the area of the slip plane. The applied force creating the shear stress is $\tau A$. The work done when the increment of slip occurs is

$$dW = \tau A \left( \frac{ds\, dl}{A} \right) b$$

The force on a dislocation is always defined as a force $F$ *per unit length* of dislocation line. Since $F = dW/dl$ and remembering that this is a force per unit length $(ds)$, we have

$$F = \frac{dW}{dl\, ds} = \tau b \qquad (5\text{-}16)$$

This force is normal to the dislocation line at every point along its length and is directed toward the unslipped part of the glide plane. Since the Burgers vector is constant along a curved dislocation line, if $\tau$ is constant, the value of $F$ will be the same at any point along the dislocation line but its direction always will be normal

**Fig. 5-15**  Force acting on a dislocation line.

**Fig. 5-16**  Forces on a curved dislocation line.

to the dislocation line. As a result, the force on a dislocation is not necessarily in the same direction as the applied stress.

All of the equations for the strain energy of a dislocation in Sec. 5-7 were expressed *per unit length of dislocation*. Because the strain energy of a dislocation line is proportional to its length, work must be performed to increase its length. Therefore, a dislocation possesses a *line tension* which attempts to minimize its energy by shortening its length. For a curved dislocation line the line tension produces a restoring force which tends to straighten it out. The line tension has units of energy per unit length and is analogous to the surface energy of a soap bubble or liquid.

Consider the curved dislocation line in Fig. 5-16. The line tension $\Gamma$ will produce a force tending to straighten the line and the line only will remain curved if there is a shear stress which produces a force on the dislocation to resist the line tension. We wish to determine the value of the shear stress $\tau$ that is required to maintain the dislocation line at a radius of curvature $R$. The angle subtended by an element of arc $ds$ is $d\theta = ds/R$. The outward force on the dislocation line is $\tau b \, ds$. The opposing inward force due to the line tension is $2\Gamma \sin(d\theta/2)$, which simplifies to $\Gamma \, d\theta$ for small values of $d\theta$. For the dislocation line to remain in the curved configuration

$$\Gamma \, d\theta = \tau b \, ds$$

or

$$\tau = \frac{\Gamma}{bR} \qquad (5\text{-}17)$$

But, $\Gamma$ is an energy per unit length, and we have seen in Sec. 5-7 that a good approximation for the energy per unit length of dislocation is $Gb^2/2 \approx \Gamma$. Thus, the shear stress required to bend a dislocation to a radius $R$ is

$$\tau \approx \frac{Gb}{2R} \qquad (5\text{-}18)$$

## 5-9   FORCES BETWEEN DISLOCATIONS

Dislocations of opposite sign on the same slip plane will attract each other, run together, and annihilate each other. This can be seen readily for the case of an edge dislocation (Fig. 4-8) where the superposition of a positive and negative dislocation on the same slip plane would eliminate the extra plane of atoms and therefore the dislocation would disappear. Conversely, dislocations of like sign on the same slip plane will repel each other. We can understand this by considering the energy changes. For two dislocations separated at a large distance, the elastic strain energy for the combined situation will be

$$2 \cdot \frac{Gb^2}{4\pi(1-v)} \ln \frac{r_1}{r_0}$$

When the two dislocations are very close together the configuration can be approximated by a single dislocation of strength $2b$. For this case, the elastic strain energy will be

$$\frac{G(2b)^2}{4\pi(1-v)} \ln \left(\frac{r_1}{r_0}\right)$$

Since this is twice the energy of the dislocations when they are separated by a large distance, the dislocations will tend to repel each other to reduce the total elastic strain energy. When unlike dislocations are on closely spaced neighboring slip planes, complete annihilation cannot occur. In this situation, they combine to form a row of vacancies (for the case of $\frac{\perp}{\top}$) or an interstitial atom (for the case $\perp\top$).

The simplest situation to consider is the force between two parallel screw dislocations. Since the stress field of a screw dislocation is radially symmetrical, the force between the dislocations is a radial force which depends only on the distance of separation $r$.

$$F_r = \tau_{\theta z} b = \frac{Gb^2}{2\pi r} \qquad (5\text{-}19)$$

The force is attractive for dislocations of opposite sign (antiparallel screws) and repulsive for dislocations of the same sign (parallel screws).

Consider now the forces between two parallel edge dislocations with the same Burgers vectors. Referring to Fig. 5-14, the edge dislocations are at $P$ and $Q$, parallel to the $z$ axis, with their Burgers vectors along the $x$ axis. The force between them is not a central force, and so it is necessary to consider both a radial and a tangential component. The force per unit length is given by[1]

$$F_r = \frac{Gb^2}{2\pi(1-v)} \frac{1}{r} \qquad F_\theta = \frac{Gb^2}{2\pi(1-v)} \frac{\sin 2\theta}{r} \qquad (5\text{-}20)$$

[1] A. H. Cottrell, "Dislocations and Plastic Flow in Crystals," p. 46, Oxford University Press, New York, 1953.

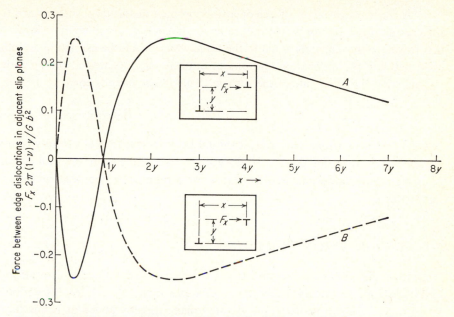

**Fig. 5-17** Graphical representation of Eq. (5-21). Solid curve $A$ is for two edge disloca-
tions of same sign. Dashed curve $B$ is for two unlike edge dislocations. (*After A. H.
Cottrell, "Dislocations and Plastic Flow in Crystals," p. 48, Oxford University Press, New
York, 1953. By permission of the publishers.*)

Because edge dislocations are mainly confined to the slip plane, the force component
along the $x$ direction, which is the slip direction, is of most interest.

$$F_x = F_r \cos \theta - F_\theta \sin \theta$$

$$= \frac{Gb^2 x(x^2 - y^2)}{2\pi(1 - v)(x^2 + y^2)^2} \qquad (5\text{-}21)$$

Figure 5-17 is a plot of the variation of $F_x$ with distance $x$, where $x$ is expressed
in units of $y$. Curve $A$ is for dislocations of the same sign; curve $B$ is for dislocations
of opposite sign. Note that dislocations of the same sign repel each other when
$x > y$ ($\theta < 45°$) and attract each other when $x < y$ ($\theta > 45°$). The reverse is true
for dislocations of opposite sign. $F_x$ is zero at $x = 0$ and $x = y$. The situation
$x = 0$, where the edge dislocations lie vertically above one another, is a condition
of equilibrium. Thus, theory predicts that a vertical array of edge dislocations of
the same sign is in stable equilibrium. This is the arrangement of dislocations
that exists in a low-angle grain boundary of the tilt variety. The calculation of
forces for more complex situations are discussed by Read[1] and Weertman and
Weertman[2].

[1] Read, *op. cit.*, p. 131.
[2] J. Weertman and J. R. Weertman, "Elementary Dislocation Theory," pp. 65–72, The Macmillan
Company, New York, 1964.

A free surface exerts a force of attraction on a dislocation, since escape from the crystal at the surface would reduce its strain energy. Koehler[1] has shown that this force is approximately equal to the force which would be exerted in an infinite solid between the dislocation and one of opposite sign located at the position of its image on the other side of the surface. This *image force* is equal to

$$F = \frac{Gb^2}{4\pi(1-v)} \frac{1}{r} \qquad (5\text{-}22)$$

for an edge dislocation. However, it should be noted that metal surfaces are often covered with thin oxide films. A dislocation approaching a surface with a coating of an elastically harder material will encounter a repulsive rather than an attractive image force.

## 5-10  DISLOCATION CLIMB

An edge dislocation can glide only in the slip plane containing the dislocation line and its Burgers vector. However, under certain conditions an edge dislocation can move out of the slip plane onto a parallel plane directly above or below the slip plane. This is the process of *dislocation climb*. This type of movement is termed nonconservative, as compared with conservative movement when a dislocation glides in its slip plane.

Dislocation climb occurs by the diffusion of vacancies or interstitials to or away from the site of the dislocation. Since climb is diffusion-controlled, it is thermally activated and occurs more readily at elevated temperature. In *positive climb* atoms are removed from the extra half plane of atoms at a positive edge dislocation so that this extra half plane moves up one atom spacing. In *negative climb* a row of atoms is added below the extra half plane so that the dislocation line moves down one atom spacing. The usual mechanism for positive climb is by a vacancy diffusing to the dislocation and the extra atom moving into the vacant lattice site (Fig. 5-18). It is also possible, but less energetically favorable, for the atom to break loose from the extra half plane and become an interstitial atom.

(a)  (b)

**Fig. 5-18**  (a) Diffusion of vacancy to edge dislocation (b) dislocation climbs up one lattice spacing.

[1] J. S. Koehler, *Phys. Rev.*, vol. 60, p. 397, 1941.

To produce negative climb, atoms must be added to the extra half plane of atoms. This can occur by atoms from the surrounding lattice joining the extra half plane, which creates a flux of vacancies away from the dislocation, or less probably, by interstitial atoms diffusing to the dislocation. The accumulation of vacancies at the dislocation core causes a free-energy change which results in an *osmotic force* on the dislocation.[1]

The existence of a compressive stress in the slip direction causes a force in the upward direction of climb. Similarly, a tensile stress normal to the extra half plane causes a force in the direction of negative climb. The superposition of stress on the high temperature needed for diffusion results in an increased rate of climb.

It is highly unlikely that complete rows of atoms are removed or added at the extra half plane in the climb process. In actual fact individual vacancies or small clusters of vacancies diffuse to the dislocation and climb occurs over a short segment of the dislocation line. This results in the formation of short steps or *jogs* along the dislocation line. Climb proceeds by the nucleation and motion of jogs. There will be a thermodynamic equilibrium number of jogs per unit length of dislocation, given by

$$n_j = n_0 \, e^{-U_j/kT} \qquad (5\text{-}23)$$

where $n_0$ is the number of atom sites per unit length of dislocation and $U_j$ is the activation energy to nucleate a jog. The activation energy of a jog is about 1 eV. The activation energy for climb is given by

$$U_c = U_j + U_v + U_m = U_j + U_d \qquad (5\text{-}24)$$

where $U_v$ = the energy of formation of a vacancy
$\quad U_m$ = the energy of movement of a vacancy
$\quad U_d$ = the activation energy for self-diffusion

If the dislocations are heavily jogged due to prior plastic deformation (see Sec. 5-11), then nucleation is negligible and the activation energy for climb will be determined only by the activation energy for self-diffusion. Dislocation climb is an important mechanism in the creep of metals, where the activation energy for steady-state creep is equal to the activation energy for self-diffusion.

Climb is not concerned with screw dislocations since in this case there is no extra half plane of atoms. Because the Burgers vector of a screw dislocation is parallel to the dislocation, it is free to slip on any plane which contains the dislocation line and the Burgers vector. No diffusion of atoms is needed to allow the screw dislocation to move on to another slip plane. However, a higher stress or activation energy may be needed since the resolved shear stress may not be as high as on the original slip plane.

[1] J. P. Hirth and J. Lothe, "Theory of Dislocations," pp. 506–509, McGraw-Hill Book Company, New York, 1968.

## 5-11 INTERSECTION OF DISLOCATIONS

Since even well-annealed crystals contain many dislocations, it is not infrequent that a dislocation moving in its slip plane will intersect other dislocations crossing the slip plane. We have already seen (Sec. 4-14) that dislocation intersection mechanisms play an important role in the strain-hardening process.

The intersection of two edge dislocations with Burgers vectors at right angles to each other is illustrated in Fig. 5-19. An edge dislocation $XY$ with Burgers vector $\mathbf{b}_1$ is moving on plane $P_{XY}$. It cuts through dislocation $AD$, with Burgers vector $\mathbf{b}_2$, lying on plane $P_{AD}$. The intersection produces a *jog PP'* in dislocation $AD$. The resulting jog is parallel to $\mathbf{b}_1$, but it has a Burgers vector $\mathbf{b}_2$ since it is part of the dislocation line $APP'D$. The length of the jog is equal to $\mathbf{b}_1$. It can be seen that the jog resulting from this intersection of two edge dislocations has an edge orientation, and therefore, it readily can glide with the rest of the dislocation. To determine which dislocation forms the jog, we note that a jog forms when the Burgers vector of the intersecting dislocation is normal to the other dislocation line ($\mathbf{b}_1$ is normal to $AD$ and jogs $AD$, while $\mathbf{b}_2$ is parallel to $XY$ and no jog is formed).

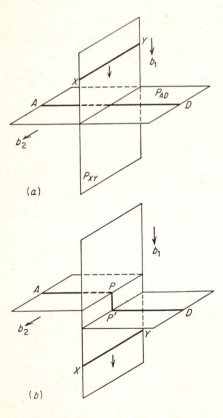

(a)

(b)

**Fig. 5-19** Intersection of two edge dislocations. (*From W. T. Read, Jr., " Dislocations in Crystals," McGraw-Hill Book Company, New York, 1953.*)

**Fig. 5-20** Intersection of edge dislocations with parallel Burgers vectors. (*a*) Before intersection; (*b*) after intersection.

The intersection of two orthogonal edge dislocations with parallel Burgers vectors is shown in Fig. 5-20. In this case both dislocations are jogged. The length of jog $PP'$ is $\mathbf{b}_1$ and the length of jog $QQ'$ is $\mathbf{b}_2$. In this case the jogs both have a screw orientation and lie in the original slip planes of the dislocations rather than on a neighboring slip plane as in the previous case. Jogs of this type which lie in the slip plane instead of normal to it are usually called *kinks*. Kinks in dislocations are unstable since during glide they can line up and annihilate the offset.

The intersection of a screw dislocation and an edge dislocation is shown in Fig. 5-21*a*. Intersection produces a jog with an edge orientation on the edge dislocation and a kink with an edge orientation on the screw dislocation. The intersection of two screw dislocations (Fig. 5-21*b*) produces jogs of edge orientation in both screw dislocations. This is the most important type of intersection from the viewpoint of plastic deformation.

The jogs produced by the intersection of two edge dislocations (of either orientation) are able to glide readily because they lie in the slip planes of the original dislocations. The only difference between the motion of the jogged dislocation and an ordinary edge dislocation is that instead of gliding along a single plane, it glides over a stepped surface. Thus, jogs in pure edge dislocations do not affect the motion of the dislocations on which they lie. However, for the screw dislocations the jogs all have an edge orientation. Since an edge dislocation can glide freely only in a

**Fig. 5-21** (*a*) Intersection of edge and screw dislocation; (*b*) intersection of two screw dislocations.

Fig. 5-22  Movement of jog on screw dislocation. The jog is constrained to move along the dislocation in plane $AA'BB'$.

plane containing its line and its Burgers vector, the only way the jog can move by slip (conservative motion) is along the axis of the screw dislocation (Fig. 5-22). The only way that the screw dislocation can slip to a new position such as $MNN'O$ and take its jog with it is by a nonconservative process such as climb. Dislocation climb is a thermally activated process and therefore the movement of jogged screw dislocations will be temperature-dependent. At temperatures where climb cannot occur, the motion of screw dislocations will be impeded by jogs. This is consistent with the experimental observation[1] that screw dislocations move more slowly through a crystal than edge dislocations.

## 5-12  JOGS

The creation of jogged segments on dislocation lines has many important implications in the theory of the plastic deformation of metals. A stable jog represents an increase in the length of the dislocation line, and therefore produces an increase in the energy of the crystal. The energy of a jog will be somewhat less than the energy of a dislocation per atom plane since its entire length lies in the distorted material of the core of the parent dislocation. Since the energy of a dislocation is $\alpha Gb^2$ (where $\alpha$ is about 0.5 to 1.0), the energy of a jog of length $b_2$ in a dislocation of Burgers vector $\mathbf{b}_1$ is

$$U_j = \alpha Gb_1{}^2 b_2 \qquad (5\text{-}25)$$

where $\alpha \approx 0.2$. The energy of a jog is about 0.5 to 1.0 eV in metals.

We have seen that the most significant dislocation intersection is the case of two screw dislocations where nonconservative jogs are produced. For all other intersections the jogs are able to move readily with their dislocations. The nonconservative motion of jogs on screw dislocations needs further discussion. Figure 5-23 shows a section of an edge jog in a screw dislocation which is gliding on a plane parallel to the paper in the direction $DB$ or $BD$. For the edge jog to move nonconservatively, it must either eliminate matter (create vacancies) or create

[1] N. K. Chen and R. B. Pond, *Trans. AIME*, vol. 194, p. 1085, 1952; W. G. Johnston and J. J. Gilman, *J. Appl. Phys.*, vol. 30, p. 121, 1957.

**Fig. 5-23** Creation of point defects by nonconservative motion of jogs of edge orientation. (*From D. Hull, "Introduction to Disloca-tions," p. 134, Pergamon Press, New York, 1965. By permission of the publishers.*)

matter (interstitials). If the jog moves nonconservatively from $A$ to $B$ and then glides along its slip plane to $C$, a row of vacancies will be created as shown in Fig. 5-23$b$. These vacancies will have a line tension and create a drag on the jog. However, if there is sufficient thermal activation, the vacancies will diffuse into the lattice and the drag will be removed. If the jog moves nonconservatively from $A$ to $D$ and then glides to $E$ (Fig. 5-23$c$), a row of interstitials will be produced along $AD$.

As a screw dislocation moves through its slip plane it will, in general, en-counter a forest of screw dislocations and many intersections will occur. Some of these will produce vacancy jogs and others will be interstitial jogs. These will tend to glide along the length of the screw dislocation and annihilate each other, leaving a net concentration of jogs all of the same sign. Then, because of their mutual repulsion, they will spread out along the dislocation line at approximately evenly spaced intervals. Figure 5-24 illustrates this situation. Under an applied shear stress $\tau$ acting in the slip direction, the jogs will act as pinning points. The dis-location will bow out between the jogs (Fig. 5-24$b$) with a radius of curvature given by Eq. (5-18). At some critical radius $R_c$ the shear stress required to further decrease $R$ is greater than the stress needed for nonconservative climb. Then the dislocation will move forward leaving a trail of vacancies (or interstitials) behind each jog. Estimates[1] of the energy required to form a vacancy or an interstitial atom at a jog in an fcc metal are 4.8 eV for an interstitial atom and 0.7 eV for a vacancy.

$$U_1 = \alpha_1 G b^3 \qquad (5\text{-}26)$$

[1] A. Seeger, "Defects in Crystalline Solids," p. 381, Physical Society, London, 1955.

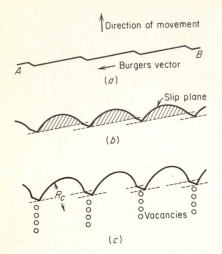

(a)

(b)

(c)

**Fig. 5-24**  Movement of jogged screw dislocation. (a) Straight dislocation under zero stress; (b) dislocation bowed out in slip plane between the jogs due to applied shear stress; (c) movement of dislocation leaving trails of vacancies behind the jogs. (*From D. Hull, "Introduction to Dislocations," p. 136, Pergamon Press, New York, 1965. By permission of the publishers.*)

where $\alpha_1 \approx 1.0$ for an interstitial atom and 0.2 for a vacancy. The work done by the applied stress in moving the jogs forward one atomic spacing by the formation of vacancies or interstitials is

$$W = (\tau b l)b = \tau b^2 l \qquad (5\text{-}27)$$

where $l$ is the spacing between jogs. By equating Eqs. (5-26) and (5-27), we obtain the shear stress required to generate a defect and move the dislocation in the absence of thermal activation.

$$\tau = \alpha_1 \frac{Gb}{l} \qquad (5\text{-}28)$$

At elevated temperatures thermal activation assists in the formation of vacancies. The activation energy for the movement of the jog or the formation of defects at the jog will be

$$U = \alpha_1 Gb^3 - \tau b^2 l \qquad (5\text{-}29)$$

Since $\alpha_1$ is much greater for interstitial atoms than for vacancies, vacancy formation will occur more readily than interstitial atom formation. Good agreement between the kinetics of annealing vacancies in quenched wires and in the annealing of cold-worked wires provides strong experimental evidence that vacancies are formed by plastic deformation.

Under certain conditions where the stress is high enough, the jogs strung out along the dislocation line can be forced together to form a *superjog* where the step height is of the order of 5 to 30$b$. As the shear stress increases, the dislocation bows out between the superjogs (Fig. 5-25a), generating long segments of edge orienta-

**Fig. 5-25**   (a) Bowing out of dislocation between superjogs to produce (b) dislocation dipoles; (c) formation of dislocation loops from dipole.

tion. As this continues, it produces elongated dislocation loops or dislocation dipoles (Fig. 5-25b). The dipole, in turn, can lower its elastic energy by forming a prismatic loop which later breaks into isolated loops (Fig. 5-25c). Loops can also form at high temperature by the collection of vacancies. The appearance of debris in the form of loops in transmission electron micrographs is a direct result of the formation of edge jogs on screw dislocations.

For very large superjogs, when the step height is greater than about 200 Å, the distance between the two dislocation segments is large enough to prevent mutual interaction. In this case the dislocations behave as separate single-ended sources (see Sec. 5-14).

Jog formation in extended dislocations is a complicated phenomena.[1] Intersection can only occur in extended dislocations by first forming a constriction in the region of the intersection.

## 5-13  DISLOCATION SOURCES

The low yield strength of pure crystals leads to the conclusion that dislocations must exist in completely annealed crystals and in crystals carefully solidified from the melt. The high strain energy of a dislocation, about 8 eV per atom plane, precludes the generation of dislocations by thermal activation and indicates that they must be produced by other processes.

There is an important difference between line defects and point defects. The density of dislocations in thermal equilibrium with a crystal is very low. There is no general relationship between dislocation density and temperature as exists for vacancies. Since dislocations are not affected by thermal fluctuations at tempera-

---

[1] M. J. Whelan, *Proc. R. Soc. London Ser. A*, vol. 249A, p. 114, 1959; P. B. Hirsch, *Philos. Mag.*, vol. 7, p. 67, 1962.

tures below which recrystallization occurs, a metal can have widely different dislocation densities depending upon prior history. Completely annealed material will contain about $10^6$ to $10^8$ dislocation lines per square centimeter, while heavily cold-worked metal will have a dislocation density of about $10^{12}$ dislocation lines per square centimeter.

All metals, with the exception of tiny whiskers, initially contain an appreciable number of dislocations produced as the result of the growth of the crystal from the melt or vapor phase. Gradients of temperature and composition may produce misalignments between neighboring dendrite arms growing from the same nucleus. These result in dislocations arranged in networks (Fig. 5-2) or in grain boundaries. Other mechanisms for introducing dislocations during growth are (1) formation of dislocation loops due to the aggregation and collapse of vacancies, and (2) heterogeneous nucleation of dislocations resulting from high local stresses at second-phase particles, grain boundaries, or as a result of phase transformations. Since a stress of about $G/30$ is required for homogeneous nucleation of dislocations, heterogeneous nucleation is preferred. A number of experimental observations of heterogeneous nucleation of dislocations have been recorded.[1]

## 5-14 MULTIPLICATION OF DISLOCATIONS

One of the original stumbling blocks in the development of dislocation theory was the formulation of a reasonable mechanism by which sources originally present in the metal could produce new dislocations by the process of slip. Such a mechanism is required when it is realized that the surface displacement at a slip band is due to the movement of about 1,000 dislocations over the slip plane. Thus, the number of dislocation sources initially present in a metal could not account for the observed slip-band spacing and displacement unless there were some way in which each source could produce large amounts of slip before it became immobilized. Moreover, if there were no source generating dislocations, cold-work should decrease, rather than increase, the density of dislocations in a single crystal. Thus. there must be a method of generating dislocations or of multiplying the number initially present to produce the high dislocation density found in cold-worked metal. The scheme by which dislocations could be generated from existing dislocations was proposed by Frank and Read[2] and is commonly called a *Frank-Read source.*

Consider a dislocation line $DD'$ lying in a slip plane (Fig. 5-26a). The plane of the figure is the slip plane. The dislocation line leaves the slip plane at points $D$ and $D'$, so that it is immobilized at these points. This could occur if $D$ and $D'$

---

[1] See for example, D. Hull, "Introduction to Dislocations," pp. 155–161, Pergamon Press, New York, 1965.

[2] F. C. Frank and W. T. Read, *Phys. Rev.*, vol. 79, pp. 722–723, 1950.

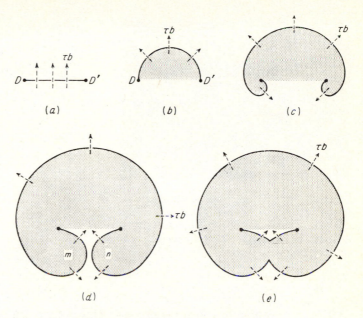

**Fig. 5-26** Schematic representation of the operation of a Frank-Read source. (*From W. T. Read, Jr., " Dislocations in Crystals," McGraw-Hill Book Company, New York, 1953.*)

were nodes where the dislocation in the plane of the paper intersects dislocations in other slip planes, or the anchoring could be caused by impurity atoms. If a shear stress $\tau$ acts in the slip plane, the dislocation line bulges out and produces slip. For a given stress the dislocation line will assume a certain radius of curvature given by Eq. (5-18). The maximum value of shear stress is required when the dislocation bulge becomes a semicircle so that $R$ has the minimum value $l/2$ (Fig. 5-26b). From the approximation that $\Gamma \approx 0.5Gb^2$ and Eq. (5-18), it readily can be seen that the stress required to produce this configuration is

$$\tau \approx \frac{Gb}{l} \qquad (5\text{-}30)$$

Beyond this point $R$ will increase and the dislocation loop will continue to expand under a decreasing stress (Fig. 5-26c). When the loop reaches Fig. 5-26d, the segments at $m$ and $n$ will meet and annihilate each other to form a large loop and a new dislocation $DD$ (Fig. 5-26e). The stage shown in Fig. 5-26d can best be understood if we assume that the original pinned length $DD'$ has a screw orientation. Then segments $m$ and $n$ are in edge orientation but of opposite sign, so that annihilation will occur. Once the loop moves into the stage shown in Fig. 5-26c, the loop can continue to expand under increased shear stress and the pinned segment

**Fig. 5-27** Frank-Read source in silicon crystal. (*From W. C. Dash, in: "Dislocations and Mechanical Properties of Crystals," John Wiley & Sons, Inc., New York, 1957. By permission of the publishers.*)

$DD'$ is in a position to repeat the process. This process can be repeated over and over again at a single source, each time producing a dislocation loop which produces slip of one Burgers vector along the slip plane. However, once the source is initiated it does not continue indefinitely. The back stress produced by the dislocations piling up along the slip plane (see Sec. 5-16) opposes the applied stress and when this equals the critical stress given by Eq. (5-30), the source ceases to operate. An excellent example of a Frank-Read source is shown in Fig. 5-27.

Although double-ended Frank-Read sources have been observed experimentally, their occurrence is not highly frequent. Several other multiplication mechanisms have been observed. A single-ended source can arise where part of a screw dislocation lies in the slip plane while part is immobilized by lying out of the slip plane. The immobilization could also arise from a superjog on a screw dislocation. Rotation around the immobilized segment of the dislocation by the dislocation segment in the slip plane results in a spiral slip step, with the height of the step proportional to the number of revolutions in the slip plane. This spiraling around the immobilized dislocation segment also results in an increase in the total length of dislocation line.

The double-ended Frank-Read source produces a large slip step in the slip plane, but it is restricted to operation in a single plane. This cannot explain the observation that slip bands widen with increased strain as well as increasing the slip offset. A *multiple cross-slip* mechanism provides a ready explanation for this behavior. In Fig. 5-8 a screw dislocation on $AB$ can cross slip into position $CD$. The edge segments $AC$ and $BD$ on the cross-slip plane can be considered as superjogs which are relatively immobile and pin the dislocations on the $AB$ and $CD$

planes. The dislocation segments lying on these planes can expand as a Frank-Read source. When cross slip can occur readily the Frank-Read sources may not complete a cycle and there will be one continuous dislocation line lying on each of many parallel glide planes connected by jogs. This produces a wide slip band.

At elevated temperature double-pinned edge segments can bow out under the influence of an osmotic driving force due to vacancy flux in the manner described for the Frank-Read source. The supersaturation of vacancies needed to operate this source is as small as 2 percent at high temperature.[1] This multiplication mechanism is known as a *Bardeen-Herring source*.[2]

## 5-15  DISLOCATION–POINT DEFECT INTERACTIONS

Isolated solute atoms and vacancies are centers of elastic distortion just as are dislocations. Therefore, point defects and dislocations will interact elastically and exert forces on each other. To a good approximation the strains around a point defect distort the lattice spherically, just as though an elastic sphere of radius $a'$ had been forced into a spherical hole of radius $a$ in an elastic continuum. The resulting strain is $\varepsilon = (a' - a)/a$. If the point defect is a vacancy, the radius $a$ is the radius of the atom normally at the lattice site, while if the defect is an interstitial atom, $a$ corresponds to the average radius of an empty interstitial site. The volume change produced by the point defect is given by

$$\Delta V = 4\pi a^3 \varepsilon \qquad (5\text{-}31)$$

Because we are concerned only with spherical distortions, the interaction only occurs with the hydrostatic component of the dislocation stress field. The elastic interaction energy between the dislocation and the point defect is

$$U_i = \sigma_m \Delta V \qquad (5\text{-}32)$$

The hydrostatic stress of a positive edge dislocation at $r$, $\theta$ (Fig. 5-14) from the dislocation is

$$\sigma_m = \frac{(1 + v)Gb \sin \theta}{3\pi(1 - v)r} \qquad (5\text{-}33)$$

so that the interaction energy is

$$U_i = \frac{4(1 + v)Gba^3 \varepsilon \sin \theta}{3(1 - v)r} \qquad (5\text{-}34)$$

[1] Hirth and Lothe, op. cit., pp. 565–568.
[2] J. Bardeen and C. Herring, in "Imperfections in Nearly Perfect Crystals," p. 261, John Wiley & Sons, Inc., New York, 1952.

However, this expression includes only the energy external to the point defect. When the strain energy due to elastic distortion of the solute atom is considered, the complete expression for interaction energy is

$$U_i = 4Gba^3\varepsilon \frac{\sin\theta}{r} = A\frac{\sin\theta}{r} \qquad (5\text{-}35)$$

Since Eq. (5-35) is derived from elasticity theory,[1] it is not strictly correct near the core of the dislocation where linear elasticity theory is no longer applicable. Since the maximum interaction energy occurs in this location, Eq. (5-35) provides only an estimate.

A negative value of interaction energy indicates attraction between the point defect and the dislocation, while a positive value denotes repulsion. A solute atom larger than the solvent atom ($\varepsilon > 1$) will be repelled from the compression side of a positive edge dislocation ($0 < \theta < \pi$) and will be attracted to the tension side ($\pi < \theta < 2\pi$). An atom smaller than the solvent atom ($\varepsilon < 1$) will be attracted to a site on the compression side of a positive edge dislocation. Similarly vacancies[2] will be attracted to regions of compression and interstitials will collect at regions of tension.

Since the strain fields around a point defect are spherically symmetrical, such a defect will produce no net force on a screw dislocation because its stress field is pure shear. Thus there is no hydrostatic stress field around a screw dislocation which can be relaxed by the presence of a point defect. However, some point defects (such as interstitial carbon atoms in a bcc lattice) produce a nonspherical distortion[3] so that there will be an interaction energy between a screw dislocation and the defect.

Even if the point defect has the same volume as the atoms in the lattice, an interaction with a dislocation can occur if the defect has different elastic constants than the matrix.[4] There will be an attraction if the point defect is elastically softer than the matrix and a repulsion if the point defect is elastically harder. For this type of interaction a binding force will exist between a screw dislocation and a spherically symmetrical point defect.

Since the energy associated with a point defect is affected by its proximity to a dislocation, Eq. (5-35), we should expect the concentration of defects to be different in the vicinity of a dislocation line. Assuming a Boltzmann distribution

[1] J. Weertman and J. R. Weertman, "Elementary Dislocation Theory," The Macmillan Company, New York, pp. 173–177; B. A. Bilby, *Proc. Phys. Soc., London*, vol. A63, p. 191, 1950.
[2] R. Bullough and R. C. Newman, *Philos. Mag.*, vol. 7, 529, 1962.
[3] A. W. Cochardt, G. Schoek, and H. Wiedersich, *Acta Metall.*, vol. 3, pp. 533–537, 1955.
[4] R. L. Fleischer, *Acta Metall.*, vol. 11, p. 203, 1963.

for the concentration of point defects, the concentration in the vicinity of a dislocation $C$ is related to the average concentration $C_0$ by the relationship

$$C = C_0 e^{-U_i/kT} \qquad (5\text{-}36)$$

The concentration of point defects around a dislocation exceeds the average value when $U_i$ is negative and is less than the average when $U_i$ is positive. An increased concentration of solute atoms around a dislocation is called an *impurity atmosphere* or an *impurity cloud*. The concentration $C$ cannot exceed one solute atom per lattice site or per interstitial site. At temperatures so low that $C_0 e^{-U_i/kT}$ exceeds 1, the sites near the dislocation core have been saturated with solute atoms. Under these conditions the impurity atmosphere is said to have "condensed" on the dislocation lines.

Interaction of dislocations with solute atoms is important for the explanation of yield-point behavior, strain aging, and solid-solution strengthening. These topics will be considered more fully in Chap. 6.

## 5-16 DISLOCATION PILE-UPS

Dislocations frequently pile up on slip planes at barriers such as grain boundaries, second phases, or sessile dislocations. The leading dislocation in the pile-up is acted on not only by the applied shear stress but also by the interaction force with the other dislocations in the pile-up. This leads to a high concentration of stress on the leading dislocation in the pile-up. When many dislocations are contained in the pile-up, the stress on the dislocation at the head of a pile-up can approach the theoretical shear stress of the crystal. This high stress either can initiate yielding on the other side of the barrier, or in other instances, it can nucleate a crack at the barrier. Dislocations piled up against a barrier produce a *back stress* acting to oppose the motion of additional dislocations along the slip plane in the slip direction.

The dislocations in the pile-up will be tightly packed together near the head of the array and more widely spaced toward the source (Fig. 5-28). The distribution of dislocations of like sign in a pile-up along a single slip plane has been studied by Eshelby, Frank, and Nabarro.[1] The number of dislocations that can occupy a distance $L$ along the slip plane between the source and the obstacle is

$$n = \frac{k\pi\tau_s L}{Gb} \qquad (5\text{-}37)$$

[1] J. D. Eshelby, F. C. Frank, and F. R. N. Nabarro, *Philos. Mag.*, vol. 42, p. 351, 1951; calculations for more complicated types of pile-ups have been given by A. K. Head, *Philos. Mag.*, vol. 4, pp. 295–302, 1959; experimental confirmation of theory has been obtained by Meakin and Wilsdorf, *op. cit.*, pp. 745–752.

**Fig. 5-28** Dislocation pile-up at an obstacle.

where $\tau_s$ is the average resolved shear stress in the slip plane and $k$ is a factor close to unity. For an edge dislocation $k = 1 - v$, while for a screw dislocation $k = 1$. When the source is located at the center of a grain of diameter $D$, the number of dislocations in the pile-up is given by

$$n = \frac{k\pi\tau_s D}{4Gb} \qquad (5\text{-}38)$$

The factor 4 is used instead of the expected factor of 2 because the back stress on the source arises from dislocations piled up on both sides of the source.

A piled-up array of $n$ dislocations can be considered for many purposes to be a giant dislocation with Burgers vector $nb$. At large distances from the array the stress due to the dislocations can be considered to be due to a dislocation of strength $nb$ located at the center of gravity three-quarters of the distance from the source to the head of the pile-up. The total slip produced by a pile-up can be considered that due to a single dislocation $nb$ moving a distance $3L/4$. Very high forces act on the dislocations at the head of the pile-up. This force is equal to $nb\tau_s$, where $\tau_s$ is the average resolved shear stress on the slip plane. Koehler[1] has pointed out that large tensile stresses of the order of $n\tau$ will be produced at the head of a pile-up. Stroh[2] has made a somewhat more detailed analysis of the stress distribution at the head of a dislocation pile-up. Using the coordinate system given in Fig. 5-28, he showed that the tensile stress normal to a line $OP$ is given by

$$\sigma = \frac{3}{2}\left(\frac{L}{r}\right)^{1/2} \tau_s \sin\theta \cos\frac{\theta}{2} \qquad (5\text{-}39)$$

The maximum value of $\sigma$ occurs at $\cos\theta = \frac{1}{3}$ or $\theta = 70.5°$. For this situation

$$\sigma_{max} = \frac{2}{\sqrt{3}}\left(\frac{L}{r}\right)^{1/2} \tau_s \qquad (5\text{-}40)$$

[1] J. S. Koehler. *Phys. Rev.*, vol. 85, p. 480, 1952.
[2] A. N. Stroh, *Proc. Roy. Soc. (London)*, vol. 223, pp. 404–414, 1954.

The shear stress acting in the plane *OP* is given by

$$\tau = \beta\tau_s \left(\frac{L}{r}\right)^{\frac{1}{2}} \qquad (5\text{-}41)$$

where $\beta$ is an orientation-dependent factor which is close to unity.

The number of dislocations which can be supported by an obstacle will depend on the type of barrier, the orientation relationship between the slip plane and the structural features at the barrier, the material, and the temperature. Breakdown of a barrier can occur by slip on a new plane, by climb of dislocations around the barrier, or by the generation of high enough tensile stresses to produce a crack.

The equations describing dislocation pile-ups are obtained using the concept of continuous dislocations. In this method of calculation[1] discrete dislocations with finite Burgers vectors are replaced by continuously distributed dislocations with infinitesimal Burgers vectors. This concept provides a link between the discrete nature of dislocations and continuum theory. It is particularly useful in dealing with cracks and fracture.

## BIBLIOGRAPHY

Cottrell, A. H.: "Dislocations and Plastic Flow in Crystals," Oxford University Press, New York, 1953.

Friedel, J.: "Dislocations," Addison-Wesley Publishing Company, Inc., Reading, Mass., 1964.

Hirth, J. P., and J. Lothe: "Theory of Dislocations," McGraw-Hill Book Company, New York, 1968.

Hull, D.: "Introduction to Dislocations," 2d ed., Pergamon Press, New York, 1975.

Nabarro, F. R. N.: "Theory of Crystal Dislocations," Oxford University Press, New York, 1967.

Read, W. T., Jr.: "Dislocations in Crystals," McGraw-Hill Book Company, New York, 1953.

Van Bueren, H. G.: "Imperfections in Crystals," Interscience Publishers, Inc., New York, 1960.

Weertman, J., and J. R. Weertman: "Elementary Dislocation Theory," The Macmillan Company, New York, 1964.

---

[1] R. Bullough, *Philos. Mag.*, vol. 9, p. 917, 1964.

# 6

## STRENGTHENING MECHANISMS

### 6-1 INTRODUCTION

Chapter 4 considered the plastic deformation of single crystals in terms of the movement of dislocations and the basic deformation mechanisms of slip and twinning. Single-crystal specimens represent the most ideal condition for study. The simplifications which result from the single-crystal condition materially assist in describing the deformation behavior in terms of crystallography and defect structure. However, with the exception of solid-state electronic devices, single crystals are rarely used for engineering applications because of limitations involving their strength, size, and production.[1] Commercial metal products invariably are made up of a tremendous number of individual crystals or grains. The individual grains of the polycrystalline aggregate do not deform in accordance with the relatively simple laws which describe plastic deformation in single crystals because of the restraining effect of the surrounding grains.

Chapter 5 was concerned with the basic relationships that govern dislocation

---

[1] An outstanding exception is the use of single-crystal turbine blades for jet engines where the elimination of grain boundaries greatly increases the resistance to thermal shock and fatigue. See F. L. Versnyder and M. E. Shank, *Mater. Sci. Eng.*, vol. 6, pp. 213–243, 1970.

behavior. It should be clear from this that strength is inversely related to dislocation mobility and that even in high-purity single crystals there are a number of possible factors that can affect the strength and mechanical behavior. Thus, the crystal structure determines the number and type of slip systems, fixes the Burgers vector, and determines the lattice friction stress (Peierls stress) which sets the base strength level and temperature dependence of strength. In close-packed structures the stacking-fault energy determines the extent of dislocation dissociation, which influences the ease of cross slip and the subsequent strain-hardening rate. The purity and method of preparation determine the initial dislocation density and substructure. These limited variables introduce enough complexity that mechanical behavior cannot in general be predicted with high precision as a function of strain, strain rate, temperature, and stress rate.

However, the introduction of still greater complexity is needed to produce materials of highest strength and usefulness. Thus, fine grain size is often desired for high strength, large additions of solute atoms are added to increase strength and bring about new phase relationships, fine particles may be added to increase strength and phase transformations may be utilized to increase strength. These various strengthening mechanisms and other aspects of the deformation of polycrystalline materials are considered in this chapter.

## 6-2 GRAIN BOUNDARIES AND DEFORMATION

The boundaries between grains in a polycrystalline aggregate are a region of disturbed lattice only a few atomic diameters wide. In the general case, the crystal-lographic orientation changes abruptly in passing from one grain to the next across the grain boundary. The ordinary high-angle grain boundary represents a region of random misfit between the adjoining crystal lattices.[1] As the difference in orientation between the grains on each side of the boundary decreases, the state of order in the boundary increases. For the limiting case of a low-angle boundary where the orientation difference across the boundary may be less than 1° (see Sec. 6-4), the boundary is composed of a regular array of dislocations.

Ordinary high-angle grain boundaries are boundaries of rather high surface energy. For example, a grain boundary in copper has an interfacial surface energy of about 600 erg/cm², while the energy of a twin boundary is only about 25 erg/cm². Because of their high energy, grain boundaries serve as preferential sites for solid-state reactions such as diffusion, phase transformations, and precipitation reactions. An important point to consider is that the high energy of a grain bound-ary usually results in a higher concentration of solute atoms at the boundary than

---

[1] For a review of the proposed models of grain boundaries see D. McLean, "Grain Boundaries in Metals," chap. 2, Oxford University Press, New York, 1957.

in the interior of the grain. This makes it difficult to separate the pure mechanical effect of grain boundaries on properties from an effect due to impurity segregation.

When a single crystal is deformed in tension, it is usually free to deform on a single slip system for a good part of the deformation and to change its orientation by lattice rotation as extension takes place. However, individual grains in a polycrystalline specimen are not subjected to a single uniaxial stress system when the specimen is deformed in tension. In a polycrystal, continuity must be maintained, so that the boundaries between the deforming crystals remain intact. Although each grain tries to deform homogeneously in conformity with the deformation of the specimen as a whole, the constraints imposed by continuity cause considerable differences in the deformation between neighboring grains and within each grain. Studies of the deformation in coarse-grain aluminum[1] showed that the strain in the vicinity of a grain boundary usually differs markedly from the strain at the center of the grain. Although the strain is continuous across boundaries, there may be a steep strain gradient in this region. As the grain size decreases and strain increases, the deformation becomes more homogeneous. Because of the constraints imposed by the grain boundaries slip occurs on several systems, even at low strains. Also, this causes slip to occur on non-close-packed planes in regions near the grain boundary. Slip in polycrystalline aluminum has been observed on the {100}, {110}, and {113} planes. The fact that different slip systems can operate in adjacent regions of the same grain results in complex lattice rotations which result in the formation of deformation bands. Since more slip systems are usually operative near the grain boundary, the hardness usually will be higher near the boundary than in the center of a grain. As the grain diameter is reduced more of the effects of grain boundaries will be felt at the grain center. Thus, the strain hardening of a fine grain size metal will be greater than in a coarse-grain polycrystalline aggregate.

Von Mises[2] showed that for a crystal to undergo a general change of shape by slip requires the operation of five independent slip systems. This arises from the fact that an arbitrary deformation is specified by the six components of the strain tensor, but because of the requirement of constant volume ($\Delta V = 0 = \varepsilon_{11} + \varepsilon_{22} + \varepsilon_{33}$), there are only five independent strain components. Crystals which do not possess five independent slip systems are never ductile in polycrystalline form, although small plastic elongation may be obtained if there is twinning or a favorable preferred orientation. Cubic metals easily satisfy this requirement, which accounts for their general high ductility. Hexagonal close-packed and other low-symmetry metals do not satisfy this requirement and have low ductility at room temperature in polycrystalline form. Polycrystalline Zn and Mg become ductile at some elevated temperature at which npnbasal slip can become operative and increase the number of slip systems to at least five.

[1] W. Boas and M. E. Hargreaves, *Proc. R. Soc. London Ser. A*, vol. A193, p. 89, 1948; V. M. Urie and H. L. Wain, *J. Inst. Met.*, vol. 81, p. 153, 1952.
[2] R. Von Mises, *Z. Angew. Math. Mech.*, vol. 8, p. 161. 1928.

Kelly[1] has pointed out that it appears as if the possession of five independent slip systems is a necessary but not sufficient condition for polycrystalline ductility. The other requirement is *slip flexibility*, the ability for the five independent slip systems to operate simultaneously in a small volume of the metal of the order of a cubic micron. The basic requirement for slip flexibility is that the dislocations must be able to cross slip easily and the slip bands must be able to interpenetrate each other. An example where slip flexibility is lacking is in bcc FeCo-2% V alloys. Although there are more than five independent slip systems, when the alloy becomes ordered cross slip cannot occur and the ductility is reduced greatly. Magnesium oxide represents a good example of how many requirements need to be met to achieve ductility in polycrystalline aggregate of limited ductility. Below 350°C slip occurs on $\{110\}\langle 1\bar{1}0\rangle$, but since this provides only two independent slip systems the material is brittle. Above 350°C slip also occurs on $\{001\}$ and there are five independent slip systems. However, the shear stress to cause slip on $\{001\}$ is much larger than the stress required to produce $\{110\}$ slip, so that the fracture stress is reached easily and only about 1 per cent extension is achieved before fracture. The ratio $\tau_{001}/\tau_{110}$ decreases with increasing temperature and becomes unity at 1500°C. Now all conditions for ductility are satisfied with the exception that the slip bands cannot interpenetrate during cross slip. Only at 1700°C can the slip bands interpenetrate and the polycrystalline MgO becomes fully ductile.

At temperatures above about one-half of the melting point, deformation can occur by sliding along the grain boundaries. *Grain-boundary sliding* becomes more prominent with increased temperature and decreasing strain rate, as in creep. The restriction of deformation to the grain-boundary region is one of the primary sources of high-temperature fracture. Because impurities tend to segregate to grain boundaries, intergrannular fracture is strongly influenced by composition. A rough way of distinguishing when grain-boundary sliding becomes prominent is with the *equicohesive temperature*. Above this temperature the grain-boundary region is weaker than the grain interior and strength increases with increasing grain size. Below the equicohesive temperature the grain-boundary region is stronger than the grain interior and strength increases with decreasing grain size (increasing grain-boundary area).

## 6-3 STRENGTHENING FROM GRAIN BOUNDARIES

Direct evidence for the mechanical strengthening of grain boundaries is provided by experiments[2] on bicrystals in which the orientation difference between a longitudinal grain boundary was varied in a systematic manner. The yield stress of the

[1] A. Kelly, "Strong Solids," pp. 82–85, Oxford University Press, New York, 1966.
[2] B. Chalmers, *Proc. R. Soc. London Ser. A*, vol. A193, p. 89, 1948; R. Clark and B. Chalmers, *Acta. Metall.*, vol. 2, p. 80, 1954.

bicrystals increased linearly with increasing misorientation across the grain boundary, and extrapolation to zero misorientation angle gave a value close to that of the yield stress of a single crystal. These results imply that a simple grain boundary has little inherent strength and that the strengthening due to grain boundaries results from mutual interference to slip within the grains.

Several attempts have been made to calculate the stress-strain curve for a polycrystal from stress-strain curves for single crystals. In Chap. 4 we saw that the resolved shear stress in a single crystal was given by

$$\tau = \sigma \sin \chi \cos \lambda = \frac{\sigma}{M} \qquad (6\text{-}1)$$

where $M$ is an orientation factor (the reciprocal of the Schmid factor). For a polycrystal the orientation factor $M$ varies from grain to grain and it is necessary to determine some average orientation factor $\overline{M}$. The best estimate for an fcc lattice is $\overline{M} = 3.1$, obtained by G. I. Taylor[1] based on the use of the von Mises compatibility condition and assuming that all grains undergo the same deformation as the overall deformation. The energy expended in deforming a polycrystal must be equal to the sum of the increments of work performed on each of the $n$ slip systems.

$$\sigma \, d\varepsilon = \sum_{i=1}^{n} \tau_i \, d\gamma_i \qquad (6\text{-}2)$$

If we assume that the critical shear stress is the same on each slip system, then

$$\frac{\sigma}{\tau} = \frac{\displaystyle\sum_{i=1}^{n} |d\gamma_i|}{d\varepsilon} = \overline{M} \qquad (6\text{-}3)$$

The value of $\overline{M}$ was determined[2] by finding the combination of slip systems that minimized the value of $\Sigma |d\gamma_i|$ but still satisfied the required continuity at the grain boundaries. To get reasonable agreement, single-crystal curves must be used which involve the same slip mechanisms as the polycrystalline specimen. Since polycrystals involve multiple slip, the single-crystal curves for fcc metals should have the $\langle 111 \rangle$ or $\langle 100 \rangle$ orientation for which easy glide is a minimum. The polycrystalline curve is calculated from the relations

$$\sigma = \overline{M}\tau \qquad \text{and} \qquad \varepsilon = \frac{\gamma}{\overline{M}} \qquad (6\text{-}4)$$

---

[1] G. J. Taylor, *J. Inst. Met.*, vol. 62, p. 307, 1938.

[2] Essentially the same value of $\overline{M} = 3.1$ was found by a more rigorous treatment. J. F. W. Bishop and R. Hill, *Philos. Mag.*, vol. 42, pp. 414–427, 1298–1307, 1951. The Taylor and Bishop-Hill approaches have been shown to be equivalent and have been generalized by G. Y. Chin and W. L. Mammel, *Trans. Metall. Soc. AIME*, vol. 245, pp. 1211–1214, 1969.

Quite reasonable agreement has been found.[1] By combining these equations we find that the strain-hardening rate for an fcc polycrystal should be about 9.5 times that for a single crystal.

$$\frac{d\sigma}{d\varepsilon} = \overline{M}^2 \frac{d\tau}{d\gamma} \qquad (6\text{-}5)$$

Up to this point we primarily have emphasized the role of grain boundaries in causing nonhomogeneous deformation and in introducing multiple slip. It was empirically established by Hall[2] and Petch[3] that the tensile yield stress was related to grain size by

$$\sigma_0 = \sigma_i + k'D^{-\frac{1}{2}} \qquad (6\text{-}6)$$

when

$\sigma_0 =$ the yield stress

$\sigma_i =$ friction stress opposing motion of dislocation

$k' =$ "'unpinning constant," measuring the extent to which dislocations are piled up at barriers

$D =$ grain diameter

The *Hall-Petch equation* has been found to express the grain-size dependence of the flow stress at any plastic strain out to ductile fracture. It also has been found to express the variation of brittle fracture stress with grain size and the dependence of fatigue strength on grain size.[4]

Two models have evolved for explaining the dependence of resistance to plastic deformation on grain size.[5] The first is based on the concept that grain boundaries act as barriers to dislocation motion (Sec. 5-16). Figure 6-1 shows a transmission electron micrograph of dislocations piled up at a grain boundary. In order for the shear displacements associated with the impinging slip bands to be accommodated plastically in the adjacent grain, the stress concentration at or near the grain boundary must be sufficient to nucleate slip in the next grain. From Eq. (5-41) the shear stress on a slip plane at a distance $r$ on the other side of the barrier (Fig. 5-28) is $\tau = \tau_s(L/r)^{1/2}$. We can take $r$ to be the distance from the head of the pile-up to the nearest dislocation source in the next grain, and $L$ to be equal to the grain diameter $D$. The applied shear stress $\tau_s$ must overcome the frictional resistance to slip due to interaction with fine precipitates, impurity atoms, or

[1] U. F. Kocks, *Acta Metall.*, vol. 6, p. 85, 1958; for a detailed analysis see U. F. Kocks, *Metall. Trans.*, vol. 1, pp. 1121–1143, 1970.

[2] E. O. Hall, *Proc. Phys. Soc. London*, vol. 643, p. 747, 1951.

[3] N. J. Petch, *J. Iron Steel Inst. London*, vol. 173, p. 25, 1953.

[4] R. W. Armstrong, *Metall. Trans.*, vol. 1, pp. 1169–1176, 1970.

[5] For a detailed review of the development of these models see J. C. M. Li and Y. T. Chou, *Metall. Trans.*, vol. 1, pp. 1145–1159, 1970; J. P. Hirth, *Metall, Trans.*, vol. 3, pp. 3047–3067, 1972.

**Fig. 6-1** Dislocations piled up against a grain boundary as observed with the electron microscope in a thin foil of stainless steel (17,500X). (*From M. J. Whelan, P. B. Hirsch, R. W. Horne, and W. Bollman, Proc. R. Soc., London, vol. 240A, p. 524, 1957.*)

lattice defects on the slip plane in the first grain. Thus, when yielding occurs $\tau_s = \tau_0$ and

$$(\tau_0 - \tau_i)\left(\frac{D}{r}\right)^{1/2} = \tau_d$$

where $\tau_d$ is the shear stress required to nucleate slip in the adjacent grain. Expressing in terms of normal stresses, and assuming $\tau = \sigma/2$

$$\sigma_0 = \sigma_i + 2\tau_d\left(\frac{r}{D}\right)^{1/2} = \sigma_i + k'D^{-1/2} \qquad (6\text{-}7)$$

While the Hall-Petch equation is a very general relationship, it must be used with some caution. For example, if Eq. (6-7) were extrapolated to the smallest grain size imaginable (approximately 40 Å), it would predict strength levels close to the theoretical shear strength. Such an extrapolation is in error because the equations for the stresses in a pile-up on which Eq. (6-7) is based were derived for large pile-ups containing more than 50 dislocations. For small pile-ups other equations must be considered.[1]

R. W. Amstrong, Y. T. Chou, R. A. Fisher, and N. Louat, *Philos. Mag.*, vol. 14, p. 943, 1966.

The factor $k'$ is the slope of the straight line that is obtained when $\sigma_0$ is plotted against $D^{-1/2}$. It was originally considered that $k'$ was related to the unpinning of a dislocation source that was locked by interaction with solute atoms. However, for this mechanism $\tau_d$ should vary with temperature, but experimentally it is found that $k'$ is independent of temperature. Also, $k'$ varies considerably for different bcc metals, which would not be expected if it measured the pinning of dislocation sources. An alternative mechanism for yield propagation[1] is that the stress concentration at the head of the pile-up is high enough so that dislocations are created at the boundary in the new grain. Transmission electron microscopy supports the presence of grain-boundary sources and theory[2] shows that it is possible to create dislocations at atomic ledges in grain boundaries. Such a mechanism will not be strongly temperature-dependent.

The term $\sigma_i$ is the intercept[3] along the ordinate in a plot of $\sigma_0$ versus $D^{-1/2}$. It is usually interpreted as the friction stress needed to move unlocked dislocations along the slip plane. This term depends strongly on temperature, strain, and alloy (impurity) content.

The second model for the grain-size dependence of flow stress does not require the presence of dislocation pile-ups at grain boundaries. This model avoids the description of the stresses at grain boundaries and instead concentrates on the influence of grain size on the dislocation density, and hence, on the yield or flow stress. The flow stress is given in terms of dislocation density by

$$\sigma_0 = \sigma_i + \alpha Gb\rho^{1/2} \qquad (6\text{-}8)$$

where $\sigma_i$ has the same meaning as in Eq. (6-6), $\alpha$ is a numerical constant generally between 0.3 and 0.6, and $\rho$ is the dislocation density. The justification for this equation was given in Sec. 4-14. The tie-in with grain size is based on the experimental observation[4] that $\rho$ is an inverse function of the grain size. Thus, $\rho = 1/D$

$$\sigma_0 = \sigma_i + \alpha GbD^{-1/2} = \sigma_i + k'D^{-1/2} \qquad (6\text{-}9)$$

Grain size is measured with a light microscope by counting the number of grains within a given area, by determining the number of grains (or grain boundaries) that intersect a given length of random line, or by comparing with standard-grain-size charts. Most grain-size measurements invoke assumptions relative to the shape and size distribution of the grains and, therefore, must be interpreted

[1] A. H. Cottrell, "The Relation Between Structure and Mechanical Properties of Metals," vol. II, p. 455, Her Majesty's Stationery Office, London, 1963.

[2] J. C. M. Li, *Trans. Metall. Soc. AIME*, vol. 227, pp. 239–247, 1963.

[3] The various techniques used to determine $\sigma_i$, are discussed by R. Phillips and J. F. Chapman, *J. Iron Steel Inst. London*, vol. 203, pp. 511–513, 1965.

[4] D. G. Brandon and J. Nutting, *J. Iron Steel Inst.*, vol. 196, p. 160, 1960; A. H. Keh and S. Weissman, in G. Thomas and J. Washburn (eds.), " Electron Microscopy and Strength of Crystals," p. 231, John Wiley & Sons, Inc., New York, 1963.

with some degree of caution. As pointed out by DeHoff and Rhines,[1] the most applicable technique is that which provides structural information which may be correlated with property data and which may be accomplished by relatively simple measurements on the polished surface.

Most grain-size measurements seek to correlate the interaction of grain boundaries with a specific mechanical property. Thus, a measurement of the grain-boundary area per unit volume $S_v$ is a useful parameter. Smith and Guttman[2] have shown that $S_v$ can be calculated without assumptions concerning grain shape and size distribution from measurements of the mean number of intercepts of random test lines with grain boundaries per unit length of test line $N_L$.

$$S_v = 2N_L \qquad (6\text{-}10)$$

If a mean grain diameter $D$ is required from $S_v$, this may be obtained by assuming constant-size spherical grains and noting that each boundary is shared by two adjacent grains

$$2S_v = \frac{4\pi(D/2)^2}{4\pi/3(D/2)^3}$$

or

$$D = \frac{3}{S_v} = \frac{3}{2N_L} \qquad (6\text{-}11)$$

An average grain size may also be obtained from measurements of the number of grains per unit area on a polished surface $N_A$. Fullman[3] has shown that the mean area on a plane of polish passing through spheres of constant size $A$ is

$$A = \frac{2}{3}\pi\left(\frac{D}{2}\right)^2 = \frac{\pi}{6}D^2 \qquad (6\text{-}12)$$

Thus, measurements of $N_A$, the reciprocal of $A$, give a grain size defined as

$$D = \sqrt{\frac{6}{\pi N_A}} \qquad (6\text{-}13)$$

Many studies have employed the mean intercept length of random test lines as a measure of grain size. This determination is made by dividing the total length of test line by the number of grains traversed. By comparison with Eq. (6-11), it can be seen that the "mean intercept length" grain size will be somewhat smaller than the actual grain size.

A very common method of measuring grain size in the United States is to compare the grains at a fixed magnification with the American Society for Testing

[1] R. T. DeHoff and F. N. Rhines, "Quantitative Microscopy," pp. 201–266, McGraw-Hill Book Company, 1968.
[2] C. S. Smith and L. Guttman, *Trans. AIME*, vol. 197, p. 81, 1953.
[3] R. L. Fullman, *Trans. AIME*, vol. 197, pp. 447 and 1267, 1953.

Table 6-1  COMPARISON OF GRAIN-SIZE MEASURING SYSTEMS†

| ASTM no. | Grains/in.$^2$ at 100× | Grains/mm$^2$ | Grains/mm$^3$ | Average Grain Diameter, mm |
|---|---|---|---|---|
| −3 | 0.06 | 1 | 0.7 | 1.00 |
| −2 | 0.12 | 2 | 2 | 0.75 |
| −1 | 0.25 | 4 | 5.6 | 0.50 |
| 0 | 0.5 | 8 | 16 | 0.35 |
| 1 | 1 | 16 | 45 | 0.25 |
| 2 | 2 | 32 | 128 | 0.18 |
| 3 | 4 | 64 | 360 | 0.125 |
| 4 | 8 | 128 | 1,020 | 0.091 |
| 5 | 16 | 256 | 2,900 | 0.062 |
| 6 | 32 | 512 | 8,200 | 0.044 |
| 7 | 64 | 1,024 | 23,000 | 0.032 |
| 8 | 128 | 2,048 | 65,000 | 0.022 |
| 9 | 256 | 4,096 | 185,000 | 0.016 |
| 10 | 512 | 8,200 | 520,000 | 0.011 |
| 11 | 1,024 | 16,400 | 1,500,000 | 0.008 |
| 12 | 2,048 | 32,800 | 4,200,000 | 0.006 |

† "Metals Handbook, American Society for Metals, Metals Park, Ohio, 1948.

and Materials (ASTM) grain-size charts. The ASTM grain-size number $n$ is related to $N^*$, the number of grains per square inch at a magnification of 100 × by the relationship

$$N^* = 2^{n-1} \qquad (6\text{-}14)$$

Table 6-1 compares the ASTM grain-size numbers with several other useful measures of grain size.

## 6-4  LOW-ANGLE GRAIN BOUNDARIES

A definite substructure can exist within the grains surrounded by high-energy grain boundaries. The subgrains are low-angle boundaries in which the difference in orientation across the boundary may be only a few minutes of arc or, at most, a few degrees. Because of this small orientation difference, special x-ray techniques are required to detect the existence of a substructure network. Subgrain boundaries are lower-energy boundaries than grain boundaries, and therefore they etch less readily than grain boundaries. However, in many metals they can be detected in the microstructure by metallographic procedures (Fig. 6-2).

A low-angle boundary contains a relatively simple arrangement of dislocations. The simplest situation is the case of a tilt boundary. Figure 6-3a illustrates two cubic crystals with a common [001] axis. The slight difference in orientation

**Fig. 6-2** Substructure network in Fe-3% Si alloy (250X).

between the grains is indicated by the angle $\theta$. In Fig. 6-3b the two crystals have been joined to form a bicrystal containing a low-angle boundary. Along the boundary the atoms adjust their position by localized deformation to produce a smooth transition from one grain to the other. However, elastic deformation cannot accommodate all the misfit, so that some of the atom planes must end on the grain boundary. Where the atom planes end, there is an edge dislocation. Therefore, low-angle tilt boundaries can be considered to be an array of edge dislocations. From the geometry of Fig. 6-3b the relationship between $\theta$ and the spacing between dislocations is given by

$$\theta = 2\tan^{-1}\frac{b}{2D} \approx \frac{b}{D} \qquad (6\text{-}15)$$

where $b$ is the magnitude of the Burgers vector of the lattice.

The validity of the dislocation model of the low-angle boundary is found in the fact it is possible to calculate the grain-boundary energy as a function of the difference in orientation between the two grains. So long as the angle does not become greater than about 20°, good agreement is obtained between the measured values of grain-boundary energy and the values calculated on the basis of the dislocation model. Other evidence for the dislocation nature of low-angle bound-

**Fig. 6-3** Diagram of low-angle grain boundary. (*a*) Two grains having a common [001] axis and angular difference in orientation of $\theta$; (*b*) two grains joined together to form a low-angle grain boundary made up of an array of edge dislocations. (*From W. T. Read, Jr. " Dislocations in Crystals," p. 157, McGraw-Hill Book Company, New York, 1953.*)

aries comes from metallographic observations. If the angle is low, so that the spacing between dislocations is large, it is often possible to observe that the boundary is composed of a row of etch pits, with each pit corresponding to the site of an edge dislocation (Fig. 6-4).

Subboundaries or low-angle boundaries can be produced in a number of ways.[1] They may be produced during crystal growth, during high-temperature creep deformation, or as the result of a phase transformation. The veining in ferrite grains is a well-known example of a substructure resulting from the internal stresses accompanying a phase transformation. Perhaps the most general method of producing a substructure network is by introducing a small amount of deformation (from about 1 to 10 percent prestrain) and following this with an annealing treat-

[1] R. W. Cahn, "Impurities and Imperfections," American Society for Metals, Metals Park, Ohio, 1955.

**Fig. 6-4** Etch-pit structures along low-angle grain boundaries in Fe-Si alloy (1,000X).

ment to rearrange the dislocations into subgrain boundaries. The amount of deformation and temperature must be low enough to prevent the formation of new grains by recrystallization. This process has been called recrystallization *in situ*, or polygonization.

The term *polygonization* was used originally to describe the situation that occurs when a single crystal is bent to a relatively small radius of curvature and then annealed. Bending results in the introduction of an excess number of dislocations of one sign. These dislocations are distributed along the bent-glide planes as shown in Fig. 6-5a. When the crystal is heated, the dislocations group themselves into the lower-energy configuration of a low-angle boundary by dislocation climb.

(a)  (b)

**Fig. 6-5** Movement of dislocations to produce polygonization (schematic).

**Fig. 6-6** Effect of density of subboundaries on yield stress. (*From E. R. Parker and T. H. Hazlett, "Relation of Properties to Microstructure," American Society of Metals, Metals Park, Ohio, 1954. By permission of the publishers.*)

The resulting structure is a polygonlike network of low-angle grain boundaries (Fig. 6-5*b*).

Since low-angle boundaries consist of simple dislocation arrays, a study of their properties should provide valuable information on dislocation behavior. Parker and Washburn[1] demonstrated that a low-angle boundary moves as a unit when subjected to a shear stress, in complete agreement with what would be expected for a linear dislocation array. It has also been found that the boundary angle decreases with increasing distance of shear. This means that the boundary loses dislocations as it moves, a fact which would be expected if dislocations were held up at imperfections such as foreign atoms, precipitated particles, and other dislocations.

The formation of subgrains in an annealed material results in a significant increase in strength. Figure 6-6 shows the increase in the yield stress in nickel due to an increase in the density of subgrain boundaries produced by various prestraining and annealing treatments. The fact that the curves for pure nickel and alloyed nickel are nearly parallel indicates that the strengthening due to substructure is additive to that produced by solid-solution hardening. The effect of a substructure of low-angle grain boundaries on the stress-strain curve of 1020 steel is shown in Fig. 6-7. Note that the material that was cold-reduced and annealed, so as to produce a substructure, has a higher yield point and tensile strength than both the annealed material and the material which was only cold-reduced. Moreover, the ductility of the material containing a substructure is almost as good as the ductility of the annealed steel.

---

[1] E. R. Parker and J. Washburn, *Trans. Metall. Soc. AIME*, vol. 194, pp. 1076–1078, 1952.

**Fig. 6-7** Effect of a substructure of low-angle grain boundaries on the stress-strain curve of SAE 1020 steel. (*From E. R. Parker and J. Washburn, " Impurities and Imperfections," p. 155, American Society for Metals, Metals Park, Ohio, 1955. By permission of the publishers.*)

## 6-5 YIELD-POINT PHENOMENON

Many metals, particularly low-carbon steel, show a localized, heterogeneous type of transition from elastic to plastic deformation which produces a yield point in the stress-strain curve. Rather than having a flow curve with a gradual transition from elastic to plastic behavior, such as was shown in Fig. 3-1, metals with a yield point have a flow curve or, what is equivalent, a load-elongation diagram similar to Fig. 6-8. The load increases steadily with elastic strain, drops suddenly, fluctuates about some approximately constant value of load, and then rises with further strain. The load at which the sudden drop occurs is called the *upper yield point*. The constant load is called the *lower yield point*, and the elongation which occurs at constant load is called the *yield-point elongation*. The deformation occurring throughout the yield-point elongation is heterogeneous. At the upper yield point a discrete band of deformed metal, often readily visible with the eye, appears at a stress concentration such as a fillet, and coincident with the formation of the band the load drops to the lower yield point. The band then propagates along the length of the specimen, causing the yield-point elongation. In the usual case several bands will form at several points of stress concentration. These bands

**Fig. 6-8**  Typical yield-point behavior.

are generally at approximately 45° to the tensile axis. They are usually called *Lüders bands*, Hartmann lines, or stretcher strains, and this type of deformation is sometimes referred to as the Piobert effect. When several Lüders bands are formed, the flow curve during the yield-point elongation will be irregular, each jog corresponding to the formation of a new Lüders band. After the Lüders bands have propagated to cover the entire length of the specimen test section, the flow will increase with strain in the usual manner. This marks the end of the yield-point elongation.

The yield-point phenomenon was found originally in low-carbon steel. A pronounced upper and lower yield point and a yield-point elongation of over 10 percent can be obtained with this material under proper conditions. More recently the yield point has come to be accepted as a general phenomenon, since it has been observed in a number of other metals and alloys. In addition to iron and steel, yield points have been observed in polycrystalline molybdenum, titanium, and aluminum alloys and in single crystals of iron, cadmium, zinc, alpha and beta brass, and aluminum. Usually the yield point can be associated with small amounts of interstitial or substitutional impurities. For example, it has been shown[1] that almost complete removal of carbon and nitrogen from low-carbon steel by wet-hydrogen treatment will remove the yield point. However, only about 0.001 percent of either of these elements is required for a reappearance of the yield point.

A number of experimental factors affect the attainment of a sharp upper yield point. A sharp upper yield point is promoted by the use of an elastically rigid (hard) testing machine, very careful axial alignment of the specimen, the use of specimens free from stress concentrations, high rate of loading, and, frequently, testing at subambient temperatures. If, through careful avoidance of stress concentrations, the first Lüders band can be made to form at the middle of the test

[1] J. R. Low and M. Gensamer, *Trans. AIME*, vol. 158, p. 207, 1944.

specimen, the upper yield point can be roughly twice the lower yield point. However, it is more usual to obtain an upper yield point 10 to 20 percent greater than the lower yield point.

The onset of general yielding occurs at a stress where the average dislocation sources can create slip bands through a good volume of the material. Thus, the general yield stress can be expressed as

$$\sigma_0 = \sigma_s + \sigma_i \qquad (6\text{-}16)$$

where $\sigma_s$ is the stress to operate the dislocation sources and $\sigma_i$ is the friction stress representing the combining effect of all the obstacles to the motion of dislocations arising from the sources.

If the stress to operate the sources is high, then the initial yield stress is high. The explanation of the yield-point phenomenon in terms of dislocation behavior arose originally from the idea that the dislocation sources were locked or pinned by solute atom interactions (Sec. 5-15). The explanation[1] of this behavior was one of the early triumphs of dislocation theory. Carbon or nitrogen atoms in iron readily diffuse to the position of minimum energy just below the extra plane of atoms in a positive edge dislocation. The elastic interaction is so strong that the impurity atmosphere becomes completely saturated and condenses into a row of atoms along the core of the dislocation. The breakaway stress required to pull a dislocation line away from a line of solute atoms is

$$\sigma \approx \frac{A}{b^2 r_0^{\,2}} \qquad (6\text{-}17)$$

where $A$ is given by Eq. (5-35) and $r_0 \approx 2 \times 10^{-8}$ cm is the distance from the dislocation core to the line of solute atoms. When the dislocation line is pulled free from the influence of the solute atoms, slip can occur at a lower stress. Alternatively, where dislocations are strongly pinned, such as by carbon and nitrogen in iron, new dislocations must be generated to allow the flow stress to drop. This explains the origin of the upper yield stress (the drop in load after yielding has begun). The dislocations released into the slip plane pile up at grain boundaries. As discussed in Sec. 6-3, the pile-up produces a stress concentration at its tip which combines with the applied stress in the next grain to unlock sources (or create new dislocations) and in this way a Lüders band propagates over the specimen. The magnitude of the yield-point effect will depend on the interaction energy, Eq. (5-35), and the concentration of solute atoms at the dislocations, Eq. (5-36).

Although dislocation locking by interstitial atoms originally was advanced as a mechanism of the yield point, subsequent research showed that a yield-point

---

[1] A. H. Cottrell and B. A. Bilby, *Proc. Phys. Soc. London*, vol. 62A, pp. 49–62, 1949; see also E. O. Hall, "Yield Point Phenomena in Metals, Plenum Publishing Company, New York, 1970.

phenomenon was a very general behavior that was found in such diverse materials as LiF and Ge crystals and copper whiskers. In these materials the dislocation density is quite low and impurity pinning cannot explain the effect. A more general theory has been advanced[1] for all materials which exhibit a *yield drop*, i.e., where the stress falls rapidly once yielding begins. Impurity locking thus becomes a special case of yield-point behavior.

The relation between the strain rate imposed on the material by the test and the dislocation motion is given by

$$\dot{\varepsilon} = b\rho\bar{v} \qquad (6\text{-}18)$$

where $\rho$ is the density of mobile dislocations and $\bar{v}$ is the average dislocation velocity. The dislocation density increases with strain and $\bar{v}$ is a very strong function of stress

$$\bar{v} = \left(\frac{\tau}{\tau_0}\right)^{m'} \qquad (6\text{-}19)$$

where $\tau_0$ is the resolved shear stress corresponding to unit velocity. For materials with a low initial dislocation density (or with strongly pinned dislocations, as in iron) the only way that $b\rho\bar{v}$ can match the imposed strain rate is for $\bar{v}$ to be high. But, according to Eq. (6-19), this can only be achieved at a high stress. However, once some dislocations begin to move they begin to multiply and $\rho$ increases rapidly. Although this introduces some strain hardening, it is more than compensated for by the fact that $\bar{v}$ can drop and with it the stress needed to move the dislocations can drop. Thus, the stress required to deform the specimen decreases once yielding begins (yield drop). Finally, the increasing dislocation density produces increased strain hardening through dislocation interactions and the stress begins to increase with further strain.

According to this model the controlling parameters are the mobile dislocation density and the exponent describing the stress dependence of dislocation velocity, $m'$. From Eqs. (6-18) and (6-19), we can express conditions at the upper and lower yield points by

$$\frac{\tau_u}{\tau_L} = \left(\frac{\rho_L}{\rho_u}\right)^{1/m'} \qquad (6\text{-}20)$$

For small values of $m'$ ($m' < 15$), the ratio $\tau_u/\tau_L$ will be large and there will be a strong yield drop. For iron ($m' \approx 35$), the yield drop will only be substantial if $\rho_u$ is less than about $10^3$ cm$^{-2}$. Since the dislocation density of annealed iron is at least $10^6$ cm$^{-2}$, this requires that most of the dislocations must be pinned. Pinning

---

[1] W. G. Johnston and J. J. Gilman, *J. Appl. Phys.*, vol. 30, p. 129, 1959. The theory has been extended to cover yielding in iron and other bcc metals by G. T. Hahn, *Acta Metall.*, vol. 10, pp. 727–738, 1962.

can arise from the solute-dislocation interaction or by precipitation of fine carbides or nitrides along the dislocation. The yield point occurs as a result of unlocking the dislocations by a high stress, or for case of strong pinning, by creating new dislocations at the points of stress concentration.

## 6-6  STRAIN AGING

*Strain aging* is a type of behavior, usually associated with the yield-point phenomenon, in which the strength of a metal is increased and the ductility is decreased on heating at a relatively low temperature after cold-working. This behavior can best be illustrated by considering Fig. 6-9, which schematically describes the effect of strain aging on the flow curve of a low-carbon steel. Region $A$ of Fig. 6-9 shows the stress-strain curve for a low-carbon steel strained plastically through the yield-point elongation to a strain corresponding to point $X$. The specimen is then unloaded and retested without appreciable delay or any heat treatment (region $B$). Note that on reloading the yield point does not occur, since the dislocations have been torn away from the atmosphere of carbon and nitrogen atoms. Consider now that the specimen is strained to point $Y$ and unloaded. If it is reloaded after aging for several days at room temperature or several hours at an aging temperature like 300°F, the yield point will reappear. Moreover, the yield point will be increased by the aging treatment from $Y$ to $Z$. The reappearance of the yield point is due to

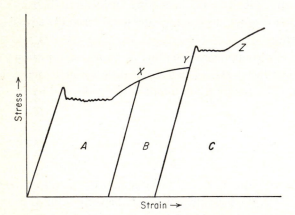

**Fig. 6-9** Stress-strain curves for low-carbon steel showing strain aging. Region $A$, original material strained through yield point. Region $B$, immediately retested after reaching point $X$. Region $C$, reappearance and increase in yield point after aging at 300°F.

the diffusion of carbon and nitrogen atoms to the dislocations during the aging period to form new atmospheres of interstitials anchoring the dislocations. Support for this mechanism is found in the fact that the activation energy for the return of the yield point on aging is in good agreement with the activation energy for the diffusion of carbon in alpha iron.

Nitrogen plays a more important role in the strain aging of iron than carbon because it has a higher solubility and diffusion coefficient and produces less complete precipitation during slow cooling. From a practical standpoint it is important to eliminate strain aging in deep-drawing steel because the reappearance of the yield point can lead to difficulties with surface markings or "stretcher strains" due to the localized heterogeneous deformation. To control strain aging, it is usually desirable to lower the amount of carbon and nitrogen in solution by adding elements which will take part of the interstitials out of solution in the form of stable carbides or nitrides. Aluminum, vanadium, titanium, columbium, and boron have been added for this purpose. While a certain amount of control over strain aging can be achieved, there is no commercial low-carbon steel which is completely non-strain aging. The usual industrial solution to this problem is to deform the metal to point $X$ by roller leveling or a skin-pass rolling operation and use it immediately before it can age. The local plastic deformation by rolling produces sufficient fresh dislocations so that subsequent plastic flow can occur without a yield point.

The occurrence of strain aging is a fairly general phenomenon in metals. In addition to the return of the yield point and an increase in the yield stress after aging, strain aging also produces a decrease in ductility and a low value of strain-rate sensitivity.[1] Strain aging also is associated with the occurrence of serrations in the stress-strain curve (discontinuous or repeated yielding). This dynamic strain-aging behavior (Fig. 6-10) is called the *Portevin-LeChatelier effect*. The serrations arise from successive yielding and aging while the specimen is tested. If the speed of a dislocation line is slow, it may be able to move by dragging its atmosphere of impurities along behind it. According to Cottrell, the maximum velocity at which this can occur is

$$v = \frac{DA}{kTr^2} \qquad (6\text{-}21)$$

where $A$ is given by Eq. (5-35) and $D$ is the diffusion coefficient.

At higher velocity the dislocation pulls away from the atmosphere and a yield drop occurs. Since solute atom mobility is high at the temperatures at which discontinuous yielding occurs, new atoms move to the dislocations and lock them.

---

[1] The strain-rate sensitivity is the change in stress required to produce a certain change in strain rate at constant temperature (see Chap. 9); also J. D. Lubahn, *Trans. Am. Soc. Met.*, vol. 44, pp. 643–666, 1952.

**Fig. 6-10**   Portevin-LeChatelier effect in iron (schematic).

The process is repeated many times, causing the serrations in the stress-strain curves.[1]

For plain carbon steel discontinuous yielding occurs in the temperature region of 450 to 700°F. This temperature region is known as the *blue brittle region* because steel heated in this temperature region shows a decreased tensile ductility and decreased notched-impact resistance. This temperature range is also the region in which steels show a minimum in strain-rate sensitivity and a maximum in the rate of strain aging. All these facts point to the realization that blue brittleness is not a separate phenomenon but is just an accelerated strain aging.

The phenomenon of strain aging should be distinguished from a process known as *quench aging*, which occurs in low-carbon steels. Quench aging is a type of true precipitation hardening that occurs on quenching from the temperature of maximum solubility of carbon and nitrogen in ferrite. Subsequent aging at room temperature, or somewhat above, produces an increase in hardness and yield stress, as in the age hardening of aluminum alloys. Plastic deformation is not necessary to produce quench aging.

## 6-7   SOLID-SOLUTION HARDENING

The introduction of solute atoms into solid solution in the solvent-atom lattice invariably produces an alloy which is stronger than the pure metal. There are two types of solid solutions. If the solute and solvent atoms are roughly similar in size, the solute atoms will occupy lattice points in the crystal lattice of the solvent

---

[1] Observations and theories of serrated yielding and the static yield point are reviewed by B. J. Brindley and P. J. Worthington, *Met. Rev.*, no. 145, *Met. Mater.*, vol. 4, no. 8, pp. 101–114, 1970.

atoms. This is called *substitutional solid solution*. If the solute atoms are much smaller than the solvent atoms, they occupy interstitial positions in the solvent lattice. Carbon, nitrogen, oxygen, hydrogen, and boron are the elements which commonly form *interstitial solid solutions*.

The factors which control the tendency for the formation of substitutional solid solutions have been uncovered chiefly through the work of Hume-Rothery. If the sizes of the two atoms, as approximately indicated by the lattice parameter, differ by less than 15 percent, the size factor is favorable for solid-solution formation. When the size factor is greater than 15 percent, the extent of solid solubility is usually restricted to less than 1 percent. Metals which do not have a strong chemical affinity for each other tend to form solid solutions, while metals which are far apart on the electromotive series tend to form intermetallic compounds. The relative valence of the solute and solvent also is important. The solubility of a metal with higher valence in a solvent of lower valence is more extensive than for the reverse situation. For example, zinc is much more soluble in copper than is copper in zinc. This relative-valence effect can be rationalized to a certain extent in terms of the electron-atom ratio.[1] For certain solvent metals, the limit of solubility occurs at approximately the same value of electron-atom ratio for solute atoms of different valence. Finally, for complete solid solubility over the entire range of composition the solute and solvent atoms must have the same crystal structure.

The acquisition of fundamental information about the causes of solid-solution hardening has been a slow process. Early studies[2] of the increase in hardness resulting from solid-solution additions showed that the hardness increase varies directly with the difference in the size of the solute and solvent atoms, or with the change in lattice parameter resulting from the solute addition. However, it is apparent that size factor alone cannot explain solid-solution hardening. An improvement in correlation of data[3] results when the relative valence of the solute and solvent are considered in addition to the lattice-parameter distortion. The importance of valence is shown in Fig. 6-11, where the yield stress of copper alloys of constant lattice parameter is plotted against the electron-atom ratio.[4] Further results[5] show that alloys with equal grain size, lattice parameter, and electron-atom ratio have the same initial yield stress, but the flow curves differ at larger strains.

Systemic studies of the effect of solid-solution alloying additions on the flow

---

[1] For example, an alloy of 30 atomic percent Zn in Cu has an electron-atom ratio of 1.3. $(3 \times 2) + (7 \times 1) = 13$ valence electrons per $3 + 7 = 10$ atoms.

[2] A. L. Norbury, *Trans. Faraday Soc.*, vol. 19, pp. 506–600, 1924; R. M. Brick, D. L. Martin, and R. P. Angier, *Trans. Am. Soc. Met.*, vol. 31, pp. 675–698, 1943; J. H. Frye and W. Hume-Rothery, *Proc. R. Soc. London*, vol. 181, pp. 1–14, 1942.

[3] J. E. Dorn, P. Pietrokowsky, and T. E. Tietz, *Trans. AIME*, vol. 188, pp. 933–943, 1950.

[4] W. R. Hibbard, Jr., *Trans. Metall. Soc. AIME*, vol. 212, pp. 1–5, 1958.

[5] N. G. Ainslie, R. W. Guard, and W. R. Hibbard, *Trans. Metall. Soc. AIME*, vol. 215, pp. 42–48, 1959.

**Fig. 6-11** Effect of electron-atom ratio on the yield stress of copper solid-solution alloys. (*From W. R. Hibbard, Jr., Trans. Metall. Soc. AIME, vol. 212, p. 3, 1958.*)

curve in tension have been made for iron,[1] copper,[2] aluminum,[3] and nickel.[4] For the case of iron the solid-solution-strengthened alloy is a power function of the alloy addition. For a given atomic percent of solute the increase in strength varies inversely with the limit of solubility.

The distribution of solute atoms in a solvent lattice is not usually completely random. There is growing evidence that solute atoms group preferentially at dislocations, stacking faults, low-angle boundaries, and grain boundaries. However, even in a perfect lattice the atoms would not be completely random. For a solid solution of $A$ and $B$ atoms, if $B$ atoms tend to group themselves preferentially around other $B$ atoms, the situation is called *clustering*. However, if a given $B$ atom is preferentially surrounded by $A$ atoms, the solid solution exhibits *short-range order*. The tendency for clustering or short-range order increases with increasing solute additions.

The usual result of solute additions is to raise the yield stress and the level of the stress-strain curve as a whole (Fig. 6-12). As was seen in Sec. 6-5, solute atoms also frequently produce a yield-point effect. Since solid-solution alloy additions affect the entire stress-strain curve, we are led to conclude that solute atoms have more influence on the frictional resistance to dislocation motion $\sigma_i$ than on the static locking of dislocations. Solute atoms fall into two broad categories[5] with respect to their relative strengthening effect. Those atoms which produce non-spherical distortions, such as most interstitial atoms, have a relative strengthening effect per unit concentration of about three times their shear modulus, while solute atoms which produce spherical distortion, such as substitutional atoms, have a relative strengthening of about $G/10$.

[1] C. E. Lacy and M. Gensamer, *Trans. Am. Soc. Met.*, vol. 32, pp. 88–110, 1944.
[2] R. S. French and W. R. Hibbard, Jr., *Trans. AIME*, vol. 188, pp. 53–58, 1950.
[3] Dorn, Pietrokowsky, and Tietz, *op. cit.*
[4] V. F. Zackay and T. H. Hazlett, *Acta Metall.*, vol. 1, pp. 624–628, 1953.
[5] R. L. Fleischer, Solid Solution Hardening in D. Peckner (ed.), "The Strengthening of Metals," Reinhold Publishing Corporation, New York, 1964.

**Fig. 6-12** Effect of solute alloy additions on stress-strain curve.

A number of factors must be considered as contributing to solid-solution strengthening. Various factors have been studied in certain systems, but as yet no coherent theory has evolved. The most important factors affecting solid-solution strengthening are:

1. *Relative size factor* $\varepsilon_a = 1/a(da/dc)$ where $a$ is the interatomic spacing of the alloy and $c$ is the atomic concentration of the solute. This factor leads to elastic interaction through Eq. (5-35).

2. *Relative modulus factor* $\varepsilon'_G = \varepsilon_G/(1 - \varepsilon_G/2)$ where $\varepsilon_G = 1/G(dG/dc)$ and $G$ is the shear modulus of the alloy. Fleischer[1] has shown that rate of change of shear stress with atomic concentration of solute plots linearly against $\varepsilon'_G - 3\varepsilon_a$ for a large number of dilute solid solutions of copper.

3. *Electrical interaction* Since the electron cloud tends to resist compression, it will tend to redistribute throughout the alloy. Electrons will tend to move from the region of compression at an edge dislocation to the region of tension, creating an electrical dipole. In a monovalent solvent the extra conduction electrons introduced by a polyvalent solute atom will tend to wander away, leaving an excess of positive charge at the impurity ion. Thus, a short-range electrostatic interaction will exist between solute atoms and dislocations. Although good calculations have not been made, it is likely[2] that the electrostatic interaction is greater than the 0.02 eV calculated for a divalent solute atom.

4. *Chemical interaction* (*Suzuki interaction*) The dissociation of dislocations in close-packed crystals into partial dislocations affects the periodic arrange-

[1] R. L. Fleischer, *Acta Metall.*, vol. 11, p. 203, 1963.
[2] P. A. Flinn, Solid Solution Strengthening in "Strengthening Mechanisms in Solids," p.21, American Society for Metals, Metals Park, Ohio, 1962.

ment of the lattice. In a fcc crystal the faulted regions have hcp stacking. Suzuki[1] observed that the change in free energy with solute concentration will not be the same in the matrix and the faulted regions, so that there will be an interaction between the extended dislocation and the solute atoms.

5. *Configurational interaction (Fisher effect)*  Solid solutions are rarely truly random and usually a preference exists for either short-range order or clustering. Fisher[2] showed that the movement of a dislocation through a region of short-range order will reduce the number of like-unlike bonds across the slip plane. This raises the energy of the system since like-unlike bonds (or like-like bonds in the case of clustering) are preferred. The increase in flow stress due to the decrease in short-range order is $\tau = \gamma/b$, where the energy $\gamma$ is estimated to be about 10 erg/cm$^2$ for brass.

The resistance to dislocation motion that constitutes solid-solution strengthening can come from one or more of these factors. At first it might appear that dislocations would not be impeded by the interactions from solute atoms since on the average as many interactions will tend to promote motion as retard it (Fig. 6-13*a*). In a random solid solution, provided the dislocation remains straight, there will be no net force on the dislocation since the algebraic sum of all interaction energies will be zero. Mott and Nabarro[3] provided the key theory by pointing out that dislocation lines generally are not straight. Rather, dislocation lines are *flexible* so that the entire line does not have to move simultaneously and can take up lower energy positions by bending around regions of high interaction energy. As was shown in Sec. 5-8, the smallest radius of curvature that the dislocation line can accommodate to under a local stress $\tau_i$ at a solute atom is

$$R \approx \frac{Gb}{2\tau_i} \qquad (6\text{-}22)$$

$(a)$ $(b)$

**Fig. 6-13**  $(a)$ Straight dislocation line in random solid solution; $(b)$ flexible dislocation line.

[1] H. Suzuki, *Sci. Rep. Res. Inst. Tohoku Univ.*, vol. 4A, p. 455, 1952.
[2] J. C. Fisher, *Acta Metall.*, vol. 2, p. 9, 1954.
[3] N. F. Mott and F. R. N. Nabarro, "Report on Conference on the Strength of Solids," p. 1, *Phys. Soc., London*, 1948. This theory has been extended by T. Stefansky and J. E. Dorn, *Trans. Metall. Soc. AIME*, vol. 245, pp. 1869–1876, 1969.

The degree of interaction that the dislocation will have will depend on the average spacing $\lambda$ of the particles. For individual solute atoms distributed through the crystal lattice, $\lambda$ is very small and is given by

$$\lambda = \frac{a}{c^{1/3}} \qquad (6\text{-}23)$$

where $a$ is the interatomic spacing and $c$ is the atomic concentration of solute. The spacing $\lambda$ of the local solute stress fields will be much smaller than the radius of curvature to which the dislocation can be bent by the local stresses. Thus, $\lambda \ll R$ and the dislocation will move in lengths much greater than $\lambda$ (Fig. 6-13$b$). The detailed calculation leads to

$$\tau_0 = 2.5 \, G\varepsilon_a^{4/3} c \qquad (6\text{-}24)$$

for very dilute solid solutions. While many studies show that the flow stress is linear with concentration for very dilute solutions, Eq. (6-24) predicts values of flow stress which are much too large.

The Mott-Nabarro theory will be important when we consider other barriers to dislocation motion such as precipitate particles.

In alloys with *long-range order* each of the constituent atoms occupies special sites in the lattice. This results in a *superlattice* with a larger unit cell and perhaps a different crystal structure. For example, in the superlattice $Cu_3Au$ the copper atoms occupy the face-centered positions of an fcc lattice while the gold atoms occupy the corner positions. A superlattice is similar to an *intermettallic compound*.[1] Usually the dislocations in a superlattice dissociate into two pairs of ordinary dislocations separated by an *antiphase domain boundary*. The width of the antiphase domain boundary is the result of a balance between the elastic repulsion of the two dislocations of the same sign and the energy of the antiphase boundary. Slip across an antiphase boundary will result in an increase in the total boundary area. From the work required to provide the energy associated with the extra area,[2] the stress required to move a dislocation is

$$\tau_0 = \gamma/t \qquad (6\text{-}25)$$

where $t$ is the antiphase boundary width and $\gamma$ is its energy. Because more antiphase boundaries are produced as slip continues, the rate of strain hardening is higher in the ordered condition than the disordered state. Ordered alloys with a fine

[1] The basic difference is that in an intermetallic compound the two atomic species may lie on lattice points that are geometrically different. For a complete discussion of intermetallic compounds see J. H. Westbrook (ed.), "Intermetallic Compounds," John Wiley & Sons, Inc., New York, 1967.

[2] P. A. Flinn, Strengthening by Superlattice Formation in D. Peckner (ed.), "The Strengthening of Metals," pp. 219–235, Reinhold Publishing Corporation, New York, 1964.

domain size (approximately 50 Å) are stronger than the disordered state but ordered alloys with a large domain size generally have a yield stress lower than the disordered state.


## 6-8 DEFORMATION OF TWO-PHASE AGGREGATES

Only a relatively small number of alloy systems permit extensive solid solubility between two or more elements, and only a relatively small hardening effect can be produced in most alloy systems by solid-solution additions. Therefore, many commercial alloys contain a heterogeneous microstructure consisting of two or more metallurgical phases. A number of different microstructures may be encountered,[1] but in general they fall into the two classes illustrated in Fig. 6-14. Figure 6-14a represents the aggregated type of two-phase structure in which the size of the second-phase particles is of the order of the grain size of the matrix. This is typified by beta brass particles in an alpha brass matrix or by pearlite colonies in a ferrite matrix in annealed steel (Fig. 15-16). The other general type of structure is the dispersed two-phase structure (Fig. 6-14b) in which each particle is completely surrounded by a matrix of a single orientation (grain). Generally the particle size of the second phase is much finer for the dispersed structure and may extend down to submicroscopic dimensions in the early stages of precipitation. The theories of strengthening in dispersion-hardened systems have been studied extensively and will be considered in Sec. 6-9.

The strengthening produced by second-phase particles is usually additive to the solid-solution strengthening produced in the matrix. For two-phase alloys produced by equilibrium methods, the existence of a second phase ensures maximum solid-solution hardening because its presence resulted from supersaturation of the matrix phase. Moreover, the presence of second-phase particles in the continuous matrix phase results in localized internal stresses which modify the plastic

(a)          (b)

**Fig. 6-14** Types of two-phase microstructures. (a) Aggregated structure; (b) dispersed structure.

[1] For a well-illustrated discussion of the relationship between phase diagrams, microstructure, and properties see R. M. Brick, R. B. Gordon, and A. Phillips, "Structure and Properties of Alloys," 3rd ed., McGraw-Hill Book Company, New York, 1965.

properties of the continuous phase. Many factors must be considered for a complete understanding of strengthening from second-phase particles. These factors include the size, shape, number, and distribution of the second-phase particles, the strength, ductility, and strain-hardening behavior of the matrix and second phase, the crystallographic fit between the phases, and the interfacial energy and interfacial bonding between the phases. It is almost impossible to vary these factors independently in experiments, and it is very difficult to measure many of these quantities with any degree of precision.

In a multiphase alloy, each phase contributes certain things to the overall properties of the aggregate. If the contributions from each phase are independent, then the properties of the multiphase alloy will be a weighted average of the properties of the individual phases. For example, the density of a two-phase alloy will be equal to the sum of the volume fraction of each phase times its density. However, for the structure-sensitive mechanical properties the properties of the aggregate are generally influenced by interaction between the two phases. Two simple hypotheses may be used to calculate the properties of a two-phase alloy from the properties of the individual ductile phases. If it is assumed that the strain in each phase is equal, the average stress in the alloy for a given strain will increase linearly with the volume fraction of the strong phase.

$$\sigma_{avg} = f_1\sigma_1 + f_2\sigma_2 \qquad (6\text{-}26)$$

The volume fraction of phase 1 is $f_1$, and $f_1 + f_2 = 1$. Figure 6-15a shows the calculation of the flow curve for an alloy with 0.5 volume fraction of phase 2 on the basis of the equal-strain hypothesis. An alternative hypothesis is to assume that the two phases are subjected to equal stresses. The average strain in the alloy at a given stress is then given by

$$\varepsilon_{avg} = f_1\varepsilon_1 + f_2\varepsilon_2 \qquad (6\text{-}27)$$

Figure 6-15b shows the flow curve for a 0.5-volume-fraction alloy on the basis of equal-stress hypothesis. Both these hypotheses are simple approximations, and the strengths of alloys containing two ductile phases usually lie somewhere between the values predicted by the two models.

The deformation of an alloy consisting of two ductile phases depends on the volume fraction of the two phases and the total deformation. Experiments have shown[1] that not all second-phase particles produce strengthening. In order for particle strengthening to occur there must be a strong particle-matrix bond. Limited experimental information is available on the deformation of two-phase alloys.[2] Slip will occur first in the weaker phase, and if very little of the stronger

[1] B. I. Edelson and W. M. Baldwin, Jr., *Trans. Am. Soc. Met.*, vol. 55, p. 230, 1962.
[2] H. Unkel, *J. Inst. Met.*, vol. 61, p. 171, 1937; L. M. Clarebrough and G. Perger, *Aust. J. Sci. Res. Ser. A*, vol. A5, p. 114, 1952.

**Fig. 6-15** Estimate of flow stress of two-phase alloy. (*a*) Equal strain; (*b*) equal stress. (*From J. E. Dorn and C. D. Starr, "Relation of Properties to Microstructure," pp. 77–78, American Society for Metals, Metals Park, Ohio, 1954. By permission of the publishers.*)

phase is present, most of the deformation will occur in the softer phase. At large deformations flow of the softer matrix will occur around the particles of the harder phase. At about 30 vol percent of the harder phase the soft phase is no longer the completely continuous phase and the two phases tend to deform more or less with equal strains. At above about 70 vol percent of the harder phase the deformation is largely controlled by the properties of this phase.

The mechanical properties of an alloy consisting of a ductile phase and a hard brittle phase will depend on how the brittle phase is distributed in the microstructure. If the brittle phase is present as a grain-boundary envelope, as in oxygen-free copper-bismuth alloys or hypereutectoid steel, the alloy is brittle. If the brittle phase is in the form of discontinuous particles at grain boundaries, as when oxygen is added to copper-bismuth alloys or with internally oxidized copper or nickel, the brittleness of the alloy is reduced somewhat. A condition of optimum strength and ductility is obtained when the brittle phase is present as a fine dispersion uniformly distributed throughout the softer matrix. This is the situation in heat-treated steel with a tempered martensitic structure.

One of the first correlations between microstructures of two-phase alloys and yield stress was made by Gensamer et al.[1] for the aggregates of cementite (iron carbide) and ferrite found in annealed, normalized, and spheroidized steel. In these coarse-dispersed second-phase aggregates they found that the flow stress at a true strain of 0.2 was inversely proportional to the logarithm of the mean inter-

---

[1] M. Gensamer, E. B. Pearsall, W. S. Pellini, and J. R. Low, Jr., *Trans., Am. Soc. Met.*, vol. 30, pp. 983–1020, 1942.

**Fig. 6-16** Flow stress vs. logarithm of mean free ferrite path in steels with pearlitic and spheroidal distribution of carbides. (*From M. Gensamer, E. S. Pearsall, W. S. Pellini, and J. R. Low, Trans. ASM, vol. 30, p. 1003, 1942.*)

particle spacing[1] (mean free ferrite path) (Fig. 6-16). This relationship also has been found to hold for the finer cementite particles in temperated martensite,[2] the coarse particles in overaged Al-Cu alloys[3] and spheroidized steel,[4] and Co-WC sintered carbides.[5] The degree of strengthening produced by dispersed particles is illustrated by Fig. 6-17. The bottom curve represents the saturated solid solution of Al-Cu. The top curve is for the solid solution containing 5 vol percent of fine $CuAl_2$ particles, while the middle curve is for a coarse dispersion of the same volume fraction.

[1] The mean free ferrite path $p$ is given by $p = (1 - f_c)/N_L$, where $f_c$ is the volume fraction of cementite and $N_L$ is the number of carbide-particles intersected per unit length of random line through the microstructure.

[2] A. M. Turkalo and J. R. Low, Jr., *Trans. Metall. Soc. AIME*, vol. 212, pp. 750–758, 1958.

[3] C. D. Starr, R. B. Shaw, and J. E. Dorn, *Trans. Am. Soc. Met.*, vol. 46, pp. 1075–1088, 1954.

[4] C. T. Liu and J. Gurland, *Trans. Metall. Soc. AIME*, vol. 242, pp. 1535–1542, 1968.

[5] J. Gurland and P. Bardzil, *Trans. AIME*, vol. 203, p. 311, 1955.

**Fig. 6-17** Variation of flow stress with temperature for Al-Cu alloy containing 5 volume percent fine and coarse second-phase particles. (*From C. D. Starr, R. B. Shaw, and J. E. Dorn, Trans. Am. Soc. Met., vol. 46, p. 1085, 1954.*)

A strong theoretical basis for the strength of coarse two-phase aggregates has not been developed. In general terms second phases block slip so that plastic deformation is not uniform in the matrix. Strain in the matrix is localized and increased over the average strain of the specimen. This leads to strengthening from local plastic constraint. A good example can be found in the relative strengthening from pearlite and spheroidite in steel (Fig. 9-22). At equal volume fraction of cementite phase, the pearlitic structure will have a greater yield stress because the ferrite matrix sandwiched between cementite platelets will be more constrained than will the ferrite surrounding the more spheroidal particles of the spheroidized microstructure. An interesting approach to this problem, just beginning, is the application of the limit design theories of continuum plasticity to the prediction of the strength of model (idealized) microstructures.[1]

---

[1] D. C. Drucker, *J. Mater.*, vol. 1, pp. 873–910, 1966; T. W. Butler and D. C. Drucker, *J. Appl. Mech.*, vol. 95, pp. 780–784, 1973.

## 6-9 STRENGTHENING FROM FINE PARTICLES

The strengthening produced by a finely dispersed insoluble second phase in a metallic matrix is known as *dispersion hardening*.[1] A very similar strengthening phenomenon, *precipitation hardening*, or *age hardening*,[2] is produced by solution treating and quenching an alloy in which a second phase is in solid solution at the elevated temperature but precipitates upon quenching and aging at a lower temperature. The age-hardening aluminum alloys and copper-beryllium alloys are common examples. For precipitation hardening to occur, the second phase must be soluble at an elevated temperature but must exhibit decreasing solubility with decreasing temperature. By constrast, the second phase in dispersion-hardening systems has very little solubility in the matrix, even at elevated temperatures. Usually there is atomic matching, or *coherency*, between the lattices of the precipitate and the matrix, while in dispersion-hardened systems there generally is no coherency between the second-phase particles and the matrix. The requirement of a decreasing solubility with temperature places a limitation on the number of useful precipitation-hardening alloy systems. On the other hand, it is at least theoretically possible to produce an almost infinite number of dispersion-hardened systems by mixing finely divided metallic powders and second-phase particles (oxides, carbides, nitrides, borides, etc.) and consolidating them by powder metallurgy techniques. Advantage has been taken of this method to produce dispersion-hardened systems which are thermally stable at very high temperatures. Because of the finely dispersed second-phase particles, these alloys are much more resistant to recrystallization and grain growth than single-phase alloys. Because there is very little solubility of the second-phase constituent in the matrix, the particles resist growth or overaging to a much greater extent than the second-phase particles in a precipitation-hardening system.

The formation of a coherent precipitate in a precipitation-hardening system, such as Al-Cu, occurs in a number of steps. After quenching from solid solution the alloy contains regions of solute segregation, or clustering. Guiner and Preston first detected this local clustering with special x-ray techniques, and therefore this structure is known as a GP zone. The clustering may produce local strain, so that the hardness of GP[1] is higher than for the solid solution. With additional aging the hardness is increased further by the ordering of larger clumps of copper atoms on the {100} planes of the matrix. This structure is known as GP[2], or $\theta''$. Next, definite precipitate platelets of $CuAl_2$, or $\theta'$, which are coherent with the matrix, form on the {100} planes of the matrix. The coherent precipitate produces an increased strain field in the matrix and a further increase in hardness. With still

[1] R. F. Decker, *Metall. Trans.*, vol. 4, pp. 2495–2518, 1973.
[2] J. W. Martin, "Precipitation Hardening," Pergamon Press, New York, 1968.

Aging time →
(Particle size →)

**Fig. 6-18** Variation of yield stress with aging time (schematic).

further aging the equilibrium phase $CuAl_2$, or $\theta$, is formed from the transition lattice $\theta'$. These particles are no longer coherent with the matrix, and therefore the hardness is lower than at the stage when coherent $\theta'$ was present. For most precipitation-hardening alloys the resolution with the light microscope of the first precipitate occurs after the particles are no longer coherent with the matrix. Continued aging beyond this stage produces particle growth and further decrease in hardness. Figure 6-18, illustrates the way in which strength varies with aging time or particle size. The sequence of events in the Al-Cu system is particularly complicated. Although other precipitation-hardening systems may not have so many stages, it is quite common for a coherent precipitate to form and then lose coherency when the particle grows to a critical size.

The deformation of alloys with fine particle strengthening is well illustrated by the stress-strain curves in Fig. 6-19. Careful studies on Al-4.5% Cu single crystals[1] have correlated observations of slip lines with the deformation behavior. When the

**Fig. 6-19** Stress-strain curves for Al-Cu single crystals in various conditions (schematic).

[1] G. Greetham and R. W. K. Honeycombe, *J. Inst. Met.*, vol. 89, p. 13, 1960–1961; J. G. Byrne, M. E. Fine, and A. Kelly, *Philos. Mag.*, vol. 6, pp. 1119–1143, 1961.

crystals are solution treated and quenched so that the alloy contains all of the copper in supersaturated solid solution, the yield stress is raised significantly over that for pure aluminum. The rate of strain hardening (slope of stress-strain curve) is low and characteristic of easy glide. Slip bands are broad and widely spaced. When the crystal is aged to form coherent GP zones, the yield stress is raised significantly and a yield drop occurs. The rate of strain hardening is still low and slip lines can be distinguished although they are finer and more closely spaced. The yield drop and low rate of strain hardening suggest that dislocations cut through the zones once the stress reaches a high enough value. Although crystals aged to peak hardness show a slight decrease in yield strength, the rate of strain hardening is significantly increased. In this condition slip lines are very short or indistinct. This suggests that dislocations are no longer cutting through particles to form well-defined slip bands but are moving around particles so as to bypass them. In the overaged conditions where particles are noncoherent and relatively coarse, the yield strength is low but the rate of strain hardening is very high. Dislocations accumulate in tangles around the particles in the process of passing between them. This promotes slip on secondary slip systems and promotes strain hardening of the matrix. The high stresses from the dislocation loops around the particles tend to make the particles conform, by elastic deformation, to the plastic deformation of the matrix. Very high elastic strains are developed in the particles and the particles support a large fraction of the total load much as in fiber-reinforced materials (Sec. 6-10). The limit of strength in overaged alloys or dispersion-strengthened alloys is the yielding, or fracture, of the particles, or the tearing of the matrix away from the particles. However, since the particles are usually very fine intermetallic compounds, they possess high strength.

The degree of strengthening resulting from second-phase particles depends on the distribution of particles in the ductile matrix. In addition to shape, the second-phase dispersion can be described by specifying the volume fraction, average particle diameter, and mean interparticle spacing. These factors are all interrelated so that one factor cannot be changed without affecting the others (see Prob. 6-8). For example, for a given volume fraction of second phase, reducing the particle size decreases the average distance between particles. For a given size particle, the distance between particles decreases with an increase in the volume fraction of second phase.

The strongest alloys are produced by combining the effects of dispersion and strain hardening. If plastic deformation precedes the aging treatment, a finer dispersion is produced when particles nucleate on the dislocations in the matrix. The strongest alloys seem to be those in which the particles are formed in dense dislocation cell structures in the deformed matrix. Extensive plastic deformation of alloys containing fine, strong, dispersed particles can result in very high strength, as in cold-drawn steel wire.

We have seen that there are several ways in which fine particles can act as barriers to dislocations. They can act as strong, impenetrable noncoherent particles through which the dislocations can move only by sharp changes in curvature of the dislocation line. On the other hand, they can act as coherent or incoherent particles through which dislocations can pass, but only at stress levels much above those required to move dislocations through the matrix phase.

Second-phase particles act in two distinct ways to retard the motion of dislocations. The particles either may be cut by the dislocations or the particles resist cutting and the dislocations are forced to bypass them. A critical parameter of the dispersion of particles is the interparticle spacing $\lambda$. Interparticle spacing has been subject to many interpretations and expressed by many parameters.[1] A simple expression for the linear mean free path is

$$\lambda = \frac{4(1-f)r}{3f} \qquad (6\text{-}28)$$

where $f$ is the volume fraction of spherical particles of radius $r$.

When particles are small and/or soft, dislocations cut and deform the particles as shown in Fig. 6-20. There are four mechanisms which can contribute to this type of strengthening. Mott and Nabarro[2] recognized that the strain field resulting from the mismatch between a particle and the matrix would be a source of strengthening.

$$\Delta\sigma \approx 2G\varepsilon f \qquad (6\text{-}29)$$

where $f$ is the volume fraction of the dispersed phase and $\varepsilon$ is the measure of the strain field. A more sophisticated estimation[3] of strengthening from coherency strains results in

$$\Delta\sigma = \frac{6G(rf)^{1/2}}{b}\varepsilon^{3/2} \qquad (6\text{-}30)$$

(a)        (b)        (c)        **Fig. 6-20**  A dislocation cutting a particle.

[1] C. W. Corti, P. Cotterill and G. Fitzpatrick, *Int. Met. Rev.*, vol. 19, pp. 77–88, June 1974.
[2] N. F. Mott and F. R. N. Nabarro, *op. cit.*
[3] V. Gerold and H. Haberkorn, *Phys. Status. Solidi*, vol. 16, p. 675, 1966.

Kelly and Nicholson[1] showed that a type of "chemical hardening" would arise when a dislocation shears through a particle. This creates a step of width $b$ on either side of the particle. Since this process increases the surface area of the particle, work must be done to shear the particle. The increment in strengthening is given by

$$\Delta\sigma = \frac{2\sqrt{6}}{\pi}\frac{f\gamma_s}{r} \qquad (6\text{-}31)$$

Many fine second-phase particles are intermetallic compounds which have ordered structures. In this case the shearing process will also produce a new interface within the particle. The *extra* increment of hardening is given by

$$\Delta\sigma = \frac{2f\gamma_p}{b} \qquad (6\text{-}32)$$

where $\gamma_s$ is the energy of the particle-matrix interface and $\gamma_p$ is the energy for disordering the particle structure.[2] This contribution to hardening is strong in nickel-based superalloys with high antiphase boundary energy.[3]

Another source of hardening by deformable particles is the difference between the elastic moduli of matrix and particle. This affects the line tension of the dislocation and an additional stress must be applied to cut through the particle.[4]

$$\Delta\sigma = \frac{0.8\,Gb}{\lambda}\left(1 - \frac{E_1^{\,2}}{E_2^{\,2}}\right)^{1/2} \qquad (6\text{-}33)$$

where $E_1$ is the modulus of elasticity of the soft phase and $E_2$ is the modulus of elasticity of the hard phase.

Finally, there will be a strengthening increment due to the difference in Peierls stress between the particle and the matrix.[3] This term will vary directly with the difference in strength between the particle $\sigma_p$ and the matrix $\sigma_m$.

$$\Delta\sigma = \frac{5.2 f^{1/3} r^{1/2}}{G^{1/2} b^2}(\sigma_p - \sigma_m) \qquad (6\text{-}34)$$

When deformation occurs by cutting of particles there is little strain hardening. Deformation is by planar coarse slip.

The summation of these strengthening mechanisms leads to an increase of strength with particle size (see left-hand side of Fig. 6-18). Eventually, a point is reached where the cutting of particles becomes very difficult, and instead the dislocations find ways of moving around the particles.

[1] A. Kelly and R. B. Nicholson, "Progress in Materials Science," vol. 10, no. 3, Pergamon Press New York, 1963.

[2] R. O. Williams, *Acta Metall.*, vol. 5, p. 241, 1957.

[3] H. Gleiter and E. Hornbogen, *Mater. Sci. Eng.*, vol. 2, pp. 285–302, 1967.

[4] K. C. Russell and L. M. Brown, *Acta Metall.*, vol. 20, p. 969, 1972.

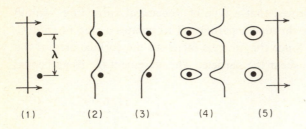

**Fig. 6-21** Schematic drawing of stages in passage of a dislocation between widely separated obstacles, based on Orowan's mechanism of dispersion hardening.

For the case of overaged noncoherent precipitates Orowan[1] proposed the mechanism illustrated in Fig. 6-21. The yield stress is determined by the shear stress required to bow a dislocation line between two particles separated by a distance $\lambda$, where $\lambda \gg R$. In Fig. 6-21, stage 1 shows a straight dislocation line approaching two particles. At stage 2 the line is beginning to bend, and at stage 3 it has reached the critical curvature. The dislocation can then move forward without further decreasing its radius of curvature. From Eq. (5-18), $R = Gb/2\tau_0$ and $\lambda = 2R$, so that the stress required to force the dislocation between the obstacles is

$$\tau_0 = \frac{Gb}{\lambda} \qquad (6\text{-}35)$$

Since the segments of dislocation that meet on the other side of the particle are of opposite sign, they can annihilate each other over part of their length, leaving a dislocation loop around each particle (stage 4). The original dislocation is then free to move on (stage 5). Every dislocation gliding over the slip plane adds one loop around the particle. These loops exert a back stress on dislocation sources which must be overcome for additional slip to take place. This requires an increase in shear stress, with the result that dispersed noncoherent particles cause the matrix to strain-harden rapidly. The strain hardening due to short-range stress can be calculated from Hart's model,[2] while that due to the average internal stress may also be determined.[3] The strain-hardening rate due to average internal stresses is

$$\frac{d\sigma}{d\varepsilon} = \frac{(7 - 5v)}{10(1 - v^2)} \frac{fE}{1 - f} \qquad (6\text{-}36)$$

[1] E. Orowan, discussion in "Symposium on Internal Stresses," p. 451, Institute of Metals, London, 1947.
[2] E. W. Hart, *Acta Metall.*, vol. 20, p. 272, 1972.
[3] T. Mori and K. Tanaka, *Acta Metall.*, vol. 21, p. 571, 1973.

The basic Orowan equation has been modified by introducing more refined estimates of the dislocation line tension,[1] by using the planar spacing $\lambda_p$, for the mean free path[2], and by adding a correction for the interaction between dislocation segments on either side of the particle.[3] These lead to a number of versions of the equation, of which the most common is the Orowan-Ashby equation[4]

$$\Delta\sigma = \frac{0.13Gb}{\lambda} \ln \frac{r}{b} \qquad (6\text{-}37)$$

The bowing of dislocations between the particles builds up dislocation loops around the particles, but it also creates a dislocation cell structure (Fig. 6-28$b$) on the particles. This arises from the generation of dislocations due to the necessity for retaining continuity between the nondeforming particles and the matrix. If the substructure is strong, it can result in Hall-Petch type of strengthening rather than Orowan strengthening.

$$\Delta\sigma = \sigma_0 + k\lambda^{-1/2} \qquad (6\text{-}38)$$

Most theory of strengthening with second-phase particles is based on idealized spherical particles. However, particle shape can be important, primarily by changing $\lambda$. At equal volume fraction, rods and plates strengthen about twice as much as spherical particles.[5]

## 6-10 FIBER STRENGTHENING

Materials of high strength, and especially high *strength-to-weight ratio,* can be produced by incorporating fine fibers in a ductile matrix. The fibers must have high strength and high elastic modulus while the matrix must be ductile and non-reactive with the fibers. Because of their very high strength, whiskers of materials such as $Al_2O_3$ have been used with good results, but most fiber-strengthened materials use fibers of boron or graphite or metal wires such as tungsten. The fibers may be long and continuous, or they may be discontinuous. Metals and polymers have been used as matrix materials. Glass-fiber-reinforced polymers are the most common fiber-strengthened materials. Fiber-reinforced materials are an important group of materials generally known as composite materials.[6]

[1] A. Kelly, and R. B. Nicholson, *op. cit.*
[2] U. F. Kochs, *Philos. Mag.*, vol. 13, p. 541, 1966.
[3] M. F. Ashby, *Acta Metall.*, vol. 14, p. 679, 1966.
[4] M. F. Ashby, in *Proc. Second Bolton Landing Conf. on Oxide Dispersion Strengthening*, 1968. *Gordon and Breach, Science Publishers, Inc., New York.*
[5] P. M. Kelly, *Scr. Metall.*, vol. 6, pp. 647–656, 1972.
[6] L. J. Broutman and R. H. Krock, "Modern Composite Materials," Addison-Wesley Publishing Co., Reading, Mass., 1967; A. Kelly and G. J. Davies, *Metall. Rev.*, vol. 10, p. 1, 1965; A. Kelly, *Metall. Trans.*, vol. 3, pp. 2313–2325, 1972.

**Fig. 6-22**  The variation of stresses along a fiber.

An important distinction between fiber-strengthened and dispersion-strengthened metals is that in fiber strengthening the high modulus fibers carry essentially all of the load.  The matrix serves to transmit the load to the fibers, to protect fibers from surface damage, and to separate the individual fibers and blunt cracks which arise from fiber breakage.  There is a fair analogy between fiber-strengthened materials and steel reinforcing bars in concrete.  The analysis of the strength of fiber-strengthened materials involves the direct application of the continuum principles of Chap. 2 to the microscopic level.[1]  There is no need to invoke dislocation theory since the material behavior is essentially elastic.

Because the fibers and the matrix have quite different elastic moduli a complex stress distribution will be developed when a composite body is loaded uniaxially in the direction of the fibers.  A fairly rigorous analysis[2] shows that shear stresses develop at the fiber-matrix interface.  The distribution of this shear stress $\tau$ and the axial tensile stress in the fiber $\sigma$ along the length of the fiber is given in Fig. 6-22.  The high shear stresses at the ends of the fiber mean that a metal matrix will flow plastically (or a polymer matrix will fail in shear) above some critical value.  In order to utilize fully the high strength of the fiber, it is necessary that the plastic zone in the matrix does not extend from the fiber ends to its midlength before the strain in the fiber reaches its failure strain.  This leads to a critical fiber length $L_c$.

$$L_c = \frac{\sigma_f d}{2\tau_0} \qquad (6\text{-}39)$$

where $\sigma_f$ is the fracture stress of a fiber of diameter $d$ in a matrix of shear yield stress $\tau_0$.  When the fiber length exceeds $L_c$, the composite ultimately fails by fiber fracture and the composite develops its full potential of strength.

The uniaxial tensile stress-strain curves for unidirectional continuous fibers or discontinuous fibers where $L > L_c$ show four stages (Fig. 6-23).  In stage 1 both fibers and matrix undergo elastic deformation.  Young's modulus for the composite

---

[1] This micromechanics approach to composite materials is well described by H. T. Corten, Micromechanics and Fracture Behavior of Composites, in L. J. Broutman and R. H. Krock, " Modern Composite Materials," see chap. 2.
[2] H. L. Cox, *Br. J. Appl. Phys.*, vol. 3, p. 72, 1952.

**Fig. 6-23** Stages in stress-strain curves of the fiber, matrix, and fiber-reinforced composite material.

$E_c$ can be determined from a simple "rule of mixtures" addition of the elastic modulus of the matrix and the fiber[1]

$$E_c = E_f f_f + E_m f_m = E_f + f_m(E_m - E_f) \qquad (6\text{-}40)$$

where $f_f$ and $f_m$ are the volume fractions of fiber and matrix. In stage 2 the matrix deforms plasticically and the fiber still is elastic. This stage begins at approximately the yield strain of the matrix material without fibers. To calculate $E_c$ in stage 2, $E_m$ must be replaced by the slope of the matrix stress-strain curve.

$$E_c = E_f f_f + \left(\frac{d\sigma}{d\varepsilon}\right)_m f_m \qquad (6\text{-}41)$$

Since the slope of the plastic part of the matrix stress-strain curve is less than $E_m$, the last term in Eq. (6-41) is small and we can express the composite modulus as

$$E_c \approx E_f f_f \qquad (6\text{-}42)$$

The composite responds in a quasi-elastic manner in stage 2. When the composite is unloaded the fibers return to their original length but the matrix is deformed into compression.[2] Stage 3 occurs when both the fibers and the matrix undergo plastic deformation. Since many of the high strength–high modulus fibers like boron are brittle, they fracture on entering stage 3, but metal wire fibers show this region. Finally, in stage 4 the fibers fracture and the composite as a whole soon fractures.

The tensile strength of a composite loaded into stage 2 is given by

$$\sigma_c = \sigma_f f_f + \sigma'_m(1 - f_f) \qquad (6\text{-}43)$$

where $\sigma_f$ is the tensile strength of the fibers (nearly equivalent to their fracture strength) and $\sigma'_m$ is the tensile stress carried by the matrix when the fibers are

[1] This simple expression is for the elastic modulus in the longitudinal (fiber) direction. The expression for $E$ in the transverse direction and at other angles to the fiber is more complex.
[2] H. P. Cheskis and R. W. Heckel, *Metall. Trans.*, vol. 1, pp. 1931–1942, 1970.

strained to their tensile strength. Since very strong fibers will have little ductility, the strain at which $\sigma'_m$ is evaluated is low and $\sigma'_m \ll \sigma_f$.

To obtain any benefit from the presence of fibers, the strength of the composite must be greater than the strength of the strain-hardened matrix, that is, $\sigma_c \gg \sigma_u$. Thus,

$$\sigma_f f_f + \sigma'_m(1 - f_f) \geq \sigma_u$$

which leads to a *critical fiber volume* which must be exceeded for fiber strengthening to occur.

$$f_{crit} = \frac{\sigma_u - \sigma'_m}{\sigma_f - \sigma'_m} \qquad (6\text{-}44)$$

For small values of $f_f$ the strength of the composite may not follow Eq. (6-43). This is because there are an insufficient number of fibers to effectively restrain the elongation of the matrix so that the fibers are rapidly stressed to their fracture point. However, the metal matrix will carry part of the load by strain hardening. Assuming when this occurs that all fibers fracture ($\sigma_f = 0$) and from Eq. (6-43), the composite strength is given by

$$\sigma_c \geq \sigma_u(1 - f_f) \qquad (6\text{-}45)$$

where $\sigma_u$ is the ultimate tensile strength of the matrix. The strength of the composite will be given by Eq. (6-43) when the volume fraction of fibers exceeds a minimum value $f_{min}$, where

$$f_{min} = \frac{\sigma_u - \sigma'_m}{\sigma_f + \sigma_u - \sigma'_m} \qquad (6\text{-}46)$$

When $f_f < f_{min}$ the strength of the composite will be given by Eq. (6-45). The relationships between composite strength and volume fraction of fibers and the location of $f_{min}$ and $f_{crit}$ are shown in Fig. 6-24.

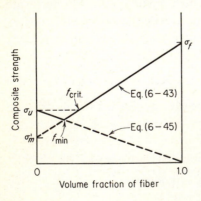

**Fig. 6-24** Theoretical variation of composite strength with volume fraction of fiber (for reinforcement with continuous fibers)

**Fig. 6-25** Variation of composite strength with angle between fibers and tensile axis.

It should be noted that the above equations apply to continuous fibers and assume no change in the strain-hardening or stress-strain behavior of the matrix due to the presence of fibers. They have been derived assuming that the fibers all have equal strength. For discontinuous fibers the ultimate strength is given by

$$\sigma_c = \sigma_f f_f \left[ 1 - (1 - \beta] \frac{L_c}{L} \right] + \sigma'_m (1 - f_f) \qquad (6\text{-}47)$$

where $\beta$ is a constant 0.5. This shows that discontinuous fibers will always produce less strengthening than continuous fibers, but if $L_c/L$ is small, the difference is unimportant.

A unidirectional array of fibers in a matrix is a highly anisotropic material. When such a composite is loaded at an angle to the fibers (Fig. 6-25), three strength parameters must be considered. The stress required to produce failure by flow parallel to the fibers is $\sigma_c$, given by Eq. (6-43). The shear stress required to produce failure by shear in the matrix or at the fiber-matrix interface is $\tau_s$, while $\sigma_s$ is the tensile stress required to produce failure of the composite in a direction normal to the fibers. The tensile stress to produce failure of the composite by fracture of the fibers is

$$\sigma = \sigma_c \sec^2 \phi \qquad (6\text{-}48)$$

If failure occurs by shear in the direction of the fibers on a plane parallel to the fibers, the failure stress is

$$\sigma = 2\tau_s \cosec 2\phi \qquad (6\text{-}49)$$

Failure by flow of the matrix transverse to the fibers or tensile failure of the interface requires a stress

$$\sigma = \sigma_s \cosec^2 \phi \qquad (6\text{-}50)$$

These failure criteria are plotted in Fig. 6-25, where it is seen that the strength of a unidirectional composite falls off significantly at small departures from the fiber orientation. This problem is solved in practice by using laminated sheets in which the fibers have a different orientation in each layer. Plywood is a familiar example of a laminated material in which this approach is used.

## 6-11 STRENGTHENING DUE TO POINT DEFECTS

In Sec. 5-12 it was shown that the movement of jogs produced by dislocation intersections can lead to the formation of point defects, either vacant lattice sites or interstitial atoms. Previously, it had been shown that quenching from temperatures near the melting point retains an excess of vacancies. Finally, appreciable concentrations of point defects can be produced by the irradiation of metals with high-energy atomic particles.

The most basic experiments[1] on the effect of vacancies on mechanical properties were performed by quenching aluminum single crystals from near the melting point. The critical resolved shear stress was increased from 50 to 500 $g/cm^2$ by the presence of quenched-in vacancies. The quench-hardened crystals showed coarse slip bands compared to the soft slow-cooled crystals. These results can be explained by the assumption that the excess vacancies migrate to dislocations and pin them in a way similar to solute atoms.

Fast-moving atomic particles create interstitials and vacancies in their collision with a solid metal. Ignoring the structural details[2] of the lattice changes brought about by high-energy radiation, it is important to realize that neutron irradiation can have marked effects on the mechanical properties of metals. In the tensile stress-strain curve the yield stress is increased by a factor of 2 to 4 for an annealed metal. Face-centered cubic metals, such as aluminum and copper, develop a sharp yield point after irradiation, but in bcc metals, like steel and molybdenum, the yield point is frequently eliminated. From an engineering point of view the most serious consequence of neutron radiation is a dramatic increase in the ductile-to-brittle transition temperature (see Sec. 14-2) in structural steels.

## 6-12 MARTENSITE STRENGTHENING

The transformation of austenite to martensite by a diffusionless shear-type transformation in quenching of steel is one of the most common strengthening processes used in engineering materials. Although martensitic transformations

---

[1] R. Maddin and A. H. Cottrell, *Philos. Mag.*, vol. 46, p. 735, 1955.
[2] For a review see G. H. Vineyard, Radiation Hardening, in "Strengthening Mechanisms in Solids," American Society for Metals, Metals Park, Ohio, 1962.

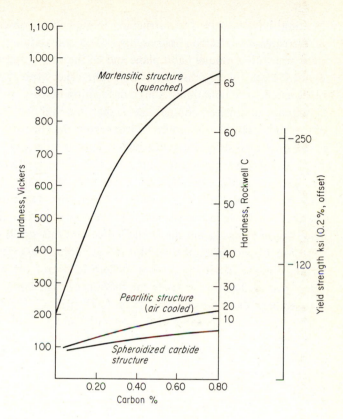

**Fig. 6-26** Hardness of various transformation products in steel. (*From E. Bain and H. W. Paxton, "Alloying Elements in Steel," 2nd ed., p. 37, American Society for Metals, Metals Park, Ohio, 1961. Copyright, American Society of Metals, 1961.*)

occur in a number of metallurgical systems,[1] only the alloys based on iron and carbon show such a pronounced strengthening effect. Figure 6-26 shows how the hardness of martensite varies with carbon content and compares this degree of strengthening with that achieved in dispersed aggregates of iron and cementite.

The high strength of martensite implies that there are many strong barriers to dislocation motion in this structure. The complexity of the system allows for considerable controversy and hardening mechanisms abound, but it appears that there are two main contributions[2] to the high strength of martensite. Kelly and

[1] C. S. Barrett and T. B. Massalski, "Structure of Metals," 3rd ed., pp. 517–531, McGraw-Hill Book Company, New York, 1966.
[2] M. Cohen, *Trans. Metall. Soc. AIME*, vol. 224, p. 638, 1962; W. Leslie and R. Sober, *Trans. Am. Soc. Met*, vol. 60, p, 459, 1967.

Nutting[1] have identified two structures in quenched iron-carbon alloys with the aid of transmission electron microscopy. The conventional martensite has a plate structure with a unique habit plane and an internal structure of parallel twins each about 0.1 $\mu$m thick within the plates. The other type of martensite structure is a block martensite containing a high dislocation density of $10^{11}$ to $10^{12}$ dislocations per square centimeter, comparable to that in a highly deformed metal. Thus, part of the high strength of martensite arises from the effective barriers to slip provided by the fine twin structure or the high dislocation density.

The second important contribution to the strength of martensite comes from the carbon atoms. Figure 6-26 shows that the hardness of martensite is very sensitive to carbon content below 0.2 percent. On rapidly transforming from austenite to ferrite in the quench, the solubility of carbon in iron is greatly reduced. The carbon atoms strain the ferrite lattice and this strain can be relieved by redistribution of carbon atoms by diffusion at room temperature. One result is that a strong binding is set up between dislocations and the carbon atoms. We have already seen that this restricts the motion of dislocations. Another result is the formation of carbon atom clusters on {100} planes. This is very similar to the GP zones discussed earlier in conjunction with the age hardening of aluminum alloys. The contribution to strength from the barriers in the martensite structure is essentially independent of carbon content, while the strengthening due to carbon atom clustering and dislocation interaction increases approximately linearly with carbon content.

An area of considerable interest has been the development of thermal-mechanical processes in which martensite is formed from an austenitic matrix which previously had been strengthened by plastic deformation.[2] This process is called *ausforming*. The plastic deformation of austenite must be accomplished without transformation to pearlite or bainite. Thus, it is necessary to work with an alloy steel which has a stable austenite region in its time-temperature-transformation (TTT) curve (Fig. 6-27). The steel is deformed in amounts in excess of 50 percent, usually by rolling, and then quenched to below the $M_s$ to form martensite. For a given alloy, the temperature of deformation and the amount of deformation are the principal variables. Highest strengths are achieved by the greatest possible deformation at the lowest temperature at which transformation does not occur. The dislocation density of ausformed martensite is very high ($10^{13}$ cm$^{-2}$) and the dislocations are usually uniformly distributed. Precipitation is more important than in ordinary quenched martensite, with the precipitates providing sites for

---

[1] P. M. Kelley and J. Nutting, *J. Iron Steel Inst.*, vol. 197, p. 199, 1961; Also J. Nutting in "Strengthening Mechanisms in Solids," chap. 4, American Society for Metals, Metals Park, Ohio, 1962.

[2] S. V. Radcliffe and E. B. Kula, in "Fundamentals of Deformation Processing," Syracuse University Press, 1964.

**Fig. 6-27** Time-temperature-transformation diagram showing steps in ausforming process.

dislocation multiplication and pinning.[1] As a result of these strengthening mechanisms ausformed steels can reach very high yield strengths of 300,000 to 400,000 psi at reductions of area ranging from 40 to 20 percent.

## 6-13 COLD-WORKED STRUCTURE

In Chap. 4 strain hardening was attributed to the interaction of dislocations with other dislocations and with other barriers to their motion through the lattice. So long as slip takes place on only a single set of parallel planes, as with single crystals of hcp metals, only a small amount of strain hardening occurs. However, even with single crystals extensive easy glide is not a general phenomenon, and with polycrystalline specimens it is not observed. Because of the mutual interference of adjacent grains in a polycrystalline specimen multiple slip occurs readily, and there is appreciable strain hardening. Plastic deformation which is carried out in a temperature region and over a time interval such that the strain hardening is not relieved is called *cold-work*.

Plastic deformation produces an increase in the number of dislocations, which by virtue of their interaction results in a higher state of internal stress. An annealed metal contains about $10^6$ to $10^8$ dislocations per square centimeter, while a severely plastically deformed metal contains about $10^{12}$ dislocations per square centimeter. Strain hardening or cold-work can be readily detected by x-ray diffraction, although detailed analysis of the x-ray patterns in terms of the structure of the cold-worked state is not usually possible. In Laue patterns cold-work produces a blurring, or *asterism*, of the spots. For Debye-Scherrer patterns the lines are broadened by cold-work. X-ray line broadening can be due to both a decrease in size of the diffraction unit, as would occur if the grains were fragmented by cold-work, and an increase in lattice strain due to dislocation interaction.

[1] O. Johari and G. Thomas, *Trans. Am. Soc. Met.*, vol. 58, pp. 563–578, 1965.

(a)                                    (b)

**Fig. 6-28** (a) Deformed to 10 percent strain. Beginning of cell formation with dislocation tangles; (b) deformed to 50 percent strain. Equilibrium cell size with heavy dislocation density in cell walls (schematic).

Techniques for analyzing the entire peak profile of x-ray lines and separating out the contribution due to lattice strain and particle size have been developed.[1]

Considerable detailed knowledge on the structure of the cold-worked state has been obtained from thin-film electron microscopy. In the early stages of plastic deformation slip is essentially on primary glide planes and the dislocations form coplanar arrays. As deformation proceeds, cross slip takes place and multiplication processes operate. The cold-worked structure forms high-dislocation-density regions or *tangles*, which soon develop into tangled networks. Thus, the characteristic structure of the cold-worked state is a *cellular substructure* in which high-density-dislocation tangles form the cell walls (Fig. 6-28). The cell structure is usually well developed at strains of around 10 percent. The cell size decreases with strain at low deformation but soon reaches a fixed size, indicating that as strain proceeds the dislocations sweep across the cells and join the tangle in the cell walls. The exact nature of the cold-worked structure will depend on the material, the strain, the strain rate, and the temperature of deformation. The development of a cell structure is less pronounced for low temperature and high strain-rate deformation and in materials with low stacking-fault energies (so that cross slip is difficult).

Most of the energy expended in deforming a metal by cold-working is converted into heat. However, roughly about 10 percent of the expended energy is stored in the lattice as an increase in internal energy. Reported values of stored energy[2] range from about 0.01 to 1.0 cal/g of metal. The magnitude of the stored

[1] B. E. Warren, in "Progress in Metal Physics," vol. 8, pp. 147–202, Pergamon Press, Ltd., London, 1959.
[2] For a comprehensive review of the stored energy of cold-work see M. B. Bever, D. L. Holt, and A. L. Titchener, "Progress in Materials Science," vol. 17, Pergamon Press, Ltd., London, 1973.

energy increases with the melting point of the metal and with solute additions. For a given metal the amount of stored energy depends on the type of deformation process, e.g., wire drawing vs. tension. The stored energy increases with strain up to a limiting value corresponding to saturation. It increases with decreasing temperature of deformation. Very careful calorimeter measurements are required to measure the small amounts of energy stored by cold-working.

The major part of the stored energy is due to the generation and interaction of dislocations during cold-working. Vacancies account for part of the stored energy for metals deformed at very low temperature. However, vacancies are so much more mobile than dislocations that they readily escape from most metals deformed at room temperature. Stacking faults and twin faults are probably responsible for a small fraction of the stored energy. A reduction in short-range order during the deformation of solid solutions may also contribute to stored energy. Elastic strain energy accounts for only a minor part of the measured stored energy.

## 6-14  STRAIN HARDENING

Strain hardening or cold-working is an important industrial process that is used to harden metals or alloys that do not respond to heat treatment. The rate of strain hardening can be gaged from the slope of the flow curve. Generally, the rate of strain hardening is lower for hcp metals than for cubic metals. Increasing temperature also lowers the rate of strain hardening. For alloys strengthened by solid-solution additions the rate of strain hardening may be either increased or decreased compared with the behavior for the pure metal. However, the final strength of a cold-worked solid-solution alloy is almost always greater than that of the pure metal cold-worked to the same extent.

Figure 6-29 shows the typical variation of strength and ductility parameters with increasing amount of cold-work. Since in most cold-working processes one or two dimensions of the metal are reduced at the expense of an increase in the other dimensions, cold-work produces elongation of the grains in the principal direction of working. Severe deformation produces a reorientation of the grains into a preferred orientation (Sec. 6-17). In addition to the changes in tensile properties shown in Fig. 6-29, cold-working produces changes in other physical properties. There is usually a small decrease in density of the order of a few tenths of a percent, an appreciable decrease in electrical conductivity due to an increased number of scattering centers, and a small increase in the thermal coefficient of expansion. Because of the increased internal energy of the cold-worked state, chemical reactivity is increased. This leads to a general decrease in corrosion resistance and in certain alloys introduces the possibility of stress-corrosion cracking.

**Fig. 6-29** Variation of tensile properties with amount of cold-work.

A high rate of strain hardening implies mutual obstruction of dislocations gliding on intersecting systems. This can come about (1) through interaction of the stress fields of the dislocations, (2) through interactions which produce sessile locks, and (3) through the interpenetration of one slip system by another (like cutting trees in a forest) which results in the formation of dislocation jogs.

The basic equation relating flow stress (strain hardening) to structure is

$$\sigma_0 = \sigma_i + \alpha G b \rho^{1/2} \qquad (6\text{-}51)$$

Much attention has been given to developing theories of strain hardening based on dislocation models. Theories based on each of the three processes listed above result[1] in equations of the form of Eq. (6-51).

Thin-film electron micrographs perhaps impart a wrong impression of the environment in which dislocations move in metals. McLean[2] gives a graphic description of the situation involving elastic interactions between dislocations and between dislocations and second-phase particles. A metal which has been plastically deformed a few percent in strain contains 50,000 km or more of dislocation line in *each* cubic centimeter. Moreover, if this cubic centimeter were enlarged to the size of a large auditorium, these dislocations would be seen to be arranged as an extremely fine irregular three-dimensional spider's web, with a mesh spacing varying from 0.1 to 1.0 mm. With this type of structure a moving dislocation can hardly avoid intersecting other dislocations and traversing the stress field of other dislocations. Because thin foils for electron microscopy sample such a small amount of material they tend to miss most of the dislocation nodes and give the impression that the dislocation network is less tightly linked than it really is.

---

[1] For a review of theories of strain hardening see D. McLean "Mechanical Properties of Metals," pp. 153–161, John Wiley & Sons, Inc., New York, 1962.
[2] D. McLean, *Trans. Metall. Soc. AIME*, vol. 242, pp. 1193–1203, 1968.

## 6-15 ANNEALING OF COLD-WORKED METAL

The cold-worked state is a condition of higher internal energy than the undeformed metal. Although the cold-worked dislocation cell structure is mechanically stable, it is not thermodynamically stable. With increasing temperature the cold-worked state becomes more and more unstable. Eventually the metal softens and reverts to a strain-free condition. The overall process by which this occurs is known as annealing.[1] Annealing is very important commercially because it restores the ductility to a metal that has been severely strain-hardened. Therefore, by interposing annealing operations after severe deformation it is possible to deform most metals to a very great extent.

The process of annealing can be divided into three fairly distinct processes: recovery, recrystallization, and grain growth. Figure 6-30 will help to distinguish between these processes. *Recovery* is usually defined as the restoration of the physical properties of the cold-worked metal without any observable change in microstructure. Electrical conductivity increases rapidly toward the annealed value during recovery, and lattice strain, as measured with x-rays, is appreciably reduced. The properties that are most affected by recovery are those which are sensitive to point defects. The strength properties, which are controlled by dislocations, are not affected at recovery temperatures. An exception to this is single crystals of hcp metals which have deformed on only one set of planes (easy glide). For this situation it is possible to recover completely the yield stress of a strain-hardened crystal without producing recrystallization. *Recrystallization* is the replacement of the cold-worked structure by a new set of strain-free grains.

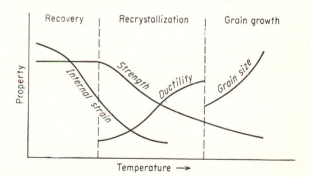

Fig. 6-30 Schematic drawing indicating recovery, recrystallization, and grain growth and chief property changes in each region.

[1] For detailed reviews of annealing, see P. A. Beck, *Adv. Phys.*, vol, 3, pp. 245–324, 1954; J. E. Burke and D. Turnbull, in "Progress in Metal Physics," vol 3, Interscience Publishers, Inc., New York, 1952.

(a)                    (b)                    (c)

**Fig. 6-31**  Changes in microstructure of cold-worked 70–30 brass with annealing. (a) Cold-worked 40 percent; (b) 440°C, 15 min; (c) 575°C, 15 min (150X). (*Courtesy L. A. Monson.*)

Recrystallization is readily detected by metallographic methods and is evidenced by a decrease in hardness or strength and an increase in ductility. The density of dislocations decreases considerably on recrystallization, and all effects of strain hardening are eliminated. The stored energy of cold-work is the driving force for both recovery and recrystallization. If the new strain-free grains are heated at a temperature greater than that required to cause recrystallization, there will be a progressive increase in grain size. The driving force for grain growth is the decrease in free energy resulting from a decreased grain-boundary area due to an increase in grain size. Figure 6-31 shows the progression from a cold-worked microstructure to a fine recrystallized grain structure, and finally to a larger grain size by grain growth.

The recrystallization process[1] consists of the nucleation of a strain-free region whose boundary can transform the strained matrix into strain-free material as it moves. In the growth of the boundary out from the nucleus the dislocations are annihilated in the region swept through. This requires that the moving boundary be a high-angle boundary so that there is a high degree of "misfit" to accommodate the dislocations. At least two distinct nucleation mechanisms have been identified for recrystallization. The first is called *strain-induced boundary migration*, where a strain-free nucleus is formed when one of the existing grain boundaries moves into its neighbor, leaving a strain-free recrystallized region behind. The boundary moves into the grain which contains the higher dislocation density in the local region. In the second nucleation mechanism new grain boundaries are formed in regions of sharp lattice curvature through subgrain growth. This mechanism

[1] For a discussion of the mechanisms and kinetics of recovery, recrystallization, and grain growth see P. G. Shewmon, "Transformations in Metals," chap. 3, McGraw-Hill Book Company, New York, 1969; also "Recrystallization Grain Growth, and Textures," American Society for Metals, Metals Park, Ohio, 1966.

seems to predominate at high strains, with nuclei appearing at grain boundaries, twin boundaries, or at inclusions or second-phase particles. The nuclei form only in regions which through inhomogeneous deformation have been rotated into an orientation appreciably different from that of the matrix.

Six main variables influence recrystallization behavior. They are (1) amount of prior deformation, (2) temperature, (3) time, (4) initial grain size, (5) composition, and (6) amount of recovery or polygonization prior to the start of recrystallization. Because the temperature at which recrystallization occurs depends on the above variables, it is not a fixed temperature in the sense of a melting temperature. For practical considerations a recrystallization temperature can be defined as the temperature at which a given alloy in a highly cold-worked state completely recrystallizes in 1 h. The relationship of the above variables to the recrystallization process can be summarized[1] as follows:

1. A minimum amount of deformation is needed to cause recrystallization.
2. The smaller the degree of deformation, the higher the temperature required to cause recrystallization.
3. Increasing the annealing time decreases the recrystallization temperature. However, temperature is far more important than time. Doubling the annealing time is approximately equivalent to increasing the annealing temperature 10°C.
4. The final grain size depends chiefly on the degree of deformation and to a lesser extent on the annealing temperature. The greater the degree of deformation and the lower the annealing temperature, the smaller the recrystallized grain size.
5. The larger the original grain size, the greater the amount of cold-work required to produce an equivalent recrystallization temperature.
6. The recrystallization temperature decreases with increasing purity of the metal. Solid-solution alloying additions always raise the recrystallization temperature.
7. The amount of deformation required to produce equivalent recrystallization behavior increases with increased temperature of working.
8. For a given reduction in cross section, different metalworking processes, such as rolling, drawing, etc., produce somewhat different effective deformations. Therefore, identical recrystallization behavior may not be obtained.

Because the driving force for grain growth is appreciably lower than the driving force for recrystallization, at a temperature at which recrystallization occurs readily grain growth will occur slowly. However, grain growth is strongly

[1] R. F. Mehl, Recrystallization, in "Metals Handbook," pp. 259–268, American Society for Metals, Metals Park, Ohio, 1948.

temperature-dependent, and a grain-coarsening region will soon be reached in which the grains increase in size very rapidly. Grain growth is inhibited considerably by the presence of a fine dispersion of second-phase particles, which restricts grain-boundary movement.

Under certain conditions, some of the grains of a fine-grained recrystallized metal will begin to grow rapidly at the expense of the other grains when heated at a higher temperature. This phenomenon is known as *exaggerated,* or *abnormal, grain growth.* The driving force for exaggerated grain growth is the decrease in surface energy, not stored energy, but because the phenomenon shows kinetics similar to those of recrystallization it is often called *secondary recrystallization.*

## 6-16 BAUSCHINGER EFFECT

In an earlier discussion of the strain hardening of single crystals it was shown that generally a lower stress is required to reverse the direction of slip on a certain slip plane than to continue slip in the original direction. The directionality of strain hardening is called the Bauschinger effect. Figure 6-32 is an example of the type of stress-strain curve that is obtained when the Bauschinger effect is considered. The Bauschinger effect is a general phenomenon in polycrystalline metals.

The initial yield stress of the material in tension is $A$. If the same ductile material were tested in compression, the yield strength would be approximately the same, point $B$ on the dashed curve. Now, consider that a new specimen is loaded in tension past the tensile yield stress to $C$ along the path $O$-$A$-$C$. If the specimen is then unloaded, it will follow the path $C$-$D$, small elastic-hysteresis effects being neglected. If now a compressive stress is applied, plastic flow will

Fig. 6-32 Bauschinger effect and hysteresis loop.

begin at the stress corresponding to point $E$, which is appreciably lower than the original compressive yield stress of the material. While the yield stress in tension was increased by strain hardening from $A$ to $C$, the yield stress in compression was decreased. This is the *Bauschinger effect*. The phenomenon is reversible, for had the specimen originally been stressed plastically in compression, the yield stress in tension would have been decreased. One way of describing the amount of Bauschinger effect is by the Bauschinger strain $\beta$ (Fig. 6-32). This is the difference in strain between the tension and compression curves at a given stress.

If the loading cycle in Fig. 6-32 is completed by loading further in compression to point $F$, then unloading, and reloading in tension, a mechanical-hysteresis loop is obtained. The area under the loop will depend on the initial overstrain beyond the yield stress and the number of times the cycle is repeated. If the cycle is repeated many times, failure by fatigue is likely to occur.

The Bauschinger effect can have important consequences in metal-forming applications. For example, it can be important in the bending of steel plates[1] and results in work-softening[2] when severely cold-worked metals are subjected to stresses of reversed sign. The best example of this is the straightening of drawn bars or rolled sheet by passing through rollers which subject the material to alternating bending stresses. Such roller-leveling operations can reduce the yield strength and increase the elongation from its cold-worked value.

The mechanism of the Bauschinger effect lies in the structure of the cold-worked state. Orowan[3] has pointed out that during plastic deformation dislocations will accumulate at barriers in tangles, and eventually form cells. Now, when the load is removed, the dislocation lines will not move appreciably because the structure is mechanically stable. However, when the direction of loading is reversed, some dislocation lines can move an appreciable distance at a low shear stress because the barriers to the rear of the dislocations are not likely to be so strong and closely spaced as those immediately in front. This gives rise to initial yielding at a lower stress level when the loading direction is reversed.

## 6-17  PREFERRED ORIENTATION (TEXTURE)

A metal which has undergone a severe amount of deformation, as in rolling or wire drawing, will develop a *preferred orientation*, or texture, in which certain crystallographic planes tend to orient themselves in a preferred manner with

[1] S. T. Rolfe, R. P. Haak, and J. H. Gross, *Trans. Amer. Soc. Mech. Eng.*, *J. Basic Eng.*, vol. 90, pp. 403–408, 1968.

[2] N. H. Polakowski, *Am. Soc. Test. Mater. Proc.*, vol. 63, p. 535, 1963.

[3] E. Orowan, Causes and Effects of Internal Stresses, in "Internal Stresses and Fatigue in Metals," Elsevier Publishing Company, New York, 1959.

respect to the direction of maximum strain. The tendency for the slip planes in a single crystal to rotate parallel to the axis of principal strain was considered previously. The same situation exists in a polycrystalline aggregate, but the complex interactions between the multiple slip systems makes analysis of the polycrystalline situation much more difficult. Since the individual grains in a polycrystalline aggregate cannot rotate freely, lattice bending and fragmentation will occur.

Preferred orientations are determined by x-ray methods. The x-ray pattern of a fine-grained randomly oriented metal will show rings corresponding to different planes where the angles satisfy the condition for Bragg reflection. If the grains are randomly oriented, the intensity of the rings will be uniform for all angles, but if a preferred orientation exists, the rings will be broken up into short arcs, or spots. The dense areas of the x-ray photograph indicate the orientation of the poles of the planes corresponding to the diffraction ring in question. The orientation of the grains of a particular crystallographic orientation with respect to the principal directions of working is best shown by means of a *pole figure*. For a description of the methods of determining pole figures and a compilation of pole figures describing the deformation textures in many metals, the reader is referred to Barrett.[1]

A preferred orientation can be detected with x-rays after about a 20 to 30 percent reduction in cross-sectional area by cold-working. At this stage of reduction there is appreciable scatter in the orientation of individual crystals about the ideal orientation. The scatter decreases with increasing reduction, until at about 80 to 90 percent reduction the preferred orientation is essentially complete. The type of preferred orientation, or deformation texture, which is developed depends primarily on the number and type of slip systems available and on the principal strains. Other factors which may be important are the temperature of deformation and the type of texture present prior to deformation.

The simplest deformation texture is produced by the drawing or rolling of a wire or rod. This is often referred to as a *fiber texture* because of its similarity to the arrangement in naturally fibrous materials. It is important to note that a distinction should be made between the *crystallographic* fibering produced by crystallographic reorientation of the grains during deformation and *mechanical fibering*, which is brought about by the alignment of inclusions, cavities, and second-phase constituents in the main direction of mechanical working. Mechanical and crystallographic fibering are important factors in producing directional mechanical properties of plastically worked metal shapes such as sheet and rods.[2] This will be discussed further in Chap. 9.

In an ideal wire texture a definite crystallographic direction lies parallel to the wire axis, and the texture is symmetrical around the wire or fiber axis. Body-

[1] C. S. Barrett and T. B. Massalski, "The Structure of Metals," 3rd ed., chaps. 20 and 21, McGraw-Hill Book Company, New York, 1966.
[2] D. V. Wilson, *Met. Technol.*, vol. 2, pp. 8–20, 1975.

centered cubic metals have a fiber texture with the $\langle 110 \rangle$ direction parallel to the wire axis. Face-centered cubic metals can have a double fiber texture with both $\langle 111 \rangle$ and $\langle 100 \rangle$ directions parallel to the wire axis. The $\langle 111 \rangle$ texture is favored by easy cross slip and predominates in high stacking-fault metals such as aluminum. Silver and brass with low stacking-fault energies have a predominantly $\langle 100 \rangle$ texture. In hcp metals the basal planes rotate so that the $\langle 10\bar{1}0 \rangle$ direction coincides with the wire axis (for magnesium).

The deformation texture of a sheet produced by rolling is described by the crystallographic planes parallel to the surface of the sheet as well as the crystallographic directions in that plane which are parallel to the direction of rolling. Two rolling textures[1] predominate in face-centered cubic metals and alloys. During initial deformation a $\{110\}\langle 112 \rangle$ texture ($\alpha$-brass-type texture) is developed, but if extensive cross slip occurs, this changes to a $\{112\}\langle 111 \rangle$ texture (copper-type texture) with more extensive plastic deformation. There is good correlation[2] between stacking-fault energy (relative ease of cross slip) and the type of texture. High stacking-fault energy and high temperature deformation favor the copper-type structure $\{112\}\langle 111 \rangle$. In body-centered cubic metals the predominant rolling texture consists of $\{100\}$ planes oriented parallel to the plane of the sheet with the $\langle 110 \rangle$ directions in the rolling direction, but other texture elements such as $\{112\}\langle 110 \rangle$, and $\{111\}\langle 112 \rangle$ may be found. For hcp metals the basal plane tends to be parallel with the rolling plane with $\langle 2\bar{1}\bar{1}0 \rangle$ aligned in the rolling direction.

The preferred orientation resulting from plastic deformation is strongly dependent on the slip and twinning systems available for deformation, but it is not generally affected by processing variables such as die angle, roll diameter, roll speed, and reduction per pass. The most important mechanical variables are the geometry of flow and the amount of deformation (reduction). Thus, the same deformation texture is produced whether a rod is made by rolling or drawing.

The recrystallization of a cold-worked metal generally produces a preferred orientation which is different from and stronger than that existing in the deformed metal. This is called an *annealing texture*, or *recrystallization texture*. An out standing example is the cube texture in copper, where the $\langle 100 \rangle$ plane lies parallel to the rolling plane with a $\langle 001 \rangle$ direction parallel to the direction of rolling. Since the existence of a recrystallization texture depends on a preferential orientation of the nuclei of the recrystallized grains, the resulting texture is strongly dependent on the texture produced by the deformation. Other important variables which affect the annealing texture are the composition, the initial grain size and grain orientation of the alloy, and the annealing temperature and time. Generally the factors which favor the formation of a fine recrystallized grain size also favor

---

[1] R. E. Smallman, *J. Inst. Met.*, vol. 84, pp. 10–18, 1955–1956.
[2] I. S. Dilamore and W. T. Roberts, *Metall. Rev.*, vol. 10. no. 39, 1965.

the formation of an essentially random orientation of recrystallized grains. Moderate cold reductions and low annealing temperatures are beneficial.

Sometimes the formation of a strong texture in a finished sheet is beneficial. One of the best examples is cube-oriented silicon-iron transformer sheet, where the energy losses are minimzed by orienting the grains in the easy direction of magnetization. The use of texture to resist yielding in titanium plate was discussed in Chap. 3, and in Chap. 20 we shall consider in detail how proper texture can greatly increase the deep-drawing quality of low-carbon steel. On the other hand, a strong preferred orientation will result in an anisotropy in mechanical properties in the plane of the sheet. This can result in uneven response of the material during forming and fabrication operations, and it must be recognized as a factor in design.

## BIBLIOGRAPHY

Cottrell, A. H: "An Introduction to Metallurgy," Edward Arnold (Publishers) Ltd., London, 1967.

Felbeck, D. F: "Introduction to Strengthening Mechanisms," Prentice-Hall, Inc., Englewood Cliffs, N.J., 1968.

Honeycombe, R. W. K: "The Plastic Deformation of Metals," Edward Arnold (Publishers) Ltd., London, 1968.

Kelly, A: "Strong Solids," 2nd ed., Oxford University Press, London, 1974.

McLean, D: "Mechanical Properties of Metals," John Wiley & Sons, Inc., New York, 1962.

Peckner, D. (ed): "The Strengthening of Metals," Reinhold Publishing Corporation, New York, 1964.

Reed-Hill, R. E: "Physical Metallurgy Principles," 2d ed., D. Van Nostrand Company, Inc., New York, 1973.

Tegart, W. J. McG: "Elements of Mechanical Metallurgy," The Macmillan Company, New York, 1966.

Wyatt, O. H. and D. Dew-Hughes: "Metals, Ceramics and Polymers," Cambridge University Press, London, 1974.

Zackay, V. F. (ed): "High Strength Materials," John Wiley & Sons, Inc., New York, 1965.

# FRACTURE

## 7-1 INTRODUCTION

Fracture is the separation, or fragmentation, of a solid body into two or more parts under the action of stress. The process of fracture can be considered to be made up of two components, crack initiation and crack propagation. Fractures can be classified into two general categories, ductile fracture and brittle fracture. A ductile fracture is characterized by appreciable plastic deformation prior to and during the propagation of the crack. An appreciable amount of gross deformation is usually present at the fracture surfaces. Brittle fracture in metals is characterized by a rapid rate of crack propagation, with no gross deformation and very little micro-deformation. It is akin to cleavage in ionic crystals. The tendency for brittle fracture is increased with decreasing temperature, increasing strain rate, and triaxial stress conditions (usually produced by a notch). Brittle fracture is to be avoided at all cost, because it occurs without warning and usually produces disastrous consequences.

This chapter will present a broad picture of the fundamentals of the fracture of metals. Since most of the research has been concentrated on the problem of brittle fracture, this topic will be given considerable prominence. The engineering aspects of brittle fracture will be considered in greater detail in Chap. 14. Fracture occurs in characteristic ways, depending on the state of stress, the rate of application of stress, and the temperature. Unless otherwise stated, it will be assumed in this chapter that fracture is produced by a single application of a uniaxial tensile stress. Fracture under more complex conditions will be considered in later chapters. Typical examples are fracture due to torsion (Chap. 10), fatigue (Chap. 12), and creep (Chap. 13), and low-temperature brittle fracture, temper embrittlement, or hydrogen embrittlement (Chap. 14).

## 7-2 TYPES OF FRACTURE IN METALS

Metals can exhibit many different types of fracture, depending on the material, temperature, state of stress, and rate of loading. The two broad categories of ductile and brittle fracture have already been considered. Figure 7-1 schematically illustrates some of the types of tensile fractures which can occur in metals. A brittle fracture (Fig. 7-1a) is characterized by separation normal to the tensile stress. Outwardly there is no evidence of deformation, although with x-ray diffraction analysis it is possible to detect a thin layer of deformed metal at the fracture surface. Brittle fractures have been observed in bcc and hcp metals, but not in fcc metals unless there are factors contributing to grain-boundary embrittlement.

Ductile fractures can take several forms. Single crystals of hcp metals may slip on successive basal planes until finally the crystal separates by shear (Fig. 7-1b). Polycrystalline specimens of very ductile metals, like gold or lead, may actually be drawn down to a point before they rupture (Fig. 7-1c). In the tensile fracture of moderately ductile metals the plastic deformation eventually produces a necked region (Fig. 7-1d). Fracture begins at the center of the specimen and then extends by a shear separation along the dashed lines in Fig. 7-1d. This results in the familiar "cup-and-cone" fracture.

Fractures are classified with respect to several characteristics, such as strain

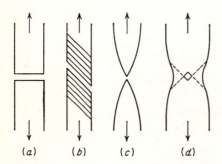

(a)    (b)    (c)    (d)

**Fig. 7-1** Types of fractures observed in metals subjected to uniaxial tension. (a) Brittle fracture of single crystals and polycrystals; (b) shearing fracture in ductile single crystals; (c) completely ductile fracture in polycrystals; (d) ductile fracture in polycrystals.

to fracture, crystallographic mode of fracture, and the appearance of the fracture. Gensamer[1] has summarized the terms commonly used to describe fractures as follows:

| Behavior described | Terms used | |
|---|---|---|
| Crystallographic mode | Shear | Cleavage |
| Appearance of fracture | Fibrous | Granular |
| Strain to fracture | Ductile | Brittle |

A shear fracture occurs as the result of extensive slip on the active slip plane. This type of fracture is promoted by shear stresses. The cleavage mode of fracture is controlled by tensile stresses acting normal to a crystallographic cleavage plane. A fracture surface which is caused by shear appears at low magnification to be gray and fibrous, while a cleavage fracture appears bright or granular, owing to reflection of light from the flat cleavage surfaces. Fracture surfaces frequently consist of a mixture of fibrous and granular fracture, and it is customary to report the percentage of the surface area represented by one of these categories. Based on metallographic examination, fractures in polycrystalline samples are classified as either *transgranular* (the crack propagates through the grains) or *intergranular* (the crack propagates along the grain boundaries). A ductile fracture is one which exhibits a considerable degree of deformation. The boundary between a ductile and brittle fracture is arbitrary and depends on the situation being considered. For example, nodular cast iron is ductile when compared with ordinary gray iron; yet it would be considered brittle when compared with mild steel. As a further example, a deeply notched tensile specimen will exhibit little gross deformation; yet the fracture could occur by a shear mode.

## 7-3 THEORETICAL COHESIVE STRENGTH OF METALS

Metals are of great technological value, primarily because of their high strength combined with a certain measure of plasticity. In the most basic terms the strength is due to the cohesive forces between atoms. In general, high cohesive forces are related to large elastic constants, high melting points, and small coefficients of thermal expansion. Figure 7-2 shows the variation of the cohesive force between two atoms as a function of the separation between these atoms. This curve is the resultant of the attractive and repulsive forces between the atoms. The interatomic spacing of the atoms in the unstrained condition is indicated by $a_0$. If the crystal

---

[1] M. Gensamer, General Survey of the Problem of Fatigue and Fracture, in "Fatigue, and Fracture of Metals," John Wiley & Sons, Inc., New York, 1952.

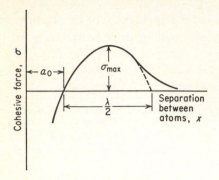

**Fig. 7-2** Cohesive force as a function of the separation between atoms.

is subjected to a tensile load, the separation between atoms will be increased. The replusive force decreases more rapidly with increased separation than the attractive force, so that a net force between atoms balances the tensile load. As the tensile load is increased still further, the repulsive force continues to decrease. A point is reached where the repulsive force is negligible and the attractive force is decreasing because of the increased separation of the atoms. This corresponds to the maximum in the curve, which is equal to the theoretical cohesive strength of the material.

A good approximation to the theoretical cohesive strength can be obtained if it is assumed that the cohesive force curve can be represented by a sine curve.

$$\sigma = \sigma_{max} \sin \frac{2\pi x}{\lambda} \qquad (7\text{-}1)$$

where $\sigma_{max}$ is the theoretical cohesive strength and $x = a - a_0$ is the displacement in atomic spacing in a lattice with wave length $\lambda$. For small displacements, $\sin x \approx x$, and

$$\sigma = \sigma_{max} \frac{2\pi x}{\lambda} \qquad (7\text{-}2)$$

Also, if we restrict consideration to a brittle elastic solid, then from Hooke's law

$$\sigma = Ee = \frac{Ex}{a_0} \qquad (7\text{-}3)$$

Eliminating $x$ from Eqs. (7-2) and (7-3), we have

$$\sigma_{max} = \frac{\lambda}{2\pi} \frac{E}{a_0} \qquad (7\text{-}4)$$

When fracture occurs in a brittle solid all of the work expended in producing the fracture goes into the creation of two new surfaces. Each of these surfaces has a *surface energy* of $\gamma_s$ ergs per square centimeter. The work done per unit area of surface in creating the fracture is the area under the stress-displacement curve.

$$U_0 = \int_0^{\lambda/2} \sigma_{\max} \sin \frac{2\pi x}{\lambda} \, dx = \frac{\lambda \sigma_{\max}}{\pi} \qquad (7\text{-}5)$$

But this energy is equal to the energy required to create the two new fracture surfaces.

$$\frac{\lambda \sigma_{\max}}{\pi} = 2\gamma_s$$

or

$$\lambda = \frac{2\pi \gamma_s}{\sigma_{\max}} \qquad (7\text{-}6)$$

and substituting into Eq. (7-4) gives

$$\sigma_{\max} = \left(\frac{E\gamma_s}{a_0}\right)^{\frac{1}{2}} \qquad (7\text{-}7)$$

It is interesting to note that the theoretical cohesive strength of a brittle solid can be expressed so simply in terms of such basic parameters. Taking typical values for these parameters and substituting into Eq. (7-7),

$$E = 10^{12} \text{ dyn/cm}^2 \ (14.5 \times 10^6 \text{ psi})$$
$$a_0 = 3 \times 10^{-8} \text{ cm}$$
$$\gamma_s = 10^3 \text{ erg/cm}^2 \ (5.7 \times 10^{-3} \text{ in-lb/in}^2)$$

yields a value of $\sigma_{\max} = 1.82 \times 10^{11}$ dyn/cm$^2$ ($2.6 \times 10^6$ psi). Measured in terms of the fraction of the elastic modulus, $\sigma_{\max} \approx E/5.5$. Using other values of the parameters and other expressions for the force-displacement curve which are more complicated than the sine-wave approximation result in estimates of $\sigma_{\max}$ varying from $E/4$ to $E/15$. A convenient choice will be to say that $\sigma_{\max} \approx E/10$.

Experience with high-strength steels shows that a fracture strength in excess of 300,000 psi is exceptional. Engineering materials typically have fracture stresses that are 10 to 1000 times lower than the theoretical value. The only materials that approach the theoretical value are tiny, defect-free metallic whiskers and very-fine-diameter silica fibers. This leads to the conclusion that flaws or cracks are responsible for the lower-than-ideal fracture strength of engineering materials.

Ignoring until later the question of where the cracks come from, it is a logical extension of the idea of stress concentration (Sec. 2-16) to show how the presence of cracks[1] will result in a reduced fracture strength. Figure 7-3 shows a thin elliptical crack in an infinitely wide plate. The crack has a length $2c$ and a radius of

---

[1] E. Orowan, *Welding J.*, vol. 34, pp. 157s–160s, 1955.

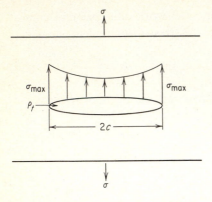

**Fig. 7-3** Elliptical crack model.

curvature at its tip of $\rho_t$. The maximum stress at the tip of the crack $\sigma_{max}$ is given by[1]

$$\sigma_{max} = \sigma\left[1 + 2\left(\frac{c}{\rho_t}\right)^{1/2}\right] \approx 2\sigma\left(\frac{c}{\rho_t}\right)^{1/2} \qquad (7\text{-}8)$$

This approach assumes that the theoretical cohesive stress $\sigma_{max}$ can be reached locally at the tip of a crack while the average tensile stress $\sigma$ is at a much lower value. Therefore, equating Eq. (7-7) and (7-8), we can solve for $\sigma$ which is the nominal fracture stress $\sigma_f$ of the material containing cracks.

$$\sigma_f \approx \left(\frac{E\gamma_s\rho_t}{4a_0 c}\right)^{1/2} \qquad (7\text{-}9)$$

The sharpest possible crack would be one where $\rho_t \approx a_0$, so that

$$\sigma_f \approx \left(\frac{E\gamma_s}{4c}\right)^{1/2} \qquad (7\text{-}10)$$

Once more substituting reasonable values into Eq. (7-9):

$E = 10^{12}$ dyn/cm$^2$, $\gamma_s = 10^3$ erg/cm$^2$, $a_0 = 2.5 \times 10^{-8}$ cm

$$\sigma_f \approx \left(10^{23}\frac{\rho_t}{c}\right)^{1/2} \approx \frac{E}{10}\left(\frac{\rho_t}{c}\right)^{1/2}$$

If $\rho_t = a_0$ and the crack length is $10^4 a_0$ (2.5 $\mu$m).

$$\sigma_f \approx \frac{E}{10}\left(\frac{a_0}{10^4 a_0}\right)^{1/2} \approx \frac{E}{1{,}000}$$

Thus, we see that in a brittle solid a very great decrease in the stress for fracture is produced by a very small crack.

---

[1] C. E. Inglis, *Trans. Inst. Nav. Archit.*, vol. 55, pt. I, pp. 219–230, 1913. Equation (7-8) is equivalent to Eq. (2-109) since for an ellipse $\rho = b^2/a$ and $a = c$ in Fig. 7-3.

## 7-4 GRIFFITH THEORY OF BRITTLE FRACTURE

The first explanation of the discrepancy between the observed fracture strength of crystals and the theoretical cohesive strength was proposed by Griffith.[1] Griffith's theory in its original form is applicable only to a perfectly brittle material such as glass. However, while it cannot be applied directly to metals, Griffith's ideas have had great influence on the thinking about the fracture of metals.

Griffith proposed that a brittle material contains a population of fine cracks which produce a stress concentration of sufficient magnitude so that the theoretical cohesive strength is reached in localized regions at a nominal stress which is well below the theoretical value. When one of the cracks spreads into a brittle fracture, it produces an increase in the surface area of the sides of the crack. This requires energy to overcome the cohesive force of the atoms, or, expressed in another way, it requires an increase in surface energy. The source of the increased surface energy is the elastic strain energy which is released as the crack spreads. Griffith established the following criterion for the propagation of a crack: *A crack will propagate when the decrease in elastic strain energy is at least equal to the energy required to create the new crack surface.* This criterion can be used to determine the magnitude of the tensile stress which will just cause a crack of a certain size to propagate as a brittle fracture.

Consider the crack model shown in Fig. 7-4. The thickness of the plate is negligible, and so the problem can be treated as one in plane stress. The cracks are

**Fig. 7-4** Griffith crack model.

[1] A. A. Griffith, *Philos. Trans. R. Soc. London*, vol. 221A, pp. 163–198, 1920; *First Int. Congr. Appl. Mech.*, Delft, 1924, p. 55; this paper has been reprinted with annotations in *Trans. Am. Soc. Met.*, vol. 61, pp. 871–906, 1968.

assumed to have an elliptical cross section . For a crack at the interior the length is
$2c$, while for an edge crack it is $c$. The effect of both types of crack on the fracture
behavior is the same. The stress distribution for an elliptical crack was determined
by Inglis.[1]  A decrease in strain energy results from the formation of a crack. The
elastic strain energy per unit of plate thickness is equal to

$$U_E = - \frac{\pi c^2 \sigma^2}{E} \qquad (7\text{-}11)$$

where $\sigma$ is the tensile stress acting normal to the crack of length $2c$.  A negative
sign is used because growth of the crack releases elastic strain energy.  The surface
energy due to the presence of the crack is

$$U_s = 4c\gamma_s \qquad (7\text{-}12)$$

The total change in potential energy resulting from the creation of the crack is

$$\Delta U = U_S + U_E \qquad (7\text{-}13)$$

According to Griffith's criterion, the crack will propagate under a constant applied
stress $\sigma$ if an incremental increase in crack length produces no change in the total
energy of the system; i.e., the increased surface energy is compensated by a decrease
in elastic strain energy.

$$\frac{d\Delta U}{dc} = 0 = \frac{d}{dc}\left(4c\gamma_s - \frac{\pi c^2 \sigma^2}{E}\right)$$

$$4\gamma_s - \frac{2\pi c \sigma^2}{E} = 0$$

$$\sigma = \left(\frac{2E\gamma_s}{\pi c}\right)^{\frac{1}{2}} \qquad (7\text{-}14)$$

Equation (7-14) gives the stress required to propagate a crack in a brittle material
as a function of the size of the microcrack.  Note that this equation indicates
that the fracture stress is inversely proportional to the square root of the crack
length.  Thus, increasing the crack length by a factor of 4 reduces the fracture stress
by one-half.

For a plate which is thick compared with the length of the crack (plane strain)
the Griffith equation is given by

$$\sigma = \left[\frac{2E\gamma_s}{(1 - v)^2 \pi c}\right]^{\frac{1}{2}} \qquad (7\text{-}15)$$

Analysis of the three-dimensional case, where the crack is a very flat oblate spheroid,[2]
results only in a modification to the constant in Griffith's equation.  Therefore,

---

[1] C. E. Inglis, *op. cit.*; Eq. (7-11) can be understood if we consider that the strain energy resides in
a circular region of radius $c$ around the crack. The strain energy per unit volume is $\sigma^2/2E$, so that $U_E$
per unit thickness is $\sigma^2(\pi c^2)/2E$. The factor $\frac{1}{2}$ drops out in the more rigorous analysis.

[2] R. A. Sack, *Proc. Phys. Soc. London*, vol. 58, p. 729, 1946.

the simplification of considering only, the two-dimensional case introduces no large error.

Let us look briefly at the equation for fracture stress derived from a stress concentration point of view, Eq. (7-9), and the Griffith equation, Eq. (7-14). Equation (7-9) can be written as

$$\sigma_f = \left( \frac{2E\gamma_s}{\pi c} \frac{\pi \rho_t}{8a_0} \right)^{\frac{1}{2}} \quad \text{or} \quad \sigma_f \approx \left( \frac{2E\gamma_s}{\pi c} \frac{\rho_t}{3a_0} \right)^{\frac{1}{2}}$$

When $\rho_t = 3a_0$ this equation reduces to the Griffith equation. Thus, $\rho_t = 3a_0$ is the lower limit of the effective radius of an elastic crack. In other words, $\sigma_f$ cannot approach zero as $\rho_t$ approaches zero. When $\rho_t < 3a_0$ the stress to produce brittle fracture is given by Eq. (7-14), but when $\rho_t > 3a_0$ the fracture stress is given by Eq. (7-9).

The Griffith's equation shows a strong dependence of fracture strength on crack length. Griffith's theory satisfactorily predicts the fracture strength of a completely brittle material such as glass. In glass, reasonable values of crack length of about 1 $\mu$m are calculated from Eq. (7-14). For zinc crystals Griffith's theory predicts a critical crack length of several millimeters. This average crack length could easily be greater than the thickness of the specimen, and therefore the theory does not apply.

The importance of the surface energy term can be demonstrated by carrying out the fracture in solutions of surface active chemicals. The fracture stress of ice when tested in bending in air is about 150 psi. If the bend specimen is sprayed with methyl cholride, to lower $\gamma_s$, the fracture stress is reduced to about 75 psi. The sensitivity of the fracture of brittle solids to surface conditions has been termed the *Joffe Effect*.[1] Use is made of surface active agents to make rock drilling more easy. In metallurgical systems the surface energy can be reduced by surface adsorption of an element in solid solution. For example, the addition of 0.5 % Sb to Cu reduces the surface energy from about 1,800 to 1,000 erg/cm$^2$. Since solute concentration often builds up in grain boundaries, this can lead to intergranular embrittlement.

## 7-5 FRACTURE MECHANICS

It is well established that even metals which fail in a completely brittle manner have undergone some plastic deformation prior to fracture. This is substantiated by x-ray diffraction studies of fracture surfaces[2] and by metallographic studies of

[1] A. F. Joffe, "The Physics of Crystals," McGraw-Hill Book Company, New York, 1928.
[2] E. P. Klier, *Trans. Am. Soc. Met.*, vol. 43, pp. 935–957, 1951; L. C. Chang, *J. Mech. Phys. Solids*, vol. 3, pp. 212–217, 1955; D. K. Felbeck and E. Orowan, *Welding J.*, vol. 34, pp. 570s–757s, 1955.

fracture (see Sec. 7-7). Therefore, Griffith's equation for the fracture stress does not apply for metals. One way of realizing that the fracture stress of a material which undergoes plastic deformation before fracture is greater than that of a truly brittle (elastic) material is to consider Eq. (7-9). Plastic deformation at the root of the crack would be expected to blunt the tip of the crack and increase $\rho_t$, thus increasing the fracture stress.

Orowan[1] suggested that the Griffith equation would be made more compatible with brittle fracture in metals by the inclusion of a term $\gamma_p$ expressing the plastic work required to extend the crack wall.

$$\sigma_f = \left[\frac{2E(\gamma_s + \gamma_p)}{\pi c}\right]^{\frac{1}{2}} \approx \left(\frac{E\gamma_p}{c}\right)^{\frac{1}{2}} \qquad (7\text{-}16)$$

The surface-energy term can be neglected since estimates of the plastic-work term are about $10^5$ to $10^6$ erg/cm$^2$ compared with values of $\gamma_s$ of about 1,000 to 2,000 erg/cm$^2$.

A similar approach by Irwin[2] laid the foundation for the important area of *fracture mechanics*. Irwin proposed that fracture occurs at a fracture stress corresponding to a critical value of the *crack-extension force* $\mathscr{G}_c$, where Eq. (7-16) is rewritten as

$$\sigma_f = \left(\frac{E\mathscr{G}_c}{\pi c}\right)^{\frac{1}{2}} \qquad (7\text{-}17)$$

Thus,

$$\mathscr{G} = \pi c \frac{\sigma^2}{E} \qquad (7\text{-}18)$$

The crack-extension force $\mathscr{G}$ has units of inch pounds per square inch (in-lb/in$^2$). $\mathscr{G}$ may also be considered the *strain-energy release rate*, i.e., the rate of loss of energy from the elastic stress field to the inelastic process of crack extension. Note that the crack sharpness $\rho/c$ does not appear in Eq. (7-17). In effect, this is part of the $\mathscr{G}_c$ value that is associated with a particular material.

Figure 7-5 shows how $\mathscr{G}$ can be measured. A single-edge notch specimen is loaded axially through pins. The sharpest possible notch is produced by introducing a fatigue crack at the root of the machined notch. The displacement of this crack as a function of the axial force is measured with a strain-gage clip gage at

[1] E. Orowan, in "Fatigue and Fracture of Metals," Symposium at Massachusetts Institute of Technology, John Wiley & Sons, Inc., New York, 1950.

[2] G. R. Irwin, Fracture in "Encyclopedia of Physics," vol. VI, Springer, Heidelberg, 1958; G. R. Irwin, J. A. Kies, and H. L. Smith, *Am. Soc. Test. Mater. Proc.*, vol. 58, pp. 640–660, 1958.

**Fig. 7-5** Determination of crack-extension force $\mathscr{G}$.

the entrance to the notch. Load vs. displacement curves are determined for different length notches, where $P = M\delta$. The elastic strain energy is given by

$$U_0 = \frac{1}{2}P\delta = \frac{P^2}{2M}$$

Since $P/M = \delta = $ constant (the fixed-grip case),

$$\left(\frac{\partial P}{\partial c}\right)\frac{1}{M} + P\frac{\partial(1/M)}{\partial c} = 0$$

$$\frac{\partial P}{\partial c} = -PM\frac{\partial(1/M)}{\partial c}$$

But the crack extension force is defined as

$$\mathscr{G} = -\frac{\partial U_0}{\partial c} = -\frac{1}{2}\frac{P}{M}\frac{\partial P}{\partial c}$$

$$\mathscr{G} = \frac{1}{2}P^2\frac{\partial(1/M)}{\partial c} \qquad (7\text{-}19)$$

Therefore, $\mathscr{G}$ is a function of the load and the slope of the compliance $(1/M)$ vs. crack-length curve.[1] The critical value of crack-extension force $\mathscr{G}_c$ is evaluated[2] at the load for which the $P$-$\delta$ curve deviates sharply from linearity.

---

[1] The same equation would be derived for $\mathscr{G}$ when the plate is subjected to a constant load $P_1$, only now $U_0$ increases with $c$, while for the fixed-grip case $U_0$ decreases with crack length.

[2] See Chap. 14 for additional details on fracture toughness testing.

**Fig. 7-6** Model for equations for stresses at a point near a crack.

The stress distribution at the crack tip in a thin plate for an elastic solid in terms of the coordinates shown in Fig. 7-6 is given by Eqs. 7-20.

$$\sigma_x = \sigma\left(\frac{c}{2r}\right)^{\frac{1}{2}}\left[\cos\frac{\theta}{2}\left(1 - \sin\frac{\theta}{2}\sin\frac{3\theta}{2}\right)\right]$$

$$\sigma_y = \sigma\left(\frac{c}{2r}\right)^{\frac{1}{2}}\left[\cos\frac{\theta}{2}\left(1 + \sin\frac{\theta}{2}\sin\frac{3\theta}{2}\right)\right] \qquad (7\text{-}20)$$

$$\tau_{xy} = \sigma\left(\frac{c}{2r}\right)^{\frac{1}{2}}\left[\sin\frac{\theta}{2}\cos\frac{\theta}{2}\cos\frac{3\theta}{2}\right]$$

where $\sigma$ = gross nominal stress = $P/wt$. These equations are valid for $c > r > \rho$. For an orientation directly ahead of the crack ($\theta = 0$)

$$\sigma_x = \sigma_y = \sigma\left(\frac{c}{2r}\right)^{\frac{1}{2}} \qquad \text{and} \qquad \tau_{xy} = 0$$

Irwin pointed out that Eqs. (7-20) indicate that the *local stresses* near a crack depend on the product of the nominal stress $\sigma$ and the square root of the half-flaw length. He called this relationship the *stress intensity factor K*, where for a sharp elastic crack in an infinitely wide plate, $K$ is defined as

$$K = \sigma\sqrt{\pi c} \qquad (7\text{-}21)$$

Note the $K$ has the unusual dimensions of $psi\sqrt{in}$ or $MN/m^{\frac{3}{2}}$ or $MPa\sqrt{mm}$. Using this definition for $K$, the equations for the stress field at the end of a crack can be written

$$\sigma_x = \frac{K}{\sqrt{2\pi r}}\left[\cos\frac{\theta}{2}\left(1 - \sin\frac{\theta}{2}\sin\frac{3\theta}{2}\right)\right]$$

$$\sigma_y = \frac{K}{\sqrt{2\pi r}}\left[\cos\frac{\theta}{2}\left(1 + \sin\frac{\theta}{2}\sin\frac{3\theta}{2}\right)\right] \qquad (7\text{-}22)$$

$$\tau_{xy} = \frac{K}{\sqrt{2\pi r}}\left(\sin\frac{\theta}{2}\cos\frac{\theta}{2}\cos\frac{3\theta}{2}\right)$$

The stress intensity factor $K$ is a convenient way of describing the stress distribution around a flaw. If two flaws of different geometry have the same value of $K$, then the stress fields around each of the flaws are identical. Values of $K$ for many geometrical cracks and many types of loading may be calculated[1] with the theory of elasticity. For the general case the stress intensity factor is given by

$$K = \alpha\sigma\sqrt{\pi c} \qquad (7\text{-}23)$$

where $\alpha$ is a parameter that depends on the specimen and crack geometry. As an example, for a plate of width $w$ loaded in tension with a centrally located crack of length $2c$

$$K = \sigma\sqrt{\pi c}\left(\frac{w}{\pi c}\tan\frac{\pi c}{w}\right)^{1/2} \qquad (7\text{-}24)$$

In dealing with the stress intensity factor there are several modes of deformation that could be applied to the crack. There have been standardized as shown in Fig. 7-7. Mode I, the crack-opening mode, refers to a tensile stress applied in the $y$ direction normal to the faces of the crack. This is the usual mode for fracture-toughness tests and a critical value of stress intensity determined for this mode would be designated $K_{Ic}$. Mode II, the forward shear mode, refers to a shear stress applied normal to the leading edge of the crack but in the plane of the crack. Mode III, the parallel shear mode, is for shearing stresses applied parallel to the leading edge of the crack. Mode I loading is the most important situation. There are two extreme cases for mode I loading. With thin plate-type specimens the stress state is plane stress, while with thick specimens there is a plane-strain condition. The plane-strain condition represents the more severe stress state and the

**Fig. 7-7** Crack-deformation modes.

[1] An excellent compendium of these relations is given by P. C. Paris and G. C. M. Sih, in J. E. Srawley and W. F. Brown (eds.), "Fracture Toughness Testing," p. 30, ASTM STP No. 381, Philadelphia, Pa. 1965.

values of $K_c$ are lower than for plane-stress specimens. Plane strain values of critical stress intensity factor $K_{Ic}$ are valid material properties, independent of specimen thickness, to describe the *fracture toughness* of strong materials like heat-treated steel, aluminum, and titanium alloys. Further details on fracture-toughness testing are given in Chap. 14.

While the crack -extension force $\mathscr{G}$ has a more direct physical significance to the fracture process, the stress intensity factor $K$ is preferred in working with fracture mechanics because it is more amenable to analytical determination. By combining Eqs. (7-18) and (7-21), we see that the two parameters are simply related.

$$K^2 = \mathscr{G}E \qquad \text{plane stress} \qquad (7\text{-}25)$$

$$K^2 = \mathscr{G}E/(1 - v^2) \qquad \text{plane strain} \qquad (7\text{-}26)$$

## 7-6 FRACTURE OF SINGLE CRYSTALS

The brittle fracture of single crystals is considered to be related to the resolved normal stress on the cleavage plane. Sohncke's law states that fracture occurs when the resolved normal stress reaches a critical value. Considering the situation used to develop the resolved shear stress for slip (Fig. 4-18), the component of the tensile force which acts normal to the cleavage plane is $P \cos \phi$, where $\phi$ is the angle between the tensile axis and the normal to the plane. The area of the cleavage plane is $A/(\cos \phi)$. Therefore, the critical normal stress for brittle fracture is

$$\sigma_c = \frac{P \cos \phi}{A/\cos \phi} = \frac{P}{A} \cos^2 \phi \qquad (7\text{-}27)$$

The cleavage planes for certain metals and values of the critical normal stress are given in Table 7-1.

Although Sohncke's law has been accepted for over 25 years, it is not based on very extensive experimental evidence. Doubt was cast on its reliability by fracture studies[1] on zinc single crystals at $-77$ and $-196°C$. The resolved normal cleavage stress was found to vary by over a factor of 10 for a large difference in orientation of the crystals. This variation from the normal-stress law may be due to plastic strain prior to fracture, although it is doubtful that this could account for the observed discrepancy.

Several modes of ductile fracture in single crystals are shown in Fig. 7-1. Under certain conditions hcp metals tested at room temperature or above will shear only on a restricted number of basal planes. Fracture will then occur by "shearing off" (Fig. 7-1b). More usually, slip will occur on systems other than

[1] A. Deruyttere and G. B. Greenough, *J. Inst. Met.*, vol. 84, pp. 337–345, 1955–1956.

**Table 7-1  CRITICAL NORMAL STRESS FOR CLEAVAGE OF SINGLE CRYSTALS†**

| Metal | Crystal lattice | Cleavage plane | Temperature, °C | Critical normal stress, $kg/mm^2$ |
|-------|-----------------|----------------|-----------------|-----------------------------------|
| Iron | bcc | (100) | −100 | 26 |
|  |  |  | −185 | 27.5 |
| Zinc (0.03% Cd) | hcp | (0001) | −185 | 0.19 |
| Zinc (0.13% Cd) | hcp | (0001) | −185 | 0.30 |
| Zinc (0.53% Cd) | hcp | (0001) | −185 | 1.20 |
| Magnesium | hcp | (0001), (10$\bar{1}$1) (10$\bar{1}$2), (10$\bar{1}$0) |  |  |
| Tellurium | Hexagonal | (10$\bar{1}$0) | 20 | 0.43 |
| Bismuth | Rhombohedral | (111) | 20 | 0.32 |
| Antimony | Rhombohedral | (11$\bar{1}$) | 20 | 0.66 |

† Data from C. S. Barrett, "Structure of Metals," 2d ed., McGraw-Hill Book Company, New York, 1952; N. J. Petch, The Fracture of Metals, in "Progress in Metal Physics," vol. 5, Pergamon Press, Ltd., London, 1954.

the basal plane, so that the crystal necks down and draws down almost to a point before rupture occurs. The usual mode of fracture in fcc crystals is the formation of a necked region due to multiple slip, followed by slip on one set of planes until fracture occurs. The crystal can draw down to a chisel edge or a point (if multiple slip continues to fracture). The best stress criterion for ductile fracture in fcc metals appears to be the resolved shear stress on the fracture plane (which is usually the slip plane).

The mode of fracture in bcc iron crystals is strongly dependent on temperature, purity, heat treatment, and crystal orientation.[1] Crystals located near the [001] corner of the stereographic triangle show no measurable ductility when tested in tension at −196°C, while crystals closer to [$\bar{1}$11] and [011] orientations may rupture by drawing down to a chisel edge when tested at the same temperature. An interesting point is that the change from brittle to ductile fracture is very sharp, occurring over a change in orientation of only about 2°.

## 7-7  METALLOGRAPHIC ASPECTS OF FRACTURE

Because of the prominence of the Griffith theory, it has been natural for metallurgists to use their microscopes in a search for Griffith cracks in metals. However, based on observations up to the magnifications available with the electron micro-

[1] N. P. Allen, B. E. Hopkins, and J. E. McLennan, *Proc. R. Soc. London*, vol. 234A, p. 221, 1956.

scope, there is no reliable evidence that Griffith cracks exist in metals in the un-stressed condition. There is, however, a considerable amount of experimental evidence to show that microcracks can be produced by plastic deformation.

Metallographic evidence of the formation of microcracks at nonmetallic inclusions in steel as a result of plastic deformation has existed for a number of years. These microcracks do not necessarily produce brittle fracture. However, they do contribute to the observed anisotropy in the ductile-fracture strength. The fact that vacuum-melted steel, which is very low in inclusions, shows a reduction in the fracture anisotropy supports the idea of microcracks being formed at second-phase particles.

An excellent correlation between plastic deformation, microcracks, and brittle fracture was made by Low.[1] He showed that for mild steel of a given grain size tested at $-196°C$ brittle fracture occurs in tension at the same value of stress that is required to produce yielding in compression. Microcracks only one or two grains long were observed. More detailed studies of the conditions for microcrack formation have been made[2] with tensile tests on mild steel at carefully controlled subzero temperatures. Figure 7-8 illustrates a typical microcrack found in a specimen before it fractured.

**Fig. 7-8** Microcracks produced in iron by tensile de-formation at $-140°C$ (250×). (*Courtesy G. T. Hahn.*)

[1] J. R. Low, I.U.T.A.M., Madrid Colloqium, "Deformation and Flow of Solids," p. 60, Springer-Verlag OHG, Berlin, 1956.
[2] G. T. Hahn, W. S. Owen, B. L. Averbach, and M. Cohen, *Welding J.*, vol. 38, pp. 367s–376s, 1959.

Detailed experiments demonstrate that the cracks responsible for brittle-cleavage-type fracture are not initially present in the material but are produced by the deformation process. The fact that at appropriate temperatures appreciable numbers of microcracks are present shows that the conditions for the initiation of a crack are not necessarily the same as conditions for the propagation of a crack The process of cleavage fracture should be considered to be made up of three steps. (1) plastic deformation to produce dislocation pile-ups, (2) crack initiation, and (3) crack propagation.

The initiation of microcracks can be greatly influenced by the presence and nature of second-phase particles.[1] A common situation is for the particle to crack during deformation. Resistance to cracking improves if the particle is well bonded to the matrix. Small particles ($r < 1$ $\mu$m) and spherical particles are more resistant to cracking. A brittle constituent at the grain boundaries, such as thin cementite films in low-carbon steel,[2] is a particularly potent source of microcracks. If the dispersion of second-phase particles is readily cut by the dislocations, then there will be planar slip and relatively large dislocation pile-ups will occur. This will lead to high stresses, easy initiation of microcracks, and brittle behavior. However, if the second-phase consists of a dispersion of fine impenetrable particles, the slip distance is greatly reduced and, correspondingly, the number of dislocations that can be sustained in a pile-up is reduced. Also, once cracks are formed they will be forced to bow between the particles, increasing the effective interfacial energy. Thus, fine dispersions of particles can lead to increased toughness under the proper circumstances. A soft, ductile phase can also impart ductility to a brittle matrix. The ductile phase must be thick enough to yield before large dislocation pile-ups are created against it.

Ductile fracture (Sec. 7-10) starts with the initiation of *voids*, most commonly at second-phase particles. The particle geometry, size, and bonding with the matrix are important parameters. Figure 7-9 shows the influence of common second phases in steel on the tensile ductility.

In Sec. 5-6 we saw that slip on intersecting (110) planes was proposed as a mechanism for brittle crack initiation in bcc metals. While this has not been strongly substantiated in metals, detailed investigations of crack nucleation by slip have been made[3] in MgO crystals. Interpenetration of the slip bands produces microcracks. In metals there is growing recognition that cross slip of screw dislocations broadens the slip band and makes it more difficult to crack a particle when the slip band impinges on an obstacle. For example,[4] the ductile-to-brittle tran-

[1] R. F. Decker, *Metall. Trans.*, vol. 4, pp. 2508–2611, 1973.
[2] C. J. McMahon, Jr., and M. Cohen, *Acta Metall.*, vol. 13, p. 591, 1965.
[3] T. L. Johnson, R. J. Stokes, and C. H. Li, *Philos. Mag.* vol. 4, p. 1361, 1959.
[4] A. R. Rosenfield, E. Votava, and G. T. Hahn, in "Ductility," pp. 75–76, American Society for Metals, Metals Park, Ohio, 1968.

**Fig. 7-9** Effect of second-phase particles on tensile ductility. (*From T. Gladman, B. Holmes, and L. D. McIvor, "Effect of Second-Phase Particles on the Mechanical Properties of Steel," p. 78, Iron and Steel Institute, London, 1971.*)

sition temperatures for the bcc metals columbium and tungsten are about $-120°C$ and $300°C$, respectively. Dislocations in columbium show profuse cross slip, but those in tungsten do not cross slip at all. Mechanical twins in bcc metals have been shown[1] to be responsible for the initiation of microcracks. The most effective barrier is the intersection of a moving twin with a stationary twin.

Most brittle fractures occur in a transgranular manner. However, if the grain boundaries contain a film of brittle constituent, as in sensitized austenitic stainless steel or molybdenum alloys containing oxygen, nitrogen, or carbon, the fracture will occur in an intergranular manner. Intergranular failure can also occur without the presence of a microscopically visible precipitate at the grain boundaries. Apparently, segregation at the grain boundaries can lower the surface energy sufficiently to cause intergranular failure. The embrittlement produced by the addition of antimony to copper and oxygen to iron and the temper embrittlement of alloy steels are good examples.

Sometimes a considerable amount of information can be obtained by examining the surfaces of the fracture at fairly high magnifications with an optical microscope. This type of examination is known as *fractography*.[2] At high magni-

[1] D. Hull, in D. C. Drucker and J. J. Gilman (eds.), " Fracture in Solids," pp. 417–453, Interscience Publishers, Inc., New York, 1963.
[2] C. A. Zappfe and C. O. Worden, *Trans. Am. Soc. Met.*, vol. 42, pp. 557–603, 1950.

**Fig. 7-10**   Cleavage steps and river pattern on a cleavage surface.

fication, transgranular-cleavage surfaces usually contain a large number of cleavage steps and a " river pattern " of branching cracks (Fig. 7-10).  These are indications of the absorption of energy by local deformation.  The surfaces of intergranular brittle fractures are much smoother, with a general absence of cleavage steps. From the appearance of the fracture surface, the energy absorbed in an inter-granular fracture is much lower than for transgranular cleavage.

A very powerful technique for studying fracture is to replicate the fracture surface with a plastic and examine its structure in the electron microscope.  Not only is the increased magnification over the optical microscope of importance, but of even greater benefit, the electron microscope has much greater depth of focus (of considerable importance on rough fracture surfaces).  *Electron Fractography*[1] has proven to be such a versatile tool that a manual[2] for fracture identification has been developed.  The need for replicating the surface has been removed by the advent of the scanning electron microscope (SEM).  The excellent three-dimensional detail provided by SEM is shown in Fig. 7-11.

---

[1] C. D. Beachem and R. M. N. Pelloux, in " Fracture Toughness Testing and Its Applications," ASTM STP-381, p. 210, 1964.
[2] " Fractography and Atlas of Fractographs," vol. 9, " Metals Handbook," American Society for Metals, Metals Park, Ohio, 1974.

**Fig. 7-11** Fracture surface of fiber-reinforced composite. Note cup-and-cone fracture of ductile stainless steel fibers and extreme ductilility of aluminum matrix which has thinned down to a knife edge. Fractograph obtained by scanning electron microscopy (300×). (*Courtesy of A. Pattnaik.*)

## 7-8 DISLOCATION THEORIES OF BRITTLE FRACTURE

The process of brittle fracture consists of three stages.

1. Plastic deformation which involves the pile-up of dislocations along their slip planes at an obstacle.
2. The buildup of shear stress at the head of the pile-up to nucleate a microcrack.
3. In some cases the stored elastic strain energy drives the microcrack to complete fracture without further dislocation movement in the pile-up. More typically in metals, a distinct growth stage is observed in which an increased stress is required to propagate the microcrack.

The idea that the high stresses produced at the head of a dislocation pile-up could produce fracture was first advanced by Zener.[1] The model is shown in

---

[1] C. Zener, The Micro-mechanism of Fracture, in "Fracturing of Metals," American Society for Metals, Metals Park, Ohio, 1948.

**Fig. 7-12** Model of microcrack formation at a pile-up of edge dislocations.

Fig. 7-12. The shear stress acting on the slip plane squeezes the dislocations together. At some critical value of stress the dislocations at the head of the pile-up are pushed so close together that they coalesce into a wedge crack or cavity dislocation of height $nb$ and length $2c$. Stroh[1] has shown that provided the stress concentration at the head of the pile-up is not relieved by plastic deformation, then the tensile stress at the pile-up is given by Eq. (5-39), which can be equated to the theoretical cohesive stress, Eq. (7-7). Thus,

$$(\tau_s - \tau_i)\left(\frac{L}{r}\right)^{1/2} = \left(\frac{E\gamma_s}{a_0}\right)^{1/2}$$

and microcrack nucleation occurs at

$$\tau_s = \tau_i + \left(\frac{Er\gamma_s}{La_0}\right)^{1/2} \qquad (7\text{-}28)$$

where $L$ is the length of the blocked slip band and $r$ is the distance from the tip of the pile-up to the point where the crack is forming. If we let $r \approx a_0$ and $E \approx 2G$, then Eq. (7-28) becomes

$$\tau_s = \tau_i + \left(\frac{2G\gamma_s}{L}\right)^{1/2} \qquad (7\text{-}29)$$

But, from Eq. (5-36), the number of dislocations in the slip band can be expressed as

$$nb \approx L\frac{\tau_s - \tau_i}{G} \qquad (7\text{-}30)$$

On eliminating $L$ from Eqs. (7-29) and (7-30) we obtain

$$(\tau_s - \tau_i)nb \approx 2\gamma_s \qquad (7\text{-}31)$$

---

[1] A. N. Stroh, *Adv. Phys.*, vol. 6, p. 418, 1957; the development given here follows that of A. S. Tetelman and A. J. McEvily, Jr., "Fracture of Structural Materials," chap. 6, John Wiley & Sons, Inc., New York, 1967.

This form of the equation for microcrack nucleation was proposed by Cottrell.[1] It has the direct physical significance that a crack will form when the work done by the applied shear stress in producing a displacement $nb$ equals the work done in moving the dislocations against the friction stress plus the work in producing the new fracture surfaces. It is interesting that Eq. (7-31) does not contain a crack-length term $2c$. Thus, the crack grows by plastic deformation so long as the dislocation source continues to force dislocations into the pile-up. It should be realized that only shearing stresses are involved in forcing the dislocations together. Tensile stresses are not involved in the microcrack nucleation process, and indeed, cleavage cracks can be formed in compression.[2] However, a tensile stress is needed to make the microcracks propagate. The fact that normal stresses (hydrostatic stress state) are not involved in microcrack nucleation leads to the conclusion that the crack-propagation stage is ordinarily more difficult than crack initiation in metals, since experience shows that fracture is strongly influenced by the hydrostatic component of stress (Sec. 7-14). Also, the occurrence of nonpropagating microcracks supports this point of view. Modification[3] of Stroh's equations to allow for a nonuniform stress field on the pile-up and dislocations of both signs in the pile-up has shown that the stress required to nucleate a microcrack is significantly lower than originally thought.

There is strong evidence that in most engineering materials the most difficult step is the propagation of deformation-produced microcracks through a strong barrier such as a grain boundary. Thus, the grain size will have a strong influence on brittle-fracture behavior. Petch[4] found that the grain-size dependence of brittle fracture in iron and steel could be expressed by

$$\sigma_f = \sigma_i + k_f D^{-\frac{1}{2}} \qquad (7\text{-}32)$$

This is analogous to Eq. (6-7) for the grain-size dependence of the yield point and the flow stress. To develop the dislocation model for Eq. (7-32), we express Eq. (7-31) in terms of normal stress

$$\sigma nb \approx 4\gamma_s \qquad (7\text{-}33)$$

and adopt a model in which the dislocation source is at the center of a grain of diameter $D$, so that $L = D/2$. Thus, substituting Eq. (7-30) into Eq. (7-31),

$$\sigma(\tau_s - \tau_i)D = 8G\gamma_s$$

But, experience shows that microcracks form when the shear stress equals the yield stress so that, from Eq. (6-7)

$$\tau_s - \tau_i = \tau_0 - \tau_i = 2k'D^{-\frac{1}{2}}$$

[1] A. H. Cottrell, *Trans. Metall. Soc. AIME*, vol. 212, pp. 192–203, 1958.
[2] A. Gilbert, et al., *Acta Metall.*, vol. 12, p. 754, 1964.
[3] E. Smith and J. T. Barnby, *Met. Sci. J.*, vol. 1, pp. 56–64, 1967.
[4] N. J. Petch, *J. Iron Steel Inst. London*, vol. 174, p. 25, 1953.

and

$$\sigma_f = \frac{4G\gamma_s}{k'D^{1/2}} \qquad (7\text{-}34)$$

This equation represents the stress required to propagate a microcrack of length $D$ in brittle fracture.

Cottrell[1] has reformulated Eq. (7-31) so that the important variables in brittle fracture are easily shown. This equation is

$$(\tau_i D^{1/2} + k')k' = G\gamma_s \beta \qquad (7\text{-}35)$$

where $\tau_i$ = the resistance of the lattice to dislocation movement

$\quad k'$ = a parameter related to the release of dislocations from a pile-up

$\quad \gamma_s$ = the effective surface energy and includes the energy of plastic deformation

$\quad \beta$ = a term which expresses the ratio of shear stress to normal stress. For torsion $\beta = 1$; for tension $\beta = \frac{1}{2}$; for a notch $\beta \approx \frac{1}{3}$.

Equation (7-35) expresses the limiting condition for the formation of a propagating crack from a pile-up of glide dislocations. If the left side of the equation is smaller than the right side, a microcrack can form but it cannot grow. This is the case of nonpropagating microcracks. When the left side of the equation is greater than the right, a propagating brittle fracture can be produced at a shear stress equal to the yield stress. Thus, this equation describes a ductile-to-brittle transition. Since many of the metallurgical parameters change with temperature, there would be a *transition temperature* at which the fracture would change from ductile to brittle.

The parameter $k'$ is important since it determines the number of dislocations that are released into a pile-up when a source is unlocked. Materials with a high value of $k'$ (e.g., iron and molybdenum) are more prone to brittle fracture than materials with lower values (e.g., columbium and tantalum). Strengthening mechanisms which depend on dislocation locking are likely to result in brittleness. The importance of grain size as a parameter in brittle fracture is shown in Fig. 7-13, where below a certain grain size there is measurable ductility at fracture. Actually, the grain size term in Eq. (7-35) should be interpreted more properly as the slip-band length. Generally this is controlled by the grain size, but in an alloy containing a fine precipitate the particle spacing will determine the slip distance. The very fine carbide precipitates in quenched and tempered steel which cause short slip distances result in the low transition temperature for this material.

A high value of frictional resistance leads to brittle fracture, since high stresses must be reached before yielding occurs. The directional bonding in ceramics results in high $\tau_i$ and high inherent hardness and brittleness. In bcc metals the frictional resistance increases rapidly as the temperature falls below room tem-

---

[1] Cottrell, op. cit.

**Fig. 7-13** Effect of grain size on the yield and fracture stresses for a low-carbon steel tested in tension at $-196°C$. (*From J. R. Low, in: "Relation of Properties to Micro-structure," American Society for Metals, Metals Park, Ohio, 1954. By permission of the publishers.*)

perature[1] and thus leads to a ductile-to-brittle transition. The $\tau_i$ term enters into Eq. (7-35) as the product of $D^{1/2}$, so that a fine-grain metal can withstand higher values of $\tau_i$ (lower temperatures) before becoming brittle. Many of the effects of the composition of steel on the ductile-to-brittle transition are due to changes in $D$, $k'$, or $\tau_i$. For example, manganese decreases the grain size and reduces $k'$, while silicon produces large grain size and increases $\tau_i$.

If the effective surface energy is large at a given temperature, then brittle fracture is suppressed. The contribution from plastic deformation will depend on the number of available slip systems and the number of mobile dislocations at the tip of the crack. Thus, zinc is brittle in large grain sizes because of its limited slip systems, while bcc metals can be brittle because impurities lock most of the mobile dislocations. Neither of these conditions exist in fcc metals and they do not ordinarily fail in brittle fracture. Various environmental factors such as corrosion or hydrogen penetration may lower the surface energy.

It is well known that the presence of a notch greatly increases the tendency for brittle fracture. The complicated effects of a notch will be considered in Sec. 7-11. The effect of a notch in decreasing the ratio of shear stress to tensile

---

[1] From Eq. (6-7) it follows that $\tau_0 D^{1/2} = \tau_i D^{1/2} + k'$, so that Eq. (7-35) can be written $\tau_0 k' D^{1/2} = G\gamma_s\beta$. The yield stress $\tau_0$ for bcc metals increases rapidly with decreasing temperature.

stress is covered in Eq. (7-35) by the constant $\beta$. Strain rate or rate of loading is not an explicit term in Eq. (7-35). However, it is well known that increasing strain rate raises both $\tau_i$ and $\tau_0$. Strain rate interacts with the notch effect in an important way. Because the deformation is localized in the vicinity of the notch, there is a strain-concentration effect and the local strain is much higher than the average value.

If fracture is to be brittle, the plastic deformation must be minimized as the crack propagates. A phenomenon which is important to the brittle-fracture process is *delayed yielding*, in which there is a certain delay time before plastic deformation occurs. Thus, a fast-moving crack can pass a spot in the metal before slip occurs. Delay times[1] can range from 1 to $10^{-6}$ s, decreasing with increasing stress and increasing with decreasing temperature.

## 7-9 CRACK PROPAGATION

Brittle fracture is not possible unless the cracks which are nucleated can propagate at a high velocity throughout the metal. Mott[2] has made an analysis of the velocity of a crack in an ideal elastic, isotropic medium. The elastic energy that is released by the movement of the crack is the driving force. This must be balanced by the surface energy of the new surface that is created and the kinetic energy associated with the rapid sidewise displacement of material on each side of the crack. The crack velocity $v$ is given by

$$v = Bv_0\left(1 - \frac{c_G}{c}\right) \qquad (7\text{-}36)$$

where $B$ is a constant and $v_0 = (E/\rho)^{1/2}$ is the velocity of sound in the material. The term $c_G$ is the length of a Griffith crack, as evaluated by Eq. (7-14), and $c$ is the actual crack length. When $c$ is large compared with $c_G$, Eq. (7-36) approaches the limiting value $Bv_0$. The constant has been evaluated[3] for the plane-stress condition and found to be $B \approx 0.38$. Table 7-2 shows that experimental values for the crack velocity in brittle materials agree quite well with the theoretical prediction that the limiting crack velocity is given by

$$v = 0.38v_0 = 0.38\left(\frac{E}{\rho}\right)^{1/2} \qquad (7\text{-}37)$$

There basically are three ways that a microcrack of half-length $c_0$ can develop into an unstable fracture.[4]

[1] D. S. Clark, *Trans. Am. Soc. Met.*, vol. 46, p. 34, 1954.
[2] N. F. Mott, *Engineering*, vol. 165, p. 16, 1948.
[3] D. K. Roberts and A. A. Wells, *Engineering*, vol. 178, p. 820, 1954.
[4] This section follows the development in Tetleman and McEvily, op. cit. pp. 58–71.

Table 7-2 VELOCITY OF PROPAGATION OF
BRITTLE FRACTURE

| Material | Observed velocity, ft/sec | $v/v_0$ | Reference |
|---|---|---|---|
| Steel | 6,000 | 0.36 | † |
| Fused quartz | 7,200 | 0.42 | ‡ |
| Lithium fluoride | 6,500 | 0.31 | § |

† T. S. Robertson, *J. Iron Steel Inst. London*, vol. 175,
p. 361, 1953.
‡ H. Schardin and W. Struth, *Glastech. Ber.*, vol. 16,
p. 219, 1958.
§ J. J. Gilman, C. Knudsen, and W. P. Walsh, *J. Appl.
Phys.*, vol. 29, p. 601, 1958.

1. *Go-No-Go Situation Under Increasing Applied Stress* Brittle materials
tested below the transition temperature undergo a rapidly propagating
brittle fracture when the applied stress reaches the fracture stress $\sigma_f$ (see
Fig. 7-14a).

2. *Slow Crack Growth Under Increasing Applied Stress* When the material
is less brittle than in case 1, the first microcrack which forms cannot propagate
unstably. The crack propagates by forming a series of microcracks or voids
which are joined together by plastic tearing. In Fig. 7-14b, crack growth begins
at $\sigma_1$ and proceeds in a series of bursts of crack growth until the crack becomes long enough and the stress level high enough to permit unstable
propagation. For example, at the lower end of the ductile and brittle transition temperature it is common for fracture to start out as fibrous rupture and
develop into unstable cleavage fracture when the crack becomes larger.

3. *Slow Crack Growth Under Constant Applied Stress* Slow crack growth
can occur at constant applied stress over a period of time when materials are
subjected to corrosion or to alternating (fatigue) loading. For this condition
the crack extends slowly until it becomes long enough to propagate unstably
at the applied stress level. The time that is required for crack growth decreases
as the stress level is increased.

Since complete brittle fracture of an engineering material (in the manner of
case 1 above) is rather rare, although not unknown, an important problem in the
fracture of materials is the development of methods for describing analytically the
formation of a plastic zone at the tip of a crack and the propagation of the crack
in the manner discussed in case 2 above. This is a rather difficult problem and only
a start has been made on developing useful engineering relationships. A suitable
fracture criterion must be related to the amount of plastic deformation that occurs

**Fig. 7-14** Three modes of crack growth. (*a*) Unstable brittle fracture without slow crack growth; (*b*) slow crack growth (ductile) followed by unstable brittle fracture; (*c*) slow growth at constant stress. (*After Tetleman and McEvily.*)

in a very small region at the crack tip. The *crack-tip displacement* concept[1] considers that the material ahead of the crack contains a series of miniature tensile specimens having a gage length $l$ and a width $w$ (Fig. 7-15). The length of the sample is determined by the root radius of the crack $\rho$, and the width is limited by those microstructural factors which control the ductility. In this model, crack growth occurs when the specimen adjacent to the crack is fractured. When the failure of the first specimen adjacent to the crack tip immediately causes the next specimen to fail, and so on, the overall fracture process is unstable and crack propagation occurs under decreasing stress (Fig. 7-14a). When each specimen does not fail immediately in turn, we have a situation of slow crack growth (Fig. 7-14b) where the applied stress must be increased for stable crack growth to continue.

If a thick plate is loaded in tension (mode I) so that plane-strain conditions prevail, the plastic deformation at the crack tip is confined to narrow bands with a thickness of the order of the diameter of the crack tip $2\rho$. The displacement (deformation) of the miniature tensile specimen at the crack tip is

$$\delta = \varepsilon l = \varepsilon 2\rho \qquad (7\text{-}38)$$

[1] A. H. Cottrell, *Proc. R. Soc. London*, vol. 285, p. 10, 1965; see also A. S. Telleman and A. J. McEvily, Jr., op. cit. pp. 60–78.

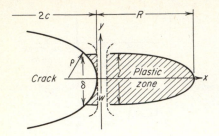

**Fig. 7-15** Model for crack-tip displacement concept.

Unstable fracture occurs when the strain in the specimen adjacent to the crack tip reaches the tensile ductility of the specimen $\varepsilon_f$. Thus, the fracture criterion is

$$\delta_c = 2\rho\varepsilon_f \qquad (7\text{-}39)$$

In *plane-stress* (thin-plate) tensile loading the strains at the crack tip are distributed over a distance of the order of the sheet thickness $t$. Thus,

$$\delta = \varepsilon l = \varepsilon t \qquad (7\text{-}40)$$

and the critical crack-opening displacement for fracture is

$$\delta_c = \varepsilon_f t \qquad (7\text{-}41)$$

The displacement of the faces of a crack can be measured with a displacement clip gage, and this can be related to the *crack-opening displacement* (COD) $\delta$ at the tip of the crack. Measures of COD are important in predicting the fracture toughness of lower yield strength, higher ductility materials (see Chap. 14).

The crack-opening displacement for a crack of length $2c$ in an infinite thin plate subjected to uniform tension $\sigma$ in a material where plastic deformation occurs at the crack tip is given by[1]

$$\delta = \frac{8\sigma_0 c}{\pi E} \ln\left(\sec\frac{\pi\sigma}{2\sigma_0}\right) \qquad (7\text{-}42)$$

Expanding the secant term and keeping only the first term results in

$$\delta = \frac{\pi\sigma^2 c}{E\sigma_0} \qquad (7\text{-}43)$$

However, remembering the definition of strain-energy release rate and dividing each by $\sigma_0$

$$\frac{\mathscr{G}}{\sigma_0} = \frac{\pi c \sigma^2}{E\sigma_0} \qquad (7\text{-}44)$$

we readily see that

$$\mathscr{G} = \sigma_0 \delta \qquad (7\text{-}45)$$

[1] G. T. Hahn and A. R. Rosenfeld, *Acta Metall.*, vol. 13, p. 293, 1965.

Since unstable crack extension occurs at a critical value of COD, the stress required to propagate this crack [from Eq. (7-17)] is

$$\sigma_f = \left(\frac{E}{\pi c}\sigma_0 \delta_c\right)^{\frac{1}{2}} \qquad (7\text{-}46)$$

Moreover, comparing this with Eq. (7-16), the plastic work required to keep a fast moving crack in motion is

$$\gamma_p = \sigma_0 \delta_c \qquad (7\text{-}47)$$

Substituting Eq. (7-39) into Eq. (7-45) gives

$$\mathcal{G}_c = 2\sigma_0 \rho \varepsilon_f \qquad (7\text{-}48)$$

This expression helps give physical significance to crack-extension force (and to stress intensity factor $K_c$). It shows that the larger the crack-tip radius and fracture strain that can be sustained prior to catastrophic fracture, the larger the energy absorption in the fracture process. A weakness of the equations just presented is that they do not contain any terms defining the microstructure of the material. Gerberich[1] has suggested that an appropriate expression is

$$\mathcal{G}_c = \pi\sigma_0 w^* \varepsilon_f \qquad (7\text{-}49)$$

where $w^*$ is the critical width from the crack tip over which the local strain reaches $\varepsilon_f$. This equation has been shown to give reasonable predictions of fracture toughness for a composite material consisting of ductile metal fibers and metal matrix.

It should be quite clear by now that the practical strength of an engineering material depends on the way the material reacts to stress concentrations such as sharp cracks. This property is called *toughness*. While toughness is defined in various ways, one approach due to Cottrell[2] is quite illuminating. The *total work* for fracture per unit area may be expressed as

$$2\gamma \approx \sigma_x \delta \qquad (7\text{-}50)$$

where $\sigma_x$ is a characteristic strength of an element ahead of the crack tip and $\delta$ is the displacement of the crack faces. For truly brittle fracture $\sigma_x \approx E/10$ and $\delta \approx b$, so that $\gamma \approx Eb/20$. In a truly ductile fracture $\gamma$ is so large ($> 10^8$ erg/cm$^2$) that general plastic deformation precedes fracture. For a material to possess both high strength and high toughness both $\sigma_x$ and $\delta$ must be large values. This is a difficult situation to achieve, and usually strong engineering materials are materials in which a compomise has been made between the two factors.

The plastic zone which develops at the crack tip in all but the most brittle materials is a factor which tends to restrict the usefulness of the fracture mechanics approach. Provided that the metal is not so ductile (like mild steel) that a large

[1] W. W. Gerberich, *J. Mater. Sci.*, vol. 5, pp. 283–294, 1970.
[2] A. H. Cottrell, *Proc. R. Soc., London*, vol. 202, pp. 2–9, 1964.

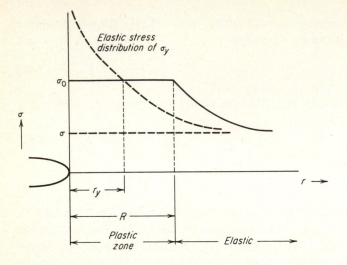

**Fig. 7-16** Distribution of longitudinal stress $\sigma_y$ ahead of a crack in plane-strain tensile loading. Normal applied stress $= \sigma$.

plastic zone extends from the crack tip, it is possible to make a correction for the plastic zone. Figure 7-16 shows the distribution of longitudinal stress $\sigma_y$ ahead of a crack in a thin plate under plane-stress tensile loading. If the material were perfectly elastic, the stress distribution would reach the high values shown in the dotted curve.

$$\sigma_y = \sigma\left(\frac{c}{2r}\right)^{1/2}$$

However, plastic yielding occurs at the crack tip out to $r = r_y$, so that $\sigma_y$ cannot exceed the yield stress $\sigma_0$.

$$\sigma_y = \sigma_0 = \sigma\left(\frac{c}{2r_y}\right)^{1/2}$$

and

$$r_y = \left(\frac{\sigma}{\sigma_0}\right)^2 \frac{c}{2} \qquad (7\text{-}51)$$

or from Eq. (7-21)

$$r_y = \frac{K^2}{2\pi\sigma_0{}^2} \qquad (7\text{-}52)$$

However, Eq. (7-52) very much underestimates the plastic zone because it does not allow for the yielding caused by the elastic stress distribution from $\sigma_y = \sigma_0$ on out to $\sigma_y = \sigma_{max}$. It turns out that this area in Fig. 7-16 is equal to the area under $\sigma_y = \sigma_0$ out to $r = r_y$. Therefore, the total depth of the plastic zone is $R \approx 2r_y$ and

$$R = \left(\frac{\sigma}{\sigma_0}\right)^2 c \qquad (7\text{-}53)$$

The stress distribution ahead of a real crack of half-length $c$ with a plastic zone of depth $R$ may be regarded as identical with the elastic stress field ahead of a fictious crack of half-length $c' = c + R/2 = c + r_y$. For example, from Eq. (7-24), the fracture toughness of a center-notched specimen that has been corrected for plastic deformation at the crack tip is

$$K = \sigma \sqrt{\pi c'} \left[ \frac{w}{\pi c'} \tan \left( \frac{c}{w} + \frac{K}{2w\sigma_0^2} \right) \right]^{\frac{1}{2}} \qquad (7\text{-}54)$$

In considering these initial attempts to include plastic deformation in the fracture criterion, it is important to note that none of the criteria discussed so far apply when fracture occurs after gross plastic yielding. Criteria for completely ductile rupture are considered in Sec. 7-10.

## 7-10  DUCTILE FRACTURE

Ductile fracture has been studied much less extensively than brittle fracture, probably because it is a much less serious problem. Up to this point ductile fracture has been defined rather ambiguously as fracture occurring with appreciable gross plastic deformation. Another important characteristic of ductile fracture, which should be apparent from previous considerations of brittle fracture, is that it occurs by a slow tearing of the metal with the expenditure of considerable energy. Many varieties of ductile fractures can occur during the processing of metals and their use in different types of service. For simplification, the discussion in this section will be limited to ductile fracture of metals produced in uniaxial tension. Other aspects of tensile fracture are considered in Chap. 9. Ductile fracture in tension is usually preceded by a localized reduction in diameter called *necking*. Very ductile metals may actually draw down to a line or a point before separation. This kind of failure is usually called *rupture*.

The stages in the development of a ductile "cup-and-cone" fracture are illustrated in Fig. 7-17. Necking begins at the point of plastic instability where the increase in strength due to strain hardening fails to compensate for the decrease in cross-sectional area (Fig. 7-17a). This occurs at maximum load or at a true strain equal to the strain-hardening coefficient (see Sec. 9-3). The formation of a neck introduces a triaxial state of stress in the region. A hydrostatic component of

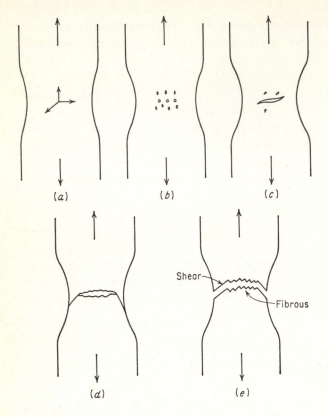

**Fig. 7-17**  Stages in the formation of a cup-and-cone fracture.

tension acts along the axis of the specimen at the center of the necked region.  Many fine cavities form in this region (Fig. 7-17$b$), and under continued straining these grow and coalesce into a central crack (Fig. 17-7$c$).  This crack grows in a direction perpendicular to the axis of the specimen until it approaches the surface of the specimen.  It then propagates along localized shear planes at roughly 45° to the axis to form the "cone" part of the fracture (Fig. 7-17$d$).

Detailed study[1] of the ductile fracture process shows that the central crack which forms early tends to concentrate the deformation at its tip in narrow bands of high shear strain.  These shear bands are at angles of 50 to 60° to the transverse direction.  Sheets of voids are nucleated in these bands, and the voids grow and coalesce into local fracture of the *void sheet*.  While the average direction of crack growth is radially outward in the direction transverse to the tensile axis,

[1] H. C. Rogers, *Trans. Metall. Soc. AIME*, vol. 218. p. 498, 1966; H. C. Rogers, in "Ductility" chap. 2, American Society for Metals, Metals Park, Ohio, 1968.

on a finer scale the crack zig-zags back and forth across the transverse plane by void-sheet formation. Thus, crack growth in ductile fracture is essentially by a process of *void coalescence.* Coalescence occurs by elongation of the voids and elongation of the bridges of material between the voids. This leads to the formation of a fracture surface[1] consisting of elongated "dimples," as if it had formed from numerous holes which were separated by thin walls until it fractures.

The voids which are the basic source of ductile fracture are nucleated heterogeneously at sites where compatibility of deformation is difficult. The preferred sites for void formation are inclusions, second-phase particles, or fine oxide particles, while in high-purity metals voids can form at grain-boundary triple points. Particles as small as 50 Å have been found to nucleate voids, so that the absence of voids on metallographic examination may not be a reliable indicator of whether void formation has occurred. In the tension test voids form prior to necking but after a neck is formed and hydrostatic tensile stresses develop, the void formation becomes much more prominent. The frequency of occurrence of nucleating particles should have a strong influence on ductile fracture.[2] It has been shown[3] that the true strain to fracture decreases rapidly with increasing volume fraction of second phase particles (Fig. 7-9). Careful metallographic study[4] of ductile fracture in carbon steels containing pearlite has shown that a combination of applied tensile stress and concentrated shear zone are required to initiate voids. The suggested mechanism is shown in Fig. 7-18. Carbides that are parallel to the applied tensile stress

$(a)$          $(b)$          $(c)$          $(d)$

**Fig. 7-18** Mechanism of ductile fracture in a pearlitic steel. (*After Miller and Smith.*)

[1] C. D. Beachem, *Trans. Am. Soc. Met.,* vol. 50, pp. 318–326, 1963.
[2] For a detailed review see A. R. Rosenfield, *Metall. Rev.* 121, *Met. Mater.,* April 1968 and J. L. Mogford, *Metall. Rev.* vol. 12, pp. 49–60, 1967.
[3] B. Edelson and W. Baldwin, *Trans. Am. Soc. Met.,* vol. 55, pp. 230–250, 1962.
[4] L. E. Miller and G. C. Smith, *J. Iron Steel Inst. London,* vol. 208, pp. 998–1005, 1970.

crack first (Fig. 7-18a).  A concentrated shear zone at about 50° to the tensile axis causes cracking of adjacent carbide plates (Fig. 7-18b).  The voids grow (Fig. 7-18c) and coalesce to form the ductile fracture (Fig. 7-18d).  Particle shape can have an important influence on ductile fracture.  When the particles are more spherical than platelike as in spheroidized pearlite, cracking of the carbides is much more difficult and the ductility is increased.  Cracking of spheroidized carbides is more difficult because dislocations in the ferrite matrix can cross slip around them more easily than for platelike carbides, and thus avoid the buildup of high stresses at pile-ups.  Also, because the of smaller contact area between spheroidized carbides and the matrix, the tensile stresses generated in the particles will be less than for lamellar carbides.  The very fine rounded carbides in quenched and tempered steels are very resistant to void formation and this accounts for the good ductility of this structure at high-strength levels.  Finally  since the second-phase particles invariably will be distorted in shape by plastic deformation processes, like rolling, it is common to find that the resistance to ductile fracture (ductility) varies greatly with orientation in a rolled sheet or plate.

An important start toward developing an analytical treatment of ductile fracture has been made by McClintock[1] using a model consisting of cylindrical holes initially of radius $b_0$ and average spacing $l_0$.  The strain to fracture is given by

$$\varepsilon_f = \frac{(1-n)\ln(l_0/2b_0)}{\sinh[(1-n)(\sigma_a+\sigma_b)/(2\bar{\sigma}/\sqrt{3})]} \tag{7-55}$$

for a material with a stress-strain curve given by $\sigma = K\varepsilon^n$.  In this equation $\sigma_a$ and $\sigma_b$ are the stresses parallel and perpendicular to the axis of the cylindrical holes, respectively, and $\bar{\sigma}$ is the true flow stress.  While Eq. (7-55) does not give close agreement with the limited data available  it certainly predicts the proper variation of fracture strain with the important variables.  Equation (7-55) indicates that ductility decreases as the void fraction increases, as the strain-hardening exponent $n$ decreases, and as the stress state changes from uniaxial tension to triaxial tension.  A somewhat different analysis[2] of ductile fracture by the internal necking of cavities leads to the same general conclusions.

## 7-11  NOTCH EFFECTS

The changes produced by the introduction of a notch have important consequences in the fracture process.  For example, the presence of a notch will increase appreciably the ductile-brittle transition temperature of a steel.  From a discussion of elastic stress concentration in Sec. 2-16, we expect that a notch creates a local

[1] F. A. McClintock, in "Ductility", chap. 9, American Society for Metals, Metals Park, Ohio, 1968; also F. A. McClintock, *J. Appl. Mech.*, vol. 90, pp. 363–371, 1968.
[2] P. F. Thomason, *J. Inst. Met.*, vol. 96, pp. 360–365, 1968.

stress peak at the root of the notch. Plastic flow begins at the notch root when this local stress reaches the yield strength of the material. The plastic flow relieves the high elastic stress and limits the peak stress to the yield stress of the material. However, the chief effect of the notch is not in introducing a stress concentration but in producing a triaxial state of stress at the notch.

The elastic stress distribution is shown in Fig. 7-19a for a notch in a thin plate. When the plate is loaded in the $y$ direction to a stress less than the elastic

**Fig. 7-19**  (a) Elastic stresses beneath a notch in a thin plate (plane stress); (b) elastic stresses beneath a notch in plane strain; (c) distribution of $\sigma_z$ with $z$ at $x = 0$ (plane strain); (d) distribution of stresses during local yielding (plane strain).

limit, the stress distribution of $\sigma_y$ is shown. At the same time a transverse elastic stress $\sigma_x$ is produced by the geometry of the notch. This can be understood physically if we imagine a series of small tensile specimens at the tip of the notch. The distribution of $\sigma_y$ is given by Fig. 7-19a. If each tensile specimen were able to deform freely, it would undergo a tensile strain $\varepsilon_y$ in response to the local value of $\sigma_y$. Because $\sigma_y$ falls off rapidly in going away from the notch root, there will be an elastic strain gradient below the notch. However, each tensile strain $\varepsilon_y$ will also have associated with it a lateral contraction $\varepsilon_x$ due to Poisson's ratio. If each tensile element is free to deform independently of its neighbors, this is given by $\varepsilon_x = -v\varepsilon_y$. Thus, all of the interfaces between the tensile specimens would pull apart. In order to maintain continuity a tensile stress $\sigma_x$ must exist across each interface. At the free surface of the notch ($x = 0$) the tensile element can undergo lateral contraction without any restraint from one side and $\sigma_x = 0$. The necessary lateral stress to maintain continuity increases with distance in from the root of the notch, but it decreases at large values of $x$ becuase the difference in longitudinal strain between adjacent elements becomes small as the $\sigma_y$ distribution flattens out at large values of $x$. Therefore, $\sigma_x$ will rise fairly steeply with $x$ and then fall more slowly as shown in Fig. 7-19a.

For the plane-stress condition of tensile loading of a thin plate the stress in the thickness direction is small and can be ignored. However, this definitely is not the case for plane-strain deformation where the thickness $B$ in the $z$ direction is large relative to the notch or crack depth. For plane-strain deformation $\varepsilon_z = 0$ and $\sigma_z = v(\sigma_y + \sigma_x)$. The elastic stress distribution along the $x$ axis for a thick notched plate loaded uniaxially in the $y$ direction is shown in Fig. 7-19b. The value of $\sigma_z$ falls to zero at both surfaces of the plate ($z = \pm B/2$) but rises rapidly with distance in from the free surfaces. The distribution of $\sigma_z$ with $z$ at the notch root ($x = 0$) is shown in Fig. 7-19c. The values of $\sigma_y$ and $\sigma_x$ are nearly independent of $z$. Figure 7-19b shows that stressing a thick plate provides a high degree of elastic triaxiality. As the thickness $B$ decreases, the values of $\sigma_y$ and $\sigma_x$ fall by less than 10 percent but the peak value of $\sigma_z$ is strongly dependent on $B$. In plane stress, when $B$ is very small, $\sigma_z = 0$.

The existence of transverse stresses raises the average value of longitudinal stress at which yielding occurs. For a Tresca yield criterion, $\sigma_0 = \sigma_1 - \sigma_3$. In an unnotched tension specimen the longitudinal stress alone measures yielding since $\sigma_0 = \sigma_y - 0$. In plane strain yielding starts at the root of the notch because $\sigma_x = 0$ at this free surface (Fig. 7-19). However, just below the notch, $\sigma_0 = \sigma_y - \sigma_x$. Since the basic material yield strength $\sigma_0$ is the same, whether notched or un-notched, it takes a higher value of longitudinal stress $\sigma_y$ to produce yielding in a notched specimen. The distribution of the principal stresses with distance in from the root of the notch is sketched in Fig. 7-19d. When local yielding occurs, the value of $\sigma_y$ drops from its high elastic value (see 7-19b) to the value of $\sigma_0$. Once the first tensile element at the notch root has yielded, it deforms plastically at constant volume with $v = 0.5$ instead of the elastic value of $v = 0.3$. Therefore, the transverse strain $\sigma_x$ is larger than for the elastic case and a larger value of $\sigma_x$ must be applied to maintain cohesion at the element interfaces. Therefore, as the plastic zone spreads from the notch root the value of $\sigma_x$ increases much more steeply with distance than in the elastic case. The maximum value of $\sigma_x$ is found at the elastic-plastic interface. Within the plastically yielding region the value of $\sigma_y$ is given by the yield criterion, $\sigma_y = \sigma_0 + \sigma_x$ and $\sigma_z$ is given by $\sigma_z = 0.5(\sigma_y + \sigma_x)$. With increasing stress the plastic zone moves inward until at some point the entire region beneath the notch becomes plastic.

As a result of the triaxial stress state produced by the notch the general yield stress of a notched specimen is greater than the uniaxial yield stress $\sigma_0$ because it is more difficult to spread the yielded zone in the presence of triaxial stresses. The ratio of notched-to-unnotched flow stress is referred to as the *plastic-constraint factor q.* Unlike elastic stress concentration, which can reach values in excess of $K_t = 10$ as the notch is made sharper and deeper, Orowan[1] has shown that the plastic-constraint factor cannot exceed a value of 2.57. Thus, the triaxial stress state of a notch results in "notch-strengthening" in a ductile metal, but in a material prone to brittle fracture the increased tensile stresses from the plastic constraint can exceed the critical value for fracture before the material undergoes general plastic yielding.

The steep gradients of stress which exist at a notch imply that there also are sharp gradients of strain. Although there are no exact methods for determining the local strain distribution in a strain-hardening material, the Neuber[2] approximation is useful for determining the magnitude of the strain concentration $K_\varepsilon$.

$$K_\varepsilon (K_\sigma) = K_t{}^2 \qquad (7\text{-}56)$$

where $K_\varepsilon$ = the plastic-strain-concentration factor

$K_\sigma$ = the plastic-stress-concentration factor

$K_t$ = the elastic-stress-concentration factor.

Therefore, another effect of the notch is to produce high, locally concentrated strain.[3] The accompanying strain hardening can lead to ductile void formation that can become converted into brittle cracks. Because the plastically strain-hardened volume beneath the notch is small, the concentration of strain leads to cracking without the expenditure of much plastic work. Still another important consequence of the plastic strain concentration at a notch is that the *local* strain rate will be much

---

[1] E. Orowan, *Trans. Inst. Eng. Shipbuild. Scot.*, vol. 89, p. 165, 1945; measurements of plastic constraint for low-temperature cleavage fracture of mild steel have been made by G. T. Hahn and A. R. Rosenfield, *ASM Trans. Q.*, vol. 59, pp. 909–199, 1966.

[2] H. Neuber, *J. Appl. Mech.*, vol. 28, p. 544, 1961.

[3] For an experimental illustration see R. Taggart D. H. Polonis, and L. A. James, *Exp. Mech.*, vol. 7, pp. 1–6, 1967.

higher than the average strain rate. Since brittle fracture depends strongly on strain rate, this can be an important, but easily overlooked factor.

In summary, a notch increases the tendency for brittle fracture in four important ways:

By producing high local stresses
By introducing a triaxial tensile state of stress
By producing high local strain hardening and cracking
By producing a local magnification to the strain rate

## 7-12 CONCEPT OF THE FRACTURE CURVE

The true-stress-true-strain curve, or flow curve, represents the stress required to cause plastic flow at any particular value of plastic strain. Plastic flow is terminated by fracture when strain hardening, triaxial stress, or high strain rate inhibit the plastic deformation sufficiently to cause a stress that is high enough to break the material. It was proposed by Ludwik[1] that a metal has a *fracture stress curve* in addition to a flow curve and that fracture occurs when the flow curve intersects the fracture curve (Fig. 7-20).

In principle, a point on the fracture curve is obtained by plastically straining a specimen to a given value of strain and then straining without gross plastic deformation until fracture occurs. In practice, a notch and/or low temperature have been used to prevent further plastic deformation before fracture. However, the fact that the embrittling effect of a notch is limited to a plastic-constraint factor of about 2.5 and the realization that even brittle fracture of metals at low temperature is preceded by plastic deformation have made it clear that good experimental measurements of the fracture stress curve cannot be made for metals.[2] However,

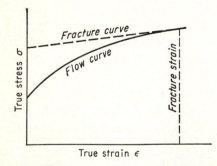

Fig. 7-20 Schematic drawing of intersection of flow curve and fracture curve according to the Ludwik theory.

[1] P. Ludwick, *Z. Ver. deut. Ing.*, vol. 71, pp. 1532–1538, 1927.
[2] Drucker has presented theoretical arguments that the fracture stress may in certain cases decrease with increasng strain; D. C. Drucker, in "Fracture of Solids," chap. 1, Interscience Publishers, Inc., New York, 1963.

**Fig. 7-21** Schematic description of transition temperature.

this does not prohibit using the concept of the fracture stress curve, in a qualitative sense, where it is useful for thinking about fracture problems.

The chief factors that influence the ductile-to-brittle transition in bcc metals are stress state, temperature, and strain rate. The transition behavior can be explained in terms of the relative resistance to shear and cleavage and how they change with temperature. Figure 7-21 shows the variation of shear resistance $\sigma_0$ and cleavage resistance $\sigma_f$ with temperature. In agreement with available data $\sigma_f$ is a less sensitive function of temperature than $\sigma_0$. For unnotched specimens the flow stress is lower than the fracture stress at all temperatures above the transition temperature. Thus, the material deforms plastically well before fracture takes place, and the material is ductile. Below the transition temperature $\sigma_f < \sigma_0$, and the material fractures before it can flow plastically.

Now, if a notch is machined in the specimen, the flow stress is raised by about a factor of 2.5 due to plastic constraint, while the fracture stress essentially is unchanged. Figure 7-21 shows how this produces a large increase in the transition temperature. Increasing the strain rate raises the flow stress in the same way as plastic constraint.

## 7-13 FRACTURE UNDER COMBINED STRESSES

The phenomenological approach to fracture is concerned with uncovering the general macroscopic laws which describe the fracture of metals under all possible states of stress. This same approach was discussed in Chap. 3 with regard to the prediction of yielding under complex states of stress. The problem of determining general laws for the fracture strength of metals is quite difficult because fracture is so sensitive to prior plastic straining and temperature. In principle we can conceive of a three-dimensional fracture surface in terms of the three principal stresses $\sigma_1$, $\sigma_2$, and $\sigma_3$. For any combination of principal stresses the metal will fracture when the limiting surface is reached.

Most experimentation in this field has been with biaxial states of stress where one of the principle stresses is zero. Tubular specimens in which an axial tensile or compressive load is superimposed on the circumferential stress produced by internal pressure are ordinarily used for this type of work. For accurate results bulging or necking during the later stages of the test must be avoided. This makes it difficult to obtain good data for very ductile metals.

Figure 7-22 illustrates the fracture criteria which have been most frequently proposed for fracture under a biaxial state of stress. The maximum-shear-stress criterion and the von Mises', or distortion-energy, criterion have already been considered previously in the discussion of yielding criteria. The maximum-normal-stress criterion proposes that fracture is controlled only by the magnitude of the greatest principal stress. Available data on ductile metals such as aluminum and magnesium alloys[1] and steel[2] indicate that the maximum-shear-stress criterion for fracture results in the best agreement. Agreement between experiment and theory is not nearly so good as for the case of yielding criteria. The fracture criterion for a brittle cast iron[3] is shown in Fig. 7-23. Note that the normal stress criterion is followed in the tension-tension region and that the fracture strength increases sig-

**Fig. 7-22** Proposed fracture criteria for biaxial state of stress in ductile metals.

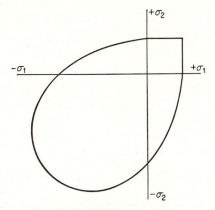

**Fig. 7-23** Biaxial fracture criterion for brittle cast iron.

[1] J. E. Dorn, " Fracturing of Metals," American Society for Metals, Metals Park, Ohio, 1948.
[2] E. A. Davis, *J. Appl. Mech.*, vol. 12, pp. A13–A24, 1945.
[3] W. R. Clough and M. E. Shank, *Trans. Am. Soc. Met.*, vol. 49, pp. 241–262, 1957.

nificantly as one of the principal stresses becomes compressive. Two theories[1,2] which consider the stress concentration of graphite flakes in cast iron are in good agreement with the fracture data. These data are also in substantial agreement with the fracture curve predicted from Griffith's theory of brittle fracture.


## 7-14  EFFECT OF HIGH HYDROSTATIC PRESSURE ON FRACTURE

The importance of triaxial tensile stresses in initiating brittle fracture was empha-sized in Sec. 7-11. It then should not be surprising to learn that triaxial compressive stress (hydrostatic pressure) resists fracture and increases the ductility. Hydro-static pressure is utilized in many metalworking operations, like wiredrawing and extrusion, to produce large plastic deformation which would not be possible in the absence of a high component of hydrostatic pressure.

The pioneering work in this field was done by Bridgman,[3] who subjected metals to tension tests with superimposed hydrostatic pressures up to 450,000 psi. More recently, mechanical tests at high pressure[4] and metalworking studies of hydrostatic extrusion have been active areas of research. The effect of superimposed hydrostatic pressure on the ductility in tension is shown in Fig. 7-24. Metals that are normally ductile at atmospheric pressure show a behavior similar to curve 1. For example, copper and aluminum reach reductions of area very close to 100 percent at a pressure of about 80 ksi. Curve 2 would be typical of brittle materials such as cast iron or zinc, while curve 3 would be more representative of very brittle materials such as tungsten or marble.

**Fig. 7-24** Effect of hydrostatic pressure on duc-tility at fracture in tension. Case 1 is a normally ductile material. Case 2 is a low-ductility material. Case 3 is a brittle material.

[1] L. F. Coffin, Jr., *J. Appl. Mech.*, vol. 17, p. 233, 1950.
[2] J. C. Fisher, *ASTM Bull*, 181, p. 74, April, 1952.
[3] P. W. Bridgman, "Studies in Large Plastic Flow and Fracture," McGraw-Hill Book Company, New York, 1952.
[4] H. L. D. Pugh (ed.), "The Mechanical Behavior of Materials Under Pressure," Elsevier Publishing Company, New York, 1970.

Since a hydrostatic component of stress exerts no shear stress, it cannot increase the number of dislocations in a pile-up or squeeze them closer together. This is the reason for the statement given earlier that hydrostatic stress influences crack propagation but not crack initiation. Compressive hydrostatic stresses act to close up small pores or separations at phase interfaces and generally make the fracture-propagation process more difficult. Studies of the fracture of iron-carbon alloys[1] have shown that high hydrostatic pressure does not suppress the fracture of carbides, but it does reduce void growth in the ferrite matrix.

## 7-15 STATISTICAL ASPECTS OF BRITTLE FRACTURE

Mechanical-strength measurements of brittle materials, like glass and ceramics, or of metals under conditions where they behave in a brittle manner, show a high variability of results which requires statistical analysis. Therefore, the mechanical properties are not expressible by a single number, but instead we must think in terms of probability of failure at a given stress. An important consequence of the statistical behavior is that the strength of brittle solids shows a pronounced *size effect* in which the strength decreases with increasing size or volume of the specimen. One of the early observations with brittle materials was that the strength of a fine glass fiber or metal wire is usually higher than that of a large-diameter rod. Moreover, if a fiber is broken into two fragments and each piece is broken successively in turn, the fracture strength will increase as the length decreases. These manifestations of the size effects are the result of the fact that the strength of a brittle material is controlled by the stress at the root of its most dangerous crack (usually the longest crack). As the specimen size decreases there is less likelihood of finding a large crack, and the strength increases.

The statistical nature of fracture in brittle materials introduces some complications not present with ductile materials. Ordinarily we consider that fracture will initiate at the most highly stressed region in a structure, usually at a discontinuity or stress concentration. However, in a brittle material whose strength is determined by the distribution of flaws, the stress distribution may be so localized at a discontinuity that the probability of failure is low because the high stress acts over a small volume of material. Failure may occur elsewhere in the structure where the peak stress is lower, but it acts over a larger volume of material.

A statistical theory of brittle fracture[2] assumes that the specimen is divided into many volume elements, each containing a single crack. The usual simplifying

[1] T. E. Davidson and G. S. Ansell, *Trans. Metall. Soc. AIME*, vol. 245, pp. 2383–2390, 1969.
[2] S. D. Volkov, " Statistical Strength Theory," Gordon and Breach, Science Publishers, Inc., New York, 1962.

**Fig. 7-25** Calculated frequency distribution of fracture stress as a function of number of cracks $N$. (*From J. C. Fisher and J. H. Hollomon, Trans. AIME, vol. 171, p. 555, 1974.*)

assumption is that there is no interaction between the cracks in the different volume elements. The strength of the specimen is determined by the element with the longest crack, for this results in the lowest value of fracture stress. Therefore, the brittle-fracture strength is determined, not by an average value of the distribution of flaws, but by the single-most dangerous flaw. This concept of brittle fracture is called the *weakest-link concept* since it uses a model in which the flaws are arranged in series like the links of a chain. Figure 7-25 shows the calculated frequency distribution of fracture stress in a brittle material as a function of the number of cracks $N$. This shows that the scatter (spread) in fracture stress decreases with an increase in the number of cracks. Also, as the number of flaws increases the mean value of the fracture stress decreases, but as the number of flaws reaches large values there is less relative decrease in the mean value of fracture stress.

The statistical theory of fracture by the weakest link was developed by Epstein.[1] Since failure results from the largest flaws, we are concerned with the distribution of the lowest strength or life or the largest flaws in a sample from some initial frequency distribution. Frequently this initial frequency distribution is not known or cannot be sampled directly, but if the number of flaws is large, we can use certain general asymptotic results to arrive at the distribution of extreme values. The statistics of extreme values[2] is important not only in fracture

[1] B. Epstein, *J. Appl. Phys.*, vol. 19, pp. 140–147, 1948.
[2] E. J. Gumbel, "Statistics of Extreme Values," Columbia University Press, New York, 1958; G. J. Hahn and S. S. Shapiro, "Statistical Models in Engineering," pp. 111–118, John Wiley & Sons, Inc., New York, 1967.

but in other engineering problems such as the prediction of the maximum wind velocity or the crest of floods. Given a random variable $x$, which can represent fracture stress, life to fracture, etc., the probability of obtaining a value less than or equal to $x$ is

$$P(x) = \int_{-\infty}^{x} p(x)\, dx \qquad (7\text{-}57)$$

where $P(x)$ = the cumulative frequency distribution of $x$
$\qquad p(x)$ = the probability distribution function of $x$
Using extreme value statistics the probability distribution function $g(x)$ of the smallest value of fracture stress $x$ in a specimen with $N$ cracks is

$$g(x) + Np(x)[1 - P(x)]^{N-1} \qquad (7\text{-}58)$$

and the cumulative frequency distribution function is

$$G(x) = 1 - [-P(x)]^{N} \qquad (7\text{-}59)$$

Therefore, if $p(x)$ is known then $g(x)$ and $G(x)$ can be obtained from Eqs. (7-58) and (7-59). The most probable of the minimum extreme values is the mode of $g(x)$ which is found by taking the most frequent value of $g(x)$ once $p(x)$ is established.

While the gaussian or normal frequency distribution is often taken as "standard," there is no theoretical or experimental justification for using it in fracture problems. In fact, since the normal frequency distribution has long tails on each side of the mean, it can be argued that this is unrealistic and does not represent the observed facts. It is more logical to expect a high frequency of occurrence of small cracks with a much lower number of large cracks. The most generally applicable frequency distribution in fracture problems is the *Weibull distribution*.[1] In its most general form this distribution may be written as

$$p(x) = b\left(\frac{x - x_0}{x_a{}^{b}}\right)^{b-1} e^{-[(x-x_0)/x_a]^{b}} \qquad (7\text{-}60)$$

$$P(x) = 1 - e^{-[(x-x_0)/x_a]^{b}} \qquad (7\text{-}61)$$

where $x_a$ = the characteristic value of $x$, corresponding to 63.2 percent failure
$\qquad$ point (scale factor)
$\quad x_0$ = the minimum value of $x$
$\quad\; b$ = the Weibull slope parameter (shape parameter)
$x_a$, $x_0$, and $b$ are constants for a given population which has a Weibull distribution. By taking natural logarithms twice, Eq. (7-61) can be transformed into

$$\ln \ln \left(\frac{1}{1 - P(x)}\right) = b \ln (x - x_0) - b \ln x_a \qquad (7\text{-}62)$$

W. Weibull, *J. Appl. Mech.*, vol. 18, pp. 293–297, 1951; pp. 109–113, 1952.

which is of the form $y = bx + c$. Special Weibull-function coordinate paper[1] permits the direct plotting of cumulative failure probability versus $x$. The three parameters may be obtained graphically from a plot of this type. The most probable value of $x$ from extreme value theory using a Weibull distribution is $x$

$$x^* = \frac{x_a}{N^{1/b}} \left(1 - \frac{1}{b}\right)^{1/b} \qquad (7\text{-}63)$$

For a gaussian distribution of the strength of the weakest element the fracture strength should decrease proportional to $(\log V)^{1/2}$, where $V$ is the volume of the specimen. For a Weibull distribution of crack sizes the strength should decrease proportional to $V^{-1/b}$.

## BILBIOGRAPHY

Averbach, B. L., D. K. Felback, G. T. Hahn, and D. A. Thomas (eds.): "Fracture," The Technology Press of The Masachussetts Institute of Technology and John Wiley & Sons, Inc., New York, 1959.

Biggs, W. D.: "The Brittle Fracture of Steel," Pitman Publishing Corporation, New York, 1960.

Drucker, D. C., and J. J. Gilman (eds.): "Fracture of Solids," *Metall. Soc. Conf.*, vol. 20, Interscience Publishers, Inc., New York, 1963.

"Fracture of Engineering Materials," American Society for Metals, Metals Park, Ohio, 1964.

Knott, J. F.: Mechanics and Mechanisms of Large-Scale Brittle Fracture in Structural Metals, *Mat. Sci. and. Eng.*, vol. 7, pp. 1–36, 1971.

Liebowitz, H. (ed.): "Fracture, An Advanced Treatise," Academic Press, Inc., New York, 1969.
    Vol. I, "Microscopic and Macroscopic Fundamentals"
    Vol. II, "Mathematical Fundamentals"
    Vol. III, "Engineering Fundamentals and Environmental Effects"
    Vol. IV, "Engineering Fracture Design"
    Vol. V, "Fracture Design of Structures"
    Vol. VI, "Fracture of Metals"
    Vol. VII, "Fracture of Nonmetals and Composites"

Pugh, S. F.: *Br. J. Appl. Phys.*, vol. 18, pp. 129–162, 1967.

Tetleman, A. S., and A. J. McEvily: "Fracture of Structural Materials," John Wiley & Sons, Inc., New York, 1967.

---

[1] This plot may be constructed rather simply according to the discussion in *Mater. Res. Stand.*, pp. 405–411, May 1962.

# 8

# MECHANICAL BEHAVIOR OF POLYMERIC MATERIALS

## 8-1  INTRODUCTION

This chapter primarily is concerned with the mechanical properties of polymeric materials, such as plastics and elastomers. Polymers are materials consisting of long-chain macromolecules. These materials are sufficiently different from metals and other crystalline solids in their mechanical behavior so that special treatment is necessary. Moreover, polymeric materials are the fastest growing class of engineering materials in volume of usage. The annual production of polymeric materials is more than 20 billion lb. This represents a fourfold increase over the past 10 years and a nearly hundredfold increase over the past 30 years. In contrast, the growth in volume of steel usage over the past 30 years has been less than tenfold.

Over the past 10 years there has been a marked shift in activity in polymer science from an emphasis on polymer chemistry and the discovery of new long-chain molecules to an understanding and control of the structure of polymer molecules and the relationship of the structure and the engineering properties of the material. The control of polymer structure is achieved in the chemical preparation steps (polymerization) or in subsequent processing into sheet, film, fiber, or other useful forms.

The chief justification for including a chapter on the mechanical properties of polymers in a book called "Mechanical Metallurgy" is that the philosophy and approach that are used today to develop materials with improved properties are essentially the same whether one deals with metals or with polymers. Of course, this unified approach to the study of structure-sensitive materials is called *materials science*. The other reason for giving attention to polymers in this book is much less noble and far more provincial. It is simply that polymers have made such deep inroads as engineering materials in applications traditionally served by metals that any progressive metallurgical engineer who is concerned with specifying or designing with materials must be aware of their properties and mechanical behavior.

## 8-2 TIME-DEPENDENT MECHANICAL BEHAVIOR

One strong characteristic of polymers compared with metals is that their mechanical properties are much more time-dependent. In our discussion of elastic behavior in Chap. 2 it was implicitly assumed that elastic strain was a single-valued function of stress alone. This is a valid assumption for the engineering analysis of metals by the theory of elasticity. However, under certain circumstances there is a time dependence to elastic strain which is called *anelasticity*. In Fig. 8-1, an elastic strain $e_1$ is applied to an anelastic material. With increasing time the strain gradually increases to a value $e_2$, the completely relaxed strain. The amount of anelastic strain is $e_2 - e_1$. If the load is suddenly removed at $t = t_1$, the material undergoes an immediate elastic contraction equal in magnitude to $e_1$ and with the passage of time the strain decays to zero. This behavior is known as an *elastic aftereffect*.

If a rod of material is loaded to an elastic stress so rapidly that there is not time for any thermal effects to equilibrate with the surroundings, the loading is

**Fig. 8-1**  Anelastic behavior and the elastic aftereffect.

done at constant entropy and under adiabatic conditions. For uniaxial loading the change in temperature of the material with strain is given by

$$\frac{\partial T}{\partial \varepsilon}\bigg|_s = -\frac{V_m \alpha E T}{c_v} \qquad (8\text{-}1)$$

where $V_m$ = the molar volume of the material
$\alpha$ = the coefficient of linear thermal expansion
$E$ = the isothermal Young's modulus
$T$ = the absolute temperature
$c_v$ = the specific heat at constant volume

Since $\alpha$ is positive for most materials and the other terms in Eq. (8-1) are also positive, it follows that an adiabatic elastic tension lowers the temperature of the material and an adiabatic compression increases the temperature. However, these temperature changes associated with the *thermoelastic effect* are usually small.

If a specimen is stretched elastically at a slow rate so that it is in thermal equilibrium with its surroundings, it will follow a path *A-B-C* along the stress-strain curve (Fig. 8-2*a*) and the path *C-B-A* on unloading. However, if the specimen is loaded rapidly to $\sigma_c$ so that there is not time for thermal equilibration with the surroundings, the temperature of the specimen will decrease a small amount and the specimen follows path *A-C'*. The strain that accumulates in the specimen will be only $e_A$ instead of the strain $e_I$ that would result from an isothermal path. Also, the elastic modulus under dynamic (rapid) loading condition is higher than the modulus under static (slow loading) conditions. With the passing of time the

Fig. 8-2  (*a*) Idealized adiabatic and isothermal stress-strain curves; (*b*) elastic hysteresis loop.

specimen will warm up and elongate due to thermal expansion along the path $C'$-$C$. If now the load is removed suddenly, the specimen will follow the adiabatic path $C$-$C''$ and the specimen temperature will rise a small amount. With the passage of time the specimen transfers heat to the surroundings and the strain decreases by thermal contraction along the path $C''$-$A$. The result is a closed hysteresis loop $A$-$C'$-$C$-$C''$-$A$. However, in an actual material where the specimen is loaded and unloaded in a continuous cycle, the elastic hysteresis loop would resemble Fig. 8-2$b$. Although the area included within the hysteresis loop may be very small, it can be an important quantity for a material subjected to rapid vibration since the total energy dissipated in a given time period is the product of the hysteresis area per cycle times the number of cycles. From an engineering point of view this elastic hysteresis leads to the generation of heat and the damping of vibrations.

The area under the hysteresis loop will be a function of the frequency at which the stress is applied and removed. If the frequency is very low, the cycle will be almost completely isothermal and the area enclosed by the hysteresis loop will be very small. At very high frequencies the loading and unloading paths are almost completely adiabatic and once again the area enclosed by the hysteresis loop is very small. However, there will be some frequency intermediate between these extremes for which the area of the hysteresis loop is a maximum.

The thermoelastic effect is but one of many mechanisms by which vibrational energy is dissipated internally by the material. Other mechanisms which produce anelastic effects in crystalline solids are stress-induced ordering of interstitial and substitutional solute atoms, grain-boundary sliding, motion of dislocations, and intercrystalline and transcrystalline thermal currents arising from the elastic anisotropy of crystals. In polymers these arise from bond rotation, moisture absorbtion, and a variety of other effects. These various energy-dissipating effects can be grouped under the generic heading of *internal friction*.[1] Precision measurements of the energy dissipation under very small strains[2] are one of the most sensitive tools for detecting changes in solid-state structure such as precipitation, diffusion, and impurity concentration.

Internal friction is usually expressed by the *logarithmic decrement, δ*. Different expressions for $\delta$ are developed depending on the type of experiment used for studying internal friction. Internal friction frequently is measured by a system which is set into motion with a certain amplitude $A_0$ and then allowed to decay freely. The amplitude at any time $A_t$ is given by

$$A_t = A_0 e^{-\beta t} \qquad (8\text{-}2)$$

C. Zener, "Elasticity and Anelasticity of Metals," University of Chicago Press, Chicago, 1948; K. M. Entwistle, *Metall. Rev.*, vol. 7, p. 175, 1962.

[2] C. Wert, "Modern Research Techniques in Physical Metallurgy," pp. 225–250, American Society for Metals, Metals Park, Ohio, 1953.

Fig. 8-3 (a) Phase lag between stress and strain; (b) resonance curve.

where $\beta$ is the attenuation coefficient. The logarithmic decrement is the logarithm of the ratio of successive amplitudes.

$$\delta = \ln \frac{A_n}{A_{n+1}} \qquad (8\text{-}3)$$

If the internal friction is independent of amplitude,[1] a plot of $\ln A$ versus the number of cycles of vibration will be linear and the slope of the curve is the decrement. With anelastic behavior the stress and strain are not *in phase* (Fig. 8-3a). The phase angle by which the strain lags behind the stress in time is $\alpha$. The logarithmic decrement is related to the lag angle by

$$\delta = \pi\alpha \qquad (8\text{-}4)$$

For a condition of forced vibration in which the specimen is driven at a constant amplitude, a measure of internal friction is the fractional decrease in vibrational energy per cycle.

$$\delta = \frac{\Delta W}{2W} \qquad (8\text{-}5)$$

where $\Delta W$ is the energy lost per cycle (area $ABCD$ in Fig. 8-2b) and $W$ is the total vibrational energy per cycle (area $ACE$ in Fig. 8-2b). If a fixed amount of energy per cycle is supplied to a specimen, a maximum amplitude of vibration occurs at the resonant frequency (Fig. 8-3b). The logarithmic decrement for a resonance curve is given approximately by

$$\delta = \frac{\pi(\text{bandwidth})}{f_r} \approx \frac{\pi(f_2 - f_1)}{f_r} \qquad (8\text{-}6)$$

[1] True anelastic behavior is independent of amplitude, although some internal friction mechanisms are not.

**Table 8-1  DAMPING CAPACITY OF SOME ENGINEERING MATERIALS†**

| Material | Specific damping capacity at various stress levels $\Delta W/W$ | | |
|---|---|---|---|
| | 4,500 psi | 6,700 psi | 11,200 psi |
| Carbon steel (0.1% C) | 2.28 | 2.78 | 4.16 |
| Ni-Cr steel—quenched and tempered | 0.38 | 0.49 | 0.70 |
| 12% Cr stainless steel | 8.0 | 8.0 | 8.0 |
| 18-8 stainless steel | 0.76 | 1.16 | 3.8 |
| Cast iron | 28.0 | 40.0 | |
| Yellow brass | 0.50 | 0.86 | |

† S. L. Hoyt, "Metal Data," rev. ed., Reinhold Publishing Corporation, New York, 1952.

A measure of internal friction that often is used is the $Q$, where $Q = \pi/\delta$. Since in electrical circuit theory the reciprocal of this value is called the $Q$ of the circuit, the symbol $Q^{-1}$ has been adopted as a measure of internal friction

$$Q^{-1} = \frac{f_2 - f_1}{f_r} \qquad (8\text{-}7)$$

Internal friction studies generally are conducted at low stress levels (10 to 100 psi) and small strains. Internal energy dissipation at larger stresses and strains generally is called *damping*.[1] High damping capacity is important in minimizing noises in machinery or in suppressing vibrations in high-speed machinery. A high damping capacity is of practical engineering importance in limiting the amplitude of vibration at resonance conditions and thereby reducing the likelihood of fatigue failure. While damping capacity is not very dependent on frequency of vibration, Table 8-1 indicates that it depends on stress level (or strain amplitude) and it varies significantly from material to material. For example, the exceptionally high damping capacity of cast iron arises from the presence of graphite flakes which do not readily transmit elastic waves.

A common and totally different kind of time-dependent mechanical behavior is *creep*. Creep is the continuing extension with time in a material subjected to a constant load (Fig. 13-1). Creep is an important engineering problem for metals at high temperature. This subject is treated in detail in Chap. 13. However, for many polymers and a few low-melting metals creep becomes significant at room temperature. *Stress relaxation* is the time-dependent decrease in stress in a body which is constrained to a fixed deformation (Fig. 13-21).

[1] Internal Friction, Damping, and Cyclic Plasticity, *ASTM Spec. Tech. Pub.* 378, 1965.

## 8-3  POLYMERIC MATERIALS

The word "polymer" comes from the Greek meaning "many membered." Organic polymers are large molecules made up of many, many repeating units, or *mers*. The simplest of polymers, polyethylene, consists of a long chain with thousands of repeating units.

$$
\begin{array}{ccccccccc}
& & & & \text{mer} & & & & \\
\text{H} & \text{H} & \text{H} & \text{H} & \text{H} & \text{H} & \text{H} & \text{H} & \text{H} \\
| & | & | & | & | & | & | & | & | \\
-\text{C}- & \text{C}- & \text{C}- & \text{C}\!+\!\text{C}\!+\!\text{C}- & \text{C}- & \text{C}- & \text{C}- \\
| & | & | & | & | & | & | & | & | \\
\text{H} & \text{H} & \text{H} & \text{H} & \text{H} & \text{H} & \text{H} & \text{H} & \text{H}
\end{array}
$$

The polyethylene molecule is formed by polymerizing the *monomer*, ethylene, under high temperature or pressure and in the presence of a catalyst, to break the double C=C bond.

$$
n\begin{bmatrix} \text{H} & \text{H} \\ | & | \\ \text{C}\!=\!\text{C} \\ | & | \\ \text{H} & \text{H} \end{bmatrix} \Longrightarrow \begin{bmatrix} \text{H} & \text{H} \\ | & | \\ -\text{C}-\text{C}- \\ | & | \\ \text{H} & \text{H} \end{bmatrix}_n
$$

Ethylene                    Polyethylene

This is a simple example of *polymerization*. The number of monomer molecules that combine to form the polyethylene molecule is from about 1,000 to 100,000. Hence, the term *macromolecule* is often applied to polymers. The molecular weight of a polymer is an important parameter which has a strong influence on properties.

There are many common polymers that form by the same addition reaction as polyethylene. The essential difference is that one or more of the hydrogen side groups has been replaced either by another atom or by a group of atoms (a radical). Table 8-2 shows some of the polymers based on the vinyl monomer, and Table 8-3 shows some addition polymers based on the vinylidene monomer.

The diene polymers have a different carbon backbone in the polymer chain than the vinyl polymers. As shown in Table 8-4, a carbon double bond remains in each repeat unit even after polymerization has taken place. The polymers listed in Table 8-4 are *elastomers*, polymers which can undergo large elongations under load and return to their original shape when the load is removed.

The second chief type of polymerization reaction is *condensation polymerization*. In this type of polymerization two monomers react chemically to produce a polymer and a reaction product. The reaction product usually is a small molecule (such as water or methanol) which is readily removed. A good example of condensation polymerization is the reaction of phenol molecules with formaldehyde to form phenol formaldehyde (Bakelite) and water (Fig. 8-4). Note that splitting

**Table 8-2  TYPICAL POLYMERS BASED ON VINYL MONOMER**

$$n\begin{bmatrix} H & H \\ | & | \\ C{=}C \\ | & | \\ H & R_1 \end{bmatrix} \Rightarrow \begin{bmatrix} H & H \\ | & | \\ -C-C- \\ | & | \\ H & R_1 \end{bmatrix}_n$$

| Side Group $R_1$ | Polymer structure | Polymer name | Typical use |
|---|---|---|---|
| Cl | (see figure) | Polyvinyl chloride (PVC) | Plastic pipe |
| Propylene H—C—H, H | (see figure) | Polypropylene | Molded Parts |
| Benzene ring | (see figure) | Polystyrene | Clear molded parts |
| Nitrile C≡N | (see figure) | Acrylonitrile (Orlon) | Fiber |
| Acetate | (see figure) | Polyvinyl acetate | Adhesive |

**Table 8-3** TYPICAL POLYMERS BASED ON THE VINYLIDENE MONOMER

$$n\begin{bmatrix} H & R_2 \\ | & | \\ C=C \\ | & | \\ H & R_3 \end{bmatrix} \Rightarrow \begin{bmatrix} H & R_2 \\ | & | \\ -C-C- \\ | & | \\ H & R_3 \end{bmatrix}_n$$

| Side groups | Polymer structure | Polymer name | Typical use |
|---|---|---|---|
| $R_2 = CH_3$ <br><br> $R_3 = -\overset{O}{\overset{\|}{C}}-O-CH_3$ | $\begin{array}{cccc} H & CH_3 & H & CH_3 \\ \| & \| & \| & \| \\ -C-C & - & C-C- \\ \| & \| & \| & \| \\ H & R_3 & H & R_3 \end{array}$ | Polymethyl methacrylate (PMMA) | Glazing |
| $R_2 = Cl$ <br><br> $R_3 = Cl$ | $\begin{array}{cccc} H & Cl & H & Cl \\ \| & \| & \| & \| \\ -C-C-C-C- \\ \| & \| & \| & \| \\ H & Cl & H & Cl \end{array}$ | Polyvinylidene chloride (Saran) | Film |

**Table 8-4** TYPICAL POLYMERS BASED ON THE DIENE MONOMER

$$n\begin{bmatrix} H & R & H & H \\ | & | & | & | \\ C=C-C=C \\ | & & & | \\ H & & & H \end{bmatrix} \Rightarrow \begin{bmatrix} H & R & H & H \\ | & | & | & | \\ -C-C=C-C- \\ | & & & | \\ H & & & H \end{bmatrix}_n$$

| Side group | Polymer structure | Polymer name |
|---|---|---|
| H | $\begin{array}{cccc} H & H & H & H \\ \| & \| & \| & \| \\ -C-C=C-C- \\ \| & & & \| \\ H & & & H \end{array}$ | Polybutadiene |
| Cl | $\begin{array}{cccc} H & Cl & H & H \\ \| & \| & \| & \| \\ -C-C=C-C- \\ \| & & & \\ H & & & \end{array}$ | Polychloroprene (neoprene) |
| $CH_3$ | $\begin{array}{cccc} H & CH_3 & H & H \\ \| & \| & \| & \| \\ -C-C= & C-C- \\ \| & & & \| \\ H & & & H \end{array}$ | Polyisoprene (natural rubber) |

**Fig. 8-4** Condensation polymerization of phenol formaldehyde.

the water molecule away leaves carbon bonds which are filled by the remains of the formaldehyde molecule. Moreover, this "methylene bridge" could form at any of the carbon atoms in the phenol molecule except the one bonded to —OH. Thus, we say that the phenol monomer is *polyfunctional*, and the polymer is not limited to forming long chains but instead polymerizes as a three-dimensional network.[1]

The functionality of a polymer is defined as the number of sites at which the monomer can attach itself to other monomers by primary bonds. The addition polymers discussed earlier are bifunctional.

All of the polymers that we have considered so far contain a single repeating unit. Therefore, they may be classified as *homopolymers*. However, useful polymeric materials may be produced by the addition polymerization of two or more monomers into a *copolymer*. A copolymer can be considered a polymer alloy. The most common synthetic rubber, SBR rubber, is a random copolymer of 75% butadiene and 25% styrene. The properties of a copolymer depend on the proportion of each monomer present and the way that they are arranged along the polymer chain. Each monomer may be arranged randomly along the chain (ABAAABBAB-AA), or they may alternate (ABABAB...), or they may be arranged in blocks (AAA...ABBB...B). Another possibility is for the main chain to be composed of

[1] Other condensation polymerization reactions may produce linear polymer molecules.

one monomer with side branches of another monomer "grafted" onto the main chain.

The picture of polymer structure that has been presented so far is misleading in one major respect. Polymer structures have been presented as flat two-dimensional structures when in fact they have an important three-dimensional aspect. The saturated carbon atom with four hydrogen atoms forms a tetrahedron in space with every pair of C—H bonds occurring at 109° 28′. Another aspect to the three-dimensional structure is rotation about the C—C bond. Still another important aspect of polymer science not considered in our brief discussion of polymers was many subtilties of the chemistry of polymerization by which the placement of side groups or the size of the polymer molecule are controlled. These points are outside of our consideration of polymers but are discussed in modern texts on polymers.[1]

## 8-4 STRUCTURE OF POLYMERS

Most aspects of the mechanical behavior of polymers depends on the ease with which the long polymer chains can slide over each other. The structure of the polymer greatly affects this behavior.

Two broad descriptions of polymer structure are found in the common description of a polymer as being either thermoplastic or thermosetting. In long-chain polymers, strong primary bonds due to electron sharing exist between the atoms in the backbone of the chain, but only relatively weak secondary bonds hold adjacent chains together. Polymers with this type of structure are said to be *thermoplastic*. They soften with increasing temperature and are easily deformed. On cooling they retain the shape into which they were molded and regain their original room-temperature properties. If primary bonds exist between molecules, as in the three-dimensional network of phenol formaldehyde, the strength does not fall off markedly with temperature until the chemical decomposition of the network occurs. The properties of such a *thermosetting* material are not reversible on cooling.

It frequently is possible for polymers with the same chemical composition to have different chain structures. This is known as *stereoisomerism* (isomers by virtue of differing geometry). There are three ways in which an unsymmetrical side group may be arranged with respect to the C—C—C backbone of the polymer chain (Fig. 8-5). A random distribution of side groups along the chain is known as an *atactic* structure. If all of the groups are arranged on the same side of the back-

[1] F. Billmeyer, "Textbook of Polymer Science," John Wiley & Sons, Inc., New York, 1962; F. Rodriguez, "Principles of Polymer Systems," McGraw-Hill Book Company, New York, 1970.

Atactic

Isotactic

**Fig. 8-5**   Stereoisomer structures (schematic).          Syndiotactic

bone chain, it is an *isotactic* structure. If the side groups alternate from side to side of the chain, it is a *syndiotactic* structure. It should be obvious from even these oversimplified "flat models" that certain stereoisomers permit closer packing of adjacent polymer chains than other arrangements.

Under certain conditions of polymerization, polyfunctional connections can be produced at random points along the backbone chain to create a branched polymer molecule (Fig. 8-6a). Branching prevents the main polymer chains from coming together as closely as linear chains, so that the secondary bonding forces between the chains are weaker than with linear chains. A branched polymer will have lower strength and be softer and more flexible than a linear polymer.

The length of the polymer chain has an important effect on properties. A long chain can form more secondary bonds with neighboring chains than a short chain. Chain length is defined by the *degree of polymerization*, which equals the average molecular weight of a chain divided by the molecular weight of the repeating unit. The degree of polymerization for natural rubber is around 3,000, while for molding polystyrene it varies from 500 to 1,000. There is usually a broad distribution of molecular weights in a polymer. However, with modern polymeri-

**Fig. 8-6**   (*a*) Branched polymer; (*b*) cross-linked polymer.

(*a*)                    (*b*)

zation techniques it is possible to produce polymers with a narrow range of polymer chain lengths.

As the frequency and length of branches on the polymer chain increases, the branches eventually reach from chain to chain to tie them together into a three-dimensional network (Fig. 8-6b). This is known as *cross-linking*. Cross-links may be formed by starting with polyfunctional monomers or by introducing a chemical substance that reacts with linear or branched molecules. The use of sulfur in the vulcanization of natural rubber is a classic example of cross-linking. Cross-links increase strength and rigidity.

A false impression that should be dispelled at this point is the idea that polymer chains are extended to full length, as we have tended to picture them in this chapter. Rather, it is more realistic to visualize polymer structure as consisting of twisted, tangled polymer chains. A crude analogy often used is that the polymer chains are like a bowl of spaghetti. However, even this analogy falls down in that it neglects the fact that segments of the polymer chain are in constant thermal motion.

An important structural consideration of thermoplastics is *crystallinity*. As in metals, crystallinity implies a highly ordered or regular arrangement of the polymer molecules. Crystalline regions are formed when sufficient chains come close together so that the secondary bonding forces can overcome the disordering effect of thermal energy. The more easily the chains can closely approach each other, the greater the probability of forming a crystalline region. Thus, linear polymers are more likely to crystallize than branched polymers. Polymers with bulky side groups, like polystyrene, do not crystallize readily. Stereoregular polymers are more likely to be crystalline than those with irregular chain structure. The syndiotactic isomer is the best for forming crystalline regions.

An important distinction of crystallinity in polymers, compared with metals, is that a completely crystalline structure never is achieved. The degree of crystallinity varies from essentially nil, (in amorphous polymers) to a maximum (in special cases) of about 98 percent. High degrees of crystallinity are achieved by drawing or stretching fibers so as to orient the polymer chains in the working direction. A conceptualization of crystallinity in polymers that has been used for a long time is the *fringed micelle model* (Fig. 8-7a). The crystalline regions are on the order of several hundred angstroms. Since the polymer chains are many times longer than the dimensions of an individual crystallite, one chain passes from a crystallite region through an amorphous area and back into another crystallite region. This model nicely explains how stretching the polymer orients the chains in the direction of stress, increasing the alignment of chains in the amorphous regions and increasing the degree of crystallinity. Also, because an individual polymer chain passes at random from one crystal line region to another, the model explains why a perfect polymer crystal can never be achieved.

**Fig. 8-7** (*a*) Fringe micelle model; (*b*) folded-chain model; (*c*) spherulite.

More recently it has been shown[1] that polymer single crystals can be grown by slow crystallization from a dilute polymer solution. These crystals are in the form of thin platelets or lamellae with a thickness of about 100 Å. Electron diffraction showed that the polymer chain in these lamellae was perpendicular to the flat faces of the platelet. Because the thickness of the platelet was only a fraction of the length of the polymer chain, the polymer chain must fold back and forth on itself within the crystal. The *folded-chain model* is shown in Fig. 8-7*b*.

Polymers may crystallize from the melt in bulk form by forming *spherulites*. Spherulites grow radially from a point of nucleation (Fig. 8-7*c*). The crystalline material is in the form of thin ribbons which move outward from the nucleus, branching as they grow radially. They eventually attain polyhedral shapes because of mutual impingement of other spherulites. In this sense they resemble the grains in a polycrystalline metal. However, there is an important distinction between spherulites and grains. Each spherulite consists of a large number of crystallites, while each grain in a metal is a single crystal.

These more recent discoveries concerning crystalline structure in polymers have shaken the hold which the fringed micelle model once held. While many modifications have been proposed,[2] none have yet been widely adopted.

The presence of crystalline regions in a polymer will strongly influence its properties. Since the polymer chains are more tightly packed in the crystalline regions, the crystallites will have a higher density than the amorphous regions. Thus, the greater the degree of crystallinity, the higher the polymer density. Increasing crystallinity raises strength and rigidity. In a sense, crystalline polymers

[1] A. Keller, *Philos. Mag.*, vol. 2, pp. 1171, 1957; P. H. Geil, "Polymer Single Crystals," John Wiley & Sons, Inc., New York, 1963.
[2] G. C. Oppenlander, *Science*, vol. 159, pp. 1311–1319, March 22, 1968.

are actually two-phase systems, a crystalline phase in an amorphous matrix. Because of the different refractive indexes of the two phases, light is scattered, and crystalline polymers appear opaque or translucent in contrast with transparent amorphous polymers.

## 8-5 DEFORMATION OF POLYMERS

It is first necessary to understand how the properties of a polymer vary over a wide range of temperature. The property most convenient to monitor is specific volume. Consider a polymer melt that is being slowly cooled (Fig. 8-8). If the polymer can form crystallite nuclei, freezing will occur at a fixed temperature, the melting point $T_m$, with the release of latent heat and a sharp decrease in specific volume. However, if crystallization cannot occur, the melt will cool as a supercooled liquid below $T_m$. With decreasing temperature the viscosity continues to increase, until below some temperature $T_g$ the material behaves like an amorphous solid. $T_g$ is called the *glass transition temperature*. It is a significant temperature for a polymeric material since there are usually important changes in mechanical properties near this temperature. Below the glass transition the polymer is a brittle elastic solid. Above the $T_g$ it is a flexible rubbery material. Values of $T_g$ vary widely from polymer to polymer. For example, $T_g$ for polyethylene is $-110°C$, while $T_g$ for polystyrene is $100°C$. In Fig. 8-8, there is a change in slope at $T_g$ for the crystalline polymer which represents a change in the amorphous phase from a supercooled liquid to a glassy solid.

A series of stress-strain curves at various temperatures for a typical amorphous polymer, PMMA, is shown in Fig. 8-9. The $T_g$ for this polymer is about 100°C. The stress-strain curves at $-40, 68,$ and $86°C$ show brittle elastic behavior. The beginning of a yield point can be seen at 104°C and it is more pronounced at higher temperatures. In contrast to metals, this amorphous polymer in a highly ductile condition does not exhibit strain hardening. It can be seen that there is a

Fig. 8-8 Change in specific volume on solidification.

**Fig. 8-9**  Tensile stress-strain curves for PMMA at various temperatures. (*From T. Alfrey, "Mechanical Behavior of High Polymers," p. 516, Interscience Publishers, a division of John Wiley & Sons, Inc., New York, 1948. By permission of the publishers.*)

fairly sharp ductile-to-brittle transition with decreasing temperature. The transition temperature is approximately $T_g$.

More extensive changes in mechanical behavior with temperature are shown in Fig. 8-10 for polystyrene in various structural conditions. The mechanical property that is measured is the relaxation modulus.[1] $E_r$. Below $T_g$ (100°C) there is little molecular motion. The polymer acts as a brittle glassy solid. The elastic deformation arises from stretching of bonds in the backbone of the polymer chain (Fig. 8-11a). With an increase in temperature just above $T_g$, sliding between short segments of the polymer chain becomes possible. For the amorphous polymers this is a self-accelerating process. Once a spot along the chain becomes free of tanglements so it can slide, it is easier for neighboring regions of the chain to become free also. Thus, the modulus falls off rapidly with temperature at a temperature corresponding roughly to $T_g$. Above $T_g$ the polymer contains many "soft regions"

---

[1] $E_r$ is a time-dependent modulus. It is measured by straining the specimen to a given strain and then allowing the stress to relax. The stress at a given relaxation time (in this case 10 s.) is observed. If a sequence of specimens are deformed to a variety of initial strains, then a curve of $\sigma(t)$ versus $e$ can be constructed and $E_r(t)$ is the slope of that curve.

**Fig. 8-10** Variation of the relaxed modulus of polystyrene with temperature. The effect of different polymer structures is shown. (*From A. V. Tobolsky, "Properties and Structure of Polymers," John Wiley & Sons, Inc., New York, 1960. By permission of the publishers.*)

in which secondary bonding is no longer localized. The polymer is said to be in a *rubbery state*. Large-scale elastic deformation occurs by the uncoiling of polymer chains (Fig. 8-11*b*), and plastic extension occurs by extensive slipping of polymer chains (Fig. 8-11*c*). Deformation in this region is both elastic and plastic and highly time-dependent. This is called the *viscoelastic region*. With increasing temperature, sliding between chains becomes progressively easier until above $T_m$ the polymer behaves as a viscous fluid.

Several important features of polymer structure are illustrated in Fig. 8-10. We note that the atactic isomers of polystyrene are amorphous because the bulky side groups randomly distributed along the backbone prevent close packing of the chains. However, when polymerized in the isotactic isomer, it results in a crystalline polymer. Referring to Fig. 8-8, which shows little change in properties at $T_g$ for the crystalline polymer, it is not surprising that the crystalline polymer does not show a marked rubbery region (Fig. 8-10). Returning to the atactic (amorphous) condition, we see that the polymer with the greater degree of polymerization is more resistant to chain sliding and is "less rubbery," while the cross-linked polymer is resistant to slippage of chains and has an $E_r$ which is almost constant over a wide temperature interval.

The yield drop phenomenon shown in Fig. 8-9 at 122 and 140°C is similar to that found in mild steel. As in mild steel, the yield drop is due to heterogeneous deformation, but in a polymer this results from the propagation of a *necked region*

**Fig. 8-11** Schematic representations of the modes of deformation of long-chain polymers. (*From M. M. Eisenstadt., "Introduction to Mechanical Properties of Materials," p. 264, The Macmillan Company, New York, 1971. By permission of the publishers.*)

along the length of the specimen (Fig. 8-12). However, if the true stress is calculated on the basis of actual cross-sectional area at the neck (see Sec. 9-2), there is no drop in stress due to yielding, only a brief plateau. In a crystalline polymer, such as nylon 6, continued straining beyond the yield point results in an increase in strength similar to that achieved by strain hardening a metal. In a polymer this strengthening arises from the orientation of the polymer chains in the direction of straining, with a corresponding increase in the degree of crystallinity.

Another type of localization of the yielding process in polymers is the formation of *deformation bands*[1] or kink bands at an angle approximately 45° to the stress axis. The bands are very thin plates a few microns in thickness which are

**Fig. 8-12**  Yield drop due to necking (cold-drawing).

[1] N. Brown, R. A. Duckeet, and I. M. Ward, *Philos. Mag.*, vol. 18, pp. 483–502, 1968.

**Fig. 8-13** Temperature dependence of yield stress.

regions of high localized shear strain. These bands form in advance of a propagating neck, and as the neck moves into their region the deformation bands coalesce with the neck and disappear, leaving behind homogeneously drawn material with no band structure.

In amorphous thermoplastics, deformation frequently occurs by *crazing* rather than by plastic flow in shear. Crazing has been defined[1] as a kind of "tortured plastic flow" which involves the production of a large number of discontinuities which are localized regions of "spongy" plastically deformed polymer. Crazes are easily mistaken for cracks. They are, however, an important mechanism of deformation which results in increased toughness and stress relief. The formation of crazes is roughly similar to the development of voids in a metal preceding ductile fracture (see Sec. 7-10).

The stress-strain curves of polymers are much more temperature and strain-rate dependent than for metals. This is a natural consequence of their irregular structure, compared to metals, which precludes a large-scale organized movement of the structure, as with a dislocation in a crystal lattice. The dependence of yield stress on $T$ and $\dot{\varepsilon}$ for an amorphous glassy polymer (PMMA) is shown in Fig. 8-13. A crystalline polymer (PTFE) does not show the linear behavior of the amorphous polymer.

## 8-6 YIELDING CRITERIA FOR POLYMERS

The von Mises' and Tresca yield criteria that are so well established for metals do not apply for polymers.[2] From Sec. 3-4 it will be recalled that the three basic assumptions concerning these yield criteria were: (1) isotropy, (2) identical yield behavior in tension and compression (no Bauschinger effect), and (3) no dependence

[1] R. P. Kambour, *J. Polymer. Sci.*, pt. D, Rev., vol. 7, pp. 1–154, 1973.
[2] I. M. Ward, *J. Mater. Sci.*, vol. 6, pp. 1397–1417, 1971.

of yielding on the hydrostatic component of the stress state. We have seen (Sec. 3-5) how the von Mises' criterion may be modified to account for anisotropy. In annealed metals the second assumption is usually reasonable, but in polymers, particularly highly oriented polymers, there can be an appreciable difference between the yield stress in tension and compression. But, most importantly, in polymers there is a strong dependence of yield stress on hydrostatic pressure. For example, the yield stress in shear of PMMA more than doubles at a pressure of 3 kbar.‡ Other influences of hydrostatic pressure on the mechanical properties of polymers are an increase in elastic modulus with pressure and a pressure-induced transition from brittle to ductile behavior in some polymers.

One approach to a yield criterion for polymers is to adopt[1] the Coulomb yield criterion for the plastic deformation of a loose soil.[2] A similar theory has been used by Mohr[3] to describe the fracture envelope of a brittle material, such as rock, in which the strength in compression exceeds the strength in tension. The Coulomb yield criterion considers that the more densely the particles of soil are packed together, the more difficult it is for them to slide over one another. Thus, the yield criteria states that the shear stress for yielding to occur on any plane in the material increases linearly with the stress applied normal to this plane.

$$\tau = \tau_c + \mu \sigma_n = \tau_c + \sigma_n \tan \phi \qquad (8\text{-}8)$$

where $\tau_c$ = the critical shear stress, known as the *cohesion* of the material

$\mu$ = a "coefficient of friction": $\mu = \tan \phi$ (see Fig. 8-14)

$\sigma_n$ = the compressive normal stress on the shear plane

The angle $\phi$ is the "internal friction angle" of the material. We can obtain a feel for this concept (which is totally different from the internal friction discussed in Sec. 8-2) by considering extreme cases. For a soft wet clay, there is good cohesion, but $\phi = 0$ and Eq. (8-8) reduces to the Tresca yield criterion $\tau = \tau_c$. For a cohesionless dry sand, $\tau_c = 0$ and $\tau = \sigma_n \tan \phi$. Now $\phi$ is the *angle of repose*, the steepest angle that a sand hill can sustain without slipping.

Figure 8-14 illustrates the Coulomb yield criterion. For uniaxial loading, if yielding occurs on a plane whose normal makes an angle of $\theta$ with the compressive stress $\sigma_1$, then from Eq. (2-2) and (2-3), we can write $\sigma_n = \sigma_1 \cos^2 \theta$ and $\tau = \sigma_1 \sin \theta \cos \theta$. The Coulomb yield criterion then becomes

$$\sigma_1 \sin \theta \cos \theta = \tau_c + \sigma_1 \tan \phi \cos^2 \theta$$

or yielding occurs when

$$\sigma_1(\cos \theta \sin \theta - \tan \phi \cos^2 \theta) \geq \tau_c$$

‡ S. Rabinowitz, I. M. Ward, and J. S. C. Parry, *J. Mater. Sci.* vol. 5, p. 29, 1970.
[1] P. B. Bowden and J. A. Jukes, *J. Mater. Sci.*, vol. 3, p. 193, 1968.
[2] C. A. Coulomb, *Mem. Math. et Phys.*, vol. 7, p. 343, 1773; A. H. Cottrell, "The Mechanical Properties of Matter," pp. 315–317, John Wiley & Sons, Inc., New York, 1964.
[3] J. C. Jaeger, "Elasticity, Fracture, and Flow," 2d ed., pp. 80–85, Methuen & Co., Ltd., London, 1962.

Fig. 8-14   Coulomb yielding criterion.

Yielding will occur at the lowest value of $\sigma_1$ when $(\cos\theta \sin\theta - \tan\phi \cos^2\theta)$ is a maximum. This occurs when $\tan\phi \tan 2\theta = -1$ or $\theta = \pi/4 + \phi/2$. Thus, the yield criterion not only defines the stress state for yielding but it also determines the direction $\theta$ in the material of the yield plane. Thus, when $\phi = 0°$, $\theta = 45°$ and if $\phi = 45°$, $\theta = 67.5°$.

Sternstein[1] has proposed a three-dimensional analog of the Coulomb yield criterion

$$\tau_{oct} = \tau_0 - \mu\sigma_m \qquad (8\text{-}9)$$

where $\tau_{oct}$ = the octahedral shear stress, Eq. (3-33), at a pressure $p$

$\tau_0$ = the octahedral shear stress at zero pressure

$\sigma_m$ = the hydrostatic or mean stress $(\sigma_1 + \sigma_2 + \sigma_3)/3$

The yield criterion may be cast into more familiar terms as

$$[(\sigma_1 - \sigma_2)^2 + (\sigma_2 - \sigma_3)^2 + (\sigma_3 - \sigma_1)^2]^{1/2} + \frac{\sqrt{2}(\sigma_{oc} - \sigma_{ot})3\sigma_m}{\sigma_{oc} + \sigma_{ot}} = \frac{2\sqrt{2}\,\sigma_{oc}\,\sigma_{ot}}{\sigma_{oc} + \sigma_{ot}}$$

$$(8\text{-}10)$$

where $\sigma_{oc}$ = the absolute value of compressive yield stress at atmospheric pressure

$\sigma_{ot}$ = the absolute value of tensile yield stress at atmospheric pressure

A related pressure modification to the von Mises' criterion has been proposed[2]

$$(\sigma_1 - \sigma_2)^2 + (\sigma_2 - \sigma_3)^2 + (\sigma_3 - \sigma_1)^2 + 6(\sigma_{oc} - \sigma_{ot})\sigma_m = 2\sigma_{oc}\sigma_{ot} \qquad (8\text{-}11)$$

[1] S. Sternstein and L. Ongchin, *Amer. Chem. Soc. Polym. Repr.*, vol. 10, no. 2, p. 1117, September, 1969.
[2] R. Raghava, R. M. Caddell, and G. S. Y. Yeh, *J. Mater. Sci.*, vol. 8, p. 225, 1973.

Comparison with rather limited data[1] shows a slight advantage for Eq. (8-11) over Eq. (8-10).

## 8-7 RHEOLOGY

Rheology[2] is the branch of science which deals with the flow and deformation of materials. Rheology treats matter very broadly and deals with the flow of such diverse materials as baker's dough, clay suspensions, wet sand, pastes, polymers, and rubbers. The concept of *viscosity* is at the heart of the science of rheology. Figure 8-15 depicts a *fluid* between two flat plates. The top plate at a distance $dy$ from the stationary plate is displaced $dx$ by the shear force $F$. The velocity of this top plate is $u = dx/dt$. The displacement gradient $dx/dy$ is the shear strain $\gamma$. The shear-strain rate, usually called just the *shear rate* is

$$\dot{\gamma} = \frac{d\gamma}{dt} = \frac{d}{dt}\left(\frac{dx}{dy}\right) = \frac{d}{dy}\left(\frac{dx}{dt}\right) = \frac{du}{dy} \qquad (8\text{-}12)$$

Therefore, the shear rate is equal to the *velocity gradient*. Newton's law of viscosity states that the shear stress acting on a fluid is *linearly proportional* to the velocity gradient at that point in the fluid.

$$\tau = \eta\dot{\gamma} = \eta\frac{d\gamma}{dt} = \eta\frac{du}{dy} \qquad (8\text{-}13)$$

where $\eta$ is the coefficient of viscosity. Most rheological work is done in the cgs system where the units of viscosity are in dyne-seconds per square centimeter

Stationary plate

**Fig. 8-15** Definition of ideal viscosity.

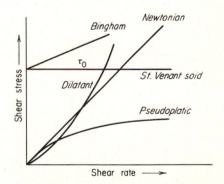

**Fig. 8-16** Flow behavior of different viscous material.

[1] R. M. Caddell, R. S. Raghava, and A. G. Atkins, *Mater. Sci. Eng.*, vol. 13, pp. 113–120, 1974.
[2] M. Reiner, "Deformation, Strain, and Flow," H. K. Lewis & Co., London 1960; R. S. Lenk, "Plastics Rheology," Interscience Publishers, Inc., New York, 1968.

(dyn-s/cm$^2$) or 1 *poise* (P). In the English engineering system the units are in pounds-force-hours per square foot (1b$_f$-h/ft$^2$) and in the SI system the units are in Pascal-seconds-(Pa-s) 1 centipoise (cP) = 1 milliPascal-second (mPa-s)]. The viscosity of materials covers a wide range. For example, the viscosity of water is around $10^{-2}$ P, heavy tar around $10^{10}$ P, and glassy polymers around $10^{15}$ to $10^{20}$ P.

Newtonian viscosity is an ideal against which real material behavior is compared (Fig. 8-16). Certain slurries and suspensions exhibit *Bingham-type* behavior in which no shear rate develops until the shear stress reaches a critical value $\tau_0$. Polymer melts exhibit *pseudoplastic behavior* in which the *apparent viscosity* decreases as the shear rate increases. This is the type of behavior that would be expected from a structure of tangled long-chain polymer molecules which would require a relatively high shear stress to get the mass in motion but once the molecules became untangled and set into motion, proportionately less stress would be needed to keep them in motion at a higher shear rate. A *dilatant* material is the opposite of pseudoplastic in that the stress required to produce the shear rate grows in a nonlinear way at a faster rate. Highly concentrated suspensions of polymers show dilatant behavior. It should be emphasized that there are varying degrees of pseudoplastic and dilatant behavior which vary from slight to marked deviation from Newtonian behavior. The flow behavior of a *St. Venant material* is one in which there is no deformation up to a critical shear stress $\tau_0$, after which the material cannot support any further stress and it will shear at any rate from zero to infinity to prevent the stress from exceeding $\tau_0$.

The flow behavior shown in Fig. 8-16 is time-independent in that as long as a constant shear rate or stress is maintained the same viscosity will be observed. However, some fluids exhibit flow behavior which is time-dependent. When a *thixotropic* fluid is sheared at a constant rate or stress the apparent viscosity will decrease with time, indicating a progressive breakdown of the structure. If the shearing is stopped for a period, the structure reforms and the process may be repeated. The opposite type of behavior is exhibited by a *rheopectic* fluid.

## 8-8 VISCOELASTIC BEHAVIOR

An amorphous polymer undergoes a complex combination of elastic deformation and slow irrecoverable viscous flow which is generally called *viscoelastic behavior*.[1] Viscoelasticity is usually discussed in terms of *linear* mechanical models of material behavior that are composed of springs (the elastic element) and dashpots (the

[1] D. R. Bland, "The Theory of Linear Elastic Viscoelasticity," Pergamon Press, London, 1960; J. D. Ferry, "Viscoelastic Properties of Polymers," 2d ed., John Wiley & Sons, Inc., New York, 1970; A. V. Tobolsky, "Properties and Structure of Polymers," John Wiley & Sons, Inc., New York, 1960; J. G. Williams, "Stress Analysis of Polymers," John Wiley & Sons, Inc., New York, 1973.

viscous element). A simple definition of linear viscoelastic behavior is that the ratio of stress to strain is a function of time only and not a function of the magnitudes of stress and strain.

The *Maxwell element* (Fig. 8-17$a$) consists of a spring and dashpot in series. The total shear strain[1] is the sum of the elastic and viscous-strain components.

$$\gamma = \gamma_E + \gamma_V$$

**Fig. 8-17** (*a*) Maxwell element; (*b*) creep of a Maxwell element; (*c*) stress relaxation of a Maxwell element; (*d*) Voigt-Kelvin element; (*e*) creep of a Voigt-Kelvin element; (*f*) four-parameter model.

[1] We shall use shear stress and shear strain in the mathematical development while recognizing that the mechanical model is visualized by a tensile deformation. The equations developed in this section may be converted to tensile deformation by making the following substitutions: $\sigma \rightarrow \tau$; $\epsilon \rightarrow \gamma$; $E \rightarrow G$; $3\eta \rightarrow \eta$.

or

$$\frac{d\gamma}{dt} = \frac{1}{G}\frac{d\tau}{dt} + \frac{\tau}{\eta} \qquad (8\text{-}14)$$

$$\tau = \eta\frac{d\gamma}{dt} - \left(\frac{\eta}{G}\right)\frac{d\tau}{dt} = \eta\dot{\gamma} - \lambda_R\dot{\tau} \qquad (8\text{-}15)$$

where $\eta$ = the coefficient of viscosity in shear

$G$ = the modulus of elasticity in shear

$\lambda_R = \eta/G$ = the relaxation time

Consider the *creep* of a Maxwell element subjected to a constant shear stress $\tau_0$. From Eq. (8-14), the strain as a function of time is given by

$$\gamma = \frac{\tau_0}{G} + \frac{\tau_0}{\eta}t \qquad (8\text{-}16)$$

Figure 8-17b shows that the application of $\tau_0$ causes an instaneous stretching of the spring to $\tau_0/G$. The dashpot extends linearly with time with a slope $\tau_0/\eta$ so long as the stress is maintained. Thus, the Maxwell element acts like a fluid because it continues to deform as long as it is stressed. When the stress is removed at $t_1$ the spring immediately undergoes an elastic recovery but the dashpot does not recover the permanent set $(\tau_0/\eta)t_1$.

The results of creep experiments often are reported in terms of a *creep compliance* $J(t)$. For a linear viscoelastic material

$$J(t) = \frac{\gamma(t)}{\tau_0} \qquad (8\text{-}17)$$

$J(t)$ is independent of stress. Using this notation the creep curve can be expressed by

$$J(t) = J_0 + \frac{t}{\eta} \qquad (8\text{-}18)$$

where $J_0$ is the instantaneous response = ($\gamma$ at $t = 0)/\tau_0$

The *stress relaxation* behavior of the Maxwell model is established by applying a strain $\gamma_0$ and following the stress as a function of time as the strain is held constant (Fig. 8-17c). When the strain is applied suddenly the elastic spring can respond instantaneously to develop a stress $\tau_0 = G\gamma_0$. The resisting force of the viscous dashpot is very high for a high rate of strain. With increasing time the extended spring begins to contract, but this contraction is resisted by the dashpot. The differential equation for stress relaxation of a Maxwell element is obtained by setting $d\gamma/dt = 0$ in Eq. (8-14).

$$\frac{d\tau}{dt} + \frac{G}{\eta}\tau = 0 \qquad (8\text{-}19)$$

A particular solution to Eq. (8-19) is

$$\tau = B\,e^{-(G/\eta)t}$$

but $\tau_0 = G\gamma_0$ at $t = 0$, so that

$$\tau = G\gamma_0\, e^{-(G/\eta)t} = G\gamma_0\, e^{-t/\lambda_R} \qquad (8\text{-}20)$$

The relaxation time is the time required for the stress to decay by a factor of $1 - (1/e)$ or 63 percent. Physically, this quantity is a measure of the resistance to viscous flow. A simple liquid like water has a $\lambda_R$ measured in microseconds, while complex solids can have relaxation times of minutes on up to years.

Another simple mechanical model is the *Voigt-Kelvin* element (Fig. 8-17d). This model consists of an elastic spring and a viscous dashpot arranged in parallel. Thus, when a stress is applied the strain in the spring and dashpot is the same, and the total stress supported by the element is the sum of the stresses in the spring and the dashpot.

$$\tau = \tau_E + \tau_V$$

$$\tau = G\gamma + \eta\frac{d\gamma}{dt} \qquad (8\text{-}21)$$

When the stress $\tau_0$ is applied suddenly in the creep test, the chief resistance is offered by the viscous dashpot which is initially very resistant. Since the strain in the spring and dashpot are equal for this model, the initial strain of the system is zero, with an initial slope $\tau_0/\eta$ (Fig. 8-17e). As the element is extended, the dashpot gradually transfers load to the spring, providing greater resistance to creep as time goes on. Eventually the element comes to equilibrium with the spring supporting all of the stress since with $d\gamma/dt = 0$, the resistance of the viscous dashpot is zero. Rearranging Eq. (8-21) into standard form

$$\frac{d\gamma}{dt} + \frac{G}{\eta}\gamma - \frac{\tau_0}{\eta} = 0$$

the solution is given by

$$\gamma = \frac{\tau_0}{G} + B\,e^{-(G/\eta)t}$$

At $t = 0$, $\gamma = 0$ and $B = -\tau_0/G$, so that the equation for creep of a Voigt-Kelvin element is

$$\gamma = \frac{\tau_0}{G}[1 - e^{-t/\lambda_R}] \qquad (8\text{-}22)$$

Voigt-Kelvin solids do not exhibit stress relaxation at constant strain. Referring to Eq. (8-21), if $d\gamma/dt = 0$, then $\tau = G\gamma_0$ and the stress will remain constant with time.

While the Maxwell and Voigt-Kelvin models show certain similarities to actual material response, by themselves they do not provide close agreement with real behavior. Considerable improvement can be achieved by combining one

Maxwell element with one Voigt-Kelvin element into the four-parameter model of Fig. 8-17f. The creep equation for this model is

$$\gamma = \frac{\tau}{G_1} + \frac{\tau}{\eta_1} t + \frac{\tau}{G_2}[1 - e^{(G_2/\eta_2)t}] \qquad (8\text{-}23)$$

$$\underbrace{\hspace{2cm}}_{\substack{\text{Hookean} \\ \text{behavior}}} \quad \underbrace{\hspace{2cm}}_{\substack{\text{Newtonian} \\ \text{viscous} \\ \text{behavior}}} \quad \underbrace{\hspace{2cm}}_{\text{Kelvin behavior}}$$

In creep, this model exhibits instantaneous elasticity, primary creep, steady-state creep, and an instantaneous elastic recovery upon the removal of load. However, it does not show sufficiently close agreement with the behavior of polymeric materials over a wide temperature range. To achieve this, it is necessary to construct an *extended Voigt-Kelvin model* in which a large number of Voigt-Kelvin elements each with its own $G$ and $\eta$ are added in series to Fig. 8-17f.

$$\gamma = \frac{\tau}{G_1} + \frac{\tau}{\eta_1} t + \sum_{n=1}^{n} \frac{\tau}{G}[1 - e^{-t/\lambda}] \qquad (8\text{-}24)$$

Although Eq. (8-24) resembles "empirical curve-fitting," there is good physical reasoning[1] to expect this kind of relationship. The first term representing the instantaneous elastic response of the material is related to the primary bonding along the polymer backbone. The second term is the viscous flow of the polymer under a constant load. The magnitude of this term depends on the resistance of individual molecules to relative slippage. For example, high cross-linking reduces this term, while high molecular symmetry increases it. The last set of terms represents the delayed elastic response of the polymer. Its physical significance comes from the uncoiling, disentangling, and temporary orientation of polymer chains. Since there will be molecular chains of varying size and varying entanglement, it is not difficult to see why the material must be characterized by a variety of retardation times.

The *Boltzmann superposition principle* may be applied to a linear viscoelastic material. Simply stated, this means that the creep strain resulting from the superposition of two or more stresses is the sum of the strains that each of these stresses would have produced had they acted independently. This is a very useful technique for analyzing mechanical behavior under complex loading conditions.

The viscoelastic properties depend strongly on temperature. Using our mechanical models, this comes about because of the temperature dependence of $G$ and $\eta$ (especially the latter). One way of comparing creep behavior of a given polymer at different temperatures is to plot $[\gamma(t)/\tau_0]T$ against logarithm of time (Fig. 8-18). The $T$ term in the ordinate accounts for the mild temperature depend-

---

[1] A. T. DiBenedetto, "The Structure and Properties of Materials," p. 371, McGraw-Hill Book Company, New York, 1967.

**Fig. 8-18**   Normalized creep curves for different temperatures.

ence of elastic modulus. The horizontal spread between temperatures arises from the strong temperature-dependence of viscosity. However, since the curves at each temperature are the same shape, the curves can all be made to superimpose on a master curve at some arbitrarily chosen reference temperature by a horizontal shift in which each value on the abscissa is corrected by $\log t - \log A_T$. In other words, a time-temperature superposition is applicable. The Williams-Landel-Ferry equation,[1] which is based on the temperature-dependence of viscosity, gives good prediction of the shift factor $A_T$ for linear amorphous polymers when the reference temperature is taken as the glass transition temperature.

$$\log A_T = \frac{-17.44\,(T - T_g)}{51.6 + T - T_g} \qquad (8\text{-}25)$$

where temperature is expressed in degrees Kelvin (°K). An important aspect of the concept of time-temperature superposition is that the shift factor applies to a particular polymer regardless of the nature of the mechanical response. Therefore, a shift factor determined for stress relaxation is applicable for predicting time-temperature behavior in creep or dynamic testing. Using this technique, it is possible to predict the properties of a polymer at a variety of temperatures when experimental data is available at only a few temperatures. More reliable results are obtained when interpolating between temperatures with known data than when extrapolating to temperatures outside of the range of the data.

## 8-9   RUBBER ELASTICITY

Elastomers and other rubberlike materials are characterized by elastically recoverable deformations of several hundred percent. In a theory of large-scale elastic deformation[2] the simple linear relationship between deformation and location of a

[1] M. L. Williams, R. F. Landel, and J. D. Ferry, *J. Am. Chem. Soc.*, vol. 77, p. 3701, 1955.
[2] A. E. Green and J. E. Adkins, "Large Elastic Deformations and Non-Linear Continuum Mechanics," Clarendon Press, Oxford, 1960.

point in a solid, Eq. (2-33), that was our earlier basis for the definition of small elastic strains must be modified. Strain is defined in terms of *extension ratio* $\lambda = L/L_0$

$$\lambda_1{}^2 = 1 + 2\bar{e}_{xx} \qquad \lambda_2{}^2 = 1 + 2\bar{e}_{yy} \qquad \lambda_3{}^2 = 1 + 2\bar{e}_{zz} \qquad (8\text{-}26)$$

$\bar{e}_{xy}$, $\bar{e}_{yy}$, $\bar{e}_{zz}$ are the components of *finite strain* defined by

$$\bar{e}_{xx} = \frac{\partial u}{\partial x} + \frac{1}{2}\left\{\left(\frac{du}{\partial x}\right)^2 + \left(\frac{\partial v}{\partial x}\right)^2 + \left(\frac{\partial w}{\partial x}\right)^2\right\}$$

$$\bar{e}_{yy} = \frac{\partial v}{\partial y} + \frac{1}{2}\left\{\left(\frac{\partial u}{\partial y}\right)^2 + \left(\frac{\partial v}{\partial y}\right)^2 + \left(\frac{\partial w}{\partial y}\right)^2\right\} \qquad (8\text{-}27)$$

$$\bar{e}_{zz} = \frac{\partial w}{\partial z} + \frac{1}{2}\left\{\left(\frac{\partial u}{\partial z}\right)^2 + \left(\frac{\partial v}{\partial z}\right)^2 + \left(\frac{\partial w}{\partial z}\right)^2\right\}$$

where $u$, $v$, and $w$ are the components of the displacement relative to the $x$, $y$, and $z$ axes, respectively. The above equations should be compared with Eq. (2-36) for small elastic strains.

The constitutive equations for small-scale elastic behavior were described by Eq. (2-24). We can rearrange these into the form

$$e_x = \frac{1+v}{E}\left[\sigma_x - \frac{v}{1+v}(\sigma_x + \sigma_y + \sigma_z)\right]$$

and can further rewrite[1] them as

$$e_x = \frac{1+v}{E}(\sigma_x - p)$$

$$e_y = \frac{1+v}{E}(\sigma_y - p)$$

$$e_z = \frac{1+v}{E}(\sigma_z - p) \qquad (8\text{-}28)$$

where $p$ is a hydrostatic pressure defined as $(v/1 + v)(\sigma_x + \sigma_y + \sigma_z)$. Summing the three equations in Eq. (8-28) and applying the incompressibility condition gives

$$e_x + e_y + e_z = 0 = \frac{1+v}{E}(\sigma_x + \sigma_y + \sigma_z - 3p)$$

and substituting for $p$ allows an evaluation of $v$.

$$\frac{1+v}{E}\left[(\sigma_x + \sigma_y + \sigma_z) - \frac{3v}{1+v}(\sigma_x + \sigma_y + \sigma_z)\right] = 0$$

$$1 - \frac{3v}{1+v} = 0 \qquad \text{and} \qquad v = \frac{1}{2}$$

---

[1] I. M. Ward, "Mechanical Properties of Solid Polymers," pp. 37–40, John Wiley & Sons, Inc., New York, 1971.

Therefore, for an isotropic, incompressible elastic solid, Hooke's law can be written

$$e_x = \frac{3}{2E}(\sigma_x - p)$$

$$e_y = \frac{3}{2E}(\sigma_y - p)$$

$$e_z = \frac{3}{2E}(\sigma_z - p) \qquad (8\text{-}29)$$

Rivlin[1] proposed that the constitutive equations for a rubber material requiring a finite-strain analysis should be written

$$1 + 2\bar{e}_x = \frac{3}{E}(\sigma_x - p)$$

$$1 + 2\bar{e}_y = \frac{3}{E}(\sigma_y - p)$$

$$1 + 2\bar{e}_z = \frac{3}{E}(\sigma_z - p) \qquad (8\text{-}30)$$

Consider the simple case of a rubber specimen loaded in tension along the $x$ axis. Then Eq. (8-30) reduce to

$$1 + 2\bar{e}_x = \lambda_1{}^2 = \frac{3}{E}(\sigma_x - p)$$

$$1 + 2\bar{e}_y = \lambda_2{}^2 = -\frac{3p}{E}$$

$$1 + 2\bar{e}_z = \lambda_3{}^2 = -\frac{3p}{E} \qquad (8\text{-}31)$$

The incompressibility condition is given by $\lambda_1 \lambda_2 \lambda_3 = 1$ so that

$$\lambda_2{}^2 = \lambda_3{}^2 = -\frac{3p}{E}\frac{1}{\lambda_1} \qquad (8\text{-}32)$$

and

$$\sigma_x = \frac{E}{3}\left(\lambda_1{}^2 - \frac{1}{\lambda_1}\right) \qquad (8\text{-}33)$$

Further insight into rubber elasticity[2] can be obtained from a brief excursion into the thermodynamics of elasticity.[3] From the first law of thermodynamics

$$dU = dQ - dW \qquad (8\text{-}34)$$

---

[1] R. S. Rivlin, *Philos. Trans. R. Soc. London,* vol. A241, p. 379, 1948.

L. R. G. Treloar, "The Physics of Rubber Elasticity," 2d ed. Oxford University Press, London, 1958.

[3] This section follows the presentation by S. L. Rosen, "Fundamental Principles of Polymeric Materials," pp. 153–163, Barnes & Noble, Inc., New York, 1971.

where $dU$ = the change in internal energy of the system,

$\quad dQ$ = the heat exchanged between the system and the surroundings

$\quad dW$ = the work done by the system on the surroundings

$$dU = TdS - pdV + PdL \qquad (8\text{-}35)$$

where $S$ = the entropy of the system at temperature $T$

$\quad pdV$ = the work done in changing the volume at pressure $p$

$\quad PdL$ = the work done by a tensile force $P$ in changing the length of the system

For applied tension at constant volume[1] and temperature

$$P = \left(\frac{\partial U}{\partial L}\right)_{T,V} - T\left(\frac{\partial S}{\partial L}\right)_{T,V} \qquad (8\text{-}36)$$

This shows that the force required to deform a polymer is due to two main effects: (1) an energy contribution (compare with Eq. 2-87) due to elastic straining of bond lengths and bond angles, and (2) an entropy contribution resulting from a decrease in entropy with deformation. The entropy of a system is concerned with the degree of disorder. Increasing disorder increases entropy. In the unstressed condition the structure of rubber is very disorganized with a random coiled and tangled structure. When rubber is stretched out under a tensile force many of these tangles are straightened and the entropy decreases.

Another interesting behavior of rubber can be established by rearranging Eq. (8-35) to establish the effect of temperature on the length of a rubberband subjected to a constant tensile force. We will assume constant volume conditions.

$$\left(\frac{\partial L}{\partial T}\right)_{P,V} = \frac{1}{P}\left(\frac{\partial U}{\partial T}\right)_{P,V} - \frac{T}{P}\left(\frac{\partial S}{\partial T}\right)_{P,V} \qquad (8\text{-}37)$$

The first term on the right represents energy elasticity of the type considered in Sec. 2-14. Ordinarily, internal energy increases with temperature, so that a temperature rise produces a thermal expansion. However, all of the factors in the entropy term of Eq. (8-37) are positive, so that an increase in temperature results in a *decrease* in length. In a rubber the entropy term predominates, so that an increase in temperature results in a contraction in length.

The entropy change in an amorphous polymer between a stretched ($L$) and unstretched ($L_0$) state can be calculated from statistical mechanics

$$S - S_0 = -\frac{1}{2}N_0 R\left[\left(\frac{L}{L_0}\right)^2 + 2\left(\frac{L_0}{L}\right) - 3\right] \qquad (8\text{-}38)$$

This equation is based on a simple structural model of flexible chainlike molecules that are pinned into a three-dimensional network by cross-links. A *network*

---

[1] It is experimentally difficult to carry out tensile tests at constant volume. Most tests are carried out at constant pressure, and since Poisson's ratio is small the volume change is small.

*chain* is the part of the polymer chain that lies between two cross-linking points. In Eq. (8-38), $N_0$ is the number of moles of network chains in the unstretched rubber, while $R$ is the universal gas constant. For an *ideal rubber*, in which there are no intermolecular forces

$$\left(\frac{\partial U}{\partial L}\right)_{T,V} = 0 \quad \text{and} \quad P = -T\left(\frac{\partial S}{\partial L}\right)_{T,V}$$

Differentiating Eq. (8-38) and substituting above gives

$$P = \frac{N_0 R T}{L_0}\left[\frac{L}{L_0} - \left(\frac{L_0}{L}\right)^2\right] \qquad (8\text{-}39)$$

However,

$$N_0 = \frac{\text{mass}}{\overline{M}_c} = \frac{\rho V}{\overline{M}_c} = \frac{\rho L_0 A_0}{\overline{M}_c} = \frac{\rho L A}{\overline{M}_c} \qquad (8\text{-}40)$$

where $\overline{M}_c$ is the number-average molecular weight of network chains, so that

$$P = \frac{\rho A_0 R T}{\overline{M}_c}\left[\frac{L}{L_0} - \left(\frac{L_0}{L}\right)^2\right] \qquad (8\text{-}41)$$

This equation shows that the force required to maintain a rubber specimen at a fixed extension *increases* with increasing temperature. Also, increasing the number of cross-links (decreasing $\overline{M}_c$) increases this force.

The engineering stress required to produce a given extension ratio $\lambda = L/L_0$ is given by

$$s = \frac{P}{A_0} = \frac{\rho R T}{\overline{M}_c}\left[\frac{L}{L_0} - \left(\frac{L_0}{L}\right)^2\right] \qquad (8\text{-}42)$$

The true tensile stress is given by

$$\sigma = \frac{P}{A} = \frac{\rho R T}{\overline{M}_c}\left[\left(\frac{L}{L_0}\right)^2 - \frac{L_0}{L}\right]$$

or

$$\sigma = \frac{\rho R T}{\overline{M}_c}\left(\lambda_1^2 - \frac{1}{\lambda_1}\right) \qquad (8\text{-}43)$$

Young's modulus of the rubber is

$$E = \left(\frac{\partial \sigma}{\partial \varepsilon}\right)_T = \frac{\partial \sigma}{\partial L/L} = = L\frac{\partial \sigma}{\partial L} = \frac{\rho R T}{\overline{M}_c}\left[2\left(\frac{L}{L_0}\right)^2 + \frac{L_0}{L}\right] \qquad (8\text{-}44)$$

The initial modulus, as $L \to L_0$, becomes

$$E_0 = \frac{3\rho R T}{\overline{M}_c} \qquad (8\text{-}45)$$

## 8-10 FRACTURE AND TOUGHNESS

We have seen several examples that the mechanical properties of polymers are strongly affected by temperature and strain rate. Figure 8-9 shows the change from ductile fracture to brittle fracture in an amorphous polymer for slow-speed tensile tests. The concepts of flow stress and fracture stress that were considered in Sec. 7-12 with metals are equally applicable to polymers in thinking about the transition from ductile to brittle fracture. The role of strain rate in increasing the transition temperature is well illustrated by the tensile deformation of nylon which cold-draws with a yield-point behavior at slow strain rate but fractures in a brittle fashion at high strain rates.

The concepts of the Griffith theory of fracture developed in Sec. 7-4 apply equally well to the brittle fracture of polymers.[1] As with metals, the surface energy measured in brittle fracture is about 1,000 times greater than would be required if fracture involved simply the breaking of a single layer of carbon-carbon bonds. This extra energy is required to initiate and propagate crazes at the tip of the crack.

The toughness of a normally brittle glassy amorphous polymer, such as polystyrene, can be increased appreciably by the dispersion of 5 to 10 vol percent of rubber particles. To be effective, the rubber particles must adhere well to the glassy polymer matrix and be in the size range 0.1 to 10 $\mu$m. The notched-impact strength of such a *polyblend* can be increased by a factor of 10 over the unmodified polymer. Deformation of a polyblend is accompanied by a noticeable increase in opacity called *stress whitening*. This is produced by a myriad of tiny crazes eminating from the rubber particles. These particles act as stress raisers to cause large numbers of crazes to initiate. Since the overall deformation of the sample is the sum of the strain produced by each craze, the resulting increase in ductility and toughness arises from this extensive craze formation.

An important constraint on the use of amorphous polymers is *solvent crazing and cracking*. This is concerned with the onset of crazing, with eventual cracking, in the presence of a wide variety of organic agents at stress levels far below the critical stress for crazing in air. It appears that liquids which wet the polymer surface and lower the surface energy and/or cause swelling of the polymer with a decrease in $T_g$ are the chief offenders. This type of environmental stress cracking (see Sec. 14-12) is less prevalent in crystalline polymers.

The toughness of amorphous polymers above $T_g$ is only moderate because although the elongation at fracture is normally several hundred percent, the tensile strength is quite low, of the order of 5,000 psi. Important inprovements in strength and toughness can be achieved by compounding the polymer with *reinforcing fillers*

---

[1] J. P. Berry, *J. Appl. Phys.*, vol. 34, 1963, p. 62; B. Rosen ed., "Fracture Processes in Polymeric Solids," Interscience Publishers, Inc., New York, 1964; E. H. Andrews, "Fracture in Polymers," Oliver and Boyd Ltd., London, 1968.

such as glass fibers, asbestos, or synthetic fibers. The theory of fiber strengthening was discussed in Sec. 6-10. The presence of glass fibers in fiber-reinforced polymers (FRP) creates many "crack stoppers" and many interfaces which can absorb energy in the fracture process.

The fracture properties of a glassy amorphous polymer are strongly influenced by the orientation of the molecular chains in the material. If polystyrene is heated above $T_g$ and stretched several hundred percent in one direction and then cooled to room temperature *while under stress*, the polymer chains will be oriented preferentially in the stretching direction. The tensile strength in the direction of orientation can be increased by a factor of 2 or 3, while the elongation at fracture may be increased from a nominal 3 percent to 50 percent or higher. Since toughness is the area under the stress-strain curve, this property can be increased up to 20 times. However, the property improvements from chain orientation are markedly anisotropic; thus a stretched material will be weaker in the direction transverse to the direction of stretching. This anisotropic behavior is not important in fibers, but in plastic films and thin sheet it may be necessary to employ biaxial molecular orientation to minimize the anisotropy. The highest strength polymers are produced by the molecular orientation of crystalline polymers.

## BIBLIOGRAPHY

Alfrey, T., and E. F. Gurney: "Organic Polymers," Prenctice-Hall Inc., Englewood Cliffs, N.J., 1967.

Baer, E. (ed.): "Engineering Design for Plastics," Reinhold Publishing Corporation, New York, 1964.

Deanin, R. D.: "Polymer Structure, Properties, and Applications," Cahners Books, Boston, 1972.

Houwink, R., and H. K. deDecker (eds.): "Elasticity, Plasticity, and Structure of Matter," 3rd ed., Cambridge University Press, London, 1971.

McKelvey, J. M.: "Polymer Processing," John Wiley & Sons, Inc., New York, 1962.

Nielsen, L. E.: "Mechanical Properties of Polymers," Reinhold Publishing Corporation, New York, 1962.

Rodriguez, F.: "Principles of Polymer Systems," McGraw-Hill Gook Company, New York, 1970.

Rosen, S. L.: "Fundamental Principles of Polymeric Materials," Barnes & Noble, Inc., New York, 1971.

Schultz, J.: "Polymer Materials Science," Prentice-Hall, Inc., Englewood Cliffs, N.J., 1974.

Ward, I. M.: "Mechanical Properties of Solid Polymers," Wiley-Interscience, New York, 1971.

# PART THREE

# Applications to Materials Testing

# THE TENSION TEST

## 9-1 ENGINEERING STRESS-STRAIN CURVE

The engineering tension test is widely used to provide basic design information on the strength of materials and as an acceptance test for the specification of materials. In the tension test[1] a specimen is subjected to a continually increasing uniaxial tensile force while simultaneous observations are made of the elongation of the specimen. An engineering stress-strain curve is constructed from the load-elongation measurements (Fig. 9-1). The significant points on the engineering stress-strain curve have already been considered in Sec. 1-5, while the appearance of a yield point in the stress-strain curve was covered in Sec. 6-5. The stress used in this stress-strain curve is the *average* longitudinal stress in the tensile specimen.[2]

---

[1] H. E. Davis, G. E. Troxell, and C. T. Wiskocil, "The Testing and Inspection of Engineering Materials," 2d ed., chaps. 2–4, McGraw-Hill Book Company, New York, 1955; Standard Methods of Tension Testing of Metallic Materials, ASTM Designation E8-69, "1970 Annual Book of ASTM Standards," pt. 31, pp. 194–213, American Society for Testing and Materials, Philadelphia, 1970.

[2] In this chapter we use $s$ to designate engineering stress and $\sigma$ to designate true stress. Elsewhere in this text we have used $\sigma$ more generally to indicate stress.

Conventional strain $e$

**Fig. 9-1**  The engineering stress-strain curve.

It is obtained by dividing the load by the *original area* of the cross section of the specimen.

$$s = \frac{P}{A_0} \qquad (9\text{-}1)$$

The strain used for the engineering stress-strain curve is the *average* linear strain, which is obtained by dividing the elongation of the gage length of the specimen, $\delta$, by its original length.

$$e = \frac{\delta}{L_0} = \frac{\Delta L}{L} = \frac{L - L_0}{L_0} \qquad (9\text{-}2)$$

Since both the stress and the strain are obtained by dividing the load and elongation by constant factors, the load-elongation curve will have the same shape as the engineering stress-strain curve. The two curves are frequently used interchangeably.

The shape and magnitude of the stress-strain curve of a metal will depend on its composition, heat treatment, prior history of plastic deformation, and the strain rate, temperature, and state of stress imposed during the testing. The parameters which are used to describe the stress-strain curve of a metal are the *tensile strength, yield strength* or *yield point, percent elongation*, and *reduction of area*. The first two are strength parameters; the last two indicate ductility.

The general shape of the engineering stress-strain curve (Fig. 9-1) requires further explanation. In the elastic region stress is linearly proportional to strain. When the load exceeds a value corresponding to the yield strength, the specimen undergoes gross plastic deformation. It is permanently deformed if the load is released to zero. The stress to produce continued plastic deformation increases with increasing plastic strain, i.e., the metal strain-hardens. The volume of the specimen remains constant during plastic deformation, $AL = A_0 L_0$, and as the specimen elongates, it decreases uniformly along the gage length in cross-sectional area. Initially the strain hardening more than compensates for this decrease in area and the engineering stress (proportional to load $P$) continues to rise with increasing strain. Eventually a point is reached where the decrease in specimen cross-sectional area is greater than the increase in deformation load arising from strain hardening. This condition will be reached first at some point in the specimen that is slightly weaker than the rest. All further plastic deformation is concentrated in this region, and the specimen begins to neck or thin down locally. Because the cross-sectional area now is decreasing far more rapidly than the deformation load is increased by strain hardening, the actual *load* required to deform the specimen falls off and the engineering stress by Eq. (9-1) likewise continues to decrease until fracture occurs.

## TENSILE STRENGTH

The tensile strength, or ultimate tensile strength (UTS), is the maximum load divided by the original cross-sectional area of the specimen.

$$s_u = \frac{P_{max}}{A_0} \qquad (9\text{-}3)$$

The tensile strength is the value most often quoted from the results of a tension test; yet in reality it is a value of little fundamental significance with regard to the strength of a metal. For ductile metals the tensile strength should be regarded as a measure of the maximum load which a metal can withstand under the very restrictive conditions of uniaxial loading. It will be shown that this value bears little relation to the useful strength of the metal under the more complex conditions of stress which are usually encountered. For many years it was customary to base the strength of members on the tensile strength, suitably reduced by a factor of safety. The current trend is to the more rational approach of basing the static design of ductile metals on the yield strength. However, because of the long practice of using the tensile strength to determine the strength of materials, it has become a very familiar property, and as such it is a very useful identification of a material in the same sense that the chemical composition serves to identify a metal or alloy. Further, because the tensile strength is easy to determine and is a quite reproducible property, it is useful for the purposes of specifications and for quality control of a product. Extensive empirical correlations between tensile strength and properties such as hardness and fatigue strength are often quite useful. For brittle materials, the tensile strength is a valid criterion for design.

## MEASURES OF YIELDING

The stress at which plastic deformation or yielding is observed to begin depends on the sensitivity of the strain measurements. With most materials there is a gradual transition from elastic to plastic behavior, and the point at which plastic deformation begins is hard to define with precision. Various criteria for the initiation of yielding are used depending on the sensitivity of the strain measurements and the intended use of the data.

1. *True elastic limit* based on microstrain measurements at strains on order of $2 \times 10^{-6}$ in|in (see Sec. 4-13). This elastic limit is a very low value and is related to the motion of a few hundred dislocations.

2. *Proportional limit* is the highest stress at which stress is directly proportional to strain. It is obtained by observing the deviation from the straight-line portion of the stress-strain curve.

3. *Elastic limit* is the greatest stress the material can withstand without any measurable permanent strain remaining on the complete release of load. With increasing sensitivity of strain measurement, the value of the elastic limit is decreased until at the limit it equals the true elastic limit determined from microstrain measurements. With the sensitivity of strain usually employed in engineering studies ($10^{-4}$ in|in), the elastic limit is greater than the proportional limit. Determination of the elastic limit requires a tedious incremental loading-unloading test procedure.

4. The *yield strength* is the stress required to produce a small specified amount of plastic deformation. The usual definition of this property is the *offset yield strength* determined by the stress corresponding to the intersection of the stress-strain curve and a line parallel to the elastic part of the curve offset by a specified strain (Fig. 9-1). In the United States the offset is usually specified as a strain of 0.2 or 0.1 percent ($e = 0.002$ or $0.001$).

$$s_0 = \frac{P_{\text{(strain offset = 0.002)}}}{A_0} \qquad (9\text{-}4)$$

A good way of looking at offset yield strength is that after a specimen has been loaded to its 0.2 percent offset yield strength and then unloaded it will be 0.2 percent longer than before the test. The offset yield strength is often referred to in Great Britain as the *proof stress*, where offset values are either 0.1 or 0.5 percent. The yield strength obtained by an offset method is commonly used for design and specification purposes because it avoids the practical difficulties of measuring the elastic limit or proportional limit.

Some materials have essentially no linear portion to their stress-strain curve, for example, soft copper or gray cast iron. For these materials the offset method cannot be used and the usual practice is to define the yield strength as the stress to produce some total strain, for example, $\varepsilon = 0.005$.

## MEASURES OF DUCTILITY

At our present degree of understanding, ductility is a qualitative, subjective property of a material. In general, measurements of ductility are of interest in three ways:[1]

1. To indicate the extent to which a metal can be deformed without fracture in metalworking operations such as rolling and extrusion.

2. To indicate to the designer, in a general way, the ability of the metal to flow plastically before fracture. A high ductility indicates that the material is "forgiving" and likely to deform locally without fracture should the designer err in the stress calculation or the prediction of severe loads.

3. To serve as an indicator of changes in impurity level or processing conditions. Ductility measurements may be specified to assess material "quality" even though no direct relationship exists between the ductility measurement and performance in service.

The conventional measures of ductility that are obtained from the tension test are the engineering strain at fracture $e_f$ (usually called the *elongation*) and the *reduction of area* at fracture $q$. Both of these properties are obtained after fracture by putting the specimen back together and taking measurements of $L_f$ and $A_f$.

$$e_f = \frac{L_f - L_0}{L_0} \qquad (9\text{-}5)$$

$$q = \frac{A_0 - A_f}{A_0} \qquad (9\text{-}6)$$

Both elongation and reduction of area usually are expressed as a percentage.

Because an appreciable fraction of the plastic deformation will be concentrated in the necked region of the tension specimen, the value of $e_f$ will depend on the gage length $L_0$ over which the measurement was taken (see Sec. 9-5). The smaller the gage length the greater will be the contribution to the overall elongation from the necked region and the higher will be the value of $e_f$. Therefore, when reporting values of *percentage elongation*, the gage length $L_0$ always should be given.

[1] G. E. Dieter, Introduction to Ductility in "Ductility," American Society for Metals, Metals Park, Ohio, 1968.

The reduction of area does not suffer from this difficulty. Reduction of area values can be converted into an equivalent *zero-gage-length elongation* $e_0$. From the constancy of volume relationship for plastic deformation $AL = A_0 L_0$, we obtain

$$\frac{L}{L_0} = \frac{A_0}{A} = \frac{1}{1-q}$$

$$e_0 = \frac{L - L_0}{L_0} = \frac{A_0}{A} - 1 = \frac{1}{1-q} - 1 = \frac{q}{1-q} \qquad (9\text{-}7)$$

This represents the elongation based on a very short gage length near the fracture.

Another way to avoid the complication from necking is to base the percentage elongation on the uniform strain out to the point at which necking begins. The uniform elongation $e_u$ correlates well with stretch-forming operations. Since the engineering stress-strain curve often is quite flat in the vicinity of necking, it may be difficult to establish the strain at maximum load without ambiguity. In this case the method suggested by Nelson and Winlock[1] is useful.

## MODULUS OF ELASTICITY

The slope of the initial linear portion of the stress-strain curve is the modulus of elasticity, or Young's modulus. The modulus of elasticity is a measure of the stiffness of the material. The greater the modulus, the smaller the elastic strain resulting from the application of a given stress. Since the modulus of elasticity is needed for computing deflections of beams and other members, it is an important design value.

The modulus of elasticity is determined by the binding forces between atoms. Since these forces cannot be changed without changing the basic nature of the material, it follows that the modulus of elasticity is one of the most structure-insensitive of the mechanical properties. It is only slightly affected by alloying additions, heat treatment, or cold-work.[2] However, increasing the temperature decreases the modulus of elasticity. The modulus is usually measured at elevated temperatures by a dynamic method.[3] Typical values[4] of the modulus of elasticity for common engineering metals at different temperatures are given in Table 9-1.

[1] P. G. Nelson and J. Winlock, *ASTM Bull.*, vol. 156, p. 53, January 1949.

[2] D. J. Mack, *Trans. AIME*, vol. 166, pp. 68–85, 1946.

[3] P. E. Armstrong, Measurement of Elastic Constants, in R. F. Bunshaw (ed.), "Techniques of Metals Research," vol. V, pt. 2, chap. 9, Interscience Publishers, Inc. New York, 1971.

[4] Standard Method of Test for Young's Modulus at Room Temperature, ASTM E111-61, op. cit., pp. 409–413.

**Table 9-1** TYPICAL VALUES OF MODULES OF ELASTICITY AT DIFFERENT TEMPERATURES

| Material | | Modulus of elasticity, psi $\times 10^{-6}$ | | | |
|---|---|---|---|---|---|
| | Room temp. | 400°F | 800°F | 1000°F | 1200°F |
| Carbon steel | 30.0 | 27.0 | 22.5 | 19.5 | 18.0 |
| Austenitic stainless steel | 28.0 | 25.5 | 23.0 | 22.5 | 21.0 |
| Titanium alloys | 16.5 | 14.0 | 10.7 | 10.1 | |
| Aluminium alloys | 10.5 | 9.5 | 7.8 | | |

## RESILIENCE

The ability of a material to absorb energy when deformed elastically and to return it when unloaded is called *resilience*. This is usually measured by the *modulus of resilience*, which is the strain energy per unit volume required to stress the material from zero stress to the yield stress $\sigma_0$. Referring to Eq. (2-80), the strain energy per unit volume for uniaxial tension is

$$U_0 = \tfrac{1}{2}\sigma_x e_x$$

From the above definition the modulus of resilience is

$$U_R = \frac{1}{2} s_0 e_0 = \frac{1}{2} s_0 \frac{s_0}{E} = \frac{s_0{}^2}{2E} \tag{9-8}$$

This equation indicates that the ideal material for resisting energy loads in applications where the material must not undergo permanent distortion, such as mechanical springs, is one having a high yield stress and a low modulus of elasticity. Table 9-2 gives some values of modulus of resilience for different materials.

**Table 9-2** MODULUS OF RESILIENCE FOR VARIOUS MATERIALS

| Material | $E$, psi | $s_0$, psi | Modulus of resilience, $U_R$ |
|---|---|---|---|
| Medium-carbon steel | $30 \times 10^6$ | 45,000 | 33.7 |
| High-carbon spring steel | $30 \times 10^6$ | 140,000 | 320 |
| Duraluminum | $10.5 \times 10^6$ | 18,000 | 17 |
| Copper | $16 \times 10^6$ | 4,000 | 5.3 |
| Rubber | 150 | 300 | 300 |
| Acrylic polymer | $0.5 \times 10^6$ | 2,000 | 4.0 |

## TOUGHNESS

The toughness of a material is its ability to absorb energy in the plastic range. The ability to withstand occasional stresses above the yield stress without fracturing is particularly desirable in parts such as freight-car couplings, gears, chains, and crane hooks. Toughness is a commonly used concept which is difficult to pin down and define. One way of looking at toughness is to consider that it is the total area under the stress-strain curve. This area is an indication of the amount of work per unit volume which can be done on the material without causing it to rupture. Figure 9-2 shows the stress-strain curves for high- and low-toughness materials. The high-carbon spring steel has a higher yield strength and tensile strength than the medium-carbon structural steel. However, the structural steel is more ductile and has a greater total elongation. The total area under the stress-strain curve is greater for the structural steel, and therefore it is a tougher material. This illustrates that toughness is a parameter which comprises *both* strength and ductility. The crosshatched regions in Fig. 9-2 indicate the modulus of resilience for each steel. Because of its higher yield strength, the spring steel has the greater resilience.

Several mathematical approximations for the area under the stress-strain curve have been suggested. For ductile metals which have a stress-strain curve like that of the structural steel, the area under the curve can be approximated by either of the following equations:

$$U_T \approx s_u e_f \tag{9-9}$$

or

$$U_T \approx \frac{s_0 + s_u}{2} e_f \tag{9-10}$$

For brittle materials the stress-strain curve is sometimes assumed to be a parabola, and the area under the curve is given by

$$U_T \approx \tfrac{2}{3} s_u e_f \tag{9-11}$$

**Fig. 9-2** Comparison of stress-strain curves for high- and low-toughness materials.

All these relations are only approximations to the area under the stress-strain curves. Further, the curves do not represent the true behavior in the plastic range, since they are all based on the original area of the specimen.


## 9-2  TRUE-STRESS–TRUE-STRAIN  CURVE

The engineering stress-strain curve does not give a true indication of the deformation characteristics of a metal because it is based entirely on the original dimensions of the specimen, and these dimensions change continuously during the test. Also, ductile metal which is pulled in tension becomes unstable and necks down during the course of the test. Because the cross-sectional area of the specimen is decreasing rapidly at this stage in the test, the load required to continue deformation falls off. The average stress based on original area likewise decreases, and this produces the fall-off in the stress-strain curve beyond the point of maximum load. Actually, the metal continues to strain-harden all the way up to fracture, so that the stress required to produce further deformation should also increase. If the *true stress*, based on the actual cross-sectional area of the specimen, is used, it is found that the stress-strain curve increases continuously up to fracture. If the strain measurement is also based on instantaneous measurements, the curve which is obtained is known as a *true-stress–true-strain* curve. This is also known as a *flow curve* (Sec. 3-2) since it represents the basic plastic-flow characteristics of the material. Any point on the flow curve can be considered the yield stress for a metal strained in tension by the amount shown on the curve. Thus, if the load is removed at this point and then reapplied, the material will behave elastically throughout the entire range of reloading.

The definitions of true stress and true strain were given in Sec. 3-3. The true stress $\sigma$ is expressed in terms of engineering stress $s$ by

$$\sigma = \frac{P}{A_0}(e+1) = s(e+1) \qquad (9\text{-}12)$$

The derivation of Eq. (9-12) assumes both constancy of volume and a homogenous distribution of strain along the gage length of the tension specimen. Thus, Eq. (9-12) should only be used until the onset of necking. Beyond maximum load the true stress should be determined from actual measurements of load and cross-sectional area.

$$\sigma = P/A \qquad (9\text{-}13)$$

The true strain $\varepsilon$ may be determined from the engineering or conventional strain $e$ by

$$\varepsilon = \ln(e+1) \qquad (9\text{-}14)$$

**Fig. 9-3** Comparison of engineering and true stress-strain curves.

This equation is applicable only to the onset of necking for the reasons discussed above. Beyond maximum load the true strain should be based on actual area or diameter measurements.

$$\varepsilon = \ln \frac{A_0}{A} = \ln \frac{(\pi/4)D_0{}^2}{(\pi/4)D^2} = 2 \ln \frac{D_0}{D} \qquad (9\text{-}15)$$

Figure 9-3 compares the true-stress–true-strain curve with its corresponding engineering stress-strain curve. Note that because of the relatively large plastic strains, the elastic region has been compressed into the $y$ axis. In agreement with Eqs. (9-12) and (9-14), the true-stress–true-strain curve is always to the left of the engineering curve until the maximum load is reached. However, beyond maximum load the high localized strains in the necked region that are used in Eq. (9-15) far exceed the engineering strain calculated from Eq. (9-2). Frequently the flow curve is linear from maximum load to fracture, while in other cases its slope continuously decreases up to fracture. The formation of a necked region or mild notch introduces triaxial stresses which make it difficult to determine accurately the longitudinal tensile stress on out to fracture (see Sec. 9-3).

The following parameters usually are determined from the true-stress–true-strain curve.

### TRUE STRESS AT MAXIMUM LOAD

The true stress at maximum load corresponds to the true tensile strength. For most materials necking begins at maximum load at a value of strain where the true stress equals the slope of the flow curve (see Sec. 9-3). Let $\sigma_u$ and $\varepsilon_u$ denote the true stress and true strain at maximum load when the cross-sectional area of the specimen is $A_u$. The ultimate tensile strength is given by

$$s_u = \frac{P_{\max}}{A_0}$$

and

$$\sigma_u = \frac{P_{\max}}{A_u} \qquad \varepsilon_u = \ln \frac{A_0}{A_u}$$

Eliminating $P_{max}$ yields

$$\sigma_u = s_u \frac{A_0}{A_u}$$

and

$$\sigma_u = s_u e^{e_u} \qquad (9\text{-}16)$$

## TRUE FRACTURE STRESS

The true fracture stress is the load at fracture divided by the cross-sectional area at fracture. This stress should be corrected for the triaxial state of stress existing in the tensile specimen at fracture. Since the data required for this correction are often not available, true-fracture-stress values are frequently in error.

## TRUE FRACTURE STRAIN

The true fracture strain $\varepsilon_f$ is the true strain based on the original area $A_0$ and the area after fracture $A_f$.

$$\varepsilon_f = \ln \frac{A_0}{A_f} \qquad (9\text{-}17)$$

This parameter represents the maximum true strain that the material can withstand before fracture and is analogous to the total strain to fracture of the engineering stress-strain curve. Since Eq. (9-14) is not valid beyond the onset of necking, it is not possible to calculate $\varepsilon_f$ from measured values of $e_f$. However, for cylindrical tensile specimens the reduction of area $q$ is related to the true fracture strain by the relationship

$$\varepsilon_f = \ln \frac{1}{1 - q} \qquad (9\text{-}18)$$

## TRUE UNIFORM STRAIN

The true uniform strain $\varepsilon_u$ is the true strain based only on the strain up to maximum load. It may be calculated from either the specimen cross-sectional area $A_u$ or the gage length $L_u$ at maximum load. Equation (9-14) may be used to convert conventional uniform strain to true uniform strain. The uniform strain is often useful in estimating the formability of metals from the results of a tension test.

$$\varepsilon_u = \ln \frac{A_0}{A_u} \qquad (9\text{-}19)$$

## TRUE LOCAL NECKING STRAIN

The local necking strain $\varepsilon_n$ is the strain required to deform the specimen from maximum load to fracture.

$$\varepsilon_n = \ln \frac{A_u}{A_f} \qquad (9\text{-}20)$$

The flow curve of many metals in the region of uniform plastic deformation can be expressed by the simple power curve relation

$$\sigma = K\varepsilon^n \qquad (9\text{-}21)$$

where $n$ is the *strain-hardening exponent* and $K$ is the *strength coefficient*. A log-log plot of true stress and true strain up to maximum load will result in a straight-line if Eq. (9-21) is satisfied by the data (Fig. 9-4). The linear slope of this line is $n$ and $K$ is the true stress at $\varepsilon = 1.0$ (corresponds to $q = 0.63$). The strain-hardening exponent[1] may have values from $n = 0$ (perfectly plastic solid) to $n = 1$ (elastic solid) (see Fig. 9-5). For most metals $n$ has values between 0.10 and 0.50 (see Table 9-3).

It is important to note that the *rate of strain hardening $d\sigma/d\varepsilon$* is not identical with the strain-hardening exponent.

From the definition of $n$

$$n = \frac{d(\log \sigma)}{d(\log \varepsilon)} = \frac{d(\ln \sigma)}{d(\ln \varepsilon)} = \frac{\varepsilon}{\sigma}\frac{d\sigma}{d\varepsilon}$$

or

$$\frac{d\sigma}{d\varepsilon} = n\frac{\sigma}{\varepsilon} \qquad (9\text{-}22)$$

**Fig. 9-4**  Log-log plot of true stress-strain curve.

**Fig. 9-5**  Various forms of power curve $\sigma = K\varepsilon^n$.

---

[1] Other techniques for measuring $n$ are discussed by J. L. Duncan, *Sheet Met. Ind.*, July 1967, pp. 483–489.

**Table 9-3  VALUES FOR $n$ AND $K$ FOR METALS AT ROOM TEMPERATURE**

| Metal | Condition | $n$ | $K$, psi | Ref. |
|---|---|---|---|---|
| 0.05% C steel | Annealed | 0.26 | 77,000 | † |
| SAE 4340 steel | Annealed | 0.15 | 93,000 | † |
| 0.6% C steel | Quenched and tempered 1000°F | 0.10 | 228,000 | ‡ |
| 0.6% C steel | Quenched and tempered 1300°F | 0.19 | 178,000 | ‡ |
| Copper | Annealed | 0.54 | 46,400 | † |
| 70/30 brass | Annealed | 0.49 | 130,000 | ‡ |

† J. R. Low and F. Garofalo, *Proc. Soc. Exp. Stress Anal.*, vol. 4, no. 2, pp. 16–25, 1947.
‡ J. R. Low, "Properties of Metals in Materials Engineering," American Society for Metals, Metals Park, Ohio, 1949.

There is nothing basic about Eq. (9-21) and deviations from this relationship frequently are observed, often at low strains ($10^{-3}$) or high strains ($\varepsilon \approx 1.0$). One common type of deviation is for a log-log plot of Eq. (9-21) to result in two straight lines with different slopes. Sometimes data which do not plot according to Eq. (9-21) will yield a straight line according to the relationship

$$\sigma = K(\varepsilon_0 + \varepsilon)^n \qquad (9\text{-}23)$$

Datsko[1] has shown how $\varepsilon_0$ can be considered to be the amount of strain hardening that the material received prior to the tension test.

Another common variation on Eq. (9-21) is the Ludwik equation

$$\sigma = \sigma_0 + K\varepsilon^n \qquad (9\text{-}24)$$

where $\sigma_0$ is the yield stress and $K$ and $n$ are the same constants as in Eq. (9-21). This equation may be more satisfying than Eq. (9-21) since the latter implies that at zero true strain the stress is zero. Morrison[2] has shown that $\sigma_0$ can be obtained from the intercept of the strain-hardening portion of the stress-strain curve and the elastic modulus line by

$$\sigma_0 = \left(\frac{K}{E^n}\right)^{1/1-n}$$

The true-stress–true-strain curve of metals such as austenitic stainless steel, which deviate markedly from Eq. (9-21) at low strains, can be expressed by[3]

$$\sigma = K\varepsilon^n + e^{K_1}\, e^{n_1 \varepsilon}$$

[1] J. Datsko, "Material Properties and Manufacturing Processes," pp. 18–20, John Wiley & Sons, Inc., New York, 1966.
[2] W. B. Morrison, *Trans. Am. Soc. Met.*, vol. 59, p. 824, 1966.
[3] D. C. Ludwigson, *Metall. Trans.*, vol. 2, pp. 2825–2828, 1971.

where $e^{K_1}$ is approximately equal to the proportional limit and $n_1$ is the slope of the deviation of stress from Eq. (9-21) plotted against $\varepsilon$. Still other expressions for the flow curve have been discussed in the literature.[1,2]

The true strain term in Eqs. (9-21) to (9-24) properly should be the *plastic strain* $\varepsilon_p = \varepsilon_{total} - \varepsilon_E = \varepsilon_{total} - \sigma/E$.

## 9-3   INSTABILITY IN TENSION

Necking generally begins at maximum load during the tensile deformation of a ductile metal.[3] An ideal plastic material in which no strain hardening occurs would become unstable in tension and begin to neck just as soon as yielding took place. However, a real metal undergoes strain hardening, which tends to increase the load-carrying capacity of the specimen as deformation increases. This effect is opposed by the gradual decrease in the cross-sectional area of the specimen as it elongates. Necking or localized deformation begins at maximum load, where the increase in stress due to decrease in the cross-sectional area of the specimen becomes greater than the increase in the load-carrying ability of the metal due to strain hardening. This condition of instability leading to localized deformation is defined by the condition $dP = 0$.

$$P = \sigma A$$

$$dP = \sigma \, dA + A \, d\sigma = 0$$

From the constancy-of-volume relationship,

$$\frac{dL}{L} = -\frac{dA}{A} = d\varepsilon$$

and from the instability condition

$$-\frac{dA}{A} = \frac{d\sigma}{\sigma}$$

so that at a point of tensile instability

$$\frac{d\sigma}{d\varepsilon} = \sigma \qquad (9\text{-}25)$$

[1] H. J. Kleemola and M. A. Nieminen, *Metall. Trans.*, vol. 5, pp. 1863–1866, 1974.
[2] C. Adams and J. G. Beese, *Trans. ASME, Ser. H*, vol. 96, pp. 123–126, 1974.
[3] An exception to this is the behavior of cold-rolled zirconium tested at 200 to 370°C, where necking occured at a strain of twice the strain at maximum load. See J. H. Keeler, *Trans. Am. Soc. Met.*, vol. 47, pp. 157–192, 1955, and discussion by A. J. Opinsky, pp. 189–190.

**Fig. 9-6** Graphical interpretation of necking criterion.

Therefore, the point of necking at maximum load can be obtained from the true-stress–true-strain curve[1] by finding the point on the curve having a subtangent of unity (Fig. 9-6a) or the point where the rate of strain hardening equals the stress (Fig. 9-6b).

The necking criterion can be expressed more explicitly if engineering strain is used. Starting with Eq. (9-25),

$$\frac{d\sigma}{d\varepsilon} = \frac{d\sigma}{de}\frac{de}{d\varepsilon} = \frac{d\sigma}{de}\frac{dL/L_0}{dL/L} = \frac{d\sigma}{de}\frac{L}{L_0} = \frac{d\sigma}{de}(1+e) = \sigma$$

$$\frac{d\sigma}{de} = \frac{\sigma}{1+e} \qquad (9\text{-}26)$$

Equation (9-26) permits an interesting geometrical construction for the determination of the point of maximum load.[2] In Fig. 9-7 the stress-strain curve is plotted in terms of true stress against conventional linear strain. Let point $A$ represent a negative strain of 1.0. A line drawn from point $A$ which is tangent to the stress-strain curve will establish the point of maximum load, for according to Eq. (9-26) the slope at this point is $\sigma/(1+e)$.

By substituting the necking criterion given in Eq. (9-25) into Eq. (9-22), we obtain a simple relationship for the strain at which necking occurs. This strain is the true uniform strain $\varepsilon_u$.

$$\varepsilon_u = n \qquad (9\text{-}27)$$

[1] An extension of Eq. (9-25) for the situation where the material transforms during tensile deformation has been given by J. R. C. Guimaraes and R. J. deAngelis, *Mater. Sci. Eng.*, vol. 13, pp. 109–111, 1974; for a more generalized discussion of stability in plastic deformation see A. S. Argon, "The Inhomogenity of Plastic Deformation," American Society for Metals, Metals Park, Ohio, 1973.
[2] A Considère, *Ann. ponts et chaussées*, vol. 9, ser. 6, pp. 574–775, 1885.

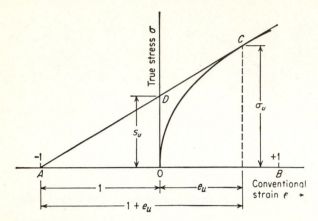

**Fig. 9-7** Considère's construction for the determination of the point of maximum load.

Although Eq. (9-22) is based on the assumption that the flow curve is given by Eq. (9-21) it has been shown[1] that $\varepsilon_u = n$ does not depend on this power law behavior.

## 9-4 STRESS DISTRIBUTION AT THE NECK

The formation of a neck in the tensile specimen introduces a complex triaxial state of stress in that region. The necked region is in effect a mild notch. As we discussed in Sec. 7-11 a notch under tension produces radial and transverse stresses which raise the value of longitudinal stress required to cause plastic flow. Therefore, the average true stress at the neck, which is determined by dividing the axial tensile load by the minimum cross-sectional area of the specimen at the neck, is higher than the stress which would be required to cause flow if simple tension prevailed. Figure 9-8 illustrates the geometry at the necked region and the stresses developed by this localized deformation. $R$ is the radius of curvature of the neck, which can be measured either by projecting the contour of the necked region on a screen or by using a tapered, conical radius gage.

   Bridgman[2] made a mathematical analysis which provides a correction to the average axial stress to compensate for the introduction of transverse stresses. This analysis was based on the following assumptions:

[1] G. W. Geil and N. L. Carwile, *J. Res. Nat. Bur. Stand.*, vol. 45, p. 129, 1950.
[2] P. W. Bridgman, *Trans. Am. Soc. Met.*, vol. 32, p. 553, 1944.

**Fig. 9-8** (*a*) Geometry of necked region; (*b*) stresses acting on element at point O.

(*a*)

(*b*)

1. The contour of the neck is approximated by the arc of a circle.
2. The cross section of the necked region remains circular throughout the test.
3. The von Mises' criterion for yielding applies.
4. The strains are constant over the cross section of the neck.

According to Bridgman's analysis, the uniaxial flow stress corresponding to that which would exist in the tension test if necking had not introduced triaxial stresses is

$$\sigma = \frac{(\sigma_x)_{\text{avg}}}{(1 + 2R/a)[\ln (1 + a/2R)]} \qquad (9\text{-}28)$$

where $(\sigma_x)_{\text{avg}}$ is the measured stress in the axial direction (load divided by minimum cross section). Figure 9-3 shows how the application of the Bridgman correction changes the true-stress–true-strain curve. A correction for the triaxial stresses in the neck of a flat tensile specimen has been considered by Aronofsky.[1] The values of $a/R$ needed for the analysis can be obtained[2] either by straining a specimen a given amount beyond necking and unloading to measure $a$ and $R$ directly or by measuring these parameters continuously past necking using photography or a tapered ring gage.

To avoid the labor of these measurements, Bridgman presented an empirical relation between $a/R$ and the true strain in the neck. Figure 9-9 shows that this gives close agreement for steel specimens, but not for other metals with widely different necking strains. Tegart[3] has shown that much better correlation is obtained between the Bridgman correction and the true strain in the neck minus the true strain at necking $\varepsilon_u$.

[1] J. Aronofsky, *J. Appl. Mech.*, vol. 18, pp. 75–84, 1951.
[2] T. A. Trozera, *Trans. Am. Soc. Met.*, vol. 56, pp. 280–282, 1963.
[3] W. J. McG. Tegart, "Elements of Mechanical Metallurgy," p. 22, The Macmillan Company, New York, 1966.

**Fig. 9-9** Relationship between Bridgman correction factor $\sigma/\sigma_{av}$ and true tensile strain. *(From E. R. Marshall and M. C. Shaw, Trans. Am. Soc. Met., vol. 44, p. 716, 1952.)*

## 9-5 DUCTILITY MEASUREMENT IN TENSION TEST

Having discussed in Sec. 9-1 the standard measurements of ductility that are obtained from the tension test, i.e., percent elongation and reduction of area, we return again to this subject armed with an understanding of the phenomenon of necking. The measured elongation from a tension specimen depends on the gage length of the specimen or the dimensions of its cross section. This is because the total extension consists of two components, the uniform extension up to necking and the localized extension once necking begins. The extent of uniform extension will depend on the metallurgical condition of the material (through $n$) and the effect of specimen size and shape on the development of the neck. Figure 9-10 illustrates the variation of the local elongation, Eq. (9-7), along the gage length of

**Fig. 9-10** Schematic drawing of variation of local elongation with position along gage length of tensile specimen.

a prominently necked tensile specimen. It readily can be seen that the shorter the gage length the greater the influence of localized deformation at the neck on the total elongation of the gage length.

The extension of a specimen at fracture can be expressed by

$$L_f - L_0 = \alpha + e_u L_0 \qquad (9\text{-}29)$$

where $\alpha$ is the local necking extension and $e_u L_0$ is the uniform extension. The tensile elongation then is given by

$$e_f = \frac{L_f - L_0}{L_0} = \frac{\alpha}{L_0} + e_u \qquad (9\text{-}30)$$

which clearly indicates that the total elongation is a function of the specimen gage length. The shorter the gage length the greater the percentage elongation.

Numerous attempts, dating back to about 1850, have been made to rationalize the strain distribution[1] in the tension test. Perhaps the most general conclusion that can be drawn is that geometrically similar specimens develop geometrically similar necked regions. According to Barba's law,[2] $\alpha = \beta \sqrt{A_0}$, and the elongation equation becomes

$$e_f = \beta \frac{\sqrt{A_0}}{L_0} + e_u \qquad (9\text{-}31)$$

It is generally recognized that in order to compare elongation measurements of different-sized specimens, the specimens must be geometrically similar. Equation (9-31) shows that the critical geometrical factor for which similitude must be maintained is $L_0/\sqrt{A_0}$ for sheet specimens or $L_0/D_0$ for round bars. In the United States the standard tensile specimen has a 0.505-in diameter and a 2-in gage length. Subsize specimens have the following diameter and (gage length): 0.357(1.4), 0.252(1.0), and 0.160(0.634). Different values of $L_0/\sqrt{A_0}$ are specified by the standardizing agencies in different countries. Table 9-4 gives some appropriate values.

To a good approximation we can say that a given elongation will be produced in a material if $\sqrt{A_0}/L_0$ is maintained constant as predcited by Eq. (9-31). Thus, at a constant value of elongation $\sqrt{A_1}/L_1 = \sqrt{A_2}/L_2$, where $A$ and $L$ are the areas and gage lengths of two different specimens, 1 and 2 of the same metal. To predict the elongation using gage length $L_2$ on a specimen with area $A_2$ by means of measurements on a specimen with area $A_1$, it only is necessary to adjust the gage length of specimen 1 to conform with $L_1 = L_2\sqrt{A_1/A_2}$. For example, suppose that $\frac{1}{8}$-in-

[1] T. C. Hsu, G. S. Littlejohn and B. M. Marchbank, *Am. Soc. Test. Mater. Proc.*, vol. 65, p. 874, 1965.
[2] M. J. Barba, *Mem. Soc. Ing. Civils*, pt. 1, p. 682, 1880.

**Table 9-4 DIMENSIONAL RELATIONSHIPS OF TENSILE SPECI-
MENS USED IN DIFFERENT COUNTRIES**

| Type Specimen | United States (ASTM) | Great Britain | | Germany |
| | | Before 1962 | Current | |
| --- | --- | --- | --- | --- |
| Sheet ($L_0/\sqrt{A_0}$) | 4.5 | 4.0 | 5.65 | 11.3 |
| Round ($L_0/D_0$) | 4.0 | 3.54 | 5.0 | 10.0 |

thick sheet is available and it is desired to predict the elongation with a 2-in gage length for the identical material but in 0.080-in thickness. Using $\frac{1}{2}$ in wide sheet specimens, we would predict that a test specimen with a gage length $L = 2\sqrt{0.125/0.080} = 2.5$ in made from the $\frac{1}{8}$-in sheet would give the same elongation as a 2-in gage length in 0.080-in-thick sheet. Experimental verification for this procedure has been shown by Kula and Fahey.[1]

Obviously, the occurrence of necking in the tension test makes any quantitative conversion between elongation and reduction of area impossible. While elongation and reduction of area usually vary in the same way, for example as a function of test temperature, tempering temperature, alloy content, etc., instances will be found where this is not the case. In the general sense, elongation and reduction of area measure different types of material behavior. Provided the gage length is not too short, percentage elongation is chiefly influenced by uniform elongation, and thus it is dependent on the strain-hardening capacity of the material. Reduction of area is more a measure of the deformation required to produce fracture and its chief contribution results from the necking process. Because of the complicated stress state in the neck, values of reduction of area are dependent on specimen geometry and deformation behavior, and they should not be taken as true material properties. However, reduction of area is the most structure-sensitive ductility parameter, and as such it is useful in detecting quality changes in the material.

## 9-6 EFFECT OF STRAIN RATE ON FLOW PROPERTIES

The rate at which strain is applied to a specimen can have an important influence on the flow stress. Strain rate is defined as $\dot{\varepsilon} = d\varepsilon/dt$, and is conventionally expressed in units of "per second." The spectrum of available strain rates is given in Table 9-5.

[1] E. G. Kula and N. N. Fahey, *Mater. Res. Stand.*, vol. 1, p. 631, 1961.

**Table 9-5  SPECTRUM OF STRAIN RATE**

| Range of Strain Rate | Condition or Type Test |
|---|---|
| $10^{-8}$ to $10^{-5}$ s$^{-1}$ | Creep tests at constant load or stress |
| $10^{-5}$ to $10^{-1}$ s$^{-1}$ | "Static" tension tests with hydraulic or screw-driven machines |
| $10^{-1}$ to $10^{2}$ s$^{-1}$ | Dynamic tension or compression tests |
| $10^{2}$ to $10^{4}$ s$^{-1}$ | High-speed testing using impact bars (must consider wave propagation effects) |
| $10^{4}$ to $10^{8}$ s$^{-1}$ | Hypervelocity impact using gas guns or explosively driven projectiles (shock-wave propagation) |

Figure 9-11 shows that increasing strain rate increases tensile strength. Moreover, the strain-rate dependence of strength increases with increasing temperature. The yield stress and flow stress at lower plastic strains are more dependent on strain rate than the tensile strength. High rates of strain cause the yield point to appear in tests on low-carbon steel that do not show a yield point under ordinary rates of loading.

Nadai[1] has presented a mathematical analysis of the conditions existing during the extension of a cylindrical specimen with one end fixed and the other

**Fig. 9-11**  Effect of strain rate on the tensile strength of copper for tests at various temperatures. (*From A. Nadai and M. J. Manjoine, J. Appl. Mech., vol. 8, p. A82, 1941.*)

[1] A. Nadai, "Theory of Flow and Fracture of Solids," vol. I, pp. 74–75, McGraw-Hill Book Company, New York, 1950.

attached to the movable crosshead of the testing machine. The crosshead velocity is $v = dL/dt$. The strain rate expressed in terms of conventional linear strain is $\dot{e}$.

$$\dot{e} = \frac{de}{dt} = \frac{d(L - L_0)/L_0}{dt} = \frac{1}{L_0}\frac{dL}{dt} = \frac{v}{L_0} \qquad (9\text{-}32)$$

Thus, the conventional strain rate is proportional to the crosshead velocity. In a modern testing machine in which the crosshead velocity can be set accurately and controlled, it is a simple matter to carry out tension tests at constant conventional strain rate.

The true strain rate $\dot{\varepsilon}$ is given by

$$\dot{\varepsilon} = \frac{d\varepsilon}{dt} = \frac{d[\ln (L/L_0)]}{dt} = \frac{1}{L}\frac{dL}{dt} = \frac{v}{L} \qquad (9\text{-}33)$$

This equation indicates that for a constant crosshead speed the true strain rate will decrease as the specimen elongates. To maintain a constant true strain rate, the crosshead velocity must increase in proportion to the increase in length of the specimen.[1] For a cylindrical specimen the true strain rate is related to the instantaneous diameter $D_i$ by

$$\dot{\varepsilon} = \frac{d\varepsilon}{dt} = \frac{d[2 \ln (D_0/D_i)]}{dt} = -\frac{2}{D_i}\frac{d(D_i)}{dt} \qquad (9\text{-}34)$$

The true strain rate is related to the conventional strain rate by the following equation:

$$\dot{\varepsilon} = \frac{v}{L} = \frac{L_0}{L}\frac{de}{dt} = \frac{1}{1 + e}\frac{de}{dt} = \frac{\dot{e}}{1 + e} \qquad (9\text{-}35)$$

Strain-rate experiments with mild steel have shown a semilogarithmic relationship between the lower yield point and the strain rate.

$$\sigma_0 = k_1 + k_2 \log \dot{\varepsilon} \qquad (9\text{-}36)$$

A more general relationship between flow stress and strain rate, at constant temperature and strain, is

$$\sigma = C(\dot{\varepsilon})^m \bigg|_{\varepsilon, T} \qquad (9\text{-}37)$$

where $m$ is known as the *strain-rate sensitivity*. The exponent $m$ can be obtained from the slope of a plot of $\log \sigma$ vs. $\log \dot{\varepsilon}$. However, a more sensitive way is a

---

[1] Modifications to a testing machine to provide for constant true strain rate are described by E. P. Lautenschlager and J. O. Brittain, *Rev. Sci. Instrum.*, vol. 39, pp. 1563–1565, 1968.

rate-change test in which $m$ is determined by measuring the change in flow stress brought about by a change in $\dot{\varepsilon}$ at constant $\varepsilon$ and $T$.

$$m = \left(\frac{\partial \log \sigma}{\partial \log \dot{\varepsilon}}\right)_{\varepsilon,T} \approx \left(\frac{\Delta \log \sigma}{\Delta \log \dot{\varepsilon}}\right)_{\varepsilon,T} = \frac{\log \sigma_2 - \log \sigma_1}{\log \dot{\varepsilon}_2 - \log \dot{\varepsilon}_1} = \frac{\log (\sigma_2/\sigma_1)}{\log (\dot{\varepsilon}_2/\dot{\varepsilon}_1)} \qquad (9\text{-}38)$$

Strain-rate sensitivity of metals is quite low ($<0.1$) at room temperature, but $m$ increases with temperature, especially at temperatures above half the absolute melting point. Strain-rate sensitivity is a good indicator of changes in deformation behavior and measurements of $m$ provide a key link[1] between dislocation concepts of plastic deformation and the more macroscopic measurements made in the tension test. In Sec. 4-6 we saw that the velocity of dislocation motion was very strongly dependent on stress according to

$$v = A\sigma^{m'} \qquad (9\text{-}39)$$

Moreover, strain rate is related to velocity of mobile dislocations by

$$\dot{\varepsilon} = \rho b v \qquad (9\text{-}40)$$

From Eqs. (9-38) and (9-40)

$$\frac{1}{m} = \frac{\partial \ln \dot{\varepsilon}}{\partial \ln \sigma} = \frac{\partial \ln v}{\partial \ln \sigma} + \frac{\partial \ln \rho}{\partial \ln \sigma}$$

and from Eq. (9-39)

$$\frac{\partial \ln v}{\partial \ln \sigma} = m'$$

Therefore,

$$m' = \frac{1}{m} - \frac{\partial \ln \rho}{\partial \ln \sigma} \qquad (9\text{-}41)$$

Thus, if there is no change in the mobile dislocation density with increasing stress, $m' = 1/m$. This hardly is a reasonable assumption. However, if $1/m$ is plotted as a function of strain, the curve extrapolated to zero gives a value close to $m'$ determined from etch-pit measurements of dislocation velocity.

While strain-rate sensitivity may be quite low in metals at room temperature, in other materials it may be quite appreciable. The extreme case is a newtonian viscous solid, where the flow stress is described by

$$\sigma = \eta \dot{\varepsilon} \qquad (9\text{-}42)$$

and by comparison with Eq. (9-37), we find that $m = 1$.

High strain-rate sensitivity is a characteristic of superplastic metals and alloys. *Superplasticity*[2] refers to extreme extensibility with elongations usually between 100 and 1,000 percent. Superplastic metals have a grain size or interphase spacing of the order of $1\mu m$. Testing at high temperature and low strain rates

[1] W. G. Johnston and D. F. Stein, *Acta Metall.*, vol. 11, pp. 317–318, 1963.
[2] W. A. Backofen, *et al.*, in "Ductility," chap. 10, American Society for Metals, Metals Park, Ohio, 1968.

accentuates superplastic behavior. While the mechanism of superplastic deformation is not yet well established, it is clear that the large elongations result from the suppression of necking in these materials with high values of $m$. An extreme case is hot glass ($m = 1$) which can be drawn from the melt into glass fibers without the fibers necking down.

In a normal metal the geometric softening that constitutes the formation of a neck is opposed by strain hardening, and so long as $d\sigma/d\varepsilon > \sigma$, the tensile specimen will not neck down. With a superplastic material the rate of strain hardening is low (because of the high temperature or structural condition) but necking is prevented by the presence of strain-rate hardening,[1] and $d\sigma/d\varepsilon > \sigma$. Consider a superplastic rod with cross-sectional area $A$ that is loaded with an axial force $P$.

$$\frac{P}{A} = \sigma = C(\dot{\varepsilon})^m$$

or

$$\dot{\varepsilon} = \left(\frac{P}{C}\right)^{1/m} \left(\frac{1}{A}\right)^{1/m} \qquad (9\text{-}43)$$

From the definition of true strain rate

$$\dot{\varepsilon} = \frac{1}{L}\frac{dL}{dt} = -\frac{1}{A}\frac{dA}{dt} \qquad (9\text{-}44)$$

Fig. 9-12 Graphical representation of Eq. (9-45). Dependence of rate of decrease of area on cross-sectional area for different values of $m$.

[1] D. H. Avery and W. A. Backofen, *Trans. Am. Soc. Met.*, vol. 58, p. 551, 1968; E. W. Hart, *Acta. Metall.*, vol. 15, p. 351, 1967; T. Y. M. Al-Naib and J. L. Duncan, *Int. J. Mech. Sci.*, vol. 12, pp. 463–477, 1970.

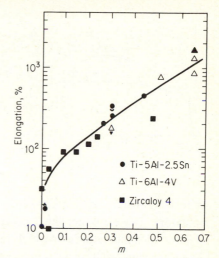

**Fig. 9-13**  Dependence of tensile elongation on strain-rate sensitivity. (*From D. Lee and W. A. Backofen, Trans. Metall. Soc. AIME, vol. 239, p. 1034, 1967.*)

and combining Eqs. (9-43) and (9-44)

$$-\frac{dA}{dt} = A\dot\varepsilon = A^{1-1/m}\left(\frac{P}{C}\right)^{1/m}$$

and

$$-\frac{dA}{dt} = \left(\frac{P}{C}\right)^{1/m}\left(\frac{1}{A^{(1-m)/m}}\right) \qquad (9\text{-}45)$$

Equation (9-45) states that so long as $m$ is $<1$ the smaller the cross-sectional area, the more rapidly the area is reduced.  Figure 9-12 shows how the area decrease varies with $m$.  When $m = 1$ the deformation is newtonian viscous and $dA/dt$ is independent of $A$ and any incipient neck is simply preserved during elongation and does not propagate inward.  As $m$ approaches unity, the rate of growth of incipient necks is drastically reduced.  Figure 9-13 shows how the tensile elongation of superplastic alloys increases with strain-rate sensitivity in agreement with the above analysis.

## 9-7  EFFECT OF TEMPERATURE ON FLOW PROPERTIES

The stress-strain curve and the flow and fracture properties derived from the tension test are strongly dependent on the temperature at which the test was conducted.[1]  In general, strength decreases and ductility increases as the test temperature is increased.  However, structural changes such as precipitation, strain aging,

[1] For test techniques see ASTM E21-69, ASTM, *op. cit.*, pp. 265–270.

**Fig. 9-14** Changes in engineering stress-strain curves of mild steel with temperature.

or recrystallization may occur in certain temperature ranges to alter this general behavior. Thermally activated processes assist deformation and reduce strength at elevated temperatures. At high temperatures and/or long exposure, structural changes occur resulting in time-dependent deformation or *creep*. These topics are discussed in greater detail in Chap. 13.

The change with temperature of the engineering stress-strain curve in mild steel is shown schematically in Fig. 9-14. Figure 9-15 shows the variation of yield strength with temperature for body-centered cubic tantalum, tungsten, molybdenum, and iron and for face-centered cubic nickel. Note that for the bcc metals the yield stress increases rapidly with decreasing temperature, while for nickel (and other fcc metals) the yield stress is only slightly temperature-dependent.

**Fig. 9-15** Effect of temperature on the yield strength of body-centered cubic Ta, W, Mo, Fe, and face-centered cubic Ni. (*From J. H. Bechtold Acta Metall., vol. 3, p. 252, 1955.*)

**Fig. 9-16** Effect of temperature on the reduction of area of Ta, W, Mo, Fe, and Ni. (*From J. H. Bechtold, Acta Metall., vol. 3, p. 253, 1955.*)

Based on the concept of fracture stress introduced in Sec. 7-12, and especially Fig. 7-21, it is easy to see why bcc metals exhibit brittle fracture at low temperatures. Figure 9-16 shows the variation of reduction of area with temperature for these same metals. Note that tungsten is brittle at 100°C, iron at −225°C, while nickel decreases little in ductility over the entire temperature interval.

In fcc metals flow stress is not strongly dependent on temperature but the strain-hardening exponent decreases with increasing temperature. This results in the stress-strain curve flattening out with increasing temperature and the tensile strength being more temperature-dependent than the yield strength. Tensile deformation at elevated temperature is complicated by the formation of more than one neck[1] in the specimen.

The best way to compare the mechanical properties of different materials at various temperatures is in terms of the ratio of the test temperature to the melting point, expressed in degrees Kelvin. This ratio is often referred to as the *homologous temperature*. When comparing the flow stress of two materials at an equivalent homologous temperature, it is advisable to correct for the effect of temperature on elastic modulus by comparing ratios of $\sigma/E$ rather than simple ratios of flow stress.

The temperature dependence of flow stress at constant strain and strain rate generally can be represented by

$$\sigma = C_2 \, e^{Q/RT} \Big|_{\varepsilon,\dot{\varepsilon}} \qquad (9\text{-}46)$$

---

[1] P. J. Wray, *J. Appl. Phys.*, vol. 41, pp. 3347–3352, 1970.

where $Q$ = an activation energy for plastic flow, cal/g mole

      $R$ = universal gas constant, 1.987 cal/(deg)(mole)

      $T$ = testing temperature, °K

If this expression is obeyed, a plot of ln $\sigma$ versus $1/T$ will give a straight line with a slope $Q/R$.

    The value of flow stress depends on the dislocation structure existing at the time the flow stress is measured. Dislocation structure will change with temperature, strain rate, and strain. One way to separate these effects is to evaluate $Q$ by a temperature change test (Fig. 9-19b). The stress-strain curve is conducted at constant strain rate, and at the desired value of plastic strain the temperature is changed from $T_2$ to $T_1$. After the temperature comes to equilibrium, the activation energy is given by

$$Q = R \ln \left(\frac{\sigma_1}{\sigma_2}\right) \frac{T_1 T_2}{T_2 - T_1} \qquad (9\text{-}47)$$

    Better than 90 percent of the energy expended in plastic deformation is converted to heat. In many metalworking operations, because of inhomogeneous flow, the deformation is localized and the temperature rises within this local region. Because the flow stress decreases with temperature, further deformation is concentrated preferentially in this zone and the process may continue until fracture occurs. At high rates of deformation there is not time for appreciable heat flow to occur and a near adiabatic condition results. This type of localized fracture is called *adiabatic shear fracture*. It is sometimes observed in the later stages of the tension test, particularly in low-temperature tests.[1] Frequently in low-temperature tests, the adiabatic heating results in a marked drop in the flow curve and serrated stress-strain curves like in Fig. 6-10. An analysis of the conditions to produce instability in the flow curve by adiabatic shear has been given by Backofen.[2] Starting with a uniaxially loaded specimen

$$P = \sigma A \qquad\qquad \sigma = f(\varepsilon, \dot{\varepsilon}, T)$$

$$dP = \sigma \, dA + A \, d\sigma \qquad \frac{d\sigma}{d\varepsilon} = \frac{\partial\sigma}{\partial\varepsilon} + \frac{\partial\sigma}{\partial T}\frac{dT}{d\varepsilon} + \frac{\partial\sigma}{\partial\dot{\varepsilon}}\frac{d\dot{\varepsilon}}{d\varepsilon}$$

and since $d\varepsilon = -dA/A$

$$\frac{dP}{d\varepsilon}\frac{1}{A} = -\sigma + \frac{d\sigma}{d\varepsilon}$$

and

$$dP = A \, d\varepsilon \left(\frac{\partial\sigma}{\partial\varepsilon} + \frac{\partial\sigma}{\partial T}\frac{dT}{d\varepsilon} + \frac{\partial\sigma}{\partial\dot{\varepsilon}}\frac{d\dot{\varepsilon}}{d\varepsilon} - \sigma\right) \qquad (9\text{-}48)$$

[1] G. Y. Chin, W. F. Hosford, Jr., and W. A. Backofen, *Trans. Metall. Soc. AIME*, vol. 230, pp. 437–449, 1964.

[2] W. A. Backofen, Metallurgical Aspects of Ductile Fracture, in "Fracture of Engineering Materials," pp. 121–124, American Society for Metals, Metals Park, Ohio, 1964.

For adiabatic heating $dT/d\varepsilon = \sigma/C\rho$, where $C$ is the specific heat and $\rho$ is the density. For low-temperature deformation the strain-rate dependence of the flow curve can be neglected and an instability (load drop) in the flow curve will occur when[1]

$$\frac{\partial \sigma}{\partial \varepsilon} - \sigma \leq \frac{\partial \sigma}{\partial T} \frac{\sigma}{C\rho} \qquad (9\text{-}49)$$

Equation (9-49) explains why load drops due to adiabatic heating are more pronounced at low temperatures: specific heat decreases at low temperatures (cryogenic) and there is also a strong temperature-dependence of flow stress in bcc metals.

## 9-8 COMBINED EFFECT OF TEMPERATURE AND STRAIN RATE

We may combine the expressions describing the temperature and strain-rate dependence of flow stress to give

$$\sigma = f(Z) = f(\dot{\varepsilon}e^{\Delta H/RT})\Big|_{\varepsilon} \qquad (9\text{-}50)$$

where $\Delta H$ is an activation energy, calorie per mole, that is related to the activation energy $Q$ in Eq. (9-46) by $Q = m\Delta H$, where $m$ is the strain-rate sensitivity. The quantity $Z$ is called the *Zener-Hollomon parameter*.[2] It may also be referred to as a *temperature-modified strain rate*.

$$Z = \dot{\varepsilon}e^{\Delta H/RT} \qquad (9\text{-}51)$$

When Eq. (9-50) was first proposed, it was interpreted much more broadly than it is today. It was suggested[3] that Eq. (9-50) represented a *mechanical equation of state*, i.e., the flow stress depends only on the instantaneous values of strain, strain rate, and temperature in the same way that the pressure of an ideal gas only depends on instantaneous values of volume and temperature. While there are a number of published experiments in which the material follows a mechanical equation of state, there are many others[4] in which the flow stress depends on previous conditions of temperature and strain rate as well as on their instantaneous values. Since flow properties fundamentally depend on dislocation structure, and this depends strongly on the temperature and strain-rate history, it is not surprising that the flow properties of a metal as usually determined in a tension test do not generally follow a mechanical equation of state law. However, Hart[5] has shown

---

[1] E. B. Kula and T. S. DeSisto, *ASTM Spec. Tech. Pub.* 387, 1966.
[2] C. Zener and J. H. Hollomon, *J. Appl. Phys.*, vol. 15, pp. 22–32, 1944.
[3] J. Hollomon, *Trans. AIME*, vol. 171, p. 355, 1947.
[4] J. E. Dorn, A. Goldberg, and T. E. Tietz, *Trans. AIME*, vol. 180, p. 205, 1949.
[5] E. W. Hart, *Acta Metall.*, vol. 18, p. 599, 1970.

theoretically how the flow stress at constant structure may be determined as a function of strain rate and temperature from stress-relaxation tests. In these experiments the specimen is first deformed rapidly in tension to a desired plastic strain level. The crosshead is then held fixed and stress relaxation results from converting the elastic strain in the specimen to plastic strain. Data on the relaxation of stress with time yield stress–strain rate relations at various values of plastic strain. Li[1] has shown how flow stress–strain rate data obtained in this way on a variety of bcc and fcc metals follow a mechanical equation law.

A relationship which is particularly useful for correlating stress, temperature, and strain rate under hot-working conditions was proposed by Sellars and Tegart.[2]

$$\dot{\varepsilon} = A(\sinh \alpha\sigma)^{n'} e^{-Q/RT} \qquad (9\text{-}52)$$

where $A$, $\alpha$, and $n'$ are experimentally determined constants and $Q$ is an activation energy. At low stresses ($\alpha\sigma < 0.8$), Eq. (9-52) reduces to a power relation such as is used to describe creep behavior

$$\dot{\varepsilon} = A_1 \sigma^{n'} e^{-Q/RT} \qquad (9\text{-}53)$$

and at high stresses ($\alpha\sigma > 1.2$) it reduces to an exponential relation

$$\dot{\varepsilon} = A_2 \exp(\beta\sigma) e^{-Q/RT} \qquad (9\text{-}54)$$

The constants $\alpha$ and $n'$ are related by $\beta = \alpha n'$, so that $\alpha$ and $n'$ can be simply determined from tests at high and low stresses. If Eq. (9-52) is followed, the data will plot as a series of parallel straight lines (one for each temperature) when $\dot{\varepsilon}$ is plotted against $\sinh \alpha\sigma$ on log-log coordinates. The activation energy $Q$ can be obtained from a plot of $\log \dot{\varepsilon}$, at constant $\sinh \alpha\sigma$, against $1/T$.

## 9-9   INFLUENCE OF TESTING MACHINE ON FLOW PROPERTIES

When a specimen is loaded in a testing machine the load is transmitted to the entire machine as well as to the specimen. While the specimen deforms first elastically and then plastically, the testing machine is designed so it only deforms elastically. For purposes of analysis this can be considered as two springs in series, the testing machine frame and fixtures having a constant $K$ and the specimen with its own spring constant.

A constant crosshead-velocity testing machine applies a constant total strain rate that is the sum of (1) the elastic strain rate in the specimen, (2) the

[1] H. Yamada and C. Y. Li, *Metall. Trans.*, vol. 4, pp. 2133–2136, 1973.
[2] C. M. Sellars and W. J. McG. Tegart, *Mem. Sci. Rev. Metall.*, vol. 63, p. 731, 1966; see also W. J. McG. Tegart, in "Ductility," chap. 5, American Society for Metals, Metals Park, Ohio, 1968.

plastic strain rate in the specimen, and (3) the strain rate resulting from the elasticity of the testing machine. At any instant of time there is some distribution of strain rate between these components.[1] If the crosshead velocity is $v$, then at a particular time $t$ the total displacement is $vt$. The force $P$ on the specimen causes an elastic machine displacement $P/K$. The elastic displacement of the specimen (from Hooke's law) is $\sigma L/E$ and the plastic displacement of the specimen is $\varepsilon_p L$. Since the total displacement is the sum of its components

$$vt = \frac{P}{K} + \frac{\sigma L}{E} + \varepsilon_p L \qquad (9\text{-}55)$$

Solving for $\varepsilon_p$, we see that the plastic strain taken from a load-time chart on a constant crosshead-velocity testing machine must be corrected for machine stiffness as well as specimen elasticity.

$$\varepsilon_p = \frac{vt}{L} - \frac{\sigma}{E} - \frac{P}{K} \qquad (9\text{-}56)$$

Testing machine–specimen interaction has a major influence on the strain rate.[2] If we eliminate time from Eq. (9-55) by treating the stress rate $\dot{\sigma}$ and the strain rate $\dot{\varepsilon}$, then after substituting $P = \sigma A$,

$$\frac{v}{L} = \frac{\dot{\sigma}}{E}\left(\frac{AE}{KL} + 1\right) + \dot{\varepsilon}_p \qquad (9\text{-}57)$$

and since

$$\dot{\varepsilon} = \dot{\varepsilon}_E + \dot{\varepsilon}_p = \frac{\dot{\sigma}}{E} + \dot{\varepsilon}_p$$

$$\dot{\varepsilon} = \frac{(vK/AE) + \dot{\varepsilon}_p}{(KL/AE) + 1} \qquad (9\text{-}58)$$

This shows that the specimen strain rate usually will differ from the preset crosshead velocity depending on the rate of plastic deformation and the relative stiffness of the testing machine and the specimen. Experimental values of machine stiffness $K$ are in the range of 40,000 to 180,000 lb/in, while a 0.505-in-diameter steel specimen with 2.5-in reduced section has a spring constant of 2,400,000 lb/in. Thus, except for wire specimens or specially designed testing machines the specimen usually has a much greater spring constant than the testing machine.[3] At the beginning of the tension test $\dot{\varepsilon}_p = 0$, and Eq. (9-58) reduces to

$$\dot{\varepsilon}_0 = \frac{v_0}{L}\left(1 + \frac{AE}{KL}\right)$$

---

[1] J. J. Gilman, "Micromechanics of Flow in Solids," p. 218, McGraw-Hill Book Company, New York, 1969.
[2] M. A. Hamstad and P. P. Gillis, *Mater. Res. Stand.*, vol. 6, pp. 569–573, 1966.
[3] J. E. Hockett and P. P. Gillis, *Int. J. Mech. Sci.*, vol. 13, pp. 251–264, 1971.

while at the *yield point* $\dot{\sigma} = 0$ so that $\dot{\varepsilon}_1 = v_1/L$. If we desire to have the strain rate constant from the start of the test to the yield point, that is, $\dot{\varepsilon}_0 = \dot{\varepsilon}_1$, we see that the crosshead velocity must be reduced from its initial value $v_0$ to its yield-point value $v_1$ by a factor of

$$\frac{v_0}{v_1} = 1 + \frac{AE}{KL} \qquad (9\text{-}59)$$

This factor typically is greater than 20 for a standard steel tension specimen. Therefore, large changes in crosshead speed usually are required to maintain a constant strain rate from the beginning of the test to the yield point.

The characteristics of the testing machine can have a strong influence on the shape of the stress-strain curve and the fracture behavior.[1] A rigid testing machine with a high spring constant is known as a "hard machine." Screw-driven mechanical machines tend to be hard machines, while hydraulically driven testing machines are "soft machines." A hard testing machine will reproduce faithfully the upper and lower yield point, but in a soft machine these will be smeared out and only the extension at constant load will be recorded.

## 9-10   THERMALLY ACTIVATED DEFORMATION

Plastic flow, especially at elevated temperature, can be considered a thermally activated process.[2] Experimental study of the activation energy of plastic deformation and its dependence on stress, temperature, deformation, strain rate, impurity concentration, etc., is one of the chief ways of determining the validity of theoretical models for dislocation processes in solids. Macroscopic tension test measurements of the temperature and strain-rate dependence of flow stress provide a direct link with dislocation behavior.

The applied shear stress[3] $\tau$ is opposed by a variety of internal stresses, whose sum is given by $\tau_i$. The effective shear stress is $\tau - \tau_i$. The internal resisting stresses can be grouped into two categories (1) *long-range obstacles* to plastic deformation, which represent barriers too high and long to be surmounted by thermal fluctuations, and (2) *short-range obstacles*, less than 10 atom diameters, for which thermal fluctuations can assist dislocations in surmounting these barriers. The first type of barriers are referred to as *athermal barriers* since they produce

---

[1] G. Y. Chin, W. F. Hosford, Jr., and W. A. Backofen, op. cit., This paper discusses fixtures for converting a machine to extra-hard or soft tension tests.

[2] H. Conrad, *J. Metals.*, vol. 16, pp. 582–885, 1964; U. F. Kochs, A. S. Argon, and M. F. Ashby, Thermodynamics and Kinetics of Slip, "Progress in Materials Science," vol. 19, Pergamon Press, London, 1975.

[3] Shear stress and shear strain are usually obtained from tension test data by the approximation $\tau = \sigma/2$ and $\gamma = 1.4\varepsilon$.

long-range internal stress fields which are not affected by temperature and strain rate except for the variation of flow stress due to the decrease in shear modulus[1] $\mu$ with increasing temperature. Accordingly, the contribution to the internal stress field from long-range obstacles is $\tau_\mu$. Short-range obstacles are called *thermal barriers*, and their contribution to the internal stress field is $\tau^*$. $\tau^*$ is a strong function of temperature and strain rate. Major contributions to $\tau_\mu$ arise from the stress fields of other dislocations and of large incoherent precipitates and massive second-phase particles. Basically, the level of $\tau_\mu$ depends on the composition, heat treatment, and dislocation structure of the material. Contributions to $\tau^*$ come from many sources, such as the Peierls-Nabarro force, stress fields of coherent precipitates and solute atoms, cross slip, dislocation climb, and intersections of dislocations.

The situation described above is illustrated schematically by Fig. 9-17. Short-range stress fields are superimposed on a fluctuating long-range stress of wave-length $\lambda$. The magnitude of this curve will depend on direction in the lattice and will vary with plastic strain and strain rate. The total internal stress is given by

$$\tau_i = \tau_\mu + \tau^* \qquad (9\text{-}60)$$

A positive value of $\tau_i$ opposes dislocation motion, while a negative $\tau_i$ assists motion of dislocations. The rate-controlling process will consist of overcoming the strongest short-range obstacle situated near the top of the opposing long-range stress field. At absolute zero, where thermal fluctuations do not exist, the applied stress $\tau$ would have to equal a value equivalent to the very top of the spike or

**Fig. 9-17** Schematic representation of long-range and short-range stress fields. (*After Conrad.*)

---

[1] Elsewhere we have used $G$ to represent shear modulus but the British usage of $\mu$ is so common to this subject that it has been retained.

**Fig. 9-18** Thermal activation barrier.

curve to cause plastic flow at a value $\tau_0$. However, at some temperature greater than 0°K, thermal fluctuations will assist the applied stress and plastic flow can occur at a stress $\tau$ which is less than $\tau_0$. The energy that must be supplied by thermal fluctuations is shown shaded in Fig. 9-17. As the temperature is increased $\tau$ can decrease as more assist is received from thermal fluctuations, until, in the limit, $\tau = \tau_\mu$, and all of the stress required to overcome the short-range stress field is supplied by thermal fluctuations. The establishment of the flat portion of the stress-temperature curve when $\tau = \tau_\mu$ was the original way of determining $\tau_\mu$, although a number of other techniques have been used since.[1]

Seeger[2] has considered the short-range barrier in detail for the common type of barrier where the activation energy is not a function of $\tau^*$. This condition applies to the intersection of dislocations and the movement of jogs. Consider the barrier shown in Fig. 9-18. On the application of a stress $\tau > \tau_\mu$ a dislocation moves up the force barrier to a level $F = \tau^* \mathbf{b} l^*$, where $\mathbf{b}$ is the Burgers vector and $l^*$ is the length of the dislocation segment involved in the thermal fluctuation. $\Delta H$ is the energy which must be supplied to overcome the barrier. The work done by the applied stress during thermal activation is

$$W = \tau^* \mathbf{b}(x_0^* - x_0) = \tau^* \mathbf{b} d^* \qquad (9\text{-}61)$$

The term $\Delta H$ is the area under the force-distance curve from $x_0$ to $x_0^*$ (designated $\Delta H^*$) minus the work done by the applied stress during the thermal activation.

$$\Delta H = \int_{x_0}^{x_0^*} [F(x) - \tau^* \mathbf{b} l^*] \, dx = \Delta H^* - v^* \tau^* \qquad (9\text{-}62)$$

The term $\Delta H^*$ represents the activation energy for zero applied stress.

The term $v^*$ is called the *activation volume*. It represents the average volume of dislocation structure involved in the deformation process. In terms of Fig. 9-18 $v^* = l^* \mathbf{b} d^*$ for a process in which $l^*$ does not vary with stress. In this relation, $l^* \mathbf{b}$ is the atomic area involved in the flow process and $d^*$ is the distance the atoms move during this process. Activation volume is an important quantity which can give valuable information on the deformation mechanism because it has a definite value and stress dependence for each atomistic process.

[1] H. Conrad, *Mater. Sci. Eng.*, vol. 6, pp. 265–273, 1970.
[2] A. Seeger, *Philos. Mag.*, vol. 1, p. 651, 1956.

One of the most basic dislocation equations is the one relating shear-strain rate with dislocation velocity, Eq. (4-12).

$$\dot{\gamma} = \rho \mathbf{b}\bar{v}$$

The strong temperature dependence of shear-strain rate can be expressed as

$$\dot{\gamma} = A e^{-\Delta G/kT} = \rho \mathbf{b} s v^* e^{-\Delta G/kT} \qquad (9\text{-}63)$$

where $\Delta G$ = the change in Gibbs free energy of the system

$A$ = the overall frequency factor

$v^*$ = the frequency of vibration of the dislocation segment involved in the thermal activation process

$s$ = the average distance a dislocation moves after every successful thermal fluctuation

Schoek[1] has shown that the activation volume $v^*$ is related to Eq. (9-63) according to

$$v^* = kT \left[ \frac{\partial \ln (\dot{\gamma}/A)}{\partial \tau} \right]_T = - \left( \frac{\partial \Delta G}{\partial \tau} \right)_T \qquad (9\text{-}64)$$

The experimentally determined activation energy $Q$ is equal to the heat of activation (enthalpy) $\Delta H$. [See Eq. (9-47) for another way of measuring $Q$.]

$$Q \equiv kT^2 \left[ \frac{\partial \ln (\dot{\gamma}/A)}{\partial T} \right]_\tau = \Delta G - T \left( \frac{\partial \Delta G}{\partial T} \right)_\tau = \Delta H \qquad (9\text{-}65)$$

The free-energy change can be expressed in terms of the measurable quantities of $Q, v^*, T, \tau,$ and $\mu$.

$$\Delta G = \frac{Q + T \dfrac{\partial \mu}{\partial T} \dfrac{\tau}{\mu} v^*}{1 - \left( \dfrac{T}{\mu} \dfrac{\partial \mu}{\partial T} \right)} \qquad (9\text{-}66)$$

Finally, we obtain another parameter from experimental data

$$p = \left( \frac{\partial \tau}{\partial T} \right)_{\dot{\varepsilon}} \qquad (9\text{-}67)$$

The parameters $Q, p,$ and $v^*$ are related according to

$$Q = -Tpv^* \qquad (9\text{-}68)$$

so long as $\dot{\gamma}$ is a function of $T$ and $\gamma$ only.

[1] G. Schoek, *Phys. Status Solidi*, vol. 8, pp. 499–507, 1965.

The main object of this theory is to identify the mechanism which controls the thermally activated deformation process by comparing experimental values of $\Delta G$, $v^*$, and $A$ with values predicted from specific dislocation models. Most of the work in the literature neglects entropy changes and uses $\Delta H$ instead of $\Delta G$. This results in a different formulation for the above equations.[1] The two approaches have been compared in detail[2] and found to be equally valid. The parameters are evaluated from macroscale measurements in tensile or creep (strain vs. time at constant stress) using incremental change tests to determine the relationships

$$\left(\frac{\partial \ln \dot{\gamma}}{\partial \tau}\right)_T \qquad \left(\frac{\partial \tau}{\partial T}\right)_{\dot{\gamma}} \qquad \text{and} \qquad \left(\frac{\partial \ln \dot{\gamma}}{\partial T}\right)_\tau$$

at nearly constant dislocation structure. The procedure for determining these functions is shown in Fig. 9-19.

**Fig. 9-19** Differential tests to determine deformation parameters at constant structure.

[1] Z. S. Basinski, *Acta Metall.*, vol. 5, p. 684, 1957; H. Conrad and H. Weidersich, *Acta Metall.*, vol. 8, p. 128, 1960.
[2] J. P. Hirth and W. D. Nix, *Phys. Status Solidi*, vol. 35, pp. 177–188, 1969.

## 9-11 NOTCH TENSILE TEST

Ductility measurements on standard smooth tensile specimens do not always reveal metallurgical or environmental changes that lead to reduced local ductility. The tendency for reduced ductility in the presence of a triaxial stress field and steep stress gradients (such as arise at a notch) is called *notch sensitivity*. A common way of evaluating notch sensitivity is a tension test using a notched specimen. The notch tensile test has been used extensively for investigating the properties of high-strength steels, for studying hydrogen embrittlement in steels and titanium, and for investigating the notch sensitivity of high-temperature alloys. More recently notched tension specimens have been used for fracture mechanics measurements (see Sec. 14-7). Notch sensitivity can also be investigated with the notched-impact test as described in Chap. 14.

The most common notch tensile specimen[1] uses a 60° notch with a root radius 0.001 in or less introduced into a round (circumferential notch) or flat (double-edge notch) tensile specimen. Usually the depth of the notch is such that the cross-sectional area at the root of the notch is one-half of the area in the unnotched section. The specimen is carefully aligned and loaded in tension until failure. The *notch strength* is defined as the maximum load divided by the original cross-

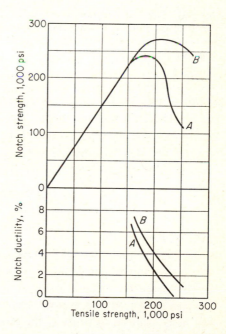

**Fig. 9-20** Notch tensile properties of two steels. Steel *A* has higher notch sensitivity than steel *B*.

[1] Standard Method of Sharp-Notch Tension Testing of High-Strength Sheet Material, ASTM E338-68 op. cit., pp. 847–854.

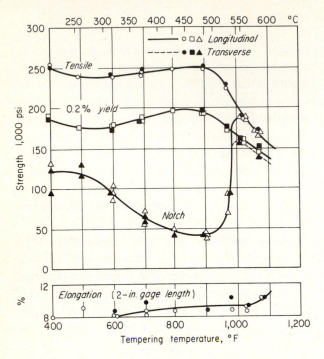

**Fig. 9-21** Notched and unnotched tensile properties of an alloy steel as a function of tempering temperature. (*From G. B. Espey, M. H. Jones, and W. F. Brown, Jr., Am. Soc. Test. Mater. Proc., vol. 59, p. 837, 1959.*)

sectional area at the notch. Because of the plastic constraint at the notch (see Sec. 7-11), this value will be higher than the tensile strength of an unnotched specimen if the material possesses some ductility. Therefore, the common way of detecting notch brittleness (or high notch sensitivity) is by determining the *notch-strength ratio* NSR.

$$\text{NSR} = \frac{s_{\text{net}} \text{ (for notched specimen at maximum load)}}{s_u \text{ (tensile strength for unnotched specimen)}} \tag{9-69}$$

If the NSR is less than unity the material is notch brittle. The other property that is measured in the notch tensile test is the reduction of area at the notch.

As strength, hardness, or some metallurgical variable restricting plastic flow increases, the metal at the root of the notch is less able to flow, and fracture becomes more likely. A typical behavior is shown in Fig. 9-20. Notch brittleness may be considered to begin at the strength level where the notch strength begins to fall off, or more conventionally, at the strength level where the NSR becomes less than unity. The sensitivity of notch strength for detecting metallurgical embrittlement is illustrated in Fig. 9-21. Note that the conventional elongation measured on a smooth specimen was unable to detect the fall off in notch strength produced by tempering in the 600 to 900°F range. For a more detailed review of notch tensile testing the reader is referred to Lubahn.[1]

## 9-12 TENSILE PROPERTIES OF STEEL

Because of the commercial importance of ferrous materials, a great deal of work has been done in correlating their tensile properties with composition and micro-structure. Moreover, from a scientific viewpoint the Fe-C system exhibits a great versatility in structure and properties. It has been clearly demonstrated that struc-ture rather than composition *per se* is the chief variable which controls the proper-ties of steel.

The tensile properties of annealed and normalized steels are controlled by the flow and fracture characteristics of the ferrite and by the amount, shape, and distribution of the cementite. The strength of the ferrite depends on the amount of alloying elements in solid solution and the ferrite grain size.[2] The carbon content has a very strong effect because it controls the amount of cementite present

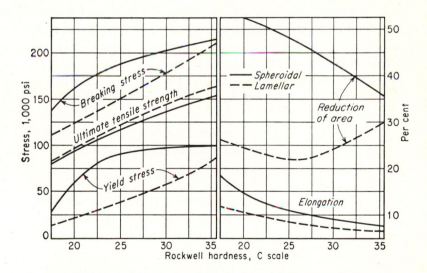

**Fig. 9-22** Tensile properties of pearlite and spheroidite in a eutectoid steel. *(From E. C. Bain, and H. W. Paxton, "Alloying Elements in Steel," 2d ed., p. 38, American Society of Metals, Metals Park, Ohio, Copyright American Society for Metals, 1961.)*

[1] J. D. Lubahn, *Trans. ASME*, vol. 79, pp. 111–115, 1957.
[2] C. E. Lacy and M. Gensamer, *Trans. Am. Soc. Met.*, vol. 32, p. 88 ,1944.

Table 9-6  EFFECT OF COLD-DRAWING ON TENSILE PROP-
ERTIES OF SAE 1016 STEEL†

| Reduction of area by drawing, % | Yield strength, psi | Tensile strength, psi | Elongation, in 2 in., % | Reduction of area, % |
|---|---|---|---|---|
| 0 | 40,000 | 66,000 | 34 | 70 |
| 10 | 72,000 | 75,000 | 20 | 65 |
| 20 | 82,000 | 84,000 | 17 | 63 |
| 40 | 86,000 | 95,000 | 16 | 60 |
| 60 | 88,000 | 102,000 | 14 | 54 |
| 80 | 96,000 | 115,000 | 7 | 26 |

† L. J. Ebert, "A Handbook on the Properties of Cold Worked Steels,"
PB 121662, Office of Technical Services, U.S. Department of Commerce, 1955.

either as pearlite or as spheroidite. The strength increases and ductility decreases with increasing carbon content because of the increased amount of cementite in the microstructure. A normalized steel will have higher strength than an annealed steel because the more rapid rate of cooling used in the normalizing treatment causes the transformation to pearlite to occur at a lower temperature, and a finer pearlite spacing results. Differences in tensile properties due to the shape of the cementite particles are shown in Fig. 9-22, where the tensile properties of a spheroidized structure are compared with a pearlitic structure for a steel with the same carbon content. Empirical correlations have been worked out[1] between composition and cooling rate for predicting the tensile properties of steels with pearlitic structures.

One of the best ways to increase the strength of annealed steel is by cold-working. Table 9-6 gives the tensile properties which result from the cold reduction of SAE 1016 steel bars by drawing through a die.

The pearlitic structure in steel can be controlled best by transforming the austenite to pearlite at a constant temperature instead of allowing it to form over a range of temperature on continuous cooling from above the critical temperature. Although isothermal transformation is not in widespread commercial use, it is a good way of isolating the effect of certain microstructures on the properties of steel. Figure 9-23 shows the variation of tensile properties for a Ni-Cr-Mo eutectoid steel with isothermal-reaction temperature.[2] This is an extension of Gensamer's work,[3] which showed that the tensile strength varied linearly with the

[1] I. R. Kramer, P. D. Gorsuch, and D. L. Newhouse, *Trans. AIME*, vol. 172, pp. 244–272, 1947.
[2] E. S. Davenport, *Trans. AIME*, vol. 209, pp. 677–688, 1957.
[3] M. Gensamer, E. B. Pearsall, W. S. Pellini, and J. R. Low, *Trans. Am. Soc. Met.*, vol. 30, pp. 983–1020, 1942.

**Fig. 9-23** Relationship of tensile properties of Ni-Cr-Mo steel to isothermal-transformation temperature. (*From E. S. Davenport, Trans. AIME, vol. 209, p. 684, 1961.*)

logarithm of the mean free ferrite path in isothermally transformed structures. In the region 1300 to 1000°F the transformation product is lamellar pearlite. The spacing between cementite platelets decreases with transformation temperature and correspondingly the strength increases. In the region 800 to 500°F the structure obtained on transformation is acicular bainite. The bainitic structure becomes finer with decreasing temperature, and the strength increases almost linearly to quite high values. Good ductility accompanies this high strength over part of the bainite temperature range. This is the temperature region used in the commercial heat-treating process known as austempering. The temperature region 1000 to 800°F is one in which mixed lamellar and acicular structures are obtained. There is a definite ductility minimum and a leveling off of strength for these structures. The sensitivity of the reduction of area to changes in microstructure is well illustrated by these results.

The best combination of strength and ductility is obtained in steel which has been quenched to a fully martensitic structure and then tempered. The best criterion for comparing the tensile properties of quenched and tempered steels is on the basis of an as-quenched structure of 100 percent martensite. However, the attainment of a completely martensitic structure may, in many cases, be commer-

**Fig. 9-24** As-quenched hardness of steel as a function of carbon content for different percentages of martensite in the microstructure. (*From Metals Handbook, American Society for Metals, Metals Park, Ohio, p. 497, 1948.*)

cially impractical. Because of the importance of obtaining a fully martensitic structure, it is desirable that the steel have adequate hardenability. *Hardenability*, the property of a steel which determines the depth and distribution of hardness induced by quenching, should be differentiated from *hardness*, which is the property of a material which represents its resistance to indentation or deformation. (This subject is discussed in Chap. 11.) Hardness is associated with strength, while hardenability is connected with the transformation characteristics of a steel. Hardenability may be increased by altering the transformation kinetics by the addition of alloying elements, while the hardness of a steel with given transformation kinetics is controlled primarily by the carbon content. Figure 9-24 shows the hardness of martensite as a function of carbon content for different total amounts of martensite in the microstructure. These curves can be used to determine whether or not complete hardening was obtained after quenching. Hardness is used as a convenient measure of the strength of quenched and tempered steels. The validity of this procedure is based on the excellent correlation which exists between tensile strength and hardness for heat-treated, annealed, and normalized steels (Fig. 9-25).

The mechanical properties of a quenched and tempered steel may be altered by changing the tempering temperature. Figure 9-26 shows how hardness and the tensile properties very with tempering temperature for an SAE 4340 steel. This is the typical behavior for heat-treated steel. Several methods for correlating and predicting the hardness change in different steels with tempering temperature

**Fig. 9-25**  Relationship between tensile strength and hardness for quenched and tempered, annealed, and normalized steels.  (*From SAE Handbook.*)

**Fig. 9-26**  Tensile properties of quenched and tempered SAE -440 steel as a function of tempering temperature (for fully hardened 1-in-diameter bars).

have been proposed.[1-3] In using tempering diagrams like Fig. 9-26, it is important to know whether or not the data were obtained on specimens quenched to essentially 100 percent martensite throughout the entire cross section of the specimen. Because of the variability in hardenability from heat to heat, there is no assurance of reproducibility of data unless this condition is fulfilled.

A great many low-alloy steels have been developed and are used in the quenched and tempered condition. A study of the tensile properties of these steels could lead to considerable confusion were it not for the fact that certain generalities can be made about their properties.[4,5] For low-alloy steels containing 0.3 to 0.5 percent carbon which are quenched to essentially 100 percent martensite and then tempered back to any given tensile strength in the range 100,000 to 200,000 psi, the other common tensile properties will have a relatively fixed value depending only on the tensile strength. In other words, the mechanical properties of this important class of steels do not depend basically on alloy content, carbon content within the above limits, or tempering temperature. It is important to note that this generalization does not say that two alloy steels given the same tempering treatment will have the same tensile properties, because different tempering temperatures would quite likely be required to bring two different alloy steels to the same tensile strength. Figure 9-27 shows this relationship between the mechanical properties of steels with tempered martensitic structures. The expected scatter in values is indicated by the shading. Because of this similarity in properties, it is logical to ask why so many different alloy steels are used. Actually, as will be seen in Chap. 14, all low-alloy steels do not have the same fracture toughness or notch sensitivity, and they may differ considerably in these respects when heat-treated to tensile strengths in excess of 200,000 psi. Further, to minimize processing difficulties such as quench cracking and weld embrittlement, it is an advantage to use a steel with the lowest carbon content consistent with the required as-quenched hardness. For this reason, steels are available with closely spaced carbon contents.

Steel sections which are too large to be quenched throughout to essentially 100 percent martensite will contain higher-temperature transformation products such as ferrite, pearlite, and bainite interspersed with the martensite. Such a situation is known as a *slack-quenched* structure. Slack quenching results in tensile properties which are somewhat poorer than those obtained with a completely tempered martensitic structure. The yield strength and the reduction of area are generally most affected, while impact strength can be very greatly reduced. The

[1] J. H. Hollomon and L. D. Jaffe, *Trans. Metall. Soc. AIME*, vol. 162, p. 223, 1945.

[2] R. A. Grange and R. W. Baughman, *Trans. Am. Soc. Met.*, vol. 48, pp. 165–197, 1956.

[3] L. D. Jaffe and E. Gordon, *Trans. Am. Soc. Met.*, vol. 49, pp. 359–371, 1957.

[4] E. J. Janitsky and M. Baeyertz, in "Metals Handbook," pp. 515–518, American Society for Metals, Metals Park, Ohio, 1939.

[5] W. G. Patton, *Met. Prog.*, vol. 43, pp. 726–733, 1943.

effect of slack quenching is greatest at high hardness levels. As the tempering temperature is increased, the deviation in the properties of slack-quenched steel from those of tempered martensite becomes smaller. In steels with sufficient hardenability to form 100 percent martensite it is frequently found that not all the austenite is transformed to martensite on quenching. Studies[1] have shown that the greatest effect of retained austenite on tensile properties is in decreasing the yield strength.

**Fig. 9-27** Relationships between tensile properties of quenched and tempered low-alloy steels. (*From W. G. Patton, Met. Progr., vol. 43, p. 726, 1943.*)

[1] L. S. Castleman, B. L. Averbach, and M. Cohen, *Trans. Am. Soc. Met.*, vol. 44, pp. 240–263, 1952.

## 9-12 ANISOTROPY OF TENSILE PROPERTIES

It is frequently found that the tensile properties of wrought-metal products are not the same in all directions. The dependence of properties on orientation is called *anisotropy*. Two general types of anisotropy are found in metals. *Crystallographic anisotropy* results from the preferred orientation of the grains which is produced by severe deformation. Since the strength of a single crystal is highly anisotropic, a severe plastic deformation which produces a strong preferred orientation will cause a polycrystalline specimen to approach the anisotropy of a single crystal. The yield strength, and to a lesser extent the tensile strength, are the properties most affected. The yield strength in the direction perpendicular to the main (longitudinal) direction of working may be greater or less than the yield strength in the longitudinal direction, depending on the type of preferred orientation which exists. This type of anisotropy is most frequently found in nonferrous metals, especially when they have been severely worked into sheet. Crystallographic anisotropy can be eliminated by recrystallization, although the formation of a recrystallization texture can cause the reappearance of a different type of anisotropy. A practical manifestation of crystallographic anisotropy is the formation of "ears," or nonuniform deformation in deep-drawn cups. Crystallographic anisotropy may also result in the elliptical deformation of a tensile specimen.

*Mechanical fibering* is due to the preferred alignment of structural discontinuities such as inclusions, voids, segregation, and second phases in the direction of working. This type of anisotropy is important in forgings and plates. The principal direction of working is defined as the *longitudinal direction*. This is the long axis of a bar or the rolling direction in a sheet or plate. Two transverse directions must be considered. The *short-transverse direction* is the minimum dimension of the product, for example, the thickness of a plate. The *long-transverse direction* is perpendicular to both the longitudinal and short-transverse directions. In a round or square, both these transverse directions are equivalent, while in a sheet the properties in the short-transverse direction cannot be measured. In wrought-steel products mechanical fibering is the principal cause of directional properties. Measures of ductility like reduction of area are most affected. In general, reduction of area is lowest in the short-transverse direcion, intermediate in the long-transverse direction, and highest in the longitudinal direction.

Transverse properties are particularly important in thick-walled tubes, like guns and pressure vessels, which are subjected to high internal pressures. In these applications the greatest principal stress acts in the circumferential direction, which corresponds to the transverse direction of a cylindrical forging. While there is no direct method for incorporating the reduction of area into the design of such a member, it is known that the transverse reduction of area (RAT) is a good index of steel quality for these types of applications. For this reason the RAT may be

**Fig. 9-28** Relationship between reduction of area and angle between the longitudinal direction in forging and the specimen axis. (*From C. Wells and R. F. Mehl, Trans. Am. Soc. Met., vol. 41, p. 753, 1949.*)

the limiting value in the design of a part. A great deal of work[1-3] on the transverse properties of gun tubes and large forgings has provided data in this field. Figure 9-28 shows the variation of reduction of area with the angle between the axis of the tensile specimen and the longitudinal direction in a forging of SAE 4340 steel. No similar variation with orientation is found for the yield strength or the tensile strength. This figure shows both the maximum and minimum values of reduction of area obtained for different specimen orientations. Because of the large scatter in measurements of RAT it is necessary to use statistical methods. The degree of anisotropy in reduction of area increases with strength level. In the region of tensile strength between 80,000 and 180,000 psi the RAT decreases by about 1.5 percent for each 5,000 psi increase in tensile strength. Figure 9-29 shows the way in which the longitudinal and transverse reduction of area varies with reduction by forging. The forging ratio is the ratio of the original to the final cross-sectional area of the forging. It is usually found that the optimum properties are obtained with a forging ratio of 2 to 3 : 1. Nonmetallic inclusions are considered to be a major source of low transverse ductility. This is based on the fact that vacuum-melted steels give higher RAT values than air-melted steel and on correlations which have been made[4] between inclusion content and RAT. Other factors

[1] C. Wells and R. F. Mehl, *Trans. Am. Soc. Met.,* vol. 41, pp. 715–818, 1949.
[2] A. H. Grobe, C. Wells, and R. F. Mehl, *Trans. Am. Soc. Met.,* vol. 45, pp. 1080–1122, 1953.
[3] E. A. Loria, *Trans. Am. Soc. Met.,* vol. 42, pp. 486–498, 1950.
[4] J. Welchner and W. G. Hildorf, *Trans. Am. Soc. Met.,* vol. 42, pp. 455–485, 1950; R. A. Cellitti and C. J. Carter, *Met. Prog.,* pp. 215–230, October 1969.

**Fig. 9-29** Effect of forging reduction on longitudinal and transverse reduction of area. Tensile strength 118,000 psi. (*From C. Wells and R. F. Mehl, Trans. Am. Soc. Met., vol. 41, p. 755, 1949.*)

such as microsegregation and dendritic structure may also be responsible for low transverse ductility in forgings.

An interesting graphic method for correlating the amount and direction of deformation with the resulting directionality of tensile properties has been presented by Hunsicker.[1] The procedure has been well documented for directional properties in aluminum alloys, but it has not yet been widely applied to other systems.

An interesting aspect of the anisotropic strength of metals concerns the effect of prior torsional deformation on tensile properties. Swift[2] twisted mild-steel bars in torsion and then determined the tensile properties of the bar. If the torsional shear strain at the surface exceeded unity, it was found that the tensile-fracture stress and reduction of area were greatly reduced. At the same time the tensile fracture changed from a cup-and-cone fracture to a fracture on a 45° plane. If the specimens were twisted to this strain and then untwisted, there was little effect of the torsional deformation on the fracture stress, ductility, or type of fracture. In interpreting these results it was suggested[3] that the twisting produced a preferred orientation of initially randomly oriented microcracks. The cracks were presumed to become oriented along the helical surface, which is in compression during twisting (see Fig. 10-4). Separation occurs along this 45° plane when axial tension is applied. The cracks were assumed to become reoriented in the longitudinal direction of the bar when it was untwisted, and with this orientation they would have little effect on the tensile properties. Although there was no real experimental evidence for the existence of microcracks, it was considered that they could be initiated at inclusions and second-phase particles. However, similar experiments[4] with OFHC copper, in which

[1] H. Y. Hunsicker, *Trans. Metall. Soc. AIME*, vol. 245, pp. 29–42, 1969.
[2] H. W. Swift, *J. Iron Steel Inst. London*, vol. 140, p. 181, 1939.
[3] C. Zener and J. H. Hollomon, *Trans. Am. Soc. Met.*, vol. 33, p. 163, 1944.
[4] W. A. Backofen, A. J. Shaler, and B. B. Hundy, *Trans. Am. Soc. Met.*, vol. 46, pp. 655–680, 1954.

there were no second-phase particles and no preferred orientation, confirmed and extended Swift's observations. The mechanical anisotropy which was observed was explained on the assumption that the metal contained a fibrous flaw structure, which has the characteristics of submicroscopic cracks. There is some indication that these cracks originate during solidification of the ingot and perhaps during plastic deformation, when they are oriented in the principal working direction.

## BIBLIOGRAPHY

Fenner, A. J.: " Mechanical Testing of Materials," Philosophical Library, Inc., New York, 1965.

Marin, Joseph: " Mechanical Behavior of Engineering Materials," chap. 1, Prentice-Hall, Inc., Englewood Cliffs, N.J., 1962.

Nadai, A.: "Theory of Flow and Fracture of Solids," vol. I, chap. 8, McGraw-Hill Book Company, New York, 1950.

Polakowski, N. H., and E. J. Ripling: " Strength and Structure of Engineering Materials," chap. 10, Prentice-Hall, Inc., Englewood Cliffs, N.J., 1966.

Symposium on Significance of the Tension Test of Metals in Relation to Design, *Am. Soc. Test. Mater. Proc.*, vol. 40, pp. 501–609, 1940.

Tegart, W. J. McG.: "Elements of Mechanical Metallurgy," chap. 1, The Macmillan Company, New York, 1966.

# 10
# THE TORSION TEST

## 10-1  INTRODUCTION

The torsion test has not met with the wide acceptance and the use that have been given the tension test. However, it is useful in many engineering applications and also in theoretical studies of plastic flow. Torsion tests are made on materials to determine such properties as the modulus of elasticity in shear, the torsional yield strength, and the modulus of rupture. Torsion tests also may be carried out on full-sized parts, such as shafts, axles, and twist drills, which are subjected to torsional loading in service. It is frequently used for testing brittle materials, such as tool steels, and has been used in the form of a high-temperature twist test to evaluate the forgeability of materials. The torsion test has not been standardized to the same extent as the tension test and is rarely required in materials specifications.

Torsion-testing equipment consists of a twisting head, with a chuck for gripping the specimen and for applying the twisting moment to the specimen, and a weighing head, which grips the other end of the specimen and measures the twisting moment, or torque. The deformation of the specimen is measured by a twist-measuring device called a troptometer. Determination is made of the angular displacement of a point near one end of the test section of the specimen with respect

to a point on the same longitudinal element at the opposite end. A torsion specimen generally has a circular cross section, since this represents the simplest geometry for the calculation of the stress. Since in the elastic range the shear stress
varies linearly from a value of zero at the center of the bar to a maximum value at
the surface, it is frequently desirable to test a thin-walled tubular specimen. This
results in a nearly uniform shear stress over the cross section of the specimen.

## 10-2  MECHANICAL PROPERTIES IN TORSION

Consider a cylindrical bar which is subjected to a torsional moment at one end
(Fig. 10-1). The twisting moment is resisted by shear stresses set up in the cross
section of the bar. The shear stress is zero at the center of the bar and increases
linearly with the radius. Equating the twisting moment to the internal resisting
moment,

$$M_T = \int_{r=0}^{r=a} \tau r \, dA = \frac{\tau}{r} \int_0^a r^2 \, dA \qquad (10\text{-}1)$$

But $\int r^2 \, dA$ is the polar moment of inertia of the area with respect to the axis of
the bar. Thus,

$$M_T = \frac{\tau J}{r}$$

or

$$\tau = \frac{M_T r}{J} \qquad (10\text{-}2)$$

where $\tau$ = shear stress, psi
$\quad M_T$ = torsional moment, lb-in
$\quad\quad r$ = radial distance measured from center of bar, in
$\quad\quad J$ = polar moment of inertia, $in^4$

Fig. 10-1  Torsion of a solid bar.

Since the shear stress is a maximum at the surface of the bar, for a solid cylindrical specimen where $J = \pi D^4/32$, the maximum shear stress is

$$\tau_{max} = \frac{M_T D/2}{\pi D^4/32} = \frac{16 M_T}{\pi D^3} \qquad (10\text{-}3)$$

For a tubular specimen the shear stress on the outer surface is

$$\tau = \frac{16 M_T D_1}{\pi (D_1{}^4 - D_2{}^4)} \qquad (10\text{-}4)$$

where $D_1$ = outside diameter of tube

$\quad\;\; D_2$ = inside diameter of tube

The troptometer is used to determine the angle of twist $\theta$, usually expressed in radians. If $L$ is the test length of the specimen, from Fig. 10-1 it will be seen that the shear strain is given by

$$\gamma = \tan \phi = \frac{r\theta}{L} \qquad (10\text{-}5)$$

During a torsion test measurements are made of the twisting moment $M_T$ and the angle of twist $\theta$. A torque-twist diagram is usually obtained, as shown in Fig. 10-2.

The elastic properties in torsion may be obtained by using the torque at the proportional limit or the torque at some offset angle of twist, frequently 0.001 rad/in of gage length, and calculating the shear stress corresponding to the twisting moment from the appropriate equations given above. A tubular specimen is usually required for a precision measurement of the torsional elastic limit or yield strength. Because of the stress gradient across the diameter of a solid bar, the surface fibers are restrained from yielding by the less highly stressed inner fibers. Thus, the first onset of yielding is generally not readily apparent with the instruments ordinarily used for measuring the angle of twist. The use of a thin-walled tubular specimen minimizes this effect because the stress gradient is practically

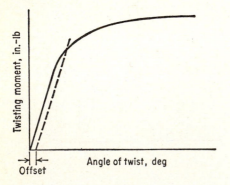

Fig. 10-2   Torque-twist diagram.

eliminated. Care should be taken, however, that the wall thickness is not reduced too greatly, or the specimen will fail by buckling rather than torsion. Experience has shown that for determinations of the shearing yield strength and modulus of elasticity the ratio of the length of the reduced test section to the outside diameter should be about 10 and the diameter-thickness ratio should be about 8 to 10.

Once the torsional yield strength has been exceeded the shear-stress distribution from the center to the surface of the specimen is no longer linear and Eq. (10-3) or (10-4) does not strictly apply. However, an ultimate torsional shearing strength, or *modulus of rupture*, is frequently determined by substituting the maximum measured torque into these equations. The results obtained by this procedure overestimate the ultimate shear stress. A more precise method of calculating this value will be discussed in the next section. Although the procedure just described results in considerable error, for the purpose of comparing and selecting materials it is generally sufficiently accurate. For the determination of the modulus of rupture with tubular specimens, the ratio of gage length to diameter should be about 0.5 and the diameter-thickness ratio about 10 to 12.

With the elastic range the shear stress can be considered proportional to the shear strain. The constant of proportionality $G$ is the *modulus of elasticity in shear*, or the *modulus of rigidity*.

$$\tau = G\gamma \qquad (10\text{-}6)$$

Substituting Eqs. (10-2) and (10-5) into Eq. (10-6) gives an expression for the shear modulus in terms of the geometry of the specimen, the torque, and the angle of twist.

$$G = \frac{M_T L}{J\theta} \qquad (10\text{-}7)$$

## 10-3 TORSIONAL STRESSES FOR LARGE PLASTIC STRAINS

Beyond the torsional yield strength the shear stress over a cross section of the bar is no longer a linear function of the distance from the axis, and Eqs. (10-3) and (10-4) do not apply. Nadai[1] has presented a method for calculating the shear stress in the plastic range if the torque-twist curve is known. To simplify the analysis, we shall consider the angle of twist per unit length $\theta'$, where $\theta' = \theta/L$. Referring to Eq. (10-5), the shear strain will be

$$\gamma = r\theta' \qquad (10\text{-}8)$$

[1] A. Nadai, "Theory of Flow and Fracture of Solids," 2d ed., vol. I, pp. 347–349, McGraw-Hill Book Company, New York, 1950.

Equation (10-1), for the resisting torque in a cross section of the bar, can be expressed as follows:

$$M_T = 2\pi \int_0^a \tau r^2 \, dr \qquad (10\text{-}9)$$

Now the shear stress is related to the shear strain by the stress-strain curve in shear.

$$\tau = f(\gamma)$$

Introducing this equation into Eq. (10-9) and changing the variable from $r$ to $\gamma$ by means of Eq. (10-8) gives

$$M_T = 2\pi \int_0^{\gamma_a} f(\gamma) \frac{(\gamma^2)}{(\theta')^2} \frac{d\gamma}{\theta'}$$

$$M_T(\theta')^3 = 2\pi \int_0^{\gamma_a} f(\gamma)\gamma^2 \, d\gamma \qquad (10\text{-}10)$$

where $\gamma_a = a\theta'$. Differentiating Eq. (10-10) with respect to $\theta'$

$$\frac{d}{d\theta'}(M_T \theta'^3) = 2\pi a f(a\theta') a^2 (\theta')^2 = 2\pi a^3 (\theta')^2 f(a\theta')$$

But, the maximum shear stress in the bar at the outer fiber is $\tau_a = f(a\theta')$. Therefore,

$$\frac{d(M_T \theta'^3)}{d\theta'} = 2\pi a^3 (\theta')^2 \tau_a$$

$$3M_T(\theta')^2 + (\theta')^3 \frac{dM_T}{d\theta'} = 2\pi a^3 (\theta')^2 \tau_a$$

Therefore,

$$\tau_a = \frac{1}{2\pi a^3} \left( \theta' \frac{dM_T}{d\theta'} + 3M_T \right) \qquad (10\text{-}11)$$

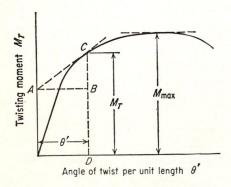

Fig. 10-3   Method of calculating shear stress from torque-twist diagram.

If a torque-twist curve is available, the shear stress can be calculated with the above equation. Figure 10-3 illustrates how this is done. Examination of Eq. (10-11) shows that it can be written in terms of the geometry of Fig. 10-3 as follows:

$$\tau_a = \frac{1}{2\pi a^3}(BC + 3CD) \qquad (10\text{-}12)$$

It will also be noticed from Fig. 10-3 that at the maximum value of torque $dM_T/d\theta' = 0$. Therefore, the ultimate torsional shear strength, or modulus of rupture, can be expressed by

$$\tau_u = \frac{3M_{max}}{2\pi a^3} \qquad (10\text{-}13)$$

Large plastic strains in torsion can result in considerable changes in the length of the specimen. If both ends of the specimen are fixed, these length changes lead to the superposition of longitudinal stresses of unknown magnitude on the torsional shearing stresses. While these longitudinal stresses usually are small compared to the shearing stresses and can be ignored, they may be important in influencing the torsional strain to fracture. Extraneous stresses due to length change can be greatly minimized, at some increase in experimental complexity, by using a torsion tester in which one end of the specimen is free to move.

## 10-4 TYPES OF TORSION FAILURES

Figure 10-4 illustrates the state of stress at a point on the surface of a bar subjected to torsion. The maximum shear stress occurs on two mutually perpendicular planes, perpendicular to the longitudinal axis $yy$ and parallel with the longitudinal axis $xx$. The principal stesses $\sigma_1$ and $\sigma_3$ make an angle of $45°$ with the longitudinal axis and are equal in magnitude to the shear stresses. $\sigma_1$ is a tensile stress, and $\sigma_3$ is an equal compressive stress. The intermediate stress $\sigma_2$ is zero.

**Fig. 10-4**  State of stress in torsion.

**Fig. 10-5** Typical torsion failures. (*a*) Shear (ductile) failure;
(*b*) tensile (brittle) failure.

Torsion failures are different from tensile failures in that there is little localized reduction of area or elongation. A ductile metal fails by shear along one of the planes of maximum shear stress. Generally the plane of the fracture is normal to the longitudinal axis (see Fig. 10-5*a*). A brittle material fails in torsion along a plane perpendicular to the direction of the maximum tensile stress. Since this plane bisects the angle between the two planes of maximum shear stress and makes an angle of 45° with the longitudinal and transverse directions, it results in a helical fracture (Fig. 10-5*b*). Fractures are sometimes observed in which the test section of the specimen breaks into a large number of fairly small pieces. In these cases it can usually be determined that the fracture started on a plane of maximum shear stress parallel with the axis of the specimen. A study of torsion failures in a tool steel as a function of hardness[1] showed that fracture started on planes of maximum shear stress up to a Vickers hardness of 720 and that above this hardness tensile stresses were responsible for starting fracture.

## 10-5 TORSION TEST VS. TENSION TEST

A good case can be made for the position advanced by Sauveur[2] that the torsion test provides a more fundamental measure of the plasticity of a metal than the tension test. For one thing, the torsion test yields directly a shear-stress–shear-strain curve. This type of curve has more fundamental significance in characterizing plastic behavior than a stress-strain curve determined in tension. Large values of strain can be obtained in torsion without complications such as necking in tension or barreling due to frictional end effects in compression. Moreover, in torsion, tests can be made fairly easily at constant or high strain rates. On the other hand, considerable labor is involved in converting torque–angle-of-twist data into shear-stress–strain curves. Furthermore, unless a tubular specimen is used, there will be a steep stress gradient across the specimen. This will make it difficult to make accurate measurements of the yield strength.

[1] R. D. Olleman, E. T. Wessel, and F. C. Hull, *Trans. Am. Soc. Met.*, vol. 46, pp. 87–99, 1954.
[2] A. Sauveur, *Am. Soc. Test. Mater. Proc.*, vol. 38, pt. 2, pp. 3–20, 1938.

The tension test and the torsion test are compared below in terms of the state of stress and strain developed in each test:

<table>
<tr><td align="center">Tension test</td><td align="center">Torsion test</td></tr>
<tr><td align="center">$\sigma_1 = \sigma_{max}; \sigma_2 = \sigma_3 = 0$</td><td align="center">$\sigma_1 = -\sigma_3; \sigma_2 = 0$</td></tr>
<tr><td align="center">$\tau_{max} = \dfrac{\sigma_1}{2} = \dfrac{\sigma_{max}}{2}$</td><td align="center">$\tau_{max} = \dfrac{2\sigma_1}{2} = \sigma_{max}$</td></tr>
<tr><td align="center">$\varepsilon_{max} = \varepsilon_1; \varepsilon_2 = \varepsilon_3 = -\dfrac{\varepsilon_1}{2}$</td><td align="center">$\varepsilon_{max} = \varepsilon_1 = -\varepsilon_3; \varepsilon_2 = 0$</td></tr>
<tr><td align="center">$\gamma_{max} = \dfrac{3\varepsilon_1}{2}$</td><td align="center">$\gamma_{max} = \varepsilon_1 - \varepsilon_3 = 2\varepsilon_1$</td></tr>
</table>

$$\bar{\sigma} = \frac{\sqrt{2}}{2}[(\sigma_1 - \sigma_2)^2 + (\sigma_2 - \sigma_3)^2 + (\sigma_3 - \sigma_1)^2]^{\frac{1}{2}}$$

$$\bar{\varepsilon} = \frac{\sqrt{2}}{3}[(\varepsilon_1 - \varepsilon_2)^2 + (\varepsilon_2 - \varepsilon_3)^2 + (\varepsilon_3 - \varepsilon_1)^2]^{\frac{1}{2}}$$

<table>
<tr><td align="center">$\bar{\sigma} = \sigma_1$</td><td align="center">$\bar{\sigma} = \sqrt{3}\sigma_1$</td></tr>
<tr><td align="center">$\bar{\varepsilon} = \sigma_1$</td><td align="center">$\bar{\varepsilon} = \dfrac{2}{\sqrt{3}}\varepsilon_1 = \dfrac{\gamma}{\sqrt{3}}$</td></tr>
</table>

This comparison shows that $\tau_{max}$ will be twice as great in torsion as in tension for a given value of $\sigma_{max}$. Since as a first approximation it can be considered that plastic deformation occurs on reaching a critical value of $\tau_{max}$ and brittle fracture occurs on reaching a critical value of $\sigma_{max}$, the opportunity for ductile behavior is greater in torsion than in tension, This is illustrated schematically in Fig. 10-6, which can

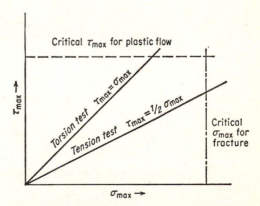

**Fig. 10-6**  Effect of ratio $\tau_{max}/\sigma_{max}$ in determining ductility. (*After Gensamer.*)

be considered representative of the condition for a brittle material such as hardened tool steel. In the torsion test the critical shear stress for plastic flow is reached before the critical normal stress for fracture, while in tension the critical normal stress is reached before the shear stress reaches the shear stress for plastic flow. Even for a metal which is ductile in the tension test, where the critical normal stress is pushed far to the right in Fig. 10-6, the figure shows that the amount of plastic deformation is greater in torsion than in tension.

   The tensile stress-strain curve can be derived from the curve for torsion when the stress-strain curve is plotted in terms of significant stress and strain or the octahedral shear stress and strain (see Prob. 10-4). Figure 10-7 shows a true-stress–true-strain curve from a tension test and the shear-stress–shear-strain curve for the same material in torsion. When both curves are plotted in terms of significant stress and significant strain (the tension curve is unchanged), the two curves superimpose within fairly close limits.

   Good agreement between flow curves determined in tension and compression and in torsion has been obtained for tests at both room temperature[1] and elevated

**Fig. 10-7**  Tension and torsion true-stress–true-strain curves for low-carbon steel.

[1] J. H. Faupel and J. Marin, *Trans. Am. Soc. Met.*, vol. 43, pp. 993–1012, 1951.

temperature.[1] For large plastic shearing strains the effective strain should be given by[2]

$$\bar{\varepsilon} = \frac{2}{\sqrt{3}} \ln \left[ \left( 1 + \frac{\gamma^2}{4} \right)^{1/2} + \frac{\gamma}{2} \right] \qquad (10\text{-}14)$$

## 10-6  HOT TORSION TESTING

The torsion test is well adapted to obtaining data on the flow properties[3] and fracture[4] of metals under hot-working conditions, that is, $T > 0.6\, T_m$ and $\bar{\varepsilon}$ up to $10^3\ \mathrm{s}^{-1}$. Since it is easy to vary and control the speed of rotation, tests may be carried out over a wide range of strain rates. Moreover, by proper control of temperature and strain rate it is possible to simulate the metallurgical structures[5] produced in multiple-pass processes such as rolling. Because the torsion test specimen is not subject to necking, as in tension, or barreling, as in compression, it is possible to carry out the test to large plastic strains. Figure 10-8 shows typical torque-twist curves obtained with a hot torsion tester. The top curve (800°F) is for an essentially cold deformation condition in which the torque (flow stress) increases continuously up to fracture. The bottom curve (1800°F) is characteristic of hot-working deformation. The flow stress peaks at a relatively small plastic strain and then falls to an essentially constant value which is maintained until fracture occurs at a large plastic strain. The curve at 1600°F is characteristic of the deformation conditions known as warm-working.

The analysis discussed in Sec. 11-3 for converting torque-twist curves to shear-stress–shear-strain curves requires modification for use with hot torsion tests because it assumed that the stress is independent of strain rate. An analysis which considers the strong strain-rate dependence of flow stress at high temperature was proposed by Fields and Backofen[6]. It results in the relation

$$\frac{dM_T}{d\theta'} = \frac{M_T}{\theta'}\,(m + n)$$

[1] F. A. Hodierne, *J. Inst. Met.*, vol. 91, pp. 267–273, 1962–1963.

[2] N. H. Polakowski and E. J. Ripling, "Strength and Structure of Engineering Materials," pp. 382–383, Prentice–Hall, Inc., Englewood Cliffs, N. J., 1966.

[3] W. J. McG. Tegart, The Role of Ductility in Hot Working, in "Ductility," chap. 5, American Society for Metals, Metals Park, Ohio, 1968.

[4] E. Shapiro and G. E. Dieter, *Metall. Trans.*, vol. 1, pp. 1711–1719, 1970; *Metall. Trans.*, vol. 2, pp. 1385–1391, 1971.

[5] M. M. Farag, C. M. Sellars, and W. J. McG. Tegart, in "Deformation Under Hot Working Conditions," pp. 60–67, Pub. No. 108, Iron and Steel Institute, London, 1968.

[6] D. S. Fields and W. A. Backofen, *Am. Soc. Test. Mater. Proc.*, vol. 57, pp. 1259–1272, 1957.

**Fig. 10-8** Torque-twist curves at different temperatures for a nickel-based alloy (Inconel 600) tested in torsion at a constant shear-strain rate of 2.5 s⁻¹. (*From E. Shapiro and G. E. Dieter, Metall. Trans., vol. 1, p. 1712, 1970.*)

where $m$ is the strain-rate sensitivity and $n$ is the strain-hardening exponent. At hot-working temperatutes $m \gg n$. Substitution of the above expression into Eq. (10-11) results in

$$\tau_a = \frac{M_T}{2\pi a^3}(3 + m + n) \qquad (10\text{-}15)$$

## BIBLIOGRAPHY

Davis, H. E., G. E. Troxell, and C. T. Wiskocil: "The Testing and Inspection of Engineering Materials," 2d ed., chap. 5, McGraw-Hill Book Company, New York, 1955.

Gensamer, M.: "Strength of Metals Under Combined Stresses," American Society for Metals, Metals Park, Ohio, 1941.

Marin, Joseph: "Mechanical Behavior of Engineering Materials" chap. 2, Prentice-Hall, Inc., Englewood Cliffs, N.J., 1962.

Tegart W. J. McG.: "Elements of Mechanical Metallurgy," chap. 3, The Macmillan Company, New York, 1966.

# 11

# THE HARDNESS TEST

## 11-1 INTRODUCTION

The hardness of a material is a poorly defined term which has many meanings depending upon the experience of the person involved. In general, hardness usually implies a resistance to deformation, and for metals the property is a measure of their resistance to permanent or plastic deformation. To a person concerned with the mechanics of materials testing, hardness is most likely to mean the resistance to indentation, and to the design engineer it often means an easily measured and specified quantity which indicates something about the strength and heat treatment of the metal. There are three general types of hardness measurements depending on the manner in which the test is conducted. These are (1) scratch hardness, (2) indentation hardness, and (3) rebound, or dynamic, hardness. Only indentation hardness is of major engineering interest for metals.

Scratch hardness is of primary interest to mineralogists. With this measure of hardness, various minerals and other materials are rated on their ability to scratch one another. Scratch hardness is measured according to the Mohs' scale. This consists of 10 standard minerals arranged in the order of their ability to be scratched. The softest mineral in this scale is talc (scratch hardness 1), while diamond has a hardness of 10. A fingernail has a value of about 2, annealed copper has a value of 3, and martensite a hardness of 7. The Mohs' scale is not well suited for metals since the intervals are not widely spaced in the high-hardness range. Most hard metals fall in the Mohs' hardness range of 4 to 8. A different type of scratch-hardness test[1] measures the depth or width of a scratch made by drawing a diamond stylus across the surface under a definite load. This is a useful tool for measuring the relative hardness of microconstituents, but it does not lend itself to high repro-ducibility or extreme accuracy.

In dynamic-hardness measurements the indenter is usually dropped onto the metal surface, and the hardness is expressed as the energy of impact. The Shore sceleroscope, which is the commonest example of a dynamic-hardness tester, measures the hardness in terms of the height of rebound of the indenter.

## 11-2 BRINELL HARDNESS

The first widely accepted and standardized indentation-hardness test was proposed by J. A. Brinell in 1900. The Brinell hardness test consists in indenting the metal surface with a 10-mm-diameter steel ball at a load of 3,000 kg. For soft metals the load is reduced to 500 kg to avoid too deep an impression, and for very hard metals a tungsten carbide ball is used to minimize distortion of the indenter. The load is applied for a standard time, usually 30 s, and the diameter of the indentation is measured with a low-power microscope after removal of the load. The average of two readings of the diameter of the impression at right angles should be made. The surface on which the indentation is made should be relatively smooth and free from dirt or scale. The Brinell hardness number (BHN) is expressed as the load $P$ divided by the *surface area* of the indentation. This is expressed by the formula[2]

$$\text{BHN} = \frac{P}{(\pi D/2)\left(D - \sqrt{D^2 - d^2}\right)} = \frac{P}{\pi D t} \qquad (11\text{-}1)$$

where $P$ = applied load, kg
$\quad D$ = diameter of ball, mm
$\quad d$ = diameter of indentation, mm
$\quad t$ = depth of the impression, mm

It will be noticed that the units of the BHN are kilograms per square millimeter. However, the BHN is not a satisfactory physical concept since Eq. (11-1) does not give the mean pressure over the surface of the indentation.

---

[1] E. B. Bergsman, *ASTM Bull.* 176, pp. 37–43, September, 1951.
[2] Tables giving BHN as a function of $d$ for standard loads may be found in most of the references in the Bibliography at the end of this chapter. See *ASTM* Standard E10-66.

**Fig. 11-1**  Basic parameters in Brinell test.

From Fig. 11-1 it can be seen that $d = D \sin \phi$. Substitution into Eq. (11-1) gives an alternate expression for Brinell hardness number.

$$\text{BHN} = \frac{P}{(\pi/2)D^2(1 - \cos \phi)} \qquad (11\text{-}2)$$

In order to obtain the same BHN with a nonstandard load or ball diameter it is necessary to produce geometrically similar indentations. Geometric similitude is achieved so long as the included angle $2\phi$ remains constant. Equation (11-2) shows that for $\phi$ and BHN to remain constant the load and ball diameter must be varied in the ratio

$$\frac{P_1}{D_1{}^2} = \frac{P_2}{D_2{}^2} = \frac{P_3}{D_3{}^2} \qquad (11\text{-}3)$$

Unless precautions are taken to maintain $P/D^2$ constant, which may be experimentally inconvenient, the BHN generally will vary with load. Over a range of loads the BHN reaches a maximum at some intermediate load. Therefore, it is not possible to cover with a single load the entire range of hardnesses encountered in commercial metals. The relatively large size of the Brinell impression may be an advantage in averaging out local heterogeneities. Moreover, the Brinell test is less influenced by surface scratches and roughness than other hardness tests. On the other hand, the large size of the Brinell impression may preclude the use of this test with small objects or in critically stressed parts where the indentation could be a potential site of failure.

## 11-3  MEYER HARDNESS

Meyer[1] suggested that a more rational definition of hardness than that proposed by Brinell would be one based on the *projected area* of the impression rather than the surface area. The mean pressure between the surface of the indenter and the

[1] E. Meyer, *Z. ver. Deut. Ing*, vol. 52, pp. 645–654, 1908.

indentation is equal to the load divided by the projected area of the indentation.

$$p_m = \frac{P}{\pi r^2}$$

Meyer proposed that this mean pressure should be taken as the measure of hardness. It is referred to as the *Meyer hardness*.

$$\text{Meyer hardness} = \frac{4P}{\pi d^2} \qquad (11\text{-}4)$$

Like the Brinell hardness, Meyer hardness has units of kilograms per square millimeter. The Meyer hardness is less sensitive to the applied load than the Brinell hardness. For a cold-worked material the Meyer hardness is essentially constant and independent of load, while the Brinell hardness decreases as the load increases. For an annealed metal the Meyer hardness increases continuously with the load because of strain hardening produced by the indentation. The Brinell hardness, however, first increases with load and then decreases for still higher loads. The Meyer hardness is a more fundamental measure of indentation hardness; yet it is rarely used for practical hardness measurements.

Meyer proposed an empirical relation between the load and the size of the indentation. This relationship is usually called *Meyer's law*.

$$P = kd^{n'} \qquad (11\text{-}5)$$

where $P$ = applied load, kg
$\qquad d$ = diameter of indentation, mm
$\qquad n'$ = a material constant related to strain hardening of metal
$\qquad k$ = a material constant expressing resistance of metal to penetration

The parameter $n'$ is the slope of the straight line obtained when $\log P$ is plotted against $\log d$, and $k$ is the value of $P$ at $d = 1$. Fully annealed metals have a value of $n'$ of about 2.5, while $n'$ is approximately 2 for fully strain-hardened metals. This parameter is roughly related to the strain-hardening coefficient in the exponential equation for the true-stress–true-strain curve. The exponent in Meyer's law is approximately equal to the strain-hardening coefficient plus 2.

There is a lower limit of load below which Meyer's law is not valid. If the load is too small, the deformation around the indentation is not fully plastic and Eq. (11-5) is not obeyed. This load will depend on the hardness of the metal. For a 10-mm-diameter ball the load should exceed 50 kg for copper with a BHN of 100, and for steel with a BHN of 400 the load should exceed 1,500 kg. For balls of different diameter the critical loads will be proportional to the square of the diameter.

## 11-4 ANALYSIS OF INDENTATION BY AN INDENTER

The plastic zone beneath a hardness indentation is surrounded with elastic material which acts to hinder plastic flow in a manner similar to the die constraint forces in a closed-die forging. Therefore, the mean compressive stress required to cause plastic flow in the hardness test exceeds that in simple compression because of this constraint. The prediction of the load required to indent a solid is one of the classic problems in plasticity. Prandtl applied slip-line field theory to show that the constraint factor for plane-strain compression was 2.57 (see Sec. 3-9 and Fig. 3-12).

$$\frac{p_m}{\sigma} = 1 + \frac{\pi}{2} = 2.57$$

In this idealized model the material surrounding the deformed zone is rigid and upward flow of material compensates for the material displaced by the punch. However, the geometry of the Brinell test is axisymmetric as opposed to plane strain. Shaw and DeSalvo have shown[1] that the plastic region under this blunt indenter bears little resemblance to the slip-line field, but instead is very similar to an elastic-plastic boundary resembling the line of constant maximum shear stress beneath a sphere pressed against a flat surface (Fig. 11-2). This curve may be derived by applying Hertz's theory of contact stresses.[2] Using this plastic-elastic model the material displaced by the indenter is completely accounted for by the decrease in volume of the elastic material. This removes the need for an upward flow of material around the indenter, which agrees with observations that show only a small amount of upward flow. The plastic-elastic analysis results in a constraint factor of $C = 3.0$ for indentation with a spherical ball.

$$p_m = C\sigma_0 = 3\sigma_0 \qquad (11\text{-}6)$$

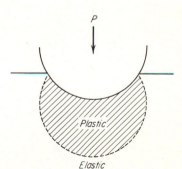

**Fig. 11-2** Plastic zone under a Brinell indenter. (*After Shaw and DeSalvo.*)

[1] M. C. Shaw and G. J. DeSalvo, *Trans. ASME, Ser. B; J. Eng. Ind.*, vol. 92, pp. 469–479, 1970; M. C. Shaw and G. J. DeSalvo, *Met. Eng. Q.*, vol. 12, pp. 1–7, May 1972.
[2] See for example S. Timoshenko and J. N. Goodier, "Theory of Elasticity," 2d ed., pp. 372–382, McGraw-Hill Book Company, New York, 1951.

## 11-5  RELATIONSHIP BETWEEN HARDNESS AND THE FLOW CURVE

Tabor[1] has suggested a method by which the plastic region of the true-stress–true-strain curve may be determined from indentation hardness measurements. The method is based on the fact that there is a similarity in the shape of the flow curve and the curve obtained when the Meyer hardness is measured on a number of specimens subjected to increasing amounts of plastic strain. The method is basically empirical, since the complex stress distribution at the hardness indentation precludes a straightforward relationship with the stress distribution in the tension or compression test. However, the method has been shown to give good agreement for several metals, and thus should be of interest as a means of obtaining flow data in situations where it is not possible to measure tensile properties. The true stress (flow stress) is obtained from Eq. (11-6), where $\sigma_0$ is to be considered the flow stress at a given value of true strain. From a study of the deformation at indentations, Tabor concluded that the true strain was proportional to the ratio $d/D$ and could be expressed as

$$\varepsilon = 0.2\,\frac{d}{D} \qquad (11\text{-}7)$$

Thus, if the Meyer hardness is measured under conditions such that $d/D$ varies from the smallest value for full plasticity up to large values and Eqs. (11-6) and (11-7) are used, it is possible at least to approximate the tensile-flow curve. Figure 11-3 shows the agreement which has been obtained by Tabor between the flow curve and hardness versus $d/D$ curve for mild steel and annealed copper. Tabor's results have

Fig. 11-3  Comparison of flow curve determined from hardness measurements (circles, and crosses) with flow curve determined from compression test (solid lines). (*From D. Tabor, "The Hardness of Metals," p. 74, Oxford University Press, New York, 1951.*)

---

[1] D. Tabor, "The Hardness of Metals," pp. 67–76, Oxford University Press, New York 1951; *J. Inst. Met.*, vol. 79, p. 1, 1951.

been verified by Lenhart[1] for duralumin and OFHC copper. However, Tabor's analysis did not predict the flow curve for magnesium, which was attributed by Lenhart to the high anisotropy of deformation in this metal. This work should not detract from the usefulness of this correlation but, rather, should serve to emphasize that its limitations should be investigated for new applications.

The 0.2 percent offset yield strength can be determined[2] with good precision from Vickers hardness measurements (see Sec. 11-6) according to the relation

$$\sigma_0 = \frac{DPH}{3}(0.1)^{n'-2} \qquad (11\text{-}8)$$

where $\sigma_0$ = the 0.2 percent offset yield strength, kg/mm$^2$

$DPH$ = the Vickers hardness number

$n' = n + 2$ = the exponent in Meyer's law

There is a very useful engineering correlation between the Brinell hardness and the ultimate tensile strength of heat-treated plain-carbon and medium-alloy steels:

Ultimate tensile strength, in pounds per square inch = 500(BHN)

A brief consideration will show that this is in agreement with Tabor's results. If we make the simplifying assumption that this class of materials does not strain-harden, then the tensile strength is equal to the yield stress and Eq. (11-6) applies.

$$s_u = \tfrac{1}{3} p_m = 0.33 p_m \qquad \text{kg/mm}^2$$

The Brinell hardness will be only a few percent less than the value of Meyer hardness $p_m$. Upon converting to engineering units the expression becomes

$$s_u = 470(\text{BHN})$$

It should now be apparent why the same relationship does not hold for other metals. For example, for annealed copper the assumption that strain hardening can be neglected will be grossly in error. For a metal with greater capability for strain hardening the "constant" of proportionality will be greater than that used for heat-treated steel.

## 11-6 VICKERS HARDNESS

The Vickers hardness test uses a square-base diamond pyramid as the indenter. The included angle between opposite faces of the pyranid is 136°. This angle was chosen because it approximates the most desirable ratio of indentation diameter

---

[1] R. E. Lenhart, *WADC Tech. Rept.*, 55–114, June, 1955.
[2] J. B. Cahoon, W. H. Broughton, and A. R. Kutzak, *Metall. Trans.*, vol. 2, pp. 1979–1983, 1971.

to ball diameter in the Brinell hardness test.[1]  Because of the shape of the indenter, this is frequently called the *diamond-pyramid hardness test*.  The diamond-pyramid hardness number (DPH), or Vickers hardness number (VHN, or VPH), is defined as the load divided by the surface area of the indentation.  In practice, this area is calculated from microscopic measurements of the lengths of the diagonals of the impression.  The DPH may be determined from the following equation

$$\text{DPH} = \frac{2P \sin (\theta/2)}{L^2} = \frac{1.854P}{L^2} \qquad (11\text{-}9)$$

where $P$ = applied load, kg

$\quad L$ = average length of diagonals, mm

$\quad \theta$ = angle between opposite faces of diamond = $136°$

The Vickers hardness test has received fairly wide acceptance for research work because it provides a continuous scale of hardness, for a given load, from very soft metals with a DPH of 5 to extremely hard materials with a DPH of 1,500.  With the Rockwell hardness test, described in Sec. 11-7, or the Brinell hardness test, it is usually necessary to change either the load or the indenter at some point in the hardness scale, so that measurements at one extreme of the scale cannot be strictly compared with those at the other end.  Because the impressions made by the pyramid indenter are geometrically similar no matter what their size, the DPH should be independent of load.  This is generally found to the case, except at very light loads.  The loads ordinarily used with this test range from 1 to 120 kg, depending on the hardness of the metal to be tested.  In spite of these advantages, the Vickers hardness test has not been widely accepted for routine testing because it is slow, requires careful surface preparation of the specimen, and allows greater chance for personal error in the determination of the diagonal length.

A perfect indentation made with a perfect diamond-pyramid indenter would be a square.  However, anomalies corresponding to those described earlier for Brinell impressions are frequently observed with a pyramid indenter (Fig. 11-4).  The pincushion indentation in Fig. 11-4b is the result of sinking in of the metal around the flat faces of the pyramid.  This condition is observed with annealed metals and results in an overestimate of the diagonal length.  The barrel-shaped indentation in Fig. 11-4c is found in cold-worked metals.  It results from ridging or piling up of the metal around the faces of the indenter.  The diagonal measurement in this case produces a low value of the contact area so that the hardness numbers are erroneously high.  Empirical corrections for this effect have been proposed.[2]

---

[1] In most Brinell tests $d/D$ lies between 0.25 and 0.50.  For the diamond-pyramid indenter a value of $d = 0.375 D$ was used, which results in cone angle of $136°$.  As a result, DPH and BHN hardnesses are nearly identical so long as the Brinell impressions are of normal depth.

[2] T. B. Crowe and J. F. Hinsely, *J. Inst. Met.*, vol. 72, p. 14, 1946.

**Fig. 11-4** Types of diamond-pyramid indentations. (*a*) Perfect indentation; (*b*) pin-cushion indentation due to sinking in; (*c*) barreled indentation due to ridging.

(*a*)

(*b*)

(*c*)

## 11-7 ROCKWELL HARDNESS TEST

The most widely used hardness test in the United States is the Rockwell hardness test. Its general acceptance is due to its speed, freedom from personal error, ability to distinguish small hardness differences in hardened steel, and the small size of the indentation, so that finished heat-treated parts can be tested without damage. This test utilizes the depth of indentation, under constant load, as a measure of hardness. A minor load of 10 kg is first applied to seat the specimen. This minimizes the amount of surface preparation needed and reduces the tendency for ridging or sinking in by the indenter. The major load is then applied, and the depth of indentation is automatically recorded on a dial gage in terms of arbitrary hardness numbers. The dial contains 100 divisions, each division representing a penetration of 0.00008 in. The dial is reversed so that a high hardness, which corresponds to a small penetration, results in a high hardness number. This is in agreement with the other hardness numbers described previously, but unlike the Brinell and Vickers hardness designations, which have units of kilograms per square millimeter, the Rockwell hardness numbers are purely arbitrary.

One combination of load and indenter will not produce satisfactory results for materials with a wide range of hardness. A 120° diamond cone with a slightly rounded point, called a *Brale indenter*, and $\frac{1}{16}$- and $\frac{1}{8}$-in-diameter steel balls are generally used as indenters. Major loads of 60, 100, and 150 kg are used. Since the Rockwell hardness is dependent on the load and indenter, it is necessary to specify the combination which is used. This is done by prefixing the hardness number with a letter indicating the particular combination of load and indenter for the hardness scale employed. A Rockwell hardness number without the letter prefix is meaningless. Hardened steel is tested on the C scale with the diamond indenter and a 150-kg major load. The useful range for this scale is from about $R_C$ 20 to $R_C$ 70. Softer materials are usually tested on the B scale with a $\frac{1}{16}$-in-diameter steel ball and a 100-kg major load. The range of this scale is from $R_B$ 0 to $R_B$ 100. The A scale (diamond penetrator, 60-kg major load) provides the most extended Rockwell hardness scale, which is usable for materials from annealed brass to cemented carbides. Many other scales are available for special purposes.[1]

---

[1] See ASTM Standard E18-67.

The Rockwell hardness test is a very useful and reproducible one provided that a number of simple precautions are observed. Most of the points listed below apply equally well to the other hardness tests:

1. The indenter and anvil should be clean and well seated.
2. The surface to be tested should be clean and dry, smooth, and free from oxide. A rough-ground surface is usually adequate for the Rockwell test.
3. The surface should be flat and perpendicular to the indenter.
4. Tests on cylindrical surfaces will give low readings, the error depending on the curvature, load, indenter, and hardness of the material. Theoretical[1] and empirical[2] corrections for this effect have been published.
5. The thickness of the specimen should be such that a mark or bulge is not produced on the reverse side of the piece. It is recommended that the thickness be at least 10 times the depth of the indentation. Tests should be made on only a single thickness of material.
6. The spacing between indentations should be three to five times the diameter of the indentation.
7. The speed of application of the load should be standardized. This is done by adjusting the dashpot on the Rockwell tester. Variations in hardness can be appreciable in very soft materials unless the rate of load application is carefully controlled. For such materials the operating handle of the Rockwell tester should be brought back as soon as the major load has been fully applied.

## 11-8 MICROHARDNESS TESTS

Many metallurgical problems require the determination of hardness over very small areas.[3] The measurement of the hardness gradient at a carburized surface, the determination of the hardness of individual constituents of a microstructure, or the checking of the hardness of a delicate watch gear might be typical problems. The use of a scratch-hardness test for these purposes was mentioned earlier, but an indentation-hardness test has been found to be more useful.[4] The development of the Knoop indenter by the National Bureau of Standards and the introduction of the Tukon tester for the controlled application of loads down to 25 g have made microhardness testing a routine laboratory procedure.

The Knoop indenter is a diamond ground to a pyramidal form that produces a diamond-shaped indentation with the long and short diagonals in the approximate

---

[1] W. E. Ingerson, *Am. Soc. Test. Mater. Proc.*, vol. 39, pp. 1281–1291, 1939.
[2] R. S. Sutton and R. H. Heyer, *ASTM Bull.* 193, pp. 40–41, October, 1953.
[3] See *ASTM* Standard E334-69.
[4] For a review of microhardness testing see H. Bückle, *Metall. Rev.*, vol. 4, no. 3, pp. 49–100, 1959.

ratio of 7 : 1. The Knoop hardness number (KHN) is the applied load divided by the unrecovered projected area of the indentation.

$$\text{KHN} = \frac{P}{A_p} = \frac{P}{L^2 C} \qquad (11\text{-}10)$$

where $P$ = applied load, kg

$\quad A_p$ = unrecovered projected area of indentation, $\text{mm}^2$

$\quad L$ = length of long diagonal, mm

$\quad C$ = a constant for each indenter supplied by manufacturer.

The special shape of the Knoop indenter makes it possible to place indentations much closer together than with a square Vickers indentation, e.g., to measure a steep hardness gradient. Its other advantage is that for a given long diagonal length the depth and area of the Knoop indentation are only about 15 percent of what they would be for a Vickers indentation with the same diagonal length. This is particularly useful when measuring the hardness of a thin layer (such as an electro-plated layer), or when testing brittle materials where the tendency for fracture is proportional to the volume of stressed material.

The low load used with microhardness tests requires that extreme care be taken in all stages of testing. The surface of the specimen must be carefully prepared. Metallographic polishing is usually required. Work hardening of the surface during polishing can influence the results. The long diagonal of the Knoop impression is essentially unaffected by elastic recovery for loads greater than about 300 g. However, for lighter loads the small amount of elastic recovery becomes appreciable. Further, with the very small indentations produced at light loads the error in locating the actual ends of the indentation become greater. Both these factors have the effect of giving a high hardness reading, so that it is usually observed that the Knoop hardness number increases as the load is decreased below about 300 g. Tarasov and Thibault[1] have shown that if corrections are made for elastic recovery and visual acuity the Knoop hardness number is constant with load down to 100 g.

## 11-9 HARDNESS-CONVERSION RELATIONSHIPS

From a practical standpoint it is important to be able to convert the results of one type of hardness test into those of a different test. Since a hardness test does not measure a well-defined property of a material and since all the tests in common use are not based on the same type of measurements, it is not surprising that no universal hardness-conversion relationships have been developed. It is important to

---

[1] L. P. Tarasov and N. W. Thibault, *Trans. Am. Soc. Met.*, vol. 38, pp. 331–353, 1947.

realize that hardness conversions are empirical relationships. The most reliable hardness-conversion data exist for steel which is harder than 240 Brinell. The ASTM, ASM, and SAE (Society of Automotive Engineers) have agreed on a table[1] for conversion between Rockwell, Brinell, and diamond-pyramid hardness which is applicable to heat-treated carbon and alloy steel and to almost all alloy constructional steels and tool steels in the as-forged, annealed, normalized, and quenched and tempered conditions. However, different conversion tables are required for materials, with greatly different elastic moduli, such as tungsten carbide, or with greater strain-hardening capacity. Heyer[2] has shown that the indentation hardness of soft metals depends on the strain-hardening behavior of the material during the test, which in turn is dependent on the previous degree of strain hardening of the material before the test. As an extreme example of the care which is required in using conversion charts for soft metals, it is possible for Armco iron and cold-rolled aluminum each to have a Brinell hardness of 66; yet the former has a Rockwell B hardness of 31 compared with a hardness of $R_B$ 7 for the cold-worked aluminum. On the other hand, metals, such as yellow brass and low-carbon sheet steel have a well-behaved Brinell-Rockwell conversion[3] relationship for all degrees of strain hardening. Special hardness-conversion tables for cold-worked aluminum, copper, and 18-8 stainless steel are given in the ASM Metals Handbook.

## 11-10  HARDNESS AT ELEVATED TEMPERATURES

Interest in measuring the hardness of metals at elevated temperatures has been accelerated by the great effort which has gone into developing alloys with improved high-temperature strength. Hot hardness gives a good indication of the potential usefulness of an alloy for high-temperature strength applications. Some degree of success has been obtained in correlating hot hardness with high-temperature strength properties. This will be discussed in Chap. 13. Hot-hardness testers using a Vickers indenter made of sapphire and with provisions for testing in either vacuum or an inert atmosphere have been developed,[4] and a high-temperature microhardness test has been described.[5]

---

[1] This table may be found in ASTM Standard E-140-67, SAE Handbook, ASM Metals Handbook, and many other standard references.

[2] R. H. Heyer, *Am. Soc. Test. Mater. Proc.*, vol. 44, pp. 1027, 1944.

[3] The Wilson Mechanical Instrument Co. Chart 38 for metals softer than BHN 240 (see ASM Hand book, 1948 ed., p. 101) is based on tests on these metals.

[4] F. Garofalo, P. R. Malenock and G. V. Smith, *Trans. Am. Soc. Met.*, vol. 45, pp. 377–396, 1953; M. Semchyshen and C. S. Torgerson, *Trans. Am. Soc. Met.*, vol. 50, pp. 830–837, 1958.

[5] J. H. Westbrook, *Am. Soc. Test. Mater. Proc.*, vol. 57, pp. 873–897, 1957; *ASTM Bull.* 246, pp. 53–58, 1960.

   In an extensive review of hardness data at different temperatures Westbrook[1] showed that the temperature dependence of hardness could be expressed by

$$H = A \, e^{-BT} \qquad (11\text{-}11)$$

where $H$ = hardness, kg/mm$^2$

$T$ = test temperature, °K

$A, B$ = constants

Plots of log $H$ versus temperature for pure metals generally yield two straight lines of different slope. The change in slope occurs at a temperature which is about one-half the melting point of the metal being tested. Similar behavior is found in plots of the logarithm of the tensile strength against temperature. Figure 11-5 shows this behavior for copper. It is likely that this change in slope is due to a change in the deformation mechanism at higher temperature. The constant $A$ derived from the

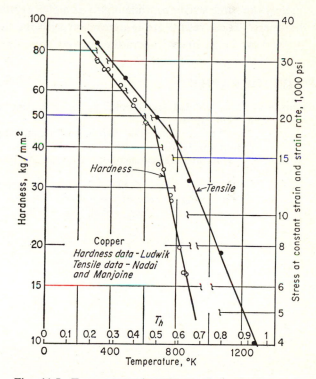

**Fig. 11-5** Temperature dependence of the hardness of copper. (*From J. H. Westbrook, Trans. Am. Soc. Met., vol. 45, p. 233, 1953.*)

---

[1] J. H. Westbrook, *Trans. Am. Soc. Met.*, vol. 45, pp. 221–248, 1953.

low-temperature branch of the curve can be considered to be the intrinsic hardness of the metal, that is, $H$ at $0°K$. This value would be expected to be a measure of the inherent strength of the binding forces of the lattice. Westbrook correlated values of $A$ for different metals with the heat content of the liquid metal at the melting point and with the melting point. This correlation was sensitive to crystal structure. The constant $B$, derived from the slope of the curve, is the temperature coefficient of hardness. This constant was related in a rather complex way to the rate of change of heat content with increasing temperature. With these correlations it is possible to calculate fairly well the hardness of a pure metal as a function of temperature up to about one-half its melting point.

Hardness measurements as a function of temperature will show an abrupt change at the temperature at which an allotropic transformation occurs. Hot-hardness tests on Co, Fe, Ti, U, and Zr have shown[1] that the body-centered cubic lattice is always the softer structure when it is involved in an allotropic transformation. The face-centered cubic and hexagonal close-packed lattices have approximately the same strength, while highly complex crystal structures give even higher hardness. These results are in agreement with the fact that austenitic iron-based alloys have better high-temperature strength than ferritic alloys.

## BIBLIOGRAPHY

Hardness Tests, "Metals Handbook," pp. 93–105, American Society for Metals, Metals Park, Ohio, 1948.

Mott, B. W.: "Micro-indentation Hardness Testing," Butterworth & Co. (Publishers), Ltd., London, 1956.

O'Neil, H.: "Hardness Measurement of Metals and Alloys," Chapman and Hall, London, 1967.

Tabor, D.: "The Hardness of Metals," Oxford University Press, New York, 1951.

Symposium on the Significance of the Hardness Test of Metals in Relation to Design, *Am. Soc. Test. Mater. Proc.*, vol. 43, pp. 803–856, 1943.

[1] W. Chubb, *Trans. AIME*, vol. 203, pp. 189–192, 1955.

# FATIGUE OF METALS

## 12-1 INTRODUCTION

It has been recognized since 1850 that a metal subjected to a repetitive or fluc-tuating stress will fail at a stress much lower than that required to cause fracture on a single application of load. Failures occurring under conditions of dynamic loading are called *fatigue failures*, presumably because it is generally observed that these failures occur only after a considerable period of service. There is no obvious change in the structure of a metal which has failed in fatigue which can serve as a clue to our understanding of the reasons for fatigue failure. Fatigue has become progressively more prevalent as technology has developed a greater amount of equipment, such as automobiles, aircraft, compressors, pumps, turbines, etc., sub-ject to repeated loading and vibration, until today it is often stated that fatigue accounts for at least 90 percent of all service failures due to mechanical causes.[1]

A fatigue failure is particularly insidious because it occurs without any ob-vious warning. Fatigue results in a brittle-appearing fracture, with no gross defor-

---

[1] Numerous examples of fatigue failures are given in Failure Analysis and Prevention, "Metals Hand-book," vol. 10, 8th ed., American Society for Metals, Metals Park, Ohio, 1975.

**Fig. 12-1** Fracture surface of fatigue failure which started at sharp corner of a keyway in a shaft (1×).

mation at the fracture. On a macroscopic scale the fracture surface is usually normal to the direction of the principal tensile stress. A fatigue failure can usually be recognized from the appearance of the fracture surface, which shows a smooth region, due to the rubbing action as the crack propagated through the section (top portion of Fig. 12-1), and a rough region, where the member has failed in a ductile manner when the cross section was no longer able to carry the load. Frequently the progress of the fracture is indicated by a series of rings, or "beach marks," progressing inward from the point of initiation of the failure. Figure 12-1 also illustrates another characteristic of fatigue, namely, that a failure usually occurs at a point of stress concentration such as a sharp corner or notch or at a metallurgical stress concentration like an inclusion.

Three basic factors are necessary to cause fatigue failure. These are (1) a maximum tensile stress of sufficiently high value, (2) a large enough variation or fluctuation in the applied stress, and (3) a sufficiently large number of cycles of the applied stress. In addition, there are a host of other variables, such as stress concentration, corrosion, temperature, overload, metallurgical structure, residual stresses, and combined stresses, which tend to alter the conditions for fatigue. Since we have not yet gained a basic understanding of what causes fatigue in metals, it will be necessary to discuss each of these factors from an essentially empirical standpoint. Because of the mass of data of this type, it will be possible to describe only the highlights of the relationship between these factors and fatigue. For more complete details the reader is referred to the number of excellent publications listed at the end of this chapter.

## 12-2 STRESS CYCLES

At the outset it will be advantageous to define briefly the general types of fluctuating stresses which can cause fatigue. Figure 12-2 serves to illustrate typical fatigue stress cycles. Figure 12-2a illustrates a *completely reversed cycle of stress* of sinusoidal form. This is an idealized situation which is produced by an R. R. Moore rotating-beam fatigue machine[1] and which is approached in service by a rotating

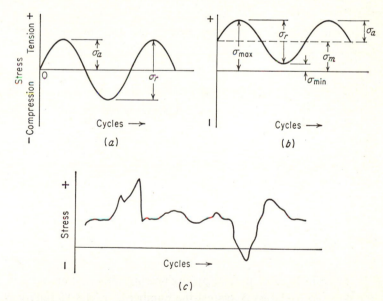

**Fig. 12-2** Typical fatigue stress cycles. (*a*) Reversed stress; (*b*) repeated stress; (*c*) irregular or random stress cycle.

[1] Common types of fatigue machines are described in the references listed at the end of this chapter and in the Manual on Fatigue Testing, *ASTM Spec. Tech. Publ.* 91, 1949.

shaft operating at constant speed without overloads. For this type of stress cycle the maximum and minimum stresses are equal. In keeping with the conventions established in Chap. 2, the minimum stress is the lowest algebraic stress in the cycle. Tensile stress is considered positive, and compressive stress is negative. Figure 12-2b illustrates a *repeated stress cycle* in which the maximum stress $\sigma_{max}$ and minimum stress $\sigma_{min}$ are not equal. In this illustration they are both tension, but a repeated stress cycle could just as well contain maximum and minimum stresses of opposite signs or both in compression. Figure 12-2c illustrates a complicated stress cycle which might be encountered in a part such as an aircraft wing which is subjected to periodic unpredictable overloads due to gusts.

A fluctuating stress cycle can be considered to be made up of two components, a *mean*, or steady, stress $\sigma_m$, and an *alternating*, or variable, stress $\sigma_a$. We must also consider the *range of stress $\sigma_r$*. As can be seen from Fig. 12-2b, the range of stress is the algebraic difference between the maximum and minimum stress in a cycle.

$$\sigma_r = \sigma_{max} - \sigma_{min} \qquad (12\text{-}1)$$

The alternating stress, then, is one-half the range of stress.

$$\sigma_a = \frac{\sigma_r}{2} \qquad (12\text{-}2)$$

The mean stress is the algebraic mean of the maximum and minimum stress in the cycle.

$$\sigma_m = \frac{\sigma_{max} + \sigma_{min}}{2} \qquad (12\text{-}3)$$

Two quantities are used in presenting fatigue data in terms of the stress ratio:

$$R = \frac{\sigma_{max}}{\sigma_{min}} \qquad (12\text{-}4)$$

$$A = \frac{\sigma_a}{\sigma_m} \qquad (12\text{-}5)$$

## 12-3 THE S-N CURVE

The basic method of presenting engineering fatigue data is by means of the *S-N* curve, a plot of stress *S* against the number of cycles to failure *N*. A log scale is almost always used for *N*. The value of stress that is plotted can be $\sigma_a$, $\sigma_{max}$, or $\sigma_{min}$. The stress values are usually nominal stresses, i.e., there is no adjustment for stress concentration. The *S-N* relationship is determined for a specified value of $\sigma_m$, *R*, or *A*. Most determinations of the fatigue properties of materials have

**Fig. 12-3**  Typical fatigue curves for ferrous and nonferrous metals.

been made in completed reversed bending, where the mean stress is zero. Figure 12-3 gives typical $S$-$N$ curves from rotating-beam tests. Cases where the mean stress is not zero are of considerable engineering importance and will be considered later in this chapter.

It will be noted that this $S$-$N$ curve is concerned chiefly with fatigue failure at high numbers of cycles ($N > 10^5$ cycles). Under these conditions the stress, on a gross scale, is elastic, but as we shall see shortly the metal deforms plastically in a highly localized way. At higher stresses the fatigue life is progressively decreased, but the gross plastic deformation makes interpretation difficult in terms of stress. For the low-cycle fatigue region ($N < 10^4$ cycles) tests are conducted with controlled cycles of elastic plus plastic strain instead of controlled load or stress cycles. *Low-cycle fatigue* will be considered in Sec. 12-5.

As can be seen from Fig. 12-3, the number of cycles of stress which a metal can endure before failure increases with decreasing stress. Unless otherwise indicated, $N$ is taken as the number of cycles of stress to cause complete fracture of the specimen. Fatigue tests at low stresses are usually carried out for $10^7$ cycles and sometimes to $5 \times 10^8$ cycles for nonferrous metals. For a few important engineering materials such as steel and titanium, the $S$-$N$ curve becomes horizontal at a certain limiting stress. Below this limiting stress, which is called the *fatigue limit*, or endurance limit, the material presumably can endure an infinite number of cycles without failure. Most nonferrous metals, like aluminum, magnesium, and copper alloys, have an $S$-$N$ curve which slopes gradually downward with increasing number of cycles. These materials do not have a true fatigue limit because the $S$-$N$ curve never becomes horizontal. In such cases it is common practice to characterize the fatigue properties of the material by giving the fatigue strength at an arbitrary number of cycles, for example, $10^8$ cycles.

The usual procedure for determining an *S-N* curve is to test the first specimen at a high stress where failure is expected in a fairly short number of cycles, e.g., at about two-thirds the static tensile strength of the material. The test stress is decreased for each succeeding specimen until one or two specimens do not fail in the specified number of cycles, which is usually at least $10^7$ cycles. The highest stress at which a runout (nonfailure) is obtained is taken as the fatigue limit. For materials without a fatigue limit the test is usually terminated for practical considerations at a low stress where the life is about $10^8$ or $5 \times 10^8$ cycles. The *S-N* curve is usually determined with about 8 to 12 specimens. It will generally be found that there is a considerable amount of scatter in the results, although a smooth curve can usually be drawn through the points without too much difficulty. However, if several specimens are tested at the same stress, there is a great amount of scatter in the observed values of number of cycles to failure, frequently as much as one log cycle between the minimum and maximum value. Further, it has been shown[1] that the fatigue limit of steel is subject to considerable variation and that a fatigue limit determined in the manner just described can be considerably in error. The statistical nature of fatigue will be discussed in the next section.

## 12-4   STATISTICAL NATURE OF FATIGUE

A considerable amount of interest has been shown in the statistical analysis of fatigue data and in the reasons for the variability in fatigue-test results.[2] Since fatigue life and fatigue limit are statistical quantities, it must be realized that considerable deviation from an average curve determined with only a few specimens is to be expected. It is necessary to think in terms of the probability of a specimen attaining a certain life at a given stress or the probability of failure at a given stress in the vicinity of the fatigue limit. To do this requires the testing of considerably more specimens than in the past so that the statistical parameters[3] for estimating these probabilities can be determined. The basic method for expressing fatigue data should then be a three-dimensional surface representing the relationship between stress, number of cycles to failure, and probability of failure. Figure 12-4 shows how this can be presented in a two-dimensional plot.

---

[1] J. T. Ransom and R. F. Mehl. *Trans., AIME*, vol. 185, pp. 364–365, 1949.

[2] P. H. Armitage, *Metall. Rev.*, vol. 6, pp. 353–385, 1964; R. E. Little and E. H. Jebe, "Statistical Design of Fatigue Experiments," John Wiley & Sons, Inc., New York, 1975.

[3] The chief statistical parameters to be considered are the estimate of the mean (average) and standard deviation (measure of scatter) of the population.

**Fig. 12-4** Representation of fatigue data on a probability basis.

A distribution of fatigue life at constant stress is illustrated schematically in this figure, and based on this, curves of constant probability of failure are drawn. Thus, at $\sigma_1$, 1 percent of the specimens would be expected to fail at $N_1$ cycles, 50 percent at $N_2$ cycles, etc. The figure indicates a decreasing scatter in fatigue life with increasing stress, which is usually found to be the case. The statistical distribution function which describes the distribution of fatigue life at constant stress is not accurately known, for this would require the testing of over 1,000 identical specimens under identical conditions at a constant stress. Muller-Stock[1] tested 200 steel specimens at a single stress and found that the frequency distribution of $N$ followed the gaussian, or normal, distribution if the fatigue life was expressed as log $N$. For engineering purposes it is sufficiently accurate to assume a logarithmic normal distribution of fatigue life at constant life in the region of the probability of failure of $P = 0.10$ to $P = 0.90$. However, it is frequently important to be able to predict the fatigue life corresponding to a probability of failure of 1 percent or less. At this extreme limit of the distribution the assumption of a lognormal distribution of life is no longer justified, although it is frequently used. Alternative approaches have been the use of the extreme-value distribution[2] or Weibull's distribution.[3]

[1] H. Muller-Stock, *Mitt. Kohle Eisenforsch. GmbH*, vol. 8, pp. 83–107, 1938.
[2] A. M. Freudenthal and E. J. Gumbel, *J. Am. Stat. Assoc.*, vol. 49, pp. 575–597, 1954.
[3] W. Weibull, *J. Appl. Mech.*, vol. 18, no. 3, pp. 293–297, 1951; W. Weibull, "Fatigue Testing and the Anaiysis of Results," Pergamon Press, New York, 1961.

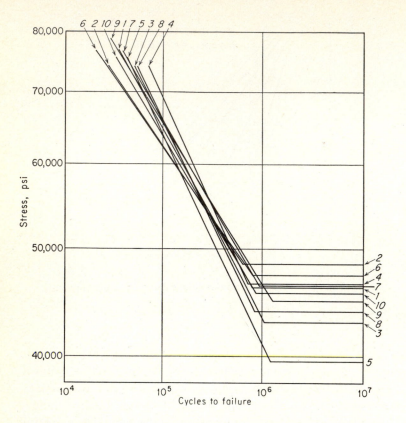

**Fig. 12-5** Summary of *S-N* curves, each based on 10 specimens, drawn from the same bar of steel. (*From J. T. Ransom, ASTM Spec. Tech. Publ. 121, p. 61, 1952.*)

For the statistical interpretation of the fatigue limit we are concerned with the distribution of stress at a constant fatigue life. The fatigue limit of steel was formerly considered to be a sharp threshold value, below which all specimens would presumably have infinite lives. However, it is now recognized that the fatigue limit is really a statistical quantity which requires special techniques for an accurate determination. For example, in a heat-treated alloy forging steel the stress range which would include the fatigue limits of 95 percent of the specimens could easily be from 40,000 to 52,000 psi. An example of the errors which can be introduced by ordinary testing with a few specimens is illustrated in Fig. 12-5. This figure summarizes[1] 10 *S-N* curves determined in the conventional manner for the *same* bar of alloy steel, each curve being based on 10 specimens. The

[1] J. T. Ransom, discussion in *ASTM Spec. Tech. Publ.* 121, pp. 59–63, 1952.

specimens were as identical as it was possible to make them, and there was no excessive scatter or uncertainty as to how to draw the *S-N* curves. Yet, as can be seen from the figure, there is considerable difference in the measured values of the fatigue limit for the steel due to the fact that the curves were based on insufficient data.

In determining the fatigue limit of a material, it should be recognized that each specimen has its own fatigue limit, a stress above which it will fail but below which it will not fail, and that this critical stress varies from specimen to specimen for very obscure reasons. It is known that inclusions in steel have an important effect on the fatigue limit and its variability, but even vacuum-melted steel shows appreciable scatter in fatigue limit. The statistical problem of accurately determining the fatigue limit is complicated by the fact that we cannot measure the individual value of the fatigue limit for any given specimen. We can only test a specimen at a particular stress, and if the specimen fails, then the stress was somewhere above the fatigue limit of the specimen. Since the specimen cannot be retested, even if it did not fail at the test stress, we have to estimate the statistics of the fatigue limit by testing groups of specimens at several stresses to see how many fail at each stress. Thus, near the fatigue limit fatigue is a "go–no-go" proposition, and all that we can do is to estimate the behavior of a universe of specimens by means of a suitable sample. The two statistical methods which are used for making a statistical estimate of the fatigue limit are called *probit analysis* and the *staircase method*. The procedures for applying these methods of analysis to the determination of the fatigue limit have been well established.[1]

## 12-5 LOW-CYCLE FATIGUE

Although historically fatigue studies have been concerned with conditions of service in which failure occurred at more than $10^4$ cycles of stress, there is growing recognition of engineering failures which occur at relatively high stress and low numbers of cycles to failure.[2] This type of fatigue failure must be considered in the design of nuclear pressure vessels, steam turbines, and most other types of power machinery. Low-cycle fatigue conditions frequently are created where the repeated stresses are of thermal origin.[3] Since thermal stresses arise from the thermal expansion of the material, it is easy to see that in this case fatigue results from cyclic strain rather than from cyclic stress.

---

[1] "A Guide for Fatigue Testing and the Statistical Analysis of Fatigue Data," *ASTM Spec. Tech. Publ.*, 91-A, 2d ed., 1963.

[2] L. F. Coffin, Jr., *Met. Eng. Q.*, vol. 3, pp. 15–24, 1963.

[3] S. S. Manson, "Thermal Stress and Low-Cycle Fatigue," McGraw-Hill Book Company, New York, 1966.

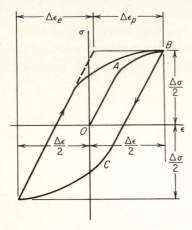

**Fig. 12-6** Stress-strain loop for constant strain cycling.

Figure 12-6 illustrates a stress-strain loop under controlled constant strain cycling in a low-cycle fatigue test.[1] During initial loading the stress-strain curve is $O$-$A$-$B$. On unloading yielding begins in compression at a lower stress $C$ due to the Bauschinger effect. In reloading in tension a hysteresis loop develops. The dimensions of this loop are described by its width $\Delta\varepsilon$, the total strain range, and its height $\Delta\sigma$, the stress range. The total strain range $\Delta\varepsilon$ consists of an elastic strain component $\Delta\varepsilon_e = \Delta\sigma/E$ plus a plastic strain component $\Delta\varepsilon_p$. The width of the hysteresis loop will depend on the level of cyclic strain. When the level of cyclic strain is small, the hysteresis loop becomes very narrow. For tests conducted under constant $\Delta\varepsilon$ the stress range $\Delta\sigma$ will usually change with increasing number of cycles. Annealed materials undergo cyclic strain hardening so that $\Delta\sigma$ increases with cycles and then levels off after about 100 strain cycles. The increase in stress range is greater the larger the value of $\Delta\varepsilon$. Materials which are initially cold-worked undergo cyclic strain softening, so that $\Delta\sigma$ decreases with increasing number of strain cycles.

The usual way of presenting low-cycle fatigue test results is to plot the plastic strain range $\Delta\varepsilon_p$ against $N$. Figure 12-7 shows that a straight line is obtained on log-log coordinates. The slope of this line shows little variation between materials and has an average value of about $-0.5$. This relationship, often called the *Coffin-Manson law*, has the form

$$\Delta\varepsilon_p N^b = C \qquad (12\text{-}6)$$

[1] "Manual on Low-Cycle Fatigue Testing," *ASTM Spec. Tech. Publ.*, 465, 1969.

**Fig. 12-7**  Low-cycle fatigue curve ($\Delta\varepsilon_p$ versus $N$) for Type 347 stainless steel. (*From L. F. Coffin, Jr., Met. Eng. Q., vol. 3, p. 22, 1963; Copyright American Society for Metals, 1963.*)

The constant $C$ can be evaluated by considering that the upper limit of the low-cycle fatigue test is failure in the tension test, where $\Delta\varepsilon_p \approx \varepsilon_f = \ln(1/1 - q)$ and $N = \frac{1}{4}$ cycle. Taking $b = \frac{1}{2}$, we find that $C = \varepsilon_f \frac{1}{4}^{1/2} = \varepsilon_f/2$, and

$$\Delta\varepsilon_p N^b = \frac{\varepsilon_f}{2} = \frac{1}{2}\ln\frac{1}{1-q} \qquad (12\text{-}7)$$

The elastic stress range is related to number of cycles to failure by the relation

$$\Delta\varepsilon_e = \frac{s_u}{E}N^c \qquad (12\text{-}8)$$

where $c \approx -0.08$ and $s_u$ is the ultimate tensile strength.

Since the total strain range is $\Delta\varepsilon = \Delta\varepsilon_e + \Delta\varepsilon_p$,

$$\Delta\varepsilon = \frac{s_u}{E}N^c + \frac{\varepsilon_f}{2}N^{-b} \qquad (12\text{-}9)$$

By multiplying through by $E$, we obtain a nominal alternating stress $s_a$ that is useful in comparing with stresses calculated from elastic stress analysis

$$s_a = s_u N^c + \frac{E\varepsilon_f}{2}N^{-b} \qquad (12\text{-}10)$$

An interesting aspect of Eqs. (12-9) and (12-10) is that they depend on only two structure-sensitive material properties: the tensile strength and the reduction of area.

## 12-6  STRUCTURAL FEATURES OF FATIGUE

Studies of the basic structural changes[1] that occur when a metal is subjected to cyclic stress have found it convenient to divide the fatigue process into the following stages:

1. *Crack initiation*—includes the early development of fatigue damage which can be removed by a suitable thermal anneal.
2. *Slip-band crack growth*—involves the deepening of the initial crack on planes of high shear stress. This frequently is called *stage I crack growth*.
3. *Crack growth on planes of high tensile stress*—involves growth of well-defined crack in direction normal to maximum tensile stress. Usually called *stage II crack growth*.
4. *Ultimate ductile failure*—occurs when the crack reaches sufficient length so that the remaining cross section cannot support the applied load.

The relative proportion of the total cycles to failure that are involved with each stage depends on the test conditions and the material. However, it is well established that a fatigue crack can be formed before 10 percent of the total life of the specimen has elapsed. There is, of course, considerable ambiguity in deciding when a deepened slip band should be called a crack. In general, larger proportions of the total cycles to failure are involved with the propagation of stage II cracks in low-cycle fatigue than in long-life fatigue, while stage I crack growth comprises the largest segment for low-stress, high-cycle fatigue. If the tensile stress is high, as in the fatigue of sharply notched specimens, stage I crack growth may not be observed at all.

An overpowering structural consideration in fatigue is the fact that fatigue cracks usually are initiated at a free surface. In those rare instances where fatigue cracks initiate in the interior there is always an interface involved, such as the interface of a carburized surface layer and the base metal. Fatigue has certain things in common with plastic flow and fracture under static or unidirectional deformation. The work of Gough[2] has shown that a metal deforms under cyclic strain by slip on the same atomic planes and in the same crystallographic directions as in unidirectional strain. Whereas with unidirectional deformation slip is usually widespread throughout all the grains, in fatigue some grains will show slip lines while other grains will give no evidence of slip. Slip lines are generally formed during the first few thousand cycles of stress. Successive cycles produce additional slip bands, but the number of slip bands is not directly proportional to the number of cycles of stress. In many metals the increase in visible slip soon reaches a saturation value, which is observed as distorted regions of heavy slip. Cracks are

[1] W. J. Plumbridge and D. A. Ryder, *Metall. Rev.*, vol. 14, no. 136, 1969.
[2] H. J. Gough, *Am. Soc. Test. Mater. Proc.*, vol. 33, pt, 2, pp. 3–114, 1933

usually found to occur in the regions of heavy deformation parallel to what was originally a slip band. Slip bands have been observed at stresses below the fatigue limit of ferrous materials. Therefore, the occurrence of slip during fatigue does not in itself mean that a crack will form.

A study of crack formation in fatigue can be facilitated by interrupting the fatigue test to remove the deformed surface by electropolishing. There will generally be several slip bands which are more persistent than the rest and which will remain visible when the other slip lines have been polished away. Such slip bands have been observed after only 5 percent of the total life of the specimen.[1] These persistent slip bands are embryonic fatigue cracks, since they open into wide cracks on the application of small tensile strains. Once formed, fatigue cracks tend to propagate initially along slip planes, although they later take a direction normal to the maximum applied tensile stress. Fatigue-crack propagation is ordinarily transgranular.

An important structural feature which appears to be unique to fatigue deformation is the formation on the surface of ridges and grooves called *slip-band extrusions* and *slip-band intrusions*.[2] Extremely careful metallography on tapered sections through the surface of the specimen has shown that fatigue cracks initiate at intrusions and extrusions.[3] Considerable experimental evidence indicates that cross slip is important to the slip-band extrusion process. For example, fatigue failure is difficult to produce in certain ionic crystals which do not easily cross slip, and it is very difficult to produce fatigue failure in zinc crystals which are oriented to deform only in easy glide. On the other hand, pure aluminum, in which cross slip is extremely easy, does not produce slip-band extrusions (in contrast to most aluminum alloys).

A mechanism for the formation of extrusions and intrusions has been suggested by Cottrell and Hull.[4] Figure 12-8 shows that it depends on the existence of slip on two slip systems. During the tensile part of the stress cycle both slip systems operate sequentially to produce two surface steps (Fig. 12-8b and c). When the stress becomes compressive, the intrusion forms by slip on the first system to operate (Fig. 12-8d) and an extrusion forms when the other slip system operates (Fig. 12-8e).

W. A. Wood,[5] who has made many basic contributions to the understanding of the mechanism of fatigue, has suggested a mechanism for producing slip-band extrusions and intrusions. He interprets microscopic observations of slip produced

[1] G. C. Smith, *Proc. R. Soc. London*, vol. 242A, pp. 189–196, 1957.
[2] P. J. E. Forsyth and C. A. Stubbington, *J. Inst. Met.*, vol. 83, p. 395, 1955–1956.
[3] W. A. Wood, Some Basic Studies of Fatigue in Metals, in "Fracture," John Wiley & Sons, Inc., New York, 1959.
[4] A. H. Cottrell and D. Hull, *Proc. R. Soc. London*, vol. 242A, pp. 211–217, 1957.
[5] W. A. Wood, *Bull. Inst. Met.*, vol. 3, pp. 5–6, September, 1955.

**Fig. 12-8** Mechanism for the formation of slip-band extrusions and intrusions. (*From A. H. Cottrell and D. Hull, Proc. R. Soc. London, vol. 242A, pp. 211–213, 1957.*)

by fatigue as indicating that the slip bands are the result of a systematic buildup of fine slip movements, corresponding to movements of the order of $10^{-7}$ cm rather than steps of $10^{-5}$ to $10^{-4}$ cm, which are observed for static slip bands. Such a mechanism is believed to allow for the accommodation of the large total strain (summation of the microstrain in each cycle) without causing appreciable strain hardening. Figure 12-9 illustrates Wood's concept of how continued deformation by fine slip might lead to a fatigue crack. The figures illustrate schematically the fine structure of a slip band at magnifications obtainable with the electron microscope. Slip produced by static deformation would produce a contour at the metal surface similar to that shown in Fig. 12-9a. In contrast, the back-and-forth fine slip movements of fatigue could build up notches (Fig. 12-9b) or ridges (Fig. 12-9c) at the surface. The notch would be a stress raiser with a notch root of atomic dimensions. Such a situation might well be the start of a fatigue crack. This mechanism for the initiation of a fatigue crack is in agreement with the facts that fatigue cracks start at surfaces and that cracks have been found to initiate at slip-band intrusions and extrusions.

(a)    (b)    (c)

**Fig. 12-9** W. A. Wood's concept of microdeformation leading to formation of fatigue crack. (a) Static deformation; (b) fatigue deformation leading to surface notch (intrusion); (c) fatigue deformation leading to slip-band extrusion.

There is a broad similarity between the dislocation structure produced by fatigue and static deformation. The formation of a dislocation cell structure is favored by a large strain amplitude and a high stacking-fault energy, both of which are factors favoring cross slip. At low strain amplitudes dislocation loops and dipoles are prevalent. There is strong evidence that cell formation is related to the development of persistent slip bands, but the exact details are not well established.[1]

The stage I crack propagates initially along the persistent slip bands. In a polycrystalline metal the crack may extend for only a few grain diameters before the crack propagation changes to stage II. The rate of crack propagation in stage I is generally very low, on the order of angstroms per cycle, compared with crack propagation rates of microns per cycle for stage II. The fracture surface of stage I fractures is practically featureless.

By marked contrast the fracture surface of stage II crack propagation frequently shows a pattern of ripples or fatigue fracture striations (Fig. 12-10). Each striation represents the successive position of an advancing crack front that is normal to the greatest tensile stress. Each striation was produced by a single cycle of stress.

**Fig. 12-10**   Fatigue striations in cold-worked copper. (*Courtesy C. Laird.*)

---

[1] For extensive studies of fatigue mechanisms see *Acta Metall.*, vol. 11, no. 7, July 1963.

The presence of these striations unambiguously defines that failure was produced by fatigue, but their absence does not preclude the possibility of fatigue fracture. Failure to observe striations on a fatigue surface may be due to a very small spacing that cannot be resolved with the observational method used, insufficient ductility at the crack tip to produce a ripple by plastic deformation that is large enough to be observed, or obliteration of the striations by some sort of damage to the surface. Since stage II cracking does not occur for the entire fatigue life, it does not follow that counting striations will give the complete history of cycles to failure.

Stage II crack propagation occurs by a plastic blunting process[1] that is illustrated in Fig. 12-11. At the start of the loading cycle the crack tip is sharp (Fig. 12-11a). As the tensile load is applied the small double notch at the crack tip concentrates the slip along planes at 45° to the plane of the crack (Fig. 12-11b). As the crack widens to its maximum extension (Fig. 12-11c) it grows longer by plastic shearing and at the same time its tip becomes blunter. When the load is changed to compression the slip direction in the end zones is reversed (Fig. 12-11d). The crack faces are crushed together and the new crack surface created in tension is forced into the plane of the crack (Fig. 12-11e) where it partly folds by buckling to form a resharpened crack tip. The resharpened crack is then ready to advance and be blunted in the next stress cycle.

There are a number of indications that cyclic deformation results in a higher concentration of vacancies than cold-working by unidirectional deformation. The

**Fig. 12-11** Plastic blunting process for growth of stage II fatigue crack. (*From C. Laird, ASTM Spec. Tech. Publ. 415, 1967, p. 136.*)

---

[1] C. Laird, in "Fatigue Crack Propagation," ASTM *Spec. Tech. Publ.* 415, pp. 131–168, 1967.

difference in the release of stored energy between fatigued and cold-worked copper is in line with what would be expected from a large concentration of point defects. The fact that initially cold-worked copper becomes softer as a result of fatigue[1] can be explained by the generation of point defects which allow the metal partly to recover by permitting dislocations to climb out of the slip plane. Age-hardening aluminum alloys in the precipitation-hardened condition can be overaged by fatigue deformation at room temperature. This suggests that vacancies produced by fatigue are available to accomplish the diffusion required for the overaging process.[2] Moreover, the fatigue strength increases markedly on going from 20 to $-190°C$, where vacancy movement is negligible. However, the fact that fatigue fracture can be produced at $4°K$ indicates that a temperature-activated process such as the diffusion of vacancies is not essential for fatigue failure.[3]

## 12-7  FATIGUE CRACK PROPAGATION

Considerable research has gone into determining the laws of fatigue crack propagation for stage II growth.[4] Reliable crack propagation relations will permit the implementation of a fail-safe design philosophy which recognizes the inevitability of cracks in engineering structures and aims at determining the safe load and crack length which will preclude failure in a conservatively estimated service life. The crack propagation rate $da/dN$ is found to follow an equation

$$\frac{da}{dN} = C\sigma_a^m a^n \qquad (12\text{-}11)$$

where $C$ = a constant
$\sigma_a$ = the alternating stress
$a$ = the crack length

In different investigations $m$ ranges from 2 to 4 and $n$ varies from 1 to 2. Crack propagation can also be expressed in terms of total strain[5] by a single power-law expression which extends from elastic to plastic strain region.

$$\frac{da}{dN} = C_1 \varepsilon^{m_1} \qquad (12\text{-}12)$$

---

[1] N. H. Polakowski and A. Palchoudhuri, *Am. Soc. Test. Mater. Proc.*, vol. 54, p. 701, 1954.
[2] T. Broom, J. H. Molineux, and V. N. Whittaker, *J. Inst. Met.*, vol. 84, pp. 357–363, 1955–1956.
[3] R. D. McCammon and H. M. Rosenberg, *Proc. R. Soc. London*, vol. 242A, p. 203, 1957.
[4] "Fatigue Crack Propagation," ASTM *Spec. Tech. Publ.* 415, 1967; D. Walton and E. G. Ellison, *Int. Metall. Rev.*, vol. 17, pp. 100–116, 1972.
[5] T. W. Crooker and E. A. Lange, *op. cit.* p. 94.

Fatigue crack propagation can be expressed in terms of the stress intensity factor $K$ from fracture mechanics (Sec. 7-5) according to

$$\frac{da}{dN} = A\Delta K^{\rho} \qquad (12\text{-}13)$$

where $\rho$ can vary from 1 to 6 depending on the material and stress level and $A$ is a constant. $K = \sigma\sqrt{\pi a}$ and $\Delta K$ is the range in stress intensity calculated from the $\sigma_{max}$ of the fatigue cycle minus $K$ calculated from $\sigma_{min}$. Figure 12-12 shows a schematic representation[1] of the relationship between fatigue crack growth and

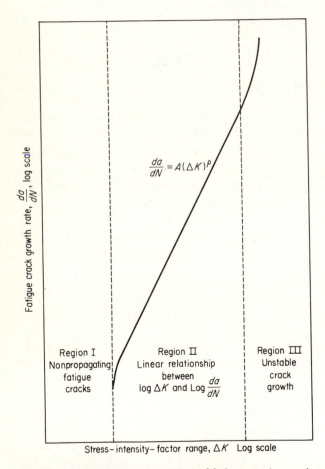

**Fig. 12-12** Schematic representation of fatigue–crack growth behavior in a nonaggressive environment. (*From W. G. Clark Jr., Met. Eng. Q., vol. 14, p. 17, August, 1974; Copyright American Society for Metals, 1974.*)

[1] W. G. Clark Jr., *Met. Eng. Q.*, vol. 14, pp. 16–22, 1974.

$\Delta K$ for a structural alloy tested in a nonaggressive environment such as air. At low stress levels, $\Delta K < 10$ ksi$\sqrt{\text{in}}$, there is a region where preexisting cracks do not grow with cyclic loading. The relationship given in Eq. (12-13) holds in region II. Region III is unstable crack growth just before the specimen fails. For a given class of alloys (steels, aluminum alloys, etc.) the relationship in region II between $da/dN$ and $\Delta K$ is essentially the same, independent of strength level and micro-structure. For steel, a conservative evaluation of the data gives $da/dN = 6.6 \times 10^{-9}$ $(\Delta K)^{2 \cdot 5}$. However, fatigue tests carried out in aggressive environments (saltwater, hydrogen, etc.) result in large changes in the crack-propagation relationship.[1]

## 12-8 DESIGN FOR FATIGUE

A considerable literature has developed regarding methods and procedures for designing against fatigue failure. The remaining sections of this chapter consider the various elements which go into a fatigue design. However, space limitations prevent inclusion of sufficient design examples. In lieu of this, we have listed below appropriate references which contain extensive examples.

Ruiz, C., and F. Koenigsberger: "Design for Strength and Production," Gordon and Breach Science Publishers, Inc., New York, 1970. Pages 106–120 give a concise discussion of the general fatigue design procedure.

Juvinall, R. C.: "Engineering Considerations of Stress, Strain, and Strength," McGraw-Hill Book Company, New York, 1967. Chapters 11 to 16 cover in considerable detail the machine design aspects of fatigue design.

Graham, J. A. (ed.): "Fatigue Design Handbook," Society of Automotive Engineers, New York, 1968.

Heywood, R. B.: "Designing Against Fatigue of Metals," Reinhold Publishing Corporation, New York, 1962. Heavily oriented toward stress-concentration calculations.

Osgood, C. C.: "Fatigue Design," John Wiley & Sons, Inc., New York, 1970. An encyclopedic collection of data and design examples.

Fatigue data on various materials are given in the references listed above. In addition, the following are useful sources of fatigue data.

MIL-HDBK-5A, "Metallic Materials and Elements for Flight Vehicle Structures," Department of Defense, December 1968.

Grover, H. J., S. A. Gordon, and L. R. Jackson: "The Fatigue of Metals and Structures," U.S. Government Printing Office, 1960 (revised).

Properties and Selection of Metals, "Metals Handbook," vol. 1, 8th ed., American Society for Metals, Metals Park, Ohio, 1961.

Horger, O. J. (ed.): "ASME Handbook: Metals Engineering-Design, 2d ed. McGraw-Hill Book Company, New York, 1965.

[1] C. J. McMahon, Jr., *Trans. ASME, Ser. H: J. Eng. Mater. Technol.* vol. 95, pp. 133–149 1973.

## 12-9 EFFECT OF STRESS CONCENTRATION ON FATIGUE

Fatigue strength is seriously reduced by the introduction of a stress raiser such as a notch or hole. Since actual machine elements invariably contain stress raisers like fillets, keyways, screw threads, press fits, and holes, it is not surprising to find that fatigue cracks in structural parts usually start at such geometrical irregularities. One of the best ways of minimizing fatigue failure is by the reduction of avoidable stress raisers through careful design and the prevention of accidental stress raisers by careful machining and fabrication. While this section is concerned with stress concentration resulting from geometrical discontinuities, stress concentration can also arise from surface roughness and metallurgical stress raisers such as porosity, inclusions, local overheating in grinding, and decarburization.

The effect of stress raisers on fatigue is generally studied by testing specimens containing a notch, usually a V notch or a circular notch. It has been shown in Chap. 7 that the presence of a notch in a specimen under uniaxial load introduces three effects: (1) there is an increase or concentration of stress at the root of the notch; (2) a stress gradient is set up from the root of the notch in toward the center of the specimen; (3) a triaxial state of stress is produced.

The ratio of the maximum stress to the nominal stress is the *theoretical stress-concentration factor* $K_t$. As was discussed in Sec. 2-16, values of $K_t$ can be computed from the theory of elasticity for simple geometries and can be determined from photoelastic measurements for more complicated situations. Most of the available data on stress-concentration factors have been collected by Peterson.[1]

The effect of notches on fatigue strength is determined by comparing the S-N curves of notched and unnotched specimens. The data for notched specimens are usually plotted in terms of nominal stress based on the net section of the specimen. The effectiveness of the notch in decreasing the fatigue limit is expressed by the *fatigue-strength reduction factor*, or *fatigue-notch factor*, $K_f$. This factor is simply the ratio of the fatigue limit of unnotched specimens to the fatigue limit of notched specimens. For materials which do not exhibit a fatigue limit the fatigue-notch factor is based on the fatigue strength at a specified number of cycles. Values of $K_f$ have been found to vary with (1) severity of the notch, (2) the type of notch, (3) the material, (4) the type of loading, and (5) the stress level. The values of $K_f$ published in the literature are subject to considerable scatter and should be carefully examined for their limitations and restrictions. However, two general trends are usually observed for test conditions of completely reversed loading. First, $K_f$ is usually less than $K_t$, and, second, the ratio of $K_f/K_t$ decreases as $K_t$ increases. Thus, very sharp notches (high $K_t$) have less effect on fatigue strength than would be expected from their high value of $K_t$.

---

[1] R. E. Peterson, "Stress-Concentration Design Factors," John Wiley & Sons, Inc., New York, 1974.

The notch sensitivity of a material in fatigue is expressed by a notch-sensitivity factor $q$

$$q = \frac{K_f - 1}{K_t - 1} \qquad (12\text{-}14)$$

Equation (12-14) was chosen so that a material which experiences no reduction in fatigue due to a notch ($K_f = 1$) has a factor of $q = 0$, while a material in which the notch has its full theoretical effect ($K_f = K_t$) has a factor of $q = 1$. However, $q$ is not a true material constant since it varies with the severity and type of notch (Fig. 12-13), the size of specimen, and the type of loading. As Fig. 12-13 indicates notch sensitivity increases with tensile strength. Thus, it is possible in certain circumstances to decrease fatigue performance by increasing the hardness or tensile strength of a material.

Another approach to notch sensitivity in fatigue was advanced by Neuber.[1] He proposed that the "technical stress-concentration factor," commonly called the *Neuber factor $K_N$* was given by

$$K_f = K_N = 1 + \frac{K_t - 1}{1 + (\pi/\pi - \omega)\sqrt{a/r}} \qquad (12\text{-}15)$$

where  $r$ = notch root radius

$\omega$ = included notch flank angle, rad

$a$ = one-half the width of the *elementary block size*

**Fig. 12-13** Variation of notch-sensitivity index with notch radius for materials of different tensile strength. (*From R. E. Peterson, in G. Sines and J. L. Waisman (eds.), "Metal Fatigue," p. 301, McGraw-Hill Book Company, New York, 1959. By permission of the publishers.*)

[1] H. Neuber, "Theory of Notch Stresses," J. W. Edwards, Publisher, Incorporated, Ann Arbor, Mich. 1946.

**Table 12-1 SOME AVERAGE VALUES OF NEUBER'S CONSTANT $a$**

| Material | $\sqrt{a}$, in$^{1/2}$ |
|---|---|
| Low-alloy steel | |
| $\quad s_u = 80$ ksi | 0.08 |
| $\quad s_u = 160$ ksi | 0.03 |
| $\quad s_u = 200$ ksi | 0.01 |
| Aluminum-alloy sheet, 2024-T4 | 0.22 |

Neuber introduced the concept of the elementary block size to circumvent the fact that the ordinary elasticity analysis breaks down at the tip of a very long, sharp notch. He considered that the material was made up of small elementary blocks each $2a$ on a side and that the stresses could be considered uniform on the sides of each block. Actually, there is no physical significance to Neuber's elementary block and it should be considered an empirical but useful concept. Peterson[1] simplified Eq. (12-15) by assuming a small notch flank angle and by substituting into Eq. (12-14) derived

$$q = \frac{1}{1 + a/r} \qquad (12\text{-}16)$$

Some typical values of $a$ are given in Table 12-1.

We have already seen several examples that demonstrate that a biaxial stress state will exist at the surface of a notch when a notched specimen is subjected to bending or axial loads. Peterson[2] applied the von Mises' yield criterion to develop an expression for the theoretical stress-concentration factor $K_{t'}$, for biaxial stress conditions

$$K_{t'} = K_t(1 - C + C^2)^{1/2} \qquad (12\text{-}17)$$

where $C = \sigma_3/\sigma_1$ and $\sigma_2 = 0$. Since $\sigma_3$ is usually tensile at a notch, $K_{t'}$ is usually less than $K_t$, and this correction can be neglected in conservative design.

The equations presented above apply to low-stress fatigue conditions in which fracture occurs at greater than $10^6$ cycles under essentially elastic conditions. In high-stress, low-cycle fatigue where stresses exceed the yield stress, $K_f$ is appreciable lower than $K_t$. Neuber's approximation[3] is that

$$K_\varepsilon K_\sigma = K_t^2 \qquad (12\text{-}18)$$

[1] R. E. Peterson, "Fatigue of Metals in Engineering Design," (Edgar Marburg Lecture at American Society for Testing and Materials, Philadelphia, 1962, reprinted in *Mat. Res. Stand.*, vol. 3, January and February 1963.)

[2] R. E. Peterson, "Fatigue of Metals in Engineering Design" (Edgar Marburg Lecture at American Society for Testing and Materials, Philadelphia, 1962, reprinted in *Mater. Res. Stand.*, vol. 3, January and February, 1963.)

[3] H. Neuber, *Trans. ASME, Ser. E: J. Appl. Mech.*, vol. 28, pp. 544–550, 1961.

where $K_\varepsilon$ = plastic-strain-concentration factor = $\varepsilon_{max}/\varepsilon_{nom}$
$K_\sigma$ = plastic-stress-concentration factor = $\sigma_{max}/\sigma_{nom} \approx 1 + (K_t - 1)(E_s/E)$
$E_s$ = the secant modulus

When plastic flow starts at the root of a notch, the maximum local strain will increase rapidly and $K_\varepsilon$ increases. On the other hand, if the material is assumed to be elastic-ideal plastic, the maximum stress at the notch cannot exceed $\sigma_0$. As the average stress increases $K_\sigma$ will decrease and in the limit $K_\sigma = 1$ when the plastic flow initiated at the notch root spreads over the entire specimen so that $\sigma_{avg} = \sigma_{max} = \sigma_0$. At this limit, from Eq. (12-18), $K_\varepsilon = K_t^2$. Equation (12-18) can be rewritten[1]

$$K_t \approx K_f = \left(\frac{\varepsilon_{max}}{\varepsilon_{nom}} \frac{\sigma_{max}}{\sigma_{nom}}\right)^{\frac{1}{2}}$$

or

$$K_f(\sigma_{nom}\,\varepsilon_{nom}\,E)^{\frac{1}{2}} = (\sigma_{max}\,\varepsilon_{max}\,E)^{\frac{1}{2}} \qquad (12\text{-}19)$$

This shows that a function of nominal stress and strain $(\sigma_{nom}\,\varepsilon_{nom}\,E)^{\frac{1}{2}}$ need only be multiplied by a constant concentration factor to obtain the values of the true stress and strain at the notch root. Figure 12-14 shows the degree of correlation with Eq. (12-19). The smooth curve is the result of unnotched fatigue tests, while the data points are notched fatigue specimens. Another way of interpreting this correlation is that a notched specimen and a smooth specimen will form detectable cracks at the same life provided $K_f(\sigma_{nom}\,\varepsilon_{nom}\,E)^{\frac{1}{2}}$ for the notched specimen is equal to $(\sigma_{max}\,\varepsilon_{max}\,E)^{\frac{1}{2}}$ for the smooth specimen. Thus, unnotched fatigue results for

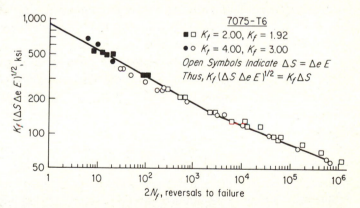

Fig. 12-14 Correlation between smooth and notched fatigue data according to Eq. (12-19). (*From T. H. Topper, R. M. Wetzel, and J. Morrow, J. Mater., vol. 4, p. 204, 1969.*)

[1] T. H. Topper, R. M. Wetzel, and J. Morrow, J. Mater., vol. 4, pp. 200–209, 1969.

completely reversed stress cycles can be used to estimate the fatigue life of notched members. Additional simplification results when the nominal stress and strain are both elastic. Since $\sigma_{avg} = \varepsilon_{avg} E$, Eq. (12-19) reduces to

$$K_f \sigma_{nom} = (\sigma_{max} \varepsilon_{max} E)^{\frac{1}{2}} \qquad (12\text{-}20)$$

## 12-10   SIZE EFFECT

An important practical problem is the prediction of the fatigue performance of large machine members from the results of laboratory tests on small specimens. Experience has shown that in most cases a *size effect* exists; i.e., the fatigue strength of large members is lower than that of small specimens. A precise study of this effect is difficult for several reasons. It is extremely difficult, if not altogether impossible, to prepare geometrically similar specimens of increasing diameter which have the same metallurgical structure and residual stress distribution throughout the cross section. The problems in fatigue testing large-sized specimens are considerable, and there are few fatigue machines which can accommodate specimens having a wide range of cross sections.

Changing the size of a fatigue specimen usually results in a variation in two factors. First, increasing the diameter increases the volume or surface area of the specimen. The change in amount of surface is of significance, since fatigue failures usually start at the surface. Second, for plain or notched specimens loaded in bending or torsion, an increase in diameter usually decreases the stress gradient across the diameter and increases the volume of material which is highly stressed.

Experimental data on the size effect in fatigue are contradictory and not very complete. For tests in reversed bending and torsion, some investigators have found no change in fatigue limit with specimen diameter, while more commonly it is observed that the fatigue limit decreases with increasing diameter. For mild steel the decrease in bending fatigue limit for diameters ranging from 0.1 to 2 in does not exceed about 10 percent. Horger's data[1] for steel shafts tested in reversed bending (Table 12-2) show that the fatigue limit can be appreciably reduced in large section sizes.

No size effect has been found[2] for smooth plain-carbon-steel fatigue specimens with diameters ranging from 0.2 to 1.4 in when tested in axial tension-compression loading. However, when a notch is introduced into the specimen, so that a stress gradient is produced, a definite size effect is observed. These important experiments support the idea that a size effect in fatigue is due to the existence of a stress gradient. The fact that large specimens with shallow stress gradients have lower fatigue limits

[1] O. J. Horger, Fatigue Characteristics of Large Sections, in "Fatigue," American Society for Metals, Metals Park, Ohio, 1953.
[2] C. E. Phillips and R. B. Heywood, *Proc. Inst. Mech. Eng. (London)*, vol. 165, pp. 113–124, 1951.

**Table 12-2    FATIGUE LIMIT OF NORMALIZED PLAIN CARBON STEEL IN REVERSED BENDING**

| Specimen diam, in | Fatigue limit, psi |
|---|---|
| 0.30 | 36,000 |
| 1.50 | 29,000 |
| 6.00 | 21,000 |

is consistent with the idea that a critical value of stress must be exceeded over a certain finite depth of material for failure to occur. This appears to be a more realistic criterion of size effect than simply the ratio of the change in surface area to the change in specimen diameter. The importance of stress gradients in size effect helps explain why correlation between laboratory results and service failure is often rather poor. Actual failures in large parts are usually directly attributable to stress concentrations, either intentional or accidental, and it is usually impossible to duplicate the same stress concentration and stress gradient in a small-sized laboratory specimen.

Analysis of considerable data for steels has shown[1] a size-effect relationship between fatigue limit and the critically stressed volume of the material.

$$\sigma_{fl} = \sigma_{f0} \left( \frac{V}{V_0} \right)^{-0.034} \tag{12-21}$$

where $\sigma_{fl}$ is the fatigue limit for a critical volume $V$ and $\sigma_{f0}$ is the known fatigue limit for a specimen with volume $V_0$. Critically stressed volume is defined as the volume near the surface of the specimen that is stressed to at least 95 percent of $\sigma_{max}$.

## 12-11    SURFACE EFFECTS AND FATIGUE

Practically all fatigue failures start at the surface. For many common types of loading, like bending and torsion, the maximum stress occurs at the surface so that it is logical that failure should start there. However, in axial loading the fatigue failure nearly always begins at the surface. There is ample evidence that fatigue properties are very sensitive to surface condition. The factors which affect the surface of a fatigue specimen can be divided roughly into three categories, (1) surface roughness or stress raisers at the surface, (2) changes in the fatigue strength of the surface metal, and (3) changes in the residual stress condition of the surface. In addition, the surface is subjected to oxidation and corrosion.

[1] R. Kuguel, *Am. Soc. Test. Mater. Proc.*, vol. 61, pp. 732–748, 1961.

## SURFACE ROUGHNESS

Since the early days of fatigue investigations, it has been recognized that different surface finishes produced by different machining procedures can appreciably affect fatigue performance. Smoothly polished specimens, in which the fine scratches (stress raisers) are oriented parallel with the direction of the principal tensile stress, give the highest values in fatigue tests. Such carefully polished specimens are usually used in laboratory fatigue tests and are known as "par bars." Table 12-3 indicates how the fatigue life of cantilever-beam specimens varies with the type of surface preparation. Extensive data on this subject have been published by Siebel and Gaier.[1]

Figure 12-15 shows the influence of various surface finishes on steel in reducing the fatigue limit of carefully polished "par bars." Note that the surface finish is characterized by the process used to form the surface. The extreme sensitivity of high-strength steel to surface conditions is well illustrated.

## CHANGES IN SURFACE PROPERTIES

Since fatigue failure is so dependent on the condition of the surface, anything that changes the fatigue strength of the surface material will greatly alter the fatigue properties. Decarburization of the surface of heat-treated steel is particularly detrimental to fatigue performance. Similarly, the fatigue strength of aluminum-alloy sheet is reduced when a soft aluminum coating is applied to the stronger age-hardenable aluminum-alloy sheet. Marked improvements in fatigue properties can result from the formation of harder and stronger surfaces on steel parts

Table 12-3 FATIGUE LIFE OF SAE 3130 STEEL SPECIMENS TESTED UNDER COMPLETELY REVERSED STRESS AT 95,000 PSI[†]

| Type of finish | Surface roughness, $\mu$in | Median fatigue life, cycles |
|---|---|---|
| Lathe-formed | 105 | 24,000 |
| Partly hand-polished | 6 | 91,000 |
| Hand-polished | 5 | 137,000 |
| Ground | 7 | 217,000 |
| Ground and polished | 2 | 234,000 |
| Superfinished | 7 | 212,000 |

[†] P. G. Fluck, *Am. Soc. Test. Mater. Proc.*, vol. 51, pp. 584–592, 1951.

[1] E. Siebel and M. Gaier, *VDI Z.*, vol. 98, pp. 1715–1723, 1956; abstracted in *Engineer's Digest*, vol. 18, pp. 109–112, 1957.

**Fig. 12-15** Reduction factor for fatigue limit of steel due to various surface treatments. (*From R. C. Juvinall, "Stress, Strain, and Strength," p. 234, McGraw-Hill Book Company, New York, 1967. By permission of the publishers.*)

by carburizing and nitriding.[1]  However, since favorable compressive residual stresses are produced in the surface by these processes, it cannot be considered that the higher fatigue properties are due exclusively to the formation of higher-strength material on the surface.  The effectiveness of carburizing and nitriding in improving fatigue performance is greater for cases where a high stress gradient exists, as in bending or torsion, than in an axial fatigue test.  The greatest percentage increase in fatigue performance is found when notched fatigue specimens are nitrided.  The amount of strengthening depends on the diameter of the part and the depth of surface hardening.  Improvements in fatigue properties similar to those caused by carburizing and nitriding may also be produced by flame hardening and induction hardening.  It is a general characteristic of fatigue in surface-hardened parts that the failure initiates at the interface between the hard case and the softer case, rather than at the surface.

Electroplating of the surface generally decreases the fatigue limit of steel. Chromium plating is particularly difficult to accomplish without impairment of

[1] "Fatigue Durability of Carburized Steel, "American Society for Metals, Metals Park, Ohio, 1957.

fatigue properties, while a softer cadmium plating is believed to have little effect on fatigue strength. The particular plating conditions used to produce an electro-plated surface can have an appreciable effect on the fatigue properties, since large changes in the residual stress, adhesion, porosity, and hardness of the plate can be produced.[1]

## SURFACE RESIDUAL STRESS

The formation of a favorable compressive residual-stress pattern at the surface is probably the most effective method of increasing fatigue performance. It can be considered that residual stresses are locked-in stresses which are present in a part which is not subjected to an external force. Only macrostresses, which act over regions which are large compared with the grain size, are considered here. They can be measured by x-ray methods or by noting the changes in dimensions when a thin layer of material is removed from the surface. Residual stresses arise when plastic deformation is not uniform throughout the entire cross section of the part being deformed. Consider a metal specimen where the surface has been deformed in tension by bending so that part of it has undergone plastic deformation. When the external force is removed, the regions which have been plastically deformed prevent the adjacent elastic regions from undergoing complete elastic recovery to the unstrained condition. Thus, the elastically deformed regions are left in residual tension, and the regions which were plastically deformed must be in a state of residual compression to balance the stresses over the cross section of the specimen. In general, for a situation where part of the cross section is deformed plastically while the rest undergoes elastic deformation, the region which was plastically deformed in tension will have a compressive residual stress after unloading, while the region which was deformed plastically in compression will have a tensile residual stress when the external force is removed. The maximum value of residual stress which can be produced is equal to the elastic limit of the metal.

For many purposes residual stresses can be considered identical to the stresses produced by an external force. Thus, the addition of a compressive residual stress, which exists at a point on the surface, to an externally applied tensile stress on that surface decreases the likelihood of fatigue failure at that point. Figure 12-16 illustrates this effect. Figure 12-16*a* shows the elastic stress distribution in a beam with no residual stress. A typical residual-stress distribution, such as would be produced by shot peening, is shown in Fig. 12-16*b*. Note that the high compressive residual stresses at the surface must be balanced by tensile residual stresses over the interior of the cross section. In Fig. 12-16*c* the stress distribution due to the

---

[1] A detailed review of the effect of electroplating on fatigue strength is given by R. A. R. Hammond and C. Williams, *Metall. Rev.*, vol. 5, pp. 165–223, 1960.

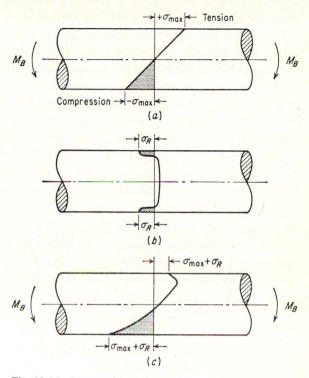

**Fig. 12-16**  Superposition of applied and residual stresses.

algebraic summation of the external bending stresses and the residual stresses is shown. Note that the maximum tensile stress at the surface is reduced by an amount equal to the surface compressive residual stress. The peak tensile stress is displaced to a point in the interior of the specimen. The magnitude of this stress depends on the gradient of applied stress and the residual-stress distribution. Thus, subsurface initiation of failure is possible under these conditions. It should also be apparent that the improvements in fatigue performance which result from the introduction of surface compressive residual stress will be greater when the loading is one in which a stress gradient exists than when no stress gradient is present. However, some improvement in the fatigue performance of axial-loaded fatigue specimens results from surface compressive residual stresses, presumably because the surface is such a potential source of weakness.

The chief commercial methods of introducing favorable compressive residual stresses in the surface are by surface rolling with contoured rollers and by shot peening.[1] Although some changes in the strength of the metal due to strain hard-

---

[1] J. O. Almen and P. H. Black, "Residual Stresses and Fatigue in Metals," McGraw-Hill Book Company, New York, 1963.

ening occur during these processes, the improvement in fatigue performance is due chiefly to the formation of surface compressive residual stress. Surface rolling is particularly adapted to large parts. It is frequently used in critical regions such as the fillets of crankshafts and the bearing surface of railroad axles. Shot peening consists in projecting fine steel or cast-iron shot against the surface at high velocity. It is particularly adapted to mass-produced parts of fairly small size. The severity of the stress produced by shot peening is frequently controlled by measuring the residual deformation of shot-peened beams called Almen strips. The principal variables in this process are the shot velocity and the size, shape, and hardness of the shot. Care must be taken to ensure uniform coverage over the area to be treated. Frequently an additional improvement in fatigue properties can be obtained by carefully polishing the shot-peened surface to reduce the surface roughness. Other methods of introducing surface compressive residual stresses are by means of thermal stresses produced by quenching steel from the tempering temperature and from stresses arising from the volume changes accompanying the metallurgical changes resulting from carburizing, nitriding, and induction hardening.

It is important to recognize that improvements in fatigue properties do not automatically result from the use of shot peening or surface rolling. It is possible to damage the surface by excessive peening or rolling. Experience and testing are required to establish the proper conditions which produce the optimum residual-stress distribution. Further, certain metallurgical processes yield surface tensile residual stresses. Thus, surface tensile stresses are produced by quenching deep-hardening steel, and this stress pattern may persist at low tempering temperatures. Grinding of hardened steel requires particular care to prevent a large decrease in fatigue properties. It has been shown[1] that either tensile or compressive surface residual stresses can be produced, depending on the grinding conditions. Further, the polishing methods ordinarily used for preparing fatigue specimens can result in appreciable surface residual stress.

Residual-stress distributions may be modified by plastic deformation or by thermal activation. When gross plastic deformation occurs, the stress drops to the yield stress. Thus, periods of overloading in the fatigue cycle or testing at high stresses in the low-cycle region may alter the residual-stress distribution by plastic deformation. This is referred to as *fading of residual stress*. Residual stresses have their greatest influence near the fatigue limit, where little fading occurs. Conversely, the fatigue life at high applied stresses depends little on residual stresses.

---

[1] L. P. Tarasov, W. S. Hyler, and H. R. Letner, *Am. Soc. Test. Mater. Proc.*, vol. 57, pp. 601–622, 1957.

## 12-12 CORROSION FATIGUE

The simultaneous action of cyclic stress and chemical attack is known as corrosion fatigue.[1] Corrosive attack without superimposed stress often produces pitting of metal surfaces. The pits act as notches and produce a reduction in fatigue strength. However, when corrosive attack occurs simultaneously with fatigue loading, a very pronounced reduction in fatigue properties results which is greater than that produced by prior corrosion of the surface. When corrosion and fatigue occur simultaneously, the chemical attack greatly accelerates the rate at which fatigue cracks propagate. Materials which show a definite fatigue limit when tested in air at room temperature show no indication of a fatigue limit when the test is carried out in a corrosive environment. While ordinary fatigue tests in air are not affected by the speed of testing, over a range from about 1,000 to 12,000 cycles/min, when tests are made in a corrosive environment there is a definite dependence on testing speed. Since corrosive attack is a time-dependent phenomenon, the higher the testing speed, the smaller the damage due to corrosion. Corrosion-fatigue tests may be carried out in two ways. In the usual method the specimen is continuously subjected to the combined influences of corrosion and cyclic stress until failure occurs. In the two-stage test the corrosion fatigue test is interrupted after a certain period and the damage which was produced is evaluated by determining the remaining life in air. Tests of the last type have helped to establish the mechanism of corrosion fatigue.[2] The action of the cyclic stress causes localized disruption of the surface oxide film so that corrosion pits can be produced. Many more small pits occur in corrosion fatigue than in corrosive attack in the absence of stress. The cyclic stress will also tend to remove or dislodge any corrosion products which might otherwise stifle the corrosion. The bottoms of the pits are more anodic than the rest of the metal so that corrosion proceeds inward, aided by the disruption of the oxide film by cyclic strain. Cracking will occur when the pit becomes sharp enough to produce a high stress concentration.

There is evidence to indicate that even fatigue tests in air at room temperature are influenced by corrosion fatigue. Fatigue tests on copper showed that the fatigue strength was higher in a partial vacuum than in air.[3] Separate tests in oxygen and water vapor showed little decrease over the fatigue strength in vacuum. It was concluded that water vapor acts as a catalyst to reduce the fatigue strength in air, indicating that the relative humidity may be a variable to consider in fatigue testing. Subsequent work with copper[4] showed that the fatigue life was much longer in

[1] A. J. McEvily and R. W. Staehle, eds., "Corrosion Fatigue," Nat. Assoc. Corrosion Eng., Houston, 1972.
[2] U. R. Evans and M. T. Simnad, *Proc. R. Soc. London*, vol. 188A, p. 372, 1947.
[3] J. J. Gough and D. G. Sopwith, *J. Inst. Met.*, vol. 72, pp. 415–421, 1946.
[4] N. Thompson, N. Wadsworth, and N. Louat, *Phil. Mag.*, vol. 1, pp. 113–126, 1956.

oxygen-free nitrogen than in air. Metallographic observation showed that the development of persistent slip bands was slowed down when tests were made in nitrogen. Additional studies on one effect of environment on fatigue cracking have been given by Achter.[1]

A number of methods are available for minimizing corrosion-fatigue damage. In general, the choice of a material for this type of service should be based on its corrosion-resistant properties rather than the conventional fatigue properties. Thus, stainless steel, bronze, or beryllium copper would probably give better service than heat-treated steel. Protection of the metal from contact with the corrosive environment by metallic or nonmetallic coatings is successful provided that the coating does not become ruptured from the cyclic strain. Zinc and cadmium coatings on steel and aluminum coatings on Alclad aluminum alloys are successful for many corrosion-fatigue applications, even though these coatings may cause a reduction in fatigue strength when tests are conducted in air. The formation of surface compressive residual stresses tends to keep surface notches from opening up and giving ready access to the corrosive medium. Nitriding is particularly effective in combating corrosion fatigue, and shot peening has been used with success under certain conditions. In closed systems it is possible to reduce the corrosive attack by the addition of a corrosion inhibitor. Finally, the elimination of stress concentrators by careful design is very important when corrosion fatigue must be considered.

## FRETTING

Fretting is the surface damage which results when two surfaces in contact experience slight periodic relative motion. The phenomenon is more related to wear than to corrosion fatigue. However, it differs from wear by the facts that the relative velocity of the two surfaces is much lower than is usually encountered in wear and that since the two surfaces are never brought out of contact there is no chance for the corrosion products to be removed. Fretting is frequently found on the surface of a shaft with a press-fitted hub or bearing. Surface pitting and deterioration occur, usually accompanied by an oxide debris (reddish for steel and black for aluminum). Fatigue cracks often start in the damaged area, although they may be obscured from observation by the surface debris. Fretting is caused by a combination of mechanical and chemical effects. Metal is removed from the surface either by a grinding action or by the alternate welding and tearing away of the high spots. The removed particles become oxidized and form an abrasive powder which continues the destructive process. Oxidation of the metal surface occurs and the oxide

---

[1] M. R. Achter, "Fatigue Crack Propagation," *ASTM Spec. Tech. Publ.*, 415, pp. 181–204, 1967.

film is destroyed by the relative motion of the surfaces. Although oxidation is not essential to fretting, as is demonstrated by relative motion between two nonoxidizing gold surfaces, when conditions are such that oxidation can occur fretting damage is many times more severe.

There are no completely satisfactory methods of preventing fretting. If all relative motion is prevented, then fretting will not occur. Increasing the force normal to the surfaces may accomplish this, but the damage increases with the normal force up to the point where relative motion is stopped. If relative motion cannot be completely eliminated, then reduction of the coefficient of friction between the mating parts may be beneficial. Solid lubricants such as MoS are most successful, since the chief problem is maintaining a lubricating film for a long period of time. Increasing the wear resistance of the surfaces so as to reduce surface welding is another approach. Exclusion of the atmosphere from the two surfaces will reduce fretting, but this is frequently difficult to do with a high degree of effectiveness. Several excellent reviews of this subject have been published.[1,2]

## 12-13 EFFECT OF MEAN STRESS ON FATIGUE

Most of the fatigue data in the literature have been determined for conditions of completely reversed cycles of stress, $\sigma_m = 0$. However, conditions are frequently met in engineering practice where the stress situation consists of an alternating stress and a superimposed mean, or steady, stress. The possibility of this stress situation has already been considered in Sec. 12-2, where various relationships between $\sigma_m$ and $\sigma_a$ have been given.

There are several possible methods of determining an $S$-$N$ diagram for a situation where the mean stress is not equal to zero. Figure 12-17 shows the two most common methods of presenting the data. In Fig. 12-17$a$ the maximum stress is plotted against log $N$ for constant values of the stress ratio $R = \sigma_{min}/\sigma_{max}$. This is achieved by applying a series of stress cycles with decreasing maximum stress and adjusting the minimum stress in each case so that it is a constant fraction of the maximum stress. The case of completely reversed stress is given at $R = -1.0$. Note that as $R$ becomes more positive, which is equivalent to increasing the mean stress, the measured fatigue limit becomes greater. Figure 12-17$b$ shows the same data plotted in terms of the alternating stress vs. cycles to failure at constant values of mean stress. Note that as the mean stress becomes more positive the allowable alternating stress decreases. Other ways of plotting these data are maximum stress vs. cycles to failure at constant mean stress and maximum stress vs. cycles to failure at constant minimum stress.

---

[1] R. B. Waterhouse, *Proc. Inst. Mech. Eng. London*, vol. 169, pp. 1157–1172, 1955.
[2] P. L. Teed, *Metall. Rev.*, vol. 5, pp. 267–295, 1960.

Fig. 12-17 Two methods of plotting fatigue data when the mean stress is not zero.

For each value of mean stress there is a different value of the limiting range of stress, $\sigma_{max} - \sigma_{min}$, which can be withstood without failure. Early contributions to this problem were made by Goodman,[1] so that curves which show the dependence of limiting range of stress on mean stress are frequently called *Goodman diagrams*. Figure 12-18 shows one common type of Goodman diagram which can be constructed from fatigue data of the type illustrated in Fig. 12-17. Basically, this diagram shows the variation of the limiting range of stress, $\sigma_{max} - \sigma_{min}$ with mean stress. Note that as the mean stress becomes more tensile the allowable range of stress is reduced, until at the tensile strength $\sigma_u$ the stress range is zero. However, for practical purposes testing is usually stopped when the yield stress $\sigma_0$ is exceeded. The test data usually lie somewhat above and below the $\sigma_{max}$ and $\sigma_{min}$ lines, respectively, so that these lines shown on Fig. 12-18 may actually be curves. A conservative approximation of the Goodman diagram may be obtained, in lieu of actual test data, by drawing straight lines from the

[1] John Goodman, "Mechanics Applied to Engineering," Longmans, Green & Co., Ltd., London, 1899.

**Fig. 12-18**   Goodman diagram.

fatigue limit for completely reversed stress (which is usually available from the literature) to the tensile strength. A diagram similar to Fig. 12-18 may be constructed for the fatigue strength at any given number of cycles. Very few test data exist for conditions where the mean stress is compressive. Data[1] for SAE 4340 steel tested in axial fatigue indicate that the allowable stress range increases with increasing compressive mean stress up to the yield stress in compression. This is in agreement with the fact that compressive residual stress increases the fatigue limit.

An alternative method of presenting mean-stress data is shown in Fig. 12-19. This is sometimes known as the Haig-Soderberg diagram.[2] The alternating stress is plotted against the mean stress. A straight-line relationship follows the suggestion of Goodman, while the parabolic curve was proposed by Gerber. Test data for ductile metals generally fall closer to the parabolic curve. However, because of the scatter in the results and the fact that tests on notched specimens fall closer to the Goodman line, the linear relationship is usually preferred in engineering design. These relationships may be expressed by the following equation,

$$\sigma_a = \sigma_e \left[ 1 - \left( \frac{\sigma_m}{\sigma_u} \right)^x \right] \qquad (12\text{-}22)$$

where $x = 1$ for the Goodman line, $x = 2$ for the Gerber parabola, and $\sigma_e$ is the fatigue limit for completely reversed loading. If the design is based on the yield

[1] J. T. Ransom, discussion in *Am. Soc. Test. Mater. Proc.*, vol. 54, pp. 847–848, 1954.
[2] C. R. Soderberg, *Trans. ASME*, vol. 52, APM-52-2, 1930, pp. 13–28.

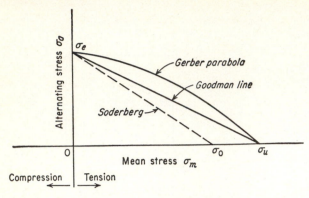

**Fig. 12-19**   Alternative method of plotting the Goodman diagram.

strength, as indicated by the dashed Soderberg line in Fig. 12-19, then $\sigma_0$ should be substituted for $\sigma_u$ in Eq. (12-22).

Figure 12-19 is obtained for alternating axial or bending stresses with static tension or compression or alternating torsion with static tension. However, for alternating torsion with static torsion or alternating bending with static torsion there is no effect of the static stress on the allowable range of alternating stress provided that the static yield strength is not exceeded.[1]

There is no general agreement regarding the proper method of applying stress-concentration factors in situations of combined mean and alternating stress. While all authorities apply a $K_f$ factor to the alternating stress component there is disagreement as to whether a factor should be applied to the mean stress. The problem of how to handle residual stresses resulting from peak stresses beyond yielding is also a question. These points are treated by Juvinall[2] in a series of design examples.

## 12-14   ENGINEERING ANALYSIS OF FATIGUE STRENGTH

An analytical theory of fatigue failure in ductile metals has been proposed by Fuchs.[3] Although the theory is not exact in every detail, it considers the influence of mean stress and alternating stress, residual stresses, and of combined loading, e.g., bending plus torsion. The theory is applicable to the long-life fatigue region

[1] G. Sines, Failure of Materials under Combined Repeated Stresses with Superimposed Static Stresses, *NACA Tech. Note* 3495, 1955.

[2] R. C. Juvinall, "Engineering Considerations of Stress, Strain, and Strength," pp. 279–297, McGraw-Hill Book Company, New York, 1967.

[3] H. O. Fuchs, *Trans. ASME, Ser. D: J. Basic Eng.*, vol. 87, pp. 333–343, 1965; also discussed in "Fatigue Design Handbook," pp. 30–36, Society of Automotive Engineers, New York, 1968.

where the gross strains are elastic. It assumes constant values of mean stress and that the loads are in phase, i.e., all alternating load components reach their peak values at the same time. To determine whether a given nominal cyclic stress state will result in failure in a given number of cycles, the stress state is compared to three criteria of failure: crack initiation, crack propagation, and gross yielding.

Although the data on combined stress fatigue failure are fewer and less reliable than for static yielding, certain generalizations can be made. Fatigue tests with different combinations of bending and torsion[1] show that for ductile metals the distortion-energy criterion (von Mises) provides the best overall fit. For brittle materials a maximum principal-stress theory provides the best criterion of failure. Sines[2] has proposed a combined stress fatigue failure criterion based on the distortion-energy theory.

$$[(\sigma_{a1} - \sigma_{a2})^2 + (\sigma_{a2} - \sigma_{a3})^2 + (\sigma_{a3} - \sigma_{a1})^2]^{1/2} + C_2(\sigma_{m1} + \sigma_{m2} + \sigma_{m3}) \geq \frac{\sqrt{2}\sigma_e}{K_f}$$

$$(12\text{-}23)$$

where $\sigma_{a1}$ = alternating component of principal stress in "1" direction, etc.

$\sigma_{m1}$ = static component of principal stress in "1" direction, etc.

$\sigma_e$ = fatigue strength (or fatigue limit) for completely reversed stress

$C_2$ = material constant giving influence of $\sigma_m$ on $\sigma_a$, that is, the negative slope of Goodman line on Fig. 12-19. As first approximations $C_2 = 0.5$.

The effect of residual stresses is included by adding their value to the appropriate $\sigma_m$ term. Since compressive residual stresses will substract from the $\sigma_m$ term, they will allow a greater alternating stress for the same fatigue life. For plane-stress conditions Eq. (12-23) will plot as a symmetrical ellipse similar to that shown in Fig. 3-5. However, the introduction of mean (static) stress terms will move the ellipse away from the center of the coordinate system.

Fuchs assumed that the general criterion for crack initiation is given by Eq. (12-23). For the simplified situation of a uniaxial stress state it reduces to

$$\sigma_a + C_2 \sigma_m \geq \frac{\sqrt{2}\sigma_e}{K_f} \qquad (12\text{-}24)$$

A maximum-tensile-stress criterion is used as the criterion of crack propagation, i.e., fatigue cracks will propagate if the alternating tensile stress is equal to or greater than some critical value $\sigma_c$. Mathematically, this criterion is simply

$$\sigma_{ta} \geq \sigma_c \qquad (12\text{-}25)$$

[1] H. J. Gough, *Proc. Inst. Mech. Eng. London*, vol. 160, pp. 417–440, 1949; W. N. Findley and P. N. Mathur, *Proc. Soc. Exp. Stress Anal.*, vol. 14, no. 1, pp. 35–46, 1956.
[2] Sines, op. cit.

where $\sigma_{ta}$ is the alternating tensile stress.

$$\sigma_{ta} = \frac{(\sigma_{\text{max tensile}} - \sigma_{\text{min tensile}})}{2} \qquad (12\text{-}26)$$

If $\sigma_{\min}$ of the stress cycle is compressive, then the minimum tensile stress is zero. No factor for $K_f$ is used in this failure criterion and, accordingly, the values taken for $\sigma_c$ are conservatively low:

$$\sigma_c \approx 3 \text{ ksi} \qquad \text{for high-strength aluminum alloy}$$
$$\sigma_c \approx 4 \text{ ksi} \qquad \text{for mild steel}$$
$$\sigma_c \approx 10 \text{ ksi} \qquad \text{for quenched and tempered steel}$$

The yielding criterion for uniaxial loading is simply that gross yielding occurs when the sum of the alternating stress and mean stress equal or exceed the yield strength of the material.

$$\sigma_a + \sigma_m \geq \sigma_0 \qquad (12\text{-}27)$$

For uniaxial loading these three failure criteria can be shown on a $\sigma_a$ versus $\sigma_m$ plot, such as Fig. 12-20. The yield criterion, Eq. (12-27), plots as the outer dashed triangle. The crack-initiation criterion, Eq. (12-24), is plotted as line $AB$ for a value of $K_f = 1$, and as line $CD$ for $K_f = 3$. The dot-dash line represents the crack-propagation criterion, Eq. (12-25). Since compressive values of $\sigma_{\min}$ are considered as zero in this criterion, it plots as a 45° line (constant $\sigma_{\max}$, variable negative $\sigma_{\min}$) in the compressive mean-stress region. The failure line is the heavy solid line in Fig. 12-20. Combinations of mean applied stress, residual stress, and alternating stress that fall inside the line will not produce failure in the number of cycles for which the diagram is constructed. Combinations of stress that fall outside the line should be considered unsafe. At the extreme tensile limit of mean stress the material fails by yielding. In the crosshatched regions fatigue cracks will initiate but they will not propagate to failure. In the middle region, cracks will initiate and propagate to failure.

**Fig. 12-20** Fatigue-failure diagram. (*After Fuchs.*)

## 12-15  CUMULATIVE FATIGUE DAMAGE

The conventional fatigue test subjects a specimen to a fixed amplitude until the specimen fails. Tests may be made at a number of different values of stress to determine the $S$-$N$ curve, but each time the maximum stress is held constant until the test is completed. However, there are many practical applications where the cyclic stress does not remain constant, but instead there are periods when the stress either is above or below some average design level. Further, there are applications involving complex loading conditions where it is difficult to arrive at an average stress level and the loading cannot be assumed to vary sinusoidally. Special fatigue tests which apply random loads have been developed for these applications.[1]

*Overstressing* is the process of testing a virgin specimen for some number of cycles less than failure at a stress above the fatigue limit, and subsequently running the specimen to failure at another test stress. The ratio of the cycles of overstress at the prestress to the virgin fatigue life at this same stress is called the *cycle ratio*. The damage produced by a cycle ratio of overstress may be evaluated by the reduction in fatigue life at the test stress. Alternatively, the damage due to overstressing may be measured by subjecting a number of specimens to a certain cycle ratio at the prestress and then determining the fatigue limit of the damaged specimens. Bennett[2] has shown that increasing the cycle ratio at the prestress produces a greater decrease in the fatigue limit of damaged specimens, while similar experiments[3] which employed a statistical determination of the fatigue limit showed a much greater reduction in the fatigue limit due to the overstressing. In line with this point, it is important to note that, because of the statistical nature of fatigue, it is very difficult to reach reliable conclusions by overstressing tests, unless statistical methods are used.

If a specimen is tested below the fatigue limit, so that it remains unbroken after a long number of cycles, and if then it is tested at a higher stress, the specimen is said to have been *understressed*. Understressing frequently results in either an increase in the fatigue limit or an increase in the number of cycles of stress to fracture over what would be expected for virgin specimens. It has frequently been considered that the improvements in fatigue properties due to understressing are due to localized strain hardening at sites of possible crack initiation. A different interpretation of the understressing effect has resulted from experiments on the statistical determination of the fatigue limit.[4] Specimens which did not fail during the determination of the fatigue limit showed greater than normal fatigue lives when retested at a higher stress. By means of statistical analysis it was possible

[1] S. R. Swanson, *Mater. Res. Stand.*, vol. 8, pp. 10–44, 1968.
[2] J. A. Bennett, *Am. Soc. Test. Mater. Proc.*, vol. 46, pp. 693–714, 1946.
[3] G. E. Dieter, G. T. Horne, and R. F. Mehl, *NACA Tech. Note* 3211, 1954.
[4] E. Epremian and R. F. Mehl, *ASTM Spec. Tech. Publ.*, 137, 1952.

to show that the observed lives at the higher stress were to be expected owing to the elimination of the weaker specimens during the prior testing below the fatigue limit. Thus, it was concluded that understressing was at least partly due to a statistical selectivity effect.

If a specimen is tested without failure for a large number of cycles below the fatigue limit and the stress is increased in small increments after allowing a large number of cycles to occur at each stress level, it is found that the resulting fatigue limit may be as much as 50 percent greater than the initial fatigue limit. This procedure is known as *coaxing*. An extensive investigation of coaxing[1] showed a direct correlation between a strong coaxing effect and the ability for the material to undergo strain aging. Thus, mild steel and ingot iron show a strong coaxing effect, while brass, aluminum alloys, and heat-treated low-alloy steels show little improvement in properties from coaxing.

Considerable data indicates that the percentage of life consumed by operation at one overstress level depends on the magnitude of subsequent stress levels. However, the linear cumulative damage rule,[2] also called *Miner's rule*, assumes that the total life of a part can be estimated by adding up the percentage of life consumed by each overstress cycle. If $n_1, n_2, \ldots, n_k$ represent the number of cycles of operation at specific overstress levels and $N_1, N_2, \ldots, N_k$ represent the life (in cycles) at these same overstress levels, then

$$\frac{n_1}{N_1} + \frac{n_2}{N_2} + \cdots + \frac{n_k}{N_k} = 1 \quad \text{or} \quad \sum_{j=1}^{j=k} \frac{n_j}{N_j} = 1 \qquad (12\text{-}28)$$

While many deviations from Miner's rule have been observed, and numerous modifications to this relationship have been proposed, none has been proven better or gained wide acceptance.

## 12-16   EFFECT OF METALLURGICAL VARIABLES ON FATIGUE

The fatigue properties of metals are quite structure-sensitive. However, at the present time there are only a limited number of ways in which the fatigue properties can be improved by metallurgical means. By far the greatest improvements in fatigue performance result from design changes which reduce stress concentration and from the intelligent use of beneficial compressive residual stress, rather than from a change in material. Nevertheless, there are certain metallurgical factors which must be considered to ensure the best fatigue performance from a particular metal or alloy. Fatigue tests designed to measure the effect of some metallurgical variable, such as special heat treatments, on fatigue performance are usually made

[1] G. M. Sinclair, *Am. Soc. Test. Mater. Proc.*, vol. 52, pp. 743–758, 1952.
[2] M. A. Miner, *J. Appl. Mech.*, vol., 12, pp. A159–A164, 1945.

with smooth, polished specimens under completely reversed stress conditions. It is usually assumed that any changes in fatigue properties due to metallurgical factors will also occur to about the same extent under more complex fatigue conditions, as with notched specimens under combined stresses. That this is not always the case is shown by the notch-sensitivity results discussed previously.

Fatigue properties are frequently correlated with tensile properties. In general, the fatigue limit of cast and wrought steels is approximately 50 percent of the ultimate tensile strength. The ratio of the fatigue limit (or the fatigue strength at $10^8$ cycles) to the tensile strength is called the *fatigue ratio*. Several nonferrous metals such as nickel, copper, and magnesium have a fatigue ratio of about 0.35. While the use of correlations of this type is convenient, it should be clearly understood that these constant factors between fatigue limit and tensile strength are only approximations and hold only for the restricted condition of smooth, polished specimens which have been tested under zero mean stress at room temperature. For notched fatigue specimens the fatigue ratio for steel will be around 0.20 to 0.30. However, as yield strength is increased by the various strengthening mechanisms, the fatigue limit usually does not increase proportionately. Most high-strength materials are fatigue-limited.

Several parallels can be drawn between the effect of certain metallurgical variables on fatigue properties and the effect of these same variables on tensile properties. The effect of solid-solution alloying additions on the fatigue properties of iron[1] and aluminum[2] parallels nearly exactly their effect on the tensile properties. Gensamer[3] showed that the fatigue limit of a eutectoid steel increased with decreasing isothermal-reaction temperature in the same fashion as did the yield strength and the tensile strength. However, the greater structure sensitivity of fatigue properties, compared with tensile properties, is shown in tests comparing the fatigue limit of a plain carbon eutectoid steel heat-treated to coarse pearlite and to spheroidite of the same tensile strength.[4] Even though the steel in the two structural conditions had the same tensile strength, the pearlitic structure resulted in a significantly lower fatigue limit due to the higher notch effects of the carbide lamellae in pearlite.

There is good evidence[5] that high fatigue resistance can be achieved by homogenizing slip deformation so that local concentrations of plastic deformation are avoided. This is in agreement with the observation that fatigue strength is directly

[1] E. Epremian, and E. F. Nippes, *Trans. Am. Soc. Met.*, vol. 40, pp. 870–896, 1948.
[2] J. W. Riches, O. D. Sherby, and J. E. Dorn, *Trans. Am. Soc. Met.*, vol. 44, pp. 852–895, 1952.
[3] M. Gensamer, E. B. Pearsall, W. S. Pellini, and J. R. Low, Jr., *Trans. Am. Soc. Met.*, vol. 30, pp. 983–1020, 1942.
[4] G. E. Dieter, R. F. Mehl, and G. T. Horne, *Trans. Am. Soc. Met.*, vol. 47, pp. 423–439, 1955.
[5] J. C. Grosskreutz, *Metall. Trans.*, vol. 3, pp. 1255–1262, 1972.

proportional to the difficulty of dislocation cross slip. Materials with high stacking-fault energy permit dislocations to cross slip easily around obstacles, which promotes slip-band formation and large plastic zones at the tips of cracks. Both of these phenomena promote the initiation and propagation of fatigue cracks. In materials with low stacking-fault energy, cross slip is difficult and dislocations are constrained to move in a more planar fashion. This limits local concentrations of plastic deformation and suppresses fatigue damage. Feltner and Laird[1] have referred to these two extremes of deformation as "wavy" and "planar" slip.

While the concept has been useful in understanding fatigue mechanisms, the ability to control fatigue strength by altering stacking-fault energy has practical limitations. A more promising approach to increasing fatigue strength appears to be the control of microstructure through thermomechanical processing to promote homogeneous slip with many small regions of plastic deformation as opposed to a smaller number of regions of extensive slip.

The dependence of fatigue life on grain size varies also depending on the deformation mode.[2] Grain size has its greatest effect on fatigue life in the low-stress, high-cycle regime in which stage I cracking predominates. In high stacking-fault-energy materials (such as aluminum and copper) cell structures develop readily and these control the stage I crack propagation. Thus, the dislocation cell structure masks the influence of grain size, and fatigue life at constant stress is insensitive to grain size. However, in a low stacking-fault-energy material (such as alpha brass) the absence of cell structure because of planar slip causes the grain boundaries to control the rate of cracking. In this case, fatigue life is proportional to $(\text{grain diameter})^{-\frac{1}{2}}$.

In general, quenched and tempered microstructures result in the optimum fatigue properties in heat-treated low-alloy steels. However, at a hardness level above about $R_C$ 40 a bainitic structure produced by austempering results in better fatigue properties than a quenched and tempered structure with the same hardness.[3] Electron micrographs indicate that the poor performance of the quenched and tempered structure is the result of the stress-concentration effects of the thin carbide films that are formed during the tempering of martensite. For quenched and tempered steels the fatigue limit increases with decreasing tempering temperature up to a hardness of $R_C$ 45 to $R_C$ 55, depending on the steel.[4] Figure 12-21 shows the results obtained for completely reversed stress tests on smooth specimens. The fatigue properties at high hardness levels are extremely sensitive to surface preparation, residual stresses, and inclusions. The presence of only a trace of

[1] C. E. Feltner and C. Laird, *Acta Metall.*, vol. 15, p. 1621, 1967.
[2] A. W. Thompson and W. A. Backofen, *Acta Metall.*, vol. 19, pp. 597–606, 1971.
[3] F. Borik and R. D. Chapman, *Trans. Am. Soc. Met.*, vol. 53, 1961.
[4] M. F. Garwood, H. H. Zurburg, and M. A. Erickson, "Interpretation of Tests and Correlation with Service," American Society for Metals, Metals Park, Ohio, 1951.

**Fig. 12-21**  Fatigue limit of alloy steels as a function of Rockwell hardness. (*From M. F. Garwood, H. H. Zurburg, and M. A. Erickson, in "Interpretation of Tests and Correlation with Service," p. 12, American Society for Metals, Metals Park, Ohio, 1951.*]

decarburization on the surface may drastically reduce the fatigue properties. Only a small amount of nonmartensitic transformation products can cause an appreciable reduction in the fatigue limit.[1] The influence of small amounts of retained austenite on the fatigue properties of quenched and tempered steels has not been well established.

The results, which are summarized in Fig. 12-21, indicate that below a tensile strength of about 200,000 psi the fatigue limits of quenched and tempered low-alloy steels of different chemical composition are about equivalent when the steels are tempered to the same tensile strength. This generalization holds for steels made by the open-hearth or electric-furnace processes and for fatigue properties determined in the longitudinal direction of wrought products. However, tests have shown[2] that the fatigue limit in the transverse direction of steel forgings may be only 60 to 70 percent of the longitudinal fatigue limit. It has been established[3]

[1] F. Borik, R. D. Chapman, and W. E. Jominy, *Trans. Am. Soc. Met.*, vol. 50, pp. 242–257, 1958.
[2] J. T. Ransom and R. F. Mehl, *Am. Soc. Test. Mater. Proc.*, vol. 52, pp. 779–790, 1952.
[3] J. T. Ransom, *Trans. Am. Soc. Met.*, vol. 46, pp. 1254–1269, 1954.

**Table 12-4  INFLUENCE OF INCLUSIONS ON FATIGUE LIMIT OF SAE 4340 STEEL†**

|  | Electric-furnace-melted | Vacuum-melted |
|---|---|---|
| Longitudinal fatigue limit, psi | 116,000 | 139,000 |
| Transverse fatigue limit, psi | 79,000 | 120,000 |
| Ratio transverse/longitudinal | 0.68 | 0.86 |
| Hardness, $R_C$ | 27 | 29 |

† Determined in repeated-bending fatigue test ($R = 0$). Data from J. T. Ransom, *Trans. Am. Soc. Met.*, vol 46, pp. 1254–1269, 1954.

that practically all the fatigue failures in transverse specimens start at nonmetallic inclusions. Nearly complete elimination of inclusions by vacuum melting produces a considerable increase in the transverse fatigue limit (Table 12-4). The low transverse fatigue limit in steels containing inclusions is generally attributed to stress concentration at the inclusions, which can be quite high when an elongated inclusion stringer is oriented transverse to the principal tensile stress. However, the fact that nearly complete elimination of inclusions by vacuum melting still results in appreciable anisotropy of the fatigue limit indicates that other factors may be important. Further investigations[1] of this subject have shown that appreciable changes in the transverse fatigue limit which cannot be correlated with changes in the type, number, or size of inclusions are produced by different deoxidation practices. Transverse fatigue properties appear to be one of the most structure-sensitive engineering properties.

The existence of a fatigue limit in certain materials, especially iron and titanium alloys, has been shown to depend on the presence of interstitial elements.[2] The mechanism by which this occurs is shown schematically in Fig. 12-22. The S-N curve for a pure metal (A) is a monotonic function with N increasing as stress decreases. The introduction of a solute element raises the yield strength and since it is more difficult to initiate a slip band, the S-N curve is shifted upward and to the right to curve B. If the alloy has suitable interstitial content so it undergoes strain aging, there is an additional strengthening mechanism. Since strain aging will not be a strong function of applied stress, there will be some limiting stress at which a balance occurs between fatigue damage and localized strengthening due

[1] G. E. Dieter, D. L. Macleary, and J. T. Ransom, Factors Affecting Ductility and Fatigue in Forgings, "Metallurgical Society Conferences," vol. 3, pp. 101–142, Interscience Publishers, Inc., New York, 1959.

[2] J. C. Levy and G. M. Sinclair, *Am. Soc. Test. Mater. Proc.*, vol. 55, p. 866, 1955; H. A. Lipsitt and G. T. Horne, *Am. Soc. Test. Mater. Proc.*, vol. 57, pp. 587–600, 1957; J. C. Levy and S. L. Kanitkar, *J. Iron Steel Inst. London*, vol. 197, pp. 296–300, 1961; H. A. Lipsitt and D. Y. Wang, *Trans. Metall. Soc. AIME*, vol. 221, p. 918, 1961.

**Fig. 12-22** Steps in the development of a material with a fatigue limit (schematic): *A* (pure metal). *B* (effect of solid solution elements on *A*), *C* (fatigue limit due to strain aging from interstitials), *D* (increased fatigue limit from enhanced strain aging.) (*After Tegart.*)

to strain aging. This results in the fatigue limit of curve *C*. With enhanced strain aging, brought about by higher interstitial content or elevated temperature the fatigue limit is raised and the break in the curve occurs at a lower number of cycles, curve *D*. In quenched and tempered steels, which do not normally exhibit strain aging in the tension test, the existence of a pronounced fatigue limit presumably is the result of localized strain aging at the tip of the crack.

## 12-17 EFFECT OF TEMPERATURE ON FATIGUE

### LOW-TEMPERATURE FATIGUE

Fatigue tests on metals at temperatures below room temperature show that the fatigue strength increases with decreasing temperature. Although steels become more notch-sensitive in fatigue at low temperatures, there is no evidence to indicate any sudden change in fatigue properties at temperatures below the ductile-to-brittle transition temperature. The fact that fatigue strength exhibits a proportionately greater increase than tensile strength with decreasing temperature has been interpreted as an indication that fatigue failure at room temperature is associated with vacancy formation and condensation.

### HIGH-TEMPERATURE FATIGUE

In general, the fatigue strength of metals decreases with increasing temperature above room temperature. An exception is mild steel, which shows a maximum in fatigue strength at 400 to 600°F. The existence of a maximum in the tensile strength in this temperature range due to strain aging has been discussed previously. As the temperature is increased well above room temperature, creep will become important and at high temperatures (roughly at temperatures greater than half the melting point) it will be the principal cause of failure. The transition from fatigue failure to creep failure with increasing temperature will result in a change in the type of failure from the usual transcrystalline fatigue failure to the intercrystalline creep failure. Local grain-boundary oxidation can contribute significantly to crack initiation. At any given temperature the amount of creep will increase with increasing mean stress.

Ferrous materials, which ordinarily exhibit a sharp fatigue limit in room-temperature tests, will no longer have a fatigue limit when tested at temperatures above approximately 800°F. Also, fatigue tests at high temperature will depend on the frequency of stress application. It is customary to report the total time to failure as well as the number of cycles to failure.

In general, the higher the creep strength of a material, the higher its high-temperature fatigue strength. However, the metallurgical treatment which produces the best high-temperature fatigue properties does not necessarily result in the best creep of stress-rupture properties. This has been shown by Toolin and Mochel[1] in high-temperature tests of a number of superalloys. Fine grain size results in better fatigue properties at lower temperatures. As the test temperature is increased the difference in fatigue properties between coarse- and fine-grain material decreases until at high temperatures, where creep predominates, coarse-grain material has higher strength. In general, wrought alloys show somewhat superior fatigue resistance, while castings are often more resistant to creep. Procedures which are successful in reducing fatigue failures at room temperature may not be effective in high-temperature fatigue. For example, compressive residual stresses may be annealed out before the operating temperature is reached.

Coffin[2] has extended the analysis for low-cycle fatigue to consider the frequency dependence of high-temperature fatigue. Thus, Eq. (12-6) becomes

$$\Delta\varepsilon_p = C_2 N^{-b} \nu^{(1-k)b} \qquad (12\text{-}29)$$

where $\nu$ is the frequency of stress application in cycles/min, and $k$ is an exponent which measures the effect of frequency on fatigue life. Stress amplitude vs. fatigue life may be correlated for different cyclic frequencies by

$$\frac{\Delta\sigma}{2} = A' N^{-b'} \nu^{k'} \qquad (12\text{-}30)$$

where the constants $A'$, $b'$, and $k'$ are obtained by regression analysis.

[1] P. R. Toolin and N. L. Mochel, *Am. Soc. Test. Mater. Proc.*, vol. 47, pp. 677–694, 1947; "Fatigue at Elevated Temperatures," *ASTM Spec. Tech. Publ.* 520, 1973.

[2] L. F. Coffin, Jr., *Proc. Air Force Conf. Fatigue and Fracture Aircraft Structures and Materials*, AFFDL TR 70-144, pp. 301–312, September 1970; also *Metall. Trans.*, vol. 2, pp. 3105–3113, 1971.

## THERMAL FATIGUE

The stresses which produce fatigue failure at high temperature do not necessarily need to come from mechanical sources. Fatigue failure can be produced by fluctuating thermal stresses under conditions where no stresses are produced by mechanical causes. Thermal stresses result when the change in dimensions of a member as the result of a temperature change is prevented by some kind of constraint. For the simple case of a bar with fixed end supports, the thermal stress developed by a temperature change $\Delta T$ is

$$\sigma = \alpha E \, \Delta T \qquad (12\text{-}31)$$

where $\alpha$ = linear thermal coefficient of expansion
$\qquad E$ = elastic modulus

If failure occurs by one application of thermal stress, the condition is called *thermal shock*. However, if failure occurs after repeated applications of thermal stress, of a lower magnitude, it is called *thermal fatigue*.[1] Conditions for thermal-fatigue failure are frequently present in high-temperature equipment. Austenitic stainless steel is particularly sensitive to this phenomenon because of its low thermal conductivity and high thermal expansion. Extensive studies of thermal fatigue in this material have been reported.[2] The tendency for thermal-fatigue failure appears related to the parameter $\sigma_f k/E\alpha$, where $\sigma_f$ is the fatigue strength at the mean temperature and $k$ is the thermal conductivity. A high value of this parameter indicates good resistance to thermal fatigue. An excellent review of the entire subject of high-temperature fatigue has been prepared by Allen and Forrest[3] and Ellison.[4]

## BIBLIOGRAPHY

American Society for Testing and Materials: "Metal Fatigue Damage-Mechanism, Detection, Avoidance and Repair," *ASTM Spec. Tech. Publ.* 495, 1971.

Battelle Memorial Institute: "Prevention of Fatigue of Metals," John Wiley & Sons, Inc., New York, 1941. (Contains very complete bibliography up to 1941.)

Burke, J. J., N. Reed, and V. Weiss (eds.): "Fatigue—An Interdisciplinary Approach," Syracuse University Press, 1964.

Forrest, P. G.: "Fatigue of Metals," Pergamon Press, New York, 1962.

---

[1] Failure of metals like uranium which have highly anisotropic thermal-expansion coefficients under repeated heating and cooling is also called thermal fatigue.
[2] L. F. Coffin, Jr. *Trans. ASME*, vol. 76, pp. 931–950, 1954.
[3] N. P. Allen and P. G. Forrest, *Proc. Int. Conf. Fatigue of Metals*, London, 1956, pp. 327–340.
[4] E. G. Ellison, *J. Mech. Eng. Sci.*, vol. 11, pp. 318–339, 1969.

Frost, N. E., Marsh, K. J., and L. P. Pook: "Metal Fatigue," Oxford University Press, London, 1974.

Harris, W. J.: "Metallic Fatigue," Pergamon Press, New York, 1961.

Kennedy, A. J.: "Processes of Creep and Fatigue in Metals," John Wiley & Sons, Inc., New York, 1963.

Kravchenko, P. E.: "Fatigue Resistance," Pergamon Press, New York, 1964.

Madayez, A. F. (ed.): "Metal Fatigue: Theory and Design," John Wiley & Sons, Inc., New York, 1969.

Sandor, B. I.: "Fundamentals of Cyclic Stress and Strain," The University of Wisconsin Press, Madison, 1972.

Sines, G., and J. L. Waisman (eds.): "Metal Fatigue," McGraw-Hill Book Company, New York, 1959.

# CREEP AND STRESS RUPTURE

## 13-1  THE HIGH-TEMPERATURE MATERIALS PROBLEM

In several previous chapters it has been mentioned that the strength of metals decreases with increasing temperature. Since the mobility of atoms increases rapidly with temperature, it can be appreciated that diffusion-controlled processes can have a very significant effect on high-temperature mechanical properties. High temperature will also result in greater mobility of dislocations by the mechanism of climb. The equilibrium concentration of vacancies likewise increases with temperature. New deformation mechanisms may come into play at elevated temperatures. In some metals the slip system changes, or additional slip systems are introduced with increasing temperature. Deformation at grain boundaries becomes an added possibility in the high-temperature deformation of metals. Another important factor to consider is the effect of prolonged exposure at elevated temperature on the metallurgical stability of metals and alloys. For example, cold-worked metals will recrystallize and undergo grain coarsening, while age-hardening alloys may overage and lose strength as the second-phase particles coarsen. Another important consideration is the interaction of the metal with its environment at high temperature. Catastrophic oxidation and intergranular penetration of oxide must be avoided.

Thus, it should be apparent that the successful use of metals at high temperatures involves a number of problems. Greatly accelerated alloy-development programs have produced a number of materials with improved high-temperature properties, but the ever-increasing demands of modern technology require materials with even better high-temperature strength and oxidation resistance. For a long time the principal high-temperature applications were associated with steam power plants, oil refineries, and chemical plants. The operating temperature in equipment such as boilers, steam turbines, and cracking units seldom exceeded 1000°F. With the introduction of the gas-turbine engine, requirements developed for materials to operate in critically stressed parts, like turbine buckets, at temperatures around 1500°F. The design of more powerful engines has pushed this limit to around 1800°F. Rocket engines and ballistic-missile nose cones present much greater problems, which can be met only by the most ingenious use of the available high-temperature materials and the development of still better ones. There is no question that the available materials of construction limit rapid advancement in high-temperature technology.

An important characteristic of high-temperature strength is that it must always be considered with respect to some time scale. The tensile properties of most engineering metals at room temperature are independent of time, for practical purposes. It makes little difference in the results if the loading rate of a tension test is such that it requires 2 h or 2 min to complete the test. Further, in room-temperature tests the anelastic behavior of the material is of little practical consequence. However, at elevated temperature the strength becomes very dependent on both strain rate and time of exposure. A number of metals under these conditions behave in many respects like viscoelastic materials. A metal subjected to a constant tensile load at an elevated temperature will *creep* and undergo a time-dependent increase in length.

A strong time dependence of strength becomes important in different materials at different temperatures. What is high temperature for one material may not be so high for another. To compensate for this, temperature often is expressed as a homologous temperature, i.e., the ratio of the test temperature to the melting temperature on an absolute temperature scale. Generally, creep becomes of engineering significance at a homologous temperature greater than 0.5.

The tests which are used to measure elevated-temperature strength must be selected on the basis of the time scale of the service which the material must withstand. Thus, an elevated-temperature tension test can provide useful information about the high-temperature performance of a short-lived item, such as a rocket engine or missile nose cone, but it will give only the most meager information about the high-temperature performance of a steam pipeline which is required to withstand 100,000 h of elevated-temperature service. Therefore, special tests are required to evaluate the performance of materials in different kinds of high-temperature service. The *creep test* measures the dimensional changes which occur from elevated-temperature exposure, while the *stress-rupture test* measures the effect of temperature on the long-time load-bearing characteristics. Other tests may be used to measure special properties such as thermal-shock resistance and stress relaxation. These high-temperature tests will be discussed in this chapter from two points of view. The engineering significance of the information obtained from the tests will be discussed, and information which is leading to a better understanding of the mechanism of high-temperature deformation will be considered.

## 13-2  THE CREEP CURVE

The progressive deformation of a material at constant stress is called *creep*. To determine the engineering creep curve of a metal, a constant load is applied to a tensile specimen maintained at a constant temperature, and the strain (extension) of the specimen is determined as a function of time. Although the measurement of creep resistance is quite simple in principle, in practice it requires considerable laboratory equipment.[1,2] The elapsed time of such tests may extend to several months, while some tests have been run for more than 10 years. The general procedures for creep testing are covered in ASTM Specification E139-69.

**Fig. 13-1**  Typical creep curve showing the three stages of creep. Curve *A*, constant-load test; curve *B*, constant-stress test.

[1] Creep and Creep Rupture Tests, ASM Metals Handbook, Supplement, August, 1955, *Met. Prog.*, vol. 68, no. 2A, pp. 175–184, Aug. 15, 1955.
[2] W. V. Green, Creep and Stress Rupture Testing, in R. F. Bunshaw (ed.), "Techniques of Metals Research," vol. V, pt. 1, chap. 6, Interscience Publishers, Inc., New York, 1971.

Curve *A* in Fig. 13-1 illustrates the idealized shape of a creep curve. The slope of this curve ($d\varepsilon/dt$ or $\dot{\varepsilon}$) is referred to as the *creep rate*. Following an initial rapid elongation of the specimen, $\varepsilon_0$, the creep rate decreases with time, then reaches essentially a steady state in which the creep rate changes little with time, and finally the creep rate increases rapidly with time until fracture occurs. Thus, it is natural to discuss the creep curve in terms of its three stages. It should be noted, however, that the degree to which these three stages are readily distinguishable depends strongly on the applied stress and temperature.

In making an engineering creep test, it is usual practice to maintain the load constant throughout the test. Thus, as the specimen elongates and decreases in cross-sectional area, the axial stress increases. The initial stress which was applied to the specimen is usually the reported value of stress. Methods of compensating for the change in dimensions of the specimen so as to carry out the creep test under constant-stress conditions have been developed.[1,2] When constant-stress tests are made, it is frequently found that no region of accelerated creep rate occurs (region III, Fig. 13-1) and a creep curve similar to *B* in Fig. 13-1 is obtained. Accelerated creep is found, however, in constant-stress tests when metallurgical changes occur in the metal. Curve *B* should be considered representative of the basic creep curve for a metal.

Andrade's pioneering work[3] on creep has had considerable influence on the thinking on this subject. He considered that the constant-stress creep curve represents the superposition of two separate creep processes which occur after the sudden strain which results from applying the load. The first component of the creep curve is a *transient creep* with a creep rate decreasing with time. Added to this is a constant-rate *viscous creep* component. The superposition of these creep processes is shown in Fig. 13-2. Andrade found that the creep curve could be represented by the following empirical equation:

$$\varepsilon = \varepsilon_0(1 + \beta t^{1/3})e^{(\kappa t)} \qquad (13\text{-}1)$$

**Fig. 13-2** Andrade's analysis of the competing processes which determine the creep curve.

[1] E. N. da C. Andrade and B. Chalmers, *Proc. R. Soc. London*, vol. 138A, p. 348, 1932.
[2] R. L. Fullman, R. P. Carreker, and J. C. Fisher, *Trans. AIME*, vol. 197, pp. 657–659, 1953.
[3] E. N. da C. Andrade, *Proc. R. Soc. London*, vol. 90A, pp. 329–342, 1914; "Creep and Recovery," pp. 176–198, American Society for Metals, Metals Park, Ohio, 1957.

where $\varepsilon$ is the strain in time $t$ and $\beta$ and $\kappa$ are constants.[1] The transient creep is represented by $\beta$ and Eq. (13-1) reverts to this form when $\kappa = 0$. The constant $\kappa$ describes an extension per unit length which proceeds at a constant rate. An equation which gives better fit than Andrade's equation, although it has been tested on a limited number of materials, was proposed by Garofalo.[2]

$$\varepsilon = \varepsilon_0 + \varepsilon_t(1 - e^{-rt}) + \dot{\varepsilon}_s t \qquad (13\text{-}2)$$

where $\varepsilon_0$ = the instantaneous strain on loading

$\varepsilon_t$ = the limit for transient creep

$r$ = the ratio of transient creep rate to the transient creep strain

$\dot{\varepsilon}_s$ = the steady-state creep rate

The various stages of the creep curve shown in Fig. 13-1 require further explanation. It is generally considered in this country that the creep curve has three stages. In British terminology the instantaneous strain designated by $\varepsilon_0$ in Fig. 13-1 is often called the first stage of creep, so that with this nomenclature the creep curve is considered to have four stages. The strain represented by $\varepsilon_0$ occurs practically instantaneously on the application of the load. Even though the applied stress is below the yield stress, not all the instantaneous strain is elastic. Most of this strain is instantly recoverable upon the release of the load (elastic), while part is recoverable with time (anelastic) and the rest is nonrecoverable (plastic). Although the instantaneous strain is not really creep, it is important because it may constitute a considerable fraction of the allowable total strain in machine parts. Sometimes the instantaneous strain is subtracted from the total strain in the creep specimen to give the strain due only to creep. This type of creep curve starts at the origin of coordinates.

The first stage of creep, known as *primary creep*, represents a region of decreasing creep rate. Primary creep is a period of predominantly transient creep in which the creep resistance of the material increases by virtue of its own deformation. For low temperatures and stresses, as in the creep of lead at room temperature, primary creep is the predominant creep process. The second stage of creep, known also as *secondary creep*, is a period of nearly constant creep rate which results from a balance between the competing processes of strain hardening and recovery. For this reason, secondary creep is usually referred to as *steady-state creep*. The average value of the creep rate during secondary creep is called the *minimum creep rate*. Third-stage or *tertiary creep* mainly occurs in constant-load creep tests at high stresses at high temperatures. Tertiary creep occurs when there is an effective reduction in cross-sectional area either because of necking or internal void formation.

---

[1] For a least squares method for evaluating $\beta$ and $\kappa$ see J. B. Conway, *Trans. Metall. Soc. AIME*, vol. 223, pp. 2018–2019, 1965.

[2] F. Garofalo, "Properties of Crystalline Solids," *ASTM Spec. Tech. Publ.*, 283, p. 82, 1960.

**Fig. 13-3** Schematic representation of the effect of stress on creep curves at constant temperature.

Third-stage creep is often associated with metallurgical changes such as coarsening of precipitate particles, recrystallization, or diffusional changes in the phases that are present.

Figure 13-3 shows the effect of applied stress on the creep curve at constant temperature. It is apparent that a creep curve with three well-defined stages will be found for only certain combinations of stress and temperature. A similar family of curves is obtained for creep at constant stress for different temperatures. The higher the temperature, the greater the creep rate.

The minimum creep rate is the most important design parameter derived from the creep curve. Two standards of this parameter are commonly used in this country: (1) the stress to produce a creep rate of 0.0001 percent per hour or 1 percent/10,000 h or (2) the stress for a creep rate of 0.00001 percent per hour or 1 percent/100,000 h (about $11\frac{1}{2}$ years). The first criterion is more typical of the requirements for jet-engine alloys, while the last criterion is used for steam turbines and similar equipment. A log-log plot of stress vs. minimum creep rate frequently results in a straight line. This type of plot is very useful for design purposes, and its use will be discussed more fully in a later part of this chapter.

## 13-3  THE STRESS-RUPTURE TEST

The stress-rupture test is basically similar to the creep test except that the test is always carried out to the failure of the material. Higher loads are used with the stress-rupture test than in a creep test, and therefore the creep rates are higher. Ordinarily the creep test is carried out at relatively low stresses so as to avoid tertiary creep. Emphasis in the creep test is on precision determination of strain, particularly as to the determination of the minimum creep rate. Creep tests are frequently conducted for periods of 2,000 h and often to 10,000 h. In the creep test the total strain is often less than 0.5 percent, while in the stress-rupture test the

**Fig. 13-4** Method of plotting stress-rupture data (schematic).

total strain may be around 50 percent. Thus, simpler strain-measuring devices, such as dial gages, can be used. Stress-rupture equipment is simpler to build, maintain, and operate than creep-testing equipment, and therefore it lends itself more readily to multiple testing units. The higher stresses and creep rates of the stress-rupture test cause structural changes to occur in metals at shorter times than would be observed ordinarily in the creep test, and therefore stress-rupture tests can usually be terminated in 1,000 h. These factors have contributed to the increased use of the stress-rupture test. It is particularly well suited to determining the relative high-temperature strength of new alloys for jet-engine applications. Further, for applications where creep deformation can be tolerated but fracture must be prevented, it has direct application in design.

The basic information obtained from the stress-rupture test is the time to cause failure at a given nominal stress for a constant temperature. The elongation and reduction of area at fracture are also determined. If the test is of suitable duration, it is customary to make elongation measurements as a function of time and from this to determine the minimum creep rate. The stress is plotted against the rupture time on a log-log scale (Fig. 13-4). A straight line will usually be obtained for each test temperature. Changes in the slope of the stress-rupture line are due to structural changes occurring in the material, e.g., changes from transgranular to intergranular fracture, oxidation, recrystallization and grain growth, or other structural changes such as spheroidization, graphitization, or sigma-phase formation. It is important to know about the existence of such instabilities, since serious errors in extrapolation of the data to longer times can result if they are not detected.

## 13-4 STRUCTURAL CHANGES DURING CREEP

If we plot the slope of a creep curve (Fig. 13-1a) vs. strain, we obtain a curve of creep rate vs. total strain (Fig. 13-5). This curve dramatically illustrates the large change in creep rate which occurs during the creep test. Since the stress and temperature

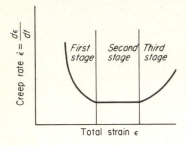

Fig. 13-5   Strain rate in creep test as function of total strain.

are constant, this variation in creep rate is the result of changes in the internal structure of the material with creep strain and time.

The principal deformation processes at elevated temperature are slip, sub-grain formation, and grain-boundary sliding. High-temperature deformation is characterized by extreme inhomogeneity. Measurements of local creep elongation[1] at various locations in a creep specimen have shown that the local strain undergoes many periodic changes with time that are not recorded in the changes in strain of the total gage length of the specimen. In large-grained specimens, local regions may undergo lattice rotations which produce areas of misorientation.

A number of secondary deformation processes have been observed in metals at elevated temperature. These include multiple slip, the formation of extremely coarse slip bands, kink bands, fold formation at grain boundaries, and grain-boundary migration. Many of the deformation studies at elevated temperature have been made with large-grain-size sheet specimens of aluminum. (Aluminum is favored for this type of study because its thin oxide skin eliminates problems from oxidation). Studies have also been made of creep deformation in iron, magnesium, and lead. It is important to remember that all the studies of high-temperature deformation have been made under conditions which give a creep rate of several percent in 100 or 1,000 h, while for many engineering applications a creep rate of less than 1 percent in 100,000 h is required. Because the deformation processes which occur at elevated temperature depend on the rate of strain as well as on the temperature, it is not always possible to extrapolate the results obtained for high strain-rate conditions to conditions of greater practical interest. Much of the work on deformation processes during creep has been reviewed by Sully[2] and Grant and Chaudhuri.[3]

[1] H. C. Chang and N. J. Grant, *Trans. AIME*, vol. 197, p. 1175, 1953.
[2] A. H. Sully, "Progress in Metal Physics," vol. 6, pp. 135–180, Pergamon Press, Ltd., London, 1956.
[3] N. J. Grant and A. R. Chaudhuri, Creep and Fracture, in "Creep and Recovery," pp. 284–343, American Society for Metals, Metals Park, Ohio, 1957.

## DEFORMATION BY SLIP

New slip systems may become operative when metals are deformed at elevated temperature. Slip in aluminum[1] occurs on the {111}, {100}, or {211} planes above 500°F. Zinc and magnesium undergo nonbasal slip at high temperature. The ease with which fresh slip systems operate in hcp metals with increasing temperature is shown by some data on zirconium.[2] The critical resolved shear stress for slip on the $(10\bar{1}0)[\bar{1}2\bar{1}0]$ system is 1.0 kg/mm$^2$ at 77°K, 0.2 kg/mm$^2$ at 575°K, and 0.02 kg/mm$^2$ at 1075°K. The slip bands produced at high temperature are coarser and more widely spaced than for room-temperature deformation. At high temperature and low strain rates it may be difficult to detect any slip lines at all. However, McLean[3] was able to detect very fine slip bands within the creeping grains of aluminum using phase contrast microscopy. Slip under high-temperature creep conditions occurs on many slip planes for small slip distances. Weertman[4] has suggested that this arises from the operation of many dislocation sources which under low-temperature conditions would stop operating because the dislocations from adjacent loops would repel each other. Since at high temperature the loops can climb and annihilate each other, there can be a steady stream of new dislocations from many sources.

## SUBGRAIN FORMATION

Creep deformation is quite inhomogeneous and many opportunities for lattice bending occur, especially near grain boundaries. Bending results in the formation of excess dislocations of one sign, and since dislocation climb can occur readily at high temperature, the dislocations arrange themselves into a low-angle grain boundary as was shown in Fig. 6-5. The formation of a subgrain structure, or cell structure during primary creep has been studied by x-rays, metallography, and thin-film electron microscopy. The dislocation density of the subgrain network increases during primary creep to a level that remains essentially constant during steady-state creep. The size of the subgrain depends on the stress and the temperature. Large subgrains are produced by high temperature and a low stress or creep rate.

The formation of a subgrain structure occurs most readily in metals of high-stacking-fault energy. There is less experimental information for metals of low stacking-fault energy which tend to recrystallize rather than form a cell structure during primary creep, but the cell's walls are less regular and well-defined. Thus, they serve as less effective barriers to moving dislocations.

[1] I. S. Servi, J. T. Norton, and N. J. Grant, *Trans. Metall. Soc. AIME*, vol. 194, p. 965, 1952.
[2] E. J. Rapperport and C. S. Hartley, *Trans. Metall. Soc. AIME*, vol. 218, p. 869, 1960.
[3] D. McLean, *J. Inst. Met.*, vol. 81, p. 133, 1952–1953.
[4] J. Weertman, *Trans. Metall. Soc. AIME*, vol. 218, p. 207, 1960.

## GRAIN-BOUNDARY SLIDING

At elevated temperature the grains in polycrystalline metals are able to move relative to each other. *Grain-boundary sliding* is a shear process which occurs in the
direction of the grain boundary. It is promoted by increasing the temperature
and/or decreasing the strain rate. Although most investigations indicate that sliding
occurs along the grain boundary as a bulk movement of the two grains,[1] other
observations[2] indicate that flow occurs in a softened area a finite distance into one
grain away from the grain boundary. Grain-boundary sliding occurs discontinuously with time, and the amount of shear displacement is not uniform along the
grain boundary.

   Grain-boundary sliding may be demonstrated by scribing a line on a polished
and etched surface and observing the shear offset where the line crosses the grain
boundary. More sophisticated techniques have been developed[3] for measuring
grain-boundary shear strain, but the measurements are not at all routine and there
is considerable doubt whether surface measurements represent the bulk behavior

**Fig. 13-6** Creep curves for total elongation and elongation
due to grain-boundary sliding in a Cu-0.37 wt% Be alloy.
(*From D. McLean and M. H. Farmer, J. Inst. Met., vol. 85,
p. 41, 1956–1957.*)

[1] H. C. Chang, and N. J. Grant, *Trans. AIME*, vol. 206, p. 169, 1956.
[2] R. N. Rhines, W. E. Bond, and M. A. Kissel, *Trans. Am. Soc. Met.*, vol. 48, p. 919, 1956.
[3] R. N. Stevens, *Metall. Rev.*, vol. 11, pp. 129–142, 1966.

of the specimen. Studies indicate that the strain due to grain-boundary sliding represents from only a few percent to as high as 50 percent of the total strain, depending on the metal and test conditions. However, a linear relation between the total grain-boundary sliding distance and the total elongation has been found for most systems studied. Since the total strain is the sum of the strain due to slip within the grains and grain-boundary sliding, the close relationship between sliding distance and total strain means that there is an equally close relationship between crystallographic slip and grain-boundary sliding. This was demonstrated by experiments on Cu-Be alloy[1] shown in Fig. 13-6. Strain aging occurred after 16 h and produced a decrease in the creep rate because slip in the grains became more difficult. The rate of grain-boundary sliding was reduced by the same ratio. One way that slip within the grains could influence grain-boundary sliding would be if sliding were impeded by ledges or protuberances in the grain boundaries which must be sheared through. The rate of sliding would then be controlled by the rate at which the grain could be deformed. Whatever the precise mechanism by which slip inside the grains controls grain-boundary sliding, from a practical point of view grain-boundary sliding is not critical to creep deformation since alloying elements added to prevent slip also prevent sliding. While complex high-temperature alloys often have alloying elements added specifically to alter the grain-boundary properties, usually these are more concerned with fracture initiation and low ductility.

The main import of grain-boundary sliding is on the initiation of grain-boundary fracture (Sec. 13-9). For grain-boundary deformation to occur without producing cracks at the grain boundaries, there must be a deformation mechanism available to achieve continuity of strain along the grain boundary. One way of accommodating grain-boundary strain at high temperature is by the formation of folds at the end of a grain boundary[2] (Fig. 13-7). Another recovery process is *grain-boundary migration*, in which the grain boundary moves normal to itself

**Fig. 13-7**   Fold formation at a triple point (schematic).

[1] D. McLean and M. H. Farmer, *J. Inst. Met.*, vol. 85, p. 41, 1956–1957.
[2] H. C. Chang and N. J. Grant, *Trans. Metall. Soc. AIME*, vol. 194, p. 619, 1952.

under the influence of shear stress and relieves the strain concentration. The wavy grain boundaries which frequently are observed during high-temperature creep are a result of inhomogeneous grain-boundary deformation and grain-boundary migration.

## 13-5  CREEP AT LOWER TEMPERATURES

Creep is possible only because obstacles to deformation can be overcome by the combined action of thermal fluctuations and stress. Diffusion-controlled processes are important chiefly at temperatures greater than about one-half the melting point. At lower temperatures recovery processes which are not dependent on diffusion, such as cross slip, play important roles in the creep process. High-temperature creep is predominantly steady-state or viscous creep, while below $T_m/2$ transient, or primary, creep predominates.

Andrade's equation for describing transient and steady-state creep was discussed in Sec. 13-2. An alternative general equation for the time laws of creep was suggested by Cottrell.[1]

$$\dot{\varepsilon} = At^{-n'} \qquad (13\text{-}3)$$

where $A$ and $n'$ are empirical constants. Different types of creep behavior are described by Eq. (13-3) depending on the value of $n'$. If $n' = 0$, the creep rate is constant and Eq. (13-3) represents steady-state creep. When $n' = 1$, Eq. (13-3) becomes

$$\varepsilon = \alpha \ln t \qquad (13\text{-}4)$$

where $\alpha$ is a constant. This is the logarithmic creep law found at low temperatures.[2] When $n' = \frac{2}{3}$, Eq. (13-3) becomes Andrade's equation for transient creep,

$$\varepsilon = \beta t^{\frac{1}{3}} \qquad (13\text{-}5)$$

Logarithmic creep occurs at low temperatures and low stresses, where recovery cannot occur. It is believed to be a true exhaustion process in which the rate-determining step is the activation energy to move a dislocation. On the initial application of stress, the dislocations with the lowest activation energy move first to produce the initial creep strain. As these easy-to-move dislocations are exhausted, creep can continue only by the movement of dislocations of higher activation energy. Therefore, the activation energy for the process continuously increases,

[1] A. H. Cottrell, *J. Mech. Phys. Solids*, vol. 1, pp. 53–63, 1952.
[2] Logarithmic creep has been observed for copper below 200°K. (O. H. Wyatt, *Proc. Phys. Soc. London*, vol. 66B, p. 495, 1953).

and the creep rate decreases. Theoretical treatments of exhaustion creep that result in a logarithmic equation have been proposed by Mott and Nabarro[1] and Cottrell.[2]

Low-temperature logarithmic creep obeys a mechanical equation of state; i.e., the rate of strain at a given time depends only on the instantaneous values of stress and strain and not on the previous strain history. However, creep at higher temperatures is strongly dependent on prior strain and thermal history and hence does not follow a mechanical equation of state.

## 13-6  ACTIVATION ENERGY FOR STEADY-STATE CREEP

Steady-state or secondary creep predominates at temperatures above about $T_m/2$. Since creep clearly depends on thermally activated processes (Sec. 9-10), it is important to examine the role of temperature on creep mechanisms.

The simplest assumption is that creep is a singly activated process which can be expressed by an Arrhenius-type rate equation

$$\dot{\varepsilon} = A \, e^{-\Delta H/RT} \qquad (13\text{-}6)$$

where $\Delta H$ = the activation energy for the rate-controlling process

$A$ = the preexponential complex constant containing the frequency of vibration of the flow unit, the entropy change, and a factor that depends on the structure of the material

$T$ = the absolute temperature

$R$ = the universal gas constant,[3] 1.987 cal/(mol)(°K)

Dorn[4] was able to develop a method for determining the activation energy $\Delta H$ by assuming that the $A$ term in Eq. (13-6) remained constant with small temperature changes. Since dislocations are not easily formed by thermal activation, it is reasonable to assume that the dislocation structure remains constant during the temperature change. By using a differential temperature change in the creep test, similar to that shown in Fig. 9-19c, we can write

$$\dot{\varepsilon}_1 \, e^{\Delta H/RT_1} = A = \dot{\varepsilon}_2 \, e^{\Delta H/RT_2}$$

and
$$\Delta H = R\left(\ln \frac{\dot{\varepsilon}_1}{\dot{\varepsilon}_2}\right)\left(\frac{T_1 T_2}{T_1 - T_2}\right) \qquad (13\text{-}7)$$

[1] N. F. Mott and F. R. N. Nabarro, "Report on Strength of Solids," p. 1, Physical Society, London, 1948.

[2] Cottrell, op. cit.

[3] Note that $R = kN$, where $k$ is Boltzmann's constant and $N$ is Avogadro's number ($6.02 \times 10^{23}$ atoms/mol).

[4] J. E. Dorn, The Spectrum of Activation Energies for Creep in "Creep and Recovery," pp. 255–283, American Society for Metals, Metals Park, Ohio, 1957.

**Fig. 13-8**   Creep strain vs. $\theta = t\,e^{-\Delta H/RT}$. (*From J. E. Dorn, in "Creep and Recovery," American Society for Metals, Metals Park, Ohio, 1957.*)

An alternative method is based on Dorn's observation that creep data can be correlated in terms of a temperature-compensated time parameter $\theta$

$$\theta = t\,e^{-\Delta H/RT} \qquad (13\text{-}8)$$

where $t$ is the time of creep exposure. Creep curves expressed as $\varepsilon$ versus $\theta$ determined at different temperatures but at the same stress superimpose as one common creep curve if a single rate-controlling process prevails (Fig. 13-8). An important characteristic of this correlation is that equivalent structure is obtained for the same values of $\varepsilon$ and $\theta$. Thus the activation energy for creep can be determined from two creep curves conducted with the same applied stress at two temperatures not too far apart so that the structures are the same at equal values of $\theta$.

$$\theta_1 = \theta_2 = t_1\,e^{-\Delta H/RT_1} = t_2\,e^{-\Delta H/RT_2}$$

and

$$\Delta H = R\left(\ln\frac{t_2}{t_1}\right)\left(\frac{T_1 T_2}{T_1 - T_2}\right) \qquad (13\text{-}9)$$

An extensive correlation[1] of creep and diffusion data for pure metals shows that the activation energy for high-temperature creep is equal to the activation energy for self-diffusion (Fig. 13-9). Since the activation energy for self-diffusion is the sum of the energies for the formation and movement of vacancies, this gives strong support to the view that dislocation climb is the rate-controlling step in high-temperature creep. The formation of a dislocation subgrain structure is

[1] O. D. Sherby, R. L. Orr, and J. E. Dorn, *Trans. AIME*, vol. 200, pp. 71–80, 1954.

**Fig. 13-9** Correlation between activation energies for high-temperature creep and self-diffusion. (*From J. E. Dorn, in "Creep and Recovery," p. 274, American Society for Metals, Metals Park, Ohio, 1957.*)

another factor in support of this view. We would expect that metals in which vacancies move rapidly would have low creep resistance. Sherby[1] has pointed out that the creep rates for alpha and gamma iron at the transition temperature differ by $\dot{\varepsilon}_\alpha/\dot{\varepsilon}_\gamma = 200$, which correlates with the much higher self-diffusion in alpha iron in the ratio $D_\alpha/D_\gamma = 350$. As another example, the self-diffusion coefficient in alpha iron increases with carbon content and so does the steady-state creep rate.

One of the important ways that dislocation climb is involved in creep deformation is to help dislocations overcome obstacles to slip. In addition to climb, there are a number of other thermally activated creep mechanisms. Dislocation intersection with other dislocations in the network is an important mechanism. The work to drive one dislocation through another varies from a fraction of an electron volt to several electron volts depending on the metal. Thermal energy corresponds to $\frac{1}{40}$ eV at 300°K and $\frac{1}{10}$ eV at 1200°K, and this will assist the applied stress in creating dislocation intersections. Intersection creates jogs (Sec. 5-11). In screw dislocations the jogs will have to move nonconservatively (Sec. 5-12), which re-

---

[1] O. D. Sherby, *Acta Metall.*, vol. 10, pp. 135–147, 1962.

quires the creation of vacancies. These vacancies will create a drag on the jog unless there is sufficient thermal activation for the vacancies to diffuse into the lattice. Therefore, the rate-controlling process is the rate of diffusion of vacancies and the activation energy of the creep mechanism is that for self-diffusion. Yet another important mechanism for creep in fcc metals is thermal activation of cross slip. Cross slip cannot occur when a screw dislocation is dissociated into a pair of partial dislocations. For an extended dislocation to cross slip it is necessary for it to form a constriction over part of its length in which the partial dislocations come together. The length of the constriction has a critical value. If it is too small, the dislocation will collapse and return to its original slip plane without cross slipping. The higher the resolved shear stress on the cross-slip plane, the more difficult it will be for the dislocation to collapse. The constriction and cross slip mechanism is a thermally activated process. Since the length of the constriction required for cross slip becomes smaller the larger the stress on the cross slip plane, the work to form the constriction decreases with increased stress. Thus, the activation energy for cross slip is stress-dependent and decreases with increasing stress.

Because of the variety of possible creep mechanisms, it is unlikely that only a single mechanism operates under a particular set of creep conditions. If these mechanisms operate independently of each other, the one with the fastest rate (or lower activation energy) will be the dominant mechanism. However, it is more likely that the various mechanisms will depend on each other. In this case the mechanism with the slowest rate that requires the highest activation energy will be

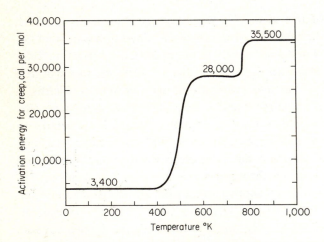

**Fig. 13-10**  Activation energy for creep in aluminum single crystals. (*From J. L. Lytton, L. A. Shepard, and J. E. Dorn, Trans. Metall. Soc. AIME, vol. 212, p. 220, 1958.*)

the controlling mechanism.  In using activation energy measurements to identify the predominant creep mechanism it is important to realize that changes in temperature, stress, or structure can produce changes in the relative importance of competing mechanisms.[1]  Figure 13-10 shows that three different activation energies have been observed[2] in aluminum single crystals.  The largest activation energy, 35,500 cal/mol, is associated with dislocation climb.  The intermediate value, 28,000 cal/mol is close to the calculated activation energy for thermally activated cross slip.  No clear mechanism has been associated with the low activation energy of 3,400 cal/mol, although mechanisms based on the Peierls stress and dislocation intersection have been suggested.  A similar type of behavior in which the activation energy is different in different temperature regions has been found in zinc, copper, and steel.

## 13-7  HIGH-TEMPERATURE CREEP

A steady-state creep condition occurs when the rate of recovery is fast enough and the rate of strain hardening slow enough so that a balance can be reached between these competing factors.  This condition is sometimes referred to as *recovery creep*.  If the rate of strain hardening is $h = \partial\sigma/\partial\varepsilon$ and the rate of recovery is $r = -\partial\sigma/\partial t$, then for a steady-state creep rate $\dot{\varepsilon}_s$ the flow stress must remain constant

$$d\sigma = \frac{\partial\sigma}{\partial t}\, dt + \frac{\partial\sigma}{\partial\varepsilon}\, d\varepsilon = 0$$

and

$$\dot{\varepsilon}_s = \frac{d\varepsilon}{dt} = -\frac{\partial\sigma/\partial t}{\partial\sigma/\partial\varepsilon} = \frac{r}{h} \qquad (13\text{-}10)$$

One of the important problems in high-temperature creep is the establishment of the stress dependence of the steady-state creep rate.  Well-tested empirical relationships have been established, but there is still difficulty in developing valid $\dot{\varepsilon}_s$ vs. $\sigma$ equations from first principles using dislocation models.  This situation is not surprising, since unlike a rapid change in temperature, a rapid change in stress can be expected to cause a significant change in the dislocation structure.  Since creep rate is determined by temperature, stress, and dislocation structure, and a change in stress changes structure, it is not easy to determine the effect of stress alone on the creep rate.

---

[1] For a simple example see R. E. Reed-Hill, "Physical Metallurgy Principles," pp. 589–592, D. Van Nostrand Company, Inc., Princeton, N.J., 1964.

[2] J. L. Lytton, L. A. Shepard, and J. E. Dorn, *Trans. Metall. AIME*, vol. 212, p. 220, 1958.

Sherby[1] has shown that the steady-state creep rates for a number of pure metals can be correlated (Fig. 13-11), according to the relation

$$\dot{\varepsilon}_s = SL^2 D \left(\frac{\sigma}{E}\right)^5 \qquad (13\text{-}11)$$

where $\dot{\varepsilon}_s$ = the steady-state creep rate, $h^{-1}$
$\qquad S$ = a constant, $10^{29}$ $cm^{-4}$
$\qquad L$ = the grain diameter, cm
$\qquad D$ = the self-diffusion coefficient, $cm^2/s$
$\qquad \sigma$ = the creep stress, psi
$\qquad E$ = the elastic modulus, psi

The general form of the empirical creep equations can be developed from a model of the effect of stress on an activated creep process.[2] In order for the creep mechanism to operate, an energy barrier of the type shown in Fig. 13-12a must be overcome. For the material to undergo a unit of strain, an atom must be moved from the free-energy minimum position at $A$ over the energy barrier at $B$ to another free-energy minimum at $C$. The height of the energy barrier[3] is $Q$. Now if a stress acts in the direction from $A$ to $C$, positive work is performed, and the height of the barrier is lowered for deformation from $A$ to $C$ and raised for deformation from $C$ to $A$. If the energy barrier now is given by Fig. 13-12b, the frequencies with which the mechanism operates from $A$ to $C$ and $C$ to $A$ are:

$A$ to $C$: $\quad v_1 = v_0 \, e^{-(Q-W_1)/kT}$

$C$ to $A$: $\quad v_2 = v_0 \, e^{-(Q+W_2)/kT}$

In general, the frequency corresponding to strain in the direction of the stress will be larger than the frequency in the direction opposite to the stress. If a large number of identical mechanisms are capable of operating in this fashion, then the rate of creep strain $\dot{\varepsilon}_s$ will be proportional to the difference between the forward and reverse frequencies

$$\dot{\varepsilon}_s \approx v_1 - v_2 \approx v_0[e^{-(Q-W_1)/kT} - e^{-(Q+W_2)/kT}]$$

or $\qquad\qquad \dot{\varepsilon}_s = Ae^{-Q/kT}(e^{W_1/kT} - e^{W_2/kT})$ $\qquad (13\text{-}12)$

If we recall the definition of the hyperbolic sine function

$$\sinh x = \frac{e^x - e^{-x}}{2}$$

[1] O. D. Sherby, *Acta Metall.*, vol. 10, pp. 135–147, 1962.
[2] R. E. Reed-Hill, "Physical Metallurgy Principles," pp. 575–577, D. Van Nostrand Company, Inc., Princeton, N.J., 1964.
[3] In strict thermodynamic terms the energy barrier should be expressed in terms of free energy. For simplicity, and because little is known about the entropy functions of activation processes, the energy barrier is expressed by enthalpy $\Delta H = Q$.

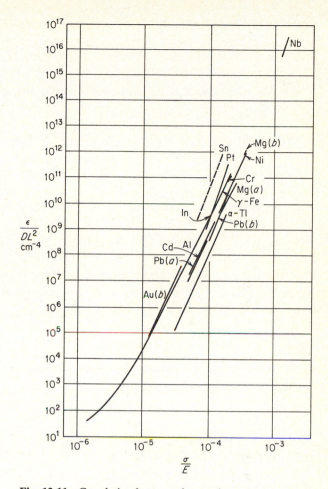

**Fig. 13-11** Correlation between the steady-state creep rate $\dot{\varepsilon}$ and the creep stress $\sigma$ for pure metals. (*From O. D. Sherby, Acta Metall., vol. 10, p. 142, 1962.*)

**Fig. 13-12** Effect of stress on thermally activated creep mechanism.

and simplify by assuming $W_1 = W_2 = W$, then Eq. (13-12) becomes

$$\dot{\varepsilon}_s = Ae^{-Q/kT}2\sinh\frac{W}{kT} \qquad (13\text{-}13)$$

The value of $W$ depends on the applied stress, and as the simplest assumption

$$W = v\sigma \qquad (13\text{-}14)$$

where $v$ is the activation volume. Thus, Eq. (13-13) becomes

$$\dot{\varepsilon}_s = Ae^{-Q/kT}2\sinh\frac{v\sigma}{kT} \qquad (13\text{-}15)$$

For very small applied stresses $\sinh \sigma \approx \sigma$ and Eq. (13-15) becomes

$$\dot{\varepsilon}_s = Ae^{-Q/kT}\frac{2v\sigma}{kT} \qquad (13\text{-}16)$$

On the other hand, if $Q$ is large compared to $kT$ and the stress is large, the frequency of the backward deformation mechanism $v_2$ will be low compared with $v_1$, and

$$\dot{\varepsilon}_s = Ae^{-(Q-W)/kT} = Ae^{-Q/kT}e^{v\sigma/kT} \qquad (13\text{-}17)$$

or

$$\dot{\varepsilon}_s = Ae^{-(Qv\sigma/kT)} \qquad (13\text{-}18)$$

The most generally applicable empirical equation for relating the steady-state creep rate to stress and temperature is

$$\dot{\varepsilon}_s = A(\sinh \alpha\sigma)^{n'}e^{-Q/RT} \qquad (13\text{-}19)$$

This was first proposed by Garofalo[1] and adopted by Sellers and Tegart for hot-working studies. Equation (13-19) reduces to an exponential dependence of stress like Eq. (13-18) at high stresses ($\alpha\sigma > 1.2$). At low stresses it reduces to a power dependence of stress

$$\dot{\varepsilon}_s = A_1\sigma^{n'}e^{-Q/RT} \qquad (13\text{-}20)$$

Experimental data that is summarized by Eq. (13-11) shows that $n' \approx 5$. The linear dependence on stress given by Eq. (13-16) rarely is observed. Further information on Eq. (13-19) is given in Sec. 9-8.

Theories of high-temperature creep[2] are based on three main dislocation models: (1) the jogged screw-dislocation model, (2) climb models, and (3) network models. In the jogged screw-dislocation model[3] dislocations move with the aid of

[1] F. Garofalo, *Trans. Metall. Soc. AIME*, vol. 227, p. 351, 1963.
[2] For reviews see J. J. Jonas, C. M. Sellers, and W. J. McG. Tegart, *Metall. Rev.*, vol. 14, pp. 1–24, 1969; G. Schoeck, in "Mechanical Behavior of Materials at Elevated Temperatures," chap. 5, McGraw-Hill Book Company, New York, 1961.
[3] P. B. Hirsch and D. H. Warrington, *Philos. Mag.*, vol. 6, p. 735, 1961; C. R. Barret and W. D. Nix, *Acta Metall.*, vol. 13, p. 1247, 1965.

thermal activation by a process in which vacancies are created and move away from jogs in the same jump. The activation energy is that for self-diffusion and the stress dependence is given by Eq. (13-18). The chief model based on dislocation climb is due to Weertman.[1] This model explains the presence of fine slip in creep as due to many dislocations emanating from many sources. In low-temperature deformation these dislocations would interfere with each other and block the sources, but at high-temperature dislocation climb prevents this and enables the sources to continue to operate. Weertman's model predicts that stress is proportional to the 4.5 power, in close agreement with Eq. (13-11). In the network model of McLean[2] the creep stress is determined by the three-dimensional dislocation network. Deformation occurs by bowing out of the longest link in the network, which decreases the average length of the remaining links and raises the flow stress. However, climb-controlled recovery takes place at elevated temperature, which increases the average link length. This model leads to Eq. (13-10).

## 13-8  HOT-WORKING

Hot-working refers to deformation carried out under conditions of temperature and strain rate such that recovery processes occur substantially during the deformation process so that large strains can be achieved with essentially no strain harden-ing. Hot-working processes such as rolling, extrusion, or forging typically are used in the first step of converting a cast ingot into a wrought product. The strain in hot-working is large ($\varepsilon \approx 2$ to 4) compared with tension or creep tests. Hot-working usually is carried out at temperatures above $0.6T_m$ and at high strain rates in the range 0.5 to 500 s$^{-1}$. The hot torsion test (Sec. 10-6) is a convenient laboratory method for studying metallurgical changes in hot-working.

The temperature and stress dependence of the steady-state hot-working rate is given by Eq. (13-19). The stress used in this equation is the horizontal steady-state value in Fig. 10-8. There are two mechanisms[3] responsible for softening in hot-working, depending on the metal. In aluminum and alpha iron *dynamic recovery* is the softening mechanism. This occurs by the formation of a well-developed subgrain structure by cross slip and climb, as occurs in creep deformation. The activation energy for hot-working is equal to that for creep and for self-diffusion. In metals with a lower stacking-fault energy (copper, nickel, and austenitic stainless steel), the activation energy for softening in hot-work is higher than for creep. For these metals the softening process is *dynamic recrystallization*. Although data are

[1] J. Weertman, op. cit.
[2] D. McLean, *Trans. Metall. Soc. AIME*, vol. 242, pp. 1195–1203, 1968.
[3] J. J. Jonas, C. M. Sellers, and W. J. McG. Tegart, op. cit.

limited, it has been shown[1] that the time to produce recrystallization in nickel decreases as (stress)³, and times of less than 1 s might be obtained at the relatively high stresses that act in hot-working.

## 13-9 FRACTURE AT ELEVATED TEMPERATURE

It has been known since the early work of Rosenhain and Ewen[2] that metals undergo a transition from transgranular fracture to intergranular fracture as the temperature is increased. When transgranular fracture occurs, the slip planes are weaker than the grain boundaries, while for intergranular fracture the grain boundary is the weaker component. Jeffries[3] introduced the concept of the *equicohesive*

Fig. 13-13 The equicohesive temperature. (*After N. J. Grant, in " Utilization of Heat Resistant Alloys," American Society for Metals, Metals Park, Ohio, 1954.*)

[1] G. J. Richardson, C. M. Sellers, and W. J. McG. Tegart, *Acta Metall.*, vol. 14, p. 1225, 1967.
[2] W. Rosenhain and D. Ewen, *J. Inst. Met.*, vol. 10, p. 119, 1913.
[3] Z. Jeffries, *Trans. AIME*, vol. 60, pp. 474–576, 1919.

*temperature* (ECT), which was defined as that temperature at which the grains and grain boundaries have equal strength (Fig. 13-13a). Like the recrystallization temperature, the ECT is not a fixed one. In addition to the effect of stress and temperature on the ECT, the strain rate has an important influence. Figure 13-13b shows that decreasing the strain rate lowers the ECT and therefore increases the tendency for intergranular fracture. The effect of strain rate on the strength-temperature relationship is believed to be much larger for the grain-boundary strength than for the strength of the grains. Since the amount of grain-boundary area decreases with increasing grain size, a material with a large grain size will have higher strength above the ECT than a fine-grain material. Below the ECT the reverse is true. For metals and alloys of commercial purity the ECT will occur within a fairly sharp temperature interval. However, for high-purity material there is a wide range of temperature where the strengths of the grains and the grain boundaries are not very different[1] (Fig. 13-13c) so that transgranular fracture can occur up to rather high temperatures.

The third stage of creep is involved with the development of intergranular cracks and cavities. Two types of intergranular cracking occur in creep[2] and hot-working.[3] These are wedge-type cracking and the formation of roughly spherical cavities at the grain boundaries. Wedge-shaped cracks (*w* type) initiate mostly at grain-boundary triple points (Fig. 13-14a) and propagate along the grain boundaries which are roughly normal to the applied stress. They initiate by grain-boundary sliding at a point of stress concentration in the manner illustrated in Fig. 13-15. *w*-type cracks are associated with creep conditions involving moderate creep temperatures and higher stress levels.

Round or elliptical cavities in the grain boundaries (*r*-type cracks) form at the higher creep temperatures under lower stresses (Fig. 13-14b). Grain-boundary sliding is necessary to initiate *r*-type cavities. As the temperature is raised, the number of cavities per unit length of grain boundary increases in the same way that grain-boundary sliding increases with temperature. The initiation of cavities in the grain boundary is believed to occur at grain-boundary ledges and at fine second-phase particles in the grain boundary. McLean[4] has shown that a spherical cavity will be stable if

$$r > \frac{2\gamma}{\sigma} \qquad (13\text{-}21)$$

[1] J. S. Servi and N. J. Grant, *Trans. AIME*, vol. 191, p. 917, 1951.
[2] F. Garofalo, in "Ductility," chap. 4, American Society for Metals, Metals Park, Ohio, 1968.
[3] W. J. McG. Tegart, in "Ductility," chap. 5, American Society for Metals, Metals Park, Ohio, 1968.
[4] D. McLean, "Mechanical Properties of Metals," pp. 327–334, John Wiley & Sons, Inc., New York, 1962.

(a)                    (b)

**Fig. 13-14**    (a) w-type cracks; (b) r-type crack.

**Fig. 13-15** Schematic drawings of the way intergranular cracks form owing to grain-boundary sliding. (*From H. C. Chang and N. J. Grant, Trans. Metall. Soc. AIME, vol. 206, p. 545, 1956.*)

where $r$ = the cavity radius

  $\gamma$ = the surface energy

  $\sigma$ = the tensile stress at the grain boundary

Grain-boundary cracks initiated as $r$-type cavities propagate by the growth of the cavities by the joint action of vacancy diffusion and grain-boundary sliding and their link up to form a continuous crack.

The shear stress required to nucleate a $w$-type crack has been calculated by Smith and Barnby[1] based on grain-boundary sliding at triple points.

$$\tau = \left[ \frac{3\pi\gamma_b G}{8(1 - v)L} \right]^{1/2} \qquad (13\text{-}22)$$

where $\gamma_b$ = the grain-boundary surface energy

  $G$ = the shear modulus

  $L$ = the grain-boundary sliding distance (assumed equal to the grain diameter)

Williams[2] used the same model to calculate the rupture time $t_r$ in terms of the steady-state creep rate $\dot{\varepsilon}_s$

$$t_r = \frac{1}{\dot{\varepsilon}_s} \left[ \frac{L\gamma 8\pi(1 - v)}{Gd^2} \right]^{1/2} \qquad (13\text{-}23)$$

where $L$ = the grain edge length ($\sim d$)

  $d$ = the grain diameter

  $\gamma$ = the effective fracture surface energy

The shear stress to initiate $r$-type cavities at grain-boundary particles or ledges of width $2c$ and spacing $2d$ was estimated by Smith and Barnby[3] as

$$\tau = \frac{\pi}{2} \left( \frac{c}{d} \right)^{1/2} \left[ \frac{4\gamma_b G}{(1 - v)d} \right]^{1/2} \qquad (13\text{-}24)$$

Equations (13-22) and (13-24) predict crack initiation at values of shear stress much lower than previous models. The calculated stresses are in general agreement with experimental observations.

While distinct examples of $w$- and $r$-type cracks have been observed under the proper conditions, in the usual case intergranular creep fracture is not so easily classified. A fractographic study of creep cavities in tungsten[4] suggests that all cavities are initially polyhedral, but they change their shape depending on whether grain-boundary sliding or vacancy diffusion predominate according to the conditions of temperature and stress.

[1] E. Smith and J. T. Barnby, *Met. Sci. J.*, vol. 1, p. 56, 1967.
[2] J. A. Williams, *Philos. Mag.*, vol. 15, p. 1289, 1967; *ibid.*, vol. 20, p. 635, 1969.
[3] E. Smith and J. T. Barnby, *Met. Sci. J.*, vol. 1, p. 1, 1967.
[4] J. O. Stiegler, K. Farrell, B. T. M. Loh, and H. E. McCoy, *Trans. Am. Soc. Met.*, vol. 60, p. 494, 1967.

High-temperature fracture frequently is sensitive to changes in microstructure brought about by heat treatment or mechanical processing. It is often stated simply that high-temperature alloys are sensitive to *thermomechanical processing*. A good example[1] occurs in the nickel-based superalloy Udimet 700. This alloy gains much of its strength from $\gamma'$ precipitate particles, essentially $Ni_3(Ti, Al)$. At temperatures above the $\gamma'$ solvus ($\sim 2070°F$) intergranular fracture occurs readily by *w*-type crack formation. On the other hand, if the alloy first is aged to form a fine $\gamma'$ precipitate, it can be rolled extensively at 1950°F where the precipitate is a stable phase. Careful thin-film electron microscopy has shown clearly the role of microstructure on fracture. At temperatures above the $\gamma'$ solvus, the grain boundaries are planar and free of second-phase particles. Deformation is concentrated along the grain boundaries and failure occurs by grain-boundary sliding. Since the sliding length is a significant fraction of the grain diameter, according to Eq. (13-22) cracks will initiate at a low shear stress. On the other hand, in the age-hardened condition the coherent $\gamma'$ precipitates are dispersed in the grain boundaries as well as within the grains. For this microstructure deformation occurs uniformly within the grains with no appreciable strain concentration at the grain boundaries. Moreover, since the sliding distance is reduced considerably by the grain-boundary precipitate, the shear stress to produce a *w* crack is increased considerably, as evidenced by the good ductility at 1950°F.

If the elongation or reduction of area is measured with increasing temperature from room temperature, there is usually an intermediate temperature region of low ductility.[2] For aluminum the ductility minimum falls near room temperature but for many alloys, especially nickel-based, it falls within the hot-working region. The region of minimum ductility typically occurs just below the recrystallization temperature in a region where grain-boundary sliding can occur to produce intergranular cracking. When the temperature is raised beyond the ductility minimum into a region where recrystallization can occur readily, these new grains isolate the grain-boundary cracks and the ductility is increased. Especially large losses in ductility can occur in certain materials when they are subjected both to elevated temperature and nuclear irradiation.[3]

## 13-10   HIGH-TEMPERATURE ALLOYS

High-temperature alloys are a class of complex materials developed for a specific application. Some appreciation of the metallurgical principles behind the development of these alloys is important to an understanding of how metallurgical vari-

[1] M. Oblak, W. A. Owczarski, and D. S. Duvall, *Metall. Trans.*, vol. 2, pp. 1499–1501, 1971.
[2] F. N. Rhines and P. J. Wray, *Trans. Am. Soc. Met.*, vol. 54, p. 117, 1961.
[3] J. O. Stiegler and J. R. Weir, Jr., in "Ductility," chap. 11, American Society for Metals, Metals Park, Ohio, 1968.

ables influence creep behavior. The development of high-temperature alloys has, in the main, been the result of painstaking empirical investigation, and it is really only in retrospect that principles underlying these developments have become evident.

The nominal composition of a number of high-temperature alloys is given in Table 13-1. Only a few of the many available alloys[1] could be included in this table. A number of general guidelines have developed for increasing high-temperature strength. In general, the creep resistance is higher the greater the melting point of the metal since the rate of self-diffusion is lower in metals with a high $T_m$. Since

**Table 13-1  COMPOSITIONS OF SOME HIGH-TEMPERATURE ALLOYS**

| Alloy | C | Cr | Ni | Mo | Co | W | Cb | Ti | Al | Fe | Other |
|---|---|---|---|---|---|---|---|---|---|---|---|
| | | | | | Ferritic steels | | | | | | |
| 1.25 Cr-Mo | 0.10 | 1.25 | — | 0.50 | | | | | | Bal. | |
| 5 Cr-Mo | 0.20 | 5.00 | — | 0.50 | | | | | | Bal. | |
| Greek Ascoloy | 0.12 | 13.0 | 2.0 | | | 3.0 | | | | Bal. | |
| | | | | | Austenitic steels | | | | | | |
| 316 | 0.08 | 17.0 | 12.0 | 2.50 | | | | | | Bal. | |
| 16-25-6 | 0.10 | 16.0 | 25.0 | 6.00 | | | | | | Bal. | |
| A-286 | 0.05 | 15.0 | 26.0 | 1.25 | | | | 1.95 | 0.2 | Bal. | |
| | | | | | Nickel-based alloys | | | | | | |
| Inconel | 0.04 | 15.5 | 76.0 | | | | | | | 7.0 | |
| Inconel 718 | 0.04 | 19.0 | Bal. | 3.0 | | | 5.0 | 0.80 | 0.60 | 18.0 | |
| René 41 | 0.10 | 19.0 | Bal. | 10.0 | 11.0 | | | 3.2 | 1.6 | 2.0 | |
| Mar-M-200 | 0.15 | 9.0 | Bal. | — | 10.0 | 12.5 | 1.0 | 2.0 | 5.0 | | |
| TRW 1900 | 0.11 | 10.3 | Bal. | — | 10.0 | 9.0 | 1.5 | 1.0 | 6.3 | | |
| Udimet 700 | 0.15 | 15.0 | Bal. | 5.2 | 18.5 | | | 3.5 | 4.25 | 1.0 | |
| In-100 | 0.15 | 10.0 | Bal. | 3.0 | 15.0 | | | 4.7 | 5.5 | | 1.0 V |
| TD Nickel | — | — | Bal. | | | | | | | | 2.0 $ThO_2$ |
| | | | | | Cobalt-based alloys | | | | | | |
| Vitallium (HS-21) | 0.25 | 27.0 | 3.0 | 5.0 | Bal. | | | | | 1.0 | |
| S-816 | 0.40 | 20.0 | 20.0 | 4.0 | Bal. | 4.0 | | 4.0 | | 3.0 | |

[1] W. F. Simmons, Rupture Strengths of Selected High-Iron, Nickel-Base, Cobalt-Base and Refractory Metal Alloys, *DMIC memo* 234, Battelle Memorial Institute, Columbus, Ohio, May 1, 1968; *Met. Progr.*, March 1973, pp. 56–59.

dislocation cross slip is an important step in dislocations climbing to avoid obstacles, metals with a low stacking-fault energy have higher creep resistance because extended partial dislocations have difficulty in cross slipping. Solid-solution alloy additions of high valence are the most effective because they produce a strong decrease in the stacking-fault energy. Solid-solution additions can increase the strength by a variety of mechanisms, such as:

1. Segregation to stacking faults (Suzuki interaction)
2. Elastic interactions of solute atoms with moving dislocations to increase the Peierls-Nabarro stress or the friction stress
3. Interaction with vacancies and dislocation jogs
4. Segregation to grain boundaries so as to influence grain-boundary sliding and migration

Fine dispersed precipitates are necessary for high creep resistance. Many nickel-based superalloys contain small amounts of Al and/or Ti which combine with the matrix to form fine precipitates of the intermetallic compounds $Ni_3Al$, $Ni_3Ti$, or $Ni_3(Al, Ti)$. In creep-resistant steels the precipitates are carbides (such as VC, TiC, NbC, $Mo_2C$, or $Cr_{23}C_6$) which can be precipitated by heat treatment prior to creep service or may be precipitated preferentially at dislocations during creep deformation. *Thermal stability* of the precipitate or strengthening phases is important. Since the greatest strengthening is achieved with the finest particles which are the most unstable, there is usually a critical heat treatment and thermo-mechanical processing for optimum high-temperature strength. Coarsening of the precipitate is minimized by using a dispersed phase which is nearly insoluble in the matrix so that the re-solution of fine particles and the growth of coarser particles is slow. This situation is achieved best in dispersion-strengthened alloys in which inert oxide particles (such as $Al_2O_3$, $ThO_2$, or $SiO_2$) are mixed with metal matrix powders and the mixture is consolidated by powder metallurgy techniques. TD-nickel and sintered aluminum powder (SAP) are examples of this type of alloy.

While emphasis has been given to strengthening processes in high-temperature alloys, other considerations are important. An alloy with high-temperature strength may be difficult to fabricate by hot-working, cold-shaping, or welding. Some of the strongest alloys are so highly alloyed that they can only be produced by precision casting. In addition to strength, high-temperature materials must be resistant to the thermal environment in which they operate. Some of the strongest alloys, such as molybdenum, have inadequate oxidation resistance and must be coated with an oxidation-resistant material if used at elevated temperature. Unfortunately, this introduces problems with reliability, quality control, and fabrication. In other cases the alloy composition must be adjusted to produce a balance between high strength and oxidation resistance.

## 13-11  EFFECT OF SOME METALLURGICAL VARIABLES

Equation (13-11) indicates that creep resistance increases (steady-state creep rate decreases) as the grain size is decreased. Actually, the influence of grain size on creep resistance is not clear cut. In many situations coarse-grain material has lower creep rates and higher stress-rupture strengths than a fine-grain-size material. The simplistic explanation is that a coarse-grain material has a smaller total grain-boundary area available for grain-boundary sliding than a fine-grain material. In several alloys an optimum grain size for maximum creep resistance has been observed. Figure 13-16 shows that increasing or decreasing the grain size beyond this critical value increases the steady-state creep rate.

Parker[1] has suggested that the effect of grain size on creep rate is due primarily to the change in grain-boundary structure with different grain-coarsening treatments. Since high-temperature creep depends on dislocation climb, the rate at which vacancies can diffuse to edge dislocations is the controlling factor. Vacancy diffusion is more rapid along high-energy grain boundaries than through the bulk lattice. In a fine-grain-size material with many high-angle grain boundaries vacancy diffusion is rapid and the creep rate is high. When the same material is heated to a

**Fig. 13-16**  Influence of grain size on the minimum creep rate for Monel at 595°C. (*From P. Shahinian and J. R. Lane, Trans. Am. Soc. Met., vol 45, p. 177, 1953. Copyright American Society for Metals, 1953.*)

[1] E. R. Parker, *Trans. Am. Soc. Met.*, vol. 50, pp. 52–104, 1958. A theoretical interpretation of the effect of grain size in creep is given by D. K. Matlock and W. D. Nix, *Metall. Trans.*, vol. 5, pp. 961–963, 1974.

high temperature to coarsen the grain size, most of the high-energy grain boundaries disappear due to grain growth. The boundaries which remain are fewer in number and are mostly lower-energy grain boundaries for which vacancy diffusion is relatively slow. Therefore, dislocation climb will be slower in the coarse-grain material and the creep rate will be lower.

Because grain boundaries are the nucleation sites for high-temperature fracture, the control, or elimination, of grain boundaries will suppress fracture and increase rupture life. The best example is the control of grain orientation by directional solidification.[1] In a directionally solidified turbine blade the grain boundaries predominantly lie parallel to the longitudinal axis of the blade so that the bending stresses on the grain boundaries are low. This material exhibits the same creep curve as a conventionally cast specimen with random grain orientation with the exception that third-stage creep is considerably delayed in time and there is appreciably greater elongation at fracture. Even greater improvements in rupture time and ductility can be obtained by using a monocrystal in which grain boundaries have been eliminated.

High-temperature properties can show considerable scatter, and frequently measurably different results are obtained between different heats of the same material or even between different bars from the same heat. The creep properties of steels are particularly subject to variations in properties, which are related in complex ways to the composition, melting practice, type of mechanical working, and microstructure. Aluminum deoxidized steels generally have poorer creep properties than silicon deoxidized steels. Not only does aluminum refine the grain size, but it also accelerates spheroidization and graphitization. Considerable improvement in the high-temperature properties of superalloys results from vacuum melting. Fabricability is also improved, presumably because of the decrease in the number and size of inclusions. In general, life to rupture and ductility at fracture are both increased by vacuum melting. Better control of composition, and therefore more uniform response to heat treatment, is obtained with vacuum melting.

The environment surrounding the specimen can have an important influence on high-temperature strength. Creep tests on zinc single crystals showed that creep practically stopped when copper was plated on the surface of the specimens.[2] When the copper was electrolytically removed, creep began again at nearly the original rate. Stress-rupture tests on nickel and nickel-chromium alloys show a complex dependence on atmosphere.[3] At high temperatures and low strain rates these materials are stronger in air than in vacuum, while the reverse is true at low temperatures and high strain rates. This behavior is attributed to the competing effects of strengthening resulting from oxidation and weakening due to lowering

[1] F. L. Ver Snyder and M. E. Shank, *Mater. Sci. Eng.*, vol. 6, pp. 213–247, 1970.
[2] M. R. Pickus and E. R. Parker, *Trans. AIME*, vol. 191, pp. 792–796, 1951.
[3] P. Shahinian and M. R. Achter, *Trans. Am. Soc. Met.*, vol. 51, pp. 244–255, 1959.

of the surface energy by absorbed gases. The nature of the oxidation can have an important influence on the high-temperature properties. A thin oxide layer or a finely dispersed oxide will usually lead to strengthening, but intergranular penetration of oxide will usually lead to decreased rupture life and intergranular fracture. When materials must operate in an atmosphere of hot combustion gases or in corrosive environments, the service life is materially reduced.

## 13-12  PRESENTATION OF ENGINEERING CREEP DATA

From the previous sections it should be apparent that understanding of the high-temperature strength of materials has not advanced to the point where creep and stress-rupture behavior can be predicted reliably from theoretical models. For design purposes it is essential that decisions be based on reliable experimental data. In reporting high-temperature strength data it is common practice to speak of *creep strength* or *rupture strength*. Creep strength is defined as the stress at a given temperature which produces a steady-state creep rate[1] of a fixed amount, usually taken as 0.00001 or 0.001 percent per hour. Alternatively, the creep strength may be defined as the stress to cause a creep strain of 1 percent at the given temperature. Rupture strength refers to the stress at a given temperature to produce a life to rupture of a certain amount, usually 1,000, 10,000, or 100,000 h.

It frequently is important to be able to extrapolate creep or stress-rupture data into regions where data are not available. Therefore, the common methods of plotting creep data are based on plots which yield reasonable straight lines. Figure 13-17 shows the common method of presenting the influence of stress on the steady-state or minimum creep rate. Note that a log-log plot is used, so that extrapolation of one log-cycle represents a tenfold change. A change in slope of the line is not unusual. It has been shown[2] that the value of the minimum creep rate depends on the length of time the creep test has been carried out. For precise determination of $\dot{\varepsilon}_s$, creep curves which have entered the third stage of creep should be used. When $\dot{\varepsilon}_s$ is based on creep curves of shorter duration, the value of minimum creep rate obtained is larger than the true value. Thus, the error is on the conservative side. It has also been shown for long-time creep tests ($t > 10,000$ h) that the creep strength based on 1 percent creep strain is essentially equal to the creep strength based on true minimum creep rate. This may provide an economy of time in determining design values of $\dot{\varepsilon}_s$.

Another method of presenting creep data is a plot of stress vs. the time to produce different amounts of total strain (instantaneous strain plus creep strain)

---

[1] In engineering usage $\varepsilon_s$ is often called the *minimum creep rate*. The reason for the terminology is evident from Fig. 13-5.

[2] R. F. Gill and R. M. Goldhoff. *Met. Eng. Q.*, vol. 10, pp. 30–39, August 1970.

**Fig. 13-17**   Log-log plot of stress vs. minimum creep rate.   For 16-25-6 alloy.   (*Courtesy C. L. Clark, Timken Roller Bearing Co.*)

(Fig. 13-18).   The uppermost curve in Fig. 13-18 is the stress-rupture curve.   The percentages beside each data point are the percentage reduction of area at failure.

For short-time high-temperature applications (such as in missiles and high-speed aircraft), data are needed at higher temperatures and stresses and at shorter times than are usually determined for creep tests.   From a set of creep curves at constant temperature and various stresses (Fig. 13-3) it is possible to construct stress-strain curves by drawing lines at fixed times, for example, $t = 0, 1, 10, 100$ h.   The resulting fictitious stress-strain curves are called *isochronus stress-strain curves*.

It often is cheaper and more convenient to conduct stress-rupture tests than creep tests, and it would be very useful if creep strength could be estimated from rupture strength with sufficient accuracy for design purposes.   Monkman and Grant[1] showed empirically that a log-log plot of minimum creep rate vs. time to rupture results in a straight line

$$\log t_r + C \log \dot{\varepsilon}_s = K \qquad (13\text{-}25)$$

where $C$ and $K$ are constants for a given alloy.   Goldhoff and Gill[2] have shown a correlation between the rupture strength and the creep strength which should prove useful for predicting creep data from stress-rupture tests.

[1] F. C. Monkman and N. J. Grant, *Am. Soc. Test. Mater. Proc.*, vol. 56, pp. 593–620, 1956.
[2] R. M. Goldhoff and R. F. Gill, *Trans. ASME, Ser. D: J. Basic Eng.*, vol. 94, pp. 1–6, March 1972.

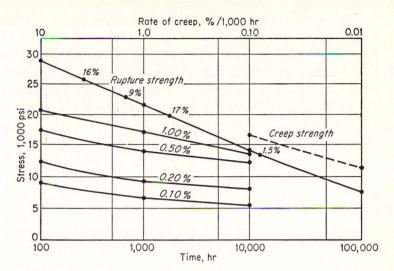

**Fig. 13-18** Deformation-time curves at 1300°F for 16-25-6 alloy. (*From C. L. Clark, in " Utilization of Heat Resistant Alloys," p. 40, American Society for Metals, Metals Park, Ohio, 1954.*)

## 13-13  PREDICTION OF LONG-TIME PROPERTIES

Frequently high-temperature strength data are needed for conditions for which there is no experimental information. This is particularly true of long-time creep and stress-rupture data, where it is quite possible to find that the creep strength to give 1 percent deformation in 100,000 h (11.4 years) is required, although the alloy has been in existence for only 2 years. Obviously, in such situations extrapolation[1] of the data to long times is required. Reliable extrapolation of creep and stress-rupture curves to longer times can be made only when it is certain that no structural changes occur in the region of extrapolation which would produce a change in the slope of the curve. Since structural changes generally occur at shorter times for higher temperatures, one way of checking on this point is to examine the log-stress–log-rupture life plot at a temperature several hundred degrees above the required temperature. For example, if in 1,000 h no change in slope occurs in the curve at 200°F above the required temperature, extrapolation of the lower temperature curve as a straight line to 10,000 h is probably safe and extrapolation even to 100,000 h may be possible. A logical method of extrapolating stress-rupture curves which takes into consideration the changes in slope due to structural changes has been proposed by Grant and Bucklin.[2]

[1] A. Mendelson and S. S. Manson, *Trans. ASME, Ser. D: J. Basic Eng.*, vol. 82, pp. 839–847, 1960.
[2] N. J. Grant and A. G. Bucklin, *Trans. Am. Soc. Met.*, vol. 42, pp. 720–761, 1950.

As an aid in extrapolation of stress-rupture data several time-temperature parameters have been proposed for trading off temperature for time.[1] The basic idea of these parameters is that they permit the prediction of long-time rupture behavior from the results of shorter time tests at higher temperatures at the same stress. The temperature-compensated time parameter $\theta$ provides a rational basis for the development of these parameters. Taking the logarithm of both sides of Eq. (13-8)

$$\log \theta = \log t - M\frac{\Delta H}{RT}$$

$$\log t = \log \theta + \frac{M}{R}\frac{\Delta H}{T} \qquad (13\text{-}26)$$

where $M = \log e = 0.434$. If $\theta$ and $\Delta H/R$ are functions of stress only, then Eq. (13-26) is linear in $\log t$ and $1/T$. While emphasis is given to time to rupture in parameter methods these techniques also can be used with creep data by using the time to reach a given strain.

Larson and Miller[2] showed that experimental stress-rupture data plotted in accordance with Eq. (13-26). See Fig. 13-19. Since the line for each constant stress converges to a common point on the $\log t$ axis this plot indicates that $\Delta H$ varies with stress but $\theta$ does not. The point of convergence is $\log \theta = -C_1$. For most alloys $C_1$ varies from about 15 to 25 depending on material. Figure 13-19 also in-

**Fig. 13-19** Stress-rupture data at various stresses plotted according to Eq. (13-26).

[1] The development of time-temperature parameters from reaction-rate theory is considered by M. Grounes, *Trans. ASME, Ser. D: J. Basic Eng.*, vol. 91, pp. 59–62, March 1969.
[2] F. R. Larson and J. Miller, *Trans. ASME*, vol. 74, pp. 765–771, 1952.

Table 13-2  TIME COMPRESSION OF
OPERATING CONDITIONS
BASED ON LARSON-MILLER
PARAMETER $C_1 = 20$

| Operating conditions | Larson-Miller test conditions |
|---|---|
| 10,000 h at 1000°F | 13 h at 1200°F |
| 1,000 h at 1200°F | 12 h at 1350°F |
| 1,000 h at 1350° | 17 h at 1500°F |

dicates that the slope $b = (M/R)\Delta H$ is a function of stress. From Eq. (13-26) the Larson-Miller parameter[1] can be formulated as

$$T(\log t + C_1) = b = f(\sigma) = P_1 \qquad (13\text{-}27)$$

or

$$T(\log t + C_1) = P_1$$

where $T$ = the test temperature  °R = °F + 460

$t$ = the time to rupture  h

$C_1$ = the Larson-Miller constant  often assumed to be 20

Table 13-2 illustrates how Eq. (13-27) provides a time compression at the expense of increased temperature.  When "trading-up" in temperature through the Larson-Miller parameter, it is important to be sure that the higher-temperature does not cause a change in metallurgical structure.

If $P_1$ is evaluated for pairs of $t$ and $T$ obtained over a range of stress, a single master curve is obtained for the material (Fig. 13-20). A plot such as this is a good

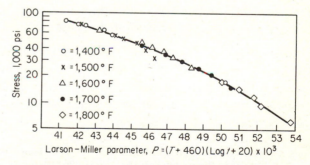

Fig. **13-20**  Master curve for Larson-Miller parameter for Astroloy.

[1] The Larson-Miller parameter expressed in terms of minimum creep rate instead of time to rupture is $T(C_1 - \log \dot{\varepsilon}_s) = P_1$.

way of summarizing for comparison the stress-rupture characteristics of several materials. To test that the value of $C_1$ is appropriate for the material, the values of $P_1$ should all fit on a smooth-curve over the useful range of stress. The master curve allows the Larson-Miller parameter to be used for time-temperature trade-off at a variety of stresses. It is also used in extrapolation. For example, if we want to know the 50,000-h rupture stress for Astroloy at 1300°F, we first evaluate $P_1$.

$$P_1 = (1,300 + 460)(\log 50,000 + 20) = 43,000$$

and from Fig. 13-20 we find a creep-rupture stress of about 60,000 psi.

While the Larson-Miller parameter has found wide application in high-temperature materials design, other parameters have been proposed for the same purpose. The simplest is the temperature-compensated time $\theta$ (sometimes called the *Dorn parameter*) given by Eq. (13-8). Time to rupture may be normalized with respect to temperature by the parameter $\theta$, and $\theta$ plots vs. stress in a master curve for the material. For the Dorn parameter to fit the data, a plot of $\log t$ versus $1/T$ yields a series of parallel lines, one line for each value of stress. This indicates that $\Delta H$ is not a function of stress. In general, high-temperature materials do not correlate well with this parameter.

In some materials the data do not fall on a good single curve of $\log \sigma$ versus $T(\log t + C_1)$ even when $C_1$ is adjusted for the best fit. This can lead to considerable error in the extrapolation to long rupture times from short-time data. The Manson-Haferd parameter[1] has additional flexibility which may be useful in such instances. In general, the Manson-Haferd parameter[2] gives the best agreement in extrapolation of stress-rupture data.[3] It has the form

$$\frac{T - T_a}{\log t - \log t_a} = P_2 \qquad (13\text{-}28)$$

where   $T$ = the test temperature, °F

$t$ = the rupture or creep time, h

$T_a$ and $t_a$ = constants derived from test data

This is often called a linear parameter because $\log t$ is linear with $T$ at a constant stress. The lines obtained at different stresses converge toward a point with coordinates $T_a, \log t_a$ (although for best results these constants should be obtained from a least-squares analysis[4]). A master plot, similar to that in Fig. 13-20, is needed to take different stress levels into account.

[1] S. S. Manson and A. M. Haferd, *NACA Tech. Note* 2890, March 1953; S. S. Manson G. Succop, and W. R. Brown, Jr., *Trans. Am. Soc. Met.*, vol. 51, pp. 911–934, 1959.

[2] A detailed example of the use of the Larson-Miller and Manson-Haferd parameters is given by J. D. Lubahn and R. P. Felgar, "Plasticity and Creep in Metals," pp. 204–217, John Wiley & Sons, Inc. New York, 1961.

[3] R. M. Goldhoff, *Trans. ASME, Ser. D: J. Basic Eng.*, vol. 81, pp. 629–644, 1959; W. Betteridge, *J. Inst. Met.*, vol. 86, pp. 232–237, 1957–1958.

[4] J. B. Conway, "Stress-Rupture Parameters: Origin, Calculation, and Use," Gordon and Breach, Science Publishers, Inc., New York, 1969.

## 13-14 CREEP UNDER COMBINED STRESSES

Considerable attention has been given to the problem of design for combined stress conditions during steady-state creep.[1] In the absence of metallurgical changes, the basic simplifying assumptions of plasticity theory hold reasonably well for these conditions. The assumption of incompressible material leads to the familiar relationship $\dot{\varepsilon}_1 + \dot{\varepsilon}_2 + \dot{\varepsilon}_3 = 0$. The assumption that principal shear-strain rates are proportional to principal shear stresses gives

$$\frac{\dot{\varepsilon}_1 - \dot{\varepsilon}_2}{\sigma_1 - \sigma_2} = \frac{\dot{\varepsilon}_2 - \dot{\varepsilon}_3}{\sigma_2 - \sigma_3} = \frac{\dot{\varepsilon}_3 - \dot{\varepsilon}_1}{\sigma_3 - \sigma_1} = C \qquad (13\text{-}29)$$

Combining these equations results in

$$\dot{\varepsilon}_1 + \frac{2}{3C}[\sigma_1 - \tfrac{1}{2}(\sigma_2 + \sigma_3)] \qquad (13\text{-}30)$$

Similar expressions are obtained for $\dot{\varepsilon}_2$ and $\dot{\varepsilon}_3$.

For engineering purposes the stress dependence of the creep rate can be expressed by Eq. (13-20). For combined stress conditions $\dot{\varepsilon}$ and $\sigma$ must be replaced by the effective strain rate $\dot{\bar{\varepsilon}}$ and the effective stress $\bar{\sigma}$. Thus we can write

$$\dot{\bar{\varepsilon}} = B\bar{\sigma}^{n'} \qquad (13\text{-}31)$$

Combining Eqs. (13-30) and (13-31) results in

$$\dot{\varepsilon}_1 = B\bar{\sigma}^{n'-1}[\sigma_1 - \tfrac{1}{2}(\sigma_2 + \sigma_3)] \qquad (13\text{-}32)$$

The effective stress and the effective strain rate are useful parameters for correlating steady-state creep data. When plotted on log-log coordinates, they give a straight-line relationship. Correlation has been obtained between uniaxial creep tests, creep of thick-walled tubes under internal pressure, and tubes stressed in biaxial tension.[2]

## 13-15 STRESS RELAXATION

Stress relaxation is the time-dependent decrease in stress in a body which is constrained to a certain fixed deformation. Important practical examples are the relaxation of stress in bolted joints[3] and press-fitted assemblies. Consider a tension

[1] The original analysis of this problem was given by C. R. Soderberg, *Trans. ASME*, vol. 58, p. 733, 1936. Subsequent analysis has been made by I. Finnie, *Trans. ASME*, Ser. D; *J. Basic Eng.*, vol. 82, pp. 462–464, 1960. For a critical review see A. E. Johnson, *Met. Rev.*, vol. 5, pp. 447–506, 1960.

[2] E. A. Davis, *Trans. ASME*, Ser. D: *J. Basic Eng.*, vol. 82, pp. 453–461, 1960; A.I. Smith and A. M. Nicholson, (eds.): "Advances in Creep Design," John Wiley & Sons, Inc., New York, 1972; R. K. Penny, "Design for Creep," McGraw-Hill Book Company, New York, 1971.

[3] E. L. Robinson, *Am. Soc. Test. Mater. Proc.*, vol. 48, p. 214, 1948.

specimen which is subjected to a total strain $\varepsilon$ at an elevated temperature where creep can occur.

$$\varepsilon = \varepsilon_e + \varepsilon_p = \frac{\sigma}{E} + \varepsilon_p \qquad (13\text{-}33)$$

where $\varepsilon$ = total strain

$\varepsilon_e$ = elastic strain

$\varepsilon_p$ = plastic (creep) strain

As the material creeps (lengthens) the total strain only can remain constant if the elastic strain decreases. Differentiating Eq. (13-33) with respect to time, and remembering that $d\varepsilon/dt = 0$,

$$\frac{d\varepsilon_e}{dt} = -\frac{d\varepsilon_p}{dt}$$

But $\varepsilon_e = \sigma/E$, and if $d\varepsilon_p/dt = B\sigma^{n'}$

$$\frac{1}{E}\frac{d\sigma}{dt} = -B\sigma^{n'} \qquad (13\text{-}34)$$

$$\int \frac{d\sigma}{\sigma^{n'}} = -BE \int dt$$

$$-\frac{1}{(n'-1)\sigma^{n'-1}} = -BEt + C$$

At $t = 0$, $\sigma = \sigma_i$, so that

$$C = -\frac{1}{(n'-1)\sigma_i^{n'-1}}$$

Therefore, the relation between stress and time in stress relaxation is given by

$$\frac{1}{\sigma^{n'-1}} = \frac{1}{\sigma_i^{n'-1}} + BE(n'-1)t \qquad (13\text{-}35)$$

$T_1 < T_2 < T_3$

$T_1$

$T_2$

$T_3$

Stress

Time

**Fig. 13-21** Stress-relaxation curves.

Figure 13-21 shows the type of relaxation curves that are obtained. The initial rate of decrease of stress is high, but it levels off because as the stress leve decreases, the creep rate decreases. These types of curves also can be used to estimate the time required to relieve residual stresses by thermal treatments. Lee and Hart[1] have shown how to convert stress-relaxation data into curves of stress vs. strain rate.

## BIBLIOGRAPHY

Clauss, F. J.: "Engineer's Guide to High Temperature Materials," Addison-Wesley, Reading, Mass., 1969.

Conway, S. B.: "Numerical Methods for Creep and Rupture," Gordon and Breach, Science Publishers, Inc., New York, 1967.

Dorn, J. E. (ed.): "Mechanical Behavior of Materials at Elevated Temperatures," McGraw-Hill Book Company, New York, 1961.

Garofalo, F.: "Fundamentals of Creep and Creep-Rupture in Metals," The Macmillan Company, New York, 1965.

*Joint Int. Conf. Creep*, Institution of Mechanical Engineers, London, 1963.

Kennedy, A. J.: "Processes of Creep and Fatigue in Metals," John Wiley & Sons, Inc., New York, 1963.

McLean, D.: "Deformation at High Temperatures," *Metall. Rev.*, vol. 7, no. 28, 1962.

Oding, I. A., U. S. Ivanova, V. V. Burdukskii, and V. N. Geminov: "Creep and Stress Relation in Metals," (English translation), Oliver & Boyd, Ltd., London, 1965.

[1] D. Lee and E. W. Hart, *Metall. Trans.* vol. 2, pp. 1245–1248, 1972.

# 14

# BRITTLE FRACTURE AND IMPACT TESTING

## 14-1 THE BRITTLE-FRACTURE PROBLEM

During World War II a great deal of attention was directed to the brittle failure of welded Liberty ships and T-2 tankers.[1] Some of these ships broke completely in two, while, in other instances, the fracture did not completely disable the ship. Most of the failures occurred during the winter months. Failures occurred both when the ships were in heavy seas and when they were anchored at dock. These calamities focused attention on the fact that normally ductile mild steel can become brittle under certain conditions. A broad research program was undertaken to find the causes of these failures and to prescribe the remedies for their future prevention. In addition to research designed to find answers to a pressing problem, other research was aimed at gaining a better understanding of the mechanism of brittle fracture and fracture in general. Many of the results of this basic work are described in Chap. 7, which should be reviewed before proceeding further with this chapter. While the brittle failure of ships concentrated great attention to

---

[1] M. L. Williams, Analysis of Brittle Behavior in Ship Plates, Symposium on Effect of Temperature on the Brittle Behavior of Metals with Particular Reference to Low Temperatures, *ASTM Spec. Tech. Publ.*, 158, pp. 11–44, 1954.

brittle failure in mild steel, it is important to understand that this is not the only application where brittle fracture is a problem. Brittle failures in tanks, pressure vessels, pipelines, and bridges have been documented[1] as far back as the year 1886.

Three basic factors contribute to a brittle-cleavage type of fracture. They are (1) a triaxial state of stress, (2) a low temperature, and (3) a high strain rate or rapid rate of loading. All three of these factors do not have to be present at the same time to produce brittle fracture. A triaxial state of stress, such as exists at a notch, and low temperature are responsible for most service failures of the brittle type. However, since these effects are accentuated at a high rate of loading, many types of impact tests have been used to determine the susceptibility of materials to brittle fracture. Steels which have identical properties when tested in tension or torsion at slow strain rates can show pronounced differences in their tendency for brittle fracture when tested in a notched-impact test. However, there are certain disadvantages to this type of test, so that much work has been devoted to the development of additional tests for defining the tendency for brittle fracture, and much effort has been expended in correlating the results of different brittle-fracture tests.

The greatest success has been achieved in applying the concepts of fracture mechanics (Sec. 7-5) to brittle fracture. Quantitative, predictive ability has been achieved for high-stength, low-plasticity materials. However, for the more usual structural steel which shows some plasticity prior to brittle fracture modifications to the fracture mechanics analysis are needed. For purposes of fracture analysis a low-yield-strength material is one with $\sigma_0 < E/300$ ($\sigma_0 < 100,000$ psi for steels) while a high-yield-strength material is one with $\sigma_0 > E/150$ ($\sigma_0 > 200,000$ psi for steels).

Since the ship failures occurred primarily in structures of welded construction, it was considered for a time that this method of fabrication was not suitable for service where brittle fracture might be encountered. A great deal of research has since demonstrated that welding, per se, is not inferior in this respect to other types of construction. However, strict quality control is needed to prevent weld defects which can act as stress raisers or notches. New electrodes have been developed that make it possible to make a weld with better properties than the mild-steel plate. The design of a welded structure is more critical than the design of an equivalent riveted structure, and much effort has gone into the development of safe designs for welded structures. It is important to eliminate all stress raisers and to avoid making the structure too rigid. To this end, riveted sections, known as crack arresters, were incorporated in some of the wartime ships so that, if a brittle failure did occur, it would not propagate completely through the structure.

---

[1] M. E. Shank, A Critical Survey of Brittle Failure in Carbon Plate Steel Structures Other than Ships, *ASTM Spec. Tech. Publ.*, 158, pp. 45–110, 1954.

## 14-2 NOTCHED-BAR IMPACT TESTS

Various types of notched-bar impact tests are used to determine the tendency of a material to behave in a brittle manner. This type of test will detect differences between materials which are not observable in a tension test. The results obtained from notched-bar tests are not readily expressed in terms of design requirements, since it is not possible to measure the components of the triaxial stress condition at the notch. Furthermore, there is no general agreement on the interpretation or significance of results obtained with this type of test.

A large number of notched-bar test specimens of different design have been used by investigators of the brittle fracture of metals. Two classes of specimens have been standardized[1] for notched-impact testing. Charpy bar specimens are used most commonly in the United States, while the Izod specimen is favored in Great Britain. The Charpy specimen has a square cross section (10 × 10 mm) and contains a 45° V notch, 2 mm deep with a 0.25-mm root radius. The specimen is supported as a beam in a horizontal position and loaded behind the notch by the impact of a heavy swinging pendulum (the impact velocity is approximately 16 ft/s). The specimen is forced to bend and fracture at a high strain rate on the order of $10^3$ s$^{-1}$. The Izod specimen, which is used rarely today, has either a circular or square cross section and contains a V notch near the clamped end. The difference in loading between the Charpy and Izod tests is shown in Fig. 14-1.

Plastic constraint at the notch produces a triaxial state of stress similar to that shown in Fig. 7-19. The maximum plastic stress concentration is given by

$$K_\sigma = \left(1 + \frac{\pi}{2} - \frac{\omega}{2}\right) \qquad (14\text{-}1)$$

where $\omega$ is the included flank angle of the notch. The relative values of the three principal stresses depend strongly on the dimensions of the bar and the details of

Charpy V-notch (top view)          Izod (side view)

**Fig. 14-1** Sketch showing method of loading in Charpy and Izod impact tests.

[1] ASTM Standards, pt. 31, 1969, Designation E23-566.

the notch. The standard specimen is thick enough to ensure a high degree of plane-strain loading and triaxiality across almost all of the notched cross section. Thus, the standard Charpy V-notch specimen provides a severe test for brittle fracture. Therefore, nonstandard specimens[1] should be used with great care.

The principal measurement from the impact test is the energy absorbed in fracturing the specimen. After breaking the test bar, the pendulum rebounds to a height which decreases as the energy absorbed in fracture increases. The energy absorbed in fracture, usually expressed in foot-pounds, is read directly from a calibrated dial on the impact tester. The energy required for fracture of a Charpy specimen is often designed $C_V$, for example, $C_V$ 25 ft-lb. In Europe impact test results are frequently expressed in energy absorbed per unit cross-sectional area of the specimen. It is important to realize that fracture energy measured by the Charpy test is only a relative energy and cannot be used directly in design equations.

Another common measurement obtained from the Charpy test results from the examination of the fracture surface to determine whether the fracture is fibrous (shear fracture), granular (cleavage fracture), or a mixture of both. These different modes of failure are readily distinguishable even without magnification. The flat facets of cleavage fracture provide a high reflectivity and bright appearance, while the dimpled surface of a ductile fibrous fracture provides a light-absorptive surface and dull appearance. Usually an estimate is made of the percentage of the fracture surface that is cleavage (or fibrous) fracture. Figure 14-2 shows how the fracture appearance changes from 100 percent flat cleavage (left) to 100 percent fibrous fracture (right) as the test temperature is increased. Note that the fibrous

**Fig. 14-2** Fracture surfaces of Charpy specimens tested at different temperatures. *Left*, 40°F *center*, 100°F; *right*, 212°F. Note gradual decrease in the granular region and increase in lateral contraction at the notch with increasing temperature.

[1] R. C. McNicol, *Weld. J.*, vol. 44, pp. 1s–9s, 1965.

**Fig. 14-3** Transition-temperature curves for two steels, showing fallacy of depending on room-temperature results.

fracture appears first around the outer surface of the specimen (shear lip) where the triaxial constraint is the least. A third measurement that is sometimes made in the Charpy test is the ductility, as indicated by the percent contraction of the specimen at the notch.

The notched-bar impact test is most meaningful when conducted over a range of temperature so that the temperature at which the ductile-to-brittle transition takes place can be determined. Figure 14-3 illustrates the type of curves which are obtained. Note that the energy absorbed decreases with decreasing temperature but that for most cases the decrease does not occur sharply at a certain temperature. This makes it difficult to determine accurately the transition temperature. In selecting a material from the standpoint of notch toughness or tendency for brittle failure the important factor is the transition temperature. Figure 14-3 illustrates how reliance on impact resistance at only one temperature can be misleading. Steel *A* shows higher notch toughness at room temperature; yet its transition temperature is higher than that of steel *B*. The material with the lowest transition temperature is to be preferred.

Notched-bar impact tests are subject to considerable scatter,[1] particularly in the region of the transition temperature. Most of this scatter is due to local variations in the properties of the steel, while some is due to difficulties in preparing perfectly reproducible notches. Both notch shape and depth are critical variables, as is the proper placement of the specimen in the impact machine.

The principal advantage of the Charpy V-notch impact test is that it is a relatively simple test that utilizes a relatively cheap, small test specimen. Tests can readily be carried out over a range of subambient temperatures. Moreover, the design of the test specimen is well suited for measuring differences in notch toughness in low-strength materials such as structural steels. The test is used for comparing the influence of alloy studies and heat treatment on notch toughness. It

---

[1] N. H. Fahey, Impact Testing of Metals, *ASTM Spec. Tech. Publ.*, 466, pp. 76–92, 1970.

frequently is used for quality control and material acceptance purposes. The chief difficulty is that the results of the Charpy test are difficult to use in design. Since there is no measurement in terms of stress level,[1] it is difficult to correlate $C_V$ data with service performance. Moreover, there is no correlation of Charpy data with flaw size. In addition, the large scatter inherent in the test may make it difficult to determine well-defined transition-temperature curves.

## 14-3   SIGNIFICANCE OF TRANSITION-TEMPERATURE CURVE

The chief engineering use of the Charpy test is in selecting materials which are resistant to brittle fracture by means of transition-temperature curves. The design philosophy is to select a material which has sufficient notch toughness when subjected to severe service conditions so that the load-carrying ability of the structural member can be calculated by standard strength of materials methods without considering the fracture properties of the material or stress-concentration effects of cracks or flaws.

The transition-temperature behavior of a wide spectrum of materials falls into the three categories shown in Fig. 14-4. Medium- and low-strength fcc metals and most hcp metals have such high notch toughness that brittle fracture is not a problem unless there is some special reactive chemical environment. High-strength materials ($\sigma_0 > E/150$) have such low notch toughness that brittle fracture can occur at nominal stresses in the elastic range at all temperatures and strain rates when flaws are present. High-strength steel, aluminum and titanium alloys fall into this category. At low temperature fracture occurs by brittle cleavage,

**Fig. 14-4**   Effect of temperature on notch toughness (schematic).

---

[1] An instrumented Charpy test which measures the dynamic load on the specimen can give quantitative stress measurements; see T. R. Wilshaw, *J. Iron Steel Inst. London*, vol. 204, p. 935, 1966; R. A. Wullaert, *ASTM Spec. Tech. Publ.* No. 466, pp. 148–164, 1970.

while at higher temperatures fracture occurs by low-energy rupture. It is under these conditions that fracture mechanics analysis is useful and appropriate. The notch toughness of low- and medium-strength bcc metals, as well as Be, Zn, and ceramic materials is strongly dependent on temperature. At low temperature the fracture occurs by cleavage while at high temperature the fracture occurs by ductile rupture. Thus, there is a transition from notch brittle to notch tough behavior with increasing temperature. In metals this transition occurs at 0.1 to 0.2 of the absolute melting temperature $T_m$, while in ceramics the transition occurs at about 0.5 to $0.7T_m$.

The design philosophy using transition-temperature curves centers about the determination of a temperature above which brittle fracture will not occur at elastic stress levels. Obviously, the lower this transition temperature, the greater the fracture toughness of the material. The shape of the typical $C_V$ versus temperature curve (Fig. 14-3) shows that there is no single criterion that defines the transition temperature. The various definitions of transition temperature obtained from an energy vs. temperature curve or a fracture appearance vs. temperature curve are illustrated in Fig. 14-5. The most conservative criterion for transition temperature is to select $T_1$, corresponding to the upper shelf in fracture energy and the temperature above which the fracture is 100 percent fibrous (0 percent cleavage). This transition temperature criterion is called the *fracture transition plastic* (FTP). The FTP is the temperature at which the fracture changes from totally ductile to substantially brittle. The probability of brittle fracture is negligible above the FTP.

The use of the FTP is very conservative and in many applications it may be impractical. An arbitrary, but less conservative criterion is to base the transition

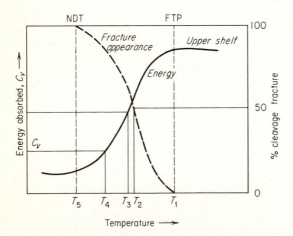

**Fig. 14-5** Various criteria of transition temperature obtained from Charpy test.

temperature on 50 percent cleavage–50 percent shear, $T_2$. This is called a *fracture-appearance transition temperature* (FATT). Correlations between Charpy impact tests and service failures indicate[1] that less than 70 percent cleavage fracture in the Charpy bar indicates a high probability that failure will not occur at or above the temperature if the stress does not exceed about one-half of the yield stress. Roughly similar results are obtained by defining the transition temperature as the average of the upper and lower shelf values, $T_3$.

A common criterion is to define the transition temperature $T_4$ on the basis of an arbitrary low value of energy absorbed $C_V$. This is often called the *ductility transition temperature*. As a result of extensive tests on World War II steel ship plates, it was established that brittle fracture would not initiate if $C_V$ was equal to 15 ft-lb at the test temperature. A 15 ft-lb $C_V$ transition temperature has become an accepted criterion for low-strength ship steels. However, it is important to realize that for other materials $C_V 15$ bears no special significance. For higher strength steel the corresponding value would exceed $C_V 15$, but in general this value is not known.

A well-defined criterion is to base the transition temperature on the temperature at which the fracture becomes 100 percent cleavage, $T_5$. This point is known as *nil ductility temperature* (NDT). The NDT is the temperature at which fracture initiates with essentially no prior plastic deformation. Below the NDT the probability of ductile fracture is negligible.

### 14-4 METALLURGICAL FACTORS AFFECTING TRANSITION TEMPERATURE

Changes in transition temperature of over 100°F can be produced by changes in the chemical composition or microstructure of mild steel. The largest changes in transition temperature result from changes in the amount of carbon and manganese.[2] The 15 ft-lb transition temperature for V-notch Charpy specimens (ductility transition) is raised about 25°F for each increase of 0.1 percent carbon. This transition temperature is lowered about 10°F for each increase of 0.1 percent manganese. Increasing the carbon content also has a pronounced effect on the maximum energy and the shape of the energy transition-temperature curves (Fig. 14-6). The Mn:C ratio should be at least 3:1 for satisfactory notch toughness. A maximum decrease of about 100°F in transition temperature appears possible by

[1] T. J. Hodgson and G. M. Boyd, *Trans. Inst. Nav. Archit. London*, vol. 100, p. 141, 1958.
[2] J. A. Rinebolt and W. J. Harris, Jr., *Trans. Am. Soc. Met.*, vol. 43, pp. 1175–1214, 1951; for a broad review see A. S. Tetelman and A. J. McEvily, Jr., "Fracture of Structural Materials," chap. 10, John Wiley & Sons, Inc., New York, 1967.

**Fig. 14-6** Effect of carbon content on the energy-transition-temperature curves for steel. (*From J. A. Rinebolt and W. J. Harris, Jr., Trans. Am. Soc. Met., vol. 43, p. 1197, 1951.*)

going to higher Mn:C ratios. The practical limitations to extending this beyond 7:1 are that manganese contents above about 1.4 percent lead to trouble with retained austenite, while about 0.2 percent carbon is needed to maintain the required tensile properties.

Phosphorus also has a strong effect in raising the transition temperature. The 15 ft-lb V-notch Charpy transition temperature is raised about 13°F for each 0.01 percent phosphorus. Since it is necessary to control phosphorus, it is not generally advisable to use steel made by the Bessemer process for low-temperature applications. The role of nitrogen is difficult to assess because of its interaction with other elements. It is, however, generally considered to be detrimental to notch toughness. Nickel is generally accepted to be beneficial to notch toughness in amounts up to 2 percent and seems to be particularly effective in lowering the ductility transition temperature. Silicon, in amounts over 0.25 percent, appears to raise the transition temperature. Molybdenum raises the transition almost as rapidly as carbon, while chromium has little effect.

Notch toughness is particularly influenced by oxygen. For high-purity iron[1] it was found that oxygen contents above 0.003 percent produced intergranular fracture and corresponding low energy absorption. When the oxygen content was raised from 0.001 percent to the high value of 0.057 percent, the transition temperature was raised from 5 to 650°F. In view of these results, it is not surprising

---

[1] W. P. Rees, B. E. Hopkins, and H. R. Tipler, *J. Iron Steel Inst. London,* vol. 172, pp. 403–409, 1952

that deoxidation practice has an important effect on the transition temperature. Rimmed steel, with its high iron oxide content, generally shows a transition temperature above room temperature. Semikilled steels, which are deoxidized with silicon, have a lower transition temperature, while for steels which are fully killed with silicon plus aluminum the 15 ft-lb transition temperature will be around $-75°F$. Aluminum also has the beneficial effect of combining with nitrogen to form insoluble aluminum nitride.

Grain size has a strong effect on transition temperature. An increase of one ASTM number in the ferrite grain size (actually a decrease in grain diameter) can result in a decrease in transition temperature of 30°F for mild steel. Decreasing the grain diameter from ASTM grain size 5 to ASTM grain size 10 can change the 10 ft-lb Charpy V-notch transition temperature[1] from about 70 to $-60°F$. A similar effect of decreasing transition temperature with decreasing austenite grain size is observed with higher alloyed heat-treated steels. Many of the variables concerned with processing mild steel affect the ferrite grain size and therefore affect the transition temperature. Since normalizing after hot rolling results in a grain refinement, if not carried out at too high a temperature, this treatment results in reduced transition temperature. The cooling rate from the normalizing treatment and the deoxidation practice are variables which also must be considered. Air cooling and aluminum deoxidation result in a lower transition temperature. Using the lowest possible finishing temperature for hot rolling of plate is also beneficial. Spray cooling from the rolling temperature before coiling[2] can lower the transition temperature 100°F.

An important development which has resulted in high-strength, low-alloy ferritic steels ($\sigma_0 = 60,000$ psi) with good impact properties (FATT below $-20°F$) is the addition of small amounts of Cb or V. This causes CbC precipitates to form during transformation. The dispersion strengthening raises the yield strength, while at the same time retarding grain growth and improving the impact resistance.

For a given chemical composition and deoxidation practice, the transition temperature will be appreciably higher in thick hot-rolled plates than in thin plates. This is due to the difficulty of obtaining uniformly fine pearlite and grain size in a thick section. Generally speaking, allowance for this effect must be made in plates greater than $\frac{1}{2}$in in thickness.

The notched-impact properties of rolled or forged products vary with orientation in the plate or bar. Figure 14-7 shows the typical form of the energy-temperature curves for specimens cut in the longitudinal and transverse direction of a rolled plate. Specimens A and B are oriented in the longitudinal direction of the plate. In specimen A the notch is perpendicular to the plate, while in B the notch

[1] W. S. Owen, D. H. Whitmore, M. Cohen, and B. L. Averbach, *Weld. J.*, vol. 36, pp. 503s–511s, 1957.
[2] J. H. Bucher and J. D. Grozier, *Met. Eng. Q.*, vol. 5, no. 1, 1965.

**Fig. 14-7** Effect of specimen orientation of Charpy transition-temperature curves.

lies parallel to the plate surface. The orientation of the notch in specimen *A* is generally preferred. In specimen *C* this notch orientation is used, but the specimen is oriented transverse to the rolling direction. Transverse specimens are used in cases where the stress distribution is such that the crack would propagate parallel to the rolling direction. Reference to Fig. 14-7 shows that quite large differences can be expected for different specimen orientations at high energy levels, but the differences become much less at energy levels below 20 ft-lb. Since ductility transition temperatures are evaluated in this region of energy, it seems that specimen and notch orientation are not a very important variable for this criterion. If, however, materials are compared on the basis of room-temperature impact properties, orientation can greatly affect the results.

Low-carbon steels can exhibit two types of aging phenomena which produce an increase in transition temperature. *Quench aging* is caused by carbide precipitation in a low-carbon steel which has been quenched from around 1300°F. *Strain aging* occurs in low-carbon steel which has been cold-worked. Cold-working by itself will increase the transition temperature, but strain aging results in a greater increase, usually around 40 to 60°F. Quench aging results in less loss of impact properties than strain aging. The phenomenon of *blue brittleness*, in which a decrease in impact resistance occurs on heating to around 400°F, is due to strain aging.

It has been demonstrated many times that a tempered martensitic structure produces the best combination of strength and impact resistance of any micro-structure that can be produced in steel. In Chap. 9 it was shown that the tensile properties of tempered martensites of the same hardness and carbon content are alike, irrespective of the amount of other alloy additions. This generalization holds approximately for the room-temperature impact resistance of heat-treated steels, but it is not valid for the variation of impact resistance with temperature. Figure 14-8 shows the temperature dependence of impact resistance for a number of different alloy steels, all having about 0.4 percent carbon and all with a tempered martensite structure produced by quenching and tempering to a hardness of $R_C$ 35. Note that a maximum variation of about 200°F in the transition temperature at the 20 ft-lb level is possible. Even greater spread in transition temperature would be obtained if the tempering temperature were adjusted to give a higher hardness.[1]

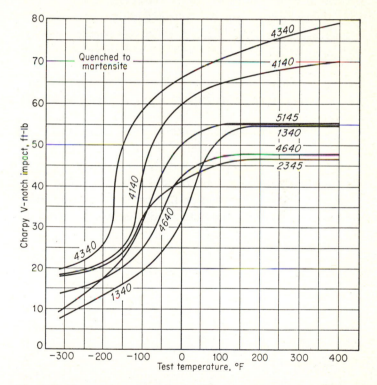

**Fig. 14-8** Temperature dependence of impact resistance for different alloy steels of same carbon content, quenched and tempered to $R_c$ 35. (*From H. J. French, Trans. Metall. Soc. AIME, vol. 206, p. 770, 1956.*)

[1] H. J. French, *Trans. AIME*, vol. 206, pp. 770–782, 1956.

Slack quenching so that the microstructure consists of a mixture of tempered martensite, bainite, and pearlite results in even greater differences between alloy steels and in a general increase in the transition temperature.

The energy absorbed in the impact test of an alloy steel at a given test temperature generally increases with increasing tempering temperature. However, there is a minimum in the curve in the general region of 400 to 600°F. This has been called 500°F *embrittlement*, but because the temperature at which it occurs depends on both the composition of the steel and the tempering time, a more appropriate name is *tempered-martensite embrittlement*. Embrittlement of steel in this tempering region is one of the chief deterrents to using steels at strength levels much above 200,000 psi. Studies of this embrittlement phenomenon have shown[1] that it is due to the precipitation of platelets of cementite from $\varepsilon$-carbide during the second stage of tempering. These platelets have no effect on the reduction of area of a tensile specimen, but they severely reduce the impact resistance. They can be formed at temperatures as low as 212°F and as high as 800°F, depending on the time allowed for the reaction. Silicon additions of around 2.25 percent are effective in increasing the temperature at which the platelets precipitate, and this permits tempering in the region 500 to 600°F, without severe embrittlement.

## 14-5 DROP-WEIGHT TEST AND OTHER LARGE-SCALE TESTS

Probably the chief deficiency of the Charpy impact test is that the small specimen is not always a realistic model of the actual situation. Not only does the small specimen lead to considerable scatter, but a specimen with a thickness of 0.394 in cannot provide the same constraint as would be found in a structure with a much greater thickness. The situation that can result is shown in Fig. 14-9. At a particular service temperature the standard Charpy specimen shows a high shelf energy,

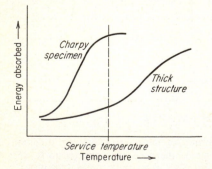

Fig. 14-9  Effect of section thickness on transition-temperature curves.

[1] L. J. Klingler, W. J. Barnett, R. P. Frohmberg, and A. R. Troiano, *Trans. Am. Soc. Met.*, vol. 46, pp. 1557–1598, 1954.

while actually the same material in a thick-section structure has low toughness at the same temperature.

The most logical approach to this problem is the development of tests that are capable of handling specimens at least several inches thick. The development of such tests and their rational method of analysis has been chiefly the work of Pellini[1] and his coworkers at the Naval Research Laboratory. The basic need for large specimens resulted from the inability to produce fracture in small laboratory specimens at stresses below the gross yield stress, whereas brittle fractures in ship structures occur at service temperatures at elastic stress levels.

The first development[2] was the explosion-crack-starter test which featured a short, brittle weld bead deposited on the surface of a 14 × 14 × 1 in steel plate. The plate was placed over a circular die and dynamically loaded with an explosive charge. The brittle weld bead introduces a small natural crack in the test plate similar to a weld-defect crack. Tests are carried out over a range of temperature and the appearance of the fracture determines the various transition temperatures (Fig. 14-10). Below the NDT the fracture is a flat (elastic) fracture running completely to the edges of the test plate. Above the nil ductility temperature a plastic bulge forms in the center of the plate, but the fracture is still a flat elastic fracture out to the plate edge. At still higher temperature the fracture does not propagate outside of the bulged region. The temperature at which elastic fracture no longer propagates to the edge of the plate is called the *fracture transition elastic* (FTE). The FTE marks the highest temperature of fracture propagation by purely elastic stresses. At yet higher temperature the extensive plasticity results in a helmet-type bulge. The temperature above which this fully ductile tearing occurs is the *fracture transition plastic* (FTP).

**Fig. 14-10** Fracture appearance vs. temperature for explosion-crack-starter test.

[1] W. S. Pellini, *Weld. J.*, vol. 50, pp. 915–1095, 147s–162s, 1971.
[2] P. P. Puzak, M. E. Shuster ,and W. S. Pellini, *Weld. J.*, vol. 33, p. 481s, 1954.

**Fig. 14-11** Drop-weight test (DWT).

*The drop-weight test* (DWT) was developed[1] specifically for the determination of the NDT on full thickness plates. A short bead of brittle weld metal is deposited on the surface of a plate, typically $3\frac{1}{2} \times 14 \times \frac{5}{8}$ to 1 in thick. A small notch is introduced in the weld bead and the specimen supported as a simple beam in a constant temperature bath (Fig. 14-11). The brittle weld bead is fractured at near yield-stress levels as a result of dynamic loading from a falling weight. The anvil stop restricts the deflection of the test specimen. Since the specimen is a wide beam loaded in three-point bending, this restriction limits the stress on the tension face of the plate to a value that does not exceed the yield stress. If the starter-crack propagates across the width of the plate on the tension surface to the edges, the test temperature is below the NDT. Complete separation on the compression side of the specimen is not required. The NDT is the highest temperature at which a nil ductility break is produced. The test is quite reproducible and the NDT can be determined to the nearest 10°F.

The *dynamic tear test*[2] (DT) is in effect a giant Charpy test (Fig. 14-12). While specimens are usually $\frac{5}{8}$ and 1 in thick, DT tests have been made on speci-

**Fig. 14-12** Dynamic-tear test (DT).

---

[1] P. P. Puzak, and W. S. Pellini, *NRL Rept.* 5831, Aug. 21, 1962; ASTM Standards, pt. 31, 1969, pp. 582–601, Designation E208-69.
[2] Early versions of this test were known as the drop-weight tear test (DWTT).

mens up to 12 in thick. The notch is an electron-beam weld which is embrittled metallurgically by alloying (Ti is added to produce a brittle Fe-Ti alloy). The narrow weld is fractured easily providing a reproducible sharp crack. As in the Charpy test, specimens are fractured over a range of temperature in a pendulum-type machine[1] and the energy absorbed in fracture is measured. However, while the maximum energy capacity of a standard Charpy impact tester is 240 ft-lb, the pendulum tester used with the DT test has a capacity of 10,000 ft-lb. The use of this test in fracture analysis will be discussed in Sec. 14-6.

Another important type of test is the *crack-arrest test*[2] which provides a relationship between the stress level and the ability of the material to arrest a rapidly propagating crack. Figure 14-13 illustrates the Robertson crack-arrest test. A uniform elastic tensile stress is applied to a plate specimen 6 in wide. A rapidly moving brittle fracture is initiated by impact loading at a starter crack on the cold side of the specimen. The crack propagates up a temperature gradient toward the hot side. The point across the specimen width at which the temperature is high enough to give enough ductility to blunt the crack is called the *crack-arrest temperature* (CAT). In an alternative form of the test, the temperature across the specimen is constant and tests are carried out with successive specimens at

**Fig. 14-13**  Robertson crack-arrest test.

[1] P. P. Puzak, and E. A. Lange, *NRL Rept.* 6851, Feb. 13, 1969; E. A. Lange and F. J. Loss, "Impact Testing of Metals," pp. 241–258, ASTM *Spec. Tech. Publ.* 466, 1970.
[2] T. S Robertson, *Engineering*, vol. 172, pp. 445–448; *J. Iron Steel Inst.* London, vol. 175, p. 361, 1953; F. J. Feely, M. S. Northrup, S. R. Kleepe, and M. Gensamer, *Weld. J.*, vol. 34, pp. 596s–607s, 1955.

increasing temperature until the CAT is reached. Crack-arrest tests on mild steel below the NDT show that the CAT is independent of temperature but the stress level for crack arrest is very low. If the stress is greater than 5,000 to 8,000 psi, brittle fracture will occur. Obviously, this stress level is too low for practical engineering design, so that steels cannot be used below the NDT. While the crack-arrest tests are among the most qualitative of brittle-fracture tests, they are not used extensively because they require large testing machines and large specimens.

## 14-6 FRACTURE ANALYSIS DIAGRAM

The previous sections have introduced a number of terms dealing with brittle fracture, such as NDT, FTE, FTP, etc. The tests for determining these transition temperatures have been described. Before seeing how they are used in engineering design through the fracture analysis diagram, we redefine these transition points through reference to basic properties of the tension test. The sub-ambient temperature dependence of yield strength $\sigma_0$ and ultimate tensile strength $\sigma_u$ in a bcc metal are shown in Fig. 14-14. For an unnotched specimen without flaws the material is ductile until a very low temperature, point $A$, where $\sigma_0 = \sigma_u$. Point $A$ represents the NDT temperature for a flaw-free material. The curve $BCD$ represents the fracture strength of a specimen containing a small flaw ($c < 0.1$). The temperature corresponding to point $C$ is the highest temperature at which the fracture strength $\sigma_f \approx \sigma_0$. Point $C$ represents the NDT for a specimen with a small crack or flaw. The presence of a small flaw raises the NDT of a steel by about

**Fig. 14-14** Temperature dependence of yield strength ($\sigma_0$), tensile strength ($\sigma_u$), and fracture strength for a steel containing flaws of different sizes.

200°F.  Increasing the flaw size decreases the fracture stress curve, as in curve
*EF*, until with increasing flaw size a limiting curve of fracture stress *HJKL* is
reached.  Below the NDT the limiting safe stress is 5,000 to 8,000 psi.  Above the
NDT the stress required for the unstable propagation of a long flaw (*JKL*) rises
sharply with increasing temperature.  This is the crack-arrest temperature curve
(CAT).  The CAT defines the highest temperature at which unstable crack propagation can occur at any stress level.  Fracture will not occur for any point to the right
of the CAT curve.

The temperature above which *elastic* stresses cannot propagate a crack is the
*fracture transition elastic* (FTE).  This is defined by the temperature when the
CAT curve crosses the yield-strength curve (point *K*).  The *fracture transition
plastic* (FTP) is the temperature where the CAT curve crosses the tensile-strength
curve (pont *L*).  Above this temperature the material behaves as if it were flaw-free,
for any crack, no matter how large, cannot propagate as an unstable fracture.

Data obtained from the DWT and other large-scale fracture tests have been
assembled by Pellini and coworkers[1] into a useful design procedure called the
*fracture analysis diagram* (FAD).  The NDT as determined by the DWT provides a
key data point to start construction of the fracture analysis diagram (Fig. 14-15).
For mild steel below the NDT the CAT curve is flat.  A stress level in excess of
5,000 to 8,000 psi causes brittle fracture regardless of the size of the initial flaw.
Extensive correlation between the NDT and Robertson CAT tests for a variety of
structural steels have shown that the CAT curve bears a fixed relationship to the
NDT temperature.  Thus, the NDT + 30°F provides a conservative estimate of

**Fig. 14-15**  Fracture-analysis diagram showing influence of
various initial flaw sizes.

[1] W. S. Pellini, and P. P. Puzak, *NRL Rept.* 5920, Mar. 15, 1963 and *Trans. ASME, Ser. A: J. Eng.
Power*, vol. 86, pp. 429–443, 1964.

the CAT curve at stress of $\sigma_0/2$. NDT + 60°F provides an estimate of the CAT at $\sigma = \sigma_0$, that is, the FTE and NDT + 120°F provides an estimate of the FTP. Therefore, for structural steels, once the NDT has been determined, the entire scope of the CAT curve can be established well enough for engineering design.[2]

The curve that has been traced out on Fig. 14-15 represents the worse possible case for large flaws in excess of 24 in. One can envision a spectrum of curves translated upward and to the left for smaller, less severe flaws. Correlation with service failures and other tests has allowed the approximate determination of curves for a variety of initial flaw sizes. Thus, the FAD provides a generalized relationship of flaw size, stress, and temperature for low-carbon structural steels of the type used in ship construction.

The fracture analysis diagram can be used several ways in design. One simple approach would be to use the FAD to select a steel which had an FTE that was lower than the lowest expected service temperature. With this criterion the worst expected flaw would not propagate so long as the stress remained elastic. Since the assumption of elastic behavior is basic in structural design, this design philosophy would be tantamount to being able to ignore the presence of flaws and brittle fracture. However, this procedure may prove to be too expensive and over-conservative. A slightly less conservative design against brittle fracture, but still a practical approach, would be to design on the basis of an allowable stress level not exceeding $\sigma_0/2$. From Fig. 14-15 we see that any crack will not propagate under this stress so long as the temperature is not below NDT + 30°F. If for example, the service temperature is not expected to be below 10°F, we would select a steel whose NDT is $10° - 30° = -20°F$.

The dynamic tear test (DT) can be used to construct the FAD as shown in Fig. 14-16. Below the NDT the fracture is brittle and has a flat, featureless surface devoid of any shear lips. At temperatures above the NDT there is a sharp rise in

Slant fracture

Shear lip

DT energy ⟶

NDT

Temperature

**Fig. 14-16**   Test results from DT test.

<hr>

[2] Tests on very thick sections (6 to 12 in.) show that the transition region is expanded from NDT + 120°F to NDT + 180°F to NDT + 220°F.

energy for fracture and the fracture surfaces begin to develop shear lips. The shear lips become progressively more prominent as the temperature is increased to the FTE. Above the FTE the fracture is ductile, void coalescence-type fracture. The fracture surface is a fibrous slant fracture. The upper shelf of energy represents the FTP. The lower half of the DT energy curve traces the temperature course of the CAT curve from NDT to FTE. The DT test is a highly versatile test because it is equally useful with low-strength ductile materials which show a high upper energy shelf and with high-strength low-toughness materials which have a low value of upper shelf energy. The large size of the DT specimen provides a high degree of triaxial constraint and results in a minimum of scatter. Extensive correlations are being developed between DT results and fracture toughness and $C_V$ test data.

## 14-7  $K_{Ic}$ PLANE-STRAIN TOUGHNESS TESTING

The concepts of crack-extension force and stress intensity factor have been introduced and thoroughly explored in Chap. 7. In this section we consider the testing procedures by which the fracture mechanics approach can be used to measure meaningful material properties. Since the methods of analysis are based on linear elastic fracture mechanics, these testing procedures are restricted to materials with limited ductility. Typical materials are high-strength steel, titanium, and aluminum alloys.

We have seen that the elastic stress field near a crack tip can be described by a single parameter called the *stress intensity factor K*. The magnitude of this stress intensity factor depends on the geometry of the solid containing the crack, the size and location of the crack, and the magnitude and distribution of the loads imposed on the solid. In Chap. 7 we saw that the criterion for brittle fracture in the presence of a crack-like defect was that unstable rapid failure would occur when the stresses at the crack tip exceeded a critical value. Since the crack-tip stresses can be described by the stress intensity factor $K$, a critical value of $K$ can be used to define the conditions for brittle failure. As the usual test involves the opening mode of loading (mode I) the critical value of $K$ is called $K_{Ic}$, *the plane-strain fracture toughness*. $K_{Ic}$ can be considered a material property which describes the inherent resistance of the material to failure in the presence of a crack-like defect. For a given type of loading and geometry the relation is

$$K_{Ic} = \alpha\sigma\sqrt{\pi a_c} \qquad (14\text{-}2)$$

where $\alpha$ is a parameter which depends on specimen and crack geometry and $a_c$ is the critical crack length. If $K_{Ic}$ is known, then it is possible to compute the maximum allowable stress for a given flaw size. While $K_{Ic}$ is a basic material property, in the same sense as yield strength, it changes with important variables such as temperature and strain rate. For materials with a strong temperature and strain-rate dependence $K_{Ic}$ usually decreases with decreased temperature and increased strain rate. For a given alloy, $K_{Ic}$ is strongly dependent on such metallurgical variables as heat treatment, texture, melting practice, impurities, inclusions, etc.

There has been so much research activity and rapid development[1] in the field of fracture-toughness testing that, in a period of about 10 years, it has evolved from a research activity to a standardized procedure.[2] In the discussion of the influence of a notch on fracture (Sec. 7-11), we saw that a notch in a thick plate is far more damaging than in a thin plate because it leads to a *plane-strain* state of stress with a high degree of triaxiality. The fracture toughness measured under plane-strain conditions is obtained under maximum constraint or material brittleness. The plane-strain fracture toughness is designated $K_{Ic}$ and is a true material property. Figure 14-17 shows how the measured fracture stress varies with specimen thickness $B$. A mixed-mode, ductile brittle fracture with 45° shear lips is obtained for thin specimens. Once the specimen has the critical thickness for the toughness of the material, the fracture surface is flat and the fracture stress is constant with increasing specimen thickness. The minimum thickness to achieve plane-strain conditions and valid $K_{Ic}$ measurements is

$$B = 2.5 \left( \frac{K_{Ic}}{\sigma_0} \right)^2 \qquad (14\text{-}3)$$

where $\sigma_0$ is the 0.2 percent offset yield strength.

Fig. 14-17 Effect of specimen thickness on stress and mode of fracture. (*From C. C. Osgood, Machine Design, August 5, 1971, p. 91.*)

[1] "Fracture Toughness Testing and its Applications," ASTM STP 381, 1965, "Plane Strain Crack Toughness Testing of High Strength Metallic Materials," ASTM *Spec. Tech. Publ.* 410, 1966; "Review of Developments in Plane Strain Fracture, Toughness Testing," ASTM *Spec. Tech. Publ.* 463, 1970.

[2] ASTM Standards, Designations E399-70T; A. S. Kobayashi (ed.), "Experimental Techniques in Fracture Mechanics," Society for Experimental Stress Analysis, Westport, Conn., 1973.

$a = B$
$W = 2B$

*Compact tension specimen*

$a = B$
$W = 2B$
$S = 8B$

*Bend specimen*

*Notched round specimen*

$$K_I = \frac{P}{D^{3/2}}\left[1.72\left(\frac{D}{d}\right) - 1.27\right]$$

**Fig. 14-18**  Common specimens for $K_{1c}$ testing.

A variety of specimens have been proposed for measuring $K_{1c}$ plane-strain fracture toughness. The three specimens shown in Fig. 14-18 represent the most common specimen designs. The compact tension specimen and the three-point loaded bend specimen have been standardized by ASTM.[1]  Other useful specimen designs are the single-edge-cracked-plate specimen (Fig. 14-17), the center-cracked and double-edge-cracked plate.[2]  After the notch is machined in the specimen, the sharpest possible crack is produced at the notch root by fatiguing the specimen in a low-cycle, high-strain mode (typically 1,000 cycles with a strain of 0.03 in/in.).  The

[1] ASTM Standards, pt. 31, Designation E399-70T.
[2] W. F. Brown and J. E. Srawley, *ASTM, Spec. Tech. Publ.* 410, 1966; J. M. Kraft, "Techniques of Metals Research," vol. V, pt. 2, chap. 7, John Wiley & Sons, Inc., New York, 1971.

initial crack length $a_i$ includes both the depth of the notch and the length of the fatigue crack and is approximately equal to the specimen width. Plane-strain toughness testing is unusual because there can be no advance assurance that a valid $K_{Ic}$ will be determined in a particular test. Equation (14-3) should be used with an estimate of the expected $K_{Ic}$ to determine the specimen thickness.

The test must be carried out in a testing machine which provides for a continuous autographic record of load $P$ and relative displacement across the open end of the notch (proportional to crack displacement). The three types of load–crack-displacement curves that are obtained for materials are shown in Fig. 14-19. The type I load-displacement curve represents the behavior for a wide variety of ductile metals in which the crack propagates by a tearing mode with increasing load. This curve contains no characteristic features to indicate the load corresponding to the onset of unstable fracture. The ASTM procedure is to first draw the secant line $OP_s$ from the origin with a slope that is 5 percent less than the tangent $OA$. This determines $P_s$. Next draw a horizontal line at a load equal to 80 percent of $P_s$ and measure the distance $x_1$ along this line from the tangent $OA$ to the actual curve. If $x_1$ exceeds one-fourth of the corresponding distance $x_s$ at $P_s$, the material is too ductile to obtain a valid $K_{Ic}$ value. If the material is not too ductile, then the load $P_s$ is designated $P_Q$ and used in the calculations described below.

The type II load-displacement curve has a point where there is a sharp drop in load followed by a recovery of load. The load drop represents a " pop in " which arises from sudden unstable, rapid crack propagation before the crack slows-down to a tearing mode of propagation. The same criteria for excessive ductility is applied to type II curves, but in this case $P_Q$ is the maximum recorded load.

Fig. 14-19 Load-displacement curves. (Note that slope $OP_s$ is exaggerated for clarity.)

The type III curve shows complete "pop in" instability where the initial crack movement propagates rapidly to complete failure. This type of curve is characteristic of a very brittle "elastic material."

The value of $P_Q$ determined from the load-displacement curve is used to calculate a conditional value of fracture toughness denoted $K_Q$. The equations relating specimen geometry (see Fig. 14-18), crack length $a$, and critical load $P_Q$ are Eqs. (14-4) and (14-5).

For the compact tension specimen:

$$K_Q = \frac{P_Q}{BW^{1/2}} \left[ 29.6 \left( \frac{a}{W} \right)^{1/2} - 185.5 \left( \frac{a}{W} \right)^{3/2} + 655.7 \left( \frac{a}{W} \right)^{5/2} \right.$$
$$\left. - 1017.0 \left( \frac{a}{W} \right)^{7/2} + 638.9 \left( \frac{a}{W} \right)^{9/2} \right] \qquad (14\text{-}4)$$

For the bend specimen:

$$K_Q = \frac{P_Q S}{BW^{3/2}} \left[ 2.9 \left( \frac{a}{W} \right)^{1/2} - 4.6 \left( \frac{a}{W} \right)^{3/2} + 21.8 \left( \frac{a}{W} \right)^{5/2} \right.$$
$$\left. - 37.6 \left( \frac{a}{W} \right)^{7/2} + 38.7 \left( \frac{a}{W} \right)^{9/2} \right] \qquad (14\text{-}5)$$

The crack length $a$ used in the equations is measured after fracture. Next calculate the factor $2.5 \, (K_Q/\sigma_0)^2$. If this quantity is less than both the thickness and crack length of the specimen, then $K_Q$ is equal to $K_{Ic}$ and the test is valid. Otherwise it is necessary to use a thicker specimen to determine $K_{Ic}$. The measured value of $K_Q$ can be used to estimate the new specimen thickness through Eq. (14-3).

## 14-8  DESIGN WITH FRACTURE MECHANICS

The criterion for failure in plane-strain fracture toughness is that crack growth to failure will occur whenever the magnitude of the crack-tip stress exceeds some critical condition $K_{Ic}$. While there is little question about this failure criterion for the type III load-displacement curves (Fig. 14-19), it could be argued that type I and II curves, where fracture resistance increases with displacement, do not satisfy this criterion. Nevertheless, it is usual practice to apply fracture mechanics equally to materials which exhibit the differing behavior shown in Fig. 14-19. If the relationship is known[1] between the stress intensity factor and the loading conditions and the geometry of the structure and the crack, it is possible to compute either the maximum allowable stress or flaw size which will not result in failure of the structure made from a material having a fracture toughness $K_{Ic}$.

---

[1] An exhaustive compendium will be found in P. C. Paris and G. C. Sih, "Stress Analysis of Cracks," *ASTM Spec. Tech. Publ.*, 381, p. 30–83, 1965.

$$\sigma = \frac{PD}{2t}$$

$$K_I^2 = \frac{1.21a\pi\sigma^2}{Q}$$

Sec. A-A

**Fig. 14-20**   Flaw geometry and design of cylindrical pressure vessel.

The procedure is best understood with an example.[1]  Consider a cylindrical pressure vessel (Fig. 14-20).  The dimensions and mechanical properties are as follows.

| | |
|---|---|
| $D = 20$ in | Material:   7079-T6 aluminum alloy |
| $t = 1$ in | $\sigma_0 = 65,000$ psi |
| | $K_{Ic} = 35,000$ psi $\sqrt{\text{in}}$ |
| Internal pressure fluctuation | $P = 0$ to 4,000 psi |
| Maximum life | 1,000 pressure cycles |

The stress analysis for this example is quite simple.  The largest stress in the vessel is the circumferential hoop stress, which can be computed from $\sigma = PD/2t$.  Substitution from above results in a maximum hoop stress of 40,000 psi, and since the applied pressure varies from 0 to 4,000 psi, the alternating hoop stress $\Delta\sigma$ is also 40,000 psi.  The crack is a semielliptical surface crack oriented with the major plane of the crack perpendicular to the uniform tensile hoop stress.  For this type of loading and geometry the stress intensity[2] is given by

$$K_I^2 = \frac{1.21a\pi\sigma^2}{Q} \qquad (14\text{-}6)$$

where $a =$ the surface crack depth, in

$\sigma =$ the applied nominal stress, psi

$Q = \phi^2 - 0.212\,(\sigma/\sigma_0)^2$, where $\phi$ is an elliptic integral of the second kind

Figure 14-21 shows the flaw-shape parameter $Q$ plotted against the crack-shape ratio $a/2c$.  These curves can be used for surface cracks and elliptical internal cracks.[3]  The critical crack length to cause failure in a single cycle of loading can be obtained from Eq. (14-6).

$$a_{cr} = \frac{K_{Ic}^2 Q}{1.21\pi\sigma^2} \qquad (14\text{-}7)$$

[1] W. G. Clark, Jr., *Trans. ASME, Ser. B: J. Eng. Ind.*, vol. 94, pp. 291–298, 1972.
[2] G. R. Irwin, *J. Appl. Mech.*, vol. 24, pp. 109–114, 1957.
[3] Internal cracks are less severe than surface cracks.  For internal elliptical cracks Eq. (14-6) becomes $K_I^2 = a\pi\sigma^2/Q$.

**Fig. 14-21**  Flaw shape parameter $Q$ for surface and internal elliptical cracks.

To evaluate the flaw shape parameter $Q$, we shall investigate a range of surface cracks from long shallow cracks to short deep flaws.

| Flaw length to depth ratio $2c/a$ | $Q$ | Critical flaw size, depth × length |
|---|---|---|
| 20:1 | 0.95 | 0.19 × 3.8 in |
| 4:1 | 1.40 | 0.28 × 1.12 in |
| 2:1 | 2.30 | 0.46 × 0.92 in |

Calculations of this type are commonly made in conjunction with *proof testing*, where the pressure vessel is subjected to a single prolonged pressure in excess of the operating pressure, e.g., 150 percent of the operating. For example, if the operating pressure were 2,670 psi, then a single pressurization to 4,000 psi would constitute a 150 percent proof test. If the vessel survives this test, we could state that any defects in the pressure vessel are smaller than those given in the above table.

Since the pressure vessel in our example must withstand 1,000 cycles of alternating pressure, 0 to 4,000 psi, the next problem is to determine the flaw sizes that will just grow to the critical sizes after 1,000 pressure cycles. Thus, we must consider fatigue crack propagation. In Sec. 12-7 we saw that fatigue crack growth could be expressed as

$$\frac{da}{dN} = A \, \Delta K^p \qquad (14\text{-}8)$$

where $p$ is the slope of the log $da/dN$ versus log $\Delta K$ curve (Fig. 12-12). The number of cycles required for a crack to grow from its initial length $a_i$ to a critical length $a_{cr}$ can be determined by numerically integrating the crack-growth rate data between appropriate limits or it can be obtained[1] from Eq. (14-9)

$$N = \left[\frac{2}{(p-2)AM^{p/2}\,\Delta\sigma^p}\right]\left[\frac{1}{a_i^{(p-2)/2}} - \frac{1}{a_{cr}^{(p-2)/2}}\right] \qquad (14\text{-}9)$$

and $M$ is a component geometry and flaw-shape parameter described by

$$K_I = \sigma\sqrt{Ma}$$

Values of $A$ and $p$, based on plots like Fig. 12-12, are given as $A = 4 \times 10^{-18}$ and $p = 3$. We could now solve Eq. (14-9) for $a_i$ when $N = 1,000$. However, much more information is obtained by solving Eq. (14-9) for $N$ starting with the smallest detectable crack (approximately 0.05 in) and working on up to the critical crack size. Figure 14-22 shows the cyclic life curves for the three crack geometries considered in this example. The maximum initial allowable flaw sizes for a 1,000-cycle life are $0.09 \times 1.80$ in (20:1), $0.17 \times 0.68$ in (4:1) and $0.33 \times 0.66$ in (2:1). If we wish to provide for a safety factor of two on cyclic life ($N = 2,000$) the allowable flaw sizes are $0.05 \times 1$ in $0.12 \times 0.48$ in, and $0.24 \times 0.48$ in.

**Fig. 14-22** Effect of crack geometry and size on fatigue life of 7379-T6 aluminum. (*From W. G. Clark, Jr., Trans. ASME, Ser. E: J. Appl. Mech., vol. 94, p. 296, 1972.*)

[1] E. T. Wessel, W. G. Clark, Jr., and W. K. Wilson, Engineering Methods for the Design and Selection of Materials Against Fracture, *DDC Rept.* AD801001, 1966.

## 14-9  FRACTURE TOUGHNESS OF LOW-STRENGTH MATERIALS

While linear elastic fracture mechanics works well for high-strength materials ($\sigma_0 > 200{,}000$ psi) it is less universally applicable for low- and medium-strength structural materials. Valid quantitative results can be obtained for these materials only at very low temperature, or in very thick sections, or at high strain rates. The need for a correction to crack length for the small plastic zone at the crack tip was recognized at an early stage in the development of linear elastic fracture mechanics (see Sec. 7-9). When the plastic zone becomes an appreciable fraction of the crack length other approaches are necessary.

Widespread plasticity at the crack tip enables the crack surfaces to move apart at the crack tip without an increase in crack length. This relative movement of the two crack faces at the crack tip is termed the *crack-opening displacement* (COD). It has been shown (Sec. 7-9) that the crack-extension force is related to COD by

$$\mathscr{G} = \sigma_0 \delta \qquad (14\text{-}10)$$

where $\mathscr{G}$ is the crack-extension force and $\delta$ is the crack-opening displacement. Since in linear elastic fracture it is well established that the material can be characterized by a critical value of $K$ or $\mathscr{G}$, it is logical to expect that a critical value of $\delta$ describes the material fracture behavior outside the range of validity of linear elastic analysis. While COD tests are not yet as plentiful as $K_{Ic}$ tests, there is strong encouragement that this approach will be a useful method of fracture analysis for low-strength materials.

Tentative standards for COD testing have been proposed.[1] Inasmuch as possible, the test techniques have been designed to be an extension of $K_{Ic}$ test techniques. A notched and precracked square specimen is loaded in three-point bending. The specimen thickness should be the same as the structure on which a fracture analysis is to be made. During the test the applied force is plotted autographically against the displacement of a clip gage located across the outer face of the notch. These plots will generally correspond to one of the three curves shown in Fig. 14-19. The critical value of gage displacement $V_c$ must be converted to crack-opening displacement using an equation of the type

$$\delta_c = V_c \frac{1}{1 + k[(a + z)/(w + a)]} \qquad (14\text{-}11)$$

where $z = $ the distance between the clip gage and the specimen surface
    $a = $ the crack length
    $w = $ the specimen width
    $k = $ a constant for a specimen of a particular geometry

---

[1] R. W. Nichols, et al., M. O. Dobson (ed.), "Practical Fracture Mechanics for Structural Steel," chap. F, Chapman and Hall, Ltd., London, 1969.

A different approach to describing the toughness of a material having appreciable plasticity before fracture is with the *J integral*.[1] The *J* integral eliminates the need to describe the behavior of the material near the crack tip by considering the average deformation energy over an arbitrary area around the crack. There is reason to believe that the critical value of the *J* integral is a material property in the same way that $K_{Ic}$ is a material for a linear elastic solid. Values of *J* integral can be found experimentally by using a series of specimens with slightly different crack lengths.[2]

## 14-10  TEMPER EMBRITTLEMENT

Temper embrittlement[3] refers to the decrease in notch toughness which often occurs in alloy steels when heated in, or cooled slowly through, the temperature range 800 to 1100°F. This can become a particularly important problem with heavy sections that cannot be cooled through this temperature region rapidly enough to suppress embrittlement. Temper embrittlement also can be produced by isothermal treatments in this temperature region. A time-temperature-embrittlement diagram exhibits a set of *C* curves. The embrittlement is reversible. Even severely embrittled steels can be deembrittled in a few minutes by heating above the nose of the *C* curve and cooling rapidly to room temperature.

The presence of temper embrittlement usually is determined by measuring the transition temperature with the notched-bar impact test. The hardness and tensile properties are not sensitive to the embrittlement, except for very extreme cases, but the transition temperature can be raised around 200°F by ordinary embrittling heat treatments. The fracture in a temper-embrittled steel is intergranular and propagates along prior austenitic grain boundaries. The embrittlement occurs only in the presence of specific impurities and increases in severity with the concentration of these impurities. The chief embrittling elements are antimony, phosphorus, tin, and arsenic (in decreasing importance). Plain carbon steels with less than 0.5 percent manganese are not susceptible to temper embrittlement. Susceptibility is enhanced greatly by Cr and Mn and less strongly by Ni. Small amounts of Mo and W inhibit embrittlement, but larger amounts enhance it.

The above facts strongly support a mechanism in which the embrittling elements segregate to austenite grain boundaries during austenitization. This

---

[1] J. R. Rice, *Trans. ASME, Ser. E: Appl. J. Mech.*, vol. 90, pp. 379–386, 1968.

[2] J. A. Begley and J. D. Landes, *ASTM Spec. Tech. Publ.* 514, pp. 40–69, 1972.

[3] The extensive literature in this field has been reviewed by B. C. Woodfine, *J. Iron Steel Inst. London*, vol. 173, pp. 229–240, 1953; J. R. Low, Jr., in "Fracture of Engineering Materials," American Society for Metals, Metals Park, Ohio, 1964; C. J. McMahon, Jr., in "Temper Embrittlement in Steel," *ASTM Spec. Tech. Publ.* 407, 1967.

segregation has been verified for Sb and P by Auger electron spectroscopy. The segregation is retained on quenching, and on heating to the tempering temperature the segregates weaken the cohesion at the ferrite-carbide interfaces. Fracture initiates along the ferrite-cementite interface[1] but propagates along prior austenite grain boundaries because they have a high embrittling element content and they contain a large density of carbide platelets.

## 14-11  HYDROGEN EMBRITTLEMENT

Severe embrittlement can be produced in many metals by very small amounts of hydrogen. Body-centered cubic and hexagonal close-packed metals are most susceptible to hydrogen embrittlement. As little as 0.0001 weight percent of hydrogen can cause cracking in steel. Face-centered cubic metals are not generally susceptible to hydrogen embrittlement.[2] Hydrogen may be introduced during melting and entrapped during solidification, or it may be picked up during heat treatment, electroplating, acid pickling, or welding.

The chief characteristics of hydrogen embrittlement are its strain-rate sensitivity, its temperature dependence, and its susceptibility to delayed fracture. Unlike most embrittling phenomena, hydrogen embrittlement is enhanced by slow strain rates. At low temperatures and high temperatures hydrogen embrittlement is negligible, but it is most severe in some intermediate temperature region. For steel the region of greatest susceptibility to hydrogen embrittlement is in the vicinity of room temperature. Slow bend tests and notched and unnotched tension tests will detect hydrogen embrittlement by a drastic decrease in ductility, but notched-impact tests are of no use for detecting the phenomenon.

A common method of studying hydrogen embrittlement is to charge notched tensile specimens with known amounts of hydrogen, load them to different stresses in a dead-weight machine, and observe the time to failure. A typical delayed-fracture curve is shown in Fig. 14-23. Note that the notched tensile strength of a charged specimen may be much lower than the strength of a hydrogen-free specimen. There is a region in which the time to fracture depends only slightly on the applied stress. There is also a minimum critical value below which delayed fracture will not occur. The similarity of the delayed fracture curve to the fatigue *S-N* curve has led to the use of the term "static fatigue" for the delayed-fracture phenomenon. The minimum critical stress, or "static fatigue limit," increases with a decrease in hydrogen content or a decrease in the severity of the notch. The

---

[1] C. J. McMahon, op. cit.

[2] The familiar example of the embrittlement of copper by hydrogen at elevated temperature is due to the reaction of hydrogen with oxygen to form internal pockets of steam.

**Fig. 14-23** Delayed-fracture curve.

hydrogen content of steel may be reduced by "baking," or heating at around 300 to 500°F.

Hydrogen is present in solution as monatomic hydrogen. Because it is a small interstitial atom, it diffuses very rapidly above room temperature. A commonly held concept of hydrogen embrittlement is that monatomic hydrogen precipitates at internal voids as molecular hydrogen. These voids might be true voids, microcracks, or perhaps simply regions of high dislocation density. As hydrogen diffuses into the voids, the pressure builds up and produces fracture. While this concept explains the general idea of hydrogen embrittlement, it is not in agreement with all the experimental facts. Further insight into the mechanism has resulted from the work of Troiano and coworkers.[1] By determining the rate of crack propagation by means of resistivity measurements, they were able to show that the crack propagates discontinuously. This indicates that the rate of crack propagation is controlled by the diffusion of hydrogen to the region of high triaxial stress just ahead of the crack tip. When a critical hydrogen concentration is obtained, a small crack forms and grows, to link up with the previous main crack. The fact that the time for the initiation of the first crack has been found to be insensitive to applied stress supports the idea that the process depends on the attainment of a critical hydrogen concentration. The main effect of stress is to assist in the accumulation of this concentration. The minimum critical stress in Fig. 14-23 can be interpreted as the stress needed to cause a critical accumulation of hydrogen. The higher the average hydrogen content, the lower the necessary critical stress.

The formation of hairline cracks, or flakes, in large ingots and forgings during cooling or room-temperature aging has long been attributed to the presence

[1] A. R. Troiano, *Trans. Am. Soc. Met.*, vol. 52, pp. 54–80, 1960. I. M. Bernstein and A. W. Thompson. eds., "Hydrogen in Metals," Amer. Soc. for Metals, Metals Park, Ohio, 1974.

of hydrogen. Studies[1] of flake formation have shown that in addition to containing hydrogen the steel must be subjected to transformation stresses for flaking to occur. The hydrogen content necessary for flaking varies widely with composition, size, and segregation. Flakes have been observed in steel with as low as 3 ppm of hydrogen. On the other hand, very high amounts of hydrogen can be tolerated without causing flakes if the transformation stresses are minimized by decomposing the austenite above the $M_s$ temperature before cooling to room temperature.

## 14-12 STRESS-CORROSION CRACKING

Stress-corrosion cracking is the failure of an alloy from the combined effects of a corrosive environment and a static tensile stress. The stress may be a result of applied loads or "locked-in" residual stress. Only specific combinations of alloys and chemical environment lead to stress-corrosion cracking. However, many are of common occurrence, such as aluminum alloys and seawater, copper alloys and $NH_3$, mild steel and NaOH, and austenitic stainless steel and saltwater. Over 80 combinations of alloys and corrosive environments have been reported to cause stress-corrosion cracking.[2] The static fatigue of glass and crystalline ceramics and the stress cracking of polymers are analogous processes occurring in nonmetallic materials.

The dependence of stress on the time to failure is very similar to the curve for a hydrogen-charged steel in Fig. 14-23. The threshold stress invariably is less than the yield strength of the alloy. The fracture mechanics approach can be used to good advantage in the analysis of stress-corrosion cracking.[3] The basic stress-time curve can be expressed in terms of the *initial level* of $K_I$ (Fig. 14-24). The threshold value of stress intensity is designated $K_{Iscc}$. Crack growth does not occur

**Fig. 14-24** Fracture mechanics plot for stress-corrosion test data.

[1] A. W. Dana, F. J. Shortsleeve, and A. R. Troiano, *Trans. Metall. Soc., AIME*, vol. 203, pp. 895–905, 1955.

[2] H. L. Logan, "The Stress Corrosion of Metals," John Wiley & Sons, Inc., New York, 1966.

[3] B. F. Brown, *Metall. Rev.*, vol. 13, pp. 171–183, 1968; "Stress Corrosion Testing," ASTM *Spec. Tech. Publ.*, 425, 1966.

**Fig. 14-25** Effect of initial stress intensity on cracking behavior in stress corrosion.

for stress intensities below $K_{Iscc}$. If the initial value of $K$ is above $K_{Iscc}$ the existing cracks grow with time. Thus, the stress intensity increases with time (Fig. 14-25) until fracture occurs when $K$ equals the fracture toughness $K_{Ic}$. For safe design in a material containing flaws the greatest initial value of $K$ must be less than $K_{Iscc}$. In general, the value of $K_{Iscc}$ decreases with an increase in yield strength.

Stress-corrosion cracks result in a nonductile type of failure. Depending on the alloy and corrosive environment, the fracture path may be either intergranular or transgranular, although the later is the most common. Because of the specific nature of stress corrosion, it is difficult to propose a general mechanism. It is widely believed that electrochemical dissolution plays a major role. A crack may initiate at a surface pit, and the crack tip acts as a local anode. Metal ions go into solution at the anode and electrons flow from the anode to a cathodic region near the surface. The intense plastic deformation at the crack tip breaks down protective surface films and aids in the rapid dissolution of metal ions. In addition, local plastic tearing at the crack tip is part of the crack-propagation process. Stress-corrosion cracks propagate intermittently and more slowly than a low-ductility brittle fracture. Because hydrogen is evolved at the anode, embrittlement may be caused in some systems. Also, the corrosion products that are formed have a large volume and cause a wedging action in the narrow crack.

## 14-13 FLOW AND FRACTURE UNDER VERY RAPID RATES OF LOADING

The flow and fracture of metals generally is found to be very sensitive to the rate of deformation. The spectrum of available strain rates is shown in Fig. 14-26. Also shown is a characteristic time, corresponding to the time required to produce

**Fig. 14-26** Spectrum of strain rates for mechanical testing. (*After Lindholm.*)

1 percent strain at the corresponding strain rate.[1] Creep and quasi-static testing is done with dead-weight, hydraulic, or screw testing machines. Inertia effects may be neglected and tests are carried out under essentially isothermal conditions. In the intermediate strain-rate range fast pneumatic or mechanical testing machines are used, and the mechanical resonance of the specimen and the machine must be considered. In the range $10^2$ to $10^4$ s$^{-1}$ impact bars are usually used. Elastic and elastic-plastic wave propagation must be considered in detail. At the highest strain rates testing usually involves explosively driven plate impact where measurements are made of the propagation and reflection of shock waves. From a strain rate of about $10^0$ s$^{-1}$ on up, the deformation becomes essentially adiabatic because the internal heat generated during plastic deformation does not have time to dissipate. The influence of the resulting temperature rise on the flow stress must be considered when comparing properties as a function of strain rate.

Material behavior under high strain-rate conditions is of practical importance in such problems as the response of structures to blast and impulsive loads, contact stresses under high-speed bearings, high-speed machining, explosive forming,[2] and ballistics.

A rapidly applied load is not instantaneously transmitted to all parts of the solid body. At a brief instant after the load has been applied the remote portions of the body remain undisturbed. The deformation and stress produced by the load move through the body in the form of a wave.[3] Since, except for the highest strain rates, most dynamic testing is done with circular cylindrical specimens of small lateral dimensions, it is permissible to consider only the case of one-dimensional wave propagation. An element of the bar is subjected to an axial stress $\sigma_x$ giving

[1] U. S. Lindholm, High Strain Rate Tests, chap. 4 in R. F. Bunshah (ed.), "Techniques of Metals Research," vol. V, pt. I, Interscience Publishers, Inc., New York, 1971.

[2] J. S. Rinehart and J. Pearson, "Explosive Working of Metals," The Macmillan Company, New York, 1963.

[3] H. Kolsky, "Stress Waves in Solids," Oxford University Press, New York, 1953.

rise to a particle velocity of $v_x$ in the axial direction. The governing first-order partial differential equations are

Equation of motion:
$$\frac{\partial \sigma_x}{\partial x} = \rho \frac{\partial v_x}{\partial t} \qquad (14\text{-}12)$$

Equation of continuity:
$$\frac{\partial v_x}{\partial x} = \frac{\partial \varepsilon_x}{\partial t} \qquad (14\text{-}13)$$

where $v_x = \dfrac{\partial u_x}{\partial t}$

$\quad\quad \varepsilon_x = \dfrac{\partial u_x}{\partial x}$

$\quad\quad u_x =$ the axial particle displacement

$\quad\quad \rho =$ the mass density

For elastic waves the constitutive equation is

$$\sigma_x = E\varepsilon_x \qquad (14\text{-}14)$$

If Eqs. (14-12) to (14-14) are expressed in terms of $u_x$, they can be combined to form the well-known wave equation

$$\frac{\partial^2 u_x}{\partial x^2} = \frac{1}{c_0{}^2} \frac{\partial^2 u_x}{\partial t^2} \qquad (14\text{-}15)$$

where $c_0 = (E/\rho)^{1/2}$ is the *elastic wave velocity*. The simplest solution of Eq. (14-15) yields

$$\frac{\partial u_x}{\partial t} = c_0 \left( \frac{\partial u_x}{\partial x} \right)$$

or
$$v_x = c_0 \varepsilon_x \qquad (14\text{-}16)$$

Substituting Eq. (14-14) into Eq. (14-16) gives

$$\sigma_x = \rho c_0 v_x \qquad (14\text{-}17)$$

The propagation of plastic stress waves has been studied by von Kármán[1] and Wood.[2] For the simplifying assumption that the stress-strain curve is independent of strain rate, the velocity of plastic wave propagation is given by

$$c_p = \left( \frac{1}{\rho} \frac{ds}{de} \right)^{1/2} \qquad (14\text{-}18)$$

where $ds/de$ is the slope of the engineering stress-strain curve. Thus each increment of plastic strain propagates at a different velocity since the slope of the curve

[1] T. von Kàrmàn and P. E. Duwez, *J. Appl. Phys.*, vol. 21, pp. 987–994, 1950.
[2] D. S. Wood, *J. Appl. Mech.*, vol. 19, p. 521, 1952.

$(ds/de)$ depends on the magnitude of plastic strain. Therefore, the shape of the plastic wave changes as it propagates along the bar. The particle velocity is related to the plastic wave velocity by

$$v_x = \int_0^e c_p \, de = \frac{1}{\rho} \int_0^s \frac{ds}{c_p} = \int_0^e \left(\frac{ds/de}{\rho}\right)^{1/2} de \qquad (14\text{-}19)$$

The first of these equations is used to determine the strain or the stress in a dynamically loaded material. The last version of Eq. (14-19) leads to the important conclusion that at the tensile strength of the material $ds/de = 0$ so that $v_x = 0$. Thus, there is a *critical impact velocity* $v_c$. If a bar is struck with an impact velocity greater than $v_c$, it ruptures at the impacted end at the instant of impact with a weak plastic strain propagating along the bar. The value of critical impact velocity for most metals lies in the range 100 to 500 ft/s.

We have already seen that if $ds/de$ decreases with strain (the normal case) the wave front becomes diffuse. However, if $ds/de$ increases with strain, large strains travel faster than small strains and the wave front sharpens into a plastic *shock wave*. This can occur in materials which show a pronounced yield point[1] or in metal cutting with a localized shear plane.

The highest strain rates are produced by impulsive loading with explosive charges. Particle velocities are very high and stresses can easily exceed 100 kbar $(1.47 \times 10^6$ psi). Much work has been done under these conditions in determining thermodynamic equations of state for metals. Because of high shock pressures, the shear strength of the metal is neglected and the problem is treated by compressible hydrodynamic theory.

The impact of explosively driven plates has been developed into a fairly well-understood test. It is useful for studying the hardening and transformations in metals[2] subjected to shock pressures in excess of 500 kbar. Shock-wave propagation through a thin plate is essentially one dimensional in the direction of the shock front. For small plastic strains this produces a state of near triaxial loading.[3] Thus shock loading provides an environment for subjecting materials to high stresses without corresponding large plastic deformations.

There are marked differences between fracture under impulsive loads and under static loads. With impulsive loads there is not time for the stress to be distributed throughout the entire body, so that fracture can occur in one part of the solid independently of what happens in another part. The velocity of propagation of stress waves in solids is in the range 3,000 to 20,000 ft/s, while the velocity

---

[1] A. H. Cottrell, *Proc. Conf. Properties of Materials at High Rates of Strain*, Institute of Mechanical Engineers, London, 1957.

[2] G. E. Dieter, Hardening Effect of Shock Waves, in "Strengthening Mechanisms in Solids," American Society for Metals, Metals Park, Ohio, 1962.

[3] J. L. O'Brien and R. S. Davis, "Fracture in Engineering Materials," chap. 12, American Society for Metals, Metals Park, Ohio, 1964.

**Fig. 14-27** Reflection of a shock wave from a free surface and development of a spall (fracture).

of crack propagation is about 6,000 ft/s. Therefore, with impulsive loads it often is found that cracks have formed but did not have time to propagate before the stress state changed.

Reflections of stress waves occur at free surfaces and fixed ends, at changes in cross section, and at discontinuities within the solid. Figure 14-27a shows a compressive shock wave approaching the free surface of a plate. The wave moves from left to right. At the free surface the wave is reflected back as a tensile wave from right to left. The heavy lines indicate the net stress after summing the incident (compressive) and reflected (tensile) waves. The dashed line corresponds to the portion of the pulse that has already been reflected. There is a rapid buildup of tensile stress as the wave is reflected back from the free surface. When the tensile stress reaches a high enough value, fracture occurs and the plate is said to have *spalled*. If the stress distribution within the pulse is known from velocity measurements, then the fracture stress can be determined[1] by a measurement of the position of the fracture relative to the free surface.

## BIBLIOGRAPHY

"A Review of Engineering Approaches to Design Against Fracture," American Society of Mechanical Engineers, New York, 1964.
Broek, D.: "Elementary Engineering Fracture Mechanics," Noordhoft International Publishing, Leyden, 1974.
Knott, J. F.: "Fundamentals of Fracture Mechanics," John Wiley & Sons, Inc., New York, 1973.

[1] J. S. Rinehart, *J. Appl. Phys.*, vol. 22, p. 555, 1951.

Parker, E. R.: "Brittle Behavior of Engineering Structures," John Wiley & Sons, Inc., New York, 1957.

Shank, M. E.: "Control of Steel Construction to Avoid Brittle Failure," Welding Research Council, New York, 1957.

Späth, W.: "Impact Testing of Materials," Thames-Hudson, London, 1962.

Tetelman, A. S., and A. J. McEvily, Jr.: "Fracture of Structural Materials," John Wiley & Sons, Inc., New York, 1967.

Tipper, C. F.: "The Brittle Fracture Story," Cambridge University Press, London, 1962.

# Plastic Forming
of Metals

# FUNDAMENTALS OF METALWORKING

## 15-1  CLASSIFICATION OF FORMING PROCESSES

The importance of metals in modern technology is due, in large part, to the ease with which they may be formed into useful shapes such as tubes, rods, and sheets. Useful shapes may be generated in two basic ways:

1. By *plastic deformation processes* in which the volume and mass of metal are conserved and the metal is displaced from one location to another
2. By *metal removal* or *machining processes* in which material is removed in order to give it the required shape.

Major emphasis in Part Four is given to plastic deformation processes. However, some of the fundamentals of metal removal are considered in the last chapter of this book.

Equal in importance to the creation of useful shapes by plastic forming processes is the control of mechanical properties by the forming process. For example, blowholes and porosity in a cast ingot may be eliminated by hot-forging or hot-rolling, to the improvement of ductility and fracture toughness. In many products the mechanical properties depend on the control of strain hardening during processing, while in other instances precise control of deformation, temperature, and strain rate during processing is required to develop the optimum structure and properties. The *thermomechanical processing* in ausforming (Sec. 6-12) is a good example.

Hundreds of processes have been developed for specific metalworking applications. However, these processes may be classified into only a few categories on the basis of the type of forces applied to the workpiece as it is formed into shape. These categories are:

1. Direct-compression-type processes
2. Indirect-compression processes
3. Tension-type processes
4. Bending processes
5. Shearing processes

In direct-compression processes the force is applied to the surface of the workpiece, and the metal flows at right angles to the direction of the compression. The chief examples of this type of process are forging and rolling (Fig. 15-1). Indirect-compression processes include wiredrawing and tube drawing, extrusion, and the deep drawing of a cup. The primary applied forces are frequently tensile, but the indirect compressive forces developed by the reaction of the workpiece with the die reach high values. Therefore, the metal flows under the action of a combined stress state which includes high compressive forces in at least one of the principal directions. The best example of a tension-type forming process is stretch forming, where a metal sheet is wrapped to the contour of a die under the application of tensile forces. Bending involves the application of bending moments to the sheet, while shearing involves the application of shearing forces of sufficient magnitude to rupture the metal in the plane of shear. Figure 15-1 illustrates these processes in a very simplified way.

Plastic working processes which are designed to reduce an ingot or billet to a standard mill product of simple shape, such as sheet, plate, and bar, are called *primary mechanical working processes*. Forming methods which produce a part to a final finished shape are called *secondary mechanical working processes*. Most sheet-metal forming operations, wire drawing, and tube drawing are secondary processes. The terminology in this area is not very precise. Frequently the first category is referred to as *processing operations*, and the second is called *fabrication*.

The deformation processing situation is best viewed as a total system (Fig. 15-2). The deformation zone is concerned with the distribution of stress, strain, and particle velocities, and with the overall pressure required to perform the operation. Obviously, the applied forces must develop yielding in the material but the stresses must not locally create fracture. Such metallurgical phenomena as strain hardening, recrystallization and fracture are important, but often under

Forging

Rolling

Wire drawing

Extrusion

Deep drawing

Die

Stretch forming

Bending

Shearing

**Fig. 15-1** Typical forming operations.

Heat flow

Tools

Lubrication system

Undeformed Metal

Deformation Zone

Product

Tools

Friction

**Fig. 15-2** Deformation processing system.

specialized conditions of high strain rates and/or high temperature. The flow stress of the material will be a strong function of strain, strain rate, and temperature, often at values of these variables that are difficult to simulate in the laboratory. The workpiece will be in contact with nondeforming (elastic) tools or dies. The friction along this interface and the heat transfer from the workpiece to the die are important considerations, as are such practical concerns as tool wear and surface finish of the product.

## 15-2   MECHANICS OF METALWORKING

The analysis of the stresses in metalworking processes has been an active area of applied plasticity for the past 40 years.[1] Since the forces and deformations generally are quite complex, it is usually necessary to use simplifying assumptions to obtain a tractable solution. Because the strains involved in deformation processes are large, it is usually possible to neglect elastic strains and consider only the plastic strains (plastic-rigid solid). As a first approximation, strain hardening is often neglected. In hot-working this may not be a serious assumption. Practically all analyses consider the material to be isotropic and homogeneous. Usually the deformation of the metal between the dies is not uniform, Fig. 15-7. To accurately predict the nonuniform deformation and calculate the local stresses is the chief analytical problem.

The principal use of analytical studies of metalworking processes is for determining the forces required to produce a given deformation for a certain geometry prescribed by the process. Such calculations are useful for selecting or designing the equipment to do a particular job. In this area existing theory is generally adequate for the task. A mechanics analysis of a process may also be used to develop information on the frictional conditions in the process. An important problem area is predicting the limiting deformation at which fracture will occur. In general, theory concerning formability limits and fracture criteria are not well developed.

Before proceeding further it would be desirable to review Chap. 3, Elements of the Theory of Plasticity, especially Secs. 3-1 to 3-4. The most basic relationship for plastic deformation is the *constant-volume relationship*. Because of the large deformations, it is important to express strain in terms of true strain or natural strain. From Eq. (3-5) we recall that the constant-volume relationship is expressed by

$$\varepsilon_1 + \varepsilon_2 + \varepsilon_3 = 0 \qquad (15\text{-}1)$$

[1] W. Johnson, *Appl. Mech. Rev.*, vol. 24, pp. 977–989, 1971.

Compressive stresses and strains often predominate in metalworking processes. If a block of initial height $h_0$ is compressed to $h_1$, the axial compressive strain (using the usual sign convention) will be

True strain:

$$\varepsilon = \int_{h_0}^{h_1} \frac{dh}{h} = \ln \frac{h_1}{h_0} = -\ln \frac{h_0}{h_1} \qquad h_0 > h_1 \qquad (15\text{-}2)$$

Conventional strain:

$$e = \frac{h_1 - h_0}{h_0} = \frac{h_1}{h_0} - 1 \qquad (15\text{-}3)$$

The calculated strains will be negative, indicating compressive strains. However, in metalworking problems it is common practice to reverse the convention so that compressive stresses and strains are defined as *positive*. When using this convention we shall use the subscript $c$. Thus, the strains described above would be calculated from

$$\varepsilon_c = \ln \frac{h_0}{h_1} \qquad e_c = \frac{h_0 - h_1}{h_0} = 1 - \frac{h_1}{h_0}$$

Metalworking deformations frequently are expressed as reduction in cross-sectional area. The fractional reduction is given by

$$r = \frac{A_0 - A_1}{A_0} \qquad (15\text{-}4)$$

From the constancy-of-volume relation,

$$A_1 L_1 = A_0 L_0$$

$$r = 1 - \frac{A_1}{A_0} \qquad \text{or} \qquad \frac{A_1}{A_0} = 1 - r$$

$$\varepsilon = \ln \frac{L_1}{L_0} = \ln \frac{A_0}{A_1} = \ln \frac{1}{1 - r} \qquad (15\text{-}5)$$

The requirement of a theory describing the mechanics of a metalworking process is the ability to make an accurate prediction of the stresses, strains, and velocities at every point in the deformed region of the workpiece. The various approaches differ in complexity and in the degree to which they meet this requirement. In general, such a theory consists of three sets of equations:

1. The static equilibrium of force equations
2. The Levy-Mises equations (see Sec. 3-8) expressing a relationship between stress and strain rate
3. The yield criterion (see Sec. 3-4)

In the general case there are nine independent equations containing nine unknowns; six stress components and three velocity (strain-rate) components. While an analytical solution is possible if a sufficient number of boundary conditions are specified, the mathematical difficulties in a general solution are formidable. Thus, most analyses of actual metalworking processes are limited to two-dimensional or symmetric three-dimensional problems.

The methods of analysis, in increasing order of complexity and ability to predict fine detail, are:

1. *The slab method*—assumes homogeneous deformation
2. *Uniform-deformation energy method*—calculates average forming stress from the work of plastic deformation
3. *Slip-line field theory*—permits point-by-point calculation of stress for plane-strain conditions only
4. *Upper- and lower-bound solutions*—based on the theory of limit analysis, uses reasonable stress and velocity fields to calculate the bounds within which the actual forming load must lie

The general framework of each of these approaches will be illustrated below using the problem of the drawing of a strip through wedge dies (Fig. 15-3). While strip drawing is not a common production process it is probably the most often studied problem in the mechanics of metalworking theory. At this stage in the presentation we will ignore the vital practical consideration of friction between the die and workpiece interface. Methods of considering friction will be discussed in Sec. 15-7.

**Fig. 15-3** Stresses acting on an element during strip drawing of a wide sheet.

The *slab method* assumes that the metal deforms uniformly in the deformation zone. A square grid placed in the deformation zone would be distorted uniformly into rectangular elements. This was the earliest approach developed in the 1920s by von Kármán, Hencky, Siebel, and later by Sachs.[1] The approach essentially is that used in strength of materials.

In Fig. 15-3, a wide strip is being drawn through a frictionless die with a total included angle of $2\alpha$. Because the original thickness of the strip is assumed to be much less than the width, the conditions approximate plane-strain deformation where there is no strain in the width direction. An element under the die of length $dx$ is subjected to a longitudinal stress $\sigma_x$ and a vertical stress $\sigma_y$ due to the die pressure $p$. A characteristic of the slab method is that these stresses are assumed to be principal stresses. The equilibrium of forces in the $x$ direction is made up of two components

1. Due to the change in longitudinal stress with $x$ increasing positively to the left

$$(\sigma_x + d\sigma_x)(h + dh)w - \sigma_x hw$$

2. Due to the die pressure at the two interfaces

$$2p\left(w \frac{dx}{\cos \alpha}\right) \sin \alpha$$

Taking the equilibrium of forces in the $x$ direction and neglecting the product $d\sigma_x \, dh$, we have

$$\sigma_x \, dh + h \, d\sigma_x + 2p \tan \alpha \, dx = 0 \qquad (15\text{-}6)$$

From the geometry, $h$ may be expressed in terms of $x$ as

$$h = 2x \tan \alpha \qquad \text{and} \qquad dh = 2 \, dx \tan \alpha$$

so that Eq. (15-6) simplifies to

$$\sigma_x \, dh + h \, d\sigma_x + p \, dh = 0 \qquad (15\text{-}7)$$

To find a relationship between $\sigma_x$ and $p$, we turn to the equilibrium of forces in the $y$ direction (vertical).

$$\sigma_y \, dxw + p \cos \alpha \left(\frac{w \, dx}{\cos \alpha}\right) = 0$$

$$\sigma_y = -p \qquad (15\text{-}8)$$

Since up to this point we have considered $p$ to be a positive quantity, we see from Eq. (15-8) that $\sigma_y$ is compressive. We now relate the two principal stresses by means

[1] O. Hoffman and G. Sachs, "Introduction to the Theory of Plasticity for Engineers," McGraw-Hill Book Company, New York, 1953.

of the yield criterion. For plane-strain conditions the von Mises' and Tresca criteria are equivalent, (Eq. 3-54).

$$\sigma_1 - \sigma_3 = 2k = \frac{2\sigma_0}{\sqrt{3}} = \sigma_0'$$

$$\sigma_x + p = \sigma_0' \qquad \text{or} \qquad p = \sigma_0' - \sigma_x \qquad (15\text{-}9)$$

Substituting into Eq. (15-7), we get

$$\sigma_0' \, dh + h \, d\sigma_x = 0$$

$$\frac{d\sigma_x}{\sigma_0'} = -\frac{dh}{h}$$

On integrating

$$\frac{\sigma_x}{\sigma_0'} = -\ln h + \text{constant}$$

The boundary condition is that at the entry to the die, $x = b$; $h = h_b$, the axial stress equals zero. Therefore, constant $= \ln h_b$.

$$\sigma_x = \sigma_0' \ln \frac{h_b}{h} = \frac{2}{\sqrt{3}} \sigma_0 \ln \frac{h_b}{h} \qquad (15\text{-}10)$$

The axial stress at the die exit needed to cause deformation under this ideal frictionless condition is

$$\sigma_{xa} = \frac{2}{\sqrt{3}} \sigma_0 \ln \frac{h_b}{h_a} = \frac{2}{\sqrt{3}} \sigma_0 \ln \frac{1}{1-r} \qquad (15\text{-}11)$$

The above result can be obtained much more simply by means of the *uniform-deformation energy method*. Consider the simplest case of a cylinder loaded in tension. The increment of work for an increment in length $\delta L$ is

$$\delta W = \delta P \, \delta L = \bar{\sigma} A \, \delta L$$

where $\bar{\sigma}$ is the mean yield stress. The deformation energy per unit volume is

$$\frac{\delta W}{V} = \frac{\bar{\sigma} A \, \delta L}{A \, L} = \bar{\sigma} \frac{\delta L}{L}$$

The plastic work of deformation per unit volume is

$$U_p = \bar{\sigma} \int d\varepsilon = \bar{\sigma} \int \frac{\delta L}{L} = \bar{\sigma} \ln \frac{L_1}{L_0} \qquad (15\text{-}12)$$

If we apply this to drawing a cylindrical wire from area $A_b$ down to area $A_a$,

$$W = P_a L_a = U_p V = A_a L_a \bar{\sigma} \ln \frac{L_b}{L_a}$$

$$P_a = A_a \bar{\sigma} \ln \frac{L_b}{L_a} = A_a \bar{\sigma} \ln \frac{A_a}{A_b}$$

The draw stress

$$\sigma_{xa} = \frac{P_a}{A_a} = \bar{\sigma} \ln \frac{A_a}{A_b}$$

or

$$\sigma_{xa} = \bar{\sigma}_0 \ln \frac{1}{1-r} \qquad (15\text{-}13)$$

Equation (15-13) not only neglects friction, but it neglects the influence of transverse stresses (plastic constraint). Notice that it differs from Eq. (15-11) by the magnitude of the average yield-stress term.

The total work per unit volume in an actual deformation process is appreciably in excess of that given by Eq. (15-12). In addition to the ideal work of deformation $U_p$, we have the work to overcome the friction at the metal-tool interface $U_f$ and the redundant work $U_r$. The redundant work is the work involved in internal shearing processes due to nonuniform deformation that does not contribute to the change in shape of the body. The amount of $U_r$ will depend on the geometry of the process and the friction. Thus, the total work is given by

$$U_T = U_p + U_f + U_r \qquad (15\text{-}14)$$

The efficiency of a process can be expressed as

$$\eta = \frac{U_p}{U_T} \qquad (15\text{-}15)$$

Typical values of efficiency are 30 to 60 percent for extrusion and 75 to 95 percent for rolling.

The first approach to the analysis of metalworking processes that did not assume homogeneous uniform deformation was *slip-line field theory*. The mathematics of slip-line fields was introduced in Sec. 3-9 where the technique was used to analyze the quite simple problem of the pressure under the indentation of a narrow frictionless punch. The presentation given there was incomplete since it did not discuss the Geiringer velocity equations[1] and the construction of a *hodograph* describing the admissible velocity fields[2] corresponding to the slip-line field. These topics are aptly discussed in a monograph[3] which presents complete bibliographic information on nearly all available slip-line fields.

The reason for using slip-line fields is that they account for inhomogeneous deformation in the calculation of the overall forming loads. Figure 15-4 shows the slip-line field proposed by Hill and Tupper[4] for frictionless strip drawing through

[1] J. Halling, *Engineer*, vol. 207, p. 250, 1959.
[2] A. P. Green, *J. Mech. Phys. Solids*, vol. 2, p. 73, 1954.
[3] W. Johnson, R. Sowerby, and J. B. Haddon, " Plane-Strain Slip-Line Fields: Theory and Bibliography," American Elsevier Publishing Company, New York, 1970.
[4] R. Hill and S. J. Tupper, *J. Iron Steel Inst. London*, vol. 159, p. 353, 1948.

**Fig. 15-4** Slip-line field for strip drawing through a frictionless wedge-shaped die at a small reduction. (*After Hill and Tupper.*)

wedge dies of included angle $2\alpha$. This solution is valid for reductions up to $r = 2 \sin \alpha/(1 + 2 \sin \alpha)$. It will be recalled that the shear stress $k$ is everywhere constant within the slip-line field but that the hydrostatic pressure normal to the slip line changes from point to point as the line curves. The Hencky equations, Eq. (3-55), are used to find the pressure $p$ at any point along $ADF$. Integrating over the pressure distribution gives the average draw stress for the strip

$$\sigma_{xa} = \frac{2}{\sqrt{3}} \sigma_0 \left[ \frac{2(1 + \alpha) \sin \alpha}{1 + 2 \sin \alpha} \right] \qquad (15\text{-}16)$$

For many metalworking processes there are no slip-line fields to allow prediction of stresses. Moreover, slip-line fields are valid only for plane-strain conditions. *Upper- and lower-bound techniques*[1] have been developed that have general applicability. An upper-bound solution provides an overestimation of the required deformation force while a lower-bound solution provides an underestimate of the force. The degree of agreement between the upper- and lower-bound predictions is an indication of how close the prediction is to the exact value. A lower-bound analysis requires that a statically admissible stress field be found throughout the entire material without making an attempt to ensure that the velocity conditions are satisfied at every point in the material. The term *statically*

[1] W. Johnson and H. Kudo, "The Mechanics of the Metal Extrusion," Manchester University Press, 1962; B. Avitzur, "Metal Forming: Processes and Analysis," McGraw-Hill Book Company, New York, 1968.

*admissible stress field* means that the assumed stress field satisfies the equilibrium equations and the stress boundary conditions and that it does not violate the yield criterion.

From a practical viewpoint the upper-bound technique is more important than the lower bound since calculations based on upper bound will always result in an overestimate of the load that the press or machine will be called upon to deliver. The upper-bound technique[1] has been used widely by civil engineers for the plastic design of steel structures (theory of limit design). In the upper-bound analysis a *kinematically admissible velocity field* is constructed, i.e., the hodograph must be satisfied, and the loads are calculated to cause the velocity field to operate. No attempt is made to satisfy the stress equilibrium conditions at any point in the field. In a useful form of the upper-bound analysis[2] the deformation is assumed to take place by rigid-body movements of triangular elements in which all particles in a given element move with the same velocity. Figure 15-5 shows an admissible velocity field and hodograph for the strip-drawing problem.[3] Material enters the die with an initial velocity $U$ and leaves the die with an increased velocity $u$. In constructing the velocity field, the number of triangular elements $n$ and the angle of inclination of the first velocity discontinuity $\theta$ are the independent variables. These have been selected so as to minimize the average drawing stress. The calculation of stress is based on Drucker's upper-bound theorem. This theorem

*Velocity field*

*Hodograph*

**Fig. 15-5**    An upper-bound solution for strip drawing.

[1] D. C. Drucker, H. J. Greenberg, and W. Prager, *J. Appl. Mech.*, vol. 18, p. 371, 1951; J. Baker, M. R. Horne, and J. Heyman, "The Steel Skeleton," vols. I and II, Cambridge University Press, London, 1956.

[2] H. Kudo, *Int. J. Mech. Sci.*, vol. 1, pp. 57–83, 229–252, 336–368, 1960.

[3] S. Kobayashi and E. G. Thomsen, Methods of Solution of Metal-forming Problems, in W. A. Backofen, et al. (eds.), "Fundamentals of Deformation Processing," Syracuse University Press, 1964.

states that the calculated load is above the critical upper-bound load if a velocity field can be found for which the rate of work due to the external load exceeds the rate of internal energy dissipation. Mathematically stated the theorem is

$$Pu = \sigma_{\text{avg}} uA \leq \int_V \bar{\sigma} \dot{\bar{\varepsilon}} \, dV + \int_S \tau_s v^* \, dS \qquad (15\text{-}17)$$

where $P$ = the upper-bound load at a forming velocity $u$

$\dot{\bar{\varepsilon}}$ = the effective strain rate in an element of volume $dV$

$v^*$ = the rate of relative slip (velocity discontinuity) along the surface $S$

$\tau_s$ = the shear stress due to friction along $S$

The average draw stress for the velocity field in Fig. 15-5 as developed by the upper-bound theorem is

$$\sigma_{xa} = \frac{2\sigma_0}{\sqrt{3}} \frac{n \sin \alpha}{2} \left[ \frac{1}{\sin \theta \sin (\theta - \alpha)} + \frac{1}{\sin (\psi + a) \sin \psi} \right] \qquad (15\text{-}18)$$

The geometry of the net is given, for each reduction $r$, by the following equation:

$$(1 - r)^{1/n} = \frac{\sin (\theta - \alpha) \sin \psi}{\sin \theta \sin (\psi + \alpha)} \qquad (15\text{-}19)$$

Figure 15-6 compares the average draw stress for the upper-bound solution with that obtained from uniform-deformation energy, Eq. (15-13), and with the draw stress obtained from slip-line field theory.

A useful technique, particularly for axisymmetric problems, is *visioplasticity*.[1] The flow field is determined experimentally from measurements on a gridded workpiece. By measuring the strain-rate distribution throughout the deformation zone, it is possible to calculate the stress distribution[2] without resorting to assumptions regarding the nature of the deformation.

The potentially most powerful calculation technique is the *finite element method*.[3] Originally developed for problems in structural analysis it has been applied to metalworking problems[4] involving elastic-plastic conditions. Briefly the technique consists of dividing the material into a number of idealized elements. The stiffness matrix of a typical element is calculated after choosing an appropriate displacement function which satisfies the compatibility conditions with adjacent elements. The individual stiffness matrices are combined to form the

[1] E. G. Thomsen, C. T. Yang, and S. Kobayashi, "Mechanics of Plastic Deformation in Metal Processing," The Macmillan Company, New York, 1965.

[2] A. Shabaik and S. Kobayashi, *Trans. ASME, Ser. B: J. Eng. Ind.*, vol. 89, pp.339–346, 1967.

[3] O. C. Zienkiewicz and Y. K. Cheung, "The Finite Element Method in Structural and Continuum Mechanics," McGraw-Hill Book Company, New York, 1967.

[4] C. H. Lee and S. Kobayashi, *Int. J. Mech. Sci.*, vol. 12, pp. 349–370, 1970.

**Fig. 15-6**  Comparison of the average drawing stresses calculated by various methods. (*From S. Kobayashi and E. G. Thomsen, in "Fundamentals of Deformation Processing," p. 61, Syracuse University Press, 1964.*)

matrix equation for the entire material. The large-scale algebraic equations involved require solution by high-speed digital computers. The method is applicable to complex geometries and can take into account realistic friction conditions and material properties.

## 15-3  FLOW-STRESS DETERMINATION

The various expressions that will be developed in succeeding chapters to describe the forming stress, or pressure, in a particular metalworking process invariably consist of three terms:

$$p = \bar{\sigma}_0 g(f) h(c)$$

where  $\bar{\sigma}_0$ = the flow resistance of the material for the appropriate stress state, i.e., uniaxial, plane strain, etc. It is a function of strain, temperature, and strain rate.

$g(f)$ = an expression for the friction at the tool-workpiece interface.

$h(c)$ = a function of the geometry of the tooling and the geometry of the deformation. This term may or may not include a contribution from redundant deformation.

It is obvious from the above relationship that if we are to make accurate predictions of forming loads and stresses, we need accurate values of flow resistance (flow stress). The experimental problems in measuring the flow curve under metalworking conditions are more severe than in the usual stress-strain test determined for structural or mechanical design applications. Since metalworking processes involve large plastic strains, it is desirable to measure the flow curve out to a true strain of 2.0 to 4.0. In addition, many of these processes involve high strain rates ($\dot{\varepsilon} \approx 100 \, s^{-1}$), which may not be obtained easily with ordinary test facilities. Further, many metalworking processes are carried out at elevated temperatures where the flow stress is strongly strain-rate sensitive but nearly independent of strain. Thus, tests for determining flow stress must be carried out under controlled conditions of temperature and constant true-strain rate.[1]

The true-stress–true-strain curve determined from the *tension test* is of limited usefulness because necking limits uniform deformation to true strains less than 0.5 (see Sec. 9-3). This is particularly severe in hot-working, where the low rate of strain hardening allows necking to occur at $\varepsilon \approx 0.1$. The formation of a necked region in the tension specimen introduces a complex stress state and locally raises the strain rate by a factor of 10 to 20.

The *compression* of a short cylinder between anvils is a much better test for measuring the flow stress in metalworking applications. There is no problem with necking and the test can be carried out to strains in excess of 2.0 if the material is ductile. However, the friction between the specimen and anvils can lead to difficulties unless it is controlled. In the *homogeneous upset test* a cylinder of diameter $D_0$ and initial height $h_0$ would be compressed in height to $h$ and spread out in diameter to $D$ according to the law of constancy of volume:

$$D_0{}^2 h_0 = D^2 h$$

During deformation, as the metal spreads over the compression anvils to increase its diameter, frictional forces will oppose the outward flow of metal. This frictional resistance occurs in that part of the specimen in contact with the anvils, while the metal at specimen midheight can flow outward undisturbed. This leads to a *barreled* specimen profile, and internally a region of undeformed metal is created near the anvil surfaces (Fig. 15-7). As these cone-shaped zones approach and overlap, they cause an increase in force for a given increment of deformation and the load-deformation curve bends upward (Fig. 15-8). For a fixed diameter, a shorter specimen will require a greater axial force to produce the same percentage reduction in height because of the relatively larger undeformed region (Fig. 15-7). Thus, one way to minimize the barreling and nonuniform deformation is to use a

---

[1] H. J. McQueen and J. J. Jonas, Hot Workability Testing Techniques, in A. L. Hoffmanner (ed.), "Metal Forming: Interrelation Between Theory and Practice," Plenum Publishing Corporation, New York, 1971.

**Fig. 15-7** Undeformed regions (shaded) due to friction at ends of a compression specimen.

low value of $D_0/h_0$. However, there is a practical limit of $D_0/h_0 \approx 0.5$, for below this value the specimen buckles instead of barreling. The true flow stress in compression without friction can be obtained[1] by plotting load versus $D_0/h_0$ for several values of reduction and extrapolating each curve to $D_0/h_0 = 0$.

The friction at the specimen-platen interface can be minimized[2] by using smooth, hardened platens, grooving the ends of the specimen to retain lubricant, and carrying out the test in increments so that the lubricant can be replaced at intervals.[3] Teflon sheet for cold deformation and glass for hot deformation are

**Fig. 15-8** Load-deformation curves for compression tests with different values of $D_0/h_0$.

[1] M. Cooke and E. C. Larke, *J. Inst. Met.*, vol. 71, pp. 371–390, 1945.
[2] G. T. van Rooyen and W. A. Backofen, *Int. J. Mech. Sci.*, vol. 1, pp. 1–27, 1960; G. W. Pearsall and W. A. Backofen, *Trans. ASME, Ser. D: J. Basic Eng.*, vol. 85, p. 68, 1963.
[3] Standard Methods of Compression Testing of Metallic Materials at Room Temperature, ASTM Standards, pt. 31, Designation E9-70; T. C. Hsu, *Mater. Res. Stand.*, vol. 9, pp.20–25, 47–53, 1969.

especially effective lubricants. With these techniques it is possible to reach a strain of about $\varepsilon = 1.0$ with only slight barreling. When friction is not present the uniaxial compressive force required to produce yielding is

$$P = \sigma_0 A$$

The true compressive stress $p$ produced by this force $P$ is

$$p = \frac{4P}{\pi D^2}$$

and using the constancy-of-volume relationship

$$p = \frac{4Ph}{\pi D_0{}^2 h_0} \qquad (15\text{-}20)$$

where $D_0$ and $h_0$ are the initial diameter and height, and $h$ is the height of the cylindrical sample at any instant during compression. The true *compressive* strain is given by

$$\varepsilon_c = \ln \frac{h_0}{h} \qquad (15\text{-}21)$$

In a compression test at constant crosshead speed, the true-strain rate continuously increases with deformation [see Eq. (9-33)]. At strain rates below $1 \text{ s}^{-1}$ a servocontrolled testing machine can be modified to maintain a constant true-strain rate. For metalworking strain rates of 1 to $10^3 \text{ s}^{-1}$ the only equipment capable of providing a constant true-strain rate is the cam plastometer.[1]

A compression test, such as that shown in Fig. 15-7, would be very difficult to conduct on a thin sheet, since it might be impossible to machine the specimens. A much more suitable test for sheet metal is the *plane-strain compression test*.[2] In this test a narrow band across the width of a strip is compressed by narrow platens which are wider than the strip (Fig. 15-9). The constraints of the undeformed shoulders of material on each side of the platens prevent extension of the sheet in the width dimension. There is deformation in the direction of platen motion and in the direction normal to the length of the platen, as occurs in the rolling process. In addition to its suitability with thin sheet, other advantages of this test are that it simulates the stress state in rolling, eliminates problems with barreling, and because the area under the platen is constant, the total deformation force does not rise as rapidly as in the compression of a cylinder. On the other hand, unless good lubrication is maintained, a dead metal zone will form in the specimen next to the face of the platens. Therefore, the test usually is carried out incrementally, measur-

[1] J. F. Alder and V. A. Phillips, *J. Inst. Met.*, vol. 83, pp. 80–86, 1954–1955; J. E. Hockett, *Am. Soc. Test. Mater. Proc.*, vol. 59, pp. 1309–1319, 1959.
[2] A. B. Watts and H. Ford, *Proc. Inst. Mech. Eng.*, vol. 169, pp. 1141–1156, 1955.

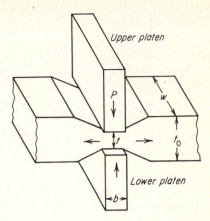

**Fig. 15-9** Plane-strain compression test.

ing the sheet thickness after each increment and relubricating for the next higher load. Slip-line field theory has been used to analyze the plane-strain compression test. It has been found that the axial compressive stress simply is the applied load divided by the contact area of the platens so long as the ratio of indentation thickness to platen width $t/b$ is in the range $\frac{1}{4}$ to $\frac{1}{2}$. By changing platens to maintain this $t/b$ range, it is possible to achieve deformations of about 90 percent. An additional requirement to maintain the plane-strain condition is that $w/b$ must be greater than 5.

True stress and true strain determined in the plane strain compression test may be expressed by

$$p = \frac{P}{wb} \qquad \varepsilon_{pc} = \ln \frac{t_0}{t}$$

The mean pressure on the platens is 15.5 percent higher than it would be in the corresponding uniaxial compression test (see Prob. 15-8). The true-stress–true-strain curve in uniaxial compression ($\sigma_0$ versus $\varepsilon_c$) may be obtained from the corresponding plane-strain compression curve ($p$ versus $\varepsilon_{pc}$) by the following relations:

$$\sigma_0 = \frac{p}{\sqrt{3/2}} = \frac{p}{1.155} \qquad (15\text{-}22)$$

$$\varepsilon_c = \frac{\sqrt{3}}{2}\,\varepsilon_{pc} = 1.155\,\varepsilon_{pc} \qquad (15\text{-}23)$$

The *hot torsion test* (see Sec. 10-6) is capable of producing very large strains of the order of 20. Since the dimensions of the specimen do not change, the strain rate remains constant for a constant rpm.[1] Strain rates varying from $10^{-5}$

---

[1] J. A. Bailey and S. L. Haas, *J Mater.*, vol. 7, pp. 8–13, 1972.

**Fig. 15-10**  Definition of mean-flow stress.

to $10^3$ s$^{-1}$ are readily achieved.  The chief difficulty with this test is the fact that stress and strain vary with radial distance in the specimen.  Usually the maximum values at the surface are reported, but this may lead to problems in interpretation if the surface strain-hardens more than the core.  The problem of nonuniform stress and strain is largely eliminated by using a tubular specimen,[1] but care must be exercised to use a short specimen or buckling will occur.  The uniaxial stress and strain are obtained from the torsional shear stress $\tau$ and shear strain $\gamma$ by the following relations based on the von Mises' criterion.

$$\sigma_0 = \sqrt{3}\,\tau \qquad \varepsilon = \frac{\gamma}{\sqrt{3}}$$

The variation of flow stress with strain can be neglected for true hot-working (see Fig. 10-8) or for a highly cold-worked metal.  In cases where strain hardening is present, the best approach to selecting a flow stress for use in forming load calculations is to use the mean flow stress as given by

$$\bar{\sigma}_0 = \frac{1}{\varepsilon_a - \varepsilon_b} \int_{\varepsilon_b}^{\varepsilon_a} \sigma_0 \, d\varepsilon \qquad (15\text{-}24)$$

Figure 15-10 shows the interpretation of this equation.  This is considered a better choice than the flow stress based on $\varepsilon = (\varepsilon_a + \varepsilon_b)/2$.  When analytical expressions for the flow curve are required, it is usually possible to fit the data to either Eq. (9-21) or Eq. (9-24).

## 15-4  TEMPERATURE IN METALWORKING

Forming processes are commonly classified into *hot-working* and *cold-working* operations.  Hot-working is defined as deformation under conditions of temperature and strain rate such that recovery processes take place simultaneously with

---

[1] F. Hodierne, *J. Inst. Met.*, vol. 91, pp. 267–273, 1963.

the deformation. On the other hand, cold-working is deformation carried out under conditions where recovery processes are not effective. In hot-working the strain hardening and distorted grain structure produced by deformation are very rapidly eliminated by the formation of new strain-free grains as the result of recrystallization. Very large deformations are possible in hot-working because the recovery processes keep pace with the deformation. Hot-working occurs at an essentially constant flow stress, and because the flow stress decreases with increasing temperature, the energy required for deformation is generally much less for hot-working than for cold-working. Since strain hardening is not relieved in cold-working, the flow stress increases with deformation. Therefore, the total deformation that is possible without causing fracture is less for cold-working than for hot-working, unless the effects of cold-work are relieved by annealing.

It is important to realize that the distinction between cold-working and hot-working does not depend on any arbitrary temperature of deformation. For most commercial alloys a hot-working operation must be carried out at a relatively high temperature in order that a rapid rate of recrystallization be obtained. However, lead and tin recrystallize rapidly at room temperature after large deformations, so that the working of these metals at room temperature constitutes hot-working. Similarly, working tungsten at 2000°F, in the hot-working range for steel, constitutes cold-working because this high-melting metal has a recrystallization temperature above this working temperature.

About 95 percent of the mechanical work of deformation is converted into heat. Some of this heat is conducted away by the tools or lost to the atmosphere, but a portion remains to increase the temperature of the workpiece. For a simple frictionless process the maximum theoretical increase in temperature is given by

$$\Delta T = \frac{U_p}{\rho c J} \qquad (15\text{-}25)$$

where $U_p$ = the work of plastic deformation per unit volume
$\rho$ = the density of the workpiece
$c$ = the specific heat of the workpiece
$J$ = the mechanical equivalent of heat, 778 ft-lb/Btu

The theoretical temperature increase (assuming no heat losses) for $\varepsilon = 1.0$ is 165°F for aluminum, 530°F for steel, and 1060°F for titanium. If the deformation process is carried out at a rapid rate, heat losses will be low and the temperature increase will approach the values given by Eq. (15-25). For slow-speed processes the heat losses may exceed the temperature rise due to deformation and the temperature will fall with increasing deformation.

## HOT WORKING

Hot-working is the initial step in the mechanical working of most metals and alloys. Not only does hot-working result in a decrease in the energy required to deform the metal and an increased ability to flow without cracking, but the rapid diffusion at hot-working temperatures aids in decreasing the chemical inhomogeneities of the cast-ingot structure. Blowholes and porosity are eliminated by the welding together of these cavities, and the coarse columnar grains of the casting are broken down and refined into smaller equiaxed recrystallized grains. These changes in structure from hot-working result in an increase in ductility and toughness over the cast state.

However, there are certain disadvantages to hot-working. Because high temperatures are usually involved, surface reactions between the metal and the furnace atmosphere become a problem. Ordinarily hot-working is done in air, oxidation results, and a considerable amount of metal may be lost. Reactive metals like titanium are severely embrittled by oxygen, and therefore they must be hot-worked in an inert atmosphere or protected from the air by a suitable barrier. Surface decarburization of hot-worked steel can be a serious problem, and frequently extensive surface finishing is required to remove the decarburized layer. Rolled-in oxide makes it difficult to produce good surface finishes on hot-rolled products, and because allowance must be made for expansion and contraction, the dimensional tolerances for hot-worked mill products are greater than for cold-worked products. Further, the structure and properties of hot-worked metals are generally not so uniform over the cross section as in metals which have been cold-worked and annealed. Since the deformation is always greater in the surface layers, the metal will have a finer recrystallized grain size in this region. Because the interior will be at higher temperatures for longer times during cooling than will be the external surfaces, grain growth can occur in the interior of large pieces, which cool slowly from the working temperature.

The lower temperature limit for the hot-working of a metal is the lowest temperature at which the rate of recrystallization is rapid enough to eliminate strain hardening in the time when the metal is at temperature. For a given metal or alloy the lower hot-working temperature will depend on such factors as the amount of deformation and the time that the metal is at temperature. Since the greater the amount of deformation the lower the recrystallization temperature, the lower temperature limit for hot-working is decreased for large deformations. Metal which is rapidly deformed and cooled rapidly from temperature will require a higher hot-working temperature for the same degree of deformation than will metal slowly deformed and slowly cooled.

The upper limit for hot-working is determined by the temperature at which either melting or excessive oxidation occurs. Generally the maximum working temperature is limited to 100°F below the melting point. This is to allow for the possibility of segregated regions of lower-melting-point material. Only a very small amount of a grain-boundary film of a lower-melting constituent is needed to make a material crumble into pieces when it is deformed. Such a condition is known as *hot shortness*, or *burning*.

Most hot-working operations are carried out in a number of multiple passes, or steps. Generally the working temperature for the intermediate passes is kept well above the minimum working temperature in order to take advantage of the economies offered by the lower flow stress. It is likely that some grain growth will occur subsequent to the recrystallization at these temperatures. Since a fine-grain-sized product is usually desired, common practice is to lower the working temperature for the last pass to the point where grain growth during cooling from the working temperature will be negligible. This *finishing temperature* is usually just above the minimum recrystallization temperature. In order to ensure a fine recrystallized grain size, the amount of deformation in the last pass should be relatively large.

The temperature range over which an alloy can be worked can be visualized with the aid of Fig. 15-11. For a given working pressure and temperature there will be a maximum amount of deformation that can be imparted to the workpiece. We are assuming here that this limitation is based on the flow resistance of the material and not on its ductility. As the preheat temperature of the billet increases the flow stress falls off and the amount of possible deformation increases. This gives the left-hand series of curves that increase to the right. The obvious limit

**Fig. 15-11** Schematic effect of temperature, pressure, and strain rate on the allowable working range. (*After S. Hirst and D. H. Ursell, Metal Treatment and Drop Forging, vol. 25, p. 409, 1958.*)

on the temperature scale is the solidus temperature of the workpiece. Hot short-ness will also be a limitation at somewhat lower temperatures. As the strain rate of deformation increases more heat will be retained in the workpiece and the workpiece temperature will have to be decreased to keep its final temperature from reaching the hot-shortness temperature.

## COLD-WORKING

As was shown in Sec. 6-13, cold-working of a metal results in an increase in strength or hardness and a decrease in ductility. When cold-working is excessive, the metal will fracture before reaching the desired size and shape. Therefore, in order to avoid such difficulties, cold-working operations are usually carried out in several steps, with intermediate annealing operations introduced to soften the cold-worked metal and restore the ductility. This sequence of repeated cold-working and annealing is frequently called the *cold-work–anneal cycle*. Figure 15-12 schematically illustrates the property changes involved in this cycle.

Although the need for annealing operations increases the cost of forming by cold-working, particularly for reactive metals which must be annealed in vacuum or inert atmospheres, it provides a degree of versatility which is not possible in hot-working operations. By suitably adjusting the cold-work–anneal cycle the part can be produced with any desired degree of strain hardening. If the finished product must be stronger than the fully annealed material, then the final operation must be a cold-working step with the proper degree of deformation to produce the desired strength. This would probably be followed by a stress relief to remove residual stresses. Such a procedure to develop a certain combination of strength and ductility in the final product is more successful than trying to achieve the same combinations of properties by partially softening a fully cold-worked material,

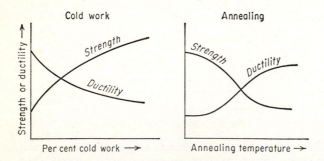

**Fig. 15-12** Typical variation of strength and ductility in the cold-work-anneal cycle.

because the recrystallization process proceeds relatively rapidly and is quite sensitive to small temperature fluctuations in the furnace. If it is desired to have the final part in the fully softened condition, then an anneal follows the last cold-working step.

It is customary to produce cold-worked products like strip and wire in different *tempers*, depending on the degree of cold reduction following the last anneal. The cold-worked condition is described as the annealed (soft) temper: quarter-hard, half-hard, three-quarter-hard, full-hard, and spring temper. Each temper condition indicates a different percentage of cold reduction following the annealing treatment.

## 15-5   STRAIN-RATE EFFECTS

Strain-rate, or deformation velocity, has three principal effects in metalworking: (1) the flow stress of the metal increases with strain rate, (2) the temperature of the workpiece is increased because of adiabatic heating, and (3) there is improved lubrication at the tool-metal interface, so long as the lubricant film can be maintained.

The usual definition of true-strain rate for a cylinder upset in compression is

$$\dot{\varepsilon} = \frac{d\varepsilon}{dt} = \frac{1}{h}\frac{dh}{dt} = \frac{v}{h} \qquad (15\text{-}26)$$

where $h$ = the instantaneous height
$\quad\ v$ = the deformation velocity

Since in many processes the workpiece is deformed between converging dies where $h$ varies with axial distance (as in Fig. 15-3), it is convenient to define a mean strain rate by

$$\dot{\bar{\varepsilon}}_x = \frac{1}{L}\int_0^L \dot{\varepsilon}\, dx \qquad (15\text{-}27)$$

where $L$ is the length of contact between the tool and the workpiece (see Fig, 15-23).

It is more usual to evaluate the mean strain rate in terms of the time for an element to travel through the die $t_f$.

$$\dot{\bar{\varepsilon}}_t = \frac{1}{t_f}\int_0^{t_f} \dot{\varepsilon}\, dt \qquad (15\text{-}28)$$

However, Chandra and Jonas[1] have shown that the appropriate measure of strain rate for strain-rate sensitive materials at large reductions (as in hot extrusion) is the root-mean-power strain rate

$$\dot{\varepsilon}_{rmp} = \left[ \frac{1}{\ln R} \int_0^{\ln R} (\dot{\varepsilon})^m \, d\varepsilon \right]^{1/m} \qquad (15\text{-}29)$$

where $R = A_0/A$ is the deformation ratio
$m$ = strain-rate sensitivity [see Eq. (9-38)]

Table 15-1 list typical values of deformation velocity for different testing and forming operations. It is important to note that the forming velocity of most commercial equipment is appreciably faster than the crosshead velocity used in the standard tension test. However, they are not high enough that stress-wave effects become important. When these forming velocities are combined with a situation in which the deformation zone is small, it is possible to produce very high local strain rates. The drawing of a fine wire at a speed of 120 ft/s can result in a strain rate in excess of $10^5 \text{ s}^{-1}$. A mean strain rate of around $2 \times 10^3 \text{ s}^{-1}$ is achieved in rolling thin tinplate. These high strain rates are attained by concentrating deformation into a narrow zone rather than by producing very high particle velocities.

However, there is a group of newer metalworking processes which utilize velocities as high as 700 ft/s to carry out forging, extrusion, sheet forming, etc. These processes are known as *high-energy-rate forming* (*HERF*) since the energy of deformation is delivered at a much higher rate than in conventional practice. These high-velocity forming processes[2] utilize the energy from exploding gas or a

**Table 15-1** TYPICAL VALUES OF VELOCITY ENCOUNTERED IN DIFFERENT TESTING AND FORMING OPERATIONS

| Operation | Velocity, ft/s |
|---|---|
| Tension test | $2 \times 10^{-6}$ to $2 \times 10^{-2}$ |
| Hydraulic extrusion press | 0.01 to 10 |
| Mechanical press | 0.5 to 5 |
| Charpy impact test | 10 to 20 |
| Forging hammer | 10 to 30 |
| Explosive forming | 100 to 400 |

[1] T. Chandra and J. J. Jonas, *Metall. Trans.*, vol. 2, pp. 877–881, 1971.
[2] G. E. Dieter, Strain Rate Effects in Deformation Processing, in W. A. Backofen (ed.), "Fundamentals of Deformation Processing," chap. 7, Syracuse University Press, 1964; "High-Velocity Forming of Metals," Prentice-Hall, Inc., Englewood Cliffs, N.J., 1964.

conventional explosive to produce high particle velocities. For many materials the elongation to fracture increases with strain rate beyond the usual metalworking range until a critical strain rate is reached where the ductility falls off sharply. Explosive forming has developed as a useful process for producing large parts in limited numbers where the force from the explosive energy is substituted for a large hydraulic press. Explosive forming usually is conducted with water filling the "stand-off distance" between the workpiece and the explosive charge. As such there are no unusual hardening effects or structural changes in the metal. However, when the explosive charge is placed in direct contact with the metal, very high transient stresses are produced in the metal and unusual hardening and structural changes may be produced depending on the metal.[1] A remarkable aspect of explosive hardening is that quite high hardness can be produced with essentially no grain distortion or gross distortion of the workpiece.

At the other extreme of the strain-rate spectrum is *superplasticity forming*. In Sec. 9-6 we saw that materials with a high strain-rate sensitivity ($0.3 \leq m \leq 1.0$) exhibit pronounced resistance to necking. Generally this occurs in material with a very fine grain size, of the order of 1 $\mu$m, and at deformation temperatures above $0.4T_m$. Also, for any superplastic material there is a limiting strain rate above which it is no longer superplastic. Generally, superplastic forming must be done at strain rates below 0.01 s$^{-1}$. The high resistance to plastic instability of a super-plastic metal means that forming processes generally restricted to hot polymers are possible. Biaxial stretching of sheet under small pressure differentials (vacuum forming) and blow molding are two examples. However, the chief advantage of superplasticity probably is the low flow stress, on the order of 1,000 to 5,000 psi, that exists at superplastic conditions. This has been utilized in the forging of difficult-to-work superalloys (the gatorizing process) or in embossing fine details in other applications.

## 15-6   METALLURGICAL STRUCTURE

The marked strengthening that results from cold-drawing an iron wire is shown in Fig. 15-13. Here the stress-strain curves determined after the wire had been drawn to a specific true strain are superimposed to show the overall strain-hardening effect. It is quite remarkable that the rate of strain hardening does not diminish significantly even at $\varepsilon > 6.0$.

The dislocation structure of strain-hardened metal was shown (in Sec. 6-13) to consist of a cellular substructure with the cell walls composed of tight-packed

[1] G. E. Dieter, Hardening Effect Produced With Shock Waves, in "Strengthening Mechanisms in Solids," American Society for Metals, Metals Park, Ohio, 1962.

**Fig. 15-13** True-stress–true-strain curve for iron wire deformed by wiredrawing at room temperature. (*From G. Lankford and M. Cohen, Trans. Am. Soc. Met., vol. 62, p. 623, 1969, Copyright American Society for Metals, 1969.*)

tangles of dislocations (Fig. 6-28). For the large plastic strains characteristic of wiredrawing and rolling, the cold-worked structure consists of highly elongated grains containing a relatively equiaxed dislocation cell structure.[1] The smaller strain-hardening ability of fcc based metals compared with iron-based alloys is explained by their development of an almost stable cell size, while with iron the cell size continues to decrease with plastic strain. The relationship between flow stress and cell size $d$ for cold-drawn iron wire has been found to be[2]

$$\sigma_0 = a + md^{-1} \qquad (15\text{-}30)$$

On a more macroscopic scale, the structure of severely cold-worked metal is characterized by the development of a strong crystallographic texture (see Sec. 6-17). The presence of preferred orientation causes anistropy of mechanical properties. This is of particular importance in determining the deep-drawing properties of rolled sheet (see Sec. 20-6). Closely allied with the development of

[1] J. D. Embury, A. S. Keh, and R. M. Fisher, *Trans. Metall. Soc. AIME*, vol. 236, pp. 1252–1260, 1966.
[2] G. Lankford and M. Cohen, *Trans. Am. Soc. Met.*, vol. 62, pp. 623–638, 1969.

texture is the formation of *deformation bands* and shear bands.[1] Deformation bands are regions of distortion where a portion of a grain has rotated towards another orientation to accommodate the applied strain. When these regions extend across many grains they are called shear bands. Exhaustive shear along these bands is a common type of failure in worked products.

Considerable work in recent years has led to increased understanding of the mechanism of hot-working.[2] The predominant softening mechanism in hot-working is *dynamic recovery.* When dynamic recovery has occurred, the microstructure consists of elongated grains with a fine substructure. Transmission electron microscopy shows the grains divided by dislocation subboundaries into subgrains with diameters of the order of 1 to 10 $\mu$m. Because this substructure resembles that found in cold-worked aluminum which has been annealed to a static recovery structure, the process has been called *dynamic recovery.* Dynamic recovery is the sole softening mechanism in metals where climb and cross slip of the dislocations are relatively easy. As a result, the misorientation of the subboundaries is maintained sufficiently low so that they never become capable of migrating as grain boundaries of recrystallization nuclei. This occurs in metals of high stacking-fault energy such as aluminum, alpha iron, and most bcc metals. During hot-working, all metals initially undergo dynamic recovery, but when climb and cross slip are not easy processes another mechanism intrudes at larger strains. This occurs in metals with low stacking-fault energy such as copper, nickel, austenitic iron, and their alloys. For these metals, when deformed at the same strain rate and fraction of the melting point, the subgrains are smaller and the dislocations in the cell walls more densely tangled than if climb and cross slip could occur easily. When the misorientation between subgrains reaches approximately 10°, the subgrains act as recrystallization nuclei and *dynamic recrystallization* occurs as the softening mechanism. These two hot-working–softening mechanisms are reflected in the hot-working stress-strain curve Fig. 15-14. A metal which softens by dynamic recovery rises to a plateau without a peak (curve *a*)

**Fig. 15-14** (*a*) Hot-working stress-strain curve for a metal which shows dynamic recovery; (*b*) metal which undergoes dynamic recrystallization after initial period of dynamic recovery.

[1] K. Brown, *J. Inst. Met.*, vol. 100, pp. 341–345, 1972.
[2] J. J. Jonas, C. M. Sellars, and W. J. McG. Tegart, *Metall. Rev.*, vol. 14, pp. 1–24, 1969.

while in curve *b* the onset of dynamic recrystallization causes the stress-strain curve to peak.

In metals which soften only by dynamic recovery, it is often possible to cool them sufficiently rapidly so that static recrystallization is prevented and the dynamic recovery structure is retained. This results in improved strength for non-heat-treatable Al-Mg alloys. For metals which soften by dynamic recrystallization it usually is not possible to retain the as-worked structure because static recrystallization occurs very rapidly when the deformation has been completed.

Hot-working greatly accelerates diffusional processes. Two examples of practical importance are the elimination of compositional inhomogeneities, such as a cored structure, by hot-working and the coarsening of a second-phase structure, such as the spheroidization of a pearlitic steel. For example,[1] a degree of spheroidization at 700°C which would take 660 h of annealing can be achieved in 210 s by deformation at $0.016 \text{ s}^{-1}$. Diffusion is greatly enhanced by the formation during deformation of a dynamic recovery substructure in which the subboundaries provide paths for pipe diffusion which is considerably more rapid than volume diffusion.

The plastic working characteristics of two-phase alloys depend on the microscopic distribution of the two phases. The presence of a high volume fraction of hard uniformly dispersed particles, such as are found in many superalloys, greatly increases the flow stress and makes working difficult. If the second-phase particles are hard and more massive, they will tend to fracture on deformation with the softer matrix extruding into the voids created by this fracturing. If the second-phase particles are ductile, failure usually occurs by the matrix pulling apart between the particles. Alloys which contain a hard second phase located primarily in the grain boundaries are difficult to work because of the tendency for fracture to occur along the grain boundaries. Cast ingots usually contain voids or porosity that arise because of the shrinkage associated with solidification or because of the liberation of gas. If these cavities have clean surfaces, they may be closed shut by pressure welding from mechanical working. Even if the cavity contains a gas, it may be closed if the gas is readily soluble in the metal at the working temperature.

As the result of a mechanical working operation second-phase particles will tend to assume a shape and distribution which roughly correspond to the deformation of the body as a whole. Second-phase particles or inclusions which are originally spheroidal will be distorted in the principal working direction into an ellipsoidal shape if they are softer and more ductile than the matrix. If the inclusions or particles are brittle, they will be broken into fragments which will be oriented parallel to the working direction, while if the particles are harder and

---

[1] E. A. Chojnowski and W. J. McG. Tegart, *Met. Sci. J.*, vol. 2, pp. 14–18, 1968.

**Fig. 15-15**   Fiber structure in wrought product revealed by macro-etching (5×).

stronger than the matrix, they will be essentially undeformed.[1] The orientation of second-phase particles during hot- or cold-working and the preferred fragmentation of the grains during cold-working are responsible for the fibrous structure which is typical of wrought products. The fiber structure can be observed after macroetching (Fig. 15-15). Microscopic examination of wrought products frequently shows the results of this *mechanical fibering* (Fig. 15-16). An important consequence of mechanical fibering is that the mechanical properties may be different for different orientations of the test specimen with respect to the fiber (working) direction. In general, the tensile ductility, fatigue properties, and impact properties will be lower in the transverse direction (normal to the fiber) than in the longitudinal direction.

The forming characteristics of an alloy can be affected if it undergoes a strain-induced precipitation or strain-induced phase transformation. If a *precipitation reaction* occurs in a metal while it is being formed, it will produce an increase in the flow stress but, more important, there will be an appreciable decrease in ductility, which can result in cracking. When brittleness is caused by precipitation, it usually results when the working is carried out at a temperature just below the solvus line[2] or from cold-working after the alloy had been heated to the same temperature region. Since precipitation is a diffusion-controlled process, difficulty

---

[1] H. Unkel, *J. Inst. Met.*, vol. 61, pp. 171–196, 1937; F. B. Pickering, *J. Iron Steel Inst. London*, vol. 189, pp. 148–159, 1958.

[2] The boundary on the phase diagram of the limit of solid solubility of a solid solution.

**Fig. 15-16** Fibered and banded structure in longitudinal direction of a hot-rolled mild-steel plate (*100×*).

from this factor is more likely when forming is carried out at a slow speed at an elevated temperature. To facilitate forming, age-hardenable aluminum alloys are frequently refrigerated just before forming in order to suppress the precipitation reaction.

A most outstanding practical example of a *strain-induced phase transformation* occurs in certain austenitic stainless steels where the Cr : Ni ratio results in an unstable austenite phase. When these alloys are cold-worked, the austenite transforms to ferrite along the slip lines and produces an abnormal increase in the flow stress for the amount of deformation received. While this phase transformation is often used to increase the yield strength by cold rolling, it can also result in cracking during forming if the transformation occurs in an extreme amount in regions of highly localized strain.

The importance of grain size in determining strength and fracture toughness focuses special attention on the role of processing in creating a fine grain size. Although every stage in the processing chain from ingot casting to cooling from the final working or heat treatment step may exert some influence, by far the most important role is played by the temperature at which hot-working is completed or the final annealing temperature for cold-worked material. In hot-working processes fine grain size is favored by a low finishing temperature and a rapid cooling rate from the working temperature. In processing steels which undergo a structural transformation on cooling from the finishing temperature, the ferrite grain size depends on achieving a fine austenite grain size. A fine austenite grain size is promoted by taking large reductions in the final pass and working at the lowest possible temperature in the austenite phase field.

The prior discussion has emphasized that the interaction of the deformation processing conditions of strain, strain rate, and temperature with metallurgical considerations such as phase transformation, precipitation, recrystallization, or grain growth determine the final properties of the product. The optimization of mechanical properties through appropriate *thermomechanical treatments*[1] is one of the areas of greatest promise in metalworking.

## 15-7  FRICTION AND LUBRICATION

The friction forces developed between the workpiece and the forming tools are an important consideration in metalworking. In the analysis of the forces in metalworking (Sec. 15-2), we neglected the friction forces in the interest of clarity. This certainly is not a realistic assumption, for in many real metalworking processes friction is the predominant factor.

As an example of an analysis with friction included, consider the slab analysis of the homogeneous compression of a flat circular disk (Fig. 15-17). We shall invoke the simplifying assumptions that there is no barreling of the edges of the disk and that the thickness is small enough that the axial compressive stress $\sigma_z$ is constant through the thickness. The frictional conditions on the top and bottom faces of the disk are described by a constant coefficient of Coulomb friction

$$\mu = \frac{\tau}{p} \quad (15\text{-}31)$$

Fig. 15-17  Stresses acting in compression of sector of flat circular disk.

[1] R. J. McElroy and Z. C. Szkopiak, *Int. Metall. Rev.*, vol. 17, pp. 175–202, 1972.

where $\tau$ = the shearing stress at the interface

$\quad\quad p$ = the stress normal to the interface

Lateral flow of the metal outward as it is being compressed leads to shearing stresses at the die contact surfaces. This surface shear is directed toward the center of the disk, opposing the outward radial flow. These frictional shear stresses lead to lateral pressure in the material, which is zero at the edges of the disk and builds up to a peak value at the center.

Referring to Fig. 15-17 and taking the equilibrium of *forces* in the radial direction:

$$\sigma_r hr\,d\theta - (\sigma_r + d\sigma_r)h(r + dr)\,d\theta + 2\sigma_\theta h\,dr\sin\frac{d\theta}{2} - 2\tau r d\theta\,dr = 0 \quad\quad (15\text{-}32)$$

Using the approximation $\sin(d\theta/2) \approx d\theta/2$, this reduces to

$$\sigma_r h\,dr + d\sigma_r rh - \sigma_\theta h\,dr + 2\tau r\,dr = 0$$

From the axial symmetry of the disk, $d\varepsilon_\theta = d\varepsilon_r$, and from Eq. (3-44), $\sigma_\theta = \sigma_r$. Making this substitution, we arrive at

$$\frac{d\sigma_r}{dr} + \frac{2\tau}{h} = 0$$

and from the chosen friction law, $\tau = \mu p = \mu\sigma_z$

$$\frac{d\sigma_r}{dr} + \frac{2\mu\sigma_z}{h} = 0 \quad\quad (15\text{-}33)$$

By making the assumption that $\sigma_z$, $\sigma_\theta$, and $\sigma_r$ are principal stresses, we can utilize the von Mises' yield criterion to develop a relation between $\sigma_r$ and $\sigma_z$.

$$2\sigma_0^2 = (\sigma_1 - \sigma_2)^2 + (\sigma_2 - \sigma_3)^2 + (\sigma_3 - \sigma_1)^2$$
$$2\sigma_0^2 = (\sigma_r - \sigma_r)^2 + (\sigma_r - \sigma_z)^2 + (\sigma_z - \sigma_r)^2$$
$$\sigma_0^2 = (\sigma_r - \sigma_z)^2$$

and $\quad\quad\quad\quad\quad\quad\quad\quad\quad \sigma_0 = \sigma_r - \sigma_z \quad\quad\quad\quad\quad\quad\quad (15\text{-}34)$

If we define $p$ as a *positive* compressive stress normal to the interface, then $p = -\sigma_z$, and $\sigma_0 = \sigma_r + p$, so that $d\sigma_r = -dp$. Making these substitutions into Eq. (15-33) gives

$$\frac{dp}{p} = -\frac{2\mu\,dr}{h} \quad\quad (15\text{-}35)$$

Integration yields

$$\ln p = -\frac{2\mu r}{h} + C$$

**Fig. 15-18**  Friction hill for homogeneous compression of a disk with Coulomb friction.

At the outersurface of the disk, $r = a$, $\sigma_r = 0$, and $p = \sigma_0$, so that

$$C = \ln \sigma_0 + 2\,\frac{\mu a}{h}$$

Therefore,

$$\ln \frac{p}{\sigma_0} = \frac{2\mu}{h}(a - r) \qquad (15\text{-}36)$$

or

$$p = \sigma_0\, e^{\frac{2\mu}{h}(a-r)} \qquad (15\text{-}37)$$

Figure 15-18 shows the axial compressive stress from Eq. (15-37) plotted over the diameter of the disk. The pressure distribution is symmetrical about the centerline and rises to a sharp peak at the center of the disk. This characteristic rise in deformation pressure with distance is often called a *friction hill*. The average deformation pressure (mean height of the friction hill) is given by

$$\bar{p} = \frac{\int_0^a 2\pi p r\, dr}{\pi a^2} = \frac{\sigma_0}{2}\left(\frac{h}{\mu a}\right)^2\left[e^{2\mu a/h} - \frac{2\mu a}{h} - 1\right] \qquad (15\text{-}38)$$

Figure 15-19 plots the ratio of $\bar{p}$ to the uniaxial compressive flow stress as a function of the length-to-thickness ratio $a/h$ of the disk and the coefficient of friction. For a given $a/h$ increasing friction easily leads to more than a doubling of the deformation pressure. The role of friction in raising deformation forces should be clear from this figure. Note also that the role of friction becomes particularly important at large values of $a/h$.

The above analysis was for sliding friction at the interface, as in our first encounter with friction in elementary physics. At the other extreme, we can envision a situation where the interface has a constant film shear strength $\tau_i$. The most usual case is *sticking friction*, where there is no relative motion between the workpiece and the tools. This condition often occurs in hot-working where lubrication may be difficult. For sticking friction $\tau_i = \tau_0$, the flow stress in shear

**Fig. 15-19** Average deformation pressure in compression of a disk as function of $\mu$ and $a/h$.

($k$). With a von Mises' yield criterion, the coefficient of friction under sticking conditions is

$$\mu = \frac{\tau_i}{p} = \frac{k}{\sigma_0} = \frac{\sigma_0/\sqrt{3}}{\sigma_0} = \frac{1}{\sqrt{3}} = 0.577 \qquad (15\text{-}39)$$

Using the sticking condition $\tau = k = \sigma_0/\sqrt{3}$ in the analysis for the compression of a disk results in

$$p = \sigma_0 \left[ 1 + \frac{2}{\sqrt{3}} \frac{(a-r)}{h} \right] \qquad (15\text{-}40)$$

Thus, the pressure distribution still peaks at $r = 0$, but now the sides of the hill are straight lines. Often a condition exists where sliding occurs at the outer periphery of the disk and sticking occurs at the interior of the disk.

The analysis for the condition of *sticking friction* suggests an alternative to the concept of Coulomb sliding friction as a way of looking at friction in metalworking processes. This approach considers that the workpiece in contact with the tools can be represented as a material of constant shear strength $\tau_i$. However, instead of letting $\tau_i = k$, as for sticking friction, we consider that the interface shear strength may be some constant fraction $m$ of the yield strength in shear where $m$ is called the *interface friction factor*.

$$m = \frac{\tau_i}{k} = \frac{\text{interface shear strength}}{\text{yield stress in shear}} \qquad (15\text{-}41)$$

Values of $m$ vary from 0 (perfect sliding) to 1 (sticking).

There are basic differences between these two methods of treating friction. The interface pressure $p$ developed in most metalworking processes will at least equal the uniaxial yield stress $\sigma_0$ and may appreciably exceed this value (see Fig. 15-19). On the other hand, the shearing stress at the interface can never exceed the yield strength of the material in shear, for once $\tau_i = k$, movement along the interface is arrested (sticking friction) and deformation proceeds by subsurface shearing. Since the yield strength of the workpiece material in shear is not affected by normal pressure, the Coulomb coefficient of friction determined from Eq. (15-31) decreases in inverse proportion to interface pressure. This is contrary to physical reality and can lead to misinterpretation of reported values of $\mu$. On the other hand, $m$ is independent of the normal stress at the interface. This factor, coupled with the relative ease with which $m$ may be measured, has led to increased use of the interface friction factor for describing friction in metalworking processes. It appears to be particularly suited for hot-working processes involving large deformations, such as forging and extrusion. In processes such as cold-rolling and wiredrawing the use of coefficient of friction appears to be well established.

Most analyses of metalworking processes have been carried out using $\mu$. However, the use of $m$ not only may be more physically appropriate, but it introduces mathematical simplification. Thus, if the compression of a disk is analyzed using $m$ instead of $\mu$, we replace Eq. (15-31) by $\tau = mk = m\sigma_0/\sqrt{3}$. Equation (15-33) now becomes, after simplification,

$$h\,dp = -\frac{2m\sigma_0}{\sqrt{3}}\,dr$$

and

$$p = -\frac{2m\sigma_0}{\sqrt{3}}\frac{r}{h} + C$$

which, when the boundary conditions are applied, reduces to

$$p = \sigma_0\left(1 + \frac{2m}{\sqrt{3}}\frac{a-r}{h}\right) \qquad (15\text{-}42)$$

The similarity with Eq. (15-40) for sticking friction ($m = 1$) is to be expected.

It is usually assumed that the coefficient of friction has a constant value over the interface between the workpiece and the tools. Although there is little experimental data, it seems clear that $\mu$ varies with normal stress. Figure 15-20 shows the variation of $p$ and $\mu$ for the compression of an aluminum disk. These data were obtained by painstaking experiments in which small pressure pins were buried in the tool under the workpiece.[1] Other factors which influence the local value of $\mu$ are the strain rate and the temperature at the interface.

[1] G. T. van Rooyen and W. A. Backofen, *Int. J. Mech. Sci.*, vol. 1, pp. 1–27, 1960; G. W. Pearsall and W. A. Backofen, *Trans. ASME, Ser. D: J. Basic Eng.*, vol. 85B, pp. 68–75, 1963.

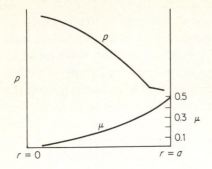

**Fig. 15-20**   Variation of normal pressure $p$ and $\mu$ with radial distance in compression of an aluminum disk without lubrication. (*After W. A. Backofen, "Deformation Processing," Addison-Wesley Publishing Company, Inc., Reading, Mass., 1972.*)

The understanding of friction between two surfaces is at the heart of the science of *tribology*, an interdisciplinary field that involves surface chemistry, materials science, and mechanics. Most basic studies of friction[2] have been concerned with light loads where there is no plastic deformation of the workpiece. The running of a shaft in a bearing would be a typical example. A basic premise of the theory of friction is that apparently flat, smooth surfaces are not smooth when viewed on a microscopic scale. Real surfaces contain hills and valleys which are large when compared on an atomic scale. Therefore, when two surfaces are brought into contact they meet where the high spots, *asperities*, come into contact (Fig. 15-21). The real area of contact $A_r$ is the sum of the area of contact at all of the asperities. Also, the surfaces are rarely clean. They contain layers of gas molecules or water molecules and often a thin oxide layer. Of course, frequently, special molecules are added to these surfaces. We call these lubricants.

As the load normal to the interface increases, the metal at the asperities deforms plastically, increasing the area of contact. Cold welding may also occur at these points. The situation at an asperity is very similar to indentation in a hardness test, where the hardness is the normal load $P$ divided by the projected area of the indentation. $H = P/A_p \approx P/A_r$. Since hardness will be independent of the sum of the area of contact of the asperities and $A_r = P/H$, the real area of contact varies directly with the normal load. We shall let $F$ be the shearing force parallel to the interface that tears apart the welded junctions. Thus, the coefficient of friction is given by

$$\mu = \frac{F}{P} = \frac{\tau_i A_r}{H A_r} = \frac{\tau_i}{H} \qquad (15\text{-}43)$$

where $\tau_i$ is the shear stress to shear through the asperities. For light normal loads $\tau_i$ and $H$ are independent of $P$. The coefficient of friction is thus independent of normal load. This is known as *Amonton's law* and is the criteria for Coulomb

---

[1] F. P. Bowden and D. Tabor, "The Frictional Lubrication of Solids," Oxford University Press, London, pt. I, 1950; pt. II, 1964.

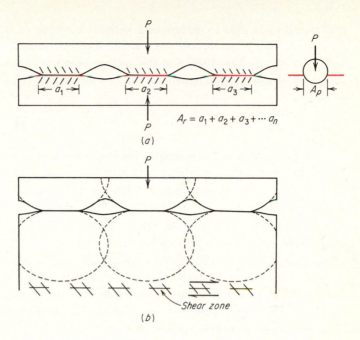

$$A_r = a_1 + a_2 + a_3 + \cdots a_n$$

(a)

(b)

**Fig. 15-21** (a) Contact at asperities; (b) overlap of deformation zones to produce subsurface shear zone.

sliding friction to exist. As the normal force increases, the real area of contact increases. At high normal loads the asperities will undergo extensive plastic deformation. The "plastic zones" below each asperity will overlap (Fig. 15-21b). Because of this plastic deformation and the accompanying plastic constraint, the bond at the workpiece-tool interface is stronger than the shear strength of the material in a subsurface zone $\tau_0$. For subsurface shear the coefficient of friction is

$$\mu = \frac{\tau_0}{H} = \tau_0 \frac{A}{P} \qquad (15\text{-}44)$$

Since $\tau_0$ and $A$ are approximately constant, this shows that $\mu$ decreases as $P$ increases, provided that bulk plastic deformation is taking place.

There are basically two approaches for measuring either $\mu$ or $m$ under metalworking conditions. Tests developed for evaluating lubricants under light load (slider) conditions are not relevant to metalworking. The most common approach is to measure the average deformation pressure and the flow stress so that $\mu$ can be calculated from the analytical relation between these factors. For example, in the compression of a disk, if $\bar{p}$ is measured and $\sigma_0$ is known for the $\dot{\varepsilon}$ and $T$ of the test conditions, then $\mu$ can be calculated with Eq. (15-38). Since information on the flow stress at the appropriate conditions may not be available, it is useful to be able to determine $\mu$ without knowing $\sigma_0$ or $k$. For example, in the plane-strain compression test,[1] if the identical reduction is made with two values of $b/t$ (see Fig. 15-9) and the average pressures are measured, then $\mu$ can be found by solving simultaneously the two sets of Eq. (15-45)

$$\frac{\bar{p}}{2k} = \frac{e^{\mu b/t} - 1}{\mu b/t} \qquad (15\text{-}45)$$

A particularly useful technique for measuring $m$ (or $\mu$) is the *ring-compression test*.[2] The test uses a thin ring with the dimensions OD : ID : thickness in the ratio 6 : 3 : 1. The friction factor is determined by measuring the percentage of change in the inside diameter of the ring. For low values of $m$ the ID increases, while for higher values of $m$ the ID of the ring decreases. Calibration curves between $\Delta$ID, percent reduction, and $m$ can be calculated from the analytical analysis of Avitzur.[3] Figure 15-22 shows a typical calibration curve.[4] It is important to note that it is not necessary to make a direct measurement of the deformation force or to know the flow stress in order to measure $m$ with the ring-compression test. Thus, $m$ may conveniently be measured at the high temperature and strain rate typical of hot-working processes.

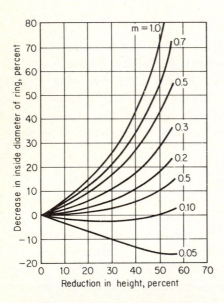

Fig. 15-22 Calibration curve for upset ring test with outside diameter, inside diameter, and thickness in a ratio of 6:3:1. (*From C. H. Lee and T. Altan, Trans. ASME, Ser. B: J. Eng. Ind., vol. 94, p. 775, 1972.*)

[1] H. Takahashi and J. M. Alexander, *J. Inst. Met.*, vol. 90, pp. 72–79, 1961–1962.
[2] A. T. Male and M. G. Cockcroft, *J. Inst. Met.*, vol. 93, pp. 38–46, 1964–1965.
[3] B. Avitzur, "Metal Forming," McGraw-Hill Book Company, New York, 1968.
[4] C. H. Lee and T. Altan, *Trans. ASME, Ser. B: J. Eng. Ind.*, vol. 94, p. 775, 1972.

Moreover, if deformation pressure is measured during compression of the ring, it is possible to calculate[1] the flow stress as well as to determine $m$. This is particularly attractive for hot forming processes where die chilling of the workpiece makes it difficult to simulate the test conditions to measure the flow stress.

Although the emphasis in this section has been on the role of friction in increasing the deformation forces, from a practical point of view many production machines have excess load capacity. Thus, while high forming loads per se may not be of practical significance, the high stresses arising from high friction may result in various limits to the attainable deformation. Many of these will be described in subsequent chapters dealing with specific processes. It is obvious that tearing and cracking are accentuated by high friction. In many practical situations the chief effect of friction is in influencing the surface finish of the product or in affecting the wear of the dies and tooling.

When one solid surface in a sliding pair is much harder than the other (as in the typical tool-workpiece combination) the asperities of the harder surface will penetrate the softer surface and displace a volume of metal proportional to the total length of sliding and the cross-sectional area of the asperity. This is known as ploughing. Frictional resistance due to ploughing is in addition to that arising from the shearing of asperity contracts. The ploughing force[2] is related to the flow properties of the workpiece and the size and shape of the asperities. Thus, smooth dies are important in reducing the friction contribution from ploughing.

The most serious result of inadequate lubrication is the transfer of workpiece material to the tools. This is called pickup. Metal transfer occurs in two ways. If the lubricant film is depleted at the interface of a rough tool surface, the workpiece is forced into crevices in the tool surface. Subsequent tangential motion shears off the projecting soft metal and results in tool pickup and a poor surface finish on the workpiece. If the lubricant breaks down under high pressure, there is local cold welding between tool and workpiece. When a piece is torn away from the workpiece surface, it exposes fresh clean metal which is even more susceptible to cold welding. Thus, when pickup begins it often becomes progressively worse and leads to galling and seizure.

Since the tool surfaces are subjected to many sliding cycles, there is continual wear of these surfaces. The chief wear mechanism is abrasion by hard oxide particles on the workpiece. Another source of wear is surface fatigue resulting from the buildup and release of the interface pressure. Surface cracking of tools may also result from thermal stresses arising from heating and cooling of the tools.

---

[1] A. T. Male, V. dePierre, and G. Sand, *ASLE Trans.*, vol. 16, pp. 177–184, 1973; J. R. Douglas and T. Altan, *ASME*, Paper No. 73–WA/Prod-13, 1973.
[2] J. Goddard and H. Wilman, *Wear*, vol. 5, pp. 114–135, 1962.

In selecting a lubricant, the workpiece, the die, and the lubricant should be considered a total system.[1] The functions of a metalworking lubricant are many:

1. Reduces deformation load
2. Increases limit of deformation before fracture
3. Controls surface finish
4. Minimizes metal pickup on tools
5. Minimizes tool wear
6. Thermally insulates the workpiece and the tools
7. Cools the workpiece and/or the tools

There are many, often contradictory, attributes required of a good metalworking lubricant. It must be capable of functioning over a wide range of pressure, temperature, and sliding velocities. Since one of the characteristics of most plastic working processes is the generation of a large amount of new surface area, the lubricant must have favorable spreading and wetting characteristics. It must be compatible with both the die and workpiece material with regard to wetting and chemical attack. It should have good thermal stability and resistance to bacteriological attack and minor contaminants. A good lubricant produces a harmless residue that does not cause staining on subsequent heat treatment or welding, and which is easily removed. Finally, a lubricant should be nontoxic, free of fire hazard, and inexpensive.

There are two general categories of lubrication: boundary lubrication and hydrodynamic lubrication. In *boundary lubrication* the surfaces are not fully separated by a lubricant film. This is the most common situation for metalworking lubrication. In *hydrodynamic lubrication* the two surfaces are completely separated by a lubricant film. This condition exists in bearings but is not common in metalworking. It does occur, for example, in hydrostatic extrusion and wiredrawing with pressurized lubrication. Thick-film lubrication, such as the extrusion of steel with a viscous glass lubricant, may be considered quasi-hydrodynamic lubrication.

Lubricants reduce the friction by introducing an interface that is easily sheared. Typical *boundary lubricants* are long-chain organic molecules. These polar molecules attach themselves by one end to the metal surface. Typical examples are fatty acids, such as oleic acid, which forms a metallic soap when mixed with water or oil. Organic lubricants have poor temperature stability and cannot be used above 200 to 300°C. *Extreme pressure (EP)* additives react locally when the boundary film is broken down under heat and pressure. When a new surface is formed due to boundary film breakdown, the EP additives react chemically to form a compound with a low interfacial shear strength. Examples of EP additives

---

[1] J. A. Schey (ed.), " Metal Deformation Processes: Friction and Lubrication," Marcel Dekker, Inc., New York, 1970.

are halogen compounds and colloidal sulfur. *Conversion coatings* (oxides, phosphates, or chromates) are often applied to the workpiece to serve as a base for the retention of the lubricant. Some of these coatings also have lubricating properties.

Quasi-hydrostatic lubrication is provided by *continuous lubricant films*. These may consist of soft ductile solids (such as lead, copper, Teflon), polyethylene, or solids with a lamellar structure (such as $MoS_2$, graphite, boron nitride, and certain oxides). Continuous-film lubrication may also be obtained with fluids such as molten glass or with graphite dispersions. With complete fluid film lubrication the interface shear strength is a product of the film viscosity and the shear-strain rate.

The surface finish of a wrought product depends on the frictional conditions. In cold-working the surface asperities are flattened by a smooth polished tool to produce a burnished or shiny surface. A rough tool will not produce a finish better than the finish of the tool. With thick-film lubrication and smooth tools a dull or matte finish is produced. When the workpiece is formed out of contact with the tools, as in certain sheet-forming operations, then the surface finish is controlled by the properties of the workpiece. Defects such as orange peel and stretcher strains may result.

## 15-8 DEFORMATION-ZONE GEOMETRY

Most deformation processes require the material to flow through a converging channel, which we call dies. Figure 15-23 shows some typical situations. The basic feature of each channel is the ratio of the mean thickness to the length of the deformation zone.[1] This ratio is designated $\Delta$, and for the simple case of parallel (nonconverging) dies is $\Delta = h/L$. Values of $\Delta$ for more complex situations are given in Fig. 15-23. $\Delta_s$ is based on a plane-strain reduction $r_s = 1 - (h_1/h_0)$ for strip. $\Delta_w$ is based on an axisymmetric reduction $r_w = 1 - (d_1/d_0)^2$ for wire or rod. To decrease $\Delta$, we can reduce the semidie angle $\alpha$, thereby increasing $L$, or we can increase the reduction $r$ which decreases $h$.

The importance of deformation-zone geometry $\Delta$ on the yield pressure for frictionless plane-strain indentation[2] is shown in Fig. 15-24. Examples of various deformation-zone geometries and typical processes which tend to have that value of $\Delta$ are given in this useful illustration. The increase in $p/2k$ with $\Delta$ is chiefly the result of increasing redundant work.[3]

[1] W. A. Backofen, "Deformation Processing," Addison-Wesley Publishing Company, Inc., Reading, Mass., 1972.
[2] R. Hill, "The Mathematical Theory of Plasticity," p. 257, Oxford University Press, London, 1950.
[3] W. A. Backofen, op. cit., p. 137.

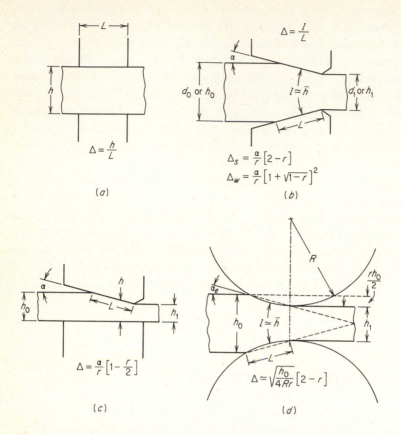

**Fig. 15-23** Converging channels for different deformation processes showing calculation of $\Delta$ (a) parallel indenters (plane-strain compression); (b) drawing or extrusion of strip or wire; (c) tube drawing over a mandrel; (d) rolling. (*From W. A. Backofen. "Deformation Processing", Addison-Wesley Publishing Company, Inc., Reading, Mass., 1972.*)

## 15-9 HYDROSTATIC PRESSURE

An important factor in achieving a successful forming operation without fracture is the level of hydrostatic pressure achieved in the process. The presence of a high hydrostatic pressure reduces the tensile stresses below the critical value for cracking, while at the same time the flow stress is unaffected. As an extra bonus, deformation carried out under high hydrostatic pressure creates less damage to the material during deformation. The effect of hydrostatic pressure in increasing ductility in the tension test has already been discussed in Sec. 7-14.

The hydrostatic pressure may arise from the hydrostatic component of the stress state, chiefly by the interaction of the workpiece and the tooling as in extru-

**Fig. 15-24** Dependence of yield pressure on deformation-zone geometry for the frictionless plane-strain indentation of a rigid-ideal plastic material. (*From W. A. Backofen, " Deformation Processing," p. 135, Addison-Wesley Publishing Company, Inc., Reading, Mass., 1972.*)

sion, or it may be introduced externally to the process, as in hydrostatic extrusion. Figure 15-25 shows the elements of the hydrostatic extrusion process in which the billet is forced through the die by a high hydrostatic fluid pressure. The chief advantages of this process are the elimination of the large drag friction force between the billet and the container wall and the achievement of hydrodynamic lubrication in the die.

The influence of hydrostatic stress state on the structural damage produced in cold-drawing of strip was studied by Coffin and Rogers.[1] Using the slip-line field for frictionless[2] strip drawing (Fig. 15-4), it is possible to map out the hydrostatic pressure at each point in the deformation zone. Figure 15-26 shows the hydrostatic pressure developed at the centerline (midplane) of the sheet as a function of semidie angle and reduction per pass. The damage produced by this deformation was measured by precision density measurements. Extreme damage also results in reduced ductility in the tension test. Decreases in density could be

[1] L. F. Coffin, Jr., and H. C. Rogers, *Trans. Am. Soc. Met.*, vol. 60, pp. 672–686, 1967.
[2] The influence of friction was subsequently covered in H. C. Rogers and L. F. Coffin, Jr., *Int. J. Mech. Sci.*, vol. 13, pp. 141–155, 1971.

**Fig. 15-25**  Typical arrangement for hydrostatic extrusion.

measured before any void formation or cracking was observed in the microstructure. Figure 15-27 shows that large decreases in density correlate with drawing conditions in which the hydrostatic stress at the centerline is most tensile. The strong influence of external hydrostatic pressure in greatly reducing damage from cold-drawing is shown in Fig. 15-28.

Hydrostatic pressure may be used to advantage in working brittle materials such as bismuth or cast iron. For brittle materials it is necessary to extrude into a pressurized chamber in order to avoid cracking.[1] Figure 15-29 shows that the

**Fig. 15-26**  Ratio of hydrostatic pressure $p$ to tensile yield stress $2k$ at centerline of cold-drawn strip. Each curve for dies with different semiangle die. Positive value of $p/2k$ denotes compressive hydrostatic stress. (*From L. F. Coffin and H. C. Rogers, Trans, Am. Soc. Met., vol. 60, p. 674, 1967. Copyright American Society for Metals, 1967.*)

[1] H. L. D. Pugh and D. Gunn, *Symp. Physics and Chemistry of High Pressure*, Society of Chemical Industry, London, 1963, pp. 157–162.

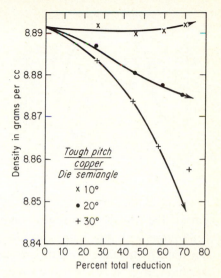

Fig. 15-27 Structural damage produced in strip drawing of tough pitch copper, as measured by density. Increasing die semiangle corresponds to greater hydrostatic tension. (Compare Fig. 15-26.) (*L. F. Coffin and H. C. Rogers, Trans. Am. Soc. Met., vol. 60, p 678, 1967. Copyright American Society for Metals, 1967.*)

Fig. 15-28 Influence of external hydrostatic pressure on the density change in 6061 aluminum alloy deformed in strip drawing under highly tensile conditions. (*From L. F. Coffin and H. C. Rogers, Trans, Am. Soc. Met., vol. 60, p. 681, 1967, Copyright American Society for Metals, 1967.*)

**Fig. 15-29** Effect of reduction on critical pressure to inhibit cracking (schematic).

hydrostatic pressure required to avoid cracking varies with extrusion ratio. This reflects the way the hydrostatic component of stress produced in the deformation zone varies with reduction, since the pressure plotted is the sum of this stress plus the fluid hydrostatic pressure. While the curve given is specific to a particular situation, it generally is found that cracking problems are less severe at very low or very large reductions and that there is an intermediate range of reduction for which cracking is more prevalent. The beneficial effects of hydrostatic pressure may be achieved without the difficulty of employing high-pressure equipment if the workpiece is encased with a metal of low flow stress compared with the workpiece. For example, brittle berylium billets are "canned" in mild steel to produce hydrostatic pressure support during forging.

While the presence of a high hydrostatic compressive stress enhances the limit of workability in processes which are limited by cracking, it will not improve workability in processes which are limited by the formation of a local necked region. Examples of metalworking processes where local plastic instability limits the extent of deformation are wiredrawing and deep drawing of sheet. The reason for this is that hydrostatic pressure does not affect the tensile instability of metals,[1] and therefore, the initiation of local necking in a metal under tension is independent hydrostatic state of stress.

## 15-10 WORKABILITY

Workability is concerned with the extent to which a material can be deformed in a specific metalworking process without the formation of cracks. In some processes the limit of workability is determined by the formation of local necking (plastic instability) rather than by the occurrence of fracture. Workability is a complex technological concept that depends not only on the fracture resistance (ductility)

[1] H. L. D. Pugh, *ASTM Spec. Tech. Publ.* 374, p. 68, 1965; J. M. Alexander, *J. Inst. Met.*, vol. 93, p. 366, 1965.

**Fig. 15-30** Examples of cracks in metalworking processes. (*a*) Free surface crack in upsetting; (*b*) surface cracks from heavy die friction in extrusion; (*c*) center burst or chevron cracks in a drawn rod.

(*a*)          (*b*)

(*c*)

of the material but also on the specific details of the process such as the reduction, friction, temperature, and strain rate. It is not surprising that widely accepted tests for workability are not available,[1] and that very specialized and empirical tests often are used in industry for rating the workability of materials.

The cracks that occur in metalworking processes can be grouped into three broad categories (see Fig. 15-30):

1. Cracks at a free surface, such as at the bulge in upsetting a cylinder or in edge cracking in rolling
2. Cracks that develop in a surface where interface friction is high, such as in extrusion
3. Internal cracks, such as chevron cracks in drawn bars

Internal cracks develop as a result of "secondary tensile stresses" which typically occur with large values of $\Delta = h/L$. Temperature gradients often occur in the workpiece, and this produces a situation very conducive to fracture. A particularly common situation is where the workpiece is chilled by the die, and since the flow stress is usually strongly temperature-dependent, a chilled region produces a local nondeforming zone. The presence of regions of "hard and easy flow" leads to the development of shear bands, and the localization of flow into these bands results in very high shear strains and often shear fracture.

Cracks that form in metalworking occur by ductile fracture (see Sec. 7-10). Ductile fracture is characterized by the formation and growth of holes around second-phase particles and the localization of shear strain in sharply defined bands. Void formation in ductile fracture in the tension test has been associated with the plastic instability of necking, and it appears that tensile cracking due to bulging in upsetting is similarly related to plastic instability.[2]

To predict workability, a fracture criterion for ductile fracture must be established. Perhaps the most generally applicable ductile-fracture criterion is that proposed by Cockcroft and Latham.[3] They reasoned that a criterion of

[1] A. L. Hoffmanner (ed.), "Metal Forming: Interrelation Between Theory and Practice," Plenum Publishing Corporation, New York, 1971.
[2] H. A. Kuhn and P. W. Lee, *Metall. Trans.*, vol. 2, pp. 3197–3202, 1971.
[3] M. G. Cockcroft and D. J. Latham, *J. Inst. Met.*, vol. 96, pp. 33–39, 1968.

ductile fracture will be based on some combination of stress *and* strain and that only the highest local tensile stress σ* will be important. They proposed that ductile fracture will occur when

$$\int_{0}^{\varepsilon_f} \sigma^* \, d\varepsilon = \text{constant} \qquad (15\text{-}46)$$

for a given temperature and strain rate. This criterion was shown to provide good correlation between fracture strain in tension and torsion tests and helped predict safe ranges of reduction in cold extrusion. However, more extensive studies[1] of free surface fracture in room-temperature deformation have shown consistent fracture criteria in the form of a unique correlation for each material between the tensile and compressive strains at fracture (Fig. 15-31). The ratio of these strains is varied by changing the *D/h* ratio and the frictional conditions for upset cylinders and by using other tests such as bending and edge cracking in rolling. The dashed curve shows the strain ratio which would be obtained for ideal (nonbarreling) compression of the cylinders. Fracture occurs when the tensile and compressive strains at fracture place the strain state above the line with a slope of ½. Other studies[2] have shown that this fracture criterion is sensitive to the

**Fig. 15-31** Fracture criterion expressed in terms of tensile and compressive strains at fracture for upset and bend tests at room temperature. (*From P. W. Lee and H. A. Kuhn, Metall. Trans., vol. 4, pp. 969–74, 1973.*)

[1] P. W. Lee and H. A. Kuhn, *Metall. Trans.*, vol. 4, pp. 969–974, 1973.
[2] T. Erturk, W. L. Otto, and H. A. Kuhn, *Metall. Trans.*, vol. 5, pp. 1883–1886, 1974.

anisotropy in rolled steel plate. The local surface strains at fracture in upset tests correlate well with the true zero-gage-length fracture strain in the tension test provided that proper consideration is given to fiber orientation. The constant slope of $\frac{1}{2}$ and the use of a strain correlation of this type have been shown[1] to be consistent with the Cockcroft and Latham fracture criterion.

The complete analysis of workability involves first establishing the ductile-fracture criterion for the material. Then complete analysis of the deformation mechanics in terms of the process geometry, reduction, friction, etc., is needed to compute the stress and strain histories at the critical points of fracture. This cannot yet be done for the general problem, but using a powerful numerical technique, Kobayashi[2] was able to calculate these parameters for the slow-speed upsetting of steel cylinders at room temperature. The resulting workability diagram is shown in Fig. 15-32.

The chief contribution to predicting internal cracks has been made by Avitzur[3] in a study of the formation of center bursts or chevron cracking in extrusion and wiredrawing. This study predicts the regions of die angle and reduction which must be avoided to prevent this type of fracture.

Extensive work has been done in developing fracture diagrams for sheet-forming operations. This work will be considered in detail in Chap. 20.

**Fig. 15-32** Workability of AISI 1040 steel in upsetting at room temperature. (*After Kobayashi.*)

[1] H. A. Kuhn, P. W. Lee, and T. Erturk, *Trans. ASME, Ser. H: J. Eng. Mater. Technol.* vol. 95, pp. 213–218, 1973; S. Kobayashi, C. H. Lee, and S. I. Oh, AFML-TR-73-192, May 1973.
[2] S. Kobayashi, et al., *op. cit.*
[3] B. Avitzur, *Trans. ASME, Ser. B: J. Eng. Ind.*, vol. 90, pp. 79–91, 1968.

## 15-11  RESIDUAL STRESSES

Residual stresses are a system of stresses which can exist in a body when it is free from external forces. Since residual stresses are generated by nonuniform plastic deformation, it is important to consider the residual stresses developed in each metalworking process. The general method by which residual stresses are produced in a metalworking process is shown in Fig. 15-33. In this example the rolling conditions are such that plastic flow occurs only near the surfaces of the sheet. The surface grains in the sheet are deformed and tend to elongate, while the grains in the center of the sheet are unaffected. Since the sheet must remain a continuous whole, the surface and center regions of the sheet must undergo a strain accommodation. The center fibers tend to restrain the surface fibers from elongating, while the surface fibers seek to stretch the central fibers of the sheet. The result is a residual-stress pattern in the sheet which consists of a high compressive stress at the surface and a tensile residual stress at the center of the sheet (Fig. 15-33*b*). In general, the sign of the residual stress which is produced by inhomogeneous deformation will be opposite to the sign of the plastic strain which produced the residual stress. For the case of the rolled sheet, the surface fibers which were elongated in the longitudinal direction by rolling are left in a state of compressive residual stress when the external rolling load is removed.

The residual-stress system existing in a body must be in static equilibrium. Thus, the total force acting on any plane through the body and the total moment

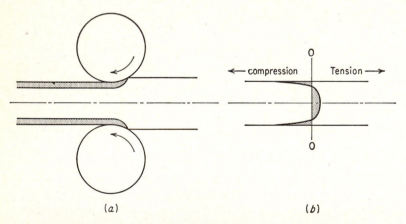

(*a*)                                         (*b*)

**Fig. 15-33**  (*a*) Inhomogeneous deformation in rolling of sheet; (*b*) resulting distribution of longitudinal residual stress over thickness of sheet (schematic).

of forces on any plane must be zero. For the longitudinal residual-stress pattern in Fig. 15-33*b*, the area under the curve subjected to compressive residual stresses must balance the area subjected to tensile residual stresses. However, it should be kept in mind that the actual situation is made complex by the presence of a three-dimensional state of residual stress.

Residual stresses are only elastic stresses. The maximum value which a residual stress can reach is the yield stress of the material. For purposes of analysis, residual stresses can be considered the same as ordinary applied stresses. Thus, a compressive residual stress effectively subtracts from the applied tensile stress, and a tensile residual stress adds to an applied tensile stress.

Metals containing residual stresses can be *stress relieved* by heating to a temperature where the yield strength of the material is the same or less than the value of the residual stress. Thus, the material can deform and release the stress. Creep deformation is important in thermal stress relief. It is important to realize that nonuniform thermal expansion or contraction due to nonuniform heating or cooling can produce residual stresses in the same way as nonuniform plastic deformation. Therefore, slow cooling from the stress-relief temperature may be an important consideration.

The differential strains that produce residual stresses also can be reduced substantially at room temperature by plastic deformation. For example, products such as sheet, plate, and extrusions are often stretched several percent beyond the yield stress to relieve differential strains by plastic deformation. Residual stresses in cold-drawn rod and tube are relieved by roller straightening.

Residual stresses are difficult to calculate with precision by analytical methods, and therefore, they are usually determined by a variety of experimental techniques.[1] Most methods of measuring residual stresses are destructive because they involve removing part of the stressed material to cause a redistribution of stress in the remaining body. This may be done by machining layers from a cylinder or drilling a small hole and measuring the redistribution of strain with suitably placed strain gages. The chief nondestructive method of measuring residual stresses is x-ray analysis.[2] In this method x-rays are used to determine the interatomic spacing of a particular set of lattice planes in the strained material. The residual stresses can be calculated by comparing the spacing measured on the material with residual stress with the spacing in a stress-free sample.

[1] A. A. Denton, *Metall. Rev.*, vol. 11, pp. 1–23, 1966; W. M. Baldwin, *Am. Soc. Test. Mater. Proc.*, vol. 49, pp. 539–583, 1949; "Residual Stress Measurements," American Society for Metals, Metals Park, Ohio, 1952.

[2] C. S. Barrett and T. B. Massalski, "Structure of Metals," 3rd ed., chap. 17, McGraw Hlil Book Company, New York, 1966; E. Macherauch, *Exp. Mech.*, vol. 6, pp. 140–153, 1966.

## 15-12 EXPERIMENTAL TECHNIQUES FOR METALWORKING PROCESSES

Experimental studies of metalworking processes require techniques for measuring the forming loads and the deformations. Measurement of the forces presents no fundamental problems, although considerable ingenuity may be required to instrument many pieces of production equipment., Force usually is measured with strain-gage load cells. The upsetting of standardized copper cylinders[1] permits presses to be calibrated so that loads can be measured by strain gages attached to the press frame. Other instruments[2] frequently needed are transducers to measure ram displacement, velocity, and acceleration. High-speed photography may be useful for measuring free surface displacement and many other kinematic variables.

Most studies[3] of the deformation in forming operations have employed grid networks. A number of identical specimens are deformed by different amounts, and the progress of the deformation is determined from the distortion of the network. Either rectangular or polar-coordinate grids are used, depending on the application. The grid network may be applied either by scribing or by photographic methods. However, only a limited amount of information can be obtained from grids placed on the surface of the workpiece. The flow in the interior of the workpiece can be studied by cutting the billet in half, affixing a grid network to the two faces, fastening them back together, and then machining to a symmetrical shape. Under certain circumstances the interior deformation can be determined by embedding a lead grid in the casting. The distortion of the lead grid is obtained by radiographing the deformed billet. Sometimes plugs are inserted in the workpiece, and their distortion is measured by sectioning and examining under the microscope.

Metallographic techniques are useful for determining regions of heavy deformation. The direction of flow can be determined from the grain distortion and the preferred alignment of second-phase particles and inclusions. An initially banded structure provides a unidirectional internal grid system. Etches are available for most metals which selectively attack the plastically deformed regions. Since recrystallization will begin first in the most heavily deformed grains, the examination of the microstructure of a deformed metal after annealing will indicate the presence of nonuniform deformation.

[1] T. Altan and D. E. Nichols, *Trans. ASME, Ser. B: J. Eng. Ind.*, vol. 94, pp. 769–774, 1972.
[2] H. N. Norton, "Handbook of Transducers for Electronic Measuring Systems," Prentice-Hall, Inc., Englewood Cliffs, N.J., 1969.
[3] H. P. Tardiff, *Steel Process. Conver.*, vol. 43, pp. 626–632, 643–644, 650, 1957; G. L. Baraya, J. Parker and, J. W. Flowett, *J. Mech. Sci.*, vol. 5, pp. 365–367, 1963; R. L. Bell and T. G. Langdon, *J. Sci. Instrum.*, vol. 42, p. 896, 1965.

Experimentation in metalworking processes may be expensive and time-consuming when dealing with large production equipment. When many trial-and-error experiments are required, as may be necessary in optimizing tooling geometry, it may be advantageous to conduct a preliminary series of experiments in which the real workpiece material is simulated with a model material. Model materials (such as plasticine, wax, lead, or clay) have low flow stresses compared with metals so that low-cost plastic or wooden tooling can be used in the experiments. Thus, it is possible to study a wide range of tooling geometries or to construct a model of a complex piece of production equipment. Strict similarity between the real and model experiments is very difficult to achieve,[1] but useful results can be obtained if the friction conditions are similar and if appropriate temperatures and strain rates are used for the model material. Plasticine[2] is a popular model material because its room-temperature flow curve approximates that of metals under hot-working conditions. Plasticine is often used to check theoretical analysis of plastic forming because it deforms similar to an ideal plastic material. Grid networks can easily be developed by working with a billet built up from layers of black and white plasticine. It is a simple matter to slice through the deformed material to observe the deformation pattern in any desired direction.

## BIBLIOGRAPHY

Alexander, J. M., and R. C. Brewer: "Manufacturing Properties of Materials," D. Van Nostrand Company, Inc., Princeton, N.J., 1963.

Avitzur, B.: "Metal Forming: Processes and Analysis," McGraw-Hill Book Company, New York, 1968.

Backofen, W. A.: "Deformation Processing," Addison-Wesley Publishing Company, Inc., Reading, Mass., 1972.

Cook, Nathan H.,: "Manufacturing Analysis," Addison-Wesley Publishing Company, Inc., Reading, Mass., 1966.

Datsko, Joseph: "Material Properties and Manufacturing Processes," John Wiley & Sons, Inc., New York, 1966.

Kalpakjian, Serope: "Mechanical Processing of Materials," D. Van Nostrand Company, Inc., Princeton, N.J., 1967.

Parkins, R. N.: "Mechanical Treatment of Metals," American Elsevier Publishing Company, New York, 1968.

Rowe, Geoffrey W.: "An Introduction to the Principles of Metalworking," St Martin's Press, Inc., New York, 1965.

Schey, John A.: The More Common Fabrication Processes, in R. F. Bunshah (ed.), "Techniques of Metals Research," vol. I, pt. 3, John Wiley & Sons, Inc., New York, 1965.

[1] T. Altan, H. J. Henning, and A. M. Sabroff, *Trans. ASME, Ser. B: J. Eng. Ind.*, vol. x, pp. 444–452, 1970.

[2] A. P. Green, *Philos. Mag.*, vol. 42, pp. 365–373, 1951.

# 16

# FORGING

## 16-1 CLASSIFICATION OF FORGING PROCESSES

Forging is the working of metal into a useful shape by hammering or pressing. It is the oldest of the metalworking arts, having its origin with the primitive black-smith of Biblical times. The development of machinery to replace the arm of the smith occurred early during the Industrial Revolution. Today there is a wide variety of forging machinery which is capable of making parts ranging in size from a bolt to a turbine rotor or an entire airplane wing.

Most forging operations are carried out hot, although certain metals may be cold-forged. Two major classes of equipment are used for forging operations. The forging hammer, or drop hammer, delivers rapid impact blows to the surface of the metal, while the forging press subjects the metal to a slow-speed compressive force.

The two broad categories of forging processes are *open-die forging* and *closed-die forging*. Open-die forging is carried out between flat dies or dies of very simple shape. The process is used mostly for large objects or when the number of parts produced is small. Often open-die forging is used to preform the workpiece for closed-die forging. In closed-die forging the workpiece is deformed between two die halves which carry the impressions of the desired final shape. The workpiece is deformed under high pressure in a closed cavity, and thus precision forgings with close dimensional tolerances can be produced.

The simplest open-die forging operation is the *upsetting* of a cylindrical billet between two flat dies. The compression test (Fig. 15-7) is a small-scale prototype of this process. As the metal flows laterally between the advancing die surfaces, there is less deformation at the die interfaces because of the friction forces than at the midheight plane. Thus, the sides of the upset cylinder becomes barreled.[1] As a general rule, metal will flow most easily toward the nearest free surface because this represents the lowest frictional path.

**Fig. 16-1** Forging operations. (*a, b*) Edging; (*c*) fullering; (*d*) drawing; (*e*) swaging; (*f*) piercing; (*g*) punching.

[1] K. M. Kulkarni and S. Kalpakjian, *Trans. ASME, Ser. B: J. Eng. Ind.*, vol. 91, pp. 743–754, 1969.

The effect of friction in restraining metal flow is used to produce shapes with simple dies. *Edging* dies are used to shape the ends of the bars and to gather metal. As is shown in Fig. 16-1a and b, the metal is confined by the die from flowing in the horizontal direction but it is free to flow laterally to fill the die. *Fullering* is used to reduce the cross-sectional area of a portion of the stock. The metal flow is outward and away from the center of the fullering die (Fig. 16-1c). An example of the use of this type of operation would be in the forging of a connecting rod for an internal-combustion engine. The reduction in cross section of the work with concurrent increase is length is called *drawing down,* or drawing out (Fig. 16-1d). If the drawing-down operation is carried out with concave dies (Fig. 16-1e) so as to produce a bar of smaller diameter, it is called *swaging.* Other operations which can be achieved by forging are bending, twisting, extrusion, piercing (Fig. 16-1f), punching (Fig. 16-1g), and indenting.

*Closed-die forging* uses carefully machined matching die blocks to produce forgings to close dimensional tolerances. Large production runs are generally required to justify the expensive dies. In closed-die forging the forging billet is usually first fullered and edged to place the metal in the correct places for subsequent forging. The preshaped billet is then placed in the cavity of the *blocking die* and rough-forged to close to the final shape. The greatest change in the shape of the metal usually occurs in this step. It is then transferred to the *finishing die,* where it is forged to final shape and dimensions. Usually the blocking cavity and the finishing cavity are machined into the same die block. Fullering and edging impressions are often placed on the edges of the die block.

It is important to use enough metal in the forging blank so that the die cavity is completely filled. Because it is difficult to put just the right amount of metal in the correct places during fullering and edging, it is customary to use a slight excess of metal. When the dies come together for the finishing step, the excess metal squirts out of the cavity as a thin ribbon of metal called *flash.* In order to prevent the formation of a very wide flash, a ridge, known as a *flash gutter,* is usually provided (Fig. 16-2). The final step in making a closed-die forging is the removal of the flash with a *trimming die.*

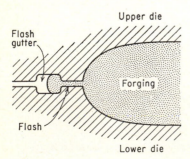

Fig. 16-2  Sectional view through closed-die forging.

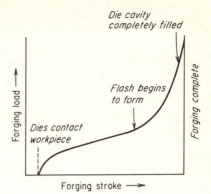

Fig. 16-3  Typical curve of forging load vs. stroke for closed-die forging.

Because of the flash, the term *closed-die forging* is a bit of a misnomer, and a better description for the process would be *impression-die forging*. The flash serves two purposes. As described above, it acts as a "safety valve" for excess metal in the closed-die cavity. Of more importance, the flash regulates the escape of metal, and thus the thin flash greatly increases the flow resistance of the system so that the pressure builds up to high values to ensure that metal fills all recesses of the die cavity. Figure 16-3 shows a typical curve of forging load vs. die advance (press stroke) for a closed-die forging process. The trick in designing the flash is to adjust its dimensions so that the extrusion of metal through the narrow flash opening is more difficult than the filling of the most intricate detail in the die. But, this must not be done to excess so as to create very high forging loads with attendent problems of die wear and breakage. The ideal is to design for the minimum flash needed to do the job.

In closed-die forging it is particularly difficult to produce parts with sharp fillets, wide thin webs, and high ribs (Fig. 16-4). Moreover, forging dies must be tapered to facilitate removal of the finished piece. This *draft allowance* is approximately 5° for steel forgings.

Fig. 16-4  Some typical forging nomenclature.

## 16-2   FORGING EQUIPMENT

Forging equipment may be classified with respect to the principle of operation.[1]   In *forging hammers* the force is supplied by a falling weight or *ram*.   These are energy-restricted machines since the deformation results from dissipating the kinetic energy of the ram.   *Mechanical forging presses* are stroke-restricted machines since the length of the press stroke and the available load at various positions of the stroke represent their capability.   *Hydraulic presses* are load-restricted machines since their capability for carrying out a forming operation is limited chiefly by the maximum load capacity.   Each of these classes of forging equipment needs to be examined with respect to its load and energy characteristics, its time-dependent characteristics, and its capability for producing parts to dimension with high accuracy.

To successfully complete a forging operation, the available machine load must exceed the required load at any point in the process and the available machine energy must exceed the energy required by the process for the entire stroke.   The most important characteristic of any machine is the number of strokes per minute, for this determines the production rate.   The velocity under pressure $v_p$ is the velocity of the machine slide under load.   This variable determines the strain rate (which influences the flow stress) and the contact time under pressure $t_p$.   Typical values of $v_p$ are given in Table 16-1.   The contact time is the time that the workpiece remains in the die under load.   Since heat transfer between the hotter forging and the cooler dies is so effective when the interface is under high pressure, die wear increases with $t_p$.   Dimensional accuracy of the parts produced by a forging machine is directly related to the stiffness of the equipment.   In general terms, stiffness is increased by

**Table 16-1   TYPICAL VALUES OF VELOCITY FOR DIFFERENT FORGING EQUIPMENT[†]**

| Forging Machine | Velocity range | |
| --- | --- | --- |
| | ft/s | m/s |
| Gravity drop hammer | 12–16 | 3.6–4.8 |
| Power drop hammer | 10–30 | 3.0–9.0 |
| HERF machines | 20–80 | 6.0–24.0 |
| Mechanical press | 0.2–5 | 0.06–1.5 |
| Hydraulic press | 0.2–1.0 | 0.06–0.30 |

[†] T. Altan, "Forging Equipment, Materials, and Practices," Metals and Ceramics Information Center, MCIC-HB-03, 1973.

[1] T. Altan, "Forging Equipment, Materials, and Practice", chap. 1, Metals and Ceramics Information Center, MCIC-HB-03, Battelle Memorial Laboratories, Columbus, Ohio, 1973.

using larger components in the construction of the machine, and this is directly related to increased equipment cost. A particularly critical problem is the tilting of the ram under load.

The most commonly used piece of forging equipment is the forging hammer. The two basic types of hammers are the *board hammer* and the *power hammer* (Fig. 16-5). In the board hammer the upper die and ram are raised by friction rolls gripping the board. When the board is released, the ram falls under the in-

**Fig. 16-5**  Schematic drawings of forging equipment.

fluence of gravity to produce the blow energy. The board is immediately raised again for another blow. Forging under a hammer usually is done with repeated blows. Hammers can strike between 60 and 150 blows per minute depending on size and capacity. The energy supplied by the blow is equal to the potential energy due to the weight of the ram and the height of the fall. Forging hammers are rated[1] by the weight of the ram. However, since the hammer is an energy-restricted machine, in which the deformation proceeds until the total kinetic energy is dissipated by plastic deformation of the workpiece or elastic deformation of the dies and machine, it is more correct to rate these machines in terms of energy delivered.

Greater forging capacity is achieved with the power hammer in which the ram is accelerated on the downstroke by steam or air pressure in addition to gravity. Steam or air also is used to raise the ram on the upstroke. The total energy supplied to the blow in a power drop hammer is given by

$$W = \frac{1}{2}\frac{mv^2}{g} + pAH = (m + pA)H \qquad (16\text{-}1)$$

where $m$ = weight of ram, lb
$\quad v$ = velocity of ram at start of deformation
$\quad g$ = acceleration of gravity, 32.2 ft/s$^2$
$\quad p$ = air or steam pressure acting on ram cylinder on downstroke
$\quad A$ = area of ram cylinder
$\quad H$ = height of the ramp drop

An important feature of the power hammer is that the energy of the blow can be controlled, whereas in the board hammer the mass and height of fall are fixed. Power hammers are preferred over board hammers for closed-die forging. This equipment ranges in size from 1,000 to 50,000 lb and can produce forgings ranging in weight from a few pounds to several tons. The forging hammer is the cheapest source of a high forging load. For example, (see Prob. 16-2) a 3,000 lb power hammer can produce a forging load in excess of 600 tons. The forging hammer also has the shortest contact time under pressure, ranging from 1 to 10 ms. However, hammers generally do not provide the forging accuracy obtainable in presses. Also, because of their inherent impact character, problems must be overcome with ground shock, noise, and vibration. Some of these problems are minimized with the *counterblow hammer* which uses two opposed rams which strike the workpiece at the same time so that practically all of the energy is absorbed by the work and very little energy is lost as vibration in the foundation and the environment. A new class of forging equipment is the *high-energy-rate forging* (*HERF*) machine in

---

[1] " Metals Handbook," 8th ed., vol. 5, pp. 12–13, American Society for Metals, Metals Park, Ohio, 1970.

which unusually high velocities (30 to 80 ft/s) are substituted for mass to achieve the required energy level.[1]

Forging presses are of either mechanical or hydraulic design. Presses are rated on the basis of the force developed at the end of the stroke. Next to hammers, mechanical presses are the most widely used equipment for closed-die forging in the United States. Most mechanical presses utilize an eccentric crank (Fig. 16-5) to translate rotary motion into reciprocating linear motion of the press slide. The ram stroke is shorter than in a hammer or hydraulic press, so that mechanical presses are best suited for low-profile forgings. The maximum load is attained when the ram is about $\frac{1}{8}$ in of the bottom dead center position. Mechanical presses with load ratings from 300 to 12,000 tons are available. The blow of a press is more like a squeeze than like the impact of a hammer. Because of this, dies can be less massive and die life is longer than with a hammer. However, the initial cost of a press is much higher than that of a hammer, so that large production runs are needed. The production rate is comparable to that of a hammer, but since each blow is of equal force, a press may be less suitable for carrying out preliminary shaping and finishing operations in the same piece of equipment.

The total energy supplied during the stoke of a press is given by

$$W = \frac{1}{2}I(\omega_0^2 - \omega_f^2) = \frac{1}{2}I\left(\frac{\pi}{30}\right)^2(n_0^2 - n_f^2) \qquad (16\text{-}2)$$

where $I$ = moment of inertia of the flywheel
$\omega$ = angular velocity, rads
$n_0$ = initial speed of flywheel, rpm
$n_f$ = speed of flywheel after deformation, rpm

Hydraulic presses are load-restricted machines in which hydraulic pressure moves a piston in a cylinder. A chief feature is that the full press load is available at any point during the full stroke of the ram. This feature makes the hydraulic press ideally suited for extrusion-type forging operations. The ram velocity can be controlled and even varied during the stroke. The hydraulic press is a relatively slow-speed machine. This results in longer contact time, which may lead to problems with heat loss from the workpiece and die deterioration. On the other hand, the slow squeezing action of a hydraulic press results in close-tolerance forgings. Hydraulic presses are available in ratings from 500 to 18,000 tons, although several presses with ratings of 50,000 tons have been built. The initial cost of a hydraulic press is higher than that of a mechanical press of equal capacity. Factors for converting between the capacity of presses and hammers are available.[2]

[1] T. Altan, op. cit., chap. 2.
[2] "Metals Handbook," 8th ed., vol. 5, pp. 14–18, American Society for Metals, Metals Park, Ohio, 1970.

*Screw presses* are widely used in Europe for both hot and cold closed-die forging. In a screw press the ram is connected by a rotary joint to a spindle, which is in effect a large screw. The rotary motion of a flywheel is transformed into linear motion by the multiple thread on the spindle and its nut.

*Forging machines*, also known as *upsetters* or *headers*, are horizontal mechanical presses which are used for the high-production forging of symmetrical shapes from bar stock, such as bolts, rivets, and gear blanks.[1] *Forging rolls*[2] are used for initial forming prior to closed-die forging and for producing tapered or long slender sections like leaf springs and long bolts. Forging rolls have matching grooves over part of their circumference to produce the shape change.

## 16-3   FORGING IN PLANE STRAIN

Figure 16-6 shows the stresses in the forging of a plate of constant thickness under conditions of plane strain. The analogous problem of forging a disk was considered in Sec. 15-7. Lateral flow perpendicular to the ram travel leads to frictional shear stresses at the die contact surfaces. This surface shear is directed toward the centerline, opposing the metal flow. The presence of friction causes an imbalance of force on the element in the $x$ direction which must be accommodated by a change in lateral pressure $\sigma_x$ from one side of the element to another. It is assumed that the plate has unit width normal to the place of the paper and that this width remains constant.

**Fig. 16-6**   Stresses acting on a plate forged in plane strain.

[1] *Ibid.*, pp. 69–95.
[2] *Ibid.*, pp. 95–99.

Taking the equilibrium of forces in the $x$ direction,

$$\sigma_x h - (\sigma_x + d\sigma_x)h - 2\tau_{xy}\, dx = 0$$

$$\frac{d\sigma_x}{dx} = -\frac{2\tau_{xy}}{h} \qquad (16\text{-}3)$$

The von Mises' yield criterion for a condition of plane strain is given by Eq. (3-53).

$$\sigma_1 - \sigma_3 = \frac{2}{\sqrt{3}}\,\sigma_0 = \sigma_0' = p - \sigma_x \qquad (16\text{-}4)$$

where both compressive stresses $p$ and $\sigma_x$ are taken as positive. Since $\sigma_0'$ does not change with $x$, Eq. (16-4) gives $dp/dx = d\sigma_x/dx$ and on substituting into Eq. (16-3) the differential equation of equilibrium becomes

$$\frac{dp}{dx} = \frac{-2\tau_{xy}}{h} \qquad (16\text{-}5)$$

If the shearing stress is related to the normal pressure by *Coulomb's law of sliding friction*, $\tau_{xy} = \mu p$, then Eq. (16-5) becomes

$$dp + \frac{2\mu p}{h}\, dx = 0$$

Separating the variables results in

$$\frac{dp}{p} = -\frac{2\mu}{h}\, dx \qquad (16\text{-}6)$$

Integrating both sides gives

$$\ln p = -\frac{2\mu x}{h} + \ln C$$

*or*

$$p = Ce^{-2\mu x/h} \qquad (16\text{-}7)$$

The constant of integration $C$ is evaluated by the boundary condition that a free surface $x = a$, the lateral stress $\sigma_x = 0$. Therefore, by Eq. (16-4), $p = \sigma_0'$ and $\ln C = \ln \sigma_0' + 2\mu a/h$

$$p = \sigma_0'^{[2\mu(a-x)]/h} \qquad (16\text{-}8)$$

$$\sigma_x = \sigma_0'\{1 - e^{[2\mu(a-x)/h]}\} \qquad (16\text{-}9)$$

Since $\mu$ usually is a small number, we can use the expansion $e^{(y)} = 1 + y + y^2/2! + y^3/3! + \cdots$ to simplify the above equations.

$$p = \sigma_0'\left[1 + \frac{2\mu(a-x)}{h}\right] \qquad (16\text{-}10)$$

The mean forging pressure is

$$\bar{p} = \int_0^a \frac{p \, dx}{a} = \sigma_0' \frac{e^{(2\mu a/h)} - 1}{2\mu a/h} \qquad (16\text{-}11)$$

The total forging load $P$ can now be established, since $P = \bar{p}(2a)w$, where $w$ is the width in the direction normal to the plane of the paper.

A convenient way to write Eq. (16-8) is

$$p = \sigma_0' e^{(\mu L/h)} \left(1 - \frac{2x}{L}\right) \qquad (16\text{-}12)$$

where $L = 2a$. This equation shows that as the ratio of length to thickness $L/h$ increases, the resistance to compressive deformation increases rapidly. This fact is used to advantage in closed-die forging where the deformation resistance of the flash must be very high so that the pressure in the die will be high enough to ensure complete filling of the die cavity. Figure 16-7 shows the variation of $p$ and $\sigma_x$ over $L = 2a$. Both stresses build up to a maximum at the center of the plate, illustrating again the *friction hill*. In this simple case the centerline of the plate defines a *neutral surface*. For the vertical downward motion of the ram the metal flow is in the lateral, or horizontal, direction. The metal is stationary at the neutral surface, but the flow is outward away from the neutral surface. In forgings of more complex geometry the neutral surface may be more difficult to establish.

Another way other than Coulomb's law to describe the interface shear stress is with the *friction factor m*, where $\tau_{xy} = \tau_i = mk$. Substituting $\tau_{xy} = mk$ into Eq. (16-5) gives

$$dp = -\frac{2mk}{h} \, dx = -\frac{2\sigma_0}{3} m \frac{dx}{h} = -\sigma_0' m \frac{dx}{h} \qquad (16\text{-}13)$$

Integration is extremely straightforward

$$p = -\sigma_0' m \frac{x}{h} + C$$

Fig. 16-7 Distribution of normal stress and longitudinal stress for compression between plates.

and since $p = \sigma'_0$ at $x = a$, $C = \sigma'_0 + \sigma'_0 m \, a/h$ and

$$p = \sigma'_0 \frac{m}{h}(a - x) + \sigma'_0 \qquad (16\text{-}14)$$

The pressure distribution is linear with $x$. For the special case of *sticking friction*, $m = 1$, and Eq. (16-14) becomes

$$p = \sigma'_0 \left(\frac{a - x}{h} + 1\right) \qquad (16\text{-}15)$$

The mean forging pressure is

$$\bar{p} = \sigma'_0 \left(\frac{a}{2h} + 1\right) \qquad (16\text{-}16)$$

Frequently the frictional conditions are intermediate between full sticking and slipping so that there may be sliding friction at the edges of the plate (near $x = a$) where the pressure is lower, but at some distance closer in to the neutral surface the pressure increases to a point where $\tau_{xy} = k = \sigma_0/\sqrt{3} = \sigma'_0/2$. The distance, measured from the centerline where this transition occurs is $x_1$. At this point $\tau_{xy} = \mu p = k = \sigma'_0/2$. Substituting for $p$ with Eq. (16-8) gives a solution for the location of the boundary between slipping and sticking friction.

$$x_1 = a - \frac{h}{2\mu} \ln \frac{1}{2\mu} \qquad (16\text{-}17)$$

## 16-4 OPEN-DIE FORGING

Open-die forging typically deals with large, relatively simple shapes that are formed between simple dies in a large hydraulic press or power hammer. Examples of parts made in open-die forging are ship propeller shafts, rings, gun tubes, and pressure vessels. Since the workpiece is always larger than the tool, at any point in time deformation is confined to a small portion of the workpiece. The chief mode of deformation is compression, accompanied by considerable spreading in the lateral directions.

Probably the simplest open-die forging operation is *cogging* a billet between flat tools to reduce the cross-sectional area, usually without changing the final shape of the cross section. Figure 16-8 helps to define the nomenclature in dealing with spread in cogging. Tomlinson and Stringer[1] defined a coefficient of spread $S$

$$S = \frac{\text{width elongation}}{\text{thickness contraction}} = \frac{\ln(w_1/w_0)}{\ln(h_0/h_1)} \qquad (16\text{-}18)$$

[1] A. Tomlinson and J. D. Stringer, *J. Iron Steel Inst. London*, vol. 193, pp. 157–162, 1969.

**Fig. 16-8** Cogging operation in open-die forging. Shaded area shows where contact would occur between workpiece and upper die.

Because of barreling of the bar, it is difficult to measure the width natural strain, but the increase in length can be measured accurately. Using the constancy-of-volume relationship, we can write

$$\frac{h_1 w_1 l_1}{h_0 w_0 l_0} = 1$$

or

$$\ln (h_1/h_0) + \ln (w_1/w_0) + \ln (l_1/l_0) = 0$$

Substituting into Eq. (16-18) gives the coefficient of elongation.

$$1 - S = \frac{\text{length elongation}}{\text{thickness contraction}} = \frac{\ln (l_1/l_0)}{\ln (h_0/h_1)} \qquad (16\text{-}19)$$

If $S = 1$, then all of the deformation would manifest itself as spread, while if $S = 0$, all of the deformation would go into elongation. It was found that $S$ depended chiefly on the *bite ratio* $b/w_0$ according to

$$S = \frac{b/w_0}{1 + b/w_0} \qquad (16\text{-}20)$$

Equation (16-18) often is expressed in terms of the "spread law"

$$\beta = \left(\frac{1}{\gamma}\right)^S \qquad (16\text{-}21)$$

where $\beta$ = spread ratio = $w_1/w_0$
$\gamma$ = squeeze ratio = $h_1/h_0$

There are certain limiting ranges of these variables which must be considered. Since only that part of the surface under the bite is being deformed at any one time, there is danger of causing surface *laps* at the step separating the forged from the unforged portion of the workpiece. For a given geometry of tooling there will be a critical deformation which will produce laps. Wistreich and Shutt[1] recommend that the squeeze ratio $h_0/h_1$ should not exceed 1.3. Since open-die forging is done frequently on large sections, it is important to ensure that the billet is deformed through to the center. It is recommended that the bite ratio $b/h$ should not be less than $\frac{1}{3}$ to minimize inhomogeneous deformation. Using these criteria Wistreich and Shutt developed optimization techniques for selecting the forging schedule from the thousands of possible combinations which would require the least number of steps.

The load required to forge a flat section in open dies may be estimated[2] by

$$P = \bar{\sigma} A C \qquad (16\text{-}22)$$

where $C$ is a constraint factor to allow for inhomogeneous deformation. It will be recalled that deformation resistance increases with $\Delta = h/L$ (Fig. 15-24). Hill[3] constructed slipline fields for forging with various conditions of $\Delta$, and the results can be summarized by the relation $C = 0.8 + 0.2h/b = 0.8 + 0.2\,\Delta$.

## 16-5   CLOSED-DIE FORGING

The description of the closed-die forging process in Sec. 16-2 emphasized the important role of the flash in controlling die fill and in creating high forging loads. Usually the deformation in closed-die forging is very complex and the design of the intermediate steps to make a final precision part requires considerable experience and skill. Overall success of the forging operation requires an understanding of the flow stress of the material, the frictonal conditions, and the flow of the material in order to develop the optimum geometry for the dies. A special problem in closed-die forging is preventing rapid cooling of the workpiece by the colder dies. Toward this end active experiments are going on with *isothermal forging* in heated super-alloy dies. The elimination of die quenching results in a lower flow stress and forging loads, and permits complete die fill and closer dimensional tolerances.

An important step toward rationalizing the design of closed-die forgings is the classification of the shapes[4] commonly produced by this process. The shape

[1] J. G. Wistreich and A. Shutt, *J. Iron Steel Inst. London*, vol. 193, pp. 163–176, 1959.
[2] *Ibid.*
[3] R. Hill, "The Mathematical Theory of Plasticity," Clarendon Press, Oxford, 1950.
[4] K. Spies, "The Preforms in Closed-Die Forging and Their Preparation by Reducer Rolling," (in German), doctoral dissertation, Technical University, Hanover, Germany, 1957.

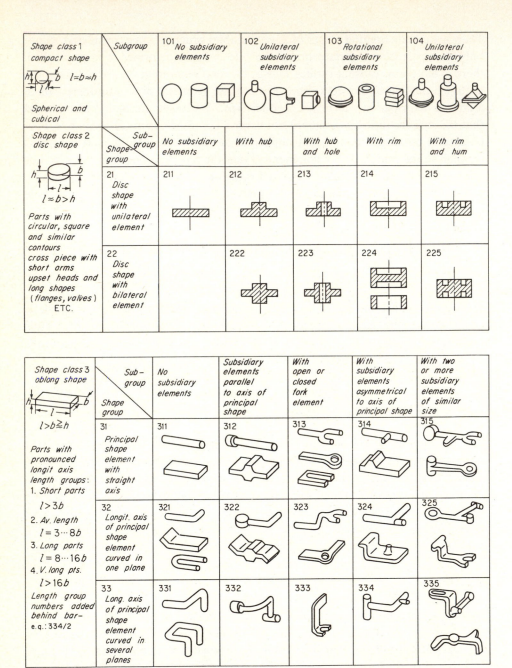

**Fig. 16-9** Shape classification for forging. (*After Spies.*) (*From "Forging Equipment, Materials, and Practices," p. 113, MCIC-HB-03, Battelle Memorial Laboratories, Columbus, Ohio, 1973.*)

classification shown in Fig. 16-9 has been widely adopted. The degree of difficulty increases as the geometry moves down and toward the right in this illustration. Roughly 70 percent of forgings fall in the third shape class with one dimension significantly longer than the other two. Although this classification system is useful in cost estimating and designing preforming steps, it is not quantitative. More quantitative methods of defining the shape difficulty factor have been developed.[1]

The design of a part for production by closed-die forging involves the prediction of:

1. Workpiece volume and weight
2. Number of preforming steps and their configuration
3. Flash dimensions in preforming and finishing dies
4. The load and energy requirements for each forging operation

Steps 1 and 3 have been discussed in detail by Altan and Henning[1] and the procedures reduced to a working computer program that is applicable to axisymmetric shapes.

Preform design is the most difficult and critical step in forging design. Proper preform design assures defect-free flow, complete die fill, and minimum flash loss. Success here depends on a thorough understanding of the metal flow during forging. Although metal flow consists only of two basic types, extrusion (flow parallel to the direction of die motion) and upsetting (flow perpendicular to the direction of die motion), in most forgings both types of flow occur simultaneously. An important step in understanding metal flow is to identify the neutral surfaces. Metal flows away from the neutral surface in a direction perpendicular to the die motion.

In designing a preform it is usual practice to take key cross sections through the forging and design the preform on the basis of the metal flow. Some general considerations[2] are:

1. The area at each cross section along the length must equal the area in the finished cross section plus the flash.
2. All concave radii on the preform should be larger than the radii on the final forged part.
3. The cross section of the preform should be higher and narrower than the final cross section, so as to accentuate upsetting flow and minimize extrusion flow.

Ideally, flow in the finishing step should be lateral toward the die cavity without additional shear at the die-workpiece interface. This type of flow minimizes fric-

---

[1] T. Altan and H. J. Henning, *Metallguria and Metal Forming*, vol. 39, pp. 83–88, March 1972.
[2] K. Lange, "Closed-Die Forging of Steel" (in German,) Springer-Verlag OHG, Berlin, 1958; *Met. Treat. Drop Forg.*, May 1965, pp. 184–195; June 1965, pp. 210–230; July 1965, pp. 264–270.

**Fig. 16-10** Flow diagram for computer-aided design (CAD) and computer-aided manufacturing (CAM) systems applied to closed-die forging. (*From "Forging Equipment Materials, and Practices," p. 140, MCIC-HB-03, Battelle Memorial Laboratories, Columbus, Ohio, 1973.*)

tion, forging load, and die wear. An example of the use of basic principles in preform design has been given by Akgerman, Becker, and Altan.[1]

A milestone in metalworking is the use of *computer-aided design (CAD)* in establishing the proper design for preforming and finishing dies in closed-die forging. CAD[2] has been applied to rib-web type airframe forgings (Fig. 16-4) and to airfoil shapes, but it can be applied to any class shape for which there is a suitable volume of parts to justify the development work. The power of the CAD system can be appreciated from Fig. 16-10. Starting with a drawing of the final part, the CAD system defines this geometry in terms of points, planes, cylinders, and other regular geometric shapes using the APT computer language. APT is a specialized computer language for describing geometric changes produced in metal cutting that is at the heart of numerical controlled (N/C) machining. Next, the coordinates of the various cross sections of the forging are determined, and these are used to perform design calculations to establish such factors as the location of the neutral surface, the shape difficulty factor, the cross-sectional area and volume, the flash geometry, and the stresses, the loads, and the center of loading. An important aspect of this system is that it takes the part geometry and flash dimensions and generates the N/C tape for machining the electrodes in the sinking of the finishing dies by electric discharge machining. Thus, this system also involves *computer-aided manufacturing (CAM)*. CAM is also used to machine the preforming dies.

---

[1] N. Akgerman, J. R. Becker, and T. Altan, *Metallurgia and Metal Forming*, vol. 40, pp. 135–138, May 1973.

[2] N. Akgerman and T. Altan, *Met. Eng. Q.*, vol. 13, pp. 26–28, February 1973.

## 16-6 CALCULATION OF FORGING LOADS IN CLOSED-DIE FORGING

The prediction of forging load and pressure in a closed-die forging operation is quite a difficult calculation. There are three general approaches to the problem. The approach used in many forge shops is to estimate the forging load required for a new part from information available from previous forgings of the same material and similar shape. Slightly more sophistication is found in what might be called the empirical approach. Schey[1] has expressed the forging load as

$$P = \bar{\sigma} A_t C_1 \qquad (16\text{-}23)$$

where   $A_t$ = cross-sectional area of the forging at the parting line, including the flash

$C_1$ = a constraint factor which depends on the complexity of the forging. $C_1$ has a value of 1.2 to 2.5 for upsetting a cylinder between flat dies. $C_1$ varies from 3 to 8 for closed-die forging of simple shapes with flash and from 8 to 12 for more complex shapes

The third approach is to use the *slab analysis*, suitably modified for the special situations found in closed-die forging. Although this level of analysis does not consider nonuniform deformation, when applied with consideration of the physical situation, it can give good agreement with experimental results.[2] The slab analysis for forging a plate in plane strain that was given in Sec. 16-3 has been extended to include forging between inclined die surfaces, and the slab analysis for upsetting a disk (Sec. 15-7) has been extended to account for lateral flow between inclined dies and extrusion (longitudinal flow) into a rib or a shaft. The basic approach is to divide the actual forging into simple geometric shapes which can be treated by the slab analysis. The total forging load is the sum of the loads found for the components parts.

As an example of this approach, consider the closed-die forging of a round part similar to shape group 211 (Fig. 16-9). The first step[3] is to establish the average thickness $h$ for the metal flowing in the die cavity toward the flash. If $\mu = 0.2$ to 0.4, the metal will slide over the die surfaces and $h = H$. For sticking friction the metal will shear internally in the die cavity as indicated by the broken lines in Fig. 16-11. Analysis shows that

$$\frac{h}{t} \approx 0.8 \left(\frac{L}{t}\right)^{0.92} \qquad \text{for } \frac{L}{t} \geq 2 \qquad (16\text{-}24)$$

[1] J. A. Schey, "Principles of Forging Design," American Iron and Steel Institute, New York, 1964.
[2] N. Akgerman and T. Altan, *Trans. ASME, Ser. B: J. Eng. Ind.*, vol. 94, pp. 1025–1034, 1972.
[3] T. Altan and R. J. Fiorentino, *Trans. ASME, Ser. B: J. Eng. Ind.*, vol. 93, pp. 477–484, 1971.

**Fig. 16-11** Closed-die forging of round forging showing critical dimensions and distribution of forging pressure. (*After Altan and Fiorentino.*)

If we simplify the equation for the compression of a disk, Eq. (15-37) with the expansion $e^y = 1 + y$

$$p = \sigma_0(1 + y) = \sigma_0 + \frac{2\mu\sigma_0(a - r)}{h}$$

Applying this equation to the ring of material in the forging flash gives

$$\sigma_f = 2\mu\bar{\sigma}\frac{w}{t} + \bar{\sigma}_F \qquad (16\text{-}25)$$

where  $\sigma_f$ = the forging pressure at the entrance to the flash from the die cavity
$\bar{\sigma}$ = the average flow stress of the metal in the cavity
$\bar{\sigma}_F$ = the average flow stress of the metal in the flash

Equation (16-25) recognizes that the die quenching and higher strain rate in the flash zone will result in a considerably different flow stress than in the cavity of the die. The forging load in the flash ring is given by

$$P_F = 2\pi\left[-\frac{2\tau_F}{t}\frac{(a^3 - L^3)}{3} + \left(\frac{2\tau_F a}{t} + \bar{\sigma}_F\right)\frac{(a^2 - L^2)}{2}\right] \qquad (16\text{-}26)$$

where $\tau_F = \mu\bar{\sigma}_F$ is the friction shear stress in the flash zone $a = L + w$ in Fig. 16-11.

The forging load $P_C$ due to deformation and friction in the die cavity is given by

$$P_C = 2\pi\left[-\frac{2\tau}{h}\frac{L^3}{3} + \left(\frac{2\tau L}{h} + \sigma_f\right)\frac{L^2}{2}\right] \qquad (16\text{-}27)$$

where $\tau = \mu\bar{\sigma}_C$ is the friction shear stress in the die cavity. The total forging load $P_T$ is the sum of the individual load for each component of the forging.

$$P_T = P_F + P_C \qquad (16\text{-}28)$$

## 16-7 FORGING DEFECTS

If the deformation during forging is limited to the surface layers, as when light, rapid hammer blows are used, the dendritic ingot structure will not be broken down at the interior of the forging. Incomplete forging penetration can readily be detected by macroetching a cross section of the forging. The examination of a deep etch disk for segregation, dendritic structure, and cracks is a standard quality-control procedure with large forgings. To minimize incomplete penetration, forgings of large cross section are usually made on a forging press.

Surface cracking can occur as a result of excessive working of the surface at too low a temperature or as a result of hot shortness. A high sulfur concentration in the furnace atmosphere can produce hot shortness in steel and nickel. Cracking at the flash of closed-die forgings is another surface defect, since the crack generally penetrates into the body of the forging when the flash is trimmed off (Fig. 16-12a). This type of cracking is more prevalent the thinner the flash in relation to the original thickness of the metal. Flash cracking can be avoided by increasing the flash thickness or by relocating the flash to a less critical region of the forging. It also may be avoided by hot trimming or stress relieving the forging prior to cold trimming of the flash.

Another common surface defect is closed-die forgings is the cold shut, or fold (Fig. 16-12b). A cold shut is a discontinuity produced when two surfaces of metal fold against each other without welding completely. This can happen when metal flows past part of the die cavity that has already been filled or that is only partly filled because the metal failed to fill in due to a sharp corner, excessive chilling, or high friction.

Loose scale or lubricant residue that accumulates in deep recesses of the die forms scale pockets and causes underfill. Incomplete descaling of the workpiece results in forged-in scale on the finished part.

Secondary tensile stresses can develop during forging, and cracking can thus be produced. Internal cracks can develop during the upsetting of a cylinder or a round (Fig. 16-12c), as a result of the circumferential tensile stresses. Proper design of the dies, however, can minimize this type of cracking. In order to minimize bulging during upsetting and the development of circumferential tensile stresses, it is usual practice to use concave dies. Internal cracking is less prevalent

**Fig. 16-12** Typical forging defects. (*a*) Cracking at the flash; (*b*) cold shut or fold; (*c*) internal cracking due to secondary tensile stresses.

in closed-die forging because lateral compressive stresses are developed by the reaction of the work with the die wall.

The deformation produced by forging results in a certain degree of directionality to the microstructure in which second phases and inclusions are oriented parallel to the direction of greatest deformation. When viewed at low magnification, this appears as *flow lines*, or *fiber structure*. The existence of a fiber structure is characteristic of all forgings and is not to be considered as a forging defect. However, as was discussed in Sec. 9-11, the fiber structure results in lower tensile ductility and fatigue properties in the direction normal to it (transverse direction). To achieve an optimum balance between the ductility in the longitudinal and transverse directions of a forging, it is often necessary to limit the amount of deformation to 50 to 70 percent reduction in cross section.

## 16-8 POWDER METALLURGY FORGING

A new and rapidly growing area is the production of closed-die forgings from powder metallurgy preforms (P/M forging). The use of sintered P/M preforms rather than bar stock as the workpiece offers advantages of improved material utilization through reduction or elimination of machining, forming to final size in one forging stroke, uniformity of structure and reduced directionality of properties relative to conventionally forged parts.

Working with a sintered powdered metal preform introduces new aspects to the mechanics and metallurgy of plastic deformation. Because it contains a dispersion of interconnected voids, the deformation of a P/M preform is much different from a conventional fully dense workpiece. With a P/M preform the workpiece decreases in volume during plastic deformation as the porosity is closed up and eliminated by the act of plastic deformation. The presence of voids causes a significant decrease in local ductility which increases the likelihood of fracture during forging. However, the forming limit concept can be applied[1] to the design of P/M preforms to prevent fracture in forging. The presence of voids increases the surface area over which unfavorable oxidation or contamination reactions can occur.

The basic plasticity mechanics of a porous powder metallurgy preform can be described[2] by the following relations. The relationship between densification and plastic deformation is achieved through relating the plastic Poisson ratio $v$ to the fraction of theoretical density $\rho/\rho_t$.

$$v = 0.5 \left(\frac{\rho}{\rho_t}\right)^2 \qquad (16\text{-}29)$$

This relationship holds for hot- and cold-working, provided the preform has been sintered. For the frictionless compression of a cylinder the relative density change is given by

$$-\frac{d\rho}{\rho} = d\varepsilon_z + d\varepsilon_r + d\varepsilon_\theta \qquad (16\text{-}30)$$

and from the definition of Poisson's ratio

$$d\varepsilon_r = d\varepsilon_\theta = -v \, d\varepsilon_z \qquad (16\text{-}31)$$

so that

$$-\frac{d\rho}{\rho} = (1 - 2v) \, d\varepsilon_z \qquad (16\text{-}32)$$

The substitution of Eq. (16-29) in Eq. (16-32) gives

$$-\frac{d\rho}{\rho} = (1 - \rho^2) \, d\varepsilon_z \qquad (16\text{-}33)$$

which can be integrated to give

$$-\varepsilon_z = \ln \left[\frac{(\rho/\rho_i)^2(1 - \rho_i^2)}{(1 - \rho^2)}\right]^{\frac{1}{2}} \qquad (16\text{-}34)$$

---

[1] C. L. Downey and H. A. Kuhn, *Trans. ASME, Ser. H., J. Eng. Mater. Technol.*, vol. 97, pp. 121–125, 1975.

[2] H. A. Kuhn and C. L. Downey, *Int. J. Powder Metall.*, vol. 7, pp. 15–25, 1971.

where $\rho_i$ is the initial density of the cylinder. The more practical problem of compression of a disk between dies with friction has been analyzed[1] using the above approach and has been shown to give excellent agreement with experiment.

The classical theory of plasticity is based on the assumption of constancy of volume, which leads to the further condition that yielding is unaffected by the hydrostatic component of the stress state. A modification of the von Mises' yield criterion is needed for dealing with porous materials which densify with plastic deformation. Kuhn has shown that a workable yield criterion is

$$\sigma_0(\rho,\varepsilon) = \left[ \frac{(\sigma_1 - \sigma_2)^2 + (\sigma_2 - \sigma_3)^2 + (\sigma_3 - \sigma_1)^2}{2} \right.$$
$$\left. + (1 - 2v)(\sigma_1\sigma_2 + \sigma_2\sigma_3 + \sigma_3\sigma_1) \right]^{\frac{1}{2}} \qquad (16\text{-}35)$$

The first term in Eq. (16-35) is the usual von Mises' criterion, and the second term accounts for the porosity through Poisson's ratio and Eq. (16-29).

## 16-9 RESIDUAL STRESSES IN FORGINGS

The residual stresses produced in forgings as a result of inhomogeneous deformation are generally quite small because the deformation is usually carried out well into the hot-working region. However, appreciable residual stresses and warping can occur on the quenching of steel forgings in heat treatment.

Special precautions must be observed during the cooling of large steel forgings from the hot-working temperature. Large forgings are subject to the formation of small cracks, or *flakes*, at the center of the cross section. Flaking is associated with the high hydrogen content usually present in steel ingots of large size, coupled with the presence of residual stresses. In order to guard against the development of high thermal or transformation residual stresses, large forgings are very slowly cooled from the working temperature. This may be accomplished by burying the forging in ashes for periods up to several weeks or, in the controlled cooling treatment which is used for hot-rolled railroad rail and certain forgings, by transferring the hot forging to an automatically controlled cooling cycle which brings the forging to a safe temperature in a number of hours. The use of vacuum-degassed steel largely eliminates problems with flaking.

[1] H. A. Kuhn and C. L. Downey, *Trans. ASME, Ser. H: J. Eng. Mater. Technol.*, vol. 95, pp. 41–46, 1973.

# BIBLIOGRAPHY

Altan, T., F. W., Boulger, J. R., Becker, N., Akgerman, and H. J. Henning: "Forging Equipment, Materials, and Practices," Metals and Ceramics Information Center, MCIC-HB-O3, October 1973. (Available through National Technical Information Service.)

"Forging Design Handbook," American Society for Metals, Metals Park, Ohio, 1972.

Geleji, A.: "Forge Equipment, Rolling Mills, and Accessories," (in English), Akademiai Kiado, Budapest, 1967.

Jensen, J. E. (ed.): "Forging Industry Handbook," Forging Industry Association, Cleveland, Ohio, 1966.

"Metals Handbook," 8th ed., vol. 5, Forging and Casting, American Society for Metals, Metals Park, Ohio, 1970.

Schey, J. A.: "Principles of Forging Design," American Iron and Steel Institute, New York, 1964.

# 17

# ROLLING OF METALS

## 17-1  CLASSIFICATION OF ROLLING PROCESSES

The process of plastically deforming metal by passing it between rolls is known as *rolling*. This is the most widely used metalworking process because it lends itself to high production and close control of the final product. In deforming metal between rolls, the work is subjected to high compressive stresses from the squeezing action of the rolls and to surface shear stresses as a result of the friction between the rolls and the metal. The frictional forces are also responsible for drawing the metal into the rolls.

The initial breakdown of ingots into blooms and billets is generally done by hot-rolling. This is followed by further hot-rolling into plate, sheet, rod, bar, pipe, rails, or structural shapes. The cold-rolling of metals has reached a position of major importance in industry. Cold-rolling produces sheet, strip, and foil with good surface finish and increased mechanical strength, at the same time maintaining close control over the dimensions of the product.

The terminology used to describe rolled products is fairly loose, and sharp limits with respect to dimensions cannot always be made for steelmaking terminology. A *bloom* is the product of the first breakdown of the ingot. Generally the width of a bloom equals its thickness, and the cross-sectional area is greater than $36$ in$^2$. A further reduction by hot-rolling results in a *billet*. The minimum cross section of a billet is about $1\frac{1}{2}$ by $1\frac{1}{2}$ in. It should be noted that in nonferrous metallurgical terminology a billet is any ingot which has received hot-working by rolling, forging, or extrusion, or the term may refer to a casting which is suitable for hot-working, as an extrusion billet. A *slab* refers to a hot-rolled ingot with cross-sectional area greater than $16$ in$^2$ and with a width that is at least twice the thickness. Blooms, billets, and slabs are known as *semifinished products* because they are subsequently formed into other mill products. The differentiation between *plate* and *sheet* is determined by the thickness of the product. In general, plate has a thickness greater than $\frac{1}{4}$ in, although there are exceptions to this limit, depending on the width. *Sheet* and *strip* refer to rolled products which generally have a thickness less than $\frac{1}{4}$ in. In general, strip refers to the rolled product with a width no greater than $24$ in, while sheet refers to the product of greater width.

Generally, rolling starts with a cast ingot or an electroplated slab. However, dense sheet may be produced by rolling directly from powder. In *powder rolling*[1] a metal powder is introduced between the rolls and compacted into a " green strip," which is subsequently sintered and subjected to further hot-working and/or cold-working and annealing cycles. A major advantage of powder rolling is the elimination of the initial hot-ingot breakdown step, with a corresponding large economy in needed capital equipment. Other advantages may be the minimization of contamination in hot-rolling, and the production of sheet with very fine grain size or with a minimum of preferred orientation.

In conventional hot- or cold-rolling the main objective is to decrease the thickness of the metal. Ordinarily little increase in width occurs, so that the decrease in thickness results in an increase in length. *Roll forming* is a special type of cold-rolling in which strip is progressively bent into complex shapes by passing it through a series of driven rolls. The thickness of the metal is not appreciably changed during this process. Roll forming is particularly suited to producing long, molded sections such as irregular-shaped channels and trim. Another specialized use of rolling is *thread rolling*, in which a blank is fed between two grooved die plates to form the threads.

## 17-2  ROLLING MILLS

A rolling mill consists basically of rolls, bearings, a housing for containing these parts, and a drive for applying power to the rolls and controlling their speed. The forces involved in rolling can easily reach many millions of pounds. Therefore, very rigid construction is needed, and very large motors are required to provide

[1] G. M. Sturgeon, " Metal Strip from Powder," Mills and Boon, London, 1972.

the necessary power. When these requirements are multiplied several times for the successive stands of a large continuous mill, it is easy to see why a modern rolling-mill installation demands many millions of dollars of capital investment and many man-hours of skilled engineering design and construction.

Rolling mills can be conveniently classified with respect to the number and arrangement of the rolls (Fig. 17-1). The simplest and most common type of rolling mill is the *two-high* mill (Fig. 17-1a). Rolls of equal size are rotated only in one direction. The stock is returned to the entrance, or rear, of the rolls for further reduction by hand carrying or by means of a platform which can be raised to pass the work above the rolls. An obvious improvement in speed results from the use of a *two-high reversing mill*, in which the work can be passed back and forth through the rolls be reversing their direction of rotation (Fig. 17-1b). Another solution is the *three-high mill* (Fig. 17-1c), consisting of an upper and lower driven roll and a middle roll which rotates by friction.

A large decrease in the power required for rolling can be achieved by the use of small-diameter rolls. However, because small-diameter rolls have less strength

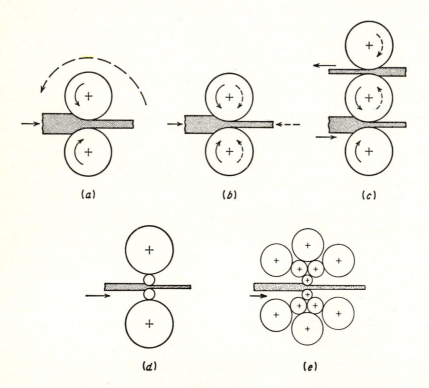

**Fig 17-1**   Typical arrangements of rolls for rolling mills. (*a*) Two-high, pullover; (*b*) two-high, reversing; (*c*) three-high; (*d*) four-high; (*e*) cluster.

**Fig. 17-2** Schematic drawing of strip rolling on a four-stand continuous mill.

and rigidity than large rolls, they must be supported by larger-diameter backup rolls. The simplest mill of this type is the *four-high mill* (Fig. 17-1*d*). Very thin sheet can be rolled to very close tolerances on a mill with small-diameter work rolls. The *cluster mill* (Fig. 17-1*e*), in which each of the work rolls is supported by two backing rolls, is a typical mill of this kind. The *Sendzimir mill* is a modification of the cluster mill which is very well adapted to rolling thin sheet or foil from high-strength alloys.

For high production it is common to install a series of rolling mills one after another in tandem (Fig. 17-2). Each set of rolls is called a *stand*. Since a different reduction is taken at each stand, the strip will be moving at different velocities at each stage in the mill. The speed of each set of rolls is synchronized so that each successive stand takes the strip at a speed equal to the delivery speed of the preceding stand. The uncoiler and windup reel not only accomplish the functions of feeding the stock to the rolls and coiling up the final product but also can be used to supply a *back tension* and a *front tension* to the strip. These added horizontal forces have several advantages that will be discussed later.

A different rolling mill design is the *planetary mill* (Fig. 17-3). This mill consists of a pair of heavy backing rolls surrounded by a large number of small planetary rolls.[1] The chief feature of the planetary mill is that it hot reduces a slab directly to strip in one pass through the mill. Each planetary roll gives an almost constant reduction to the slab as it sweeps out a circular path between the backing roll and the slab. As each pair of planetary rolls ceases to have contact with the workpiece another pair of rolls makes contact and repeats that reduction. The overall reduction is the summation of a series of small reductions by each pair of rolls in turn following each other in rapid succession. The action in the planetary mill is more like forging than rolling. It is necessary to use feed rolls to introduce the slab into the mill, and a pair of planishing rolls may be needed on the exit side to improve the surface finish.

[1] J. L. Giles and C. Gutteridge, *J. Iron Steel Inst. London*, vol. 211, pp. 9–12, 1973.

**Fig. 17-3** Arrangement of rolls in a planetary mill.

There are other innovative designs for cold-rolling. The *pendulum mill*[1] uses two small-diameter work rolls which reciprocate over the arc of contact to cold reduce a slab to a thin sheet. The *contact-bend-stretch* (*CBS*) rolling process[2] uses a four-high mill with a small-diameter floating bend roll.

## 17-3  HOT-ROLLING

The first hot-working operation for most steel products is done on the *primary roughing mill* (sometimes called blooming, slabbing, or cogging mills). These mills usually are two-high reversing mills with 24- to 54-in-diameter rolls. They are designated by the size of the rolls, as for example, a 45-in slabbing mill. The objective of this operation is the breakdown of the cast ingot into blooms or slabs for subsequent finishing into bars, plate, or sheet. The initial breakdown passes often involve only small reductions. Heavy scale is removed initially by rolling the ingot while lying on edge, while the thickness is reduced by rolling after the ingot has been turned 90° so as to be lying flat. There is appreciable spreading of the ingot width in hot-rolling of ingots. To maintain the desired width and preserve the edges, the ingot is turned 90° on intermediate passes and passed through edging grooves in the rolls. A reversing primary mill has a relatively low production rate since the workpiece may be passed back and forth and turned from 10 to 20 times. Where high production rates are of prime concern, edging passes may be eliminated by using a *universal mill*. This type of mill essentially is two rolling mills, one with two large-diameter rolls and the other with vertical rolls which control the width at the same time the thickness is reduced. The production of slabs from

[1] K. Saxl, *Engineering*, vol. 195, pp. 494–495, 1963.
[2] L. F. Coffin, *J. Met.*, vol. 15, pp. 14–22, March 1967.

cast ingots by hot-rolling can be elimintaed by using *continuous casting* to produce the slab directly from the molten steel. Another method is to produce slabs by *bottom-pressure casting*.

Plates are produced by hot-rolling, either from reheated slabs or directly from ingots. *Sheared plate* is produced by rolling between straight horizontal rolls and then trimming all edges. *Mill edge* is the normal edge produced in hot-rolling between horizontal finishing rolls. Mill-edge plates have two mill edges and two trimmed edges. *Universal-mill plates* have been rolled on a universal mill and trimmed on the ends only.

The general distinction between *strip* and *sheet* is that strip usually is less than 24 in wide. However, irrespective of width, the equipment for producing these products is known as a *continuous hot-strip mill*. In the modern wide, hot-strip mill, reheated slabs are first passed through a scalebreaker mill, then through a *roughing train* of 4 four-high mills, followed by a *finishing train* of 6 four-high finishing mills. If the sheet to be produced is wider than the width of the slab, then the first stand in the roughing train is a *broadside mill* in which the width of the slab is increased by cross rolling. The roughing mills usually are equipped with vertical edging rolls to control the width of the strip. High-pressure water jets are sprayed on the strip to remove scale. Following the last finishing stand there is either a flying shear to cut the strip to length or a coiler to produce continuous long lengths. In hot-rolling steel the slabs are heated initially at 2000 to 2400°F. The temperature in the last finishing stand varies from 1300 to 1600°F, but should be above the upper critical temperature to produce uniform equiaxed ferrite grains.

Because the nonferrous-metals industry deals with a diverse product mix, the equipment used for hot-rolling these materials is usually less specialized than equipment for hot-rolling of steel. The smaller ingot sizes and lower flow stresses found with most nonferrous alloys permit the use of smaller rolling mills. Two- and three-high mills are used for most hot-rolling of nonferrous alloys, although continuous four-high hot mills are used for aluminum alloys.

## 17-4 COLD-ROLLING

Cold-rolling is used to produce sheet and strip with superior surface finish and dimensional tolerances compared with hot-rolled strip. In addition, the strain hardening resulting from the cold reduction may be used to give increased strength. A greater percentage of rolled nonferrous metals is finished by cold-rolling compared with rolled-steel products. The starting material for cold-rolled steel sheet is pickled hot-rolled breakdown coil from the continuous hot-strip mill. Cold-rolled nonferrous sheet may be produced from hot-rolled strip, or in the case of certain copper alloys it is cold-rolled directly from the cast state.

High-speed four-high tandem mills with three to five stands are used for the cold-rolling of steel sheet, aluminum, and copper alloys. Generally, this type of mill is designed to provide both front and back tension. A continuous mill has high capacity and results in low labor costs. For example, the delivery speed of a five-stand continuous mill can reach 6,000 ft/min. However, this type of equipment requires a large capital investment and suffers further from lack of versatility. Four-high single-stand reversing mills with front and back tension are a more versatile installation. This type of mill is used often for the production of specialty items that vary widely in dimensions. However, it cannot compete with the continuous-tandem mill where large tonnages are involved.

The total reduction achieved by cold-rolling generally will vary from about 50 to 90 percent. In establishing the reduction in each pass or in each stand, it is desirable to distribute the work as uniformly as possible over the various passes without falling very much below the maximum reduction for each pass. Generally the lowest percentage reduction is taken in the last pass to permit better control of flatness, gage, and surface finish. One rational procedure[1] for developing cold-rolling schedules is to adjust the reduction in each pass so as to produce a constant rolling load.

The elimination of the yield point from annealed steel sheet is an important practical problem since the existence of a yield-point elongation results in inhomogeneous deformation (stretcher strains) during deep drawing or forming. The usual practice is to give the annealed steel a final, small cold reduction, *temper rolling*, or *skin pass*, which eliminates the yield-point elongation. Temper rolling also results in an improved surface and in improved flatness. Other methods which are used to increase the flatness of rolled sheet are *roller leveling* and *stretcher leveling*. A roller-leveling machine consists of two sets of small-diameter rolls which are arranged so that the top and bottom rows are offset. When the sheet is passed into the leveler, it is flexed up and down and the sheet is straightened as it emerges from the rolls. The stretcher leveler consists of two jaws which grip the edges of the sheet and stretch it with a pure tensile force.

## 17-5 ROLLING OF BARS AND SHAPES

Bars of circular or hexagonal cross section and structural shapes like I beams, channels, and railroad rails are produced in great quantity by hot-rolling with grooved rolls (Fig. 17-4). Actually, the hot breakdown of an ingot into a bloom falls in this category since grooved rolls are used to control the changes in shape during the blooming operation.

---

[1] B. R. Oliver and J. E. Bowers, *J. Inst. Met.*, vol. 93, pp. 218–222, 1964–1965.

**Fig. 17-4** Rolling of bars and structural shapes. (*American Iron and Steel Institute.*)

The rolling of bars and shapes differs from the rolling of sheet and strip in that the cross section of the metal is reduced in two directions. However, in any one pass the metal is usually compressed in one direction only. On the next pass it is rotated 90°. Since the metal spreads to a much greater extent in the hot-rolling of bars than in cold-rolling of sheet, an important problem in designing passes for bars and shapes is to provide allowance for the spreading. A typical method of reducing a square billet to a bar is by alternate passes through oval and square-shaped grooves. The design of roll passes for structural shapes is much more complicated and requires extensive experience.[1] Because different metals spread different amounts, it is not generally possible to roll metals of widely different rolling characteristics on the same set of bar rolls.

A rolling mill designed to roll bars is known as a *bar mill*, or *merchant mill*. Most production bar mills are equipped with guides to feed the billet into the

[1] W. Trinks, "Roll Pass Design," 2d ed., Penton Publishing Company, Cleveland, 1933; R. Stewartson, The Rolling of Rods, Bars, and Light Sections, *Metall. Rev.*, vol. 4, pp 309–379, 1959.

grooves and repeaters to reverse the direction of the bar and feed it back through the next roll pass. Mills of this type are generally either two- or three-high. A common installation consists of a roughing stand, a strand stand, and a finishing stand. It is common practice to arrange bar rolls *in train*; i.e., several mills are set close together, side by side, and the rolls in one stand are driven by connecting them to those of the adjacent stand.

## 17-6  FORCES AND GEOMETRICAL RELATIONSHIPS IN ROLLING

Figure 17-5 illustrates a number of important relationships between the geometry of the rolls and the forces involved in deforming a metal by rolling. A metal sheet with a thickness $h_0$ enters the rolls at the entrance plane $XX$ with a velocity $v_0$. It passes through the roll gap and leaves the exit plane $YY$ with a reduced thickness $h_f$. To a first approximation no increase in width results, so that the vertical compression of the metal is translated into an elongation in the rolling direction. Since equal volumes of metal must pass a given point per unit time, we can write

$$bh_0v_0 = bhv = bh_fv_f \qquad (17\text{-}1)$$

where $b$ = width of sheet
   $v$ = its velocity at any thickness $h$ intermediate between $h_0$ and $h_f$
In order that a vertical element in the sheet remain undistorted, Eq. (17-1) requires that the exit velocity $v_f$ must be greater than the entrance velocity $v_0$. Therefore,

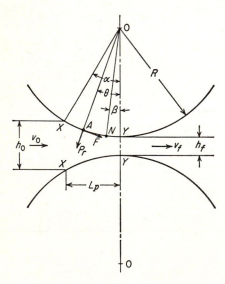

**Fig. 17-5**  Forces acting during rolling.

the velocity of the sheet must steadily increase from entrance to exit. At only one point along the surface of contact between the roll and the sheet is the surface velocity of the roll $v_r$ equal to the velocity of the sheet. This point is called the *neutral point*, or *no-slip point*. It is indicated in Fig. 17-5 by point $N$.

At any point along the surface of contact, such as point $A$ in Fig. 17-5, two forces act on the metal. These are a radial force $P_r$ and a tangential friction force $F$. Between the entrance plane and the neutral point the sheet is moving slower than the roll surface, and the frictional force acts in the direction shown in Fig. 17-5 so as to draw the metal into the rolls. On the exit side of the neutral point the sheet moves faster than the roll surface. The direction of the frictional force is then reversed so that it acts to oppose the delivery of the sheet from the rolls.

The vertical component of $P_r$ is known as the *rolling load P*. The rolling load is the force with which the rolls press against the metal. Because this is also equal to the force exerted by the metal in trying to force the rolls apart, it is frequently called the *separating force*. The *specific roll pressure p* is the rolling load divided by the contact area. The contact area between the metal and the rolls is equal to the product of the width of the sheet $b$ and the projected length of the arc of contact $L_p$.

$$L_p = \left[ R(h_0 - h_f) - \frac{(h_0 - h_f)^2}{4} \right]^{1/2} \approx [R(h_0 - h_f)]^{1/2} \qquad (17\text{-}2)$$

Therefore, the specific roll pressure is given by

$$p = \frac{P}{bL_p} \qquad (17\text{-}3)$$

The distribution of roll pressure[1] along the arc of contact is indicated in Fig. 17-6. The pressure rises to a maximum at the neutral point and then falls off. The fact that the pressure distribution does not come to a sharp peak at the neutral point, as required in theoretical treatments of rolling, indicates that the neutral point is not really a line on the roll surface but an area. The area under the curve

**Fig. 17-6** Distribution of roll pressure along the arc of contact.

[1] F. A. R. Al-Salehi, T. C. Firbank, and P. R. Lancaster, *Int. J. Mech. Sci.*, vol. 15, pp. 693–710, 1973.

is proportional to the rolling load, which for purposes of calculation acts at the center of gravity of the pressure distribution. Therefore, the shape of the pressure distribution is important because the location of the resultant rolling load with respect to the roll centers determines the torque and power required to produce the reduction. The area shown shaded in Fig. 17-6 represents the force required to overcome frictional forces between the roll and the sheet, while the area under the dashed line $AB$ represents the force required to deform the metal in plane homogeneous compression. A similarity should be noted between the pressure distribution in rolling shown in Fig. 17-6 and the pressure distribution for compression between plates (Fig. 15-18).

The angle $\alpha$ between the entrance plane and the centerline of the rolls is called the *angle of contact*, or *angle of bite*. Referring to Fig. 17-5, the horizontal component of the normal force is $P_r \sin \alpha$, and the horizontal component of the friction force is $F \cos \alpha$. For the workpiece to enter into the throat of the roll the horizontal component of the friction force, which acts toward the roll gap, must be equal to or greater than the horizontal component of the normal force, which acts away from the roll gap. The limiting condition for unaided entry of a slab into the rolls is

$$F \cos \alpha = P_r \sin \alpha$$

$$\frac{F}{P_r} = \frac{\sin \alpha}{\cos \alpha} = \tan \alpha$$

but

$$F = \mu P_r$$

so that

$$\mu = \tan \alpha \qquad (17\text{-}4)$$

The workpiece cannot be drawn into the rolls if the tangent of the contact angle exceeds the coefficient of friction. If $\mu = 0$, rolling cannot occur, but as $\mu$ increases progressively larger slabs will be drawn into the roll throat. In ingot breakdown by hot-rolling, where it is desired to achieve a large reduction in a short time, the rolls have grooves cut in them parallel to the roll axis to increase the effective value of $\mu$.

For the same friction conditions, a large-diameter roll will permit a thicker slab to enter the rolls than will a small-diameter roll. This is because the angle subtended from the center of the roll to the entrance plane will be the same in both cases ($\tan \alpha$) but the lengths of the arcs of contacts (Eq. 17-2) will be appreciably different. Referring to Fig. 17-5, we can write Eq. (17-2) as

$$L_p \approx \sqrt{R\Delta h}$$

where $\Delta h =$ the "draft" taken in rolling

$$\tan \alpha = \frac{L_p}{R - \Delta h/2} \approx \frac{\sqrt{R\Delta h}}{R - \Delta h/2} \approx \sqrt{\frac{\Delta h}{R}}$$

From Eq. (17-4), $\mu \geq \tan \alpha = \sqrt{\Delta h/R}$

or

$$(\Delta h)_{\max} = \mu^2 R \qquad (17\text{-}5)$$

The high forces generated in rolling are transmitted to the workpiece through the rolls. Under these loading conditions there are two major types of elastic distortion. First, the rolls tend to bend along their length because the workpiece tends to separate them while they are restrained at their ends. This leads to the problems with thickness variation over the width that are discussed in Sec. 17-8. Second, the rolls flatten in the region where they contact the workpiece, so that the radius of curvature is increased from $R$ to $R'$. The most widely used analysis for *roll flattening* is due to Hitchcock,[1] who represented the actual pressure distribution with elastically flattened rolls by an elliptical distribution. According to this analysis the radius of curvature increases from $R$ to $R'$

$$R' = R\left[1 + \frac{CP}{b(h_0 - h_f)}\right] \qquad (17\text{-}6)$$

where $C = 16(1 - v)/\pi E$ is evaluated for the roll material ($C = 3.34 \times 10^{-4}$ in$^2$/ton for steel rolls) and $P$ is the rolling load based on the deformed roll radius. Since $P$ is a function of $R'$, the exact solution of Eq. (17-6) requires a trial-and-error procedure.[2]

## 17-7 SIMPLIFIED ANALYSIS OF ROLLING LOAD: ROLLING VARIABLES

The main parameters in rolling are:

The roll diameter

The deformation resistance of the metal as influenced by metallurgy, temperature, and strain rate

The friction between the rolls and the workpiece

The presence of front tension and/or back tension in the plane of the sheet

A good appreciation of the role of these variables and a simple estimate of the rolling load in sheet rolling can be obtained by considering the previous analysis (Sec. 16-3) for the plane-strain homogeneous compression of a slab. Recasting[3] Eq. (16-11), the mean deformation pressure is given by

$$\frac{\bar{p}}{\bar{\sigma}_0'} = \frac{1}{Q}(e^Q - 1) \qquad (17\text{-}7)$$

[1] J. H. Hitchcock, "Roll Neck Bearings," ASME, New York, 1935; see L. R. Underwood, "The Rolling of Metals," pp. 286–296, John Wiley & Sons, Inc., New York, 1950.

[2] E. C. Larke, "The Rolling of Strip, Sheet, and Plate," 2d ed., pp. 267–273, Chapman and Hall, Ltd., London, 1963.

[3] W. A. Backofen, "Deformation Processing," p. 164, Addison-Wesley Publishing Company, Inc., Reading, Mass. 1972.

where $Q = \mu L_p / \bar{h}$

$\bar{h}$ = the mean thickness between entry and exit from the rolls.
Referring to Eq. (17-3), the rolling load is given by

$$P = \bar{p} b L_p$$

and since $L_p \approx \sqrt{R \Delta h}$

$$P = \frac{2}{\sqrt{3}} \bar{\sigma}_0 \left[ \frac{1}{Q} (e^Q - 1) b \sqrt{R \Delta h} \right] \qquad (17\text{-}8)$$

The factor $2/\sqrt{3}$ arises because flat rolling is a plane-strain situation, so that the flow stress should be the flow stress in plane strain.

Equation (17-8) shows that rolling load increases with roll diameter at a rate greater than $D^{1/2}$, depending on the contribution from the friction hill. The rolling load also increases as the sheet entering the rolls becomes thinner (due to the $e^Q$ term). Eventually a point is reached where the deformation resistance of the sheet is greater than the roll pressure which can be applied and no further reduction in thickness can be achieved. This occurs when the rolls in contact with the sheet are both severely elastically deformed. The roll diameter has an important influence on determining the minimum possible gage sheet that can be rolled with a particular mill. Both rolling load and the length of the arc of contact decrease with roll diameter. Therefore, with small-diameter rolls, properly stiffened against deflection by backup rolls, it is possible to produce a greater reduction before roll flattening becomes significant and no further reduction in sheet thickness is possible.

The mean flow stress for a rolling process can be determined directly from a plane-strain compression test. For cold-rolling it does not depend much on the strain rate or roll speed. However, as emphasized earlier, in hot-rolling changes in strain rate can produce significant changes in the flow stress of the metal.[1]

The friction between the roll and the metal surface is of great importance in rolling. We have already seen that a frictional force is needed to pull the metal into the rolls. However, Fig. 17-6 indicates that a large fraction of the rolling load comes from the frictional forces. The frictional contribution resides in the $e^Q$ term in Eq. (17-7). High friction results in high rolling load, a steep friction hill, and great tendency for edge cracking. The friction varies from point to point along the contact arc of the roll.[2] However, because it is very difficult to measure this variation in $\mu$, all theories of rolling are forced to assume a constant coefficient of friction. For cold-rolling[3] with lubricants $\mu$ varies from about 0.05 to 0.10, but for hot-rolling friction coefficients from 0.2 up to the sticking condition are common.

[1] P. M. Cook, *Proc. Conf. Properties of Materials at High Rates of Strain,* Institution of Mechanical Engineers, London, 1957, pp. 85–97.
[2] J. M. Capus and M. G. Cockcroft, *J. Inst. Met.,* vol. 90, pp. 289–297, 1961–1962.
[3] W. L. Roberts, *Blast Furn. Steel Plant,* vol. 56, pp. 382–394, 1968.

The angle of bite can be used to establish $\mu$ through Eq. (17-4). However, the method is not very accurate and other techniques have been developed. We have already seen that the neutral point is the location on the arc of contact where the direction of the friction force changes. From the entry plane to the neutral point the friction force acts in the direction of roll rotation, while on the exit side of the neutral point it reverses direction. If back tension is applied gradually to the sheet, the neutral point shifts toward the exit plane. The total rolling load $P$ and torque $M_T$ (per unit of width $b$) are given by[1]

$$\frac{P}{b} = \int_0^{L_p} p \, dx \qquad \frac{M_T}{b} = \int_0^{L_p} (\mu p \, dx)R = \mu R \int_0^{L_p} p \, dx = \mu R \frac{P}{b}$$

thus,

$$\mu = \frac{M_T}{PR} \qquad (17\text{-}9)$$

where $\mu$ is obtained by measuring the torque and the rolling load at constant roll speed and reduction with the proper back tension. The proper back tension to bring the neutral point to the exit plane is achieved when the exit velocity of the sheet $v_f$ is equal to the surface velocity of the rolls, $v_r = R\omega$. This can be expressed another way be stating that the *forward slip* $S_f$ equals zero.

$$S_f = \frac{v_f - v_r}{v_r} \qquad (17\text{-}10)$$

Measurement of forward slip for any value of back tension can be used to estimate $\mu$ through[2]

$$S_f = \frac{1}{4} \frac{r}{1-r} \left(1 - \frac{\alpha}{2\mu}\right)^2 \qquad (17\text{-}11)$$

where $r = (h_0 - h_f)/h_0$ is the reduction and $\alpha$ is the angle of bite.

The minimum-thickness sheet that can be rolled on a given mill is directly related to the coefficient of friction. Since friction coefficients are much lower for cold-rolling than hot-rolling, thinner-gage sheet is produced by cold-rolling. The thickness of the sheet produced on a cold-rolling mill can also be decreased appreciably by increasing the rolling speed. This is attributable to the decrease in friction coefficient with increasing rolling speed.[3]

The presence of tension in the plane of the sheet can materially reduce the rolling load. Back tension may be produced by controlling the speed of the uncoiler relative to the roll speed and front tension may be created by controlling the coiler.

[1] P. W. Whitton and H. Ford, *Proc. Inst. Mech. Eng.* (London), vol. 169, p. 123, 1955.
[2] M. D. Stone, *Trans. ASME, Ser. D: J. Basic Eng.*, vol. 81, pp. 681–686, 1957.
[3] R. B. Sims and D. F. Arthur, *J. Iron Steel Inst. London*, vol. 172, pp. 285–295, 1952.

The effect of sheet tension on reducing roll pressure $p$ can be shown simply from a consideration of the von Mises' criterion in plane strain.

$$\sigma_1 - \sigma_3 = \frac{2}{\sqrt{3}} \bar{\sigma}_0$$

$$p - (-\sigma_h) = \frac{2}{\sqrt{3}} \bar{\sigma}_0$$

where $\sigma_h$ is the horizontal sheet tension and compressive stresses are taken as positive.

$$p = \frac{2}{\sqrt{3}} \bar{\sigma}_0 - \sigma_h \qquad (17\text{-}12)$$

Thus, the roll pressure is reduced in direct proportion to the tension in the plane of the sheet. This results in less wear of the rolls and improved flatness and uniformity of thickness across the width of the sheet. A study[1] of the effect of sheet tension has shown that back tension is about twice as effective in reducing the rolling load as front tension. The rolling load when tension is applied $P_t$ can be calculated from the following equation:

$$P_t = P \left( 1 - \frac{\sigma_b + \sigma_f}{\bar{\sigma}_0'} \frac{\beta}{\alpha} \right) \qquad (17\text{-}13)$$

where $P$ = rolling load for same reduction but without front or back tension

$\sigma_b$ = back tension

$\sigma_f$ = front tension

$\bar{\sigma}_0'$ = mean value of plane-strain yield stress

$\alpha$ = angle of contact

$\beta$ = no-slip angle $= \dfrac{\alpha}{2} - \dfrac{1}{\mu}\left(\dfrac{\alpha}{2}\right)^2$

Nadai[2] developed a theory of rolling which permits the calculation of the effect of strip tension on the roll pressure distribution. As shown schematically in Fig. 17-7, the addition of both front and back tension materially reduces the area under the curve, although there is little shift of the neutral point. If only back tension is applied, the neutral point moves toward the roll exit. If a high enough back tension is applied, the neutral point will eventually reach the roll exit. When this happens, the rolls are moving faster than the metal and they slide over the surface. On the other hand, if only front tension is used, the neutral point will move toward the roll entrance.

[1] W. C. F. Hessenberg and R. B. Sims, *J. Iron Steel Inst. London*, vol. 168, pp. 155–164, 1951.
[2] A. Nadai, *J. Appl. Mech.*, vol. 6, pp. A54–A62, 1939.

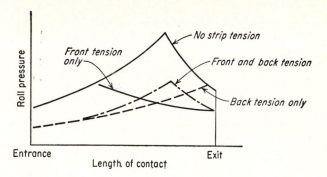

**Fig. 17-7** Effect of strip tension on the distribution of roll pressure.

## 17-8  PROBLEMS AND DEFECTS IN ROLLED PRODUCTS

A variety of problems in rolling, leading to specific defects, can arise depending on the interaction of the plastically deforming workpiece with the elastically deforming rolls and rolling mill. Under the influence of the high rolling forces the rolls flatten and bend, and the entire mill is elastically distorted. Because of *mill spring* the thickness of the sheet exiting from the rolling mill is greater than the roll gap set under no-load conditions. In order to roll to precise thickness it is necessary to know the elastic constant of the mill.[1] This is usually given in the form of a calibration curve (Fig. 17-8). The elastic constant for most screw-loaded rolling mills is 4,000 to 8,000 tons/in, while hydraulically loaded mills may exceed 10,000 tons/in.

Elastic flattening of the rolls with increasing roll pressure results in a condition where the rolls eventually deform more easily than the workpiece. Thus, for a given material and set of rolling conditions, there will be a minimum thickness below which the sheet can be reduced no further. It has already been seen that thinner-gage sheet can be produced with small-diameter rolls. More complete analysis of the problem[2] shows that the limiting thickness is nearly proportional to the coefficient of friction, the roll radius, the flow stress of the workpiece, and is inversely proportional to the elastic modulus of the rolls. For steel rolls this is given by

$$h_{\min} = \frac{\mu R \bar{\sigma}'_0}{1,860} \qquad (17\text{-}14)$$

where units are in pounds per square inch and inches.

[1] E. A. Marshall and A. Shutt, *J. Iron Steel Inst. London*, vol. 204, pp. 837–841, 1966; K. Wiedemer, *Met. Technol.*, vol. 1, pp. 181–185, 1974.
[2] H. Ford and J. M. Alexander, *J. Inst. Met.*, vol. 88, pp. 193–199, 1959–1960.

**Fig. 17-8** Typical calibration curve for the elastic constant of a rolling mill. (*From J. A. Schey, in " The Techniques of Materials Preparation and Handling," R. F. Bunshah, ed., vol. 1, pt. 3, chap. 34, p. 1452. John Wiley & Sons Inc., New York, 1965.*)

The roll gap must be perfectly parallel, otherwise one edge of the sheet will be decreased more in thickness than the other, and since volume and width remain constant, this edge of the sheet elongates more than the other, and the sheet bows. There are two aspects to the problem of the shape of a sheet. The first pertains to uniform thickness over the width and along the length. This property of a sheet can be measured accurately, and it is subject to precise control with modern automatic gage control systems. The second important property of a sheet is *flatness*. It is difficult to measure this property accurately, particularly difficult are in-process measurements taken as the sheet moves through a continuous mill at high speed. The rolling process is very sensitive to flatness. A difference in elongation of one part in 10,000 between different locations in the sheet can give rise to waviness in a thin-gage sheet. Figure 17-9 shows how waviness (lack of flatness) develops. If the rolls deflect as in Fig. 17-9a, the edges of the sheet will be elongated to a greater extent in the longitudinal direction than the center, i.e., it has "long edges." If the edges were free to move relative to the center, we would have the situation shown in Fig. 17-9b. However, the sheet remains a continuous body and the strains readjust to maintain continuity. The result is that the center portion of the sheet is stretched in tension and the edges are compressed in the rolling direction Fig. 17-9c. The usual result is a wavy edge or edge buckle (Fig. 17-9d). Under other conditions of $\Delta = h/L$ the strain distribution produced by long edge could produce short "zipper breaks" or cracks in the center of the sheet (Fig. 17-9e). The obvious solution to roll bending is to contour the roll parallel to its axis so it is larger in the center than at the ends. Then when the rolls deflect, they will present a parallel gap to the workpiece. Rolls having a ground *camber* or crown are usually used. Camber can also be produced thermally due to the thermal expansion. This corrective procedure suffers from the fact that the camber corrects for the roll deflection

**Fig. 17-9**  Consequences of roll bending to produce long edge.

at only one value of roll force, and thus it may not be an effective measure for a range of rolling conditions. A better technique is to equip the rolling mill with hydraulic jacks which permit the elastic distortion of the rolls to correct for deflection under rolling conditions. Many modern sheet mills are equipped in this way. If the rolls have excess convex camber, the center of the sheet is elongated more than the edges. The strain distribution is the opposite of Fig. 17-9c and the sheet is said to have loose middle and tight edges. Such a sheet usually will contain center buckles.

It is generally considered[1] that sheet with bad shape is created in the hot-rolling step and the cold-rolling cannot correct bad shape completely. Moreover, shape problems are greatest when rolling thin strip (less than 0.010 in) because fractional errors in roll gap profile increase with decrease in thickness, producing larger internal stresses. Also, very thin sheet is less resistant to buckling. Mild shape problems may be corrected by *stretch leveling* the sheet in tension with a bridle roll between stands, or by bend flexing the sheet in a *roller-leveler*. Suggestions for a rolling mill with "flexible rolls" which would correct bad shape have been made[2] and a good beginning has been made[3] on a mechanics analysis of shape and flatness.

---

[1] T. Sheppard and J. M. Roberts, *Int. Metall. Rev.*, vol. 18, pp. 1–18, 1973.
[2] N. H. Polakowski, D. M. Reddy, and H. N. Schmeissing, *Trans. ASME, Ser. B: J. Eng. Ind.*, vol. 91, pp. 702–709, 1969.
[3] H. W. O'Connor and A. S. Weinstein, *Trans. ASME, Ser. B: J. Eng. Ind.*, vol. 94pp. 1113–1123, 1972; D. J. McPherson, *Metall. Trans.*, vol. 5, pp. 2479–2499, 1974.

Rolling
direction

(a)                    (b)                                (c)

**Fig. 17-10**  Defects resulting from lateral spread.

Problems with shape and flatness are brought about by inhomogeneities in deformation in the rolling direction of the sheet. Other forms of inhomogeneous deformation[1] can lead to problems with *cracking*. As the workpiece passes through the rolls all elements across the width experience some tendency to expand laterally (in the transverse direction of the sheet). The tendency for lateral spread is opposed by transverse friction forces. Because of the friction hill, these are higher toward the center of the sheet so that the elements in the central region spread much less than the outer elements near the edge. Because the thickness decrease in the center of the sheet all goes into a length increase, while part of the thickness decrease at the edges goes into lateral spread, the sheet may develop a slight rounding at its ends (Fig. 17-10a). Because there is continuity between the edges and center, the edges of the sheet are strained in tension, a condition which leads to *edge cracking* (Fig. 17-10b). Under severe conditions the strain distribution shown in Fig. 17-10a can result in a *center split* of the sheet (Fig. 17-10c).

Edge cracking also can be caused by inhomogeneous deformation in the thickness direction.[2] When the rolling conditions are such that only the surface of the workpiece is deformed (as in light reductions on a thick slab), the cross section of sheet is deformed into the shape shown in Fig. 17-11a. In subsequent passes through the rolls the overhanging material is not compressed directly but is forced to elongate by the neighboring material closer to the center. This sets up high secondary tensile stresses which lead to edge cracking. This type of cracking is experienced in initial ingot breakdown in hot-rolling where it was observed[3]

[1] I. Ya Tarnovskii, A. A. Pozdeyev, and V. B. Lyashkov, "Deformation of Metals During Rolling," Pergamon Press, New York, 1965; A. Jones and B. Walker, *Met. Technol.*, vol. 1, pp. 310–315, 1974.
[2] N. H. Polakowski, *J. Inst. Met.*, vol. 76, pp. 755–757, 1949–1950.
[3] P. A. Aleksandrov, in A. M. Samarin (ed.), "Contemporary Problems of Metallurgy," pp. 416–422, Consultants Bureau, New York, 1960.

**Fig. 17-11**   Edge distribution resulting from rolling with (*a*) light reduction, (*b*) heavy reduction,
and (*c*) alligatoring.

that edge deformation like that shown in Fig. 17-11*a* occurred where $h/L_p > 2$.
With heavy reductions, so that the deformation extends through the thickness of
the sheet, the center tends to expand laterally more than the surfaces to produce
barreled edges similar to those found in upsetting a cylinder (Fig. 17-11*b*). The
secondary tensile stresses[1] created by the barreling are a ready cause of edge cracking. With this type of lateral deformation, greater spread occurs toward the center
than at the surfaces so that the surfaces are placed in tension and the center is in
compression. This stress distribution also extends in the rolling direction, and if
there is any metallurgical weakness along the center line of the slab, fracture will
occur there (Fig. 17-11*c*). This *alligatoring* type of fracture is accentuated if there
is any curling of the sheet because one roll is higher or lower than the centerline
of the roll gap.

Edge cracking is minimized in commerical rolling practice by employing
vertical edge rolls which keep the edges straight and thus prevent a cumulative
buildup of secondary tensile stresses due to barreling of the edge. Since most
laboratory mills do not have edge rolls, a simple but time-consuming procedure
to prevent edge cracking is to machine the edges square after each pass. A better
procedure is to equip the mill with edge-restraining bars.[2] Materials with low
ductility may be rolled without excessive cracking by "canning" on all sides with a
material with a flow stress similar to that of the workpiece. The canning material
minimizes thermal stresses and provides edge restraint and an increased degree of
hydrostatic compression.

Defects other than cracks can result from defects introduced during the
ingot stage of production or during rolling. Internal defects such as fissures are
due to incomplete welding of pipe and blowholes. Longitudinal stringers of
nonmetallic inclusions or pearlite banding in steels are related to melting and

---

[1] G. Cuminsky and F. Ellis, *J. Inst. Met.*, vol. 95, pp. 33–37, 1967; A. L. Hoffmanner, *Ibid.*, vol. 96, pp.
158–159, 1968.
[2] J. A. Schey, *J. Inst. Met.*, vol. 94, pp. 193–200, 1966.

solidification practices. In severe cases these defects can lead to laminations which drastically reduce the strength in the thickness direction. Because rolled products usually have a high surface-to-volume ratio, the condition of the surface during all stages of production is of importance. In order to maintain high quality, the surface of billets must be conditioned by grinding, chipping, or burning with an oxygen lance to remove surface defects such as slivers, seams, and scabs. Scratches due to defective rolls or guides must be guarded against in cold-rolled sheet. Sometimes problems arise with removing the rolling lubricant or because there is discoloration after heat treatment.

## 17-9   ROLLING-MILL CONTROL

The high throughput and production rate of modern continuous hot-strip and cold-rolling mills make it imperative that these mills be operated under automatic control. Of all the metalworking processes rolling is best suited for the adoption of automatic control because it is an essentially steady-state process in which the tooling geometry (roll gap) may be changed readily during the process. To date, the application of automatic control in rolling has been mostly concerned with control of sheet gage during the rolling of long continuous coils. This has required the development of on-line sensors to continuously measure sheet thickness. The two instruments most commonly used are the flying micrometer and x-ray, or isotope, gages which measure thickness by monitoring the amount of radiation transmitted through the sheet. More recently control procedures have been aimed at controlling strip shape[1] as well as thickness.

**Fig. 17-12**   Characteristic elastic and plastic curves for a rolling mill.

[1] A. J. Carlton, W. J. Edwards, P. W. Johnston, and N. Kuhn, *J. Aust. Inst. Met.*, vol. 18, pp. 22–36, 1973.

The problem of gage control can be understood by considering the characteristic curves for a rolling mill (Fig. 17-12). For a given set of rolling conditions, the rolling load varies with the final sheet thickness according to the *plastic curve* This is essentially what would be obtained by the solution of Eq. (17-8). The *elastic curve* for mill spring is superimposed on the figure. This indicates that a sheet of initial thickness $h_0$ would have a final thichness $h_f$ and the load on the mill would be $P$. The influence of changes in rolling variables can be visualized readily with this type of diagram. If the lubrication breaks down so that $\mu$ increases or the flow stress increases because the temperature decreases, the plastic curve will be raised (Fig. 17-13). The rolling load will be increased from $P_1$ to $P_2$ and the final thickness will be increased from $h_{f1}$ to $h_{f2}$. Figure 17-13 shows that to maintain a constant thickness $h_{f1}$ under these new conditions the roll gap would have to be decreased. Moving the elastic curve to the left, in this way, further increases the rolling load to $P_3$. If, for example, there is an increase in the incoming sheet thickness, the plastic curve would move to the right relative to the elastic curve. If there is an increase in strip tension, the plastic curve would move to the left.

In a continuous hot mill the strip thickness is measured indirectly by measuring the rolling load and using the characteristic curve of the mill to establish thickness. The error signal is fedback to the rolling mill screws to reposition them so as to minimize the error. An x-ray gage is used after the last stand to provide an absolute measurement of sheet gage. In continuous cold-strip mills thickness is measured by x-ray gages. While the error in thickness following the first stand is usually fedback to adjust the gap setting on the first stand, gage control in subsequent stands usually is achieved by controlling the strip tension through controlling the relative roll speed in successive stands or the coiler speed. Gage control through control of strip tension has faster response time than control through change in roll setting.

**Fig. 17-13**  Use of characteristic curves to show changes in rolling conditions.

## 17-10   THEORIES OF COLD ROLLING

Probably more work has been expended in developing a theoretical treatment of cold-rolling than for any other metalworking process. A theory of rolling is aimed at expressing the external forces, such as the rolling load and the rolling torque, in terms of the geometry of the deformation and the strength properties of the material being rolled.

The differential equation for the equilibrium of an element of material being deformed between rolls is common to all the theories of rolling. The derivation given below is based on the following assumptions.

1.  The arc of contact is circular—no elastic deformation of the rolls.
2.  The coefficient of friction is constant at all points on the arc of contact.
3.  There is no lateral spread, so that rolling can be considered a problem in plane strain.
4.  Plane vertical sections remain plane.
5.  The peripheral velocity of the rolls is constant.
6.  The elastic deformation of the sheet is negligible in comparison with the plastic deformation.
7.  The distortion-energy criterion of yielding, for plane strain, holds.

$$\sigma_1 - \sigma_3 = \frac{2}{\sqrt{3}}\,\sigma_0 = \sigma_0',$$

The stresses acting on an element of strip in the roll gap are shown in Fig. 17-14. At any point of contact between the strip and the roll surface, designated by the angle $\theta$, the stresses acting are the radial pressure $p_r$ and the tangential shearing stress $\tau = \mu p_r$. These stresses are resolved into their horizontal and vertical components in Fig. 17-14b. In addition, the stress $\sigma_x$ is assumed to be uniformly distributed over the vertical faces of the element. The normal stress on one end of the element is $p_r R\,d\theta$, and the horizontal component of this force is $p_r R \sin\theta\,d\theta$. The tangential friction force is $\mu p_r R d\theta$, and its horizontal component is $\mu p_r R \cos\theta\,d\theta$. Taking the summation of the horizontal forces on the element results in

$$(\sigma_x + d\sigma_x)(h + dh) + 2\mu p_r R \cos d\theta - \sigma_x h - 2p_r R \sin\theta\,d\theta = 0$$

which simplifies to

$$\frac{d(\sigma_x h)}{d\theta} = 2p_r R(\sin\theta \pm \mu \cos\theta) \qquad (17\text{-}15)$$

The positive sign applies between the exit plane and the neutral point, while the negative sign applies between the entrance plane and the neutral point. The minus and plus signs in Eq. (17-15) occur because the direction of the friction force changes at the neutral point. This equation was first derived by von Kármán[1] and is usually named for him.

[1] T. von Kármán, Z. Angew. Math. Mech., vol. 5, pp. 139–141, 1925.

**Fig. 17-14** (*a*) Geometrical relationship for element undergoing plane-strain deformation by rolling; (*b*) stresses acting on an element.

The forces acting in the vertical direction are balanced by the specific roll pressure $p$. Taking the equilibrium of forces in the vertical direction results in a relationship between the normal pressure and the radial pressure.

$$p = p_r(1 \mp \mu \tan \theta) \qquad (17\text{-}16)$$

The relationship between the normal pressure and the horizontal compressive stress $\sigma_x$ is given by the distortion energy criterion of yielding for plane strain.

$$\sigma_1 - \sigma_3 = \frac{2}{\sqrt{3}} \sigma_0 = \sigma_0'$$

or

$$p - \sigma_x = \sigma_0' \qquad (17\text{-}17)$$

where $p$ is the greater of the two compressive principal stresses.[1]

The solution of problems in cold-rolling consists in the integration of Eq. (17-15) with the aid of Eqs. (17-16) and (17-17). Unfortunately, the mathematics is rather formidable, and various approximations must be made to obtain a tractable solution. Trinks[2] provided graphical solutions to von Kármán's equation, using

---

[1] In agreement with the literature on rolling, compressive stresses are taken as positive.
[2] W. Trinks, *Blast Furn. Steel Plant*, vol. 25, pp. 617–619, 1937; see also Underwood, op. cit., pp. 210–215.

the assumptions of constant yield stress and a parabolic arc of contact. The most complete solution of the rolling equations has been obtained by Orowan.[1] In this solution, allowance was made for the fact that the flow stress changes with $\theta$ owing to strain hardening. However, the complexity of the equations makes it necessary to obtain solutions by graphical integration.[2] Even though the equations for roll pressure have been expressed[3] in algebraic form, they are still too complicated for routine calculation of rolling problems.

Some simplification to this problem has been provided by Bland and Ford.[4] By restricting the analysis to cold-rolling under conditions of low friction and for angles of contact less than 6°, they were able to put $\sin \theta \approx \theta$ and $\cos \theta \approx 1$. Thus, Eq. (17-15) can be written

$$\frac{d(\sigma_x h)}{d\theta} = 2p_r R'(\theta \pm \mu) \qquad (17\text{-}18)$$

It is also assumed that $p_r \approx p$, so that Eq. (17-17) can be written $\sigma_x = p_r - \sigma'_0$. By substituting into Eq. (17-18) and integrating,[5] relatively simple equations for the radial pressure result.

Roll entrance to neutral point:

$$p_r = \frac{\sigma'_0 h}{h_0} \left(1 - \frac{\sigma_{xb}}{\sigma'_{01}}\right) e^{\mu(H_1 - H)} \qquad (17\text{-}19)$$

Neutral point to roll exit:

$$p_r = \frac{\sigma'_0 h}{h_f} \left(1 - \frac{\sigma_{xf}}{\sigma'_{02}}\right) e^{\mu H} \qquad (17\text{-}20)$$

where

$$H = 2 \left(\frac{R'}{h_f}\right)^{1/2} \tan^{-1} \left[\left(\frac{R'}{h_f}\right)^{1/2} \theta\right]$$

and $\sigma_{xb}$ = back tension

$\sigma_{xf}$ = front tension

Subscript 1 refers to a quantity evaluated at the roll-entrance plane, and subscript 2 refers to a quantity evaluated at the roll-exit plane.

The rolling load or total force $P$ is the integral of the specific roll pressure over the arc of contact.

$$P = R'b \int_0^{\theta = \alpha} p \, d\theta \qquad (17\text{-}21)$$

[1] E. Orowan, *Proc. Inst. Mech. Eng. (London)*, vol. 150, pp. 140–167, 1943.

[2] A computer solution to Orowan's equations which also incorporates corrections for the deformation of the rolls has been given by J. E. Hockett, *Trans. Am. Soc-. Met.*, vol. 52, pp. 675–697, 1960.

[3] M. Cook and E. C. Larke, *J. Inst. Met.*, vol. 74, pp. 55–80, 1947.

[4] D. R. Bland and H. Ford, *Proc. Inst. Mech. Eng. (London)*, vol. 159, pp. 144–163, 1948.

[5] See *ibid.*, or H. Ford *Metall. Rev.*, vol. 2, pp. 5–7, 1957.

where $b$ = width of sheet

$\quad\quad\alpha$ = contact angle

This is best evaluated by graphical integration following a point-by-point calcu-lation with Eqs. (17-19) and (17-20). Computational methods based on the Bland and Ford solution have been published.[1]

The modern digital computer removes the necessity for making simplifying assumptions to obtain solutions to von Kármán's equation. Alexander[2] removed all restrictions except the assumption of homogeneous deformation, and showed that while the previous approximate theories gave reasonable prediction of rolling load, they were not capable of predicting roll torque with any degree of precision.

## 17-11  THEORIES OF HOT-ROLLING

Theories of hot-rolling have not advanced to the state of knowledge that exists for cold-rolling because of the more difficult problem of accounting for inhomo-geneous deformation and less well-defined friction conditions in hot-working. As in other hot-working processes, the flow stress for hot-rolling is a function of both temperature and strain rate (speed of rolls).

The strain rate for hot-rolling with sticking friction is given[3] by

$$\dot{\varepsilon} = \frac{v}{h} = \frac{2v_r \sin \theta}{h} = \frac{2v_r \sin \theta}{h_f + D(1 - \cos \theta)} \quad (17\text{-}22)$$

Evaluation of this equation shows that the maximum strain rate occurs near the entrance to the rolls. For equal percentage reduction, thin sheet will undergo much greater strain rates than thick slabs. The mean strain rate in rolling is given by[4]

$$\dot{\bar{\varepsilon}} = \frac{v_r}{\sqrt{Rh_0}} \sqrt{r}\left(1 + \frac{r}{4}\right) \quad (17\text{-}23)$$

where $v_r = 2\pi Rn$ and $n$ is in revolutions per second

$\quad\quad r = (h_0 - h_f)/h_0$

Ford and Alexander[5] used slip-line field analysis of hot-rolling to develop equations for the rolling load and torque in hot-rolling of nonferrous alloys and steels.[6]

$$P = kbL_p\left(\frac{\pi}{2} + \frac{L_p}{h_0 + h_f}\right) \quad (17\text{-}24)$$

where $k$ is the mean flow stress in pure shear $= \sigma_0/\sqrt{3}$

---

[1] P. W. Whitton, *J. Appl. Mech.*, vol. 23, pp. 307–311, 1956.
[2] J. M. Alexander, *Proc. R. Soc. London, Ser. A*, vol. 326, pp. 535–563, 1972.
[3] Larke, op. cit., chap. 8.
[4] H. Ford and J. M. Alexander, *J. Inst. Met.*, vol. 92, pp. 397–404, 1963–1964.
[5] *Ibid.*, pp. 397–399.
[6] E. Gupta and H. Ford, *J. Iron Steel Inst. London*, vol. 205, pp. 186–190, 1967.

The torque is given by

$$M_T = kbL_p^2 \left(1.60 + 0.91 \frac{L_p}{h_0 + h_f}\right) \qquad (17\text{-}25)$$

Denton and Crane[1] have pointed out that if Eq. (17-24) is rewritten slightly, it is similar to the equation for the force to forge a slab between perfectly rough platens. Hot-rolling:

$$P = kbL_p \left[1.31 + 0.53 \frac{L_p}{(h_0 h_f)^{1/2}}\right]$$

Hot-forging:

$$P = kbw \left[1.5 + 0.5 \frac{b}{h}\right]$$

where $(h_0 h_f)^{1/2}$ is the geometric mean thickness of the strip.

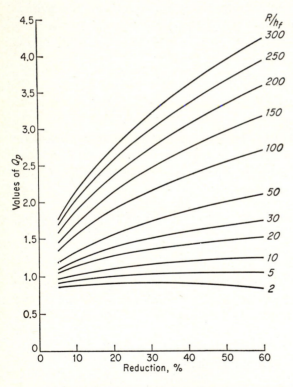

**Fig. 17-15**  Values of $Q_p$ for use with Eq. (17–26). (*From E. C. Larke, " The Rolling of Strip, Sheet, and Plate,"  Chapman & Hall, Ltd., London, 1957.*)

[1] B. K. Denton and F. A. A. Crane, *J. Iron Steel Inst. London*, vol. 210, pp. 606–616, 1972.

A frequently used analysis for the rolling load in hot-rolling is due to Sims.[1] Using a shortened form of the Orowan equation and mathematical simplifications. similar to Bland and Ford, Sims developed the relationship

$$P = \sigma_0' \, b[R(h_0 - h_f)^{1/2}]Q_p \qquad (17\text{-}26)$$

where $Q_p$ is a complex function of the reduction in thickness and the ratio $R/h_f$. Values of $Q_p$ may be obtained from Fig. 17-15.

## 17-12 TORQUE AND HORSEPOWER

Power is applied to a rolling mill by applying a torque to the rolls and by means of strip tension. The power is expended principally in four ways, (1) the energy needed to deform the metal, (2) the energy needed to overcome frictional forces in the bearings, (3) the energy lost in the pinions and power-transmission system, and (4) electrical losses in the various motors and generators. Losses in the windup reel and uncoiler must also be considered.

The total rolling load is distributed over the arc of contact in the typical friction-hill pressure distribution. However, the total rolling load can be assumed to be concentrated at a point along the arc of contact at a distance $a$ from the line of centers of the rolls. In calculating the torque the problem is mainly how to calculate this moment arm. It is usual practice[2] to consider the ratio of the moment arm $a$ to the projected length of the arc of contact.

$$\lambda = \frac{a}{L_p} = \frac{a}{[R(h_0 - h_f)]^{1/2}} \qquad (17\text{-}27)$$

For cold-rolling a correction for the change in roll radius from $R$ to $R'$ due to elastic deformation must be made in the calculation of $\lambda$. According to Ford,[3] $\lambda$ is expresssed in terms of an experimental parameter $\lambda'$ as follows,

$$\lambda = 0.5\left(\frac{R}{R'}\right)^{1/2} - (0.5 - \lambda')\left(\frac{R'}{R}\right)^{1/2} \qquad (17\text{-}28)$$

where $\lambda'$ has a value of 0.43 for cold-rolling with matte-finished rolls and $\lambda'$ is 0.48 for rolls with a smooth surface. With this information it is possible to determine the moment arm from Eq. (17-27).

The torque is equal to the total rolling load multiplied by the effective moment arm, and since there are two work rolls, the torque is given by

$$M_t = 2Pa \qquad \text{lb-ft} \qquad (17\text{-}29)$$

[1] R. B. Sims, *Proc. Inst. Mech. Eng.* (*London*), vol. 108, pp. 191–200, 1954; M. Tarokh and F. Seredynski, *J. Iron Steel Inst. London*, vol. 208, pp. 695–697, 1970.

[2] Larke, op. cit., chap. 11.

[3] H. Ford, *Proc. Inst. Mech. Eng.* (*London*), vol. 159, pp. 115, 1948.

During one revolution of the top roll the resultant rolling load $P$ moves along the circumference of a circle equal to $2\pi a$ (Fig. 17-16). Since there are two work rolls involved, the work done is equal to

$$\text{Work} = 2(2\pi a)P \qquad \text{ft-lb} \qquad (17\text{-}30)$$

Since horsepower is defined as the rate of doing work at 33,000 ft-lb/min, the horsepower needed to operate a pair of rolls revolving at $N$ rpm is given by

$$\text{hp} = \frac{4\pi a P N}{33{,}000} \qquad (17\text{-}31)$$

The above equation expresses the horsepower required in deforming the metal as it flows through the roll gap. The horsepower needed to overcome friction in the bearings and the pinions must be determined separately.

The same basic equation for horsepower holds for hot-rolling, with the important condition that the equations which were given for determining the effective moment arm in cold-rolling are not applicable in hot-rolling. A procedure for determining the moment arm in hot-rolling, based on the work of Sims,[1] is given in detail by Larke.[2] Although the calculations are straightforward, they are too detailed for inclusion here.

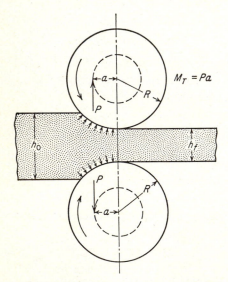

$M_T = Pa$

**Fig. 17-16** Schematic diagram illustrating roll torque.

[1] Sims, op. cit.
[2] Larke, op. cit., chap. 12.

# BIBLIOGRAPHY

"Elements of Rolling Practice," 2d ed., The United Steel Companies, Ltd., Sheffield, England, 1963.

Ford, H.: "The Theory of Rolling," *Metall. Rev.*, vol. 2, no. 5, pp. 1–28, 1957.

Larke, E. C.: "The Rolling of Strip, Sheet, and Plate," 2d ed., Chapman and Hall, Ltd., London, 1963.

"Roll Pass Design," The United Steel Companies, Ltd., Sheffield, England, 1960.

Starling, C. W.: "The Theory and Practice of Flat Rolling," University of London Press, London, 1962.

"The Making, Shaping, and Treating of Steel," 8th ed., United States Steel Corporation, Pittsburgh, 1964.

Tselikov, A. I., and V. V. Smirnov: "Rolling Mills," Pergamon Press, New York, 1965.

Underwood, L. R.: "The Rolling of Metals," John Wiley & Sons, Inc., New York, 1950.

Wusatowski, Z.: "Fundamentals of Rolling," Pergamon Press, New York, 1969.

# 18

## EXTRUSION

### 18-1 CLASSIFICATION OF EXTRUSION PROCESSES

Extrusion is the process by which a block of metal is reduced in cross section by forcing it to flow through a die orifice under high pressure. In general, extrusion is used to produce cylindrical bars or hollow tubes, but shapes of irregular cross section may be produced from the more readily extrudable metals, like aluminum. Because of the large forces required in extrusion, most metals are extruded hot under conditions where the deformation resistance of the metal is low. However, cold extrusion is possible for many metals and has become an important commercial process. The reaction of the extrusion billet with the container and die results in high compressive stresses which are effective in reducing the cracking of materials during primary breakdown from the ingot. This is an important reason for the increased utilization of extrusion in the working of metals difficult to form, like stainless steels, nickel-based alloys, and other high-temperature materials.

The two basic types of extrusion are *direct extrusion* and *indirect extrusion* (also called inverted, or back, extrusion). Figure 18-1a illustrates the process of direct extrusion. The metal billet is placed in a container and driven through the die by the ram. A dummy block, or pressure plate, is placed at the end of the ram

(a)

Die

Extrusion

Die
holder

Container

Billet

Ram

Dummy
block

Direct extrusion

(b)

Extrusion

Ram

Ram

Container

Billet

Closure
plate

Indirect extrusion

**Fig. 18-1**   Types of extrusion.

in contact with the billet. Figure 18-1*b* illustrates the indirect-extrusion process. A hollow ram carries the die, while the other end of the container is closed with a plate. Frequently, for indirect extrusion, the ram containing the die is kept stationary, and the container with the billet is caused to move. Because there is no relative motion between the wall of the container and the billet in indirect extrusion, the friction forces are lower and the power required for extrusion is less than for direct extrusion. However, there are practical limitations to indirect extrusion because the requirement for using a hollow ram limits the loads which can be applied.

Tubes can be produced by extrusion by attaching a mandrel to the end of the ram. The clearance between the mandrel and the die wall determines the wall thickness of the tube. Tubes are produced either by starting with a hollow billet

**Fig. 18-2**   Extrusion of lead sheath on electrical cable.

or by a two-step extrusion operation in which a solid billet is first pierced and then extruded.

Extrusion was originally applied to the making of lead pipe and later to the lead sheathing of cable.  Figure 18-2 illustrates the extrusion of a lead sheath on electrical cables.

*Impact extrusion* is a process used to produce short lengths of hollow shapes, such as collapsible toothpaste tubes.  It may be either indirect or direct extrusion, and it is usually performed on a high-speed mechanical press.  Although the process generally is performed cold, considerable heating results from the high-speed deformation.  Impact extrusion is restricted to the softer metals such as lead, tin, aluminum, and copper.

## 18-2   EXTRUSION EQUIPMENT

Most extrusions are made with hydraulic presses.  Hydraulic extrusion presses are classified into horizontal and vertical presses, depending upon the direction of travel of the ram.  Vertical extrusion presses are generally built with capacities of 300 to 2,000 tons.  They have the advantages of easier alignment between the press ram and the tools, higher rate of production, and the need for less floor space than horizontal presses.  However, they need considerable headroom, and to make extrusions of appreciable length, a floor pit is frequently necessary.  Vertical presses will produce uniform cooling of the billet in the container, and thus symmetrically uniform deformation will result.  In a horizontal extrusion press the bottom of the billet which lies in contact with the container will cool more rapidly than the top surface, unless the extrusion container is internally heated, and therefore the deformation will be nonuniform.  Warping of bars will result, and nonuniform wall thickness will occur in tubes. In commercial operations the chief use for vertical presses is in the production of thin-wall tubing, where uniform wall thickness and concentricity are required. Horizontal extrusion presses are used for most commercial extrusion of bars and tubes. Presses with a capacity of 1,500 to 5,000 tons are in regular operation, while a few presses of 14,000 tons capacity have been constructed.

The ram speed of the press can be an important consideration since high ram speeds are required in high-temperature extrusion where there is a problem of heat transfer from the billet to the tools. Ram speeds of 1,000 to 1,500 in/min may be used in extruding refractory metals, and this requires a hydraulic accumulator system with the press. At the other extreme, aluminum and copper alloys are prone to hot shortness so that the ram speed must be restricted to a few inches per minute. In this case, direct-drive pumping systems are adequate. Work has been done on presses equipped for preselected program control of ram speed in order to maintain a uniform finishing temperature.

The dies and tooling used in extrusion must withstand considerable abuse from the high stresses, thermal shock, and oxidation. Figure 18-3 illustrates a typical extrusion tooling assembly. This assembly is designed for easy replacement of damaged parts and for reworking and reuse of components of the tooling. The die stack consists of the die (6), made from highly alloyed tool steel, which is supported by a die holder (5) and a bolster (7), all of which are held in a die head (2). This entire assembly is sealed against the container on a conical seating surface by the pressure applied by a wedge (1). Since the extrusion container must withstand high pressures, it is usually made in two parts. A liner (4) is shrunk into the more massive container (3) to produce a compressive prestress in the inside surface of the liner. The extrusion ram or stem is usually highly loaded in compression. It is protected from the hot billet by a follower pad immediately behind the billet. Since the liner and follower pad are subjected to many cycles of thermal shock, they will need to be replaced periodically.

Fig. 18-3 Typical arrangement of extrusion tooling. (*After Schey.*)

**Fig. 18-4** Typical extrusion dies. (*a*) Flat-faced (square) die; (*b*) concical die.

There are two general types of extrusion dies. Flat-faced dies (Fig. 18-4*a*) are used when the metal entering the die forms a dead zone and shears internally to form its own die angle. A parallel land on the exit side of the die helps strengthen the die and allows for reworking of the flat face on the entrance side of the die without increasing the exit diameter. Dies with conical entrance angles (Fig. 18-4*b*) are used in extrusion with good lubrication. Decreasing the die angle increases the homogeneity of deformation and lowers the extrusion pressure, but beyond a point the friction in the die surfaces becomes too great. For most extrusion operations the optimum semidie angle ($\alpha$) is between 45 and 60°.

In addition to the extrusion press, billet-heating facilities are needed, and for production operations, there should be automatic transfer equipment for placing the heated billet in the container. Provision for heating the extrusion container may also be required. A hot saw is needed to cut off the extrusion so that the discard, or butt, can be removed from the die. Alternatively, extrusion may be done with a carbon block between the billet and the follower. This allows complete extrusion of the billet without a butt. Finally, there must be a runout table for catching the extrusion and a straightener to correct minor warpage in the extruded product.

## 18-3 HOT EXTRUSION

The principal variables which influence the force required to cause extrusion are (1) the type of extrusion (direct vs. indirect), (2) the extrusion ratio, (3) the working temperature, (4) the speed of deformation, and (5) the frictional conditions at the die and container wall.

In Fig. 18-5 the extrusion pressure is plotted against ram travel for direct and indirect intrusion. Extrusion pressure is the extrusion force divided by the cross-sectional area of the billet. The rapid rise in pressure during the initial ram travel is

**Fig. 18-5** Typical curves of extrusion pressure vs. ram travel for direct and indirect extrusion.

due to the initial compression of the billet to fill the extrusion container. For direct extrusion the metal begins to flow through the die at the maximum value of pressure, the *breakthrough pressure*. As the billet extrudes through the die the pressure required to maintain flow progressively decreases with decreasing length of the billet in the container. For indirect extrusion there is no relative motion between the billet and the container wall. Therefore, the extrusion pressure is approximately constant with increasing ram travel and represents the stress required to deform the metal through the die. While this appears to be an attractive process, in practice it is limited by the need to use a hollow ram which creates limitations on the size of the extrusion and the extrusion pressures which can be achieved. Therefore, most hot extrusion is done by the direct process. Finally, returning to Fig. 18-5, at the end of the stroke the pressure builds up rapidly and it is usual to stop the ram travel so as to leave a small discard in the container. This discard often contains defects which are unwanted in the product.

The *extrusion ratio* is the ratio of the initial cross-sectional area of the billet to the final cross-sectional area after extrusion, $R = A_0/A_f$. Extrusion ratios reach about 40:1 for hot extrusion of steel and may be as high as 400:1 for aluminum. It is important to appreciate the distinction between the fractional reduction in area, $r = 1 - A_f/A_0$, Eq. (15-4), and the extrusion ratio $R$, $R = 1/(1 - r)$. For large deformations $R$ is a more descriptive parameter. For example, the change in fractional reduction from 0.95 to 0.98 appears relatively minor, yet it corresponds to a change in area ratio from $R = 20:1$ to $R = 50:1$.

Because there is constancy of mass flow rate through the die, the velocity of the extruded product is the ram velocity $\times R$, so that quite high sliding velocities can be achieved along the die land. The extrusion pressure is directly related to the natural logarithm of the extrusion ratio, so that the extrusion force may be expressed as:

$$P = kA_0 \ln A_0/A_f \qquad (18\text{-}1)$$

where $k$ = the "extrusion constant," an overall factor which accounts for the flow stress, friction, and inhomogeneous deformation.

Most metals are extruded hot so as to take advantage of the decrease in flow stress or deformation resistance with increasing temperature. Since hot-working introduces the problems of oxidation of the billet and the extrusion tools and softening of the die and tools, as well as making it more difficult to provide adequate lubrication, it is advantageous to use the minimum temperature which will provide the metal with suitable plasticity. The upper hot-working temperature is the temperature at which hot shortness occurs, or, for pure metals, the melting point. Because of the extensive deformation produced in extrusion, considerable internal heating of the metal also results. Therefore, the top working temperature should be safely below the melting-point or hot-shortness range. For a given deformation, higher working temperatures are generally possible with extrusion than can be used in forging or rolling because the high compressive stresses minimize cracking. However, cracking in extrusions of unsymmetrical shape may occur because of unequal flow in different sections.

In the extrusion of steel[1] the billets are heated in the range 1100 to 1200°C, while the tooling is preheated to around 350°C. The extrusion pressures generally are in the range of 125,000 to 180,000 psi. The combination of high stresses and temperature place a severe demand on the glass lubrication system. The hot extrusion of lower-strength aluminum alloys is accomplished without any billet lubrication. The billet shears within itself near the container wall to create its own internal conical die surface. Metal deformation is very nonuniform, which results in a wide variation in heat-treatment response.

The interrelationship of temperature, deformation velocity, and deformation (reduction) to define conditions for successful crack-free hot-working was considered in Sec. 15-4 and illustrated with Fig. 15-11. This situation is very typical of hot extrusion.

Increasing the ram speed produces an increase in the extrusion pressure. A tenfold increase in the speed results in about a 50 percent increase in pressure. Greater cooling of the billet occurs at low extrusion speeds. When this becomes pronounced, the pressure required for direct extrusion will actually increase with increasing ram travel because of the increased flow stress as the billet cools. The higher the temperature of the billet, the greater the effect of low extrusion speed on the cooling of the billet. Therefore, high extrusion speeds are required with high-strength alloys that need high extrusion temperatures. The temperature rise due to deformation of the metal is greater at high extrusion speeds, and therefore problems with hot shortness may be accentuated.[2]

[1] R. Cockroft, *Met. Mater.*, vol. 3, pp. 351–355, September 1969.
[2] The heat balance during extrusion has been discussed by A. R. E. Singer and J. W. Coakham, *Metallurgia*, vol. 60, pp. 239–246, 1959.

The selection of the proper extrusion speed and temperature is best determined by trial and error for each alloy and billet size. The interdependence of these factors is shown schematically in Fig. 15-11. For a given extrusion pressure, the extrusion ratio which can be obtained increases with increasing temperature. For any given temperature a larger extrusion ratio can be obtained with a higher pressure. The maximum billet temperature, on the assumption that there are no limitations from the strength of the tools and die, is determined by the temperature at which incipient melting or hot shortness occurs in the extrusion. The temperature rise of the extrusions will be determined by the speed of extrusion and the amount of deformation (extrusion ratio). Therefore, the curve which represents the upper limit to the safe extrusion region slopes upward toward the left. The worst situation is for extrusion at infinite speed, where none of the heat produced by deformation is dissipated. At lower extrusion speeds there is greater heat dissipation, and the allowable extrusion ratio for a given preheat temperature increases. The allowable extrusion range is the region under the curve of constant pressure and extrusion speed.

## 18-4  DEFORMATION, LUBRICATION, AND DEFECTS IN EXTRUSION

The pressure required to produce an extrusion is dependent on the way the metal flows in the container and extrusion die, and this is largely determined by the conditions of lubrication. Certain defects which occur in extrusion are directly related to the way the metal deforms during extrusion.

Figure 18-6 shows the characteristic types of deformation in extrusion. Nearly homogeneous deformation is shown in Fig. 18-6a. This would be typical of low container friction with a well-lubricated billet, hydrostatic extrusion in which the billet is surrounded with a pressurized fluid, or indirect extrusion (Fig. 18-6d). Deformation of the billet is relatively uniform until close to the die entrance. Figure 18-6b represents a case of increased container-wall friction, as shown by the severe distortion of the grid pattern in the corners of the die to produce a *dead zone* of stagnant metal which undergoes little deformation. Grid elements at the center of the billet undergo essentially pure elongation into the extruded rod, while elements near the sides of the billet undergo extensive shear deformation. The shear distortion requires an expenditure of energy which is not related to the change in external dimensions from billet to extruded product, and this is termed redundant work. For high friction at the container-billet interface, flow is concentrated toward the center and an internal shear plane develops (Fig. 18-6c). This condition can also occur when the billet surface is chilled by a cold container. When sticking

$(a)$ $(b)$ $(c)$ $(d)$

**Fig. 18-6** Patterns of metal deformation in extrusion. (*After Schey.*)

friction prevails between the billet and container, the metal will separate internally along a shear zone and a thin skin will be left in the container. Under these deformation conditions all of the surface of the extrusion is clean new material. While the cross-sectional area of the billet decreases as (diameter)[1], the surface area increases as (diameter).[1] If the extrusion ratio is 16:1, the diameter is reduced to one-quarter of its original value and the surface area increases fourfold. Note that three-quarters of the surface area of the product is new surface that did not exist in the billet. This points out the importance of the lubricant having the ability to spread out readily and maintain a coherent film.

An effective hot-extrusion lubricant must have a low shear strength yet be stable enough to prevent breakdown at high temperature. For high-temperature extrusion of steel and nickel-based alloys the most common lubricant is molten glass, the Ugine-Sejournet process.[1] In this lubrication practice the billet is heated in an inert atmosphere and coated with glass powder before it enters the container of the press. The glass coating not only serves as a lubricant between the billet and the container wall, but it also serves as a thermal insulation to reduce heat loss to the tools. However, the main source of lubricant is a glass pad placed at the face of the die in front of the nose of the billet. During extrusion the hot billet progressively softens this glass pad and provides a lubricant film between the extrusion and the die, typically about 0.001 in thick. While it has been widely held that the viscosity of the glass is the chief criterion in selecting lubricants, more recent

---

[1] J. Sejournet and J. Delcroix, *Lubr. Eng.*, vol. 11, p. 389, 1955.

studies[1] have shown no correlation between the extrusion pressure or coating thickness and the glass viscosity. It is believed that successful lubrication depends on maintaining a reservoir of glass between the billet and the die. The coating thickness depends on the rate at which lubricant becomes available by melting or softening, rather than on its shear stress. A complex interaction exists between the optimum lubricant, the temperature, and the ram speed. If the speed is too low, the lubricant will produce thick coatings with low initial extrusion pressures, but this will rapidly exhaust the reservoir of lubricant and restrict the length that can be extruded. If the speed is too high, the coating may become dangerously thin. A modification of this technique is to apply thick films of glass only to the circumference of the billet to achieve hydrostatic lubrication.[2] In this case lubricant viscosity has a strong influence on extrusion pressure.

The lubricant film must be complete and continuous to be successful. Gaps in the film will serve to initiate shear zones which can develop into surface cracks. Also, lubricant film may be carried into the interior of the extrusion along shear bands to show up as longitudinal laminations in the product. Laminations of oxide can be created in the same way.

There are a number of other extrusion defects[3] which must be guarded against. Because of the inhomogeneous deformation in the direct extrusion of a billet, the center of the billet moves faster than the periphery. As a result, the dead metal zone extends down along the outer surface of the billet. After about two-thirds of the billet is extruded, the outer surface of the billet moves towards the centre and extrudes through the die near the axis of the rod. Since the surface of the billet often contains an oxidized skin, this type of flow results in internal oxide stringers. This defect can be considered to be an internal pipe, and it is known as the *extrusion defect*. On a transverse section through the extrusion this will appear as an annular ring of oxide. The tendency toward the formation of the extrusion defect increases as the container wall friction becomes greater. If a heated billet is placed in a cooler extrusion container, the outer layers of the billet will be chilled and the flow resistance of this region will increase. Therefore, there will be a greater tendency for the center part of the billet to extrude before the surface skin, and the tendency for formation of the extrusion defect is increased.

One way of avoiding the extrusion defect is to carry out the extrusion operation only to the point where the surface oxide begins to enter the die and then discard the remainder of the billet. This procedure may have serious economic

---

[1] J. A. Rogers and G. W. Rowe, *J. Inst., Met.*, vol. 95, pp. 257–263, 1967.
[2] K. M. Kulkarni, J. A. Schey, P. W. Wallace, and V. DePierre, *J. Inst. Met.*, vol. 100, pp. 33–39, 1972.
[3] R. N. Parkins, "Mechanical Treatment of Metals," pp. 232–239, American Elsevier Publishing Company, New York, 1968.

consequences since as much as 30 percent of the billet may remain at the point where the extrusion defect is encountered. An alternative procedure, which is frequently applied in the extrusion of brass, is to use a follower block which is slightly smaller than the inside diameter of the extrusion container. As the ram pushes the follower block forward, it scalps the billet and the oxidized surface layer remains in the container.

When extrusion is carried to the point at which the length of billet remaining in the container is about one-quarter its diameter, the rapid radial flow into the die results in the creation of an axial hole, or funnel, in the back end of the extrusion. This hole may extend for some distance into the back end of the extrusion, and therefore this metal must be discarded. The length of this defect can be reduced considerably by inclining the face of the ram at an angle to the ram axis.[1]

Surface cracking, ranging in severity from a badly roughened surface to repetitive transverse cracking called *fir-tree cracking* (Fig. 15-30b) can be produced by longtitudinal tensile stresses generated as the extrusion passes through the die. In hot extrusion this form of cracking usually is intergranular and is associated with hot shortness. The most common cause is too high a ram speed for the extrusion temperature. At lower temperatures where hot shortness cannot occur transverse cracking is believed[2] to be caused by momentary sticking in the die land and the sudden building up of pressure, and then breakaway.

Center burst, or chevron cracking (Fig. 15-30c) can occur at low extrusion ratios. It has been shown[3] that this defect is related to the influence of frictional conditions on the zone of deformation at the extrusion die. In this instance high frictional restraint at the tool-billet interface produces a sound product while center bursts occur when the friction is low.

Because of the nonuniform deformation produced in extrusion, considerable variation in hot-worked structure and properties, or in properties after heat treatment, may be expected. A common problem is variation in structure and properties from front to back end of the extrusion in both the longitudinal and transverse directions. Regions of exaggerated grain growth often are found in hot extrusions. These coarse-grain regions appear on the surface or towards the center, depending on the deformation conditions.

## 18-5   ANALYSIS OF THE EXTRUSION PROCESS

Using the uniform deformation energy approach (Sec. 15-2), the plastic work of deformation per unit volume can be expressed for direct extrusion as

[1] An analysis of this defect based on the slip-field theory has been given by W. Johnson, *Appl. Sci. Res.*, vol. 8A, pp. 52–60, 228–236, 1959.

[2] R. J. Wilcox and P. W. Whitton, *J. Inst. Met.*, vol. 88, p. 145, 1959–1960.

[3] F. J. Gurney and V. DePierre, *Trans. ASME, Ser. B: J. Eng. Ind.*, vol. 96, pp. 912–916, 1974.

$$U_p = \bar{\sigma} \int d\varepsilon = \bar{\sigma} \int_{A_0}^{A_f} d \ln A = \bar{\sigma} \ln \frac{A_f}{A_0} = -\bar{\sigma} \ln R \qquad (18\text{-}2)$$

The work involved is

$$W = U_p V = V\bar{\sigma} \ln R = pAL = \text{force} \times \text{distance} \qquad (18\text{-}3)$$

where $\bar{\sigma}$ is defined as the effective flow stress *in compression* so that

$$p = \frac{V}{AL} \bar{\sigma} \ln R = \bar{\sigma} \ln R \qquad (18\text{-}4)$$

This is an expression for the idealized extrusion pressure, since it considers neither friction nor redundant deformation. If we define the efficiency of the process $\eta$ as the ratio of the ideal to actual energy per unit volume, we can express the actual extrusion pressure $p_e$ by

$$p_e = \frac{p}{\eta} = \frac{\bar{\sigma}}{\eta} \ln R \qquad (18\text{-}5)$$

It will be noted that this is just another way of expressing Eq. (18-1).

Careful measurements of the forces in extrusion made by DePierre[1] show that the total extrusion force $P_e$ is the sum of the die force $P_d$, the friction force between the container liner and the upset billet $P_{fb}$, and the frictional force between the container liner and the follower $P_{ff}$. Generally, $P_{ff} \approx 0$. Expressed in terms of pressure,

$$p_e = p_d + \frac{4\tau_i L}{D} \qquad (18\text{-}6)$$

where $\tau_i$ = uniform interface shear stress between billet and container liner

$L$ = length of billet in the container liner

$D$ = inside diameter of the container liner

By directly measuring $P_e$ and $P_d$, it is possible to obtain good values for the frictional shear stress.

A simple approach to determining die pressure $p_d$ is to use a slab analysis to account for friction on extruding through a conical die.[2] Sachs[3] has performed this analysis for Coulomb sliding friction. The solution is directly analogous to that for wiredrawing through a conical die given in Sec. 19-3.

$$p_d = \sigma_{xb} = \sigma_0 \left( \frac{1+B}{B} \right) (1 - R^B) \qquad (18\text{-}7)$$

[1] V. DePierre, *Trans. ASME, J. Lubr. Technol.*, vol. 92, pp. 398–405, 1970.

[2] For a 180° flat die it is assumed that the material shears internally to form an effective cone angle of $2\alpha = 90°$.

[3] O. Hoffman and G. Sachs, "Introduction to the Theory of Plasticity for Engineers," pp. 176–186, McGraw-Hill Book Company, New York, 1953.

where $B = \mu \cot \alpha$

$\qquad \alpha =$ semidie angle

$\qquad R =$ extrusion ratio $= A_0/A$

While this analysis considers die friction, it does not allow for redundant deformation. However, extrusion probably is the most often studied problem using slip-line field theory.[1-5] Here the problem has been studied for plane-strain deformation without considering friction. All of these studies yield solutions of the form

$$p_d = \sigma_0(a + b \ln R) \qquad (18\text{-}8)$$

where typically $a = 0.8$ and $b = 1.5$ for axisymmetric extrusion.

Extensive work using upper-bound analysis has also contributed to an understanding of the problem. Kudo[6] found the following expression for extrusion through rough square dies ($2\alpha = 180°$):

$$p_d = \sigma_0(1.06 + 1.55 \ln R) \qquad (18\text{-}9)$$

Avitzur[7] produced a more generalized expression based on a spherical velocity field. This equation applies for lubricated extrusion through a die of semiangle $\alpha$.

$$p_d = \frac{2\sigma_0}{\sqrt{3}} \left( \frac{\alpha}{\sin^2\alpha} - \cot \alpha \right) + \sigma_0[2f(\alpha) + m \cot \alpha] \ln (r_0/r_f)$$

$$+ 2m\left[ \frac{L}{r_0} - \left(1 - \frac{r_f}{r_0}\right) \cot \alpha \right] \qquad (18\text{-}10)$$

where $m =$ interfacial friction factor

$\qquad f(\alpha) =$ complex function of semidie angle; $f(\alpha) = 1$ for small die angles

$\qquad L =$ length of land on exit from die

$\qquad r_0 =$ radius of billet

$\qquad r_i =$ radius of extruded rod

A more recent upper-bound analysis[8] has used a different velocity field which produces good agreement with experiments on hydrostatic extrusion.

Based on the above theoretical background and experimental studies of hot extrusion, DePierre[9] used the following equation to describe die pressure:

$$p_d = \sigma_0 (a + b \ln R) + mk \cot \alpha \ln R \qquad (18\text{-}11)$$

[1] R. Hill, *J. Iron Steel Inst. London*, vol. 158, pp. 177–185, 1948.
[2] W. Johnson, *J. Mech. Phys. Solids*, vol. 4, pp. 191–198, 1956.
[3] A. P. Green, *ibid.*, vol. 3, pp. 189–196, 1955.
[4] T. F. Jordan and E. G. Thomsen, *ibid.*, vol. 4, pp. 184–190, 1956.
[5] W. Johnson and H. Kudo, "The Mechanics of Metal Extrusion," Manchester University Press, 1962.
[6] H. Kudo, *Int. J. Mech. Sci.*, vol. 2, p. 102, 1960.
[7] B. Avitzur, *Trans. ASME, Sec. B: J. Eng. Ind.*, vol. 85, pp. 89–96, 1963; vol. 86, pp. 305–316, 1964.
[8] N. Ahmed, *Trans. ASME, Ser. B: J. Basic Eng.*, vol. 94, pp. 213–222, 1972.
[9] V. DePierre, *op. cit.*

where $m = \tau_i/k$ and $a$ and $b$ are evaluated as follows:

| $\alpha$ | $a$ | $b$ |
|---|---|---|
| 30 | 0.419 | 1.006 |
| 45 | 0.659 | 1.016 |
| 60 | 0.945 | 1.034 |

The flow distribution in extrusion has been measured[1] with billets gridded along the diametrial plane. Using the technique of visioplasticity,[2] it is possible to map out in detail the distribution of strain and strain rate and to calculate the variation of temperature and flow stress within the extrusion. Figure 18-7 shows the variation of *local* strain rate. Note that there are local maxima near the exit from the die on the surface, and along the centerline of the extrusion. The ambiguity of defining an average strain rate in a converging die was discussed in Sec. 15-5. The average strain rate for extrusion usually is defined by the time for material to transverse through a truncated conical volume of deformation zone defined by the billet diameter $D_b$ and the extrusion diameter $D_e$. For a 45° semicone angle,

$$V = \frac{\pi h}{3} \left( \frac{D_b^2}{4} + \frac{D_e^2}{4} + \frac{D_b D_e}{4} \right)$$

**Fig. 18-7**  Strain-rate distribution in a partially extruded steel billet.  $R = 16.5$, ram speed $= 210$ mm/s, average extrusion temperature $= 1440°$K.  (*From T. H. C. Childs, Met. Technol. N.Y. vol. 1, p. 306, 1974.*)

[1] T. H. C. Childs, *Met. Technol.*, vol. 1, pp. 305–309, 1974.
[2] A. H. Shabaik and E. G. Thomsen, *Ann. CIRP*, vol. 17, pp. 149–156, 1969.

where
$$h = (D_b - D_e)/2$$

so
$$V = \frac{\pi}{24} (D_b{}^3 - D_e{}^3)$$

For a ram velocity $v$, the volume extruded per unit time is
$$v \frac{\pi D_b{}^2}{4}$$

and the time to fill the volume of the deformation zone is
$$\frac{V}{v(\pi D_b{}^2/4)} = \frac{D_b{}^3 - D_e{}^3}{6v D_b{}^2}$$

and if $D_b \gg D_e$
$$\frac{D_b{}^3 - D_e{}^3}{D_b{}^2} \approx D_b$$

and
$$t = D_b/6v.$$

The time average mean strain rate is given by
$$\dot{\varepsilon}_t = \frac{\bar{\varepsilon}}{t} = \frac{6v \ln R}{D_b} \qquad (18\text{-}12)$$

and for the general semidie angle $\alpha$
$$\dot{\varepsilon}_t = \frac{6v D_b{}^2 \ln R \tan \alpha}{D_b{}^3 - D_e{}^3} \qquad (18\text{-}13)$$

## 18-6   COLD EXTRUSION AND COLD-FORMING

Cold extrusion is concerned with the cold forming[1] from rod and bar stock of small machine parts, such as spark plug bodies, shafts, pins, and hollow cylinders or cans. More properly, the subject should be expanded to include other cold-forming processes (such as upsetting, expanding, and coining) that are not strictly extrusion. Axisymmetric parts are particularly suited to cold-forming processes. Precision cold-forming can result in high production of parts with good dimensional control and good surface finish. Because of extensive strain hardening, it often is possible to use cheaper materials with lower alloy content. Extensive use is made of cold-formed low-alloy steels in the automotive industry.

The first major application of cold-forming of steel was the development during World War II of a process for making steel cartridge cases. The key was the

---

[1] C. H. Wick, "Chipless Machining," The Industrial Press, New York, 1961; H. D. Feldman, "The Cold Forging of Steel," Hutchinson & Co. (Publishers), Ltd., London, 1961; "Source Book on Cold Forming," American Society for Metals, Metals Park, Ohio, 1976.

development of a suitable lubricant system. For steel, a zinc phosphate conversion coating and soap is usually preferred. Other important factors are the use of a steel with high resistance to ductile fracture and the design of the tooling to minimize tensile-stress concentrations. The extensive literature on cold-forming has been reviewed by Watkins.[1]

## 18-7  HYDROSTATIC EXTRUSION

The concept of hydrostatic extrusion was introduced in Figure 15-25. Because the billet is subjected to uniform hydrostatic pressure, it does not upset to fill the bore of the container as it would in conventional extrusion. This means that the billet may have a large length-to-diameter ratio (even coils of wire can be extruded) or it may have an irregular cross section. Since there is no container-billet friction, the curve of the extrusion pressure vs. ram travel is nearly flat, like that for indirect extrusion in Fig. 18-5. Because of the pressurized fluid, lubrication is very effective, and the extruded product has good surface finish and dimensional accuracy. Since friction is nearly absent, it is possible to use dies with a very low semicone angle ($\alpha \approx 20°$) which greatly minimizes the redundant deformation.

Because hydrostatic extrusion employs a pressurized fluid, there is an inherent limitation to hot-working with this process. A practical limit on fluid pressure of around 250,000 psi currently exists because of the strength of the container and the requirement that the fluid not solidify at high pressure. This limits the obtainable extrusion ratio for mild steel to less than 20:1, while for very soft metals like aluminum, it is possible to achieve extrusion ratios in excess of 200:1. The extrusion of wire is one of the areas actively being pursued. Because of the large amount of stored energy in a pressurized fluid, the control of the extrusion on the exit from the die may be a problem. However, this is solved by *augmented* hydrostatic extrusion in which an axial force is applied either to the billet or to the extrusion. The fluid pressure is kept at less than the value required to cause extrusion, and the balance is provided by the augmenting force. In this way, much better control is obtained over the movement of the extrusion.

A number of methods have been developed to increase the rate of production of hydrostatic extrusion or to place it on a continuous basis. Thick-film hydrostatic extrusion[2] minimizes the amount of pressurized fliud, and because the billets can be precoated with the hydrostatic medium, the production rate approaches that for conventional extrusion. Fuchs[3] developed a continuous extrusion process which

[1] M. T. Watkins, *Int. Metall. Rev.*, vol. 18, September, December, 1973.
[2] R. J. Fiorentino, G. E. Meyer, and T. G. Byres, *Metallurgia and Metal Forming*, July, 1974, pp. 193–197; August 1974, pp. 210–213.
[3] F. J. Fuchs, *West. Electr. Eng.*, July–October 1971.

uses viscous drag of a flowing polymer to feed the billet rod into an extrusion chamber and through the extrusion die. Problems with pressure control have led to a modified design[1] of a continuous gear-driven extruder. Alexander[2] has discussed continuous hydrostatic extrusion from a general point of view.

## 18-8   EXTRUSION OF TUBING

With modern equipment, tubing may be produced by extrusion to tolerances as close as those obtained by cold-drawing.[3] To produce tubing by extrusion, a mandrel must be fastened to the end of the extrusion ram. The mandrel extends to the entrance of the extrusion die, and the clearance between the mandrel and the die wall determines the wall thickness of the extruded tube. Generally, a hollow billet must be used so that the mandrel can extend to the die. In order to produce concentric tubes, the ram and mandrel must move in axial alignment with the container and the die. Also, the axial hole in the billet must be concentric, and the billet should offer equal resistance to deformation over its cross section.

One method of extruding a tube is to use a hollow billet for the starting material. The hole may be produced either by casting, by machining, or by hot piercing in a separate press. Since the bore of the hole will become oxidized during

(a)

(b)

**Fig. 18-8**  Tube extrusion.
(a) Piercing; (b) extrusion.

[1] J. A. Gumprecht and I. G. Histand, *Tech. Paper* MF74-223, Society of Manufacturing Engineers, 1974.
[2] J. M. Alexander, *Mater. Sci. Eng.*, vol. 10. pp. 70–74, 1972.
[3] F. W. Morris, *J. Inst. Met.*, vol. 90, pp. 101–106, 1961–1962.

Fig. 18-9  Porthole extrusion die.

heating, the use of a hollow billet may result in a tube with an oxidized inside surface.

A more satisfactory method of extruding a tube is to use a solid billet which is pierced and extruded in one step in the extrusion press. With a modern extrusion press the piercing mandrel is actuated by a separate hydraulic system from the one which operates the ram. The piercing mandrel moves coaxially with the ram, but it is independent of its motion (Fig. 18-8). In the operation of a double-action extrusion press the first step is to upset the billet with the ram while the piercing mandrel is withdrawn. Next the billet is pierced with the pointed mandrel, ejecting a metal plug through the die. Then the ram advances and extrudes the billet over the mandrel to produce a tube.

A third method of extruding tubing, which is used with aluminum and magnesium alloys, is to use a solid billet and a porthole die with a standard extrusion ram without a mandrel. A sketch of a porthole die is shown in Fig. 18-9. The metal is forced to flow into separate streams and around the central bridge, which supports a short mandrel. The separate streams of metal which flow through the ports are brought together in a welding chamber surrounding the mandrel, and the metal exits from the die as a tube. Because the separate metal streams are joined within the die, where there is no atmospheric contamination, a perfectly sound weld is obtained. In addition to tubing, porthole extrusion is used to produce hollow unsymmetrical shapes in aluminum alloys.

## 18-9  PRODUCTION OF SEAMLESS PIPE AND TUBING

Pipe and tubing may be classified as seamless or welded, depending on the method of manufacture. Welded tubing is formed from strip and welded by hot-forming, fusion, or electric welding. As discussed in Sec. 18-7, extrusion is an excellent method of producing seamless pipe and tubing, especially for metals which are

Fig. 18-10   (a) Mannesmann mill; (b) plug rolling mill; (c) three-roll piercing mill; (d) reeling mill.

difficult to work.  However, there are other well-established processes[1] for producing seamless pipe and tubing which generally are more economical than extrusion.  Cold-working techniques for producing tubes are discussed in Chap. 19.

The Mannesmann mill (Fig. 18-10a) is used extensively for the rotary piercing of steel and copper billets.  The process employs two barrel-shaped driven rolls which are set at an angle to each other.  An axial thrust is developed as well as rotation to the billet.  Because of the low arc of contact with the billet, $\Delta$ is high and tensile stresses develop along the axis of the billet.  This assists in opening up the center of the billet as it flows around the piercing point to create the tube cavity.  Piercing is the most severe hot-working operation customarily applied to metals.

The Mannesmann mill does not provide sufficiently large wall reduction and elongation to produce finished hot-worked tubes.  Various types of plug rolling mills which drive the tube over a long mandrel containing a plug (Fig. 18-10b)

[1] T. Z. Blazynski, *Metall. Rev.*, vol. 15, no. 142, pp. 27–45, 1970.

have evolved. The Assel elongator,[2] which uses three conical driven rolls, has been widely adopted. This led to the development of three-roll piercing machines[3] (Fig. 18-10c) which produce more concentric tubes with smoother inside and outside surfaces than the older Mannesmann design. A *reeling mill* (Fig. 18-10d) which burnishes the outside and inside surfaces and removes the slight oval shape is usually one of the last steps in the production of pipe or tubing.

## BIBLIOGRAPHY

Bishop, J. F. W.: The Theory of Extrusion, *Metall. Rev.*, vol. 2, pp. 361–390, 1957.

Chadwick, Richard: The Hot Extrusion of Non-ferrous Metals, *Metall. Rev.*, vol. 4, pp. 189–255, 1959.

Haffner, E. K. L., and R. M. L. Elken: Extrusion Presses and Press Installations, *Metall. Rev.*, vol. 2, pp. 263–303, 1957.

Müller, E. "Hydraulic Extrusion Presses" (in English), Springer-Verlag, Berlin, 1961.

Pearson, C. E., and R. N. Parkins: "The Extrusion of Metals," 2d ed., John Wiley & Sons, Inc., New York, 1960.

[2] C. E. Snee, *Iron Steel Eng.*, vol. 33, p. 124, 1956.
[3] W. L. Roberts, *Ibid.*, vol. 52, pp. 56–63, 1975.

# DRAWING OF RODS, WIRES, AND TUBES

## 19-1  INTRODUCTION

Drawing operations involve pulling metal through a die by means of a tensile force applied to the exit side of the die. Most of the plastic flow is caused by compression force which arises from the reaction of the metal with the die. Usually the metal has a circular symmetry, but this is not an absolute requirement. The reduction in diameter of a solid bar or rod by successive drawing is known as *bar*, *rod*, or *wiredrawing*, depending on the diameter of the final product. When a hollow tube is drawn through a die without any mandrel to support the inside of the tube, this is known as *tube sinking*. When a mandrel or plug is used to support the inside diameter of the tube as it is drawn through a die, the process is called *tube drawing*. Bar, wire, and tube drawing are usually carried out at room temperature. However, because large deformations are usually involved, there is considerable temperature rise during the drawing operation.

## 19-2  ROD AND WIREDRAWING

The principles involved in the drawing of bars, rod, and wire are basically the same, although the equipment that is used is different for the different-sized products. Rods and tubes, which cannot be coiled, are produced on drawbenches (Fig. 19-1a). The rod is pointed with a swager, inserted through the die, and clamped to the jaws of the drawhead. The drawhead is moved either by a chain drive or by a hydraulic mechanism. Drawbenches with 300,000 lb pull and 100 ft of runout are available. Draw speeds vary from about 30 to 300 ft/min.

The cross section through a typical conical drawing die is shown in Fig. 19-1b. The entrance angle of the die is made large enough to allow room for the lubricant that adheres to the die. The *approach angle* is the section of the die where the actual reduction in diameter occurs. The *bearing* surface serves to guide the rod or wire as it exits from the die. An important characteristic of a drawing die is the *half-die angle*, denoted by α. At the present time most drawing dies are made from tungsten carbide because it provides long die life.

Wiredrawing starts with hot-rolled *wire rod*. The rod is first cleaned by pickling to remove any scale which would lead to surface defects and excessive die wear. For the production of steel wire the next step consists in coating the wire rod with lime or plating it with a thin layer of copper or tin. The lime serves as an

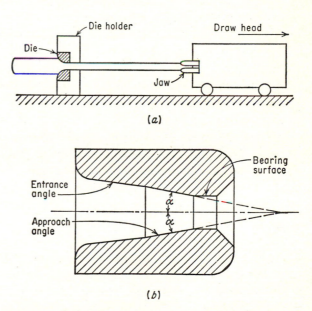

Fig. 19-1  (a) Schematic drawing of a drawbench; (b) cross section of a drawing die.

**Fig. 19-2** Wire-drawing equipment (schematic).

absorber and carrier of the lubricant during *dry drawing*, and it also serves to neutralize any acid remaining from pickling. In dry drawing the lubricant is grease or soap powder, while in *wet drawing* the entire die is immersed in a lubricating fluid.

Long lengths of thinner bar or wire (usually below $\frac{1}{2}$-in diameter) are produced by drawing with a *bull block* (Fig. 19-2). For coarse wire a single bull block is used, but for fine wire a large number of draw blocks are required, with the wire passing through a number of dies until it is reduced to its final size in one continuous operation. For fine wires reductions per pass of 15 to 25 percent are used, while for coarse wires the reduction per pass may be 20 to 50 percent. Drawing speeds of modern wiredrawing equipment may exceed 5,000 ft/min.

Nonferrous wire and low-carbon steel wire are produced in a number of tempers ranging from dead soft to full hard. Depending on the metal and the reductions involved, intermediate anneals may be required. Steel wire with a carbon content greater than 0.25 percent is given a special *patenting* heat treatment. This consists in heating above the upper critical temperature and then cooling at a controlled rate or transforming in a lead bath at a temperature around 600°F to cause the formation of fine pearlite. Patenting produces the best combination of strength and ductility for the successful drawing of high-carbon music and spring wire.

Defects in rod and wire can result from defects in the starting rod (seams, slivers, and pipe) or from the deformation process itself. The most common type of drawing defect is center burst, or chevron cracking (Fig. 15-30c). This also is called *cupping*. An upper-bound analysis[1] is capable of identifying the combinations of semidie angle and reduction for which less deformation energy is required if a hole forms at the centerline. This analysis predicts that center burst fracture will

[1] B. Avitzur, "Metal Forming: Process and Analysis," pp. 172–176, 240–241, McGraw Hill Book Company, New York, 1968.

occur for low die angles at low reductions, and as $\alpha$ increases the critical reduction for freedom from center burst increases. For a given reduction and die angle, the critical reduction to prevent fracture increases with the friction.

## 19-3 ANALYSIS OF WIREDRAWING

Although wiredrawing appears to be one of the simplest metalworking processes, a complete analysis that enables calculation of the draw force to better than $\pm 20$ percent of the observed value is a rather difficult problem. We shall use wiredrawing as an example of how to add to the model to account for realistic conditions.

We have already seen that the uniform-deformation-energy method predicts a draw stress given by Eq. (15-13).

$$\sigma_{xa} = \bar{\sigma}_0 \ln \frac{A_b}{A_a} = \bar{\sigma}_0 \frac{1}{1 - r} \qquad (19\text{-}1)$$

This equation not only neglects friction, but it neglects the influence of transverse stresses and of redundant (shearing) deformation.

As a first step we shall consider the problem of strip drawing (see Sec. 15-2 and Fig. 15-3) where a Coulomb friction coefficient $\mu$ exists between the strip and the die. The friction stress $\mu p$ opposes the motion of the strip through the die (Fig. 19-3). Referring to Eq. (15-6), the equilibrium of forces in the $x$ direction is

$$\sigma_x \, dh + h \, d\sigma_x + 2p \tan \alpha \, dx + 2\mu p \, dx = 0 \qquad (19\text{-}2)$$

and since $dh = 2 \, dx \tan \alpha$,

$$\sigma_x \, dh + h \, d\sigma_x + p(1 + \mu \cot \alpha) \, dh = 0 \qquad (19\text{-}3)$$

Since the yield condition of plane strain is $\sigma_x + p = \sigma_0'$ and $B = \mu \cot \alpha$, the differential equation for strip drawing is

$$\frac{d\sigma_x}{\sigma_x B - \sigma_0'(1 + B)} = \frac{dh}{h} \qquad (19\text{-}4)$$

**Fig. 19-3** Stresses acting on an element of wire in plane-strain strip drawing.

If $B$ and $\sigma'_0$ are both constant, Eq. (19-4) can be integrated directly to give the draw stress $\sigma_{xa}$.

$$\sigma_{xa} = \sigma'_0 \frac{1+B}{B} \left[ 1 - \left(\frac{h_a}{h_b}\right)^B \right] = \sigma'_0 \frac{1+B}{B} \left[ 1 - (1-r)^B \right] \tag{19-5}$$

However, wiredrawing is conducted with conical dies. An analysis[1] following that for strip drawing but integrating around the circumference of the die results in

$$\sigma_{xa} = \sigma_0 \frac{1+B}{B} \left[ 1 - \left(\frac{D_a}{D_b}\right)^{2B} \right] \tag{19-6}$$

The surface area of contact between the wire and the die (Fig. 19-4) is given by[2]

$$S = \frac{A_b - A_a}{\sin \alpha} \tag{19-7}$$

and the mean normal pressure on this area is $\bar{p}$. The forces acting in the axial direction are given in Fig. 19-4. The draw force $P_d$ is balanced by the horizontal component of the frictional force and the horizontal component of the normal pressure.

$$P_d = \mu \bar{p} S \cos \alpha + \bar{p} S \sin \alpha \tag{19-8}$$

$$P_d = \bar{p} S (\mu \cos \alpha + \sin \alpha) = \bar{p} \frac{A_b - A_a}{\sin \alpha} (\mu \cos \alpha + \sin \alpha)$$

$$P_d = \bar{p}(A_b - A_a)(\mu \cot \alpha + 1) = \bar{p}(A_b - A_a)(1 + B) \tag{19-9}$$

In the absence of friction, $B = 0$ and

$$P_d = \bar{p}(A_b - A_a) = \bar{\sigma}_0 A_a \ln \frac{A_b}{A_a}$$

Fig. 19-4  Forces acting on a conical element.

---

[1] O. Hoffman and G. Sachs, "Introduction to the Theory of Plasticity for Engineers," McGraw-Hill Book Company, pp. 176–180, 1953.
[2] R. W. Johnson and G. W. Rowe, *J. Inst. Met.*, vol. 96, p. 105, 1968.

from Eq. (19-1), so that the draw stress *with friction* is given by[1]

$$\sigma_{xa} = \frac{P_d}{A_a} = \bar{\sigma}_0 \ln \frac{A_b}{A_a} (1 + B) \qquad (19\text{-}10)$$

Equation (19-10) includes a term for uniform deformation energy and a term for frictional energy. The redundant deformation can be taken into consideration by including a factor $\phi$ which allows for the influence of redundant deformation in raising the flow stress of the material.

$$\sigma_{xa} = \phi \bar{\sigma}_0 \ln \frac{A_b}{A_a}(1 + B) \qquad (19\text{-}11)$$

The redundant work factor is defined by

$$\phi = f(\alpha, r) = \frac{\varepsilon^*}{\varepsilon} \qquad (19\text{-}12)$$

where $\phi$ = the redundant work factor

$\varepsilon^*$ = the "enhanced strain" corresponding to the yield stress of the metal which has been homogenously deformed to a strain $\varepsilon$

The redundant work factor may be determined in a straightforward manner for drawn wires[2] as shown in Fig. 19-5. The flow curve of a drawn wire is superimposed on the flow curve for the annealed metal. The origin of the curve for the drawn metal is displaced along the strain axis by an amount equal to the drawing reduction, $\varepsilon = \ln (A_b/A_a) = \ln [1/(1 - r)]$. The fact that the yield stress for the drawn wire is above the basic flow curve for the material is due to the redundant work that it received. To determine $\phi$, the flow curve for the drawn metal is moved to

**Fig. 19-5**  Procedure for determining redundant deformation of drawn wire.

[1] J. G. Wistreich, *Proc. Inst. Mech. Eng. (London)*, vol. 169, pp. 654–665, 1955.
[2] The use of microhardness measurements to determine $\phi$ is discussed by W. A. Backofen, "Deformation Processing," pp. 138–140, Addison-Wesley Publishing Company, Inc., Reading, Mass., 1972.

the right to $\varepsilon^*$ where the curves coincide. The area swept out in this procedure is the redundant work per unit of volume. An alternative approach[1] is to use upper-bound analysis to account for the redundant deformation by assuming simple yet reasonable deformation regions. Unfortunately, this leads to complex equations.

$$\sigma_{xa} = \sigma_0 \frac{\left\{ 2f(\alpha) \ln \frac{R_b}{R_a} + \frac{2}{\sqrt{3}} \left( \frac{\alpha}{\sin^2 \alpha} - \cot \alpha \right) + 2\mu \left( \cot \alpha \left[ 1 - \ln \frac{R_b}{R_a} \right] \ln \frac{R_b}{R_a} + \frac{L}{R_a} \right) \right\}}{1 + 2\mu(L/R_a)}$$

(19-13)

where $f(\alpha)$ = a complex function of semidie angle
$\quad L$ = the length of the die land
$\quad R_b$ = the radius of the billet
$\quad R_a$ = the wire radius

The redundant work factor has been shown[2] to be linearly related to the reduction, the semidie-angle, and the strain-hardening behavior of the material.

$$\phi = C_1 + C_2 \, \Delta = C_1 + C_2 \frac{D_b + D_a}{D_b - D_a} \sin \alpha \qquad (19\text{-}14)$$

where $C_1$ and $C_2$ are functions of $\sigma_0{}^x n^y$ for a strain-hardening relation $\sigma = \sigma_0(\varepsilon_0 + \varepsilon)^n$.

Equation (19-11) neglects strain hardening by using a mean value for the flow stress. Various equations for strain hardening have been used[3] to evaluate $\bar{\sigma}_0$, but the most direct approach is to use an experimental flow curve to evaluate $\int \sigma \, d\varepsilon$ and an analog computer[4] to arrive at the solution.

The expression for draw stress given in Eq. (19-11) contains terms for homogeneous deformation, friction, and redundant work. The last two terms are functions of semidie angle $\alpha$. The effect of die angle on these components of the total energy required to cause deformation is shown schematically in Fig. 19-6. The ideal work of plastic deformation $U_p$ is independent of die angle. For a fixed coefficient of friction, the work required to overcome friction $U_f$ decreases with increasing $\alpha$ as shown by Eqs. (19-7) and (19-8). On the other hand, the redundant work $U_r$ increases with increasing $\alpha$ as shown by Eq. (19-14). The summation of these components of total energy $U_T$ leads to a curve which has a minimum at some optimum die angle $\alpha^*$. Increasing the reduction and the friction raise the optimum die angle. Since in hot-working the reduction and friction usually are higher than in cold-working processes, the die angles tend to be greater.

[1] B. Avitzur, *Trans. ASME, Ser. B, J. Eng. Ind.*, vol. 86, pp. 305–316, 1964.
[2] R. M. Caddell and A. G. Atkins, *Trans. ASME, Ser. B: J. Eng. Ind.*, vol. 90, pp. 411–419, 1968; R. W. Johnson and G. W. Rowe, op. cit.
[3] A. G. Atkins and R. M. Caddell, *Int. J. Mech. Sci.*, vol. 10, pp. 15–28, 1968.
[4] G. W. Rowe, "An Introduction to the Principles of Metalworking," chap. 12, St Martin's Press, Inc., New York, 1965.

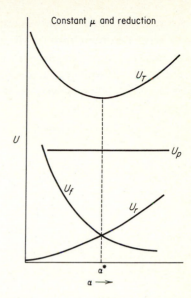

Fig. 19-6 Components of total energy of deformation.

We now shall examine the limit of drawability for steady-state wiredrawing.[1] Equation (19-11) can be expressed most simply by

$$\sigma_{xa} = \frac{1}{\eta} \int \sigma \, d\varepsilon \qquad (19\text{-}15)$$

where $\eta = U_p/U_T$ is the efficiency of the deformation process. Figure 19-7 plots the flow curve of the material, $\int \sigma \, d\varepsilon$ for an ideal process without friction or redundant work, and $1/\eta \int \sigma \, d\varepsilon$ for the actual drawing process. For a given strain $\varepsilon = \ln (A_b/A_a)$ produced by the die the values of draw stress $\sigma_{xa}$ and flow stress

Fig. 19-7 Development of limit on drawability. (*After Backofen.*)

[1] W. A. Backofen, op. cit., pp. 227–230; R. M. Caddell and A. G. Atkins, *Trans. ASME, Ser. B: J. Eng. Ind.*, vol. 91, pp. 664–672, 1969.

$\sigma_\varepsilon$ are as indicated. The weakest link is the wire which has exited from the die. On being deformed through the die the material is strain-hardened. For a severely strain-hardened material the formation of a neck is quickly followed by fracture. Therefore, the drawing limit is reached when $\sigma_d = \sigma_\varepsilon$ as shown in Fig. 19-7. If the material follows a power-law hardening relationship $\sigma_\varepsilon = K\varepsilon^n$, then Eq. (19-15) becomes

$$\sigma_d = \frac{K\varepsilon^{n+1}}{\eta(n+1)} = \frac{\sigma_\varepsilon \varepsilon}{\eta(n+1)} \qquad (19\text{-}16)$$

Substituting the criterion for the maximum drawing strain in a single pass, that is, $\sigma_d = \sigma_\varepsilon$,

$$\varepsilon_{max} = \eta(n+1) \qquad (19\text{-}17)$$

Since $\varepsilon = \ln(A_b/A_a)$,

$$\left(\frac{A_b}{A_a}\right)_{max} = e^{\eta(n+1)} \qquad (19\text{-}18)$$

and by definition the reduction is $r = 1 - (A_a/A_b)$

$$r_{max} = 1 - e^{-\eta(n+1)} \qquad (19\text{-}19)$$

For wire subjected to repeated reductions through a series of dies, $n$ will decrease toward zero and the allowable reduction will decrease. A comparison of the maximum strain in drawing with that in stretching (tension) shows the important role played by the strain-hardening exponent.

$$\left(\frac{\varepsilon(\text{drawing})}{\varepsilon(\text{stretch})}\right)_{max} = \frac{\eta(n+1)}{n} \qquad (19\text{-}20)$$

Typically the limiting strain in drawing is at least twice that in pure tension.

## 19-4 TUBE-DRAWING PROCESSES

Hollow cylinders, or tubes, which are made by hot-forming processes such as extrusion or piercing and rolling (see Chap. 18), often are cold finished by drawing. Cold-drawing is used to obtain closer dimensional tolerances, to produce better surface finishes, to increase the mechanical properties of the tube by strain hardening, to produce tubes with thinner walls or smaller diameters than can be obtained with hot-forming methods, and to produce tubes of irregular shapes.

The three basic types of tube-drawing processes are *sinking*, *plug drawing*, and *mandrel drawing*. Since the inside of the tube is not supported in tube sinking (Fig. 19-8a), the wall thickens slightly and the internal surface becomes uneven.

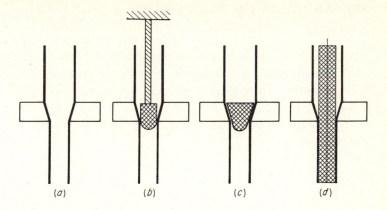

**Fig. 19-8**   Methods of tube drawing. (*a*) Sinking; (*b*) fixed plug; (*c*) floating plug; (*d*) moving mandrel.

Because the shearing at the entry and exit of the die is large, the redundant strain is higher for sinking and the limiting deformation is lower than for other tube-producing processes. Both the inner and outer diameters of the tube are controlled in drawing on a plug (Fig. 19-8*b*). The plug may be either cylindrical or conical. The plug controls the size and shape of the inside diameter and produces tubing of greater dimensional accuracy than in tube sinking. Because of the increased friction from the plug, the reduction in area seldom exceeds 30 percent. The situation where a carefully matched plug floats in the die throat is shown in Fig. 19-8*c*. Properly designed *floating plugs* can give a reduction in area of 45 percent, and for the same reduction the drawing loads are lower than for drawing with a fixed plug. An important feature of this design is that it is possible to draw and coil long lengths of tubing. However, tool design and lubrication can be very critical. Problems with friction in tube drawing are minimized in drawing with a long *mandrel* (Fig. 19-8*d*). The mandrel consists of a long hard rod or wire that extends over the entire length of the tube and is drawn through the die with the tube. In tube drawing with a moving mandrel the draw force is transmitted to the metal partly by the pull on the exit section and partly by the friction forces acting along the tube-mandrel interface. Since the mandrel is moving with a velocity equal to that of the tube on exiting from the die, and since this is higher than the velocity of the metal confined in the die channel, there is a *forward frictional drag* at the interface between the mandrel and the tube which tends to cancel the backward frictional drag between the stationary die and the tube. However, after drawing, the mandrel must be removed from the tube by rolling (reeling), which increases the tube diameter slightly and disturbs the dimensional tolerances.

## 19-5 ANALYSIS OF TUBE DRAWING

In the tube drawing with a plug or mandrel the greatest part of the deformation occurs as a reduction in wall thickness. Usually the inside diameter is reduced by a small amount equal to that needed to allow for insertion of the plug or mandrel before drawing. Thus, there is no hoop strain and the analysis[1] can be based on plane-strain conditions.

For tube drawing with a plug, the draw stress, by analogy with Eq. (19-5), can be expressed by

$$\sigma_{xa} = \sigma_0' \frac{1 + B'}{B'} \left[ 1 - \left( \frac{h_a}{h_b} \right)^{B'} \right] \qquad (19\text{-}21)$$

where $B' = \dfrac{\mu_1 + \mu_2}{\tan \alpha - \tan \beta}$

and $\mu_1$ = coefficient of friction between tube and die wall

$\mu_2$ = coefficient of friction between tube and plug

$\alpha$ = semicone angle of die

$\beta$ = semicone angle of the plug; for a cylindrical plug, $\beta = 0$

In tube drawing with a floating plug the question of proper design of the plug is more important than the magnitude of the drawing stress. This has been discussed by Blazynski.[2]

In tube drawing with a moving mandrel, the friction forces at the stationary die-tube interface are directed toward the die entrance as for wiredrawing and tube drawing with a plug. However, as discussed earlier, the friction forces at the mandrel-tube interface are directed toward the exit of the die. Therefore, for a moving mandrel $B'$ must be expressed as

$$B' = \frac{\mu_1 - \mu_2}{\tan \alpha - \tan \beta}$$

The draw stress is given by Eq. (19-21) with the proper value of $B'$. If $\mu_1 = \mu_2$, as may often be the case, then $B' = 0$. The differential equation of equilibrium for this simple case is

$$h \, d\sigma_x + (\sigma_x + p) \, dh = 0$$

which integrates directly to the equation for ideal homogeneous deformation Eq. (19-1). However, it is entirely possible for the coefficient of fricton on the mandrel $\mu_2$ to exceed that at the die $\mu_1$, so that $B$ is negative. This situation results in a draw stress less than required by frictionless ideal deformation.

---

[1] Hoffman and Sachs, op. cit., pp. 190–195.
[2] T. Z. Blazynski, *Metall. Rev.*, no. 142, April, 1970; D. J. Smith and A. N. Bramley, *Proc. 13th Int. Machine Tool Design and Research Conf.*, Birmingham, England, 1972.

Tube sinking is the process by which the inside diameter of the tube is decreased. Although we have neglected this aspect of the deformation until now, some reduction in inside diameter is part of the tube-drawing process.

The stresses involved in tube sinking have been analyzed by Sachs and Baldwin[1] on the assumption that the wall thickness of the tube remains constant. The equation for the draw stress at the die exit is analogous with the equation describing the draw stress in wiredrawing. The cross-sectional area of the tube is related to the midradius $r$ and the wall thickness $h$ by $A \approx 2\pi rh$.

$$\sigma_{xa} = \sigma_0'' \frac{1+B}{B} \left[ 1 - \left( \frac{A_f}{A_b} \right)^B \right] \qquad (19\text{-}22)$$

The yield stress $\sigma_0''$ is taken equal to $1.1\sigma_0$ to account for the complex stresses in tube sinking. A more complete analysis of tube sinking has been given by Swift.[2]

## 19-6 RESIDUAL STRESSES IN ROD, WIRE, AND TUBES

Two distinct types of residual-stress patterns are found in cold-drawn rod and wire, depending on the amount of reduction. For reductions per pass of less than about 1 percent the longitudinal residual stresses are compressive at the surface and tensile at the axis, the radial stresses are tensile at the axis and drop off to zero at the free surface, while the circumferential stresses follow the same trend as the longitudinal residual stresses. For larger reductions of commercial significance the residual-stress distribution is completely reversed from the first type of stress pattern. In this case the longitudinal stresses are tensile at the surface and compressive at the axis of the rod, the radial stresses are compressive at the axis, and the circumferential stresses follow the same pattern as the longitudinal stresses. The first type of residual-stress pattern is characteristic of forming operations where the deformation is localized in the surface layers.

The effect of die angle and the amount of reduction per pass on the longitudinal residual stress in cold-drawn brass wire was investigated by Linicus and Sachs.[3] Figure 19-9 shows that for a given reduction the longitudinal residual stress increases with the half-die angle. Maximum values of residual stress are obtained for reductions in the region of 15 to 35 percent.

For tubes produced by tube sinking, under conditions where the deformation is relatively uniform throughout the tube wall, the longitudinal residual stresses

[1] G. Sachs and W. M. Baldwin, Jr., *Trans. ASME*, vol. 68, pp. 655–662, 1946; also see Hoffman and Sachs, op. cit., pp. 252–255.
[2] H. W. Swift, *Philos. Mag.*, vol. 40, ser. 7, pp. 883–902, 1949.
[3] W. Lincius and G. Sachs, *Mitt. Dtsch. Materialprüfungsanst.*, vol. 16, pp. 38–67, 1932.

**Fig. 19-9** Longitudinal residual stresses in cold-drawn brass wire. (*After Linicus and Sachs.*)

are tensile on the outer surface and compressive on the inner surface of the tube. The residual stresses in the circumferential direction follow the same pattern, while the stresses in the radial direction are negligible. Approximate measurements[1] of the circumferential stresses on the outer surface of sunk tubes indicate that the stresses increase with increasing reduction in diameter at the same rate at which the yield stress is increased by the cold-work.

Studies[2] of residual stresses in tubes produced by drawing over a plug and mandrel showed the same distribution of residual stresses as for tube sinking. An important result was that a substantial reduction in the level of residual stress could be produced by tandem drawing, whereby a small reduction (2 percent) was produced by a second die immediately following the main reduction. A systematic approach to developing tube-drawing schedules has been proposed[3] based on the criterion of developing zero circumferential residual stress.

[1] D. K. Crampton, *Trans. Metall. Soc. AIME*, vol. 89, pp. 233–255, 1930.
[2] S. K. Misra and N. H. Polakowski, *Trans. ASME, Ser. D: J. Basic Eng.* vol. 91, pp. 810–815 1969.
[3] B. J. Meadows, and A. G. Lawrence, *J. Inst. Met.*, vol. 98, pp. 102–105, 1970.

# BIBLIOGRAPHY

Hoffman, O., and G. Sachs: "Introduction to the Theory of Plasticity for Engineers," chaps. 15–17, McGraw-Hill Book Company, New York, 1953.

Pomp, A.: "The Manufacture and Properties of Steel Wire," The Wire Industry Ltd., London, 1954.

Rowe, G. W.: "An Introduction to the Principles of Metalworking," chaps. 6 and 7 St Martin's Press, Inc., New York, 1965.

"Tube Producing Practice," American Institute of Mining and Metallurgical Engineers New York, 1951.

Wistreich, J. G.: The Fundamentals of Wire Drawing, *Metall. Rev.*, vol. 3, pp. 97–142, 1958.

# 20

## SHEET-METAL FORMING

### 20-1  INTRODUCTION

The ability to produce a variety of shapes from flat sheets of metal at high rates of production has been one of the real technological advances of the twentieth century. This transition from hand-forming operations to mass-production methods has been an important factor in the great improvement in the standard of living which occurred during the period.

In essence, a shape is produced from a flat blank by stretching and shrinking the dimensions of all its volume elements in the three mutually perpendicular principal directions. The resulting shape is then the result of the integration of all the local stretching and shrinking of the volume elements. Attempts have been made to classify the almost limitless number of shapes which are possible in metal forming into definite categories depending on the contour of the finished part. Sachs[1] has classified sheet-metal parts into five categories.

---

[1] G. Sachs, "Principles and Methods of Sheet-Metal Fabricating," pp. 9–14, Reinhold Publishing Corporation, New York, 1951.

1. Singly curved parts
2. Contoured flanged parts—including parts with stretch flanges and shrink flanges
3. Curved sections
4. Deep-recessed parts—including cups and boxes with either vertical or sloping walls
5. Shallow-recessed parts—including dish-shaped, beaded, embossed, and corrugated parts

Typical examples of these parts are shown in Fig. 20-1. Another classification system, developed in the automotive industry, groups sheet-steel parts into categories depending on the severity of the forming operation.[2] Severity of the operation is based on the maximum amount of bending or stretching in the part.

(a)

(b)

(c)

(d)

(e)

(f)

**Fig. 20-1**  Typical formed shapes. (a) Singly curved; (b) stretch flange; (c) shrink flange; (d) curved sections; (e) deep-drawn cup; (f) beaded section.

[2] "Metals Handbook," 8th ed., vol. 1, pp. 319–335, American Society for Metals, Metals Park, Ohio, 1961.

Still another way of classifying sheet-metal forming is by means of specific operations such as bending, shearing, deep drawing, stretching, ironing, etc. Most of these operations have been illustrated briefly in Fig. 15-1, and they will be discussed in considerably greater detail in this chapter.

## 20-2 FORMING METHODS

The old method of hand forming of sheet metal is today used primarily as a finishing operation to remove wrinkles left by forming machines. In the metalworking industries hand forming is primarily limited to experimental work where only a few identical pieces are required.

Most high-production-volume sheet-metal forming is done on a press, driven by either mechanical or hydraulic action. In mechanical presses energy is generally stored in a flywheel and is transferred to the movable slide on the downstroke of the press. Mechanical presses are usually quick-acting and have a short stroke, while hydraulic presses are slower-acting but can apply a longer stroke. Presses are usually classified according to the number of slides which can be operated independently of each other. In the *single-action press* there is only one slide, generally operating in the vertical direction. In the *double-action press* there are two slides. The second action ordinarily is used to operate the *hold-down*, which prevents wrinkling in deep drawing. A *triple-action press* is equipped with two actions above the die and one action below the die.

The basic tools used with a metalworking press are the *punch* and the *die*.[1] The punch is the convex tool which mates with the concave die. Generally the punch is the moving element. Because accurate alignment between the punch and die is usually required, it is common practice to mount them permanently in a *subpress*, or *die set*, which can quickly be inserted in the press. An important consideration in tooling for sheet-metal forming is the frequent requirement for a clamping pressure, or hold-down, to prevent wrinkling of the sheet as it is being formed. Hold-down can best be provided by a *hold-down ring*, which is actuated by the second action of a double-action press. However, by using mechanical springs or an auxiliary air cylinder, hold-down can be provided in a single-action press.

Frequently punches and dies are designed so that successive stages in the forming of the part are carried out in the same die on each stroke of the press. This is known as *progressive forming*. A simple example is a progressive blanking and piercing die to make a plain, flat washer (Fig. 20-2). As the strip is fed from

---

[1] "Die Design Handbook," 2d ed., McGraw-Hill Book Company, New York, 1965.

**Fig. 20-2** Progressive piercing and blanking die.

left to right, the hole for the washer is first punched and then the washer is blanked from the strip. At the same time as the washer is being blanked from the strip, the punch $A$ is piercing the hole for the next washer. The stripper plate is used to prevent the metal from separating from the die on the up stroke of the punch. The die materials depend on the severity of the operation and the required production run.[1] In aircraft work, where production runs are often small, tooling is frequently made from a zinc-base alloy called Kirksite or from wood or epoxy resins. For long die life, however, tool steel is required.

A forging hammer can be used for sheet-metal forming in place of a mechanical press. Drop-hammer forming differs from forming with a conventional press since usually many blows of varying force will be applied rather than a single stroke of fixed length or pressure. While a drop hammer is equivalent only to a single-action press, hold-down action can usually be provided by auxiliary devices. Because of the flexibility of the drop hammer, it is often used on short production runs in the aircraft industry. However, since the force cannot be controlled as accurately as in the double-action mechanical or hydraulic press, the drop hammer is not particularly well suited to the most severe forming operations.

---

[1] "Metals Handbook," 8th ed., vol. 1, pp. 685–717, American Society for Metals, Metals Park, Ohio, 1961.

**Fig. 20-3**  Rubber forming.

The *press brake* is a single-action press with a very long and narrow bed. The chief purpose of a press brake is to form long, straight bends in pieces such as channels and corrugated sheets. *Roll forming* (Chap. 17) is another common method of producing bent shapes in long lengths. The roll-forming process is also used to produce thin-wall cylinders from flat sheet.

*Rubber hydroforming* is a modification of the conventional punch and die in which a pad of rubber serves as the die. Rubber forming, or the *Guerin process*, is illustrated in Fig. 20-3. A form block (punch) is fastened to the bed of a single-action hydraulic press, and a thick blanket of rubber is placed in a retainer box on the upper platen of the press. When a blank is placed over the form block and the rubber forced down on the sheet, the rubber transmits a nearly uniform hydrostatic pressure against the sheet. A unit pressure of around 1,500 psi is sufficient for most parts, and higher local pressures can be provided by auxiliary tooling.[1] Rubber forming is used extensively in the aircraft industry. Shallow flanged parts with stretch flanges are readily produced by this method, but shrink flanges are limited because the rubber provides little resistance to wrinkling. Another limitation is that the blank tends to move on the form block unless holes for positioning pins are provided in the part.

A variety of methods are used to bend or to contour-form straight sections.[2] Cylindrical- and conical-shaped parts are produced with *bending rolls* (Fig. 20-4a). A three-roll bender is not very well suited to preventing buckling in thin-gage sheet. Often a fourth roll is placed at the exit to provide an extra adjustment in curvature. In three-point loading the maximum bending moment is at the midpoint of the span. This localization of strain can result, under certain circumstances, in the forming limit being reached at the midpoint before the rest of the part is bent to the proper contour. More uniform deformation along the length of the part is

[1] Sachs, op. cit., pp. 424–455.
[2] *Ibid.*, pp. 476–493.

**Fig. 20-4** Methods of bending and contouring. (*a*) Three-roll bender; (*b*) wiper-type benders; (*c*) wrap forming.

obtained with *wiper-type equipment*. In its simplest form this consists of a sheet which is clamped at one end against a form block; the contour is progressively formed by successive hammer blows, starting near the clamp and moving a short distance toward the free end with each blow. A wiper-type bender is sketched in Fig. 20-4*b*. In this case the form block or die has a nonuniform contour so that the wiper rolls must be pressed against the block with a uniform pressure supplied by a hydraulic cylinder. For a form block of constant radius, as in a power-driven tube bender, the clearance between the die and the wiper rolls remains constant, and there is no need for the hydraulic-actuated rolls. Still a third method of producing contours is by *wrap forming*. In wrap forming the sheet is compressed against a form block, and at the same time a longitudinal tensile stress is applied to prevent buckling and wrinkling (Fig. 20-4*c*). A simple example of wrap forming is the coiling of a spring around a mandrel. The stretch forming of curved sections is a special case of wrap forming.

A method of making tank heads, television cones, and other deep parts of circular symmetry is *spinning* (Fig. 20-5). The metal blank is clamped against a

**Fig. 20-5**   Manual spinning.

form block which is rotated at high speed. The blank is progressively formed against the block, either with a manual tool or by means of small-diameter work rolls.

   *Explosive forming* is well suited to producing large parts with a relatively low production lot size. The sheet-metal blank is placed over a die cavity and an explosive charge is detonated in water at an appropriate standoff distance from the blank. The shock wave propagating from the explosion serves as a "frictionless punch" to deform the blank.

## 20-3   SHEARING AND BLANKING

Shearing is the separation of metal by two blades moving as shown in Fig. 20-6*a*. In shearing, a narrow strip of metal is severely plastically deformed to the point where it fractures at the surfaces in contact with the blades. The fracture then propagates inward to provide complete separation. The depth to which the punch must penetrate to produce complete shearing is directly related to the ductility of the metal. The penetration is only a small fraction of the sheet thickness for brittle materials, while for very ductile materials it may be slightly greater than the thickness.

   The clearance between the blades is an important variable in shearing operations. With the proper clearance the cracks that initiate at the edges of the blades will propagate through the metal and meet near the center of the thickness to provide a clean fracture surface (Fig. 20-6*a*). Note that even with proper clearance

**Fig. 20-6** Shearing of metal. (*a*) Proper clearance; (*b*) insufficient clearance; (*c*) excessive clearance.

there is still distortion at a sheared edge. Insufficient clearance will produce a ragged fracture (Fig. 20-6*b*) and also will require more energy to shear the metal than when there is proper clearance. With excessive clearance there is greater distortion of the edge, and more energy is required because more metal must plastically deform before it fractures. Furthermore, with too large a clearance, burrs or sharp projections are likely to form on the sheared edge (Fig. 20-6*c*). A dull cutting edge also increases the tendency for the formation of burrs. The amount of plastic deformation that can be withstood in brittle metals without extensive edge cracking when the metal is sheared is small, and therefore the clearance should be small for hard and brittle metals. On the other hand, soft, ductile metals require extensive plastic deformation before fracture, and therefore the clearance must be larger.

Neglecting friction, the force required to shear a metal sheet is the product of the length cut, the sheet thickness, and the shearing strength of the metal. The shearing force can be reduced appreciably by making the edges of the cutting tool at an inclined angle so that only a short part of the total length of cut is made at one time. The inclination of the cutting edge is called *shear*.

A whole group of press operations are based on the process of shearing. The shearing of closed contours, when the metal inside the contour is the desired part, is called *blanking*. If the material inside the contour is discarded, then the operation is known as *punching*, or *piercing*. Punching indentations into the edge of the sheet is called *notching*. *Parting* is the simultaneous cutting along at least two lines which balance each other from the standpoint of side thrust on the parting tool. *Slitting* is a shearing cut which does not remove any metal from the sheet. *Trimming* is a secondary operation in which previously formed parts are finished to size, usually by shearing excess metal around the periphery. The removal of forging flash in a press is a trimming operation. When the sheared edges of a part are trimmed or squared up by removing a thin shaving of metal, the operation is called *shaving*.

*Fine blanking* is a relatively new process in which extremely close clearances (typically 0.0002 in) and very rigid presses and die sets (to prevent sheet bending) produce a blank with no tearing at the sheared edge. These parts can be used as gears, cams, levers, etc., without any machining of the edges.

## 20-4   BENDING

Bending is the process by which a straight length is transformed into a curved length. It is a very common forming process for changing sheet and plate into channel, drums, tanks, etc. In addition, bending is part of the deformation in many other forming operations. The definition of the terms used in bending are illustrated in Fig. 20-7. The bend radius $R$ is defined as the radius of curvature on the concave, or inside, surface of the bend. For elastic bending below the elastic limit the strain passes through zero halfway through the thickness of the sheet at the neutral axis.

$R$ = Bend radius
$B$ = Bend allowance
$\alpha$ = Bend angle

**Fig. 20-7**   Definition of terms used in bending.

In plastic bending beyond the elastic limit the neutral axis moves closer to the inside surface of the bend as the bending proceeds. Since the plastic strain is proportional to the distance from the neutral axis, fibers on the outer surface are strained more than fibers on the inner surface are contracted. A fiber at the mid-thickness is stretched, and since this is the average fiber, it follows that there must be a decrease in thickness (radial direction) at the bend to preserve the constancy of volume. The smaller the radius of curvature, the greater the decrease in thickness on bending.

According to the theory of bending[1] the strain increases with decreasing radius of curvature. If the change in thickness is neglected, the neutral axis will remain at the center fiber and the circumferential stretch on the top surface $e_a$ will be equal to the shrink on the bottom surface, $e_b$. The conventional strain at the outer and inner fibers is given by

$$e_a = -e_b = \frac{1}{(2R/h) + 1} \qquad (20\text{-}1)$$

Experiments show that the circumferential strain on the tension surface is considerably greater than that given by Eq. (20-1) for large values of $h/R$, while the strain on the compression surface is not very different from the strain predicted by the simplified equation.

For a given bending operation the bend radius cannot be made smaller than a certain value, or the metal will crack on the outer tensile surface. The *minimum bend radius* is usually expressed in multiples of the sheet thickness. Thus, a $3T$ bend radius indicates that the metal can be bent without cracking through a radius equal to three times the sheet thickness. Therefore, the minimum bend radius is a forming limit. It varies considerably between different metals and always increases with cold-working. Although some very ductile metals have a minimum bend radius of zero, indicating that they can be flattened upon themselves, it is general practice to use a bend radius of not less than $\frac{1}{32}$ in in order to prevent damage to punches and dies. For high-strength sheet alloys the minimum bend radius may be $5T$ or higher. The minimum bend radius is not a precise material parameter because it depends, among other things, on the geometry of the bending conditions.

The minimum bend radius for a given thickness of sheet can be predicted[2] fairly accurately from the reduction of area measured in a tension test $q$. If $q$ is less than 0.2, then the shift in the neutral axis can be neglected and $R_{\min}$ is given simply by

$$\frac{R_{\min}}{h} = \frac{1}{2q} - 1 \qquad \text{for } q < 0.2 \qquad (20\text{-}2)$$

[1] J. D. Lubahn and G. Sachs, *Trans. ASME*, vol. 72, pp. 201–208, 1950; B. W. Shaffer and E. E. Unger, *Trans. ASME.* Ser. E: *J. Appl. Mech.*, vol. 27, pp. 34–40, 1960.
[2] J. Datsko and C. T. Yang, *Trans. ASME*, Ser. B: *J. Eng. Ind.*, vol. 82, pp. 309–314, 1960.

When $q$ is greater than 0.2, the shift in the neutral axis must be taken into consideration and the minimum bend radius is given by

$$\frac{R_{min}}{h} = \frac{(1-q)^2}{2q - q^2} \qquad \text{for } q > 0.2 \qquad (20\text{-}3)$$

The ductility of the outer fiber in bending is a function of the stress state acting on the surface. It is a well-established fact that the occurrence of a biaxial state of tension produces a decrease in the ductility of the metal. The biaxiality ratio $\sigma_2/\sigma_1$ of the transverse stress to the circumferential stress increases with increasing ratio of width to thickness, $b/h$. Figure 20-8 indicates that for low values of $b/h$ the biaxiality is low because the stress state is practically pure tension, but as the width of the sheet increases relative to its thickness, the ratio $\sigma_2/\sigma_1$ increases until at approximately $b/h = 8$ the biaxiality reaches a saturation value of approximately $\frac{1}{2}$. Correspondingly, the strain to produce fracture in bending is a reverse function of the width-thickness ratio. In bending sheets with a high width-thickness ratio, the cracks will occur near the center of the sheet when the ductility is exhausted. However if the edges of the sheet are rough, edge cracking will occur. Frequently the minimum bend radius can be increased by polishing or grinding the edges of the sheet. In bending narrow sheets, the failure will occur at the edges because the biaxiality is quite low at the center of the width.

In addition to cracking on the tensile surface of a bend another common forming difficulty is springback. *Springback* is the dimensional change of the formed part after the pressure of the forming tool has been released. It results from the changes in strain produced by elastic recovery. When the load is released,

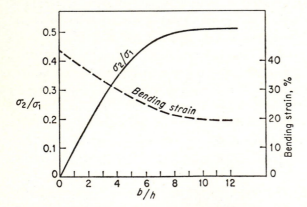

**Fig. 20-8** Effect of ratio of width to thickness on biaxiality and bend ductility in bending. (*From G. S. Sangdahl, E. L. Aul, and G. Sachs, Proc. Soc. Exp. Stress Anal., vol. 6, no. 1, p. 1, 1948.*)

Fig. 20-9   Springback in bending.

the total strain is reduced owing to the elastic recovery. The elastic recovery, and therefore the springback, will be greater the higher the yield stress, the lower the elastic modulus, and the greater the plastic strain. For a given material and strain the springback increases with the ratio between the lateral dimensions of the sheet and its thickness.

Springback is encountered in all forming operations, but it is most easily recognized and studied in bending (Fig. 20-9). The radius of curvature before release of load $R_0$ is smaller than the radius after release of the load $R_f$. Using the notation in Fig. 20-9, springback may be described by the springback ratio $K$.

$$K = \frac{\alpha_f}{\alpha_0} = \frac{R_0 + h/2}{R_f + h/2} \qquad (20\text{-}4)$$

The springback ratio defined in this way is independent of sheet thickness and depends only on the ratio of bend radius to sheet thickness. Values are available[1] for aluminum alloys and austenitic stainless steels in a number of cold-rolled tempers. Other data,[2] which include a number of high-temperature alloys, indicate that to a first approximation the springback in bending can be expressed by

$$\frac{R_0}{R_f} = 4\left(\frac{R_0 \sigma}{Eh}\right)^3 - 3\frac{R_0 \sigma}{Eh} + 1 \qquad (20\text{-}5)$$

The commonest method of compensating for springback is to bend the part to a smaller radius of curvature than is desired so that when springback occurs the part has the proper radius. The trial-and-error procedure of finding the proper die contour to correct for springback can be shortened somewhat by the use of the above equation, but the calculation is by no means a precise procedure. Furthermore, the correction to the die is valid only over a rather narrow range of yield

[1] Sachs, op. cit., p. 100.
[2] F. J. Gardiner, *Trans. ASME*, vol. 79, pp. 1–9. 1957; C. A. Queener and R. J. DeAngelis, *Trans. Am. Soc. Met.*, vol. 61, p. 757; 1968.

stress. Other methods of compensating for springback are to bottom the punch in the die so as to produce a coining action and the use of high-temperature forming so as to reduce the yield stress.

The force $P_b$ required to bend a length $L$ about a radius $R$ may be estimated from

$$P_b = \frac{\sigma_0 L h^2}{2(R + h/2)} \tan \frac{\alpha}{2} \qquad (20\text{-}6)$$

## 20-5  STRETCH FORMING

Stretch forming is the process of forming by the application of primarily tensile forces in such a way as to stretch the material over a tool or form block.[1] The process is an outgrowth of the stretcher leveling of rolled sheet. Stretch forming is used most extensively in the aircraft industry to produce parts of large radius of curvature, frequently with double curvature. An important consideration is that springback is largely eliminated in stretch forming because the stress gradient is relatively uniform. On the other hand, because tensile stresses predominate, large deformations can be obtained by this process only in materials with appreciable ductility.

Stretch-forming equipment consists basically of a hydraulically driven ram (usually vertical) which carries the punch or form block and two jaws for gripping the ends of the sheet (Fig. 20-10). No female die is used in stretch forming. The grips may be pivoted so that the tension force is always in line with the edge of the unsupported sheet, or they can be fixed, in which case a large radius is needed to prevent tearing the sheet at the jaws. In using a stretch-forming machine the sheet-metal blank is first bent or draped around the form block with relatively light tensile pull, the grips are applied, and the stretching load is increased until the blank is strained plastically to final shape. This differs from wrap forming (Sec.

Fig. 20-10  Stretch-forming operation.

[1] R. D. Edwards, *J. Inst. Met.*, vol. 84, pp. 199–209, 1956; Sachs, op. cit., pp. 456–475.

20-2) since in the latter process the blank is first gripped and then while still straight is loaded to the elastic limit before wrapping around the form block.

Stretching commonly is found as a part of many sheet-forming operations. For example, in forming a cup with a hemispherical bottom, the sheet is stretched over the punch face. Most complex automotive stampings involve a stretching component.

For a strip loaded in tension we have seen (Sec. 9-3) that the limit of uniform deformation occurs at a strain equal to the strain-hardening exponent $n$. In biaxial tension, the local necking which occurs in uniaxial tension is inhibited if $\sigma_2/\sigma_1 > \frac{1}{2}$, and instead the material develops *diffuse necking* which is not highly localized or readily visible to the eye. Eventually in the stretching of a thin sheet[1] plastic instability will occur in the form of a narrow *localized neck*. There will be a direction of zero-length increase inclined at an angle $\phi$ to the deforming axis (Fig. 20-11).

The normal strain $\varepsilon_2'$ must be zero, for if it were not, material adjoining the edges of the band would have to deform and the band would spread out along $X_1'$ and the band would grow into a diffuse neck. It can be shown (Prob. 20-3) that $\phi \approx \pm 55°$ for an isotropic material in pure tension. Moreover, the criterion for *local necking* is (Prob. 20-4 )

$$\frac{d\sigma}{d\varepsilon} = \frac{\sigma}{2} \qquad (20\text{-}7)$$

which arises from the fact that the area decreases with straining in local necking less rapidly than with diffuse necking. This means that more total strain must be accumulated in diffuse necking to cancel out strain hardening with the geometrical

**Fig. 20-11** Localized necking in a strip loaded in tension. The normal strain along $X_2'$ must be zero. *(From W. A. Backofen, " Deformation Processing," p. 205, Addison-Wesley, Publishing Company, Inc., Reading, Mass., 1972.)*

[1] S. F. Keeler and W. A. Backofen, *Trans. Am. Soc. Met.*, vol. 56, pp. 25–48, 1963.

softening. For power-law strain hardening, we have seen that for diffuse necking $\varepsilon_u = n$, but for local necking $\varepsilon_u = 2n$.

Diffuse necking usually will not constitute a limit to forming because the thinning is spread out over a fairly wide area of the sheet. Local necking is readily detected on exposed surfaces so that it represents a forming limit. Moreover, the formation of a local neck soon is followed by fracture of the sheet.

Local necking is very sensitive to the state of plastic strain,[1] usually denoted by the strain ratio $\rho = \varepsilon_2/\varepsilon_1$, where $\varepsilon_1$ is the algebraically larger of the principal strains in the plane of the sheet. If the sheet were deformed in pure shear, it would neither thicken nor thin and there would be no necking. At the other extreme, if the sheet is deformed in plane strain so that the width does not change, the local necking angle is $\phi = 90°$, and the strain for onset of both diffuse and local necking are equal. The dependence of necking strains on strain state is shown[2] in Fig. 20-12. $\varepsilon_1^*$ is the strain for necking in the direction of the largest principal strain, and $\varepsilon_u$ is the strain for diffuse necking in pure tension. If $\rho < 0$, both diffuse necking and local necking are possible, but both criteria coincide for plane-strain deformation. Theory predicts that only diffuse necking can occur if $\rho > 0$, since when both strains are tensile, there can be no zero-length change line. However, sheets tested under these conditions do thin locally and tear, generally at right angles to the $\varepsilon_1$ direction. For low-carbon steel[3] the critical strain for the onset of local necking is larger than $\varepsilon_u$ and rises with increasing strain ratio in the manner shown by the dotted line in Fig. 20-12. Other materials (such as brass and austenitic stainless steel) do not show a rising necking limit with strain ratio. The type of observations

**Fig. 20-12**  Variation of necking limits with strain ratio.

[1] R. Hill, *J. Mech. Phys. Solids*, vol. 1, p. 19, 1952.
[2] W. A. Backofen, *Metall. Trans.*, vol. 4, p. 2682, 1973.
[3] M. Azrin and W. A. Backofen, *Metall. Trans.*, vol. 1, pp. 2857–2865, 1970.

shown in Fig. 20-12 are the basis for the forming limit diagram discussed in Sec. 20-7. However, the theoretical basis for this behavior is still being developed.[1,2]

A common feature of a sheet-forming process is the existence of strain gradients. A strain gradient arises whenever the deformation is not uniform. The important role of strain hardening in modifying strain gradients can be shown by a simple example.[3] Consider a section of a sheet loaded in tension. The force is constant but the cross-sectional area increases with radial distance from some point, so that $\sigma r = $ constant. Therefore,

$$\frac{d\sigma}{dr} = -\frac{\sigma}{r}$$

and

$$\frac{d\sigma}{d\varepsilon}\frac{d\varepsilon}{dr} = \frac{d\sigma}{dr} = -\frac{\sigma}{r}$$

so the strain gradient is expressed by

$$\frac{d\varepsilon}{dr} = -\frac{\sigma}{(d\sigma/d\varepsilon)r} \qquad (20\text{-}8)$$

If the strain hardening can be expressed by the power law $\sigma = K\varepsilon^n$, $d\sigma/d\varepsilon = nK\varepsilon^{n-1} = n(\sigma/\varepsilon)$so

$$\frac{d\varepsilon}{dr} = -\left(\frac{\sigma}{r}\right)\frac{1}{n(\sigma/\varepsilon)} = -\frac{(\varepsilon/r)}{n} \qquad (20\text{-}9)$$

Equation (20-9) shows that the strain gradient is reduced by greater strain hardening (larger $n$). Since the most highly strained region will have hardened the most, the load is passed on to the neighboring elements. This forces them to strain more and in so doing the strain gradient is reduced. As a result, deeper, more complex parts or overall greater reductions can be made with material with greater strain-hardening capacity.

While the strain-hardening capacity is best measured by the strain-hardening exponent $n$ (as discussed in Sec. 9-2), in production situations it may be useful to estimate strain hardening from properties measured in the ordinary tension test.[4] The ratio of tensile strength $s_u$ to yield strength $s_0$ is a common measurement of strain hardening.

$$s_u = \frac{P_{max}}{A_0} = \frac{\sigma_u A_u}{A_0}$$

$$\varepsilon_u = \ln{(A_0/A_u)} \qquad \text{or} \qquad \frac{A_0}{A_u} = e^{\varepsilon_u} = e^n \qquad \text{if } \sigma = K\varepsilon^n$$

[1] Z. Marciniak and K. Kuczynski, *Int. J. Mech. Sci.*, vol. 9, pp. 609–620, 1967.
[2] A. Ghosh and W. A. Backofen, *Metall. Trans.*, vol. 4, pp. 1113–1123, 1973.
[3] W. A. Backofen, "Deformation Processing," p. 211, Addison-Wesley Publishing Company, Inc., Reading, Mass., 1972.
[4] W. A. Backofen, op. cit., p. 213.

Therefore,

$$s_u = \frac{\sigma_u}{e^n} = \frac{Kn^n}{e^n} \qquad (20\text{-}10)$$

The yield strength is the stress at $\varepsilon \approx 0.02$, or

$$s_0 = K(0.02)^n$$

(neglecting the small difference between engineering and true stress at small strain) so that the tensile-yield ratio is

$$\frac{s_u}{s_0} = \frac{Kn^n}{e^n} \frac{1}{K(0.02)^n} = \left(\frac{n}{0.02e}\right)^n$$

or

$$\frac{s_u}{s_0} = (18.4n)^n \qquad (20\text{-}11)$$

## 20-6  DEEP DRAWING

Deep drawing is the metalworking process used for shaping flat sheets into cup-shaped articles such as bathtubs, shell cases, and automobile fenders. This is done by placing a blank of appropriate size over a shaped die and pressing the metal into the die with a punch (Fig. 20-13). Generally a clamping or hold-down pressure is required to press the blank against the die to prevent wrinkling. This is best done by means of a *blank holder* or *hold-down ring* in a double-action press. Although the factors which control the deep-drawing process are quite evident,

**Fig. 20-13**   Deep drawing of a cylindrical cup. (*a*) Before drawing; (*b*) after drawing.

**Fig. 20-14**  Stresses and deformation in a section
from a drawn cup.

they interact in such a complex way that precise mathematical description of the
process is not possible in simple terms. The greatest amount of experimental and
analytical work has been done on the deep drawing of a flat-bottom cylindrical
cup (Swift test) from a flat circular blank. The discussion of deep drawing which
follows will be limited to this relatively simple situation.

In the deep drawing of a cup the metal is subjected to three different types of
deformations. Figure 20-14 represents the deformation and stresses developed in a
pie-shaped segment of the circular blank during deep drawing. The metal at the
center of the blank under the head of the punch is wrapped around the profile of the
punch, and in so doing it is thinned down. The metal in this region is subjected to
biaxial tensile stress due to the action of the punch. Metal in the outer portion of the
blank is drawn radially inward toward the throat of the die. As it is drawn in, the
outer circumference must continuously decrease from that of the original blank
$\pi D_0$ to that of the finished cup $\pi D_p$. This means that it is subjected to a compressive strain in the circumferential, or hoop, direction and a tensile strain in the
radial direction. As a result of these two principal strains there is a continual increase in the thickness as the metal moves inward. However, as the metal passes
over the die radius, it is first bent and then straightened while at the same time being
subjected to a tensile stress. This plastic bending under tension results in considerable thinning, which modifies the thickening due to the circumferential shrinking.
Between the inner stretched zone and the outer shrunk zone there is a narrow ring
of metal which has not been bent over either the punch or the die. The metal in
this region is subjected only to simple tensile loading throughout the drawing
operation.

If the clearance between the punch and the die is less than the thickness produced by free thickening, the metal in these regions will be squeezed, or *ironed*, between the punch and the die to produce a uniform wall thickness.  In commercial deep drawing clearances about 10 to 20 percent greater than the metal thickness are common.  Ironing operations in which appreciable uniform reductions are made in the wall thickness use much smaller clearances.

The force on the punch required to produce a cup is the summation of the ideal force of deformation, the frictional forces, and the force required to produce ironing (if present).  Figure 20-15 illustrates the way in which these components of the total punch force vary with the length of stroke of the punch.  The ideal force of deformation increases continuously with length of travel because the strain is increasing and the flow stress is increasing owing to strain hardening.  A major contribution to the friction force comes from the hold-down pressure.  This force component peaks early and decreases with increasing travel because the area of the blank under the hold-down ring is continually decreasing.  Any force required to produce ironing occurs late in the process after the cup wall has reached maximum thickness.  An additional factor is the force required to bend and unbend the metal around the radius of the die.  One measurement[1] of the work required in cupping showed that 70 percent of the work went into the radial drawing of the metal, 13 percent into overcoming friction, and 17 percent into the bending and unbending around the die radius.

From an analysis of the forces in equilibrium during the formation of a deep-drawn cup Sachs[2] has developed the following approximate equation for the

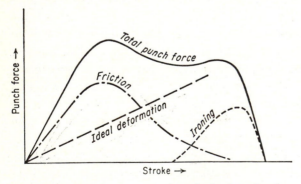

**Fig. 20-15**   Punch force vs. punch stroke for deep drawing.

[1] H. W. Swift, *J. Inst. Met.*, vol. 82, p. 119, 1952.
[2] G. Sachs, "Spanlose Formung," pp. 11–38, Springer-Verlag OHG, Berlin, 1930; G. Sachs and K. R. Van Horn, "Practical Metallurgy," pp. 430–431, American Society for Metals, Metals Park, Ohio, 1940.

total punch force as a function of the diameter of the blank, $D_0$, at any stage in the process,

$$P = \left[ \pi D_p h(1.1\sigma_0) \ln \frac{D_0}{D_p} + \mu \left( 2H \frac{D_p}{D_0} \right) \right] e^{(\mu\pi/2)} + B \quad (20\text{-}12)$$

where $P$ = total punch load

$\sigma_0$ = average flow stress

$D_p$ = punch diameter

$D_0$ = blank diameter

$H$ = hold-down force

$B$ = force required to bend and restraighten blank

$h$ = wall-thickness

$\mu$ = coefficient of friction

In Eq. (20-12) the first term expresses the ideal force required to produce the cup, and the second term is the friction force under the blank holder. The exponential term considers the friction at the die radius, and the quantity $B$ accounts for the force required to bend and unbend the sheet around this radius. A more complete and accurate analysis of the stresses and strains in the deep drawing of a cup has been presented by Chung and Swift[1] and Budiansky and Wang.[2]

The load in deep drawing is applied by the punch to the bottom of the cup and it is then transmitted to the sidewall of the cup. Usually failure occurs in the narrow band of material in the cup wall just above the radius of the punch which has undergone no radial drawing or bending but is subjected essentially to tensile straining. In this annular ring between the die wall and the punch the metal is subjected essentially to plane-strain stretching and thinning. Failure occurs by necking (followed by tearing) at a stress approximately equal to the tensile strength, increased by the plane-strain factor.

$$P_{max} = \frac{2}{\sqrt{3}} s_u \pi D_p h \quad (20\text{-}13)$$

The applied stress in deep drawing is transmitted through a "weakest-link system" to a point of incipient failure which has undergone little strengthening by strain hardening. Thus, the strain-hardening ability of the material plays very little role in deep drawing, as opposed to wire drawing where $n$ is very important. The *drawability* of a metal is measured by the ratio of the initial blank diameter to the diameter of the cup drawn from the blank (usually approximated by the punch diameter). For a given material there is a *limiting draw ratio (LDR)*, representing

[1] S. Y. Chung and H. W. Swift, *Proc. Inst. Mech. Eng. (London)*, vol. 165, pp. 199–228, 1951; this theory has been reviewed in detail by J. M. Alexander, *Metall. Rev.*, vol. 5, pp. 349–411, 1960.

[2] B. Budiansky and N. M. Wang, *J. Mech. Phys. Solids*, vol 14, p. 357, 1966.

the largest blank that can be drawn through a die $D_p$ without tearing. The theoretical upper limit on LDR is [1]

$$\text{LDR} \approx \left(\frac{D_0}{D_p}\right)_{max} \approx e^{\eta} \qquad (20\text{-}14)$$

where $\eta$ is an efficiency term to account for frictional losses. If $\eta = 1$, then LDR $\approx$ 2.7; while if $\eta = 0.7$, LDR $\approx 2$. This agrees with exerience that even with ductile metals it is difficult to draw a cup with a height much greater than its diameter.

Some of the practical considerations which affect drawability are:

Die radius—should be about 10 times sheet thickness.
Punch radius—a sharp radius leads to local thinning and tearing.
Clearance between punch and die—20 to 40 per cent greater than the sheet thickness.
Hold-down pressure—about 2 per cent of average of $s_0$ and $s_u$.
Lubricate die side to reduce friction in drawing.

Since the average maximum reduction in deep drawing is about 50 percent, to make tall slender cups (such as cartridge cases and closed-end tubes), it is necessary to use successive drawing operations. Reducing a cup or drawn part to a smaller diameter and increased height is known as *redrawing*.

The two basic methods of redrawing are *direct*, or *regular*, redrawing and *reverse*, or *indirect*, redrawing (Fig. 20-16). In direct redrawing the original outside surface of the cup remains the outside surface of the redrawn cup. Figure 20-16a illustrates direct redrawing by means of a hold-down ring. Note that the metal must bend twice and that it is bent and unbent at the punch and die radii. The high strain hardening that is encountered in the process shown in Fig. 20-16a is

$(a)$      $(b)$      $(c)$

**Fig. 20-16** Redrawing methods. (*a*) Direct redrawing; (*b*) direct redrawing with tapered die; (*c*) reverse redrawing.

---

[1] W. A. Backofen, op. cit., p. 233.

reduced somewhat by the design shown in Fig. 20-16b. Although the metal still goes through the same number of bends, the angle through which it bends is less than 90° and the punch load is reduced. The disadvantage of the design shown in Fig. 20-16b is that the first-stage drawn cup must be made with a tapered corner. This type of drawn cup cannot be produced in all metals without buckling. The thickness of the hold-down ring used in direct redrawing is determined by the percentage reduction of the redraw. For small reductions a hold-down ring cannot be used.

In reverse redrawing (Fig. 20-16c) the cup is turned inside out so that the outside surface of the drawn cup becomes the inside surface of the redrawn shell. The bending is always in the same direction, rather than in opposite directions as for direct redrawing, and therefore there is less strain hardening of the walls of the shell. Better control of wrinkling can be obtained in reverse redrawing because of the snug control of the metal around the die radius and the fact that there are no geometrical limitations to the use of a hold-down ring.

The reduction obtained by redrawing is always less than that obtainable on the initial draw because of the higher friction inherent in the redrawing process. Generally the reduction is decreased for each successive redrawing operation to allow for strain hardening. Greater reductions are, of course, possible if annealing is carried out between redraws. Most metals will permit a total reduction of 50 to 80 percent before annealing.

Redrawing operations may also be classified into drawing with appreciable decrease in wall thickness, called *ironing*, and drawing with little change in wall thickness, called *sinking*. The ironing process is basically the same as tube drawing with a moving mandrel. The predominant stress in ironing is the radial compressive stress developed by the pressure of the punch and the die. Redrawing without reduction in wall thickness is basically the same as tube sinking or tube drawing without a mandrel. The predominant stresses are an axial tensile stress from the action of the punch and a circumferential compression from the drawing in of the metal.

To improve drawability, the potential failure site near the bottom of the cup wall must be strengthened relative to the metal deforming by radial drawing near the top of the cup wall.[1] Some improvement can be obtained by shifting the failure site up the cup wall by mechanical factors such as gripping the cup more firmly around the punch by surrounding the cup with a pressurized rubber pad where the cup exits from the die. Roughening the punch or withholding lubrication to the punch may also help in this regard. It would also be possible to weaken the metal in the flange relative to the failure site by selectively heating the metal in the flange area. However, by far the greatest improvement in drawability comes

---

[1] R. T. Holcomb and W. A. Backofen, *Sheet Met. Ind.*, vol. 43, pp. 479–484, 1966.

about by the control of *crystallographic texture* in the sheet that is to be drawn. The correct texture gives the proper orientation of slip systems so that the strength in the thickness direction is greater than that in the plane of the sheet.

In Sec. 3-5 we saw that the resistance to through thickness thinning was measured by $R$, the plastic strain ratio[1] of width to thickness in a sheet. $R$ measures the *normal anisotropy*. A large value of $R$ denotes high resistance to thinning in the thickness direction (direction normal to the plane of the sheet).

$$R = \frac{\ln (w_0/w)}{\ln (h_0/h)} \qquad (20\text{-}15)$$

where $w_0$ and $w$ are the initial and final width and $h_0$ and $h$ are the initial and final thickness. Since thickness measurements are difficult to make with precision on thin sheets, the equation can be rewritten using the constancy-of-volume relationship.

$$R = \frac{\varepsilon_w}{\varepsilon_L - \varepsilon_w} = \frac{\log (w_0/w)}{\log (L/L_0) - \log (w_0/w_0)} = \frac{\log (w_0/w)}{\log (Lw/L_0 w_0)} \qquad (20\text{-}16)$$

Since most rolled sheets show a variation of elastic and plastic properties with orientation in the plane of the sheet, it is usual to allow for this planar anisotropy by $\bar{R}$ averaged over measurements taken at different angles to the rolling direction of the sheet.

$$\bar{R} = \frac{R_0 + 2R_{45} + R_{90}}{4} \qquad (20\text{-}17)$$

**Fig. 20-17** Correlation between limiting draw ratio and $\bar{R}$ for a wide range of sheet metals. (*From M. Atkinson, Sheet Met. Ind., vol. 44, p. 167, 1967.*)

[1] M. Grumbach and G. Pomey, *ibid.*, vol. 43, pp. 515–529, 1966.

Whiteley[1] first demonstrated the important of $\bar{R}$ in drawability. An example of this type of correlation is given in Fig. 20-17. The ideal crystallographic orientation[2] to maximize $\bar{R}$ in bcc metals would be a sheet texture with $\langle 111 \rangle$ in the direction normal to the sheet and $\{111\}$ oriented randomly in the plane of the sheet. Most steels fall in the range $R = 1$ to $R = 2$. The theoretical limit on $\bar{R}$ for bcc metals[3] appears to be 3. The best way to understand the role of $\bar{R}$ in deep drawing is with the use of a yield locus diagram. We need to relate the stress states at the failure site on the cup wall and in the flange to the yield locus. As Fig. 20-14 shows (neglecting the low contact pressure), in the cup wall both stresses are tensile and the deformation is essentially plane strain. This requires that $\sigma_1 = 2\sigma_2$, and on the yield locus for an isotropic sheet ($R = 1$) (Fig. 20-18), it corresponds to the loading vector **OA**. The stress state in the flange is compression in the hoop direction and tension in the radial direction. If thickening can be neglected, the stress state approaches pure shear, vector **OB** on the yield locus. The yield locus for a sheet with $R > 1$ is shown dotted in Fig. 20-18. There hardly is any difference in the length of the load path ($OB'$) for the pure shear stress state in the flange, but the load path ($OA'$) for the plane-strain tension in the cup wall is increased appreciably. This shows that the cup wall is strengthened relative to the flange by texture control, and drawability is increased.

**Fig. 20-18**  Comparison of the load paths on the yield locus for isotropic ($R = 1$) and crystallographically strengthened ($R > 1$) sheet.

[1] R. L. Whiteley, *Trans. Am. Soc. Met.*, vol. 52, pp. 154–169, 1960.
[2] D. V. Wilson, *Metall. Rev.*, no. 139, pp. 175–188, 1969.
[3] H. R. Piehler, Sc.D. thesis, Massachusetts Institute of Technology, Cambridge, 1967.

## 20-7   FORMING LIMIT CRITERIA

Because of the complexity of sheet-forming operations, simple mechanical property measurements made from the tension test are of limited value. Over the years a number of laboratory tests have been developed to evaluate the formability of sheet materials. The Swift flat-bottom cup test[1] is a standardized test for deep drawing. The drawability is expressed in terms of the *limiting draw ratio*. In the Olsen and Erichsen tests[2] the sheet is clamped between two ring dies while a punch, usually a ball, is forced against the sheet until it fractures. The depth of the bulge before the sheet fractures is measured. These tests subject the sheet primarily to stetching, while the Swift test provides nearly pure deep drawing. However, most practical sheet-forming operations provide a combination of both biaxial stretching and deep drawing.[3] The Fukui test, which produces a conical cup using a hemispherical punch, provides a combination of both stretching and drawing.

A useful technique for controlling failure in sheet-metal forming is the *forming limit diagram*.[4] The surface of the sheet is covered by a grid of circles, produced by electrochemical marking. When the sheet is deformed, the circles distort into ellipses. The major and minor axes of an ellipse represent the two principal strain directions in the stamping. The strain in these two directions is measured by the percentage change in the lengths of the major and minor axes. These strains, at any point on the surface, are then compared with the Keeler-Goodwin diagram for the material (Fig. 20-19). Strain states above the curve represent failure, those below do not cause failure. For example, point $A$ is a failure, but if the strain

**Fig. 20-19**   Keller-Goodwin forming limit diagram.

[1] O. H. Kemmis, *Sheet Met. Ind.*, vol. 34, pp. 203, 251, 1957.

[2] "Metals Handbook," 8th ed., vol. 1, pp. 322–323, American Society for Metals, Metals Park, Ohio, 1961.

[3] W. T. Lankford, S. C. Snyder, and J. A. Bauscher, *Trans. Am. Soc. Met.*, vol. 42, pp. 1192–1235, 1950.

[4] S. P. Keeler, *Met. Prog.*, October 1966, pp. 148–153.

**Fig. 20-20** Formability chart for linear stretch forming. Each material would have to be calibrated, but the form of the chart would be the same. (*After Wood et al.*)

distribution is altered (perhaps by changing a die radius), it could move to point *B*, which would not cause failure. The failure curve for the tension-tension region was determined by Keeler[1] and is nearly fixed for a variety of low-carbon steels. Other metals such as aluminum[2] have a different curve. The tension-compression region was first determined by Goodwin.[3]

Another approach to predicting sheet formability is the stretch-draw shape analysis.[4] The forming limit of the material is established with the Olsen test and the Swift cupping test. Then the part is broken down into simple shapes and the percentage of draw and stretch are calculated from the geometry. This places the part on the forming limit diagram and the degree of severity of the part can be established.

Useful formability charts have been developed[5] for such processes as brake forming, joggling, dimpling, linear stretch forming, rubber forming, beading, and spinning. Generally, these are processes used in aerospace manufacture. Critical dimensional parameters were established for each process. An example for linear stretch forming is shown in Fig. 20-20.

## 20-8 DEFECTS IN FORMED PARTS

The ultimate defect in a formed sheet-metal part is the development of a crack which destroys its structural integrity. The usefulness of the part may also be destroyed by local necking or thinning or by buckling and wrinkling in regions of

[1] S. P. Keeler, *Sheet Met. Ind.*, vol. 42, pp. 683–691, 1965.
[2] S. S. Hecker, *Trans. ASME, Ser. H: J. Eng. Mater. Technol.*, vol. 97, pp. 66–73, 1975.
[3] G. M. Goodwin, *Metall. Ital.*, vol. 60. pp. 767–774, 1968.
[4] A. S. Kasper, *Met. Prog.*, May 1971, pp. 57–60.
[5] W. W. Wood, et al., "Theoretical Formability," AFML-TR-64-411, January 1965.

compressive stress. Another troublesome defect is the failure to maintain dimensional tolerances because of springback.

In the deep drawing of a cup the most common failure is the separation of the bottom from the rest of the cup at the location of greatest thinning near the punch radius. This defect may be minimized either by reducing the thinning by using a larger punch radius or by decreasing the punch load required for the drawing operation. If radial cracks occur in the flange or the edge of the cup, this indicates that the metal does not have sufficient ductility to withstand the large amount of circumferential shrinking that is required in this region of the blank. This type of failure is more likely to occur in redrawing without annealing than on the initial draw.

Wrinkling of the flange or the edges of the cup results from buckling of the sheet as a result of the high circumferential compressive stresses. In analyzing this type of failure each element in the sheet can be considered as a column loaded in compression. If the blank diameter is too large, the punch load will rise to high values, which may exceed the critical buckling load of the column. Since column stability decreases with an increasing slenderness ratio, the critical buckling load will be achieved at lower loads for thin sheet. To prevent this defect, it is necessary to use sufficient hold-down pressure to suppress the buckling.

**Fig. 20-21**  Stretcher strains in low-carbon steel sheet. (*Courtesy E. R. Morgan and Met. Prog., June, 1958, p. 89.*)

Since sheet-metal formed parts usually present a large surface area, they are particularly susceptible to surface blemishes which detract from the appearance of the part. Pronounced surface roughness in regions of the part which have undergone appreciable deformation is usually called *orange peeling*. The orange-peel effect occurs in sheet metal of relatively large grain size. It results from the fact that the individual grains tend to deform independently of each other, and therefore grains stand out in relief on the surface. This condition is best corrected by using finer-grain-size sheet metal so that the grains deform more nearly as a whole and the individual grains are difficult to distinguish with the eye.

Another serious surface defect that is commonly found in low-carbon sheet steel is the presence of *stretcher strains*, or "worms." This defect shows up as a flamelike pattern of depressions in the surface (Fig. 20-21). The depressions first appear along planes of maximum shear stress, and then, as deformation continues, they spread and join together to produce a uniform rough surface. The existence of stretcher strains is directly associated with the presence of a yield point in the stress-strain curve and the nonuniform deformation which results from the yield-point elongation (Fig. 20-22). The metal in the stretcher strains has been strained an amount equal to $B$ in Fig. 20-22, while the remaining metal has received essentially zero strain. The elongation of the part is given by some intermediate strain $A$. As the deformation continues and the number of stretcher strains increase, the strain will increase until when the entire part is covered it has a strain equal to $B$. Beyond this strain the deformation is uniform and homogeneous. The main difficulty with stretcher strains, therefore, occurs in regions of the part where the strain is less than the yield-point elongation. The usual solution to this problem is to give the sheet steel a small cold reduction, usually $\frac{1}{2}$ to 2 percent reduction in

**Fig. 20-22**  Relation of stretcher strains to stress-strain curve.

thickness. Such a temper-rolling or skin-rolling treatment cold-works the metal sufficiently to eliminate the yield point. However, if the steel strain-ages during storage, the yield point will return and difficulties with stretcher strains will re-appear.

The directionality in mechanical properties produced by rolling and other primary working processes can have an important effect on the fabricability of the metal. Mechanical fibering has little effect on formability, whereas crystallographic fibering or preferred orientation may have a large effect. Ordinarily, bending is more difficult when the bend line is parallel to the rolling direction than when the bend is made perpendicular to the rolling direction.

One of the ways that directionality shows up in deep drawing is the phenomenon of *earing*. Earing is the formation of a wavy edge on the top of a drawn cup which necessitates extensive trimming to produce a uniform top. Usually two, four, or six ears will be formed, depending on the preferred orientation in the plane of the sheet. Earing can be directly correlated[1] with the *planar anisotropy*, measured by $\Delta R$.

$$\Delta R = \frac{R_0 + R_{90} - 2R_{45}}{2} \qquad (20\text{-}18)$$

## BIBLIOGRAPHY

Alexander, J. M.: An Appraisal of the Theory of Deep Drawing, *Metall. Rev.*, vol. 5, pp. 349–411, 1960.

Crane, E. V.: "Plastic Working of Metals and Non-metallic Materials in Presses," 3d ed., John Wiley & Sons, Inc., New York, 1944.

Eary, D. F., and E. A. Reed: "Techniques of Pressworking Sheet Metal," Prentice-Hall, Inc., Englewood Cliffs, N.J., 1958.

Jevons, J. D.: "The Metallurgy of Deep Drawing and Pressing," 2d ed., Chapman & Hall, Ltd., London. 1941.

Keeler, S. P.: Understanding Sheet Metal Formability, *Machinery*, February through July, 1968.

"Metals Handbook," 8th ed., vol. 4, Forming, American Society for Metals, Metals Park, Ohio, 1967.

Sachs, G.: "Principles and Methods of Sheet-Metal Fabricating," 2d ed., revised by H. E. Voegel, Reinhold Publishing Corporation, New York, 1966.

Willis, J.: "Deep Drawing," Butterworth & Co. (Publishers) Ltd., London, 1954.

---

[1] D. V. Wilson and R. D. Butler, *J. Inst. Met.*, vol. 90, p. 473, 1961–1962.

# 21

# MACHINING OF METALS

## 21-1 TYPES OF MACHINING OPERATIONS

In the broadest sense there are two distinct classes of solid-state manufacturing processes. *Deformation processes* produce the required shape with the necessary mechanical properties by plastic deformation in which the material is moved and its volume is conserved. *Machining processes* produce the required shape by removal of selected areas of the workpiece through a machining process. Most machining is accomplished by straining a local region of the workpiece by the relative motion of the tool and the workpiece. Although mechanical energy is the usual input to most machining processes, some of the newer metal-removal processes employ chemical, electrical, or thermal energy.

The inexperienced reader should not be misled about the importance of machining processes to the industrial world because only one chapter is devoted to this subject. This choice partly reflects the author's own areas of interest and partly the fact that deformation processes historically have been of more interest to the metallurgist, while machining processes have been of prime concern to mechanical and industrial engineers. There is no real rational for this situation since, as will be seen shortly, machining processes depend critically on the tool material and the interaction of the machining forces with the workpiece material. In the space available only the briefest introduction to the technology of machining operations is possible. Emphasis is on fundamentals of the mechanics and the material response to more generalized machining processes.

Machining usually is employed to produce shapes with high dimensional tolerance, good surface finish, and often with complex geometry. Machining is a *secondary* processing operation since it usually is conducted on a workpiece that was produced by a primary process such as hot rolling, forging, or casting. Something more than 80 percent of all manufactured parts must be machined before they are completed. The variety of machining processes and machine tools that can be utilized is very great. Since the development of machine tools paralleled the industrialization of our society, it is an old field with much specialized terminology and jargon. The student is referred to texts on machine tool practice[1] for details of machine tool construction and performance. Other reference sources[2] should be consulted for details on specific machining operations and selection of speeds, feeds, tool geometry and material, and cutting fluid.

Some generalization can be brought to this complex subject by considering how the workpiece is held in the machine and what the relative movements are between the workpiece and the tool. The ease and accuracy with which a surface can be machined depends on how well the surface can be matched to the movements of the machine and the cutting edge of the tool. Machine tools generate surfaces in two ways: (1) by using a *form tool* which matches the shape to be produced in the surface, e.g., a tool with a notch in its cutting edge produces a raised rib on the machined surface, and (2) feeding the cutting tool in and out along the length of the workpiece. Form tools lack flexibility and have high relative cost, but they are important in high-volume production. Machining by path cutting with a single-edge or multiple-edge tool requires the movement of the tool in two directions in order to generate shapes other than a simple cylinder. The *primary motion* is the main motion provided by the machine tool to cause relative motion between the tool and the workpiece so that the face of the tool approaches the workpiece. The *feed motion* is a motion provided to the tool or workpiece which when added to the primary motion leads to a repeated or continuous chip removal and the creation of a surface with the desired geometry. The feed motion absorbs a much smaller portion of the total power than the primary motion.

---

[1] C. R. Hine, "Machine Tools and Processes for Engineers," 2d ed., McGraw-Hill Book Company, New York, 1959; L. E. Doyle, et al., "Manufacturing Processes and Materials for Engineers," Prentice-Hall, Inc., Englewood Cliffs, N.J., 1961; E. P. DeGarmo, "Materials and Processes in Manufacturing," 3d ed., The Macmillan Company, New York, 1969; R. A. Lindberg, "Processes and Materials of Manufacture," Allyn and Bacon, Inc., Boston, 1968.
[2] "Metals Handbook," 8th ed., vol. 3, Machining, American Society for Metals, Metals Park, Ohio, 1967; "Machining Data Handbook," 2d ed., Metcut Research Associates Inc., Cincinnati, Ohio, 1972.

(a)        (b)        (c)        (d)

**Fig. 21-1** Schematic of some common machining operations. (*a*) Turning on a lathe; (*b*) boring; (*c*) shaping; (*d*) slab milling.

With only a few exceptions, machine tools may be grouped into two broad categories: those that generate surfaces of rotation and those that generate flat or formed surfaces by linear movement. Figure 21-1*a* shows the relative motions needed to produce a machined surface on a turning machine such as a lathe. The principal parameters are the *cutting speed v*, the *depth of cut d*, and the *feed f*. To turn a cylindrical surface of length $L_w$ requires $L_w/f$ revolutions of the workpiece. If the number of revolutions of the workpiece per minute is $n_w$, then the time for one *pass* is $t = L_w/fn_w$. If the tool operates on the inside of a hollow cylinder (Fig. 21-a*b*), the process is called *boring*. Very heavy workpieces are machined on vertical boring mills where they are rotated about the vertical axis. *Shaping* is a process in which a single-point tool reciprocates over the workpiece in a linear manner (Fig. 21-1*c*). One of the most versatile processes for producing flat surfaces is *milling* (Fig. 21-1*d*). The milling cutter is a multiple-edge tool.

## 21-2 MECHANICS OF MACHINING

A simple model of two-dimensional machining (orthogonal cutting) is shown in Fig. 21-2. The tool is a single-point tool that is characterized by the *rake angle* $\alpha$, the *clearance* (relief) angle $\theta$, and the wedge angle $\omega$. $\alpha + \omega + \theta = 90°$. The *rake face* of the tool is the surface over which the chip flows as it leaves the machined surface. The forces imposed on the tool create intense shearing action on the metal ahead of the tool. The metal in the chip is severely deformed on going from an undeformed chip of thickness $t$ to the deformed chip with thickness $t_c$. The *chip thickness ratio* (also the cutting ratio) $r = t/t_c$ is an easily measured parameter which is typically around $r \approx 0.5$. The creation of the chip subjects the metal to large plastic shear strains ($\gamma = 2$ to $4$) at strain rates above $10^3$ s$^{-1}$. There is a localized region of intense shear in the vicinity of *OA*, but for simplicity this usually is represented by a well-defined shear plane *OA* occurring at a *shear angle* $\phi$.

**Fig. 21-2** Idealized model for chip formation.

The relationship between rake angle, shear angle, and chip thickness ratio can be derived from Fig. 21-3.

$$r = \frac{t}{t_c} = \frac{OA \sin \phi}{OA \cos (\phi - \alpha)} \qquad (21\text{-}1)$$

and on solving for $\phi$

$$\tan \phi = \frac{r \cos \alpha}{1 - r \sin \alpha} \qquad (21\text{-}2)$$

Thus, the chip thickness ratio $r$ determined after the cut has been made may be used to establish the shear angle $\phi$ existing during the machining operation. Since volume is constant in plastic deformation, and chip width $b$ is essentially constant, we could also obtain $r$ from measurements of the ratio of chip length $L_c$ to the length of the workpiece from which it came, $L_w$. If $L_w$ is unknown, it can be determined by measuring the weight of chips $W_c$, and by knowing the density of the metal $\rho$, $L_w bd = W_c/\rho$.

If the deformation in cutting is assumed to be a simple block-like shearing process, we can estimate the shearing strain in machining using Fig. 21-4. An

**Fig. 21-3** Relationships expressed in Eq. (21–1).

Fig. 21-4   Average shear strain in machining.

element originally defined by $ABCD$ is sheared to $EBCF$. The shear strain $\gamma$ is the shear offset divided by the perpendicular distance between shear planes.

$$\gamma = \frac{AE}{OB} = \frac{AO}{OB} + \frac{OE}{OB} = \tan{(\phi - \alpha)} + \cot{\phi} \qquad (21\text{-}3)$$

For chips without curl there are three velocities which must be specified in cutting. The *cutting speed v* is the velocity of the tool relative to the workpiece. The velocity of the chip relative to the tool face is the *chip velocity $v_c$*. The velocity of the chip relative to the work is the *shear velocity $v_s$*. From continuity of mass, $vt = v_c t_c$, so

$$r = \frac{t}{t_c} = \frac{v_c}{v} \qquad (21\text{-}4)$$

The three velocity components give the kinematic relationship shown in Fig. 21-5, from which we see that the vector sum of the cutting velocity and the chip velocity is equal to the shear velocity vector. It can be shown from the geometry that

$$v_s = \frac{v \cos{\alpha}}{\cos{(\phi - \alpha)}} \qquad (21\text{-}5)$$

Knowing $v_s$, we can establish the shear-strain rate in cutting.

$$\dot{\gamma} = \frac{d\gamma}{dt} = \frac{v_s}{(y_s)_{\max}} \qquad (21\text{-}6)$$

Fig. 21-5   Velocity relationships in orthogonal machining.

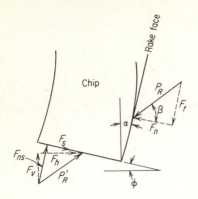

*Rake face*

**Fig. 21-6**   Force components in orthogonal cutting.

where $(y_s)_{max}$ is the estimate of the maximum value of the thickness of the shear zone, approximately $10^{-3}$ in. Using realistic values of $\phi = 20$, $\alpha = 5°$, $v = 600$ fpm and $(y_s)_{max} \approx 10^{-3}$ in, we calculate $\dot{\gamma} = 1.2 \times 10^5 \text{ s}^{-1}$, which is several orders of magnitude greater than the strain rate usually associated with high-speed metal-working operations.

We are now ready to consider the forces and stresses acting in metal cutting.[1] In Fig. 21-6, $P_R$ is the resultant force between the tool face and the chip and $P_R'$ is the equal resultant force between the workpiece and the chip along the shear plane. These resultant forces can be resolved conveniently into several sets of components. An obvious choice is the component tangential, $F_t$, and normal, $F_n$, to the rake face of the tool. However, the force in metal cutting frequently is measured by a strain-gage toolpost dynamometer[2] which independently measures the horizontal (cutting) $F_h$ and vertical (thrust) $F_v$ forces in cutting. It can be shown that

$$F_t = F_h \sin \alpha + F_v \cos \alpha$$
$$F_n = F_h \cos \alpha - F_v \sin \alpha \qquad (21\text{-}7)$$

If the components of cutting force are known, then the coefficient of friction in the tool face is given by

$$\mu = \tan \beta = \frac{F_t}{F_n} = \frac{F_v + F_h \tan \alpha}{F_h - F_v \tan \alpha} \qquad (21\text{-}8)$$

Finally, the resultant force may be resolved parallel, $F_s$, and normal, $F_{ns}$, to the shear plane. In terms of the usual measured force components

$$F_s = F_h \cos \phi - F_v \sin \phi \qquad (21\text{-}9)$$
$$F_{ns} = F_h \sin \phi + F_v \cos \phi$$

[1] M. E. Merchant, *J. Appl. Phys.*, vol. 16, pp. 207–318, 1945.
[2] E. G. Loewen, E. R. Marshall, and M. C. Shaw, *Proc. Soc. Exp. Stress Anal.*, vol. 8, p. 1, 1951; G. Boothroyd, *Engineer*, vol. 213, p. 351, 1962.

The average shear stress is $F_s$ divided by the area of the shear plane $A_s = bt/\sin\phi$.

$$\tau = \frac{F_s}{A_s} = \frac{F_s \sin\phi}{bt} \qquad (21\text{-}10)$$

and the normal stress is

$$\sigma = \frac{F_{ns}}{A_s} = \frac{F_{ns}\sin\phi}{bt} \qquad (21\text{-}11)$$

The shear stress in cutting is the main parameter affecting the energy requirement. Reasonable values of this shear stress have been predicted[1] by a dislocation model which considers high strain rates and very large strains.

Calculation of the shear stress in cutting from force measurements requires knowledge of the shear angle $\phi$. This angle can be measured experimentally by suddenly stopping the cutting process and using metallographic techniques to determine the shear zone. However, it would be convenient to be able to predict $\phi$ from other more easily determined parameters. Merchant[2] assumed that the shear plane would be at the angle which minimizes the work done in cutting. This results in the expression

$$\phi = \frac{\pi}{4} + \frac{\alpha}{2} = \frac{\beta}{2} \qquad (21\text{-}12)$$

An analysis by Lee and Shaffer[3] using slip-line field theory gave

$$\phi = \frac{\pi}{4} + \alpha - \beta \qquad (21\text{-}13)$$

Experimental data fall between both relations with marked deviations not at all unusual. However, Eq. (21-13) generally is preferred when experimental data are lacking. These relations show that for the same rake angle, less friction (lower $\beta$) results in a larger shear angle. This implies a thinner chip and less severe deformation in the chip since the shear strain is lower, Eq. (21-3).

An interesting aspect of metal cutting is that the average shear stress on the shear plane is essentially constant,[4] independent of cutting speed, depth of cut, feed, and rake angle, although the normal stress varies with cutting conditions. It has been shown[5] that shear stresses in cutting may be combined with the flow stress determined at other strain rates and temperature through a Zener-Hollomon-type parameter.

---

[1] B. F. von Turkovitch, *Trans. ASME, Ser. B: J. Eng. Ind.*, vol. 92, pp. 151–157, 1970.
[2] M. E. Merchant, op. cit.
[3] E. H. Lee and B. W. Schaffer, *Trans. ASME, Ser. E, J. Appl. Mech.*, vol. 73, p. 405, 1951.
[4] E. G. Thomsen, *Met. Eng. Q.*, May 1964, pp. 32–41.
[5] J. A. Bailey and D. G. Bhanvadia, *Trans. ASME, Ser. H*, vol. 95, pp. 94–98, 1973.

The power (energy per unit time) required for cutting is $F_h v$. The volume of metal removed per unit time (metal-removal rate) is $Z_w = btv$. Therefore, the energy per unit volume $U$ is given by

$$U = \frac{F_h v}{Z_w} = \frac{F_h v}{btv} = \frac{F_h}{bt} \qquad (21\text{-}14)$$

where $b$ is the width of the chip, and $t$ is the undeformed chip thickness. This energy involved in removing a unit volume of metal often is referred to as the *specific cutting energy*. The specific cutting energy depends on the material being machined and also on the cutting speed, feed, rake angle, and other machining parameters. However, at high cutting speeds (greater than about 600 ft/min) it is independent of speed, and thus provides an approximate but useful indication of the forces required for machining. Figure 21-7 summarizes typical values of specific cutting energy for various materials and machining processes. The total energy for cutting can be divided into (see Prob. 21-2) a number of components:

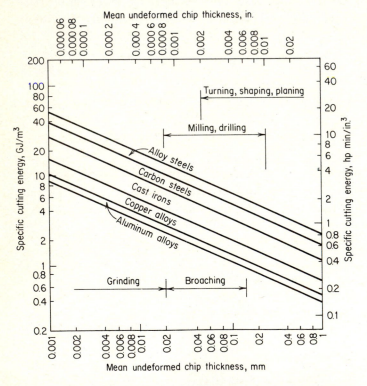

**Fig. 21-7** Typical values of specific cutting energy for various materials and machining operations. (*From G. Boothroyd, "Fundamentals of Metal Machining and Machine Tools," p. 57, McGraw-Hill Book Company, New York, 1975.*)

1. The energy required to produce the gross deformation in the shear zone,
2. The frictional energy resulting from the chip sliding over the tool face,
3. Energy required to curl the chip,
4. Momentum energy associated with the momentum change as the metal crosses the shear plane,
5. The energy required to produce the new surface area.

In a typical machining operation, the shear energy will be about 75 percent of the total, the frictional energy about 25 percent, and the energy associated with the other three components is negligible.

The model of the machining process that has been presented so far is unrealistically simplified. Basically, the model considers two-dimensional (orthogonal) cutting in which a continuous chip is produced by intense shearing over a localized shear plane. We now shall consider some of the deviations from this simplified model.

Three general classifications of chips are formed in the machining process.[1] The *continuous chip* (type 2 chip) is characteristic of cutting ductile materials under steady-state conditions. Most basic research in metal cutting utilizes this type of chip (Fig. 21-8a). However, long continuous chips present handling and removal problems in practical operations. The usual practice is to include a *chipbreaker* on the rake face of the tool to deflect the chip and cause it to break into short lengths. Under conditions where the friction between the chip and the rake face of the tool is high the chip may weld to the tool face. The accumulation of chip material (Fig. 21-8b) is known as a *builtup edge* (*BUE*). The BUE (type 3 chip) acts as a substitute cutting edge and the major part of the contact between the chip and the tool is through the BUE. Periodically the BUE breaks off, part remaining on the back side of the chip and part welding to the machined surface. Thus, BUE is a

**Fig. 21-8**  Types of machining chips. (*a*) Continuous chip; (*b*) chip with a builtup edge, BUE; (*c*) discontinuous chip.

---

[1] For a more detailed classification of chips see G. Boothroyd, "Fundamentals of Metal Machining and Machine Tools," p. 186, McGraw-Hill Book Company, New York, 1975.

source of rough surface finish. BUE is most prevalent with high feeds and low rake angle. For a set of machining parameters there usually is a cutting speed above which BUE does not occur. *Discontinuous chips* (type 1 chip) are formed in brittle materials which cannot withstand the high shear strains imposed in the machining process without fracture (Fig. 21-8c). Typical materials which machine with discontinuous chips are cast iron and cast brass, but this type of chip also may be produced in ductile materials machined at very low speeds and high feeds.

Forces other than those required to produce the sheared chip are involved in the machining process. No cutting tool is perfectly sharp, so that the tool edge "plows" its way through the work material.[1] For a large depth of cut the ploughing force is a small fraction of the total force, but for small values of depth of cut it becomes an appreciable fraction of the total cutting force. This leads to a "size effect", i.e., the specific cutting energy for processes that remove very thin chips, like grinding, is higher than for processes that produce thicker chips. Also difficult to model are the frictional forces between the chip and the rake face of the tool. Along the area of contact there is a region of sticking friction as well as an area of sliding. The best analysis of the situation is given by Zorev.[2]

Because of the complexity of practical machining operations, the machining force $F_h$ often is related empirically to the machining parameters by equations of the form

$$F_h = k d^a f^b \qquad (21\text{-}15)$$

where $d$ is the depth of cut, $f$ is the feed, and $k$ is a function of rake angle, decreasing about 1 percent per degree increase in rake angle.

## 21-3   THREE-DIMENSIONAL MACHINING

To this point we have considered only two-dimensional orthogonal machining in which the cutting edge is perpendicular to the cutting velocity vector. While a few practical machining operations (such as surface broaching, lathe cutoff operations, and plain milling) fall in this category, most practical machining operations are three dimensional. A comparison between two-dimensional and three-dimensional machining is given in Fig. 21-9. For orthogonal machining we see that rotating the tool around the $x$ axis will change the width of the cut, and rotation around the $y$ axis changes the rake angle. However, major changes in the cutting process occur if the tool is rotated about the $z$ axis by an *angle of inclination i*. Now the chip no longer flows up the tool face along a path perpendicular to the cutting edge (path $OA$), but instead it follows the path $OB$ which is inclined to the normal by

[1] P. Albrecht, *Trans. ASME, Ser B, J. Eng. Ind.*, vol. 82, pp. 348–362, 1960.
[2] N. N. Zorev, "Metal Cutting Mechanics," Pergamon Press, New York, 1966.

**Fig. 21-9** (a) Orthogonal cutting; (b) three-dimensional cutting.

the *chip flow angle* $\eta_c$. The chip flow angle may be obtained experimentally from the relation

$$\cos \eta_c = \frac{b_c}{b/\cos i} \qquad (21\text{-}16)$$

where $b_c$ is the width of the chip and $b$ is the width of the workpiece that is being cut.

Three-dimensional machining can be treated as orthogonal machining if the process is analyzed on the plane defined by the cutting velocity $v$ and the chip velocity $v_c$. The angle $zOA$ is defined as the *normal rake angle* $\alpha_n$. This is a geometric property of the tool that is independent of $i$, and for $i = 0$ it equals the rake angle for orthogonal cutting. The *effective rake angle* $\alpha_e$ is the angle between $v_c$ and the normal to $v$ in the plane through $v$ and $v_c$. From the geometry of the situation, this is given by

$$\sin \alpha_e = \sin \eta_c \sin i + \cos \eta_c \cos i \sin \alpha_n \qquad (21\text{-}17)$$

Staebler has shown[1] that to a first approximation $\eta_c = i$, so that Eq. (21-17) becomes

$$\sin \alpha_e \approx \sin^2 i + \cos^2 i \,(\sin \alpha_n) \qquad (21\text{-}18)$$

Since $i$ and $\alpha_n$ may be determined fairly readily for a tool, $\alpha_e$ can be calculated and used in the equations for orthogonal machining to make predictions about three-dimensional cutting. From Eq. (21-18) we see that increasing the inclination of the cutting edge is equivalent to increasing the rake angle. From previous discussion of orthogonal cutting we have seen that this reduces the shear strain in the chip, and reduces the cutting force, energy per unit volume, and the tendency for a BUE.

[1] G. V. Staebler, *Proc. Inst. Mech. Eng. (London)*, vol. 165, p. 14, 1951.

**Fig. 21-10** (a) Typical cutting tool terminology based on right-hand convention; (b) ANSI designation of a cutting tool. (*After S. Kalpakjian, " Mechanical Processing," p. 279, Van Nostrand Reinhold Co., New York, 1967.*)

The most common three-dimensional cutting tool is the ordinary lathe tool (Fig. 21-10). Like most practical three-dimensional tools, it has two cutting edges that cut simultaneously. The primary cutting edge is the side-cutting edge, while the secondary cutting edge is the end-cutting edge. The seven-digit code for describing the cutting tool geometry given in Fig. 21-10b is that of the American National Standards Institute (ANSI). Completely different methods for describing lathe tools are in use in Great Britain and Germany and international standards have been proposed to clarify this confusing situation.[1]

Milling and drilling are important machining operations that employ multiple-edge cutting tools.[2] The description of the tool geometry and the mechanics of the process are too detailed for discussion here.

## 21-4 TEMPERATURE IN METAL CUTTING

Although the vast majority of machining operations are conducted with the workpiece at ambient temperature, because of the large plastic strain and very high strain rate there is a significant temperature rise. This has an important bearing on

[1] G. Boothroyd, op. cit., chap. 7; A. Bhattacharyya and I. Ham, "Design of Cutting Tools," chap. 2, Society of Manufacturing Engineers, Dearborn, Mich. 1969.
[2] A. Bhattacharyya and I. Ham, op. cit., pp. 124–149; E. J. A. Armarego and R. H. Brown, "The Machining of Metals," pp. 174–200, Prentice-Hall, Inc., Englewood Cliffs, N.J., 1969.

the choice of tool materials, their useful life, and on the type of lubricant system required. The shearing energy required to form the chip is nearly all converted into heat. Heat also is generated by the rubbing under high pressure and velocity of the chip against the rake face of the tool. As a result of this rubbing action a secondary deformation zone develops in the chip adjacent to the chip-tool interface, and this also contributes to heat generation.

The heat transfer is strongly dependent on the cutting velocity. At very low cutting speeds there will be adequate time for conduction to occur, chiefly into the workpiece, and near isothermal conditions will prevail. At the other extreme, at very high cutting speeds there is little time for heat conduction and a near adiabatic condition will exist with high local temperatures in the chip. The usual machining conditions are somewhere between these two extremes.

Much data on temperature generation in metal cutting may be correlated with a dimensionless number $R_t = \kappa/vd$, where $\kappa$ is the thermal diffusivity $= k/\rho c$, $v$ is the cutting speed, and $d$ is the depth of cut. $R_t$ is called the *thermal number*. If all of the heat generated goes into the chip, the adiabatic temperature is given by

$$T_{ad} = \frac{U}{\rho c} \qquad (21\text{-}19)$$

where $U$ = the specific cutting energy
$\quad\ \rho$ = the density of the workpiece material
$\quad\ c$ = specific heat of workpiece

For lower velocities the temperature will be less than that given by Eq. (21-19). The approximate chip-tool interface temperature is given by the correlation[1]

$$\frac{T}{T_{ad}} = C \left(\frac{1}{R_t}\right)^p \qquad (21\text{-}20)$$

where $C \approx 0.4$ and the exponent $p$ ranges from $\frac{1}{3}$ to $\frac{1}{2}$. More complete analyses of the temperature in the chip have been given by Lowen and Shaw,[2] Rapier,[3] and Weiner.[4]

## 21-5 CUTTING FLUIDS

We have seen that machining processes involve high local temperatures and high friction at the chip-tool interface. Thus most practical machining operations use a cutting fluid designed to ameliorate these effects. The primary functions of a cutting fluid are:

[1] N. H. Cook, "Manufacturing Analysis," p. 47, Addison-Wesley Publishing Company, Inc., Reading, Mass., 1966.
[2] E. G. Loewen and M. C. Shaw, *Trans. ASME*, vol. 76, pp. 217–231, 1954.
[3] A. C. Rapier, *J. Appl. Phys.*, vol. 5, p. 400, 1954.
[4] J. H. Weiner, *Trans. ASME*, vol. 77, p. 1331, 1955.

1. To decrease friction and wear,
2. To reduce temperature generation in the cutting area,
3. To wash away the chips from the cutting area,
4. To protect the newly machined surface against corrosion.

While the last two factors are important practical considerations, the first two factors are the prime functions of a cutting fluid. Important practical objectives which grow out of these functions are increased tool life, improved surface finish, reduced cutting forces and power consumption, and reduced thermal distortion of the workpiece. Cutting fluids usually are liquids, but they may be gases. Solid lubricants also play a role in improving machinability.

There are two basic types of liquid cutting fluids: petroleum-based non-soluble fluids (straight cutting oils) and water-miscible fluids (soluble oils). Many additives are used in conjunction with each of these types of fluids to achieve a specific purpose. A cutting fluid of the first type may contain one or more parts of mineral oil, fatty oils, sulfur, or chlorine. Water-soluble oils may contain some combination of fatty oils, fatty acids, wetting agents, emulsifiers, sulfur, chlorine, rust inhibitors, and germicides in water. Since water has nearly twice the specific heat of any other cutting fluid, water-based fluids are better coolants than oil-based cutting fluids, and they tend to be used for high-speed machining. They also are more economical than oil-based fluids. Petroleum-based cutting fluids are better lubricants and are often preferred at lower cutting speeds. Cutting fluid selection is a field in which experience plays a major role.

The new surfaces created in cutting are clean and hot, and therefore they are active sites for chemical reaction. The sulfur and chlorine added to the cutting fluid react with the fresh metal surfaces to form sulfides or chlorides. Generally these are compounds with low shear strength and therefore they reduce friction. Iron chloride usually has a lower shear strength than iron sulfide, but the reaction time is longer to form the chloride compound. Thus, while chlorinated fluids work well at low speed and light loads, they are not as effective as sulfur compounds under severe conditions. Both sulfur and chlorine usually are added together to provide for effective lubrication over a range of machining conditions.

The cutting fluid reaches the chip-tool interface by a number of processes. For thin chips, it has been shown[1] that cutting fluid can diffuse through the highly distorted structure of the metal in the chip. In general the back of the chip is not in complete contact with the rake face of the tool so that low-surface-tension liquid can penetrate these cavities by capillary action. Since the chips are hot, the fluid often is a vapor, which can more easily penetrate in narrow cavities. Moreover, chlorine compounds have been shown[2] to prevent the welding shut of fine cracks in

---

[1] C. Cassin and G. Boothroyd, *J. Mech. Eng. Sci.*, vol. 7, p. 67, 1965.
[2] E. Usui, A. Gujral, and M. C. Shaw, *Int. J. Mach. Tool Des., Res.*, vol. 1, p. 187, 1961.

highly sheared metal, a mechanism which could introduce many paths to the chip-tool interface.

A form of solid lubrication is provided by certain *free-machining additions* to the work material. Lead added to brass or steel is present in the form of fine dispersed particles. When the leaded metal is cut, the lead spreads over the cut surface to provide a low-shear-stress interface.

## 21-6  TOOL MATERIALS AND TOOL LIFE

Cutting tools are subjected to high forces under conditions of high temperature and wear. Three main forms of wear occur in metal cutting. In *adhesion wear* the tool and the chip weld together at local asperities, and wear occurs by the fracture of the welded junctions. *Abrasive wear* occurs as a result of hard particles on the underside of the chip abrading the tool face by mechanical action as the chip passes over the rake face. The hard particles can arise from hard constituents of either the work-piece or the tool, or highly strain-hardened fragments of a BUE. Wear may also result from *solid-state diffusion* from the tool material to the workpiece at the high temperature and intimate contact existing at the interface between the chip and the rake face. Thus, a good cutting tool material[1] must possess high hot hardness and wear resistance, with sufficient toughness to resist fracture or chipping. It should have good thermal shock resistance and low reactivity with the workpiece material. It should be easy to form, grind, and sharpen to the desired tool profile, and it should be economical.

Two main types of wear are observed in a cutting tool: (1) *flank wear* is the development of a *wear land* on the tool due to abrasive rubbing between the tool flank and the newly generated surface, and (2) *crater wear* is the formation of a circular crater in the rake face of the tool (see Fig. 21-11). Other modes of wear are chipping or fracture of the tool, rounding of the cutting edge, or plastic deformation of the tool.

The primary cause of crater wear is diffusion wear due to the high temperature developed at the interface between the chip and the rake face of the tool. Typically the crater forms at about the midpoint of the tool-chip contact area at a distance from the tool tip corresponding to the location of maximum temperature. The crater is formed by wear processes which are accelerated by the local softening of the tool material. Crater wear is followed by plotting crater depth vs. cutting time. A sharp increase in rate of crater wear often is observed above a certain temperature

---

[1] E. M. Trent, *Metall. Rev.*, no. 127, pp. 129–144, 1968; "Metals Handbook," op. cit., pp. 311–332.

**Fig. 21-11**   Crater wear and flank wear.

This corresponds to a temperature at which the removal of atoms from the tool to the chip is accelerated due to a rapid rate of diffusion.  The predominant wear process depends on cutting speed.  At higher speeds crater wear is predominant, while at lower speeds it is flank wear.

Flank wear usually is monitored by plotting the length of the wear land $w$ against cutting time (Fig. 21-12).  This typical wear curve shows an initial region where the sharp cutting edge quickly is broken down to establish a wear land, a linear region of constant wear rate, and finally, a region of accelerated wear.  A similar curve could be obtained by measuring the depth of the crater.

The oldest cutting tool materials are the *high-carbon tool steels* containing from 0.7 to 1.5 percent carbon.  These are chiefly water-hardening and limited in application to low cutting speeds because they fall rapidly in hardness above 300°F.  Some improvement in hot hardness is achieved by alloying elements such as chromium, vanadium, tungsten, and cobalt.

High-speed tool steels retain their hot hardness to about 1000°F and can be used at cutting speeds about twice as fast as carbon tool steels.  The classic example of high-speed steel is 18-4-1 (W-Cr-V), but since World War II the M-series steels, in which most of the expensive tungsten is replaced by molybdenum, have been the most common grades.  The popular M-2 high-speed steel contains 6% W–4% Cr–2%V–5% Mo.

*Cast nonferrous tools* typically contain 40% Co–35% Cr–20% W.  They are called nonferrous because they contain no iron.  The structure of these materials is chiefly complex intermetallic compounds which retain useful hardness to about 1500°F.  Because the materials are brittle, they must be cast to shape and used in situations where impact and vibration are not severe.

A major achievement in tool materials was the introduction of *cemented carbide* tools, which consist of heat-resistant refractory carbides embedded in a ductile metal matrix.  They are produced by powder metallurgy using liquid-phase sintering.  The most common system consists of tungsten carbide with a

**Fig. 21-12**   Typical wear curve for cutting tool.

cobalt binder.  To minimize crater wear when machining steel, titanium and tanta-lum carbide usually are combined with the tungsten carbide.  Cemented carbides are useful at cutting temperatures up to 2000°F, and they can be used at speeds up to five times that used with high-speed steel.  Cemented carbide tools are used in the form of inserts or tips that are brazed to a steel shank, or in the form of throwaway inserts with multiple cutting edges that are mechanically fastened to the tool post.  Cemented carbides are brittle, and to prevent fracture the tool should be rigidly supported and run without vibration or chatter.  Coated carbides consisting of a very thin layer of TiC or TiN over the WC-Co tool reduce the effects of adhesion and diffusion, and reduce crater wear.

*Ceramic* or *oxide tools* of sintered $Al_2O_3$ are harder than cemented carbides, possess better wear reistance, and have less tendency for the tool to weld to the chip.  They also possess high compressive strength, but they are more brittle than WC.  Ceramic tools can operate at cutting speeds two to three times that of cemented carbides in uninterrupted cuts where shock and vibration are minimized. Tool failure can be reduced by the use of rigid tool mounts and rigid machine tools.  *Diamond* offers the highest hot hardness of any material.  Synthetic diamonds were produced in the late 1950s, and cubic boron nitride (CBN) in the early 1970s, by high-pressure high-temperature processing.  Until recently, diamond and CBN were used only for grinding, but in the midseventies "compax" tools in which these materials are sintered to a WC substrate were developed for machining.

Tool life is a very important parameter in metal cutting, but it is difficult to define without ambiguity.  In the most general sense, tool life is determined by the point at which the tool no longer makes economically satisfactory parts.  However, more specific criteria of tool life usually are used.  Complete destruction of the tool when it ceases to cut is one such criterion.  More typically, the tool life usually is defined in terms of an average or maximum allowable wear land.  A typical value would be 0.3 mm.  Less often, the tool life might be based on the degradation of the surface finish below some specified limit, or the increase in the cutting force above some value, or when the vibrational amplitude reaches a limiting value.

Cutting speed is the most important operating variable influencing tool temperature, and hence, tool life. The classic study by Taylor[1] established the relationship between cutting speed $v$ and the time $t$ to reach a wear land of certain dimensions as

$$vt^n = \text{constant} \qquad (21\text{-}21)$$

Typical values of the exponent $n$ are: high-speed steel, 0.1; cemented carbide, 0.2; ceramic tool, 0.4. Significant changes in tool geometry or machining parameters will change the value of the constant. The tool life used in this equation is the machining time between regrinding the tool, not the total life of the tool before it is discarded. While cutting speed is the major variable affecting tool life feed $f$ and depth of cut $d$ also are important parameters. Thus, the Taylor equation often is extended to the form

$$vt^n d^x f^y = \text{constant} \qquad (21\text{-}22)$$

It is important to realize that the Taylor equation is completely empirical and, as with other empirical relationships, it is dangerous to extrapolate outside of the limits over which the data extend.

To summarize briefly, tool life will depend on the tool material, the machining parameters, and on such factors as the cutting fluid and the properties of the workpiece material. The ability of a material to be machined has been termed *machinability*. A material has good machinability if the tool wear is low, the tool life is long, the cutting forces are low, the chips are well behaved and break into small ringlets instead of long snarls, and the surface finish is acceptable. This illustrates the problem in defining machinability in a scientific sense; there are too many criteria for evaluating this elusive material property. Thus, there are no standard tests for evaluating machinability.

One of the most common methods of evaluating machinability is in terms of the cutting velocity for a specified tool life, e.g., 60 or 90 min. The machinability of any material is compared with a standard material. Thus, the *relative machinability* may be defined as ($v_{60}$ material A)/($v_{60}$ standard material). However, defining machinability in this way essentially is equivalent to describing tool life. If the *tool* material changes, the relative machinability ratio would be expected to change. Moreover, this method does not consider the surface finish produced. Some success at predicting relative machinability from basic strength and ductility parameters has been obtained.[2]

[1] F. W. Taylor, *Trans. ASME*, vol. 28, pp. 31–350, 1907.
[2] A. Henkin and J. Datsko, *Trans. ASME, Ser. B, J. Eng. Ind.*, vol. 85, pp. 321–328, 1963.

## 21-7  GRINDING PROCESSES

Grinding processes employ an abrasive wheel containing many grains of hard material bonded in a matrix. The action of a grinding wheel may be considered a multiple-edge cutting tool except the cutting edges are irregularly shaped and randomly spaced around the face of the wheel. Each grain removes a short chip of gradually increasing thickness, but because of the irregular shape of the grain there is considerable ploughing action between each grain and the workpiece. This results in progressive attritious wear of the grain in which the sharp edges become dull and the force on the grain increases because of increasing friction. Eventually the force on the grain becomes high enough either to tear the grain from the bonding matrix and expose a sharp fresh grain, or to fracture the worn grain to expose new cutting edges. In either case, a grinding wheel has a self-sharpening feature. However, under other conditions a wheel can become "loaded" with dull grains and/or fine adhering chips. Then the wheel must be *dressed* by passing a diamond-tip tool across the wheel face to generate new sharp cutting surfaces.

There are several aspects of the grinding process which make it distinctly different from other metal-removal processes. These differences chiefly are related to the different nature of the cutting tool in grinding. While most cutting processes are carried out with tools having positive rake angles, the shape of the abrasive grains results in a tool with large negative rake angles. Thus, there is a tendency for the grain to slide over the work rather than cut. The depth of cut in grinding usually is very small ($10^{-4}$ in), and this results in very small chips that adhere readily to the wheel or the workpiece. The net effect is that the specific cutting energy for grinding is about 10 times greater than for turning or milling (see Fig. 21-7). In other cutting processes only about 5 percent of the energy input ends up in the finished surface, the bulk going into the chip. In grinding, greater than 70 percent of the energy goes into the finished surface. This results in considerable temperature rise and generation of residual stresses.

One of the chief applications of grinding is in *finishing operations*, where the ability to produce a highly finished surface to precise dimensions is important. The ability to grind a hard surface which cannot be machined with a conventional cutting tool also is important. Finishing processes which use abrasives more to produce a very smooth surface than to remove metal are *honing*, *lapping*, and *superfinishing*. However, there is one area of grinding processes used primarily to remove metal at high rate (abrasive machining). Examples are billet conditioning, rough grinding of castings, and abrasive cutoff operations.

A grinding wheel is a cutting tool whose properties and performance can be varied over wide limits by the selection and control of the abrasive grains[1] and the

---

[1] R. Komanduri and M. C. Shaw, *Trans. ASME, Ser. H*, vol. 96, pp. 145–156, 1974.

bonding matrix. Most commercial grinding wheels employ aluminum oxide $Al_2O_3$ or silicon carbide SiC as abrasive grain. Frequently these materials are alloyed with oxides of titanium, chromium, vanadium, zirconium, etc., to impart special properties. There is a growing use of diamond wheels for fine finishing. The performance of the wheel chiefly is controlled by the strength of the bond, which is a function of the type and amount of bonding material used. The most common bonding materials are a glassy *vitrified bond*, a *silicate bond* made from hardened water glass, a *resinoid bond* prepared from organic resins, and a *rubber bond*. The *grade* of a wheel expresses the degree to which the abrasive grains are bonded to the matrix. A soft wheel releases a worn grain easier than a hard wheel. The relative proportion of abrasive grain and binder, and the spacing between grains, is called the *structure* of the wheel. Structure is one of the variables that determines the number of cutting points per unit area on the face of the wheel and the clearance between grains for removing chips. Another variable relative to effective grain spacing is grain size. A wheel with an open structure provides for easy chip removal. but it will tend to act as a soft wheel compared with a wheel with a closed structure, The proper wheel for a grinding operation is one in which the bond permits a sharp grain to cut but allows a dull grain, which is acted on by a greater force, to be fractured or pulled from the wheel.

Figure 21-13 illustrates the geometry involved in surface grinding. The velocity of the grinding wheel is $v_g$ and the velocity of the workpiece is $v_w$. Typically, grinding wheels move at very high cutting speeds (3,000 to 16,000 ft/min) while $v_w$ might typically be 60 ft/min. An important variable in the grinding process is

(a)    (b)

**Fig. 21-13** (a) Geometry in surface grinding; (b) approximate cross section of grinding chip.

the *grain depth of cut t* (analogous to the undeformed chip thickness in ordinary metal cutting), which is distinct and much smaller than the wheel depth of cut (infeed) $d$. The metal removed by a grain is area $OAB$ times the chip width $b$. The average length of chip $L_c$ is

$$L_c \approx \frac{D}{2} \sin \theta \qquad (21\text{-}23)$$

But

$$\cos \theta = \frac{(D/2) - d}{D/2} = 1 - \frac{2d}{D} \qquad (21\text{-}24)$$

and

$$\sin^2 \theta = 1 - \cos^2 \theta = \frac{4d}{D} - \frac{4d^2}{D^2} \qquad (21\text{-}25)$$

Neglecting the second-order term $4d^2/D^2$, and substituting Eq. (21-25) into Eq. (21-23), results in

$$L_c \approx \sqrt{Dd} \qquad (21\text{-}26)$$

The total volume of chips removed per unit time $V_{\text{chip}}$ is

$$V_{\text{chip}} = v_w bd \qquad (21\text{-}27)$$

These chips tend to approximate a triangular cross section (Fig. 21-13b) where $r = b'/t$ is between 10 and 20. The average volume of a chip is $(b'tL_c)/4 = (rt^2L_c)/4$. The number of chips produced per unit time is

$$n = v_g bC \qquad (21\text{-}28)$$

where $C$ is the number of active grains on the wheel per square inch (typically 2,000).

The volume of material removed per unit time is $nV_{\text{chip}} = v_w bd$. Thus,

$$v_g bC \frac{(rt^2L_c)}{4} = v_w bd \qquad (21\text{-}29)$$

and since $L_c \approx \sqrt{Dd}$,

$$t = 2 \sqrt{\frac{v_w}{Crv_g}} \sqrt{\frac{d}{D}} \qquad (21\text{-}30)$$

The grain depth of cut $t$ given above is its maximum value. The mean value of $t$ is $\bar{t} = t/2$. This analysis was for surface grinding, where the workpiece is a flat surface. Two other common grinding processes are the grinding of the external surface of a cylinder (cylindrical grinding) and grinding the inside cylindrical surface of a hollow cylinder (internal grinding). The above analysis can be used for these processes simply by substituting the equivalent diameter $D_e$ for the wheel diameter $D$ in Eq. (21-30).

$$D_e = \frac{D_g D_w}{D_w \pm D_g} \qquad (21\text{-}31)$$

where $D_g$ is the grinding wheel diameter, $D_w$ is the workpiece diameter, and the plus sign is used for cylindrical grinding and the minus sign is used for internal grinding.

The specific cutting energy in grinding is

$$U = \frac{F_h v_g}{v_w bd} \qquad (21\text{-}32)$$

As already discussed, $U$ is appreciably higher for grinding than for turning. $U$ is strongly dependent on $t$, and to a first approximation

$$U \propto \frac{1}{t} \qquad (21\text{-}33)$$

If the grain cross section is assumed triangular, the force on a single abrasive grain $F_g$ will be

$$F_g = U(\tfrac{1}{2}b't) = \tfrac{1}{2}Urt^2 \qquad (21\text{-}34)$$

Substituting Eq. (21-33) into Eq. (21-34), we find

$$F_g \propto rt \propto \sqrt{\frac{rv_w}{Cv_g}} \sqrt{\frac{d}{D}} \qquad (21\text{-}35)$$

Equation (21-35) illustrates the important point that the hardness of a grinding wheel depends on both the strength of the bond *and* the grinding conditions, since the greater the value of $F_g$ the more easily the grains fracture or tear out. For a wheel of a given grade, the greater $F_g$ the softer the wheel appears to be. Thus, a hard wheel can be made to grind softer by increasing the workpiece velocity or depth of cut, or by decreasing the wheel speed or the number of cutting points (move to a more open structure).

It was pointed out earlier that grinding requires a much larger energy input per unit volume than cutting, and that a far larger portion of this energy goes to raising the temperature of the workpiece. While a rigorous analysis is quite difficult[1] it has been established that the surface temperature $T_w$ depends strongly on the energy per *unit surface area* that is ground.

$$T_w \propto \frac{F_g v_g}{v_w b} \propto Ud \qquad (21\text{-}36)$$

and since $U \propto 1/t$, because of the size effect,

$$T_w \propto \frac{d}{t} d^{3/4} \sqrt{\frac{v_g C}{v_w}} \sqrt{D} \qquad (21\text{-}37)$$

Temperatures of 3000°F or higher can be created in the ground surface. Since the time involved in producing a chip is of the order of microseconds, the peak tempera-

---

[1] S. Malkin, *Trans. ASME Ser. B., J. Eng. Ind.*, vol. 96, pp. 1184–1191, 1974.

ture is very transitory. However, the temperature rise in grinding can lead to melting, or metallurgical changes such as the production of untempered martensite, or grinding cracks, or a visible oxidation of the surface known as grinding burn. Grinding carried out under improper conditions can result in high tensile residual stresses in the ground surface. Residual stresses can be minimized by using the proper grinding fluid and softer wheels at lower wheel speeds.

While the picture of grinding given above adequately describes many aspects of the process, it does not explain such important observations as the large thermal input into the finished surface, or why the specific cutting energy for fine grinding is at least 10 times greater than for metal cutting. Shaw[1] has presented a new approach in which the action of an abrasive grain is modeled with the plastic deformation under a Brinell hardness indentor (see Fig. 11-2). In this model the grinding chip is removed by a special type of extrusion rather than a concentrated shear process as in ordinary metal cutting.

*Grinding ratio* (*G* ratio) is the volume of material removed from the work per unit volume of wheel wear. It is a measure of ease of grinding or grindability. The higher the *G* ratio the easier a material is to grind. The *G* ratio depends on the grinding process and grinding conditions (wheel, fluid, speed, and feed) as well as the material. Values of *G* ratio range from about 2 to over 200.

## 21-8  NONTRADITIONAL MACHINING PROCESSES

A number of new material-removal processes[2] have been developed since World War II which mostly use forms of energy other than mechanical energy. The impetus for developing most of these processes was the search for better ways of machining complex shapes in hard materials. The various techniques are listed below in terms of the major energy source. However, space permits only a brief description of a few of these processes.

| Source of energy | Name of process |
| --- | --- |
| Thermal energy processes | Electrical discharge machining, EDM<br>Laser-beam machining, LBM<br>Plasma-arc machining, PAM |
| Electrical processes | Electrochemical machining, ECM<br>Electrochemical grinding, ECG |
| Chemical process | Chemical machining, CHM |
| Mechanical process | Ultrasonic machining, USM |

[1] M. C. Shaw, *Mech. Chem. Eng. Trans.*, vol. MC8, no. 1, pp. 73–78, May 1972.
[2] R. K. Springborn (ed.), "Non-Traditional Machining Processes," Society of Manufacturing Engineers, Dearborn, Mich., 1967.

*Electrical discharge machining*[1] (*EDM*) is a method for producing holes, slots, or other cavities in an electrically conductive material by the controlled removal of material through melting or vaporization caused by a high-frequency spark discharge. The workpiece (the anode) and the tool (the cathode) are immersed in a dielectric fluid with a spark gap of 0.0005 to 0.020 in. Important advantages of EDM are that it can produce deep holes in hard material without drifting, or can machine cavities of irregular contour. The metal-removal rate is independent of the hardness of the workpiece, but it does depend on thermal properties such as heat capacity and conductivity, melting point, and latent heat of melting and vaporization. Electrode wear is a problem which requires selection of the proper electrode material for the workpiece material. Because high temperatures are attained in the spark, the metal is melted and then rapidly quenched by a mass effect when it resolidifies. This recast layer may be deleterious to fatigue properties.

*Electrochemical machining*[2] (*ECM*) is the controlled removal of metal by anodic dissolution in an electrolytic cell in which the workpiece is the anode and the tool is the cathode. It is similar to a reverse electroplating process. The electrolyte is pumped through the cutting gap while direct current is passed through the cell at low voltage to dissolve the metal from the workpiece. The rate of removal of material is proportional to the amount of current passing between the tool and the workpiece and is independent of the hardness of the workpiece. ECM is a cold process which results in no thermal damage to the workpiece. It results in a smooth burr-free surface. However, it is not suited for producing sharp corners or cavities with flat bottoms. *Electrochemical grinding* (*ECG*) is a combination of ECM and abrasive grinding in which most of the metal is removed by electrolytic action. It is used with hard carbides or difficult-to-grind alloys where wheel wear or surface damage must be minimized.

*Chemical machining*[3] (*CHM*) involves metal removal by controlled chemical attack with chemical reagents. The essential steps involve cleaning the surface, masking those areas which are not to be dissolved, attacking with chemicals, and cleaning. Chemical milling refers to chemical machining of large areas, such as aircraft structural parts. Chemical blanking is used for cutting or stamping parts from very thin sheet.

In *ultrasonic machining*[4] (*USM*) the tool is excited at around 20,000 cycles/sec with a magnetostrictive transducer while a slurry of fine abrasive particles is introduced between the tool and the workpiece. Each cycle of vibration removes minute pieces of the workpiece by fracture or erosion. USM mostly is used for machining brittle hard materials such as semiconductors, ceramics, or glass.

---

[1] Electrical Discharge Machining, " Metals Handbook," op. cit., pp. 227–233.
[2] Electrochemical Machining, " Metals Handbook," op. cit., pp. 233–240.
[3] Chemical Milling, " Metals Handbook," op. cit., pp. 240–249; J. W. Dini, *Int. Metall. Rev.*, vol. 20, pp. 29–55, March 1975.
[4] Ultrasonic Machining, " Metals Handbook," op. cit., pp. 249–251.

## 21-9  ECONOMICS OF MACHINING

The annual cost of operating machine tools in the United States (labor and over-head) exceeds \$40 billion. Thus, there is a demonstrated need for understanding the cost elements in machining. In addition, the problem of determining the cutting speed to give minimum cost is a classic problem in cost optimization using Kelvin's law. If we consider for the moment only the costs involved in actual cutting, the machining time (and thus the cost of machining labor for one piece) will decrease with increasing speed or feed. However, tool wear will increase with speed and feed, and thus, there will be increased costs for tools and tool changing. There will be an optimum speed which balances these opposing factors and results in minimum cost per piece.

The *total cost* of a machined piece is the sum of four costs:

$$C_u = C_m + C_n + C_c + C_t \qquad (21\text{-}38)$$

where $C_u$ = the total unit (per piece) cost

$C_m$ = the machining cost

$C_n$ = the cost associated with nonmachining time; including setup cost of mounting fixtures in the machine and preparing it for operation, time for loading and unloading parts in the machine, and idle machine time

$C_c$ = the cost of tool changing

$C_t$ = the tool cost per piece

*Machining cost* can be expressed by

$$C_m = t_m(L_m + O_m) \qquad (21\text{-}39)$$

where $t_m$ = the machining time per piece (including the time the feed is engaged whether or not the tool is cutting)

$L_m$ = the labor cost of a production operator per unit time

$O_m$ = the overhead charge for the machine, including depreciation, indirect labor, maintenance, etc.

For a simple turning operation, the machining time for one piece is

$$t_m = \frac{L}{fN} = \frac{LD}{12fv} \qquad (21\text{-}40)$$

where $L$ = the length of cut

$f$ = the feed, in/rev

$N$ = the rpm of the workpiece

$D$ = the diameter of the workpiece, in

$v$ = the cutting speed, ft/min

The *cost of nonmachining time* $C_n$ is usually expressed as a fixed cost in dollars per piece. The *cost of tool changing* $C_c$ can be expressed by

$$C_c = t_g\left(\frac{t_{ac}}{t}\right)(L_g + O_g) \qquad (21\text{-}41)$$

where $t_g$ = the time required to grind and change a cutting edge
$\quad\ t_{ac}$ = the actual cutting time per piece
$\quad\quad t$ = the tool life for a cutting edge
$\quad\ L_g$ = the labor rate for a toolroom operator
$\quad\ O_g$ = the overhead rate for the toolroom operation
The Taylor equation for tool life can be written

$$vt^n = \text{constant} = K$$

or

$$t = \left(\frac{K}{v}\right)^{1/n} \qquad (21\text{-}42)$$

$$t_{ac} = \frac{\pi L_a D}{12fv}$$

where $L_a$ is the actual length of cut and

$$C_c = t_g(L_g + O_g)\frac{\pi L_a Dv^{(1/n)-1}}{12fK^{1/n}} \qquad (21\text{-}43)$$

The *tool cost* per piece can be expressed by

$$C_t = C_e \frac{t_{ac}}{t} \qquad (21\text{-}44)$$

where $C_e$ is the cost of a cutting edge, and $t_{ac}/t$ is the number of tool changes required per piece.

The *total unit cost* is given by

$$C_u = C_n + C_m + C_c + C_t$$

or $\quad C_u = C_n + (L_m + O_m)Av^{-1} + (L_g + O_g)t_g\dfrac{Av^{(1/n)-1}}{K^{1/n}} + C_e\dfrac{Av^{(1/n)-1}}{K^{1/n}} \qquad (21\text{-}45)$

where $A = (\pi LD)/12f$.

The optimum value of cutting speed to give minimum unit cost can be obtained by setting $\partial C_u/\partial v = 0$. The component costs are shown in Fig. 21-14. The minimum in the total unit cost curve occurs chiefly because tool wear increases rapidly with cutting speed. Improvements in cutting tool materials would push the optimum speed to higher values. Worked examples comparing different tool materials have been presented by Boothroyd.[1] Detailed equations for costs for turning, milling, drilling, reaming, and tapping have been published.[2]

An alternative strategy to operating at the cutting speed to minimize unit cost

---

[1] G. Boothroyd, op. cit., pp. 161–164.
[2] M. Field, N. Zlatin, R. Williams, and M. Kronenberg, *Trans. ASME, Ser. B, Eng. Ind.*, vol. 90, pp. 455–466, 585–596, 1968.

**Fig. 21-14** Variation of machining costs with cutting speed.

is to operate at the cutting speed that results in maximum production rate. This occurs at a higher value of cutting speed (Fig. 21-14). Still another alternative[1] is to operate at the value of cutting speed which maximizes profit.

## BIBLIOGRAPHY

Armarego, E. J. A., and R. H. Brown: "The Machining of Metals," Prentice-Hall, Inc. Englewood Cliffs. N.J., 1969.

Bhattacharyya, A., and I. Ham: "Design of Cutting Tools," Society, of Manufacturing Engineers, Dearborn, Mich., 1969.

Boothroyd, G.: "Fundamentals of Metal Machining and Machine Tools," McGraw-Hill Book Company, New York, 1975.

Cook, N. H.: "Manufacturing Analysis," Addison-Wesley Publishing Company, Inc., Reading, Mass., 1966.

Kronenberg, M.: "Machining Science and Applications," Pergamon Press, New York, 1966.

"Metals Handbook," 8th ed., vol. 3, Machining, American Society for Metals, Metals Park, Ohio, 1967.

Shaw, M. C.: "Metal Cutting Principles," 3d ed., The Technology Press of The Massachusetts Institute of Technology, Cambridge, 1954.

---

[1] A. M. Wu and D. S. Ermer, *Trans. ASME, Ser. B, J. Eng. Ind.*, vol. 88, pp. 435–442, 1966.

# appendix A
## THE INTERNATIONAL SYSTEM OF UNITS (SI)

**BASIC SI UNITS APPROPRIATE TO MECHANICAL METALLURGY**

| Quantity | Name | Symbol |
|---|---|---|
| Length | Meter | m |
| Mass | Kilogram | kg |
| Time | Second | s |
| Thermodynamic temperature | Kelvin | K |
| Amount of substance | Mole | mol |

## SI UNITS DERIVED FROM BASIC UNITS

| Quantity | Name | Symbol | Quantity expressed in other units |
|---|---|---|---|
| Area | Square meter | $m^2$ | |
| Volume | Cubic meter | $m^3$ | |
| Velocity | Meter per second | m/s | |
| Density | Kilogram per cubic meter | $kg/m^3$ | |
| Concentration | Mole per cubic meter | $mol/m^3$ | |
| Force | Newton | N | |
| Energy (work) | Joule | J | N-m |
| Power | Watt | W | J/s |
| Stress (pressure) | Pascal | Pa | $N/m^2$ |
| Strain | Meter per meter | m/m | |
| Fracture toughness | | $MPa\sqrt{m}$ | |
| Viscosity (dynamic) | Pascal-second | Pa-s | |
| Surface tension | Newton per meter | N/m | |
| Specific heat | Joule per kilogram kelvin | J/(kg K) | |

## DECIMAL MULTIPLES AND SUBMULTIPLES OF SI UNITS

| Factor | Prefix | Symbol | Factor | Prefix | Symbol |
|--------|--------|--------|--------|--------|--------|
| $10^9$ | giga | G | $10^{-9}$ | nano | n |
| $10^6$ | mega | M | $10^{-6}$ | micro | $\mu$ |
| $10^3$ | kilo | k | $10^{-3}$ | milli | m |
| $10^2$ | hecto | h | $10^{-2}$ | centi | c |
| $10^1$ | deka | da | $10^{-1}$ | deci | d |

## CONVERSIONS FROM ENGLISH TO SI UNITS
(English unit $\times$ conversion factor = SI unit)

| Quantity | English unit | Conversion factor | SI unit |
|----------|-------------|-------------------|---------|
| Area | $in^2$ | $6.452 \times 10^{-4}$ | $m^2$ |
| Volume | $in^3$ | $1.639 \times 10^{-5}$ | $m^3$ |
| Velocity | ft/min | $5.080 \times 10^{-3}$ | m/s |
| Density | $lb_m$/cu in | $2.768 \times 10^4$ | $kg/m^3$ |
| | $g/cm^3$ | $1.000 \times 10^3$ | $kg/m^3$ |
| Force | $lb_f$ | 4.448 | N |
| | $kg_f$ | 9.806 | N |
| | dyne | $1.000 \times 10^{-5}$ | N |
| Energy | $ft\text{-}lb_f$ | 1.356 | J |
| | Btu | $1.055 \times 10^3$ | J |
| | cal | 4.190 | J |
| Power | hp | $7.460 \times 10^2$ | W |
| | Btu/h | $2.931 \times 10^{-1}$ | W |
| Stress | $lb_f/in^2$ (psi) | $6.895 \times 10^3$ | Pa |
| | ksi ($10^3$ psi) | 6.895 | MPa |
| | $dyn/cm^2$ | $1,000 \times 10^{-1}$ | Pa |
| | $kg_f/mm^2$ | $9.806 \times 10^6$ | Pa |
| Fracture toughness | $ksi\sqrt{in}$ | 1.098 | $MPa\sqrt{m}$ |
| Viscosity | centipoise | $1.000 \times 10^{-3}$ | Pa-s |
| Temperature | °F | $\frac{5}{9}(°F + 460)$ | K |

# CONSTANTS AND CONVERSION FACTORS

1 angstrom unit, $\text{Å} = 10^{-8}$ cm $= 10^{-10}$ m

1 micron, $\mu$m $= 10^{-3}$ mm $= 10^4$ Å $= 10^{-6}$ m

1 centimeter $= 0.39370$ in

1 inch $= 2.5400$ cm

1 foot $= 30.480$ cm

1 radian $= 57.29°$

1 pound $= 453.59$ g

1 kilogram $= 2.204$ lb

1 dyne $= 2.24 \times 10^{-6}$ lb

1 erg $= 1$ dyne-cm $= 10^{-7}$ joule

1 foot-pound $= 1.3549$ joules

1 calorie $= 4.182$ joules

1 kilocalorie $= 10^3$ cal

1 electron volt, eV $= 1.602 \times 10^{-12}$ erg $= 23{,}050$ cal/g-atom

Avogadro's number $n = 6.02 \times 10^{23}$ molecules per mole

Universal gas constant $R = 1.987$ cal/(deg)(mole) $= 8.34 \times 10^7$ erg/(deg)(mole)

Boltzmann's constant $k = 1.380 \times 10^{-16}$ erg/(deg)(mole)

Planck's constant $h = 6.6234 \times 10^{-27}$ erg-s

1 atmosphere $= 14.697$ psi

1 pound per square inch $= 7.04 \times 10^{-4}$ kg/mm$^2$ $= 6.93 \times 10^4$ dyn/cm$^2$

1 megapascal $(\text{MN/m}^2) = 145$ psi

1 square inch $= 6.4516$ cm$^2$

1 square foot $= 929.03$ cm$^2$

1 cubic inch $= 16.387$ cm$^3$

$\ln x = \log_e x = 2.3026 \log_{10} x$

The base of the natural system of logarithms $= 2.718$

$kT$ at room temperature (300K) $= 4.8 \times 10^{-10}$ erg $= 1/40$ eV

Lattice parameter: $a_0 \approx 3.5 \times 10^{-8}$ cm

Burgers vector: $\mathbf{b} \approx 2.5 \times 10^{-8}$ cm

# appendix B

## PROBLEMS

## CHAPTER 1

*1-1*  An annealed-steel tensile specimen ($E = 30 \times 10^6$ psi) has a 0.505-in minimum diameter and a 2in-gage length.  Maximum load is reached at 15,000 lb, and fracture occurs at 10,000 lb.
(*a*) What is the tensile strength?
(*b*) Why does fracture occur at a lower load than, maximum load?
(*c*) What is the deformation when a tensile stress of 15,000 psi is applied?

*1-2*  A wire 300 m long elongates by 3 cm when a tensile force of 200 N is applied.  What is the modulus of elasticity (in MPa) if the diameter of the wire is 5 mm?  Compare your answer with the result in units of psi.

*1-3*  Show that the deformation of a bar subjected to an axial load $P$ is $\delta = PL/AE$.

*1-4*  The deflection of a cantilever beam of length $L$ subjected to a concentrated load $P$ at one end is given by $\delta = PL^3/3EI$, where $I$ is the moment of inertia.  Compare the relative deflections of steel, titanium, and tungsten beams.  See Table 2-1 for necessary data.

*1-5*  A hollow copper cylinder (OD = 10 in  and ID = 5 in) contains a solid steel core.  Determine the compressive stresses in the steel and the copper when a force of 200,000 lb is applied with a press.

1-6  Discuss the relative factor of safety to be applied to the following situations: (1) box-car coupling; (2) pressure vessel for nuclear reactor; (3) missile nose cone; (4) flagpole (5) automobile leaf spring.  Consider each case in terms of the following factors: (a) material reliability;
(b) type of loading;
(c) reliability of stress analysis;
(d) fabrication factors;
(e) influence of time;
(f) consequences of failure.

1-7  A cylinder of cast iron 0.50 in in diameter by 2 in long is tested in compression. Failure occurs at an axial load of 50,000 lb on a plane inclined 40° to the axis of the cylinder.  Calculate the shearing stress on the plane of failure.  The same cylinder of cast iron is pulled in tension.  Give your estimate of the failure load and the plane of failure.

1-8  A bar of uniform cross section is subjected to an axial stress of 10 MN/m². $\theta$ is the angle between the transverse axis and a plane oblique to the longitudinal axis. Plot the normal stress and shear stress on this oblique plane over the range $\theta = 0°$ to $\theta = 90°$.

## CHAPTER 2

2-1  Find the principal stresses and the orientation of the axes of principal stress with the x, y axes for the following situations:

(a)   $\sigma_x = +50,000$ psi
   $\sigma_y = +5,000$ psi
   $\tau_{xy} = -8,000$ psi

(b)   $\sigma_x = -60,000$ psi
   $\sigma_y = +5,000$ psi
   $\tau_{xy} = +25,000$ psi

2-2  Construct a Mohr's circle of stress for each of the plane-stress conditions given in Prob. 2-1.

2-3  The three-dimensional state of stress is given by

$$\sigma_x = +8,000 \text{ psi} \qquad \tau_{xy} = +2,000 \text{ psi}$$
$$\sigma_y = -4,000 \text{ psi} \qquad \tau_{yz} = +3,000 \text{ psi}$$
$$\sigma_z = +6,000 \text{ psi} \qquad \tau_{xz} = -5,000 \text{ psi}$$

Determine (a) the total stress (magnitude and direction with x, y, z axes) on plane described by direction cosines $l = +1/\sqrt{2}$, $m = +\tfrac{1}{2}$, $n =$ negative; (b) magnitude of normal and shear stresses on this plane; (c) principal stresses and direction cosines of the principal planes; (d) maximum shear stress.

2-4  The state of stress at a point for a system of reference axes x, y, z is given by

$$\begin{pmatrix} 200 & 100 & 0 \\ 100 & 0 & 0 \\ 0 & 0 & 500 \end{pmatrix}$$

If a new set of axes $x', y', z'$ is formed by rotating $x, y, z$ 60° about the $z$ axis, find the matrix of the stress tensor for the new axes $x', y', z'$ through the same point. Use tensor concepts.

2-5   The strain produced by a temperature change is given by $e_{ij} = \alpha_{ij}(T - T_0)$, where $\alpha$ is the coefficient of thermal expansion.

(a) What is the rank of the thermal expansion tensor $\alpha$?

(b) Because of the symmetry of cubic crystals $\alpha_{ij}$ is isotropic. Write the components of the $\alpha$ tensor for this case.

(c) In lower symmetry crystals, such as Sn and U, $\alpha_{ij}$ is not isotropic. What practical problems will this introduce?

2-6   Write a matrix giving the components of the hydrostatic (mean) strain tensor. Evaluate the first, second, and third invariants of the hydrostatic strain tensor.

2-7   For the strain-gage arrangement shown in Fig. 2-17, $\alpha = 30°$, and $\beta = 40°$. If $e_a = 2.6 \times 10^{-3}$, $e_b = 0.4 \times 10^{-3}$, and $e_c = 1.4 \times 10^{-3}$, determine the principal strains by means of Mohr's circle.

2-8   On a plate of material ($E = 25 \times 10^6$ psi, $\nu = 0.25$) strain gages are arranged as shown. When the plate is loaded, the gages read $e_1 = 1,860 \times 10^{-6}$, $e_2 = 185 \times 10^{-6}$, and $e_3 = 1,330 \times 10^{-6}$.

(a) What is the largest normal stress?

(b) What is the smallest normal stress?

(c) What is the largest shear stress?

2-9   In a nylon part (Young's modulus = 2,000 MPa and Poisson's ratio = 0.35) the normal strains at a free surface are measured along the three directions shown in Fig. 2-17, where $\alpha = \beta = 45°$. If

$$\varepsilon_a = 0.1 \times 10^{-3}, \ \varepsilon_b = -0.2 \times 10^{-3}, \ \varepsilon_c = 0.4 \times 10^{-3},$$

determine the following:

(a) The three principal strains and their directions;

(b) The magnitude of the greatest shear strain;

(c) The three principal stresses and their directions;

(d) The magnitude of the greatest shear stress;

(e) Draw Mohr's circle for the three-dimensional state of stress.

2-10   In an isotropic solid, a hydrostatic pressure produces a dilatation in which the strain in any direction is given by $\varepsilon = p(1 - 2\nu)/E$. However, in an anisotropic solid the strain due to hydrostatic pressure is not equal in all directions.

(a) For a cubic crystal determine a relationship for strain in any direction in terms of $E_x, \nu_y, G_{xy}$.

(b) Explain whether the dilatation can be calculated from $\Delta = \varepsilon_x + \varepsilon_y + \varepsilon_z$.

(c) Under the action of a hydrostatic pressure, what will be the shape of a face of a cube whose edges are originally parallel to the cubic crystal axes?

2-11 A boron-epoxy composite material can be considered an orthotropic sheet with the following elastic properties:

$$E_{11} = 30 \times 10^6 \text{ psi} \qquad E_{22} = 3 \times 10^6 \text{ psi}$$
$$G_{12} = 1 \times 10^6 \text{ psi} \qquad \nu_{12} = 0.3$$

(a) Show by a careful plot how $E_{11}$ and $E_{22}$ vary with angle $\theta$ from the fiber axis. Let $\theta$ vary from 0 to 180°.

(b) If the sheet is loaded to 20,000 psi at an angle 30° to the fiber axis, what is the strain at 90° to the loading direction?

2-12 For a thin-wall cylindrical pressure vessel of internal diameter $D$ and wall thickness $h$ show that the stress in the circumferential direction is related to the internal pressure $p$ by the equation $\sigma_t = pD/2h$. If the ends of the vessel are welded shut, what is the average longitudinal stress in the walls? Determine the ratio of circumferential to longitudinal stress. Why would these equations not hold for a thick-wall pressure vessel?

2-13 Strain gages on the outer surface of a pressure vessel read $e_l = 0.002$ in the longitudinal direction and $e_t = 0.005$ in the circumferential direction. Compute the stresses in these two principal directions: $E = 2.07 \times 10^5$ MPa and $\nu = 0.3$. What is the error if the Poisson effect is not taken into consideration?

2-14 It is found experimentally that a certain material does not change in volume when subjected to an elastic state of stress. What is Poisson's ratio for this material?

2-15 Determine the volume of a 10-cm copper sphere that is subjected to a fluid pressure of 12 MPa.

2-16 Assuming that atoms are hard elastic spheres, show that Poisson's ratio for a close-packed array of spheres is $\frac{1}{3}$.

## CHAPTER 3

3-1 Determine the engineering strain $e$, the true strain $\varepsilon$, and the reduction in area $q$ for each of the following situations:

(a) Extension from $L$ to $1.1L$

(b) Compression from $h$ to $0.9h$

(c) Extension from $L$ to $2L$

(d) Compression from $h$ to $\frac{1}{2}h$

(e) Compression to zero thickness

3-2 A 2-in-diameter forging billet is decreased in height (upset) from 5 to 2 in

(a) Determine the average axial strain and the true strain in the direction of compression.

(b) What is the final diameter of the forging (neglect bulging)?

(c) What are the transverse plastic strains?

**3-3** A 1-in-thick plate is decreased in thickness according to the following schedule: 0.50, 0.25, 0.125 in. Compute the total strain on the basis of initial and final dimensions and the summation of the incremental strains, using
(a) conventional strain and
(b) true strain.
How does this show an advantage for the use of true strain in metal-forming work?

**3-4** Show that constancy of volume results in $e_1 + e_2 + e_3 = 0$ and $\varepsilon_1 + \varepsilon_2 + \varepsilon_3 = 0$. Why is the relationship for conventional strain valid only for small strains but the relationship for true strains is valid for all strains?

**3-5** A steel shaft ($\sigma_0 = 100,000$ psi) is subjected to static loads consisting of a bending moment of 200,000 in-lb and a torsional moment of 500,000 in-lb. Using a factor of safety of 2, determine the diameter of the shaft based on
(a) maximum-shear-stress theory and
(b) distortion-energy theory.

**3-6** A hexagonal crystal has its basal slip plane normal to the $\alpha$ axis and the slip direction parallel to the $\beta$ axis. If the stresses $\sigma_{11}$ and $\sigma_{22}$ are applied as shown in the sketch, develop and plot the two-dimensional yield locus for the 1,2 system of axes. Let $k_{\alpha\beta}$ be the shear stress for yielding on the $\alpha\beta$ slip system.

**3-7** The yield stress of a tension specimen machined from a $\frac{1}{2}$-m wide by 0.6-cm thick copper sheet is 145 MPa. The sheet is rolled further with an applied tensile force in the plane of the sheet of 0.22 MN. What roll pressure is needed to just cause yielding? Ignore friction between the metal and the rolls. For rolling, deformation occurs by plane-strain in which there is no increase in the width of the sheet.

**3-8** If the yield strength of the steel used in the vessel in Prob. 2-12 is 30,000 psi, $D$ is 12 in and $h$ is $\frac{1}{4}$ in, what will be the value of internal pressure at which the tank will yield?

**3-9** A thin-walled tube 10 in long has a wall thickness of 0.1 in and an initial outside radius of 2.0 in. The plastic stress-strain curve for the material is given by $\bar{\sigma} = 100,000 \, \bar{\varepsilon}^{0.5}$. The tube is subjected to axial load $P$ and torque $T$ so that the stress ratio $\tau_{z\theta}/\sigma_z = 1$ at all times. The deformation is continuous until the effective stress equals 25,000 psi. Determine:
(a) The effective strain at the end of the deformation process
(b) The final dimensions of the tube
(c) The final axial load and torque

*3-10* On the assumption that there is no change in width during the rolling of a sheet, derive the expression for the significant strain in terms of the change in thickness of the sheet. Also express in terms of percentage reduction.

*3-11* A steel shaft transmits 400 hp at 200 rpm. The maximum bending moment is 8,000 ft-lb. The factor of safety is 1.5. Find the minimum diameter of solid shaft on the basis of

(*a*) maximum-shear-stress theory

(*b*) distortion-energy theory. ($\sigma_0 = 40,000$ psi, $E = 30 \times 10^6$, $\nu = 0.30$.)

*3-12* A steel with a yield strength in tension of 42,000 psi is tested under a state of stress where $\sigma_2 = \sigma_1/2$, $\sigma_3 = 0$. What is the stress at which yielding will occur if it is assumed that

(*a*) the maximum-normal-stress theory holds

(*b*) the maximum-shear-stress theory holds,

(*c*) the distortion-energy theory holds?

*3-13* A material is tested under the state of stress $\sigma_1 = 3\sigma_2 = -2\sigma_3$. Yielding is observed at $\sigma_2 = 20,000$ psi.

(*a*) What is the yield stress in simple tension?

(*b*) If the material is used under conditions such that $\sigma_1 = -\sigma_3$, $\sigma_2 = 0$, at what value of $\sigma_3$ will yielding occur? (Assume distortion-energy criterion of yielding.)

*3-14* Show that slip-line fields meet at 45° to a free surface on a frictionless surface.


## CHAPTER 4

*4-1* (*a*) Interstitials in the fcc lattice occupy the tetrahedral positions ($\frac{1}{4}$, $\frac{1}{4}$, $\frac{1}{4}$) and the octahedral positions ($\frac{1}{2}$, $\frac{1}{2}$, $\frac{1}{2}$). Indicate these interstitial positions on a sketch of the structure cell

(*b*) In the bcc lattice interstitials occupy sites at ($\frac{1}{2}$, $\frac{1}{4}$, 0) and smaller sites at ($\frac{1}{2}$, $\frac{1}{2}$, 0) and ($\frac{1}{2}$, 0, 0). Show these sites on a sketch.

Compare the number of available interstitial sites in the fcc and bcc lattices.

*4-2* For what geometrical conditions can a single crystal be deformed but not show slip lines on its surface?

*4-3* Prove that the $c/a$ ratio for an ideal hcp structure is $\sqrt{8/3}$.

*4-4* How many atoms per square millimeter are there on a (100) face of a copper crystal? ($a_0 = 3.60$ Å.)

*4-5* How many atoms per square millimeter are there on a (111) face of a copper crystal?

*4-6* Verify the values for the distance between planes for fcc and bcc lattices given in Table 4-2.

*4-7* (*a*) Show by means of a model or sketch that the intersection of {111} in an fcc structure cell produces an octahedron

(*b*) How many favorable slip systems are there for tensile straining along a [001] axis?

(*c*) How many for deformation along [111]?

**4-8**  The following data for zinc single crystals were obtained by Jillson:[1]

| $90-\phi$ | $\lambda$ | P, g |
|---|---|---|
| 6.5 | 18 | 20,730 |
| 19.5 | 29 | 7,870 |
| 30 | 30.5 | 5.280 |
| 40 | 40 | 4,600 |
| 61 | 62.5 | 5,600 |
| 85 | 85.5 | 28,500 |

$$A = 122 \text{ mm}^2$$

[1] D. C. Jillson, *Trans. AIME*, vol. 188, p. 1129, 1950.

(*a*) Establish that Schmid's law is obeyed.

(*b*) Demonstrate the orientation dependence of the yield stress by plotting $\sigma_0$ versus $\cos \phi \cos \lambda$.

**4-9**  If a material has a dislocation density of $10^{12}$ cm$^{-2}$, what is the average distance between dislocations?

**4-10**  If the critical resolved shear stress for yielding in aluminum is $240 \times 10^3$ Pa, calculate the tensile stress required to cause yielding when the tensile axis is [001].

**4-11**  A bicrystal with a simple cubic crystal structure is oriented as shown below. Which crystal will slip first if the slip system is $\{100\}\langle 100\rangle$?

**4-12**  Along which crystallographic direction in an fcc crystal must the tensile axis be oriented to give multiple slip on

(*a*) four slip systems,

(*b*) six slip systems,

(*c*) eight slip systems?

**4-13**  Compare and contrast the deformation mechanisms of slip and twinning.

## CHAPTER 5

**5-1**  (*a*) Prove that the dislocation reaction $(a/2)[10\bar{1}] \rightarrow (a/6)[2\bar{1}\bar{1}] + (a/6)[11\bar{2}]$ is vectorially correct.

(*b*) Show that this reaction results in a decrease in strain energy.

5-2   Construct a two-dimensional drawing of the (110) plane of a bcc lattice. By means of vector addition show that $(a/2)[\bar{1}\bar{1}1] + (a/2)[111] \rightarrow a[001]$ and that $(a/2)[\bar{1}\bar{1}1] + (a/2)[\overline{11}1] \rightarrow a[\bar{1}\bar{1}0]$. Show that the last reaction is not energetically favorable.

5-3   Define the following:
   (a) perfect dislocation;
   (b) glissile dislocation;
   (c) cross slip;
   (d) Frank partial dislocation;
   (e) Shockley partial dislocation;
   (f) extended dislocation.

5-4   Using Fig. 5-13, show that the y components in Cottrell's equation $(a/2)[\bar{1}11] + (a/2)[111] \rightarrow a[001]$ cancel each other with a release of elastic energy. Show that this equals the energy gained by the x components of the reaction.

5-5   Express Eq. (5-4) in terms of the polar coordinates r and θ. By letting $\theta = 0$ show that the shear stress on the slip plane in the direction of the Burgers vector is equal to $\tau = \tau_0 b/r$.

5-6   (a) Derive an expression for the volume change at a point in a crystal near an edge dislocation. What is the total volume change in the region surrounding the dislocation, not including the core?
   (b) What is the volume change at a point near a screw dislocation? Explain the difference between this result and that for the edge dislocation.

5-7   The yield stress of most single crystals is $10^{-4}G$ or less. By using the concept of the Frank-Read source, develop an argument in which the yield stress is determined by the spacing of sources initially present in the crystal. Show that this leads to a reasonable estimate for the dislocation density. Give some alternative explanations for the yield stress.

5-8   (a) Estimate the strain energy of an edge dislocation where $r = 1$ cm, $r_0 = 10^{-7}$ cm, $G = 5 \times 10^{11}$ dyn/cm$^2$, $b = 2.5 \times 10^{-8}$ cm, and $\nu = \frac{1}{3}$. Express the result in electron volts per atomic plane. How much energy (electron volts) is required to produce 1 cm of dislocation line?
   (b) Show that more than one-half the strain energy resides outside the core of the dislocation in the region $r = 10^{-4}$ to $r = 1$ cm.

5-9   (a) Show that the shear stress produced on a slip plane by an edge dislocation on a parallel plane at coordinates r, θ is given by

$$\tau_{xy} = \frac{Gb}{8\pi(1-\nu)} \frac{1}{h} \sin 4\theta$$

where h is the perpendicular distance between two parallel planes.
   (b) What is the shear stress required for two edge dislocations to pass on parallel slip planes that are 1,000 Å apart in an annealed copper crystal? (Use the data in Prob. 5-8).

*5-10* Derive an expression for the stress-strain curve, using G. I. Taylor's assumption that the dislocations are arranged in the crystal in alternate rows of positive and negative edge dislocations. Explain why this concept is no longer valid.

*5-11* If the binding energy $U_i$ of carbon atoms to edge dislocations is about 0.2 eV, show that the dislocation sites become readily saturated at room temperature.

## CHAPTER 6

*6-1* Calculate the number of grains in the top surface of the head of a pin (0.1 in. diameter) if the pin is made from steel with ASTM grain size 6.

*6-2* What is the grain-boundary area in 1 $in^3$ of steel with ASTM grain size 6? Assume that the grains have a cubic shape.

*6-3* If the grain boundaries of an iron sample (ASTM grain size 6) are completely covered with a monatomic layer of oxygen, how much oxygen, in ppm, is present? (Radius of the oxygen atom is 0.6 Å.)

*6-4* For a steel with ASTM grain size 3, what reduction in grain size would double the yield stress? Express as both average grain diameter and ASTM number.

*6-5* The energy of a low-angle boundary is given by $E = E_0 \theta(A - \ln \theta)$, where $E$ is the energy per unit boundary area, $E_0$ is the elastic strain energy of a dislocation, $\theta$ is the angle of misfit, and $A$ is a constant.
(*a*) Plot $E/\theta$ versus $\ln \theta$.
(*b*) Plot $E$ versus $\theta$.

*6-6* The dislocation density $\rho$ is the total length of dislocation line in the crystal volume. The units are number per square centimeter. Show that the relationship between $\rho$ and the radius of curvature of a bent crystal, $R$, is given by $\rho = 1/Rb$. (*Hint*: Use Fig. 6-5.)

*6-7* Calculate the modulus of elasticity for a two-phase alloy composed of 50 vol percent cobalt, and 50 vol percent tungsten carbide on the basis of
(*a*) equality-of-strain and
(*b*) equality-of-stress hypotheses.
$E_{Co} = 30 \times 10^6$ psi; $E_{wc} = 100 \times 10^6$ psi. Compare the calculated values with the measured modulus of $54 \times 10^6$ psi.

*6-8* For a unit cube containing a dispersed second phase of uniform spherical particles of radius $r$, the number of particles per unit volume $N_v$ is related to the volume fraction of particles $f$ by $N_v = f/\frac{4}{3}\pi r^3$. The average distance between particles is given by $\lambda = 1/(N_v)^{1/3} - 2r$.
(*a*) Show that $\lambda = 2r[(1/1.91f)^{1/3} - 1]$.
(*b*) For $f = 0.001$, 0.01, 0.1 and $r = 10^{-6}$, $10^{-5}$, $10^{-4}$, and $10^{-3}$ cm calculate $\lambda$. Plot $f$ versus $\lambda$ on log-log paper.

*6-9* If we take the limits on matrix strength as $G/1,000$ (yield strength of the pure matrix) and $G/30$ (fracture stress of the matrix), show that the limits on interparticle spacing for effective dispersion hardening are between about 0.01 and 0.3 $\mu$m. To minimize low ductility, the volume fraction of the second phase should be less than about 15 percent. What restrictions does this place on particle size?

6-10 If the shear strength of a grain boundary in iron is assumed to be $G/10$, determine the stress required to penetrate the grain boundary for

(a) a grain size of ASTM 2

(b) a grain size of ASTM 12.

6-11 Tests on alloy single crystals, such as 70/30 brass, show that the lattice rotation "overshoots" beyond the $[001] - [1\bar{1}1]$ symmetry boundary before duplex slip occurs. On the basis of your knowledge of the effect of alloying on dislocation behavior, offer an explanation.

## CHAPTER 7

7-1 Estimate the theoretical fracture stress of iron if the surface energy is $1.2 \text{ J/m}^2$. How does this compare with the highest observed strength of heat-treated steel?

7-2 Using Griffith's equation for plane stress, determine the critical crack length for

(a) iron.

(b) zinc and

(c) NaCl if the following constants apply:

|  | $\sigma_c$, dyn/cm$^2$ | $\gamma$, erg/cm$^2$ | $E$, dyn/cm$^2$ |
|---|---|---|---|
| Iron | $9 \times 10^9$ | 1,200 | $20.5 \times 10^{11}$ |
| Zinc | $2 \times 10^7$ | 800 | $3.5 \times 10^{11}$ |
| NaCl | $2 \times 10^7$ | 150 | $5.0 \times 10^{11}$ |

7-3 What is the critical crack length in iron according to the Griffith-Orowan equation if $p \approx 10^6 \text{ erg/cm}^2$?

7-4 Show that the Cottrell mechanism for the creation of a microcrack by slip on intersecting planes is not energetically favorable in an fcc lattice.

7-5 A ship steel has a value of $\mathscr{G}_c = 200 \text{ lb/in}$.

(a) What is the fracture stress in a thin plate that is 12 in wide and that contains a central crack 0.5 in long?

(b) If the crack is 2 in long, what is the fracture stress?

(c) Increasing the plate thickness to 5 in reduces $\mathscr{G}_c$ to 100 lb/in. What is the fracture stress for a 0.5-in-long crack?

7-6 A steel plate is 12 in wide and $\frac{1}{4}$ in thick. There is a 1-in-long crack along each edge.

(a) Calculate the force required to propagate the crack the remaining 10 in across the width of the plate.

(b) Calculate the force required to break the plate in tension if there were no crack. Assume that the fracture strength is 100,000 psi.

(c) Calculate the force required to break the plate in tension if the steel had the theoretical cohesive strength.

7-7  Using Eq. (7-35), discuss how it qualitatively predicts the effect of the following factors on the ductile-to-brittle transition:
(a) material composition,
(b) impurities,
(c) melting practice,
(d) temperature,
(e) grain size,
(f) state of stress,
(g) rate of loading.

## CHAPTER 8

8-1  The internal-friction peak due to carbon in iron occurs at 35°C for $f = 0.65$ cycles/s. If $D_0 = 2 \times 10^{-2}$ cm$^2$/s and $\Delta H = 20.1$ kcal/mole, what is the relaxation time at 0, 100, and 500°C?

8-2  A steel beam 15 in long, 1.5 in wide, and 0.25 in thick is loaded as a cantilever beam. The following amplitude readings were observed at a point 12 in from the loaded end of the beam for the cases of no end load and when the beam is damped with a 3-lb weight at the free end:

| No load | | Damped with 3-lb load | |
|---|---|---|---|
| Cycles | Amplitude, $\mu$in | Cycles | Amplitude, $\mu$in |
| 0 | 563 | 0 | 464 |
| 10 | 420 | 4 | 390 |
| 20 | 300 | 8 | 360 |
| 30 | 240 | 12 | 328 |
| 40 | 185 | 16 | 270 |
| 50 | 160 | 20 | 243 |
| 60 | 135 | 24 | 205 |
| 70 | 115 | 28 | 165 |
| 80 | 100 | 32 | 125 |
| 90 | 85 | 36 | 97 |
| 100 | 78 | 40 | 63 |
| 120 | 60 | 44 | 33 |

(a) Plot ln $A$ versus cycles.
(b) Plot the logarithmic decrement vs. amplitude.

8-3  The creep rate of a chain polymer which is not cross-linked but is above $T_g$ increases with increase in applied stress. Give several brief explanations for this based on the deformation modes shown in Fig. 8-11.

8-4  Consider a Maxwell element subjected to an engineering stress-strain test at a nearly constant strain rate $\dot{\varepsilon}_c$. Develop an equation for the tensile stress $\sigma$ as a function of strain $\varepsilon$ and the polymer properties $\eta$ and $E$.

8-5  The "standard linear solid" combines a Maxwell element and an elastic element in parallel. If the components of the Maxwell element are expressed by $\sigma_1 = \eta_e \dot{\varepsilon}_3$

and $\sigma_1 = E_1\varepsilon_1$, and the elastic element is described by $\sigma_2 = E_2\varepsilon_2$, derive the equation for the strain-time (creep) relationship for this material. Use the concept of the relaxed modulus and unrelaxed modulus in formulating your answer.

8-6   Using the data for the creep compliance $J_p(t)$ given below, determine the shift factor $A_T$ necessary to make the data at different temperatures fall on a master curve using a reference temperature of 109.8°C. Give a plot of $A_T$ vs. temperature.

Creep compliance $J_0(t)$ for polystyrene. (*From D.J. Plazek, J. Polym. Sci. A-2, vol. 6, pp. 621–638, 1968.*)

8-7   For a certain amorphous polymer the modulus relaxes to $10^9$ dyn/cm² in about 0.2 h at 25°C. If the glass transition temperature is $-50$°C, estimate the time it will take to reach the same value of modulus at $-10$°C.

## CHAPTER 9

9-1   The following data were obtained during the tension test of a low-carbon steel with a specimen having a 0.505-in diameter and a 2-in gage length.

| Load, lb | Elongation, in | Load, lb | Elongation, in |
|---|---|---|---|
| 500 | 0.00016 | 6,300 | 0.020 |
| 1,000 | 0.00032 | 7,000 | 0.060 |
| 1,500 | 0.00052 | 7,500 | 0.080 |
| 2,000 | 0.00072 | 8,500 | 0.12 |
| 2,500 | 0.00089 | 9,600 | 0.18 |
| 3,000 | 0.00105 | 10,000 | 0.26 |
| 3,500 | 0.00122 | 10,100 | 0.30 |
| 4,000 | 0.00138 | 10,200 | 0.50 |
| 4,500 | 0.00154 | 10,050 | 0.58 |
| 5,000 | 0.00175 | 9,650 | 0.62 |
| 5,500 | 0.00191 | 9,100 | 0.70 |
| 6,000 | 0.00204 | 8,100 | 0.76 |

Yield point, 6,200 lb; breaking load, 6,800 lb; final gage length, 2.87 in; final diameter, 0.266 in.

(a) Plot the engineering stress-strain curve.

(b) Determine the (1) proportional limit; (2) modulus of elasticity; (3) lower yield point; (4) tensile strength; (5) fracture stress; (6) percentage elongation; (7) reduction of area.

(c) Plot the true-stress–true-strain curve up to maximum load.

9-2 Derive the relationship for the toughness of a metal whose true stress-strain curve obeys the power law $\sigma = K\varepsilon^n$.

9-3 (a) Show that $q = 1 - e^{-\varepsilon_f}$, where $q$ is the reduction of area and $\varepsilon_f$ is the true strain at fracture.

(b) If the true-stress–true-strain curve is given by $\sigma = K\varepsilon^n$, derive a relationship between the yield stress and the reduction in cross-sectional area.

9-4 A useful measure of ductility for evaluating formability is the *zero-gage-length elongation* $e_0$. This represents the maximum possible elongation on the minimum possible gage length. Show that $e_0 = q/(1 - q)$. If $q$ is measured at some point on the tensile specimen other than the neck, it gives a value of uniform reduction. Why?

9-5 The following data were obtained during the true-stress–true-strain test of a nickel specimen:

| Load, lb | Diameter, in | Load, lb | Diameter, in |
|---|---|---|---|
| 0 | 0.252 | 3,570 | 0.201 |
| 3,440 | 0.250 | 3,500 | 0.200 |
| 3,580 | 0.245 | 3,350 | 0.190 |
| 3,670 | 0.240 | 3,150 | 0.180 |
| 3,710 | 0.235 | 2,950 | 0.170 |
| 3,720 | 0.230 | 2,800 | 0.149 |

(a) Plot the true-stess–true-strain curve.

(b) For the data beyond the maximum load correct for necking by means of Bridgman's correction (Fig. 9-9).

(c) Determine the following: (1) true stress at maximum load; (2) true fracture stress; (3) true fracture strain; (4) true uniform strain; (5) true necking strain; (6) ultimate tensile strength; (7) strain-hardening coefficient.

9-6 Will the superposition of lateral hydrostatic pressure affect the strain at which necking occurs in a tensile test? Explain.

9-7 The greater the strain-hardening exponent the greater the percentage difference between the elastic limit and the yield strength. Compare the elastic limits of two metals with $n = 0.03$ and $n = 0.30$ if both have a 0.2% offset yield strength of 120,000 psi. The elastic limit may be taken as the stress at $\varepsilon = 2 \times 10^{-5}$.

9-8 For a material whose flow curve is given by $\sigma = K\varepsilon^n$, derive an expression for the tensile-yield ratio in terms of $n$ and show how $s_u/s_o$ can be used to determine $n$.

*9-9* Show that the tensile strength for a material that follows the power law is

$$\sigma_u = K \left( \frac{n}{e} \right)^n$$

where $e = 2.718$.

*9-10* The variation of percentage elongation $X$ with gage length has been found empirically to follow the relationship $X = c\sqrt{A}/L_0 + b$, where $c$ and $b$ are constants. The following data were obtained for a steel plate that is 2 in wide and $\frac{1}{2}$ in thick:

| $L_0$, in | 2 | 4 | 6 | 10 | 12 | 14 |
|---|---|---|---|---|---|---|
| $X$ | 40.0 | 30.6 | 27.5 | 25.7 | 25.0 | 24.3 |

Find $b$ and $c$, and determine the percentage elongation for a 2- by $\frac{7}{8}$-in steel plate with a 10-in gage length.

*9-11* The true-stress–true-strain curve of a material can be expressed by the equation $\sigma = 200,000\varepsilon^{0.33}$, where stress is in units of pounds per square inch. What is the ultimate tensile strength of this material?

# CHAPTER 10

*10-1* A 1-in-diameter hot-rolled steel bar was tested in torsion, with the following results:

| Torque, in-lb | Number of $\frac{1}{4}$ turns of bar | Torque, in-lb | Number of $\frac{1}{4}$ turns of bar |
|---|---|---|---|
| 6,700 | 1 | 10,600 | 12 |
| 7,400 | 2 | 11,000 | 15 |
| 8,200 | 3 | 11,400 | 18 |
| 8,700 | 4 | 11,800 | 24 |
| 9,100 | 5 | 12,400 | 32 |
| 9,700 | 7 | 12,600 | 38 |
| 10,200 | 9 | 12,800 | 39 |

If the length of the bar between chucks is 18 in, determine

(*a*) shear stress–shear-strain curve;

(*b*) modulus of rupture;

(*c*) shear-stress-shear-strain curve corrected for inelastic strain.

(*d*) If the twisting moment at the yield point was 4,500 in-lb and the angle of twist was 2.6°, determine the torsional yield stress, the modulus of elasticity, and the modulus of resilience.

(*e*) What is Poisson's ratio if $E = 29.0 \times 10^6$ psi?

*10-2* Compare the torsional strength, stiffness, and weight of two steel shafts, one solid, the other hollow. The hollow shaft has twice the diameter of the solid shaft, but the cross-sectional area is the same.

*10-3* Assume that a certain brittle material will fail at a tensile stress of 30,000 psi.

    (*a*) For a bar with a ¼-in diameter, what torsional moment will cause failure? Sketch the failure.

    (*b*) If this is an ideal brittle material ($v = \frac{1}{4}$), show that greater strain will be achieved in torsion than tension before failure occurs.

*10-4* The following data were obtained for the same low-carbon steel when tested in tension and in torsion:

| Tension | | Torsion | |
|---|---|---|---|
| True normal stress, psi | True strain, in./in, | Shear stress psi | Shear strain in./in. |
| 55,000 | 0.05 | 24,000 | 0.05 |
| 63,000 | 0.10 | 30,000 | 0.10 |
| 73,000 | 0.20 | 36,000 | 0.20 |
| 80,000 | 0.30 | 39,000 | 0.30 |
| 85,000 | 0.40 | 41,000 | 0.40 |
| 90,000 | 0.50 | 42,000 | 0.50 |
| 95,000 | 0.60 | 44,000 | 0.60 |
| 100,000 | 0.70 | 45,000 | 0.70 |
| 105,000 | 0.80 | 45,500 | 0.80 |
| 116,000 | 1.00 | 47,000 | 1.00 |
| 126,000 | 1.20 | 48,500 | 1.20 |
| 131,000 | 1.30 | 50,000 | 1.50 |

    (*a*) Plot the tension stress-strain curve and the torsion stress-strain curve on the same axis.

    (*b*) Apply Bridgman's correction for necking to the tension data.

    (*c*) Plot both tension and torsion data as significant stress and significant strain.

    (*d*) Plot both sets of data as maximum shear stress vs. maximum shear strain.

*10-5* Explain why necking is observed in the tension test of a ductile material but not in the torsion test of the same material.

# CHAPTER 11

*11-1* Estimate the Mohs' hardness for the following materials:

    (*a*) steel file,

    (*b*) chalk,

    (*c*) pine plank,

    (*d*) ball bearing,

    (*e*) sapphire.

*11-2* (*a*) Show by means of a sketch that in order to produce geometrically similar indentations in a Brinell test $d/D$ must remain constant.

(b)

| Diameter of indenter, mm | Diameter of indentation, mm | Load, kg |
|---|---|---|
| 10 | 4.75 | 3,000 |
| 7 | 3.33 | 1,470 |
| 5 | 2.35 | 750 |
| 1.2 | 0.57 | 425 |

$P/D^2$ is constant for the above data. Determine the BHN, and show that it is approximately constant with load.

*11-3* The following hardness data were obtained on copper by using a 10-mm ball indenter:

| Load, kg | Diameter of indentation, mm | | |
|---|---|---|---|
| | Annealed | $\frac{1}{4}$ hard | $\frac{1}{2}$ hard |
| 500 | 4.4 | 3.2 | 2.9 |
| 1,000 | 5.4 | 3.9 | 3.7 |
| 1,5000 | 6.2 | 4.6 | 4.5 |
| 2,000 | | 5.4 | 5.3 |
| 2,500 | | 5.9 | 5.7 |

(a) Determine whether or not Meyer's law is obeyed.
(b) Determine the Meyer's law constants $k$, $n'$, and $C$.
(c) Plot the BHN and Meyer hardness as a function of load for the $\frac{1}{4}$ hard copper.

*11-4* If a $\frac{1}{4}$-in-diameter ball is used for the annealed copper in Prob. 11-3, calculate what the indentation would be at 500-, 1,000-, and 1,500-kg loads. Compare with the observed values.

| Load, kg | Diameter of indentation, mm |
|---|---|
| 500 | 4.0 |
| 1,000 | 4.6 |
| 1,500 | 5.5 |
| 2,000 | 5.9 |

Determine $k$ and $n'$, and compare with the values for a 10-mm ball.

# CHAPTER 12

*12-1* Peterson has shown that for an average value of $K_t = 2$ the notch sensitivity index is given by $q = 1/(1 + a/r)$, where $a$ is a "particle size" dependent on the material and $r$ is the notch radius. Kuhn and Hardrath[1] have related $a$ with the tensile strength of the steel.

[1] NACA Tech. Note 2805, 1952.

| $S_u$, ksi | 200 | 160 | 100 | 50 |
|---|---|---|---|---|
| $a$ | | 0.0004 | 0.001 | 0.0042 | 0.012 |

Using these data, plot a curve of $q$ versus $r$ for steels of different strengths.

12-2 Show that increasing the size of a fatigue specimen tested in bending decreases the stress gradient. Why would this result in a size effect?

12-3 Why are the notched fatigue properties of ordinary gray cast iron about equal to the unnotched properties?

12-4 Using the ideas about stress gradient, show that the increase in fatigue limit due to nitriding the surface of a steel shaft is given approximately by $1/(D/2\delta - 1)$, where $\delta$ is the depth of the nitrided layer and $D$ is the diameter of the shaft. Would this equation be applicable for axial loading? Explain why nitriding is an effective means of counteracting stress raisers.

12-5 A steel shaft is subjected to a completely reversed bending moment of 6,000 in-lb. A transition section at one end contains a fillet with a 0.10-in radius. The steel has a tensile strength of 100,000 psi, and the fatigue limit is 45,000 psi. $K_t$ for the fillet is about 2, and the dynamic factor of safety is 2.5. What is the minimum allowable diameter of the shaft?

12-6 A 5-cm-diameter shaft is subjected to a static axial load of 100 kN. If the yield stress of the material is 350 MN/m² and the fatigue limit ($R = -1$) is 260 MN/m², what is the largest completely reversed bending moment that can be applied? The factor of safety is 2, and $K_f$ due to the presence of notches is 1.8.

12-7 A threaded bolt experiences a completely reversed axial load of 60,000 lb. The static factor of safety is 2.0, and the dynamic factor of safety is 3.0. $K_t = 2.0$ The root radius of the threads is 0.020 in.
(a) Compare the root area required for static and dynamic service for a high-strength bolt ($\sigma_0 = 180,000$ psi, $\sigma_u = 200,000$ psi).
(b) Repeat for a low-strength bolt ($\sigma_0 = 30,000$ psi, $\sigma_u = 50,000$ psi). Use data in Prob. 12-1.
(c) Determine the ratio of the areas for low-strength and high-strength bolts for the static and dynamic conditions.

12-8 A steam turbine used for peaking service experiences a thermal cycle from room temperature (70°F) to a service temperature of 1000°F several times each day. The *total* strain amplitude experienced by a turbine blade each cycle is $\varepsilon_a = \alpha(T_s - T_r)$. The properties of the blade material over the temperature interval are: 60,000-psi yield strength, 80,000-psi tensile strength, 30 percent reduction in area, elastic modulus $24 \times 10^6$ psi, and coefficient of thermal expansion $10^{-5}$ per °F. How many cycles can the blade withstand?

12-9 A pressure cylinder with a 4-in. inside diameter and a $\frac{1}{4}$-in wall thickness is subjected to internal pressure that varies from $-p/4$ to $p$. The steel has a fatigue limit of 60,000 psi in completely reversed bending. Under a repeated stress cycle of zero to a maximum stress in tension the same steel can withstand 90,000 psi without failure in $10^7$ cycles. What is the maximum internal pressure that can be withstood by the cylinder without fatigue failure before $10^7$ cycles?

*12-10* A 5-cm-diameter shaft is subjected to a static bending moment of 900 N-m. What is the value of the maximum twisting moment varying from 0 to $M_T$ to produce failure in $10^7$ cycles? The yield strength is 620 MPa and the fatigue limit at $10^7$ cycles in completely reversed stress is 400 MPa.

## CHAPTER 13

*13-1* Construct a three-dimensional plot showing the relationship between stress, strain, and time, for $T = T_m/2$.

*13-2* Assuming that the mechanical equation of state is completely valid, show how creep curves could be constructed from stress-strain curves.

*13-3* The following rupture times[1] were obtained for stress rupture tests on a steel alloy:

| Stress, psi | Temperature, °F | Rupture, time, hr |
|---|---|---|
| 80,000 | 1080 | 0.43 |
| 80,000 | 1030 | 6.1 |
| 80,000 | 1000 | 22.4 |
| 80,000 | 975 | 90.8 |
| 10,000 | 1400 | 1.95 |
| 10,000 | 1350 | 6.9 |
| 10,000 | 1300 | 26.3 |
| 10,000 | 1250 | 84.7 |

Establish the validity of the Larson-Miller and Manson-Haferd parameters.

*13-4* By means of sketches, show how isochronous stress-strain curves are derived from creep data.

*13-5* Steady-state creep rate can be expressed in dimensionless form by

$$\frac{\dot{\varepsilon}_s}{\dot{\varepsilon}_0} = \left(\frac{\sigma}{\sigma_0}\right)^{n'} e^{-H(\sigma)/kT}$$

where $\dot{\varepsilon}_0$ is an arbitrarily chosen strain rate and $\sigma_0$ is an experimentally determined constant which replaces $A_1$ in Eq. (13-20). Apply this relationship to the creep of a steel support rod in a boiler operating at 1000°F. The rod is stressed in tension to 8,000 psi and its creep elongation must not exceed 10 percent. Using the data given below, evaluate the constants in the creep equation and estimate the lifetime of the rod.

Creep rate for medium-carbon steel. (*N. P. Suh and A. P. L. Turner, "Elements of the Mechanical Behavior of Solids," p. 327, McGraw-Hill, Book Company, 1975.*)

[1] S. S. Manson, G. Succop, and W. F. Brown, Jr., *Trans. Am. Soc. Met.*, vol. 51, p. 924, 1959.

*13-6* The following data[1] were obtained in creep tests at 1500°F on an austenitic high-temperature alloy.

| Stress, psi | Minimum creep rate, %/h |
|---|---|
| 10,000 | 0.00008 |
| 15,000 | 0.0026 |
| 20,000 | 0.025 |
| 30,000 | 2.0 |
| 40,000 | 30 |
| 50,000 | 320 |

Use these data to evaluate Eq. (13-31).

*13-7* A tie rod 10 ft long is made from the austenitic alloy whose creep properties are given in Prob. 13-6. At 1500°F it is subjected to an axial load of 8,000 lb. The yield stress at 1500°F is 60,000 psi, and the stress for rupture in 1,000 h is 20,000 psi. Using a factor of safety of 3.0, establish the necessary cross-sectional area of the tie rod based on the following considerations:
(*a*) yield stress at 1500°F;
(*b*) 1,000 h rupture life at 1500°F;
(*c*) an allowable creep rate of 1 percent per 10,000 h.

*13-8* A thin-wall pressure vessel of an austenitic alloy has a 1-in wall thickness and an 18-in inside diameter. The vessel operates at 1500°F. Find the allowable internal pressure if the maximum allowable increase in diameter is 0.2 in over a 2-year period. Assume steady-state conditions, and use the values of $B$ and $n'$ derived from the data in Prob. 13-6.

*13-9* A steel bolt is loaded in tension to 10,000 psi and rigidly attached at both ends. After 3 years of service at 800°F what is the stress in the bolt? $E = 22.5 \times 10^6$ psi; $B = 3 \times 10^{-40}$ percent/(h)(psi); $n' = 8.0$.

# CHAPTER 14

*14-1* Sketch the transition-temperature curves for a plain carbon steel tested in tension, torsion, and notched impact.

*14-2* An internal crack has been detected with ultrasonics in the shell of a large cylindrical steel pressure vessel. The stresses in the region of the crack are: longitudinal tensile stress of 60,000 psi; circumferential tensile stress of 140,000 psi; radial compressive stress of 5,000 psi. The pressure vessel is so large that there is no appreciable variation of stress over the region containing the crack. The crack is oriented so that its normal makes an angle of 40° with the longitudinal direction and 60° with the circumferential direction.

[1] N. J. Grant and A. G. Bucklin, *Trans. Am. Soc. Met.* vol. 42, p. 720, 1950.

(a) What is the normal stress on the plane of the crack? What is the shear stress in the plane of the crack?

(b) If $K_{Ic}$ for the steel is 50 ksi$\sqrt{\text{in}}$, what is the smallest size crack that will cause fracture of the pressure vessel? State the assumptions made in arriving at this answer.

14-3 For the introduction of a given amount of hydrogen into a steel, would an annealed or cold-worked steel suffer the greatest embrittlement? Explain on the basis of dislocation theory.

14-4 The weight of the pendulum of an impact tester is 45 lb, and the length of the pendulum arm is 32 in. If the arm is horizontal before striking the specimen, what is the potential energy of the tester? What is the striking velocity?

14-5 A weight $W$ falls vertically downward (without friction) along a thin rod of area $A$ and length $L$ until it is brought to rest by striking a flange at the end of the rod. Using energy considerations, derive an expression for the maximum stress in the rod produced by the weight falling through a distance $h$. Assume perfectly elastic impact. Show that the sudden release of load from a height $h = 0$ produces twice the stress that would result from a gradual application of load.

14-6 A cylindrical steel pressure vessel with a yield strength of 360 MPa is subjected to a hoop stress of 140 MPa. A tensile residual stress of 80 MPa also can be assumed to be present. The lowest service temperature of the vessel will be $-25°C$, under which conditions the $K_{Ic}$ is 1,400 MPa$\sqrt{\text{mm}}$. The vessel will be designed according to a "leak-before-break" philosophy in which a detectable leak will occur before brittle fracture could occur. Determine the thickness of the pressure vessel based on brittle fracture considerations.

14-7 The critical impact velocity for a steel is 160 ft/s, and the longitudinal sound velocity is 19,500 ft/s. If $E = 30 \times 10^6$ psi and the density is 0.32 lb/in$^3$, what is the fracture stress under these impulsive-loading conditions?

14-8 A high-strength steel has a yield strength of 100 ksi and a fracture toughness ($K_{Ic}$) equal to 150 ksi $\sqrt{\text{in}}$. Based on the level of nondestructive inspection, the smallest size flaw that can be detected routinely, $a_i$, is 0.3 in. Assume that the most dangerous crack geometry in the structure is a single-edge notch so that $K_{Ic} = 1.12\,\sigma\sqrt{\pi a}$. The structure is subjected to cyclic fatigue loading in which $\sigma_{max} = 45$ ksi and $\sigma_{min} = 25$ ksi. The fatigue-crack growth rate for the steel is given by $da/dN = 0.66 \times 10^{-8}(\Delta K)^{2.25}$. With this information, estimate the fatigue life of the structure by

(a) integrating numerically using an increment of crack growth of 0.1 in for each calculation

(b) using Eq. (14-9). What changes could be made to increase the fatigue life? Which factor would be the most influential?

## CHAPTER 15

15-1 Discuss the relative ease of hot-work and cold-work for the following metals: aluminum, zirconium, lead, SAE 4340 steel, molybdenum. Consider the following factors: melting point, flow stress, oxidation behavior, brittleness.

15-2 When a compressive force of 400 tons is applied to the top surface of a well-lubricated cube it just causes plastic flow. The cube is 3.16 in on each edge. What force would be required to produce flow if the faces of the cube other than the top face were constrained by die forces of 100 and 200 tons?

15-3 A circular plate is compressed by inclined dies, as shown below. The angle of the dies $\alpha$ is small and sliding friction occurs at the surface with coefficient of friction $\mu$. Derive the differential equation for normal pressure. What is the significance of the case when $\alpha = \mu$?

15-4 Calculate the ideal work of deformation per unit volume in drawing a wire through a die. Express in terms of the strain in the longitudinal direction.

15-5 The reduction per pass is given by $q_n = (h_{n-1} - h_n)/h_{n-1}$, where $n$ is the number of the pass. Derive an expression between the total reduction $Q_n = (h_0 - h_n)/h_0$ and the summation of the reductions per pass $q_n$.

15-6 Determine $p/2k$ for a 50 percent plane-strain extrusion using the slip-line field given below.

15-7 Determine $p/2k$ for a 50 percent extrusion (frictionless) using the upper-bound field given below.

15-8  A frictionless uniaxial compression test of a copper cylinder gave the following results:

| True stress, ksi | 4 | 10 | 16 | 24 | 32 | 38 | 40 | 41.4 | 42 |
|---|---|---|---|---|---|---|---|---|---|
| Reduction in height, % | 0 | 2.5 | 5 | 10 | 20 | 30 | 40 | 50 | 60 |

(a) Plot the true-stress–true-strain curve in compression;

(b) Construct an engineering tensile stress-strain curve up to the ultimate tensile stress from these data.

15-9  Derive the expressions for the flow stress and the strain in the plane-strain compression test.

15-10  Compare the advantages and disadvantages of the tension test, the compression test, the plane-strain compression test, and the torsion test for determining the flow stress for metalworking processes.

15-11  In upsetting cylindrical specimens two types of cracks are observed:

(a) vertical cracks which are parallel to the direction of compression and run into the specimen normal to its free surface, and

(b) inclined cracks at approximately 45° to the direction of compression. Both types of fracture occur in the plane of maximum shear stress, which is a diametral plane at midheight. By applying Mohr's circle in three-dimensions show the two different stress states that must exist to produce these different fractures.

# CHAPTER 16

16-1  Plot a curve of $p_{av}/\sigma_0$ versus the parameter $2\mu a/h$ for a rectangular slab and a cylinder for values of the parameter from 0 to 5.

16-2  A 3,000-lb power forging hammer has a total nominal energy of 35,000 ft-lb. If the blow efficiency is 40 percent and the forging load builds up from $P/3$ at the beginning of the stroke to $P$ at the end of the stroke, what is the total forging load for

(a) a stroke of 0.2 in,

(b) a stroke of 0.6 in.

16-3  By means of sketches illustrate the Cook and Larke method for determining the true compression stress-strain curve.

16-4  What would be the effect of the section thickness $h$ on the forging load if the friction were reduced to zero?

16-5  Assuming that an average forging pressure of 50,000 psi is required, what is the largest closed-die forging that could be made on a 35,000-ton press (the largest available)?

16-6  Show that the strains in the transverse and radial directions are equal for a thin circular disk compressed in the $z$ direction.

16-7  Compare the average pressure needed to compress a cylinder 1 in in thickness and 2 ft in diameter with one $\frac{1}{8}$ in in thickness and 2 ft in diameter.

16-8  An SAE 1040 steel at the forging temperature has a yield stress of 6,000 psi. A right-circular cylinder 3 in high and 1 in in diameter is to be upset to half height between flat dies.

  (a) If the coefficient of friction is 0.4, what is the maximum force required for the upsetting?

  (b) How much extra force is required over what would be needed if no friction were present?

  (c) If it takes 3 s to produce the forging and the efficiency is 40 percent, how much power must be available in order to do the job?

## CHAPTER 17

17-1. Using the simplified theory of rolling, plot curves for the variation of rolling load with roll diameter, coefficient of friction, and mean strip thickness.

17-2  Sheet steel is reduced from 0.160 to 0.140 in with 20-in-diameter rolls having a coefficient of friction equal to 0.04. The mean flow stress in tension is 30,000 psi. Neglect strain hardening and roll flattening.

  (a) Calculate the roll pressure at the entrance to the rolls, the neutral point, and the roll exit.

  (b) If $\mu = 0.40$, what is the roll pressure at the neutral point?

  (c) If 5,000-psi front tension is applied in Prob. 17-2a what is the roll pressure at the neutral point?

17-3  Calculate the minimum gage possible in hot rolling low-carbon steel strip at 700°F on a 23-in mill with steel rolls. ($\sigma_0' = 35,000$ psi, $\mu = 0.4$, $E = 20 \times 10^6$ psi.) What is the minimum gage for cold rolling on the same mill? ($\sigma_0' = 60,000$ psi, $\mu = 0.10$, $E = 30 \times 10^6$ psi.)

17-4  Explain the reason for designing rolling mills with complex clusters of rolls such as in Fig. 17-1e.

17-5  In hot rolling with an 18-in rolling mill a groove was placed on the roll so as to put an imprint on the sheet every revolution. If the distance between marks on the sheet was 62.1 in, what was the forward slip? What is the coefficient of friction in rolling from 0.30 to 0.20 in in thickness?

17-6  List three basic requirements for a good rolling lubricant.

17-7  The residual stresses produced in rolling are proportional to $(h_f/L)^2$, where $L$ is the length of the arc of contact.[1] From the geometry of rolling show that $(h_f/L)^2 = (h_f/R)[(100 - q)/q]$, where $q = [(h_0 - h_f)/h_0]100$.

---

[1] R. McC. Baker, R. E. Ricksecker, and W. M. Baldwin, Jr., *Trans. AIME*, vol. 175, pp. 337–354, 1948.

*17-8* On a certain mill the rolling load for a 30 percent reduction is 19.5 tons/in of width. What is the rolling load when a back tension of 10 tons/in² and a front tension of 8 tons/in² are applied? Assume that $\sigma_0' = 35$ tons/in² and $\alpha = 2\beta$.

*17-9* Show how the mean rate of deformation varies with the reduction for 0.3-in-thick strip rolled on 28-in rolls at 500 ft/min. Determine separate curves for sticking friction and slipping friction. Use $q = 0.10, 0.20, 0.30,$ and $0.50$.

*17-10* (a) Determine the rolling schedule for cold-rolling annealed copper and mild steel strips from $16 \times \frac{1}{8}$ in to $16 \times 0.100$ in with 6-in diameter rolls on a four-high mill. The rolling load of the mill may not exceed 100 tons. Use a simplified rolling theory. The following mean flow stresses were obtained with a plane-strain compression test: copper, 20 ksi; mild steel, 35 ksi.

    (b) Determine the horsepower required to drive the mill in rolling the copper strip if it exits at 600 ft/min,

    (c) What would be the maximum temperature reached in rolling copper strip if there are no heat losses?

## CHAPTER 18

*18-1* Derive Eq. (18-6) using a slab analysis.

*18-2* Using Eq. (18-6), suggest a simple method for measuring the frictional shear stress if only breakthrough pressure, and not die pressure $p_d$, can be measured.

*18-3* In the hot extrusion of aluminum at 600°F, $\sigma_0 = 12,000$ psi.

    (a) For a 12-in-diameter billet, 36 in long, what is the breakthrough pressure required to extrude a 3-in-diameter bar if $\mu = 0.10$?

    (b) What is the required extrusion pressure at the end of the stroke?

    (c) What capacity press would be needed for this extrusion?

*18-4* For extrusion through a flat die the strain rate is given approximately by $\dot{\varepsilon} = (6v/D) \ln R$, where $v$ is the ram velocity, $D$ is the billet diameter, and $R$ is the extrusion ratio. For the conditions given in Prob. 18-3 compare the strain rates and the time the metal is in the die for a 2 in/s and 10 in/s ram speed.

*18-5* Starting with Eq. (18-8), derive an expression for the adiabatic temperature rise in extrusion as a function of the extrusion ratio. See Fig. 15-11 for the trend.

*18-6* The following equation expresses the pressure for the extrussion of aluminum bars:

$$p = \sigma_0 (0.47 + 1.2 \ln R) e^{4\mu L/D}$$

Billets 8 in in diameter and 16 in long are extruded into $\frac{3}{4}$-in-diameter bars. In order to increase the length of the product by 10 ft, would it be more economical in terms of pressure to increase the billet length or the diameter? (Assume that $\mu = 0.10$.)

## CHAPTER 19

*19-1* Plot a curve of the ratio of draw stress to flow stress vs. reduction for $B = 0, 1,$ and 2.

**19-2** Plot a curve of the ratio of draw stress to flow stress versus semiangle die for the conditions $\mu = 0.05$, $r = 0.4$, and $D_a = 0.2$ in.

**19-3** For a value of $B = 1$, plot a curve similar to Prob. 19-1 which shows the effect of back pull on draw stress. Use ratios of back stress to flow stress of 0.0, 0.4, and 0.8.

**19-4** Show that the strain rate of wire being pulled through a die can be expressed by by $\dot{\varepsilon} = \dfrac{2 \tan \alpha}{r} \dfrac{dx}{dt}$. What is the maximum strain rate for a 0.040-in-diameter wire pulled at 2,000 ft/min through a 10° semiangle die?

**19-5** Compare the force required to give a 30 percent reduction in a 0.5-in-diameter wire in a 10° semiangle die with the force required to produce the same reduction in a tube blank of 0.500-in outside diameter, 0.400-in inside diameter using a cylindrical mandrel. Assume that $\mu = 0.08$ for both cases.

**19-6** Determine the maximum reduction by drawing for
(a) a non-strain-hardening material;
(b) annealed low-carbon steel ($n = 0.26$);
(c) annealed copper ($n = 0.5$).

## CHAPTER 20

**20-1** Derive Eq. (20-1) for the fiber strain in a bent sheet. What is the basic assumption in this derivation?

**20-2** Show how minimum bend radius may be predicted by the tensile reduction of area, Eq. (20-2).

**20-3** Use Mohr's circle of strain to establish that a local neck forms at 55° to the tensile axis in the uniaxial straining of an isotropic sheet.

**20-4** Show that the criterion for *localized* necking is $d\sigma/d\varepsilon = \sigma/2$.

**20-5** The flow curves for two metals are given by

$$\text{Metal } A: \ \sigma = 89,000 \ \varepsilon^{0.5}$$
$$\text{Metal } B: \ \sigma = 27,000 \ \varepsilon^{0.1}$$

Both metals initially are loaded to a stress of 20,000 psi and a strain of 0.05. If the stress now is increased to 23,000 psi, what is the strain in each metal? What is the importance of this example to sheet-metal forming?

**20-6** Estimate the punch load required for a deep-drawn cup with a 2-in diameter and 0.040-in wall if it is made from low-carbon steel, $\sigma_0 = 30,000$ psi, $\sigma_u = 50,000$ psi. The blank diameter is 4 in, and the overall coefficient of friction is 0.08. Estimate the hold-down pressure and force required for bending and unbending, from other considerations.

## CHAPTER 21

**21-1** List 10 specific or specialized machining operations or machine tools. Give examples of the type of part or geometry produced by each.

21-2   In orthogonal cutting of a low-carbon steel the specific cutting energy is 4 GJ/m³.
       The undeformed chip thickness is 0.2 mm and the chip width is 5 mm at a cutting
       speed of 1.1 m/s. The rake angle of the tool is 5°. Determine:
       (a) the cutting force,
       (b) the average shear stress in the shear plane,
       (c) the normal stress on the shear plane,
       (d) the average shear strain in cutting,
       (e) the average shear-strain rate.

21-3   The unit power required for turning with a feed of 0.010 in/rev is given in units of
       hp/(in³)/(min). For machining gray cast iron the unit power is 1.0, and for machin-
       ing heat-treated alloy steel at $R_c40$ it is 2.0 hp/(in³)(min). Compare the time
       required for rough turning 2-in-diameter × 10 in long cylinders of cast iron and
       steel to a diameter of 1.8 in. The lathe is rated at 10 hp and has a 70 percent
       efficiency.

21-4   Describe two methods for measuring the temperature generated in metal cutting.

21-5   Specify the conditions of grinding that would produce a fine finish on the ground
       surface.

21-6   Two different cutting tool materials have the tool life relations:
       High-speed steel: $vt^{0.12} = 230$
       Cemented carbide: $vt^{0.30} = 1,500$
       Determine what the increase in tool life will be for each material as a result of a
       50 percent decrease in cutting speed.

# NAME INDEX

# SUBJECT INDEX